GROUP THEORY IN SPECTROSCOPY

With Applications to Magnetic Circular Dichroism

WILEY-INTERSCIENCE MONOGRAPHS IN CHEMICAL PHYSICS

Editors

John B. Birks, *Reader in Physics, University of Manchester*

Sean P. McGlynn, *Professor of Chemistry, Louisiana State University*

Photophysics of Aromatic Molecules:
J. B. Birks, Department of Physics,
University of Manchester

Atomic and Molecular Radiation Physics:
L. G. Christophorou, Oak Ridge National Laboratory, Tennessee

The Jahn–Teller Effect in Molecules and Crystals:
R. Englman, Soreq Nuclear
Research Centre, Yavne

Organic Molecular Photophysics: Volumes 1 and 2:
edited by J. B. Birks, Department of Physics, University of Manchester

Internal Rotation in Molecules:
edited by W. J. Orville-Thomas,
Department of Chemistry and Applied Chemistry,
University of Salford

The Dynamics of Spectroscopic Transitions: Illustrated by Magnetic Resonance and Laser Effects:
James D. Macomber, Department of
Chemistry, Louisiana State University

Principles of Ultraviolet Photoelectron Spectroscopy:
J. Wayne Rabalais, Department of Chemistry,
University of Houston

Group Theory in Spectroscopy—With Applications to Magnetic Circular Dichroism:
Susan B. Piepho, Sweet Briar College;
Paul N. Schatz, University of Virginia

GROUP THEORY IN SPECTROSCOPY

With Applications to Magnetic Circular Dichroism

Susan B. Piepho
Sweet Briar College

Paul N. Schatz
University of Virginia

A Wiley-Interscience Publication
JOHN WILEY & SONS
New York • Chichester • Brisbane • Toronto • Singapore

Library of Congress Cataloging in Publication Data:

Piepho, Susan B.
Group theory in spectroscopy.

(Wiley-Interscience monographs in chemical physics, ISSN 0277-2477)
"A Wiley-Interscience publication."
Includes bibliographical references and index.
1. Molecular spectra. 2. Magnetic circular dichroism. 3. Quantum theory. 4. Groups, Theory of. I. Schatz, Paul N. II. Title. III. Series.

QC454.M6P53 1983 539′.6 83-3665
ISBN 0-471-03302-2

Printed in the United States of America

10 9 8 7 6 5 4 3 2 1

Preface

This book has two central objectives. First, it attempts to present a unified, self-contained, and immediately usable exposition of more advanced group-theoretic techniques applicable in molecular spectroscopy and quantum mechanics. Second, it seeks to present a clear and reasonably complete discussion of the theory of magnetic circular dichroism (MCD) spectroscopy. Though these subjects need not in principle be connected, any attempt to interpret the MCD of systems of high symmetry demands the application of group-theoretic techniques. Conversely, MCD examples serve as excellent illustrations of the use of these techniques. We seek therefore to interconnect these subjects. As a result, we believe that each provides important insights into the other.

Though many books have treated the application of group theory to molecular spectroscopy, we believe the present volume has several features which make it especially useful to the practicing spectroscopist. First, it is almost completely self-contained. Group-theoretic and spectroscopic equations are developed in consistent notation from first principles. And with the use of the appendixes, virtually all problems can be carried through in their entirety. Second, while vector coupling coefficients are discussed in some detail, the emphasis is entirely on high-symmetry coefficients which are used in all applications. This allows for a smooth transition from elementary (Chapters 8–13) to more advanced (Chapters 14–23) group-theoretic methods. The irreducible tensor method has been formulated in a general way and is applicable to all common point groups (including their double-group irreducible representations). A number of important lower symmetry groups are treated explicitly, and general methods for generating the high-symmetry coefficients of any common point group are outlined. In addition, basis choice and phase conventions have been made entirely consistent with those in Butler (1981), *Point Group Symmetry Applications: Methods and Tables* (Plenum Press). This latter book uses a chain-of-groups approach based on the Racah factorization lemma and makes available extraordinarily extensive tables of high-symmetry coefficients for all the common point groups. The Butler approach is especially useful in treating low-symmetry perturbations. We provide a link to the powerful methods of this book via Chapter 15 and subsequent applications and tables. Thus we hope to make the Butler book more accessible to molecular spectroscopists. We also hope to make a persuasive case for the wide adoption of the Butler phase and basis choices, which we use exclusively. Finally, we have illustrated our methods with frequent examples and have devoted the last chapter entirely to examples from the literature.

Though we start from basic principles, it is assumed that readers have some familiarity with quantum mechanics and group theory, and in particular with the properties of (quantum-mechanical) matrix elements. It is our belief that such readers can learn quite quickly to use advanced group-theoretic techniques. If one is willing to accept all formulas (and hence the resulting prescriptions) on faith, one can correctly use even the more sophisticated and powerful techniques in a short time, and this seems to us a valid way to use this book. The many specific examples we give illustrating important formulas and procedures facilitate such an approach. Nevertheless, many readers will be uneasy with a completely "blind" approach, since prescriptions may be subject to misinterpretation and misunderstanding.

Our own experience has shown that the use of group-theoretic techniques to solve specific problems provides great stimulus and incentive to delve more deeply into the mathematical foundations of the subject. We have therefore given derivations of (or literature references to) all formulas and prescriptions used in the book; thus the reader who so desires should be able to understand in full detail the theoretical basis for any of the methods. But it would certainly require a very determined (or masochistic) reader to work through all of the formulas in Chapters 9–22 before actually trying specific examples. We recommend strongly that readers proceed quickly to actual examples—those in the book, or even better, their own—and that they then go back to examine derivations more carefully. Formulas involving vector coupling or $3jm$, $6j$, and $9j$ coefficients become vastly more meaningful after they have been used a few times. It is a curious fact that the more complicated formulas are often the easiest to use in practice. Thus the "simpler" methods in Chapters 9–13 become hopelessly laborious in more complicated problems because it is necessary to construct wave functions. Conversely, the tensor methods of Chapter 14 et seq., which can produce general formulas of horrifying appearance, are easy to use in practice. Problems of remarkable complexity can be solved by looking up a few numbers in appropriate tables.

We have tried very hard to eliminate errors and misprints in formulas and tables. However, some are inevitable, and we would heartily welcome corrections. We have also tried to avoid obscurities and errors in the text. Comments, suggestions, and criticisms in this regard are welcome.

Our debt to the late J. S. Griffith is obvious. His two powerful books, *The Theory of Transition-Metal Ions* (Cambridge University Press) and *The*

Irreducible-Tensor Method for Molecular Symmetry Groups (Prentice-Hall), are the basis for many of the group-theoretic methods we discuss. However, following Butler (1981), our choice of standard basis functions is different, and the chain-of-groups approach permits significant generalizations and simplifications of the earlier methods.

We are indebted to many colleagues for specific comments and criticisms, and wish especially to thank the following individuals: Philip Stephens, Philip H. Butler, Peter W. Atkins, Paul A. Dobosh, William L. Luken, Tim Keiderling, Sean McGlynn, Richard A. Palmer, Mike Reid, Jim Riehl, Fred S. Richardson, Asok Banerjee, Paul A. Lund, David Pullen, Kurt Neuenschwander, Eileen Stephens, Chris Misener, Debbie Smith, Mike Boyle, Anne Johnston, Mary Berry, Janna Rose, and Seth Snyder. We should particularly like to acknowledge numerous helpful conversations and much correspondence with Philip H. Butler. We owe a very special note of thanks to Philip Stephens, who pioneered the development of MCD theory and the applications of group theory to the analysis of MCD spectra. He has taught us much, and we greatly value his friendship. We also acknowledge the dedicated typing of Sarah Alcock and Mary Critzer and the skilled art work of Don Bibb and Annemarie Aigner.

This book was started while one of us (P.N.S.) was on a John Simon Guggenheim Memorial Foundation Fellowship at the University of Oxford, 1974–1975. This support is deeply appreciated, and many thanks are due Peter Day for his kind hospitality during that period. Much of the later part of the book was first formulated while the other author (S.B.P.) was a NATO postdoctoral fellow at the University of Oxford in 1975–1976. The hospitality and keen interest of Peter Atkins is much appreciated.

Finally, we express deep gratitude to the National Science Foundation for support over a period of many years. This has permitted much of the research by us and others that provided the inspiration and motivation for this book.

Last but not least, we acknowledge the support and patience of our families over the long haul.

Susan B. Piepho
Paul N. Schatz

Sweet Briar, Virginia
Charlottesville, Virginia
July 1983

Contents

1. Introduction **1**

1.1 Structure 1
1.2 User's Guide 2
Magnetic Circular Dichroism 2
Vibronic Spectroscopy 3
Irreducible-Tensor Method 3
Older Group-Theoretic Methods 4
jj Coupling 5
1.3 Relation to the Literature 5

2. Basic Spectroscopic Formulas **6**

2.1 The Beer–Lambert Law 6
2.2 The Absorption Probability 7
2.3 Lineshapes 12
2.4 Induced and Spontaneous Emission 12
2.5 Elimination of Translation and Rotation 13
2.6 Relation of the Absorption Coefficient to the Electric-Dipole Transition-Moment Integral 16
2.7 Magnetic-Dipole and Electric-Quadrupole Transition Moments 19
2.8 Demonstration That $P_{a \to j} = P_{a \leftarrow j}$ 22
2.9 Inclusion of Electron Spin 23
2.10 Relative Magnitudes of the Transition Moments 24
2.11 Solvent Effects: The Effective-Field Approximation 24

3. Vibronic Transitions **27**

3.1 Introduction 27
3.2 The Molecular Hamiltonian and the Born–Oppenheimer Approximation 28

Failure of the Born–Oppenheimer Approximation for Degenerate States: The Jahn–Teller Effect 30
Adiabatic Functions 30
Adiabatic and Diabatic Surfaces: The Adiabatic Principle 31
The Validity of the Born–Oppenheimer Approximation for Nondegenerate States and the Pseudo-Jahn–Teller Effect 32
3.3 A Perturbation Approach to the Solution of the Electronic Born–Oppenheimer Equation in the Nondegenerate Case 33
3.4 The Crude Adiabatic Approximation 34
3.5 Solution of the Nuclear Equation in the Born–Oppenheimer Approximation 35
The Harmonic Approximation 35
3.6 The Transition-Moment Integral in the Born–Oppenheimer Approximation 36
The Franck–Condon Approximation 37
Deviations from the Franck–Condon Approximation 38
Allowed Electronic Transitions: Bandshape in the Franck–Condon Approximation 39
Selection Rules When *A* and *J* Have Identical Minima and Potential Surfaces: Why Only a Single Line Is Predicted 39
Selection Rules When *A* and *J* Have Different Minima: The Appearance of Totally Symmetric Progressions 41
Hot Bands 43
3.7 Electronically Forbidden Transitions in the Born–Oppenheimer and Franck–Condon Approximations 44
How to Treat the Breakdown in the Franck–Condon Approximation: The Herzberg–Teller Approach 44
Selection Rules in the Herzberg–Teller Approximation 46
An Example: The Bandshape Predicted for $A_{1g} \rightarrow T_{1g}$ and $A_{1g} \rightarrow T_{2g}$ Electronic Transitions in Octahedral MX_6^{n-} Complexes 48
3.8 The Breakdown of the Born–Oppenheimer Approximation: How the Jahn–Teller Effect Is Treated Using Degenerate Perturbation Theory 49

A Simple Jahn–Teller Case: The $E \otimes e$ System 52
Symmetry Classification of Jahn–Teller Vibronic Eigenfunctions 53
Static and Dynamic Jahn–Teller Effects 54
Bandshapes and Vibronic Selection Rules in Jahn–Teller Systems 54
An Example of the Dispersion of a Jahn–Teller System: The $^2T_{2u}$ Band in $IrCl_6^{2-}$ 55
The Ham Effect 56
The Breakdown of the Franck–Condon Approximation in Jahn–Teller Systems 57
Nonadiabatic Coupling as an Intensity Mechanism: The Effect of Small Deviations from the Born–Oppenheimer Approximation 57

4. Magnetic Circular Dichroism (MCD) Spectroscopy 58

4.1 The Faraday Effect 58
4.2 (Magnetic) Circular Dichroism 60
Difficulties if a System is not Isotropic, Cubic, or Uniaxial 64
4.3 The Relation of $\Delta k = k_- - k_+$ to Molecular Parameters 65
4.4 A Simple Example; $\mathcal{A}$, $\mathcal{B}$, and $\mathcal{C}$ Terms 69
4.5 The MCD of Allowed Electronic Transitions between Born–Oppenheimer States in the Linear Limit 74
The Rigid-Shift Approximation 77
Definition of $\mathcal{A}_1$, $\mathcal{B}_0$, and $\mathcal{C}_0$ 79
Definition of the Dipole Strength $\mathcal{D}_0$ 80
Definition of the Bandshape Function $f(\mathcal{E})$ 81
How to Obtain $\mathcal{A}_1$, $\mathcal{B}_0$, $\mathcal{C}_0$, and $\mathcal{D}_0$ from Experimental Data 82
Limitations of the Model 82
4.6 Spatial Averaging and Basis Invariance 83
Spectroscopic Stability 83
$\mathcal{A}_1$, $\mathcal{B}_0$, $\mathcal{C}_0$ and $\mathcal{D}_0$ in Basis-Invariant Form 84
Molecule-Fixed Operators and Orientational Averaging; $\bar{\mathcal{A}}_1$, $\bar{\mathcal{B}}_0$, $\bar{\mathcal{C}}_0$, and $\bar{\mathcal{D}}_0$ 86
4.7 Emission; Magnetic Circularly Polarized Luminescence (MCPL); Photoselection 90
Spontaneous Emission 90
Magnetic Circularly Polarized Luminescence 92
MCPL with No Photoselection 93
MCPL and Total Emission with Photoselection 94

5. Properties of the MCD Parameters **98**

5.1 Properties of $\mathcal{A}_1$, $\mathcal{B}_0$, and $\mathcal{C}_0$ 98
The Parameter $\mathcal{A}_1$ 98
The Parameter $\mathcal{C}_0$ 100
The Parameter $\mathcal{B}_0$ 102
5.2 Overlapping Bands and Pseudo-$\mathcal{A}$ Terms 104
Excited-State Near-Degeneracy 104
Ground-State Near-Degeneracy 108
5.3 MCD Dispersion Forms; Quantitative Comparisons of $\mathcal{A}$, $\mathcal{B}$ and $\mathcal{C}$ Terms 110
5.4 The Utility of MCD Spectroscopy 114

6. Deviations from the MCD Linear Limit **116**

6.1 Failure of First-Order Perturbation Theory 116
6.2 Saturation: Zeeman Splittings $\gtrsim kT$ 117
6.3 Saturation Analog of the $\mathcal{C}$ Term 118
Basis Invariance 119
Molecule-Fixed Operators and Orientational Averaging 120
Saturation Analog of $\mathcal{A}$ and $\mathcal{B}$ Terms 121
6.4 Zeeman Splittings Comparable to Bandwidth 122
6.5 Saturation and Large Zeeman Splitting 122

7. Moments and Parameter Extraction **124**

7.1 Relating the Experimental Lineshape to Theoretical Parameters 124
7.2 Introduction to the Method of Moments 124
7.3 The Moment Equations in Absorption and MCD Spectroscopy 126
7.4 Moments for Franck–Condon-Allowed Transitions 131
7.5 The Moment Equations in the Born–Oppenheimer Approximation 137
Other Situations in Which $\mathcal{B}_1$ and $\mathcal{C}_1$ are Zero 138
7.6 Overlapping Bands 141
How to Deal with Excited-State Near-Degeneracy Using Moments 141
Applications of the Excited-State Near-Degeneracy Moment Equations 143
Accidental Degeneracy 145

Ground-State Near-Degeneracy 146
7.7 Higher Moments 147
7.8 Obtaining MCD Parameters from the Rigid-Shift Equations 148
Gaussian Fits of an Isolated Band 148
Moments in the RS–BO Approximation 151
Gaussian Fits of Overlapping Bands 152
7.9 The MCD of Resolved Vibronic Lines in the Franck–Condon-Allowed Case 153
Use of the RS–BO Model for Lines 154
The MCD of Vibronic Lines When the BO Approximation Breaks Down 155
7.10 Franck–Condon-Forbidden Transitions in the Rigid-Shift Approximation 156
7.11 Vibrationally Induced Lines in the Rigid-Shift Born–Oppenheimer Approximation 159
7.12 Hybrid Cases 160

8. Introduction to Group Theory **161**

8.1 Introduction 161
8.2 Vector Spaces and the Cartesian Representation 162
Symmetry Elements and Groups 162
Representations and Basis Vectors 162
The Active and Passive Conventions 166
Dot and Scalar Products and Length-Preserving Transformations 167
8.3 Change of Basis Vectors and Similarity Transformations 169
8.4 *n*-Dimensional Vector Spaces 170
8.5 Function Space 172
8.6 The Transformation Operators $\mathfrak{R}$ 173
8.7 Representations Generated by the $\mathfrak{R}$: The $\mathbf{D}(R)$ 176
Examples of the Use of the $\mathfrak{R}$ 177
8.8 Choice of Basis Functions 180
8.9 Group Theory and the Schrödinger Equation 181
The Molecular Point Group 182
Representations of the $\mathfrak{R}$ 183
8.10 Irreducible Representations (Irreps) 184
The Great Orthogonality Theorem 185
8.11 Character Analysis 186
Classes 187
Obtaining the Character-Analysis Equation 188

Significance of Character Analysis 189
Character Tables 190
Reduction of the Generalized Cartesian Representation; Symmetry Coordinates and Normal Coordinates 191
Character Analysis of Atomic Basis Sets 194
Atomic Orbitals not Centered at an Invariant Point 196
Symmetrically Equivalent Sets 197
8.12 Spin Irreps and Double Groups 197
8.13 Direct-Product Groups 198
Example: Formation of the Character Table for $D_{3h} = D_3 \otimes C_s$ 199
How to Use the Direct-Product Group to Simplify Calculations 200

9. Vector Coupling Coefficients **201**

9.1 Introduction 201
9.2 Direct-Product Representations and Coupling 202
9.3 The Wigner–Eckart Theorem 204
The Replacement Theorem 209
Repeated Representations 209
9.4 Standard Basis Functions 211
The Standard Angular-Momentum Basis for SO_3 212
Standard Basis Functions for Subgroups of SO_3: Group Generators 212
Example: The Standard Basis for the T_1 Irrep in the Real Octahedral Basis 213
9.5 Construction of Coupling Tables 214
Example: Construction of the $T_1 \otimes T_2$ Coupling Table in the Real Octahedral Basis 214
9.6 Construction of Coupling Tables in Another Basis 221
9.7 Spin Irreps and Coupling 222
9.8 Function and Operator Transformation Coefficients; Spherical Vectors 223

10. High-Symmetry Coupling Coefficients: The 3*jm* **228**

10.1 Introduction 228
Advantages of the 3*jm* 228
10.2 Definition of the 3*jm* and Related Quantities 229
The Wigner–Eckart Theorem Using the 3*jm* 229

The $3j$ Phase $\{abcr\}$ 231
Limitations on the Choice of $\{abcr\}$ Phases 232
The $2j$ Phase $\{a\}$ 233
The $2jm$ Phase $\begin{pmatrix} a \\ \alpha \end{pmatrix}$ 234
Definition of a^* and α^* 234
10.3 Useful Equations Involving the $3jm$ 234
Relation of $3jm$ Containing an A_1 Representation to the $2jm$ 234
Relation between $3jm$ and $3jm^*$ 235
The Coupling of Conjugate States 235
Orthonormality Equations for the $3jm$ 235
Inversion of the Reduced-Matrix-Element Equation 236
10.4 Example of the Use of the Wigner–Eckart Theorem: Demonstration That T_1 and T_2 States in the Complex-O Basis are Diagonal in μ_z 236

11. Elementary Uses of Coupling Tables 239

11.1 Introduction 239
11.2 Use of the Wigner–Eckart Theorem in a Simple Example: The MCD of $Fe(CN)_6^{3-}$ Neglecting Spin–Orbit Coupling 239
$\mathcal{C}$- Term Calculation 241
Parity Labels and the Relation of a Direct-Product Group G_i to a Group G 242
Wigner–Eckart-Theorem Reductions 243
Interpretation of the $Fe(CN)_6^{3-}$ MCD Spectrum 244
Calculation of $\mathcal{A}$ Terms for $Fe(CN)_6^{3-}$ 246
Selection Rules for Circularly Polarized Transitions in $Fe(CN)_6^{3-}$ 247
11.3 The Role of Spin in the Absence of Spin–Orbit Coupling 247
Calculation of $\mathcal{C}_0$ in the Absence of Spin–Orbit Coupling 250
11.4 The First Step in Matrix-Element Evaluation: The Construction of Wavefunctions 252
How to Form $|\mathcal{S}h\mathfrak{M}\theta\rangle$ Functions for a Configuration a^n 254
Hole–Particle Convention 259
Fractional-Parentage Equation and the Formation of $|\mathcal{S}h\mathfrak{M}\theta\rangle$ States for t_1^n and t_2^n Configurations 260
11.5 Formation of $|\mathcal{S}h\mathfrak{M}\theta\rangle$ Functions for Configurations with More Than One Open Shell 261

12. Spin–Orbit Coupling and Double-Group Calculations 266

12.1 The Spin–Orbit Coupling Operator 266
12.2 Spin–Orbit Basis Functions 267
12.3 Construction of Spin–Orbit States $|(\mathbb{S}S, h)rt\tau\rangle$ for Integral $\mathbb{S}$ 269
12.4 Construction of Spin–Orbit States When $\mathbb{S}$ is Half-Integer 271
12.5 Behavior of $|\mathbb{S}h\mathfrak{M}\theta\rangle$ and $|(\mathbb{S}S, h)rt\tau\rangle$ Functions under $\mathfrak{R}$ and $\mathcal{H}_{SO}$ 272
12.6 Repeated-Representation Considerations 273
Choice of the $U'U'T_1$ and $U'U'T_2$ Multiplicity Separations 273
Determination of First-Order Spin–Orbit $U'(^4T_1)$ and $U'(^4T_2)$ Functions Using $3jm$ 274
The Significance of Repeated-Representation Labels in Matrix-Element Reduction 276

13. Matrix-Element Evaluation: Elementary Concepts 277

13.1 Introduction 277
13.2 Reduction of Spin-Independent Matrix Elements to One-Electron Form: The Example of $Fe(CN)_6^{3-}$ 278
13.3 Spin–Orbit-Coupling Matrix Elements 281
Reduction of Spin–Orbit Matrix Elements to One-Electron Form: The Spin–Orbit Splittings in $Fe(CN)_6^{3-}$ 282
13.4 Evaluation of One-Electron Angular Momentum Reduced Matrix Elements 287
Single-Center One-Electron Calculations 288
Many-Center Calculations in the LCAO–MO Approximation 290
Transforming l_γ: The Octahedral Case 292
Transforming l_γ: The Tetrahedral Case 293
13.5 The $\mathcal{C}$ Terms of $Fe(CN)_6^{3-}$ with Spin–Orbit Coupling 296

14. Theory of High-Symmetry Coefficients 303

14.1 Invariance and Contragredience 303
14.2 The Relation Between Contragredient Sets for Molecular Point Groups 305
Definition of Ambivalent and Nonambivalent Groups 305

The Relation between Contragredient Sets for Ambivalent Groups 305
14.3 Use of the Time-Reversal Operation T 307
The Behavior of the $|jm\rangle$ of SO_3 in the Angular-Momentum Basis under T 308
14.4 Obtaining $\mathbf{U}(a)$ and the $2jm$ for Ambivalent Groups Using the Time-Reversal Operator T 310
Relation of $\mathbf{U}(a)$ and $\tilde{\mathbf{U}}(a)$ 312
Example: Definition of the $2jm$ and α^* for the E'' Irrep of the Group O 313
14.5 The Relation between Contragredient Sets for Nonambivalent Groups 313
Example: Definition of $\binom{a}{\alpha}$, a^*, and α^* for the Complex Irreps of D_3 314
14.6 The Invariant Product of Two Sets 315
14.7 The Invariant Product of Three Sets: The $3jm$ 316
The Even-Permutation Rule for $3jm$ 317
The Odd-Permutation Rule for $3jm$ 318
The Coupling Is Not Associative 319
A Note on the Permutation Rules for Nonambivalent Groups 319
14.8 The Relation between $3jm$ and $3jm^*$ 320
14.9 How to Find a Set of $3jm$ for a Molecular Point Group 321
An Illustration: Determination of $3jm$ for D_3 323
Groups for Which Our Formalism is Applicable 325
14.10 Adjoint Operators $A^\dagger$ and Time-Reversed Operators A^* 325
14.11 Use of Time-Reversal Symmetry to Simplify the MCD Equations 327
Behavior of Kramers Pair States and MCD Operators under T 327
Derivation of Some Results in Section 5.1 328
Simplification of MCD Equations for Use in Chapter 17 330

15. The Chain-of-Groups Approach to Symmetry Calculations 333

15.1 Introduction 333
15.2 Group–Subgroup Chains 334
How Basis States are Labeled by Specifying Irreps along the Chain 335
15.3 The Racah Factorization Lemma and the Factoring of Coupling Coefficients 336

Calculation of $2jm$ and $3jm$ Coefficients as a Product of $2jm$ and $3jm$ Factors 339
15.4 Evaluating Reduced Matrix Elements in a Group–Subgroup Scheme 340
Example: Orbital-Angular-Momentum Matrix Elements 341
Example: Spin-Angular-Momentum Matrix Elements 342
15.5 Basis Functions 344

16. $6j$ and $9j$ Coefficients 346

16.1 $6j$ Coefficients 346
Symmetry Rules for $6j$ for Simply Reducible Groups 347
Symmetry Rules for $6j$ in the General Case 347
Simple Formula for $6j$ Containing an A_1 Irrep 349
16.2 Other Equations Relating $3jm$ Products to $6j$ 349
16.3 Definition of $6j$ in Terms of Recoupling Coefficients 351
16.4 $9j$ Coefficients 353
Symmetry Rules for $9j$ for Simply Reducible Groups 353
Symmetry Rules for $9j$ in the General Case 354
How to Reduce $9j$ Containing an A_1 Irrep to $6j$ 355
The Relation of the $9j$ to Recoupling Coefficients 355

17. MCD Equations Using $3jm$ and $6j$: Allowed Transitions 356

17.1 Introduction 356
17.2 The MCD Equations for Ambivalent Groups: Oriented or Isotropic Case 357
Simplification of the $\mathcal{A}_1$, $\mathcal{B}_0$, and $\mathcal{C}_0$ Equations Using $3jm$ and $6j$ 359
Simplification of the $\mathcal{D}_0$ Equation Using $3jm$ 363
$\mathcal{A}_1/\mathcal{D}_0$ and $\mathcal{C}_0/\mathcal{D}_0$ for Simply Reducible Groups 364
17.3 The Oriented or Isotropic Case: Examples 365
$\mathcal{A}_1/\mathcal{D}_0$ and $\mathcal{C}_0/\mathcal{D}_0$ for the Groups O and T_d When A and J are Single-Valued Irreps 365
$Fe(CN)_6^{3-}$ Revisited 366
$\mathcal{A}_1/\mathcal{D}_0$ and $\mathcal{C}_0/\mathcal{D}_0$ for the Groups O and T_d in the More General Case 367

17.4 The MCD Equations for Ambivalent Groups: Space-Averaged Case 367
Comparison of MCD Equations for the Space-Averaged and Oriented Cases 372
17.5 Orientation Dependence of the MCD in Cubic Crystals 374
17.6 The MCD Equations for Nonambivalent Groups 376

18. Matrix-Element Simplification Using the Irreducible-Tensor Method: Fundamental Equations 382

18.1 Simplification of Matrix Elements of Operators Which Act on Only One Part of a Coupled System 382
Example: Simplification of Matrix Elements of Spin-Independent and Space-Independent Operators 385
18.2 Uses of the O^f and S^f Equations in MCD Calculations 386
Example: The $E''_g(^2T_{2g}) \rightarrow E''_u(^2T_{2u})$ $\mathcal{C}$ Term in $IrBr_6^{2-}$ 386
Example: The MCD of $CoCl_4^{2-}$ 388
18.3 Transformation Properties of the Dot Product of Tensor Operators 393
18.4 Matrix Elements of Dot-Product Operators: Spin–Orbit-Coupling Matrix Elements 397
18.5 The Double-Tensor Analogs of the D^d and E^e Equations of Section 18.1 400

19. Reduction of Multielectron Matrix Elements to One-Electron Form Using the Irreducible-Tensor Method 402

19.1 Permutation Properties of Determinantal Wavefunctions 403
19.2 Outline of the Method Used to Simplify Matrix Elements of a Configuration a^n 405
19.3 Coefficients of Fractional Parentage (cfp) 407
How to Calculate cfp and Use Them to Construct Octahedral $|t_2^3\,\mathbb{S}h\mathfrak{M}\theta\rangle$ States 407
General Equations for cfp 411
19.4 One Open Shell: Matrix Elements of a^n 414
Spin-Independent Matrix Elements of a^n and the $g(a^n, \mathbb{S}hfh')$ 414
Spin-Only Matrix Elements of a^n 418
Example: Magnetic-Moment Matrix Elements for an Octahedral t_{2g}^4 Configuration 419

19.5 One Open Shell: Spin–Orbit Matrix Elements of a^n and the $G(a^n, \mathbb{S}hf\mathbb{S}'h')$ 420
Example: Spin–Orbit Matrix Elements for an Octahedral t_{2g}^4 Configuration 423
Example: Calculation of the Spin–Orbit Matrix for the ${}^2T_{2g}$, ${}^2T_{1u}$, and ${}^2T_{2u}$ States of $IrCl_6^{2-}$ 424
19.6 Calculation of One-Electron Matrix Elements: Multicenter Examples 425
19.7 Two Open Shells: Expansion of the Functions of an $a^m b^n$ Configuration 429
Example: Construction of a ${}^3T_2$ Wavefunction for a $t_2 e$ Configuration 431
19.8 The Simplest Off-Diagonal Case: Matrix Elements between a^n and $a^{n-1}b$ 432
Spin-Independent Matrix Elements between a^n and $a^{n-1}b$ 432
Example: Calculation of an Orbital-Angular-Momentum Matrix Element between t_2^2 and $t_2 e$ States 433
Spin-Only Matrix Elements between a^n and $a^{n-1}b$ 434
Spin–Orbit Matrix Elements between a^n and $a^{n-1}b$ 434
19.9 Calculation of cfp, $g(a^n, \mathbb{S}hfh')$, and $G(a^n, \mathbb{S}h\mathbb{S}'h')$ for the Pseudo-p^n Configurations of the Cubic Groups 435
Calculation of Coefficients of Fractional Parentage for t_1^n and t_2^n 436
Calculation of $g(a^n, \mathbb{S}hfh')$ for t_1^n and t_2^n 437
Calculation of the $G(a^n, \mathbb{S}h\mathbb{S}'h')$ for t_1^n and t_2^n 438

20. Matrix Elements with Several Open Shells 440

20.1 Matrix Elements of $a^m b^n$ 440
Spin-Independent Matrix Elements of $a^m b^n$ 441
Spin-Only Matrix Elements of $a^m b^n$ 442
Spin–Orbit Matrix Elements of $a^m b^n$ 443
20.2 The Off-Diagonal Case: Matrix Elements between $a^m b^{n-1}$ and $a^{m-1} b^n$ 444
The Initial Procedure 444
Spin-Independent Matrix Elements between $a^m b^{n-1}$ and $a^{m-1} b^n$ 447
Spin-Only Matrix Elements between $a^m b^{n-1}$ and $a^{m-1} b^n$ 448

Spin–Orbit Matrix Elements between $a^m b^{n-1}$ and $a^{m-1} b^n$ 448
Equations Analogous to Those of Chapter 18 for Matrix Elements between $a^m b^{n-1}$ and $a^{m-1} b^n$ 450
Example: Calculation of the Spin–Orbit Matrix Elements between $U_u'(^2T_{1u})$ and $U_u'(^2T_{2u})$ for $IrCl_6^{2-}$ 451
20.3 More Than Two Open Shells 452
Configurationally Diagonal Matrix Elements for $(a^m b^n)c^l$, $a^m(b^n c^l)$, or $(a^m b^n)(c^l d^k)$ States 452
Matrix Elements between $(a^{m-1} b^n)c^l$ and $(a^m b^{n-1})c^l$ or between $a^m(b^{n-1} c^l)$ and $a^m(b^n c^{l-1})$ 453
More Complicated Cases 453

21. The jj-Coupling Approach to Molecules 454

21.1 Introduction to jj Coupling 454
21.2 jj Coupling using the OsX_6^{2-} t_{2g}^4 Configuration as an Example 455
Ground-State Basis of OsX_6^{2-} Using $\mathfrak{SL}$ Coupling 455
Ground-State Basis of OsX_6^{2-} Using jj Coupling 456
Formal Construction of States of jj Configurations with One or Two Open Shells 457
Warning About Phases of jj States 458
21.3 Calculation of the Spin–Orbit Matrix for a^n: The Example of $|t_{2g}^4 A_1 a_1\rangle$ in OsX_6^{2-} 459
The Matrix in the $\mathfrak{SL}$ Basis 459
The Matrix in the jj Basis 460
Comparison of the Methods 462
21.4 Calculations for the T_{1u} States of the OsX_6^{2-} $t_{2u}^5 t_{2g}^5$ Configuration in the jj Basis 462
Determination of the Spin–Orbit Energy of the Lowest T_{1u} State 463
The Magnetic Moment of the Lowest T_{1u} State 464
Calculation of $\mathcal{C}_1/\mathfrak{D}_0$ and $\mathfrak{D}_0$ for the Transition to the Lowest T_{1u} State 466

22. MCD Equations Using $3jm$ and $6j$ Coefficients: The Herzberg-Teller Case 469

22.1 Introduction 469
Statement of the Problem 469

Electric-Dipole and Magnetic-Dipole Matrix Elements in the Herzberg–Teller Approximation 470
22.2 The Dipole Strength of a Vibrationally Induced Transition 470
Approximation of $\mathcal{D}_0^{\text{vib}}$ 473
22.3 $\mathcal{A}_1^{\text{vib}}$, $\mathcal{B}_0^{\text{vib}}$, and $\mathcal{C}_0^{\text{vib}}$ 474
Approximation of $\mathcal{A}_1^{\text{vib}}$ and $\mathcal{C}_0^{\text{vib}}$ 476
22.4 Special Cases Where Vibration-Induced MCD Ratios Simplify 477
Octahedral Systems with A_{1g} or A_{2g} Ground States 477
Systems with Excited-State Mixing Only and Mixing States of Only One Symmetry 478

23. Analysis of Real Systems **480**

23.1 Introduction 480
Basic Considerations: Obtaining the Data 480
Data Analysis: The Initial Stages 481
23.2 A Favorite Example: $IrCl_6^{2-}$ 482
Sketching the Molecular-Orbital Diagram and Relating It to the Spectrum 482
Arriving at an Electronic State Diagram 484
Relation of the State Diagram to the Absorption Spectrum: $IrCl_6^{2-}$ in Solution 485
Use of MCD to Clarify $IrCl_6^{2-}$ Solution Assignments 486
Calculation of $\mathcal{C}_0/\mathcal{D}_0$ for Bands II and III of $IrCl_6^{2-}$ in Solution 487
23.3 The $IrCl_6^{2-}$ Spectrum in the Solid State: $Cs_2ZrCl_6 : Ir^{4+}$ 488
$Cs_2ZrCl_6 : Ir^{4+}$, Band III: The $t_{2u}(\pi) \rightarrow t_{2g}$ Band 491
$Cs_2ZrCl_6 : Ir^{4+}$, Band II: The $t_{1u}(\pi + \sigma) \rightarrow t_{2g}$ Bands 492
23.4 $OsCl_6^{2-}$ in the Solid-State Host Cs_2ZrCl_6 494
The jj Approximation in $OsCl_6^{2-}$ 494
Electrostatic Interactions in $OsCl_6^{2-}$ and $IrCl_6^{2-}$ 497
23.5 MCD and Absorption Calculations for $IrCl_6^{2-}$ and $OsCl_6^{2-}$ 499
Calculation of $IrCl_6^{2-}$ Reduced Matrix Elements: The Method of Chapters 19–20 501
Calculation of $IrCl_6^{2-}$ Reduced Matrix Elements: The jj Approach 506

Calculation of $OsCl_6^{2-}$ Reduced Matrix Elements in the jj Approximation 509
Breakdown of the jj Approximation in $OsCl_6^{2-}$ 511
23.6 Jahn–Teller Effects in a Series of Complex Ions: The General Approach 511
The Jahn–Teller Effect in $IrCl_6^{2-}$ 511
The Jahn–Teller Effect in $OsCl_6^{2-}$ 512
Analysis of the Absorption and MCD Data for the $OsCl_6^{2-}$ $t_{2u}(\pi) \rightarrow t_{2g}$ Band in $Cs_2ZrCl_6 : Os^{4+}$ 514
23.7 How to Treat Small Distortions from High Symmetry 520
Example: The $t_{2u}(\pi) \rightarrow t_{2g}$ Transitions in Trigonally Distorted $IrCl_6^{2-}$ 520
23.8 Analysis of the MCD of Vibronic Lines: The $IrBr_6^{2-}$ Spectrum 523
The Use of Hot-Band Data in $IrBr_6^{2-}$ Assignments 528
The MCD of $Cs_2ZrBr_6 : Ir^{4+}$ 528
The $\mathcal{C}$ Terms of the Broader Bands 528
The $\mathcal{A}$ Terms of the Sharp Lines 532

APPENDIXES

Appendix A. MCD Conventions and Notation 533

A.1 Circularly Polarized Light 533
A.2 (M)CD Sign 534
A.3 Magnetic-Field Sign 534
A.4 MCD Equations Using the New [Stephens (1976)] Definitions of $\mathcal{A}_1$, $\mathcal{B}_0$, $\mathcal{C}_0$, and $\mathcal{D}_0$ 534
A.5 Earlier (and Obsolete) Definitions of the Faraday Parameters 538
A.6 Some Other Useful Equations 540

Appendix B. Tables for the Group of All Rotations SO_3 541

B.1 Spherical Harmonics: The $SO_3 \supset SO_2$ Basis for Integer j 541
B.2 Real Atomic Orbitals Defined in Terms of the Y_m^j 541
B.3 High-Symmetry Coefficients and Phases for SO_3 543
Partial List of $3jm$ for $SO_3 \supset SO_2$ 544
Partial List of $6j$ for SO_3 546
B.4 Reduced Matrix Elements of SO_3: The Tables of Nielson and Koster 546
B.5 Partial List of $SO_3 \supset O$ $3jm$ Factors 549

Appendix C. Tables for the Groups O and T_d **551**

C.1 Character Table for the Groups O and T_d 551
C.2 Direct-Product Table for O and T_d 552
C.3 Table of Symmetrized $[a^2]$ and Antisymmetrized (a^2) Squares 553
C.4 Decomposition of O_h Relative to its Subgroups, T_d, D_{4h}, and D_{3d} 553
C.5 Tetragonal Bases for the Groups O and T_d 554
C.6 The Complex-O and Complex-T_d (Tetragonal) Bases Defined in Terms of the $O_3 \supset SO_3 \supset SO_2$ Basis 558
C.7 Function and Operator Transformation Coefficients 558
C.8 Functions for t_2^m and e^n Configurations 562
C.9 Strong-Field Crystal-Field and Electrostatic Matrices for d^n Configurations 564
C.10 Symmetry-Adapted Molecular Orbitals for Octahedral and Tetrahedral Complexes 568
C.11 $3j$, $2j$, and $2jm$ Phases and Related Definitions 570
The $3j$ Phase $\{abcr\}$ 570
The $2j$ Phase $\{a\}$ 571
The $2jm$ $\begin{pmatrix} a \\ \alpha \end{pmatrix}$ and Definitions of a^* and α^* 571
C.12 $3jm$ for the Groups O and T_d in the Bases of Section C.5 572
C.13 $6j$ for the Groups O and T_d 576
C.14 $9j$ for the Groups O and T_d 582
C.15 Coefficients of Fractional Parentage for O and T_d 582
C.16 Coefficients $g(a^n, \mathbb{S}hfh')$ for O and T_d 585
C.17 The Coefficients $G(a^n, \mathbb{S}hf\mathbb{S}'h')$ for O and T_d 587
C.18 Functions for $(e')^n$, $(e'')^n$, and $(u')^n$ Configurations 588
C.19 Coefficients of Fractional Parentage for $(e')^n$, $(e'')^n$, and $(u')^n$ jj Configurations 588

Appendix D. Tables for the Groups D_4 and D_{2d} **589**

D.1 Character Table for D_4 and D_{2d} 589
D.2 Direct-Product Table for D_4 and D_{2d} 590
D.3 Table of Symmetrized $[a^2]$ and Antisymmetrized (a^2) Squares 590
D.4 Bases for D_4 and D_{2d} 590
D.5 The Complex-D_4 and Complex-D_{2d} Bases Defined in Terms of the $O_3 \supset SO_3 \supset SO_2$ Basis 592
D.6 Function and Operator Transformation Coefficients 593

D.7 Symmetry-Adapted D_{4h} Molecular Orbitals for Square-Planar Compounds 594
D.8 $3j$, $2j$, and $2jm$ Phases and Related Definitions 594
D.9 $3jm$ for D_4 and D_{2d} in the Bases of Section D.4. 596
D.10 $6j$ for D_4 and D_{2d} 597
D.11 $9j$ for D_4 and D_{2d} for Single-Valued Irreps 598
D.12 $2jm$ and $3jm$ Factors for $O \supset D_4$ and $T_d \supset D_{2d}$ 598

Appendix E. Tables for the Groups D_3 and C_{3v} and for the Groups O and T_d Using Trigonal Bases **601**

E.1 Character Table for D_3 and C_{3v} 601
E.2 Direct-Product Table for D_3 and C_{3v} 602
E.3 Table of Symmetrized $[a^2]$ and Antisymmetrized (a^2) Squares 602
E.4 Bases for D_3 and C_{3v} and Trigonal Bases for O and T_d 602
E.5 Trigonal Bases Defined in Terms of the $O_3 \supset SO_3 \supset SO_2$ Basis 603
E.6 Function and Operator Transformation Coefficients 607
E.7 Symmetry-Adapted Molecular Orbitals for Trigonal Planar ML_3 608
E.8 $3j$, $2j$, and $2jm$ Phases for D_3 and C_{3v} and Related Definitions 609
E.9 $3jm$ for D_3 and C_{3v} in the Bases of Section E.4 610
E.10 $6j$ for D_3 and C_{3v} 610
E.11 $9j$ for D_3 and C_{3v} for Single-Valued Irreps 612
E.12 $2jm$, $3jm$, and Other Coefficients for O and T_d in the Trigonal Bases 612

Appendix F. Tables for the Groups D_∞, $C_{\infty v}$, and D_6 **613**

F.1 Direct Products in D_∞ and $C_{\infty v}$ 613
F.2 Table of Symmetrized $[a^2]$ and Antisymmetrized (a^2) Squares for D_∞ and $C_{\infty v}$ 613
F.3 Basis for $D_\infty \supset C_\infty$ Defined in Terms of the $SO_3 \supset SO_2$ Basis 614
F.4 Phases and Coefficients for D_∞ and $C_{\infty v}$ 615
F.5 Tables for D_6 616

References **616**

Index **623**

GROUP THEORY IN SPECTROSCOPY

With Applications to Magnetic Circular Dichroism

1 Introduction

This book covers a wide range of material in both spectroscopy and group theory. How it can be most effectively used therefore depends strongly on the background and needs of the reader. In this chapter we briefly describe the structure of the book and then suggest possible sequences of topics which might best match one or another particular need.

1.1. Structure

The book divides rather naturally into three parts, which present basic spectroscopy (Chapters 2–7), and then elementary (Chapters 8–13) and more advanced (Chapters 14–23) group-theoretic techniques and applications. We comment here briefly on each of these.

Chapters 2 and 3 develop the basics of vibronic spectroscopy, which are used in Chapters 4–7 to develop magnetic circular dichroism (MCD) spectroscopy in detail.

Chapter 8 provides a unified summary of representation theory. With perhaps a bit more detail, it covers material typical of a first course in group theory for chemists, as presented for example in Cotton (1971) or Bishop (1973). Applications of group theory to spectroscopy commence in Chapter 9 with the introduction of direct-product representations, the Wigner–Eckart theorem, and coupling tables. High-symmetry coefficients are introduced in Chapter 10 and are applied in Chapters 11–13 to a variety of examples using the "simple" methods which require explicit construction of wavefunctions. Spin–orbit coupling and double-group functions are introduced in Chapter 12, and the explicit evaluation of reduced matrix elements is the topic of Chapter 13.

Chapters 14–22 treat the theory of the irreducible-tensor method and its application to a variety of problems in molecular spectroscopy. MCD is used to illustrate the great power of this technique, which completely avoids the construction of wavefunctions. Chapter 15 introduces the Butler (1981) chain-of-groups approach, which is subsequently used in various applications. Explicit formulation of the MCD equations using high-symmetry coefficients is found in Chapters 17 and 22.

In Chapter 23 successively more complex examples from the MCD literature are presented to illustrate the methods developed earlier. Particularly noteworthy is the ease with which low-symmetry perturbations can be treated by the chain-of-groups approach (Section 23.7).

The Appendixes provide extensive sets of tables which, for the more common point groups, provide all the information required to carry out specific group-theoretic analyses of molecular systems. These tables are completely consistent with the very extensive set of Butler (1981) and with the tables of Rotenberg et al. (1959) for the full rotation group. Together the three sets of tables cover any case which is apt to be encountered. Our appendixes use a dual system of notation to define basis states; both the Mulliken representation labels familiar to chemists and their Butler (1981) equivalents are given to facilitate use of the Butler book.

1.2. User's Guide

Magnetic Circular Dichroism

For the reader primarily interested in MCD applications, the ultimate foci are Chapters 17 and 22. Equations for the Faraday (MCD) parameters ($\mathcal{A}_1$, $\mathcal{B}_0$, $\mathcal{C}_0$, $\mathcal{D}_0$) are presented there in a form which makes maximum use of symmetry for allowed transitions (Chapter 17) and for vibronic transitions (Chapter 22). With the equations given it is possible, in a matter of a few minutes, to write these parameters in terms of a minimum number of reduced matrix elements. No wave functions need be constructed, and for the more common cases all necessary quantities (high-symmetry coefficients and phase factors) are available in the appendixes. How fast one gets to these chapters depends on one's previous background and one's general philosophy. For example, to reach Chapter 17 at maximum speed, one can skim (or know) the material in Section 3.2 and Chapters 4 and 5. The crucial equations are (4.5.14), (4.6.8), and/or (4.6.14). If these are accepted at face value (or are already familiar), one can jump immediately to their "symmetrized" form in Sections 17.2–17.4. Starting with the $Fe(CN)_6^{3-}$ example (Section 17.3) and doing some backing and filling to learn the notation, the formulas of Chapter 17 can be applied with confidence to specific problems of interest.

Clearly a more leisurely approach is possible (and probably desirable), starting with a more careful study of Chapters 4 and 5. Unless specific applications demand otherwise, we recommend that Chapters 6 (saturation) and 7 (moments) be omitted initially. Whether Chapter 8 (introductory group theory) requires more than a cursory examination for notational familiarity depends on the background of the reader. Chapters 9 and 10 are important because they introduce the fundamental concepts, theorems, and notation used throughout the remainder of the book. Though the vector coupling coefficients of Chapter 9 are not used later, they are the starting point for the high-symmetry coefficients, which are introduced in Chapter 10 and are used exclusively thereafter.

Chapters 11–13 illustrate the "simpler" methods which require explicit construction of wave functions; these can be skimmed if Chapter 17 (and/or 22) is the immediate goal.

Chapter 14 (theory of high-symmetry coefficients) can be omitted entirely on a first reading, but at least the first two sections of Chapter 15 should be read. These sections are not difficult and make clear the chain-of-groups nomenclature used extensively in the Appendixes and in various examples. Chapter 16 introduces the $6j$ and $9j$ symbols used in Chapter 17 and later. Proofs can be omitted on first reading; knowledge of only a few simple properties of these quantities is required to use them with confidence.

Vibronic Spectroscopy

For readers particularly interested in vibronic spectroscopy, the starting point is Chapter 3, which discusses both Herzberg–Teller and Jahn–Teller theory in some detail. Chapter 22 presents the MCD equations for the Herzberg–Teller case, and application of this and the Jahn–Teller case are discussed in Chapter 23. Also relevant is the method of moments, discussed in Chapter 7. It is shown for example how one can obtain significant information without actually solving the Jahn–Teller problem. The chapter should also serve as a good general introduction to the method of moments.

Irreducible-Tensor Method

For readers primarily interested in the irreducible-tensor method, the relevant chapters are 9, 10, 14–16, and 18–21. The approach we use differs significantly from the earlier work of, for example, Griffith (1962), Dobosh (1972), Harnung (1973), and Silver (1976). A unified formalism is presented which is applicable to *all* molecular point groups (including double groups, groups with repeated representations, and nonambivalent groups) in *any* subgroup basis. Thus our treatment is much more complete than the treatments referred to above, which are limited to particular groups in particular bases. While our objectives parallel to a considerable extent those of Griffith (1962), we follow the conventions of Butler (1981) for high-symmetry coefficients, and use his chain-of-groups method in defining our bases. Besides offering great flexibility in the basis choice for higher groups such as O_h, T_d, or the full rotation group SO_3, the use of group-chain bases enables reduced matrix elements of a subgroup to be calculated in terms of those for a higher group (Section 15.4). Nearly all the high-symmetry coefficients required for irreducible-tensor-method calculations in the more common groups are given in this book. These may be supplemented by the very extensive tables for the point groups (all but SO_3) of Butler (1981) and those for SO_3 of Rotenberg et al. (1959), both of which are fully compatible with ours.

Chapter 15 is the obvious starting point for those interested in an introduction to the Butler (1981) chain-of-groups approach. Specific applications are found in Sections, 15.3, 15.4, 18.4, 19.4, 19.9, 23.7, and Appendix B, Section B.4.

A further feature of our presentation is the large number of examples. These provide a link between the more abstract mathematical methods and the practical world of the working spectroscopist. In this context both Chapter 15 and our use of Mulliken group-representation labels along with their Butler (1981) analogs (e.g., Appendix C, Section C.5) should help bridge the gap between the older "Griffith" methods used by most spectroscopists and those of Butler (1981). We emphasize, however, that in general all Griffith phases, conventions, bases, and reduced matrix elements, and all symmetry, g, G, and fractional-parentage coefficients have been *redefined* in this book to be consistent with the Butler (1981) conventions. This should cause no problems since our treatment [together with Butler (1981) and Rotenberg et al. (1959)] provides a self-contained system. Readers are warned, however, against "hybrid" calculations; mixing, for example, quantities calculated with Griffith (1962), Griffith (1964), Silver (1976), or Koster et al. (1963) with those calculated according to this book can produce disastrous phase errors (wrong answers).

Our formalism enables spin-orbit and double-group calculations to be handled much more easily than was possible using the C-number method of Griffith (1964) or the Ω-coefficient method of Griffith (1962)—see Sections 13.3 and 18.4. Moreover, the use of the chain-of-groups approach makes the calculation of high-spin ($\mathbb{S} \geqslant \frac{3}{2}$) matrix elements and the treatment of low-symmetry perturbations much simpler than in the Griffith or Silver formulations.

Older Group-Theoretic Methods

Chapters 9–13 should provide a detailed guide for those who desire a bridge to the earlier methods of, for example, Griffith (1964). In these chapters the construction of explicit wavefunctions and the use of vector coupling coefficients are discussed. *Remember however that our choice of standard basis functions is different*. These chapters should also serve as an introduction to the use of group-theoretic methods in spectroscopy.

jj Coupling

Finally, the use of the *jj*-coupling approximation in molecular systems is discussed in some detail in Chapter 21 and further examples of its use are given in Chapter 23.

1.3. Relation to the Literature

While our group-theoretical applications to spectroscopy are based on the pioneering work of Griffith (1962, 1964), we emphasize again that *our choice of standard basis functions is different* despite superficial resemblances. We follow Butler (1981) implicitly in all standard basis and phase choices. We hope that these choices will become standard in the future literature. Butler's group-chain bases and computational methods for calculating high-symmetry coefficients have the important advantage of allowing computer generation of a unified set of coefficients for all important molecular point groups in any desired subgroup basis. Using these methods, Butler has published by far the most extensive set of such tables available.

2 *Basic Spectroscopic Formulas*

In this chapter, we relate the experimental observable of absorption spectroscopy, the absorption coefficient, to the parameter which reflects the detailed molecular properties of the system, namely the transition moment integral. This relationship is fundamental and is the starting point for all applications of group theory to spectroscopy. We obtain the desired result using the standard semiclassical approach in which the molecular system is treated quantum-mechanically and the radiation field is treated classically. In this way, with a minimum of effort, we can obtain all required spectroscopic formulas.

2.1. The Beer–Lambert Law

Let us suppose that a monochromatic light beam of intensity I is traveling through a homogeneous, absorbing sample. If each molecule of the sample has a specific intrinsic probability of absorbing a photon of this frequency, the fractional *decrease* in intensity of the beam in traversing an infinitesimal thickness dl of the sample is given by

$$\frac{dI}{I} = -\kappa\, dl \qquad (2.1.1)$$

where $\kappa(\nu)$, commonly referred to as an absorption coefficient, is in general a strong function of frequency. Also, if we take a sample of fixed thickness, dI/I must vary in direct proportion to the concentration $\mathfrak{c}$ of absorbing molecules; thus

$$\frac{dI}{I} = -\kappa'\, d\mathfrak{c} \qquad (2.1.2)$$

Equations (2.1.1) and (2.1.2) are differential forms respectively of the Lambert and Beer laws.

If it is assumed that κ is independent of l and I, and that κ' is independent of c and I, the equations may be integrated to obtain

$$\ln \frac{I_0}{I} = \kappa l \tag{2.1.3}$$

$$\ln \frac{I_0}{I} = \kappa' c \tag{2.1.4}$$

κ should be independent of l if the sample is homogeneous. However, for κ' to be independent of c, the molecules must behave independently; that is, the absorption must be independent of intermolecular interactions. This is a reasonable assumption at low concentrations, but may break down significantly as the concentration is increased. The assumption that κ and κ' are independent of I requires a light intensity sufficiently low that the Boltzmann distribution among energy levels remains unchanged in its presence. For optical transitions, this condition is effectively satisfied in conventional spectroscopic measurements.

With the above assumptions, (2.1.3) and (2.1.4) can be combined to give the Beer–Lambert law

$$\ln \frac{I_0}{I} = \alpha c l \tag{2.1.5}$$

where α is an absorption coefficient, which is independent of concentration and sample thickness; it is characteristic of the particular absorbing molecule. In molecular spectroscopy it is conventional to convert (2.1.5) to the base 10 and to express c in moles liter^{-1} and l in centimeters. The result is then

$$\epsilon c l = \log \frac{I_0}{I} \equiv A \tag{2.1.6}$$

where ϵ is the *molar extinction coefficient*, $\log \equiv \log_{10}$, and A is the *absorbance* or *optical density*.

2.2. The Absorption Probability

Since modern spectrophotometers measure the absorbance A directly, it is a relatively simple matter to determine ϵ experimentally. One then obtains detailed molecular information by relating ϵ to appropriate molecular parameters. We summarize the derivation of such relations here.

Let us assume we have a single molecule *in a vacuum* interacting with an electromagnetic wave propagating with velocity c in the positive Z direction with respect to fixed laboratory axes. Such a wave can be represented by the real part ($\mathcal{R}e$) of the vector potential $\mathbf{A}$:

$$\mathcal{R}e\,\mathbf{A} = \mathcal{R}e\,\mathbf{A}^0\exp\left[2\pi i\nu\left(t - \frac{Z}{c}\right)\right]$$

$$= \frac{1}{2}\left[\mathbf{A}^0\exp\left\{2\pi i\nu\left(t - \frac{Z}{c}\right)\right\} + \mathbf{A}^{0*}\exp\left\{-2\pi i\nu\left(t - \frac{Z}{c}\right)\right\}\right] \tag{2.2.1}$$

The electric ($\mathbf{E}$) and magnetic ($\mathbf{B}$) field vectors of the wave are obtained from [Born and Wolf (1959), pp. 71–73][‡]

$$\mathbf{E} = -\frac{1}{c}\frac{\partial \mathbf{A}}{\partial t}, \qquad \mathbf{B} = \nabla \times \mathbf{A}, \qquad \mathbf{B} = \mu\mathbf{H}; \tag{2.2.2}$$

(where in a vacuum the magnetic permeability $\mu = \mu_0 = 1$ so in Gaussian units $\mathbf{B} = \mathbf{H}$) and with the Coulomb gauge

$$\nabla \cdot \mathbf{A} = 0 \tag{2.2.3}$$

The polarization properties of the wave are contained in the vector amplitude $\mathbf{A}^0$. It follows immediately from (2.2.1) and (2.2.2) that $\mathbf{E}$ and $\mathbf{B}$ have equal amplitudes and are perpendicular to each other as well as to the direction of propagation.

The interaction of the electromagnetic wave with a system of point charges produces a time-dependent perturbation [Griffith (1964), Appendix 4; Eyring et al. (1944), Chapter VIII],

$$\mathcal{H}' = \sum_j \frac{q_j}{m_j c}\mathcal{R}e\,\mathbf{A}_j \cdot \mathbf{P}_j \tag{2.2.4}$$

where q_j, m_j, and $\mathbf{P}_j$ are the charge, mass, and linear momentum operator respectively of the jth particle, and $\mathbf{A}_j$ is the vector potential at the jth particle. Equation (2.2.4) omits terms of the form $(q^2/2mc^2)\,\mathbf{A}\cdot\mathbf{A}$ as well

‡In our derivations we use the Gaussian system of units with the Coulomb gauge. In this system, electrical quantities are measured in electrostatic units and magnetic quantities in electromagnetic units. Tables relating Gaussian units to SI units may be found in contemporary texts on electricity and magnetism.

as all reference to spin. The term in A^2, which gives rise to nonlinear optical effects, is negligible for conventional, incoherent light sources. The omission of spin is remedied in Section 2.9.

Let us designate the molecular Hamiltonian by $\mathcal{H}^0$ and assume that the molecule is initially in stationary state $|a\rangle$. We wish to calculate the probability per unit time that the system, under the influence of the (time-dependent) perturbation (2.2.4), will absorb energy and undergo transition to the stationary state $|j\rangle$. We employ standard semiclassical radiation theory whereby the atom is treated quantum-mechanically in its interaction with a classical electromagnetic field [Eyring et al. (1944), Chapter VIII]. In a fully quantum-mechanical treatment the field is also quantized. [A concise description of a fully quantum-mechanical treatment may be found in Griffith (1964), Chapter 3. A detailed treatment may be found in Loudon (1973).] For conventional light sources $\mathcal{H}'$ is a small perturbation, and we use time-dependent perturbation theory assuming we know the eigenfunctions [$|j^0(t)\rangle$ and $|j\rangle$] of $\mathcal{H}^0$, the molecular Hamiltonian in the absence of the radiation field. Thus we write

$$|a(t)\rangle = \sum_j C_{aj}(t)|j^0(t)\rangle \tag{2.2.5}$$

where the $|j^0(t)\rangle$ satisfy the time-dependent Schrödinger equation,

$$\mathcal{H}^0|j^0(t)\rangle = -\frac{h}{2\pi i}\frac{\partial|j^0(t)\rangle}{\partial t} \tag{2.2.6}$$

It follows that

$$|j^0(t)\rangle = |j\rangle\exp(-2\pi i E_j t/h)$$

where $|j\rangle$ is a member of a complete set of time-independent eigenfunctions satisfying

$$\mathcal{H}^0|j\rangle = E_j|j\rangle \tag{2.2.7}$$

Equation (2.2.5) describes the time evolution of the state $|a(t)\rangle$ under the influence of the perturbation $\mathcal{H}'$ [Pauling and Wilson (1935), Chapter XI, Section 39]. $|a(t)\rangle$ must satisfy the time-dependent Schrödinger equation,

$$(\mathcal{H}^0 + \mathcal{H}')|a(t)\rangle = -\frac{h}{2\pi i}\frac{\partial|a(t)\rangle}{\partial t} \tag{2.2.8}$$

Substituting (2.2.5) into (2.2.8), using (2.2.6), and taking the scalar product, one obtains the familiar result,

$$\frac{dC_{aj}(t)}{dt} = -\frac{2\pi i}{h}\sum_{j'} C_{aj'}(t)\langle j|\mathcal{H}'|j'\rangle \exp\left[\frac{2\pi i(E_j - E_{j'})t}{h}\right] \tag{2.2.9}$$

Let us assume that the system is in state $|a^0(t)\rangle$ at time $t = 0$, so that

$$C_{aj}(t=0) \equiv C_{aj}(0) = \delta_{aj} \tag{2.2.10}$$

For a sufficiently small time interval $0 < t \leqslant t'$, (2.2.10) continues to apply *as a first approximation*. Using (2.2.1), we substitute (2.2.4) and (2.2.10) into (2.2.9). Integration from 0 to t' gives

$$\begin{aligned} C_{aj}(t') = \frac{\pi}{h}\Bigg[&\langle j|\sum_k \frac{q_k}{m_k c}\mathbf{A}^0 \cdot \mathbf{P}_k \exp\left(-\frac{2\pi i\nu Z_k}{c}\right)|a\rangle \\ &\times\left(\frac{1-\exp\left[2\pi i(\nu_{aj}+\nu)t'\right]}{2\pi(\nu_{aj}+\nu)}\right) \\ &+\langle j|\sum_k \frac{q_k}{m_k c}\mathbf{A}^{0*} \cdot \mathbf{P}_k \exp\left(\frac{2\pi i\nu Z_k}{c}\right)|a\rangle \\ &\times\left(\frac{1-\exp\left[2\pi i(\nu_{aj}-\nu)t'\right]}{2\pi(\nu_{aj}-\nu)}\right)\Bigg] \end{aligned} \tag{2.2.11}$$

where $\nu_{aj} \equiv (E_j - E_a)/h$.

Designating by $P_{a\to j}$ the probability per unit time that a transition takes place from $|a\rangle$ to $|j\rangle$ with absorption of a photon, we note that

$$P_{a\to j} = \frac{|C_{aj}(t')|^2}{t'} \tag{2.2.12}$$

For absorption, $\nu_{aj} > 0$ and thus C_{aj} is appreciable only when $\nu \approx \nu_{aj}$; therefore the first term in (2.2.11) may be dropped before (2.2.11) is substituted into (2.2.12). The result is

$$P_{a\to j} = \frac{\pi^2}{t'h^2}\left|\langle j|\sum_k \frac{q_k}{m_k c}\mathbf{A}^{0*}\cdot\mathbf{P}_k\exp\left(\frac{2\pi i\nu Z_k}{c}\right)|a\rangle\right|^2$$
$$\times\left|\frac{1-\exp[2\pi i(\nu_{aj}-\nu)t']}{2\pi(\nu_{aj}-\nu)}\right|^2 \tag{2.2.13}$$

where

$$\pi^2\left|\frac{1-\exp[2\pi i(\nu_{aj}-\nu)t']}{2\pi(\nu_{aj}-\nu)}\right|^2 = \frac{\sin^2[\pi(\nu_{aj}-\nu)t']}{(\nu_{aj}-\nu)^2} \tag{2.2.14}$$

Equation (2.2.13) expresses the probability of transition per unit time *under the influence of perfectly monochromatic radiation* of frequency ν_{aj}. Such radiation cannot be obtained in practice; all light sources provide a range of frequencies over which $P_{a\to j}$ must be summed. Furthermore, (2.2.14) defines a *very* sharp function of frequency whose width goes as $1/t'$ and whose area is proportional to t' [Schiff (1968), Section 35; Kauzmann (1957), p. 641]. It can be shown [Kauzmann (1957)] (1) that the transitions of interest occur over a time interval $0 \leqslant t \leqslant t'$, which is sufficiently long that the frequency distribution of any spectroscopic source is very broad compared to the width of (2.2.14), but (2) that t' is nevertheless still sufficiently small that (2.2.10) continues to apply.

Thus we may integrate (2.2.13) over a range of frequencies in the vicinity of ν_{aj}, *regarding* $\exp(2\pi i\nu Z_k/c)$ *as having the fixed value* $\exp(2\pi i\nu_{aj}Z_k/c)$. Since

$$\int_{-\infty}^{+\infty}\frac{\sin^2 ax}{x^2}\,dx = a\pi,$$

we obtain finally

$$P_{a\to j} = \frac{\pi^2}{h^2}\left|\langle j|\sum_k \frac{q_k}{m_k c}\mathbf{A}^{0*}\cdot\mathbf{P}_k\exp\left(\frac{2\pi i\nu_{aj}Z_k}{c}\right)|a\rangle\right|^2 \tag{2.2.15}$$

We see from this treatment that (2.2.14) is behaving like the Dirac delta function. In fact [Heitler (1954), pp. 69–71]

$$\delta(\nu_{aj}-\nu) = \lim_{t'\to\infty}\frac{\sin^2[\pi(\nu_{aj}-\nu)t']}{\pi^2 t'(\nu_{aj}-\nu)^2} \tag{2.2.16}$$

Using this form leads immediately from (2.2.13) to (2.2.15). Despite appearances, this does *not* contradict our assumption in obtaining (2.2.11) that t' is small.

2.3. Lineshapes

The treatment just outlined is incapable of giving finite linewidths. This is a direct consequence of the assumption that $C_{aa}(t) = 1$ for $0 \leqslant t \leqslant t'$. A finite linewidth can be introduced into the theory by dropping this assumption [Heitler (1954), p. 182]. In practice, however, it is very difficult to calculate accurate *a priori* lineshapes, since these are determined by complex intermolecular interactions. Fortunately the precise form of the lineshape function never proves a matter of crucial importance to us. We designate this function by a density-of-states function $\rho_{aj}(\nu)$, and thus write (2.2.15) as

$$P_{a\to j} = \frac{\pi^2}{h^2}\left|\langle j|\sum_k \frac{q_k}{m_k c}\mathbf{A}^{0*}\cdot\mathbf{P}_k \exp\left(\frac{2\pi i\nu Z_k}{c}\right)|a\rangle\right|^2 \rho_{aj}(\nu) \quad (2.3.1)$$

In the limit of an infinitely sharp line, $\rho_{aj}(\nu) = \delta(\nu_{aj} - \nu)$, so that

$$\int_0^\infty \rho_{aj}(\nu)\, d\nu = 1. \quad (2.3.2)$$

2.4. Induced and Spontaneous Emission

Let us consider the *induced emission* process. We may easily adapt the preceding analysis if we simply regard $|a\rangle$ as the (higher-energy) emitting state and $|j\rangle$ as the (lower-energy) ground state. However, since $\nu_{aj} = -\nu_{ja}$ is now a *negative* number [since $\nu_{aj} \equiv (E_j - E_a)/h$], the resonance denominator in (2.2.11) occurs in the first term rather than the second. Thus following the analysis through to (2.3.1), we obtain for the probability of *induced emission* $j \leftarrow a$,

$$P_{j\leftarrow a} = \frac{\pi^2}{h^2}\left|\langle j|\sum_k \frac{q_k}{m_k c}\mathbf{A}^{0}\cdot\mathbf{P}_k \exp\left(\frac{2\pi i\nu Z_k}{c}\right)|a\rangle\right|^2 \rho_{ja}(\nu) \quad (2.4.1)$$

where $\nu_{ja} > 0$. (Throughout this book, in nomenclature of the form $a \rightarrow j$, $a \leftarrow j$, and so on, we follow the convention that *the symbol on the left always designates the lower energy state, so that the direction of the arrow distinguishes absorption and emission.*) Thus the only difference in viewing a given transition from $|a\rangle$ to $|j\rangle$ as an emission rather than an absorption process is the substitution of $\mathbf{A}^0$ for $\mathbf{A}^{0*}$ and ρ_{ja} for ρ_{aj}. In cases where $\mathbf{A}^0$ is real (as it is for linearly polarized or unpolarized light), the probability of inducing a transition is the same in absorption and emission. However, in the case of circularly (or elliptically) polarized light, the polarization vector is complex, and one must take care to distinguish $\mathbf{A}^0$ and $\mathbf{A}^{0*}$ (Section 2.6). From now on we do not distinguish between ν_{aj} and ν_{ja}, and require always that $\nu_{aj} \equiv \nu_{ja} > 0$. We show in Section 2.8 that in all cases

$$P_{a\to j} = P_{a\leftarrow j} \tag{2.4.2}$$

That is, for radiation of any arbitrary polarization, the probability of inducing a transition $a \to j$ (in absorption) is exactly equal to the probability of inducing the reverse transition $a \leftarrow j$ (in emission). [Strictly speaking, (2.4.2) applies only after an integration over the density-of-states function is performed using (2.3.2), since in general $\rho_{a\to j} \not\equiv \rho_{a\leftarrow j}$. This integration is assumed to be implicit in (2.4.2), and we write $\rho_{a\to j} = \rho_{a\leftarrow j} = \rho_{aj}$ unless the distinction between them has practical significance.]

The semiclassical treatment we have presented is not able to account for spontaneous emission. However, it was shown by Einstein long ago [Pauling and Wilson (1935), Chapter XI, Section 40], using thermodynamic-type arguments that

$$\frac{P_{a\leftarrow j}(S)}{P_{a\leftarrow j}(I)} = \frac{8\pi h\nu_{aj}^3}{\rho(\nu_{aj})c^3} \tag{2.4.3}$$

where $P(S)$ and $P(I)$ are the probabilities for spontaneous and induced emission, and $\rho(\nu_{aj})$ is the radiation energy density [not to be confused with the lineshape function of (2.3.1)] present at frequency ν_{aj}. Equation (2.4.3) is an immediate consequence of the fully quantum-mechanical treatment of radiation [Loudon (1973)], the formulation of which was stimulated in the first place by the need to account for spontaneous emission.

2.5. Elimination of Translation and Rotation

Equation (2.3.1) is the fundamental relation connecting the absorption probability and the molecular eigenstates involved in the transition. In that equation Z_k and the coordinates in $\mathbf{P}_k$, $|a\rangle$, and $|j\rangle$ are defined with respect to an arbitrary laboratory-fixed coordinate system. We now introduce the new coordinates, $X_0, Y_0, Z_0, x_1, y_1, z_1, \ldots, x_l, y_l, z_l$, where X_0, Y_0, Z_0 specify the origin of the center of mass of the molecule with respect to the laboratory system and $x_1, \ldots, z_l$ describe the coordinates of the l particles (electrons and nuclei) *with respect to the molecular center of mass* in a coordinate system chosen parallel to the laboratory system. It follows that

$$Z_r = z_r + Z_0 \qquad (r = 1, 2, \ldots, l) \tag{2.5.1}$$

and

$$z_1 = -\frac{1}{m_1}\sum_{r=2}^{l} m_r z_r \tag{2.5.2}$$

Equation (2.5.2) expresses the center of mass condition $\bar{z} \equiv \Sigma_{r=1}^{l} m_r z_r / \Sigma_{r=1}^{l} m_r = 0$, and arbitrarily eliminates the coordinates of particle 1, leaving the $3l$ independent variable, $X_0, Y_0, Z_0, x_2, \ldots, z_l$. Application of the chain rule now leads to the relationships [Bunker (1979), Chapter 6],

$$\frac{\partial}{\partial Z_1} = \frac{m_1}{M}\left[\frac{\partial}{\partial Z_0} - \sum_{r=2}^{l}\frac{\partial}{\partial z_r}\right]$$

$$\frac{\partial}{\partial Z_r} = \frac{m_r}{M}\left[\frac{\partial}{\partial Z_0} - \sum_{r'=2}^{l}\frac{\partial}{\partial z_{r'}}\right] + \frac{\partial}{\partial z_r}$$

$$(r = 2, 3, \ldots, l) \tag{2.5.3}$$

where M is the total mass of the molecule. Equations exactly analogous to (2.5.1)–(2.5.3) apply if the replacements $Z \to Y$, $z \to y$, or $Z \to X$, $z \to x$ are made everywhere.

If these relations are used to form the molecular kinetic-energy operator, the molecular Hamiltonian may be written [Bunker (1979), Chapter 6]

$$\mathcal{H}(X_1, \ldots, Z_l) = \mathcal{H}_{\text{tr}}(X_0, Y_0, Z_0) + \mathcal{H}_{\text{int}}(x_2, \ldots, z_l) \tag{2.5.4}$$

and so

$$\psi(X_1, \ldots, Z_l) = \psi_{\text{tr}}(X_0, Y_0, Z_0)\psi_{\text{int}}(x_2, \ldots, z_l) \tag{2.5.5}$$

where

$$\mathcal{H}_{\text{tr}} = -\frac{h^2}{8\pi^2 M}\left(\frac{\partial^2}{\partial X_0^2} + \frac{\partial^2}{\partial Y_0^2} + \frac{\partial^2}{\partial Z_0^2}\right) \tag{2.5.6}$$

Equation (2.5.6) describes a free particle,

$$\psi_{\text{tr}} = \left(\frac{1}{8\pi^3}\right)^{1/2} e^{i\mathbf{k}\cdot\mathbf{R}_0}, \tag{2.5.7}$$

where $k^2 = 8\pi^2 M E_{\text{tr}}/h^2$ (E_{tr} is the translational energy) and $\mathbf{R}_0$ is the position vector of the center of mass.

The operators in (2.3.1) which are expressed in terms of the laboratory-fixed coordinates $X_1, \ldots, Z_l$ can now instead be expressed in terms of $X_0, Y_0, Z_0, x_2, \ldots, z_l$. By straightforward substitution from (2.5.3),

$$\begin{aligned}
&\sum_k \frac{q_k}{m_k c} \mathbf{A}^{0*} \cdot \mathbf{P}_k \exp(2\pi i\nu Z_k/c) \\
&= \left[\frac{A^{0*}}{cM} \cdot \left(\mathbf{P}_{\text{tr}} - \sum_{k=2}^{l} \mathbf{p}_k\right)\left(\sum_{r=1}^{l} q_r \exp\left(\frac{2\pi i\nu z_r}{c}\right)\right)\right. \\
&\quad \left. + \sum_{k=2}^{l} \frac{q_k}{m_k c} \mathbf{A}^{0*} \cdot \mathbf{p}_k \exp\left(\frac{2\pi i\nu z_r}{c}\right)\right] \exp\left(\frac{2\pi i\nu Z_0}{c}\right)
\end{aligned} \tag{2.5.8}$$

We drop the terms containing $1/M$ in (2.5.8). This is justified in general because $m_e/M \lesssim 10^{-4}$. Furthermore, in the electric-dipole approximation (Section 2.6) these terms vanish exactly for an electrically neutral molecule because $\Sigma_{r=1}^{l} q_r = 0$.

It is now convenient to change to a coordinate system referred to the *nuclear* center of mass. However, one finds [Bunker (1979), Chapter 7] that such coordinates are identical to $x_2, y_2, z_2, \ldots, z_l$ if terms of order m_e/M_N are neglected, where M_N is the total nuclear mass of the molecule. *We hereafter regard our coordinates as referred to the nuclear center of mass*, and the sum in (2.5.8) and subsequent equations runs over all electrons, and all but one nucleus, the excluded nucleus being chosen at the invariant point of the molecular point group when possible.

If (2.5.1), (2.5.5), (2.5.7), and (2.5.8) are substituted into (2.3.1), one obtains a form identical to (2.3.1) if the changes $\mathbf{P}_k \to \mathbf{p}_k$, $Z_k \to z_k$ are made, and if the right-hand side is multiplied by

$$\frac{1}{4\pi^2} \left| \langle e^{ik_j Z_0} | e^{ikZ_0} | e^{ik_a Z_0} \rangle \right|^2 \tag{2.5.9}$$

where $h\nu/c \equiv kh/2\pi$ is the linear momentum of the photon. The expression (2.5.9) is different from zero only if $k = k_j - k_a$, which simply expresses the fact that linear momentum is conserved when a photon is absorbed.

In summary, (2.3.1) and (2.4.1) continue to apply if it is understood that all coordinates (written in lowercase hereafter) are referred to a system chosen parallel to the laboratory axes, but fixed at the nuclear center of mass of the molecule, and that $|a\rangle$ and $|j\rangle$ are the molecular eigenstates ψ_{int} in (2.5.5) after translation has been eliminated.

Rotational motion is eliminated by fiat. We consider exclusively systems in condensed phases where it is assumed that molecules are "frozen" in a fixed orientation, any librational motion being accounted for by an empirical lineshape function (Section 2.3). Clearly, if a system is not a perfect single crystal, individual molecules may assume a variety of orientations. Thus it is necessary later to average various spectroscopic properties over orientation. If a random distribution of orientations is assumed, this is equivalent to treating rotational motion classically. [A detailed quantum group-theoretic treatment of rotation and many related matters may be found in the excellent treatise by Bunker (1979).]

As a matter of nomenclature, we refer to the coordinate system fixed at the nuclear center of mass but parallel to the laboratory axes as the *space-fixed* system, and to a coordinate system fixed in the nuclear framework, i.e., one that "rotates" with the molecule, as the *molecule-fixed* system.

2.6. Relation of the Absorption Coefficient to the Electric-Dipole Transition-Moment Integral

Suppose we have a collection of molecules at sufficiently high dilution that each molecule may be regarded as being in a vacuum with respect to the radiation field (e.g., as in a gas at low pressure). We designate the number of molecules per unit volume in states $|a\rangle$ and $|j\rangle$ by N_a and N_j, and subject this system to radiation of intensity $I(\nu)$. The decrease in intensity per unit length of penetration of the sample is then given by

$$-\frac{dI}{dl} = h\nu\left(N_a P_{a\to j} - N_j P_{a\leftarrow j}\right) \tag{2.6.1}$$

[Spontaneous emission is omitted because at ordinary temperatures ($\lesssim$ room temperature) $N_j \neq 0$ applies only if $\nu_{aj} \lesssim 1000\ \text{cm}^{-1}$, but under these circumstances, $P_{a\leftarrow j}(S)/P_{a\leftarrow j}(I) \approx 0$ for incoherent light sources. Furthermore, induced emission is in the direction of the light beam (which is why the second term appears in (2.6.1) in the first place). There is no such restriction on spontaneous emission, which will be completely isotropic, even for anisotropic molecules, if molecular reorientation is fast compared to the lifetime of the emitting state—see also Section 4.7.] Then using (2.1.1) and (2.4.2), we obtain

$$\kappa(\nu) = \frac{h\nu}{I(\nu)}(N_a - N_j)P_{a\to j} \tag{2.6.2}$$

Using electromagnetic theory, $I(\nu)$ is the time average ($\langle\ \rangle$) of the Poynting vector:

$$I(\nu) = \left|\frac{c}{4\pi}\left\langle \mathcal{R}e\,\mathbf{E}(\nu) \times \mathcal{R}e\mathbf{B}(\nu)\right\rangle\right| \tag{2.6.3}$$

where $\mathcal{R}e$ designates real part. We write the vector amplitude of the wave in the form

$$\mathbf{A}^0 = \boldsymbol{\pi} A^0 \tag{2.6.4}$$

where $\boldsymbol{\pi}$ is a (possibly complex) vector of unit magnitude which defines the polarization properties of the wave. In general, for light propagating along z, $\boldsymbol{\pi} = \mathbf{e}_x\pi_1 + \mathbf{e}_y\pi_2$, where $\mathbf{e}_x$ and $\mathbf{e}_y$ are unit vectors along x and y, and $|\pi_1|^2 + |\pi_2|^2 = 1$. Applying (2.2.2) to (2.2.1), substituting in (2.6.3), and recalling that $\langle \sin^2 2\pi\nu(t - z/c)\rangle = \langle \cos^2 2\pi\nu(t - z/c)\rangle = \frac{1}{2}$, we have

$$I(\nu) = \frac{(A^0)^2\pi\nu^2}{2c} = \frac{c}{8\pi}|\mathbf{E}^0|^2 \tag{2.6.5}$$

where the last relation follows from (2.2.1) and (2.2.2).

To obtain an explicit expression for κ from (2.6.2), we must evaluate $P_{a\to j}$ using (2.3.1). Employing the molecule-fixed coordinate system described in Section 2.5, we write

$$\begin{aligned}\exp\left(\frac{2\pi i\nu z_k}{c}\right) &= \exp\left(\frac{2\pi i z_k}{\lambda}\right) \\ &= 1 + \frac{2\pi i\nu z_k}{c} + \cdots\end{aligned} \tag{2.6.6}$$

We drop higher terms in this expansion, since for optical frequencies $\lambda \sim (2\text{–}10) \times 10^3$ Å, whereas z_k for a small molecule is $\sim$ 1–10 Å. As a first approximation we keep only the first term. Using (2.6.4) and the identity [e.g., see Schiff (1968), Chapter 11: Section 44, the discussion leading to (44.20)]

$$\langle j|\mathbf{p}_k|a\rangle = \frac{2\pi i}{h} m_k(E_j - E_a)\langle j|\mathbf{r}_k|a\rangle \tag{2.6.7}$$

which relates "dipole velocity" and "dipole length", we obtain

$$P_{a\to j} = \frac{4\pi^4 (A^0)^2 \nu^2}{h^2 c^2} |\langle j|\mathbf{m}\cdot\boldsymbol{\pi}^*|a\rangle|^2 \rho_{aj}(\nu)$$

with

$$\mathbf{m} \equiv \sum_k e_k \mathbf{r}_k \tag{2.6.8}$$

where **m** is the familiar electric dipole operator. Noting from (2.2.1) and (2.2.2) that

$$\mathbf{E} = -\frac{2\pi i\nu}{c}\mathbf{A}$$

and

$$\mathbf{E}^0 = -\frac{2\pi i\nu}{c}\mathbf{A}^0 \tag{2.6.9}$$

we see that the operator in (2.6.8) is of the form $\mathbf{m}\cdot\mathbf{E}^{0*}$, thus representing the interaction of an electric dipole with the electric field of the electromagnetic wave. Substituting (2.6.5) and (2.6.8) into (2.6.2) yields

$$\kappa(\nu) = \frac{8\pi^3 \nu}{hc}(N_a - N_j)|\langle j|\mathbf{m}\cdot\boldsymbol{\pi}^*|a\rangle|^2 \rho_{aj}(\nu) \tag{2.6.10}$$

The matrix element $\langle j|\mathbf{m}\cdot\boldsymbol{\pi}^*|a\rangle$ is referred to as an *electric-dipole transition moment* integral (often abbreviated transition moment), and transitions governed by this quantity are called *electric-dipole transitions*, or are sometimes referred to as arising from "electric dipole radiation".

We now evaluate (2.6.10) for waves in various states of polarization; those of particular interest to us are defined in Table 2.6.1 along with values of the corresponding transition moments. In evaluating $|\langle j|\mathbf{m}\cdot\boldsymbol{\pi}^*|a\rangle|^2$ for xy unpolarized light, one must average over ϕ, the angle between the electric vector of the light wave and the x axis. The cross-term involving $\cos\phi\sin\phi$ then vanishes. We define

$$m_{\pm} \equiv \frac{1}{\sqrt{2}}(m_x \pm i m_y) \tag{2.6.11}$$

Note carefully, however, that $m_{\pm}$ *differ in phase* from $m_{\pm 1}$ of (4.5.15), which have the standard vector-operator form defined in (9.8.7). Likewise, the

Table 2.6.1. The Absolute Value Squared of the Transition Moment for Different Light Polarizations [see (2.6.4) and (2.6.10)][a]

Polarization	$\boldsymbol{\pi} \equiv \mathbf{e}_x \pi_1 + \mathbf{e}_y \pi_2$	$\lvert\langle j \lvert \mathbf{m} \cdot \boldsymbol{\pi}^* \rvert a\rangle\rvert^2$
Linear along x	$\mathbf{e}_x$	$\lvert\langle j \lvert m_x \rvert a\rangle\rvert^2$
Linear along y	$\mathbf{e}_y$	$\lvert\langle j \lvert m_y \rvert a\rangle\rvert^2$
Right circular	$(\mathbf{e}_x + i\mathbf{e}_y)/\sqrt{2}$	$\lvert\langle j \lvert m_- \rvert a\rangle\rvert^2$
Left circular	$(\mathbf{e}_x - i\mathbf{e}_y)/\sqrt{2}$	$\lvert\langle j \lvert m_+ \rvert a\rangle\rvert^2$
xy unpolarized	$\mathbf{e}_x \cos\phi + \mathbf{e}_y \sin\phi$ averaged over all ϕ	$\frac{1}{2}\{\lvert\langle j \lvert m_x \rvert a\rangle\rvert^2 + \lvert\langle j \lvert m_y \rvert a\rangle\rvert^2\}$

[a]Note: $\mathbf{e}_x$, $\mathbf{e}_y$, and $\mathbf{e}_z$ are unit vectors in a right-handed coordinate system and, ϕ is the azimuthal angle in the xy plane. $m_\pm \equiv (1/\sqrt{2})(m_x \pm i m_y)$, $i \equiv \sqrt{-1}$.

expressions $(1/\sqrt{2})(\mathbf{e}_x \pm i\mathbf{e}_y)$ of Table 2.6.1 differ in phase from $\mathbf{e}_{\pm 1}$ of Section 9.8. It should also be emphasized that $m_\pm$ (and $m_{\pm 1}$) are *not* Hermitian. In particular

$$\langle a \lvert m_\pm \rvert j\rangle = \langle j \lvert m_\mp \rvert a\rangle^* \tag{2.6.12}$$

For a molecule in an isotropic radiation bath (*not* the usual situation in a spectroscopic measurement), $\kappa(\nu)$ is given by (2.6.10) with

$$\begin{aligned}\lvert\langle j \lvert \mathbf{m} \cdot \boldsymbol{\pi}^* \rvert a\rangle\rvert^2 &= \tfrac{1}{3}\left\{\lvert\langle j \lvert m_x \rvert a\rangle\rvert^2 + \lvert\langle j \lvert m_y \rvert a\rangle\rvert^2 + \lvert\langle j \lvert m_z \rvert a\rangle\rvert^2\right\} \\ &\equiv \tfrac{1}{3}\lvert\langle j \lvert \mathbf{m} \rvert a\rangle\rvert^2 \end{aligned} \tag{2.6.13}$$

This result is obtained by averaging over all directions and polarizations, whereupon again all cross-terms vanish [Griffith (1964), p. 53].

2.7. Magnetic-Dipole and Electric-Quadrupole Transition Moments

Let us now also include the second term in the expansion of (2.6.6), which gives rise to magnetic-dipole and electric-quadrupole transitions. Substituting (2.6.6) into (2.3.1) gives

$$P_{a\to j} = \frac{\pi^2(A^0)^2}{h^2}\left|\langle j|\sum_k \frac{q_k}{m_k c}\boldsymbol{\pi}^*\cdot\mathbf{p}_k\left(1+\frac{2\pi i\nu z_k}{c}\right)|a\rangle\right|^2 \rho_{aj}(\nu) \quad (2.7.1)$$

We note that (2.2.3) requires that the light wave be transverse; that is, for a wave propagating in the z direction, the polarization vector must always lie in the xy plane. Then

$$\begin{aligned} p_x z &= \tfrac{1}{2}p_x z + \tfrac{1}{2}p_x z \\ &= \tfrac{1}{2}(p_x z - xp_z) + \tfrac{1}{2}(p_x z + xp_z) \\ &= \tfrac{1}{2}l_y + \tfrac{1}{2}(p_x z + xp_z) \end{aligned} \quad (2.7.2)$$

where $\mathbf{l}$ is the angular-momentum operator. Using the same method which gave rise to (2.6.7), we obtain

$$\langle j|p_x z + xp_z|a\rangle = \frac{2\pi i m}{h}(E_j - E_a)\langle j|xz|a\rangle \quad (2.7.3)$$

Similarly,

$$p_y z = -\tfrac{1}{2}l_x + \tfrac{1}{2}(p_y z + yp_z) \quad (2.7.4)$$

and (2.7.3) applies with x replaced everywhere by y. We note from (2.2.1) and (2.2.2) that

$$B_x = \frac{2\pi i\nu}{c}A_y, \qquad B_y = -\frac{2\pi i\nu}{c}A_x$$

and thus,

$$B_x^0 = \frac{2\pi i\nu}{c}\pi_2 A^0, \qquad B_y^0 = -\frac{2\pi i\nu}{c}\pi_1 A^0 \quad (2.7.5)$$

and

$$\frac{\partial\mathbf{E}}{\partial z} = -\frac{2\pi i\nu}{c}\frac{\partial\mathbf{A}}{\partial z} = -\frac{2\pi i\nu}{c}\mathbf{E}$$

Using (2.6.8) and (2.7.5) and substituting (2.7.2) and (2.7.4) in (2.7.1), we obtain

$$P_{a\to j} = \frac{\pi^2}{h^2}\bigg|\langle j|\mathbf{m}\cdot\mathbf{E}^{0*}|a\rangle + \langle j|\boldsymbol{\mu}\cdot\mathbf{B}^{0*}|a\rangle$$

$$+ \langle j|\frac{i\pi\nu}{c}E^{0*}(\pi_1^* Q_{xz} + \pi_2^* Q_{yz})|a\rangle\bigg|^2 \rho_{aj}(\nu) \qquad (2.7.6)$$

where $\boldsymbol{\mu}$ is the magnetic dipole moment operator and Q_{xz} and Q_{yz} are components of the quadrupole-moment operator:

$$\boldsymbol{\mu} \equiv \sum_k \frac{q_k}{2m_k c}\mathbf{l}_k \approx -\frac{e}{2m_e c}\sum_k \mathbf{l}_k$$

$$Q_{\alpha\beta} \equiv \sum_k q_k\left(r_{k\alpha}r_{k\beta} - \frac{\mathbf{r}_k\cdot\mathbf{r}_k}{3}\delta_{\alpha\beta}\right) \qquad (2.7.7)$$

$\alpha, \beta = x, y, z$ and $r_{kx} \equiv x_k$, and so on. Since $m_e/M \lesssim 10^{-4}$, we drop the sum over nuclei in the expression for $\boldsymbol{\mu}$; e is the charge of the proton and m_e the mass of the electron.

The first term in (2.7.6) is the electric-dipole term already discussed. The second and third terms are the *magnetic-dipole* and *electric-quadrupole* terms, which arise respectively from the interaction of the magnetic field of the electromagnetic wave with the molecular magnetic-moment and the interaction of the electric-field gradient with the molecular quadrupole-moment.

Substituting (2.7.6) into (2.6.2) using (2.2.1), (2.2.2), and (2.6.5), and expressing the result in vector form, we obtain

$$\kappa(\nu) = \frac{8\pi^3\nu}{hc}(N_a - N_j)\bigg|\langle j|\boldsymbol{\pi}^*\cdot\mathbf{m}|a\rangle + \langle j|(\mathbf{e}_z\times\boldsymbol{\pi}^*)\cdot\boldsymbol{\mu}|a\rangle$$

$$+ \frac{i\pi\nu}{c}\langle j|\pi_1^* Q_{xz} + \pi_2^* Q_{yz}|a\rangle\bigg|^2 \rho_{aj}(\nu) \qquad (2.7.8)$$

where $\mathbf{e}_z$ is a unit vector along z, and π_1 and π_2 are the x and y components of the unit polarization vector $\boldsymbol{\pi}$ defined in (2.6.4). Results for various polarizations may be obtained by substituting appropriate values of $\boldsymbol{\pi}$ from Table 2.6.1. For xy unpolarized radiation, an average over all ϕ must be performed, with the result

$$\kappa(\nu) = \frac{8\pi^3\nu}{2hc}(N_a - N_j)\Big\{|\langle j|m_x|a\rangle|^2 + |\langle j|m_y|a\rangle|^2$$

$$+ |\langle j|\mu_x|a\rangle|^2 + |\langle j|\mu_y|a\rangle|^2$$

$$+ \frac{\pi^2\nu^2}{c^2}\left(|\langle j|Q_{xz}|a\rangle|^2 + |\langle j|Q_{yz}|a\rangle|^2\right)\Big\}\rho_{aj}(\nu) \tag{2.7.9}$$

Once again all cross terms vanish.‡

Finally, in the case of an isotropic radiation bath, averaging over all orientations and polarizations yields

$$\kappa(\nu) = \frac{8\pi^3\nu}{3hc}(N_a - N_j)\Big[|\langle j|\mathbf{m}|a\rangle|^2 + |\langle j|\boldsymbol{\mu}|a\rangle|^2$$

$$+ \frac{3\pi^2\nu^2}{10c^2}\sum_{\alpha,\beta}|\langle j|Q_{\alpha\beta}|a\rangle|^2\Big]\rho_{aj}(\nu) \tag{2.7.10}$$

2.8. Demonstration That $P_{a\to j} = P_{a\leftarrow j}$

We may now easily demonstrate the validity of (2.4.2). Comparing (2.4.1) and (2.3.1), we note that for a given transition $a \leftrightarrow j$, $P_{a\leftarrow j}$ differs from $P_{a\to j}$ only in the interchange of $|a\rangle$ and $|j\rangle$ and the substitution of $\boldsymbol{\pi}$ for $\boldsymbol{\pi}^*$. Thus consider an electric-dipole transition; in absorption ($a \to j$),

$$P_{a\to j} \sim |\langle j|\boldsymbol{\pi}^* \cdot \mathbf{m}|a\rangle|^2 \rho_{aj}(\nu) \tag{2.8.1}$$

whereas in the corresponding emission ($a \leftarrow j$),

$$P_{a\leftarrow j} \sim |\langle a|\boldsymbol{\pi} \cdot \mathbf{m}|j\rangle|^2 \rho_{aj}(\nu) \tag{2.8.2}$$

‡It must be emphasized that (2.7.9) applies with the understanding that a summation over all degenerate components of the transition ($\Sigma_{a\to j}$) is implied. In such a case, cross-terms such as $\mathcal{R}e\langle j|m_x|a\rangle\langle j|\mu_y|a\rangle^*$ always vanish. This is obvious if $|j\rangle$ and $|a\rangle$ are real (nondegenerate), since μ_y is pure imaginary and m_x is real. It also applies in degenerate cases provided one sums over all degenerate components of $|a\rangle$ and $|j\rangle$.

If $\boldsymbol{\pi}$ is real, $\boldsymbol{\pi} \cdot \mathbf{m}$ is Hermitian (since $\mathbf{m}$ is Hermitian); thus $P_{a \to j} = P_{a \leftarrow j}$ follows immediately from (2.8.1) and (2.8.2). If $\boldsymbol{\pi}$ is complex,

$$\langle a|\boldsymbol{\pi} \cdot \mathbf{m}|j\rangle = \langle j|\boldsymbol{\pi}^* \cdot \mathbf{m}|a\rangle^* \tag{2.8.3}$$

once again proving (2.4.2). Equation (2.8.3) is easily proved by letting $\boldsymbol{\pi} = \mathbf{e}_x\pi_1 + \mathbf{e}_y\pi_2$, where π_1 and π_2 may be any arbitrary complex numbers consistent with $|\boldsymbol{\pi}|^2 = 1$. Then (2.8.3) follows from the Hermitian properties of m_x and m_y. Equation (2.6.12) is a special case of (2.8.3). Clearly, the same arguments apply for magnetic dipole and electric quadrupole transitions. A complex polarization vector $\boldsymbol{\pi}$ in general represents an elliptically polarized wave. (Such waves are discussed in Section 4.1).

2.9. Inclusion of Electron Spin

We must now repair one significant omission in our results, namely that arising from the neglect of electron spin. The spin magnetic moment of each electron is given by $(-ge/2m_ec)\mathbf{s}$, where $-e$ is the electron's charge and m_e its mass, and $\mathbf{s}$ is the spin angular-momentum operator. The g factor has the value $g \approx 2.00232$, but we hereafter make the approximation $g = 2$. Interaction with the electromagnetic wave is given by the Hamiltonian,

$$\mathcal{H}_s = \frac{e}{m_ec}\sum_k \mathbf{s}_k \cdot \nabla \times \mathcal{R}e\, \mathbf{A}_k \tag{2.9.1}$$

where the sum runs over only electronic coordinates, nuclear-spin effects being ignored. One adds this extra term to the Hamiltonian in (2.2.4) but (with it) retains only the *first term* in the expansion of (2.6.6). Carrying through the previous derivation then leads to precisely the same results as before [for (2.7.7)–(2.7.9) and (2.7.10)], provided only that the definition of $\boldsymbol{\mu}$ in (2.7.7) is changed to

$$\boldsymbol{\mu} = \frac{-e}{2m_ec}\sum_k(\mathbf{l}_k + 2\mathbf{s}_k) = \frac{-e}{2m_ec}(\mathbf{L} + 2\mathbf{S}) \tag{2.9.2}$$

In these equations, the orbital angular momentum of the nuclei has been neglected, and $\mathbf{L}$ and $\mathbf{S}$ are thus total *electronic* orbital and spin angular-momentum operators respectively.

The introduction of electron spin of course also gives rise to other terms, specifically spin–orbit, spin–spin, and spin–other-orbit interactions. These may however be incorporated into $\mathcal{H}^0$ [Eq. (2.2.6)] and need not be explicitly considered at present (see Section 12.1). A more subtle point is the fact that the interaction of the spin–orbit coupling operator with the electromagnetic field should properly be included in (2.9.1) to justify the use of the dipole length formula (2.6.7) in the presence of spin–orbit coupling—see L. L. Lohr (1966).

2.10. Relative Magnitudes of the Transition Moments

Magnetic-dipole and electric-quadrupole transitions are expected to be very weak compared to allowed electric-dipole transitions. Order-of-magnitude calculations [Griffith (1964), p. 55] suggest that the intensities of such transitions are 10^{-6} to 10^{-7} times those of electric-dipole transitions. "Forbidden" (vibration-induced) electric-dipole transitions (Section 3.7), however, are typically of the order of 10^{-2} to 10^{-3} of allowed transitions, and it is not uncommon to observe magnetic-dipole transitions associated with them. Thus a typical pattern with transition-metal ions is the observation of a weak magnetic-dipole origin on which are built vibronic "sidebands" whose integrated intensities are several orders of magnitude larger (see Section 3.7).

2.11. Solvent Effects: The Effective-Field Approximation

Our results still have a severe limitation, namely, they are applicable only to isolated molecules in a vacuum. In practice this situation is approximated by a dilute gas. However, in most cases we are concerned with absorbing molecules (at low concentration) imbedded in an "inert" solvent (liquid or solid). In such circumstances, significant solute–solvent interactions occur which must be taken into account. A rigorous treatment of this difficult problem is not available, and it is customary to proceed in the following approximate manner. One takes account of the presence of the solvent by noting that the electric and magnetic fields acting on an absorbing molecule are due not simply to the electromagnetic radiation (as would be the case in a vacuum), but also to the presence of solvent molecules. Thus the fields acting on the molecules are designated as *microscopic* or *effective* or *internal fields* to distinguish them from the macroscopic fields associated with the electromagnetic radiation in the medium. We then write

$$\begin{aligned} \mathbf{E}^0_{\text{mic}} &= \alpha \mathbf{E}^0_{\text{mac}} \\ \mathbf{B}^0_{\text{mic}} &= \alpha' \mathbf{B}^0_{\text{mac}} \end{aligned} \tag{2.11.1}$$

For nonmagnetic materials, the magnetic permeability is approximately unity in Gaussian units, and in such cases we assume

$$\alpha' = 1 \tag{2.11.2}$$

We may now repeat our previous derivation. The results (2.2.1)–(2.6.2) are still applicable if we bear in mind that (2.2.1) and (2.2.2) refer to the *microscopic (effective) fields at the absorbing molecule*. However, in calculating the intensity of the incident radiation, we use the macroscopic field propagating through the medium with velocity c/n:

$$\mathcal{R}e\,\mathbf{A}_{\text{mac}} = \mathcal{R}e\,\mathbf{A}^0_{\text{mac}}\exp\left[2\pi i\nu\left(t - \frac{nZ}{c}\right)\right] \tag{2.11.3}$$

Equation (2.11.3) is written in analogy to (2.2.1), and n is the constant refractive index of the medium. Using (2.6.3) and applying (2.2.2) to (2.11.3), one obtains in place of (2.6.5)

$$I(\nu) = \frac{n\left(A^0_{\text{mac}}\right)^2\pi\nu^2}{2c} = \frac{nc}{8\pi}\left|\mathbf{E}^0_{\text{mac}}\right|^2 \tag{2.11.4}$$

Substituting (2.11.4) and (2.7.6) into (2.6.2), let us first consider the electric-dipole transition moment. Noting that E^0 in (2.7.6) is E^0_{mic} and using the first of (2.11.1), Eq. (2.6.10) applies with the right-hand side multiplied by α^2/n. Similarly, the magnetic-dipole term is handled by noting from (2.2.2) and (2.11.3) that $|\mathbf{B}_{\text{mic}}| = |\mathbf{B}_{\text{mac}}| = n|\mathbf{E}^0_{\text{mac}}|$, and the electric-quadrupole term must have the same correction factor to maintain an origin-independent absorption coefficient [Stephens (1970), Appendix]. The final result in place of (2.7.8) is

$$\kappa(\nu) = \frac{8\pi^3\nu}{hc}(N_a - N_j)\Bigg|\frac{\alpha}{\sqrt{n}}\langle j|\boldsymbol{\pi}^*\cdot\mathbf{m}|a\rangle$$
$$+\sqrt{n}\,[\langle j|(\mathbf{e}_z \times \boldsymbol{\pi}^*)\cdot\boldsymbol{\mu}|a\rangle$$
$$+\frac{i\pi\nu}{c}\langle j|(\pi_1^* Q_{xz} + \pi_2^* Q_{yz})|a\rangle]\Bigg|^2 \rho_{aj}(\nu) \tag{2.11.5}$$

The corresponding corrections to (2.7.9) and (2.7.10) are obvious.

In the simplest model, one assumes that the electric field acting on the molecule (E^0_{mic}) is the average microscopic field over the entire solvent in the presence of a field E^0_{mac} in the medium. This calculation was done by

Lorentz, with the result [Böttcher (1952), Section 33]

$$\alpha = \frac{n^2 + 2}{3} \tag{2.11.6}$$

where n is the (constant) refractive index of the (nonabsorbing) solvent. In this case $\mathbf{E}^0_{\text{mic}}$ is referred to as the *Lorentz effective field*. For many solvents, $n \approx 1.5$–2, in which case $\alpha^2/n \approx 1.3$–2, thus, for example, predicting that the electric-dipole absorption coefficient in the medium (solvent) is larger than its counterpart in vacuum by this factor. However, it is by no means clear that (2.11.6) is a good approximation in general. Many refinements have been considered, but the derivation of an appropriate expression for α remains a vexing problem. Clearly when strong specific solvent–solute interactions occur, a "dielectric approach," of which (2.11.6) is the simplest example, is suspect. Fortunately, it is often possible to bypass this entire problem in practice by working with ratios which are (at least approximately) independent of condensed-medium effects. We do this in MCD spectroscopy by considering the ratio of a "Faraday parameter" to the dipole strength—see Chapters 4 and 17. For this reason we usually ignore entirely the problem of condensed-medium effects. However, their existence must be kept in mind, for example, if one wishes to make quantitative intercomparisons of gas and condensed-phase absorption coefficients.

3 *Vibronic Transitions*

In this chapter we consider the form of the molecular Hamiltonian and show how it leads to selection rules for transitions between molecular eigenstates. These selection rules are strongly model-dependent, since at least one of the well-established approximations, namely the Born–Oppenheimer, the Franck–Condon, or the harmonic, is always made. It is important that the model upon which each of these approximations is based be well understood so that its suitability may be judged in particular cases.

3.1. Introduction

We begin the chapter by defining the full molecular Hamiltonian. The Born–Oppenheimer (BO) approximation is introduced, and its breakdown for degenerate or nearly degenerate states is discussed. We introduce the Jahn–Teller and pseudo-Jahn–Teller effects and define adiabatic functions. We next show how BO electronic functions and BO nuclear potentials may be approximated using perturbation theory. The electronic functions are known as Herzberg–Teller functions. This perturbation approach leads to a harmonic nuclear potential.

We then consider selection rules for electronic and vibronic transitions in the simplest case where the BO and harmonic approximations are supposed reasonable in both ground and excited states. Here we introduce the Franck–Condon (FC) approximation and discuss the bandshape expected for allowed transitions. Later, as an example, we show how the breakdown of the FC approximation explains the intensity of parity-forbidden *d–d* transitions in octahedral systems. Finally we discuss how the breakdown of the BO approximation leads to new selection rules in Jahn–Teller systems.

3.2 The Molecular Hamiltonian and the Born–Oppenheimer Approximation

The full vibronic Hamiltonian may be written

$$\mathcal{H}_T(q, Q) = -\frac{1}{2}\sum_{\eta}\frac{\partial^2}{\partial Q_\eta^2} - \frac{1}{2}\sum_i \nabla_i^2 + V_{nn}(Q) + V_{ee}(q) + V_{en}(q, Q) \tag{3.2.1}$$

where the Q_η (symbolized collectively by Q) are the normal coordinates of the nuclei, and the q_i (symbolized collectively by q) are electronic coordinates. We use atomic units; rotation and translation have been eliminated (Section 2.5). [For a discussion of normal coordinates see Wilson, Decius, and Cross (1955).] If we define $V(q, Q)$ as

$$V(q, Q) = V_{nn}(Q) + V_{ee}(Q) + V_{en}(q, Q) \tag{3.2.2}$$

and $T_n(Q)$ and $T_e(q)$ respectively as the nuclear and electronic kinetic energy operators, (3.2.1) may be rewritten

$$\begin{aligned}\mathcal{H}_T(q, Q) &= T_n(Q) + T_e(q) + V(q, Q)\\ &\equiv T_n(Q) + \mathcal{H}_{\text{el}}(q, Q)\end{aligned} \tag{3.2.3}$$

The last line defines the electronic Hamiltonian:

$$\mathcal{H}_{\text{el}}(q, Q) = T_e(q) + V(q, Q) \tag{3.2.4}$$

Note that $\mathcal{H}_{\text{el}}$ includes $V_{nn}(Q)$.

In order to solve the *vibronic Schrödinger equation*

$$\mathcal{H}_T(q, Q)\Psi_K(q, Q) = \mathcal{E}_K\Psi_K(q, Q) \tag{3.2.5}$$

various approximations must be made. The most central, the *Born–Oppenheimer (BO) approximation*, has as its basis the fact that nuclei are far more massive than electrons. The electrons are pictured as moving in a field of fixed nuclei. As the nuclei move, the electrons are assumed to adapt themselves more or less instantaneously to successive nuclear configurations. Thus the nuclei may be thought of as being fixed in a series of nuclear configurations Q, in each of which the *electronic Schrödinger equation* is

$$\mathcal{H}_{el}(q,Q)\phi_k(q,Q) = W_k(Q)\phi_k(q,Q) \tag{3.2.6}$$

Therefore, the $\phi_k(q, Q)$ depend parametrically on Q. In order to define the function $W_k(Q)$, (3.2.6) must be solved for each possible nuclear configuration Q.

In the BO approximation it is assumed that the solutions to (3.2.5) take the simple product form

$$\Psi_K(q,Q) = \Psi_{ki} = \phi_k(q,Q)\chi_{ki}(Q) \tag{3.2.7}$$

where $\chi_{ki}(Q)$, the vibrational function, is a function *only* of nuclear coordinates. If (3.2.7) is to be a sensible approximation, it must be possible to derive an equation for the nuclear motion. To do this we substitute (3.2.7) into (3.2.5). It is clear from (3.2.4) that $\mathcal{H}_{el}$ commutes with $\chi_{ki}(Q)$. Let us also assume that $T_n(Q)$ commutes (approximately) with $\phi_k(q, Q)$. Thus

$$T_n(Q)\phi_k(q,Q)\chi_{ki}(Q) \approx \phi_k(q,Q)T_n(Q)\chi_{ki}(Q) \tag{3.2.8}$$

Using (3.2.7), (3.2.6), and (3.2.8) in (3.2.5), we obtain

$$\begin{aligned}\mathcal{H}_T(q,Q)\phi_k(q,Q)\chi_{ki}(Q) &= \phi_k(q,Q)[T_n(Q) + W_k(Q)]\chi_{ki}(Q) \\ &= \mathcal{E}_{ki}\phi_k(q,Q)\chi_{ki}(Q) \end{aligned} \tag{3.2.9}$$

The last equality in (3.2.9) defines the BO nuclear equation:

$$[T_n(Q) + W_k(Q)]\chi_{ki}(Q) = \mathcal{E}_{ki}\chi_{ki}(Q) \tag{3.2.10}$$

Thus *the BO approximation requires that the electronic function* $\phi_k(q, Q)$ *be a slowly varying function of nuclear coordinates* so that (3.2.8) is a good approximation. Equation (3.2.10) shows that when the BO approximation is valid the nuclear motion can be pictured as taking place in a potential (or on a single potential-energy surface) $W_k(Q)$, which, by (3.2.6), is simply the electronic energy as a function of nuclear configuration. Examples of such surfaces are given in Figs. 3.6.1 and 3.6.2. The subscripts k and i in (3.2.7) et seq. take explicit account of the fact that a single vibrational function (i) is associated with each electronic state (k) in forming BO vibronic functions.

The above discussion must be modified for degenerate (or nearly degenerate) states, for reasons discussed in the next subsection; for such states it is not generally possible to obtain solutions to (3.2.5) in the simple product form (3.2.7).

Failure of the Born–Oppenheimer Approximation for Degenerate States: The Jahn–Teller Effect

Implicit in the BO discussion is the notion that each electronic state $\phi_k(q, Q)$ has its own potential surface $W_k(Q)$ which is a continuous function of each Q_η. Nuclear motion in the state is confined to that potential surface. For this to be true for degenerate eigenfunctions, the degeneracy must be maintained at all Q, not just at the potential minimum Q_0. If the degeneracy of the $\phi_k(q, Q)$ is lifted away from Q_0, (3.2.8) no longer applies and the motion is no longer confined to a single potential surface. Jahn and Teller showed that lifting of the degeneracy may occur along at least one nuclear coordinate Q_η in all orbitally degenerate electronic states in nonlinear systems. When this happens, the system is said to experience a *Jahn–Teller (JT) effect*. Spin (Kramers) doublets, which occur in odd-electron systems, are not JT-susceptible.

In JT systems the BO approximation breaks down, and it is not possible to find a set of electronic functions which obey both (3.2.6) and (3.2.8). In Section 3.8 we illustrate, following Stephens (1976), one way in which the degenerate (JT) case may be handled perturbationally. A set of basis functions which is degenerate at $Q = Q_0$ is used (where Q_0 is the potential minimum in the absence of a JT effect). These functions satisfy (3.2.8) to a good approximation but are not in general eigenfunctions of $\mathcal{H}_{\mathrm{el}}(q, Q)$, the electronic Hamiltonian defined for the degenerate case. The solution to (3.2.5) is then a sum of functions of the form (3.2.7). An alternate method uses functions which are eigenfunctions of $\mathcal{H}_{\mathrm{el}}$ at all Q as a basis. *These, however, do not commute with* T_n. We refer interested readers to Longuet-Higgins (1961).

Adiabatic Functions

Functions are termed *adiabatic* if they have the simple product form

$$\Psi(q, Q) = \phi(q, Y)\chi(Q) \tag{3.2.11}$$

where *only one* of the two functions depends on electronic coordinates. Our definition is that of Ballhausen and Hansen (1972). The electronic function may or may not depend on nuclear coordinates, depending on whether we choose $Y = Q$ or Q_0. Thus the BO functions of (3.2.7) are adiabatic. In the JT case, regardless of which of the methods above are used, vibronic eigenfunctions cannot be expressed in the simple product form of (3.2.11). Thus they are *nonadiabatic.*

Adiabatic and Diabatic Surfaces: The Adiabatic Principle

The term adiabatic is also applied to potential surfaces, and we briefly discuss the usage of this and the related term, diabatic. Suppose we start with two noninteracting, intersecting potential surfaces, as depicted by the solid curves in Fig. 3.2.1(a). The curves could for example represent two electronic states of a diatomic molecule as a function of internuclear distance, or a cut in the hyperspace of a polyatomic molecule. If these states have the same symmetry (belong to the same irreducible representation), we may imagine that a small interaction (perturbation) is now turned on so that a nonzero matrix element of $\mathcal{H}_{el}$ connects the states. A perturbational treatment [Landau and Lifshitz (1958), Section 76] shows that the two potential curves cannot cross (noncrossing rule). Near the crossing point, the two states in fact "repel" to give an avoided crossing, as shown by the dashed lines in Fig. 3.2.1(a). Thus in the presence of the perturbation, the potential surfaces are as shown in Fig. 3.2.1(b). The two surfaces in Fig.

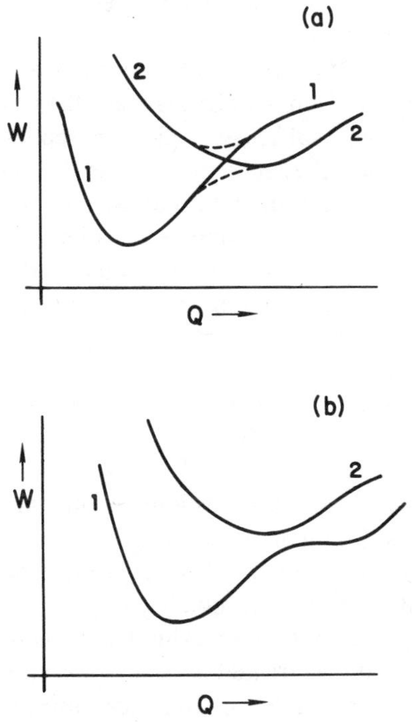

Figure 3.2.1. The solid curves 1 and 2 of (a) are diabatic surfaces. The avoided crossing produced by the perturbation is shown by the dashed lines, which give the adiabatic surfaces 1 and 2 of (b).

3.2.1(b) are called *adiabatic surfaces*, and the pair of *intersecting* surfaces in Fig. 3.2.1(a) are called *diabatic surfaces*.

This usage has its origin in the adiabatic principle [Kauzmann (1957), Chapter 14], which states: "a system will always remain in a definite quantum level if its surroundings are changed sufficiently slowly." A consequence of this principle is the following. Suppose the system started on surface 1 in Fig. 3.2.1(b), for example at the potential minimum with the nuclei held fixed ($T_n = 0$). If Q is now increased sufficiently slowly, the system will always remain on the lower surface. But as the increase in Q is accelerated, the system with increasing probability will undergo transition to the upper surface, particularly in the vicinity of the intersection point of the diabatic surfaces. Such a transition is called a *nonadiabatic* (NA) transition. If the change in Q is made sufficiently rapid, then in fact the system will move along the diabatic (the "fast") surface, namely along surface 1 in Fig. 3.2.1(a). A general expression for the probability of a NA transition in the vicinity of the intersection has been derived by Landau and Zener [Landau and Lifshitz (1958), Section 87]. A simplified treatment is given by Kauzmann (1957), Chapt. 14.

We thus see the reason for calling (3.2.11) an adiabatic function; it describes a motion confined to a single surface. In a molecule there is always nuclear motion as a consequence of the kinetic-energy operator T_n. Consequently, there is in general a finite probability of transition to another surface. If the surfaces are far apart, the NA transition probability is small and may be treated by perturbation theory (Section 3.8). In that case the BO approximation is good. If the surfaces get sufficiently close (pseudo-Jahn–Teller case) or touch (Jahn–Teller case), they mix strongly, as do the associated adiabatic functions. The BO approximation thus fails, and the stationary states of the system are described by a mixture of adiabatic functions from all the neighboring surfaces [i.e., by NA functions of the form (3.8.6) or (3.8.9)]. In such cases the nuclear motion in each vibronic state takes place over all the interacting surfaces.

The Validity of the Born–Oppenheimer Approximation for Nondegenerate States and the Pseudo-Jahn–Teller Effect

The BO approximation is good for nondegenerate states as long as they are well separated from other electronic states. Mathematically this means that the mixing of a state $\phi_k(q, Q)$ with a state $\phi_j(q, Q)$ by $\mathcal{H}_{el}(q, Q)$ must be small enough to be treated by first- and second-order nondegenerate perturbation theory. When it is not, the system is said to exhibit a *pseudo-Jahn–Teller* (*PJT*) *effect*. PJT systems may be treated by methods very similar to those used for JT systems [Englman (1972)]. Binuclear mixed-valence systems may be modeled using a one-dimensional PJT treatment which is essentially exact [Piepho, Krausz, and Schatz (1978); Wong, Schatz, and Piepho (1979); Wong and Schatz (1981)]. The model can serve as a useful introduction to the JT and PJT effects.

3.3. A Perturbation Approach to the Solution of the Electronic Born–Oppenheimer Equations in the Nondegenerate Case

Even in the absence of JT or PJT effects, solving the molecular Hamiltonian is complicated, and further approximations must be introduced. In this section we show how the electronic BO equation (3.2.6) may be simplified in the nondegenerate case using perturbation theory. The zero-order basis is chosen to be BO eigenfunctions of $\mathcal{H}_{el}(q, Q_0)$, where Q_0 symbolizes the ground-state equilibrium nuclear configuration. $\mathcal{H}_{el}(q, Q)$ is then approximated by expansion in a Taylor's series in Q where all terms cubic or higher in the nuclear coordinates are dropped. Thus

$$\mathcal{H}_{el}(q, Q) = \mathcal{H}_{el}(q, Q_0) + \Delta\mathcal{H}_{el}(q, Q), \tag{3.3.1}$$

where

$$\mathcal{H}_{el}(q, Q_0) = T_e(q) + V(q, Q_0) \tag{3.3.2}$$

and

$$\begin{aligned}\Delta\mathcal{H}_{el}(q, Q) &\equiv V(q, Q) - V(q, Q_0)\\ &\approx \sum_{\eta}\left(\frac{\partial V}{\partial Q_\eta}\right)_{Q_0} Q_\eta + \frac{1}{2}\sum_{\eta\mu}\left(\frac{\partial^2 V}{\partial Q_\eta\,\partial Q_\mu}\right)_{Q_0} Q_\eta Q_\mu\end{aligned} \tag{3.3.3}$$

Inclusion of higher terms in the Taylor expansion is inconsistent with the level of approximation already introduced in making the BO approximation [Ballhausen and Hansen (1972)]. $W_k(Q)$ is now obtained by standard perturbation theory:

$$\begin{aligned}W_k(Q) &= W_k(Q_0) + \sum_{\eta} V_\eta^{kk} Q_\eta\\ &\quad + \frac{1}{2}\sum_{\eta\mu}\left[V_{\eta\mu}^{kk} + 2\sum_{j\neq k}\frac{V_\eta^{kj}V_\mu^{jk}}{W_k(Q_0) - W_j(Q_0)}\right]Q_\eta Q_\mu\end{aligned} \tag{3.3.4}$$

where

$$V_\eta^{kj} \equiv \langle \phi_k(q, Q_0) | \left(\frac{\partial V}{\partial Q_\eta} \right)_{Q_0} | \phi_j(q, Q_0) \rangle \tag{3.3.5}$$

$$V_{\eta\mu}^{kj} \equiv \langle \phi_k(q, Q_0) | \left(\frac{\partial^2 V}{\partial Q_\eta \, \partial Q_\mu} \right)_{Q_0} | \phi_j(q, Q_0) \rangle \tag{3.3.6}$$

The first-order perturbed wavefunctions, commonly called Herzberg–Teller (HT) functions, are given by

$$\phi_k = \phi_k^0 + \sum_{j \neq k} \left[\frac{\sum_\eta \langle \phi_j^0 | \left(\frac{\partial V}{\partial Q_\eta} \right)_{Q_0} | \phi_k^0 \rangle Q_\eta}{W_k(Q_0) - W_j(Q_0)} \right] \phi_j^0 \tag{3.3.7}$$

where the notation is $\phi_k(q, Q) \equiv \phi_k$ and $\phi_k(q, Q_0) \equiv \phi_k^0$.

The $\phi_k(q, Q)$ of (3.3.7) vary slowly with Q as long as the potential surface W_k is well separated from all others. Then the BO assumption, $T_n \phi_k \chi_{ki} = \phi_k T_n \chi_{ki}$, is clearly reasonable. The equations above are obviously invalid (as is the BO approximation) in the degenerate case when $W_j(Q_0) = W_k(Q_0)$, since the denominator in (3.3.7) vanishes and nondegenerate perturbation theory fails. This is the JT case, which we examine in Section 3.8 using degenerate perturbation theory.

3.4. The Crude Adiabatic Approximation

Even the electronic functions obtained by the perturbation approach are unwieldy. Thus BO electronic functions are often further approximated as the zero-order functions of the previous section:

$$\phi_k(q, Q) \approx \phi_k(q, Q_0) \equiv \phi_k^0 \tag{3.4.1}$$

The resulting BO eigenfunctions

$$\Psi_{ki}(q, Q) = \phi_k(q, Q_0) \chi_{ki}(Q) \tag{3.4.2}$$

are termed *crude adiabatic* functions.

3.5. Solution of the Nuclear Equation in the Born–Oppenheimer Approximation

The second stage in the solution of the molecular Hamiltonian requires substitution of an explicit expression for $W_k(Q)$ into (3.2.10). The nuclear functions $\chi_{ki}(Q)$ are the vibrational eigenstates associated with the kth electronic state, and $\mathcal{E}_{ki}$ is the energy of the ith vibrational level of the kth electronic state.

The Harmonic Approximation

If no terms higher than quadratic are included in the potential, as is the case in (3.3.4), the potential $W_k(Q)$ describes a set of coupled harmonic oscillators. Thus when we use a potential of this type we are making the *harmonic approximation*. If we choose coordinates for an electronic state so that $Q = 0$ at the potential minimum (i.e., $Q_0 = 0$, which means that at the potential minimum, $Q_\eta = 0$ for all η), then all linear terms in (3.3.4) vanish. Furthermore, for any one electronic state—the usual choice is the ground state—the Q may be chosen in such a way that all cross-terms $Q_\eta Q_\mu$ ($\eta \neq \mu$) vanish. Such a set of coordinates is called *normal coordinates* [Wilson, Decius, and Cross (1955)]. For a nondegenerate ground state A in the harmonic approximation we therefore write

$$W_A(Q) = W_A(Q_0 = 0) + \frac{1}{2}\sum_\eta k_\eta^A Q_\eta^2 \tag{3.5.1}$$

The ground-state force constants k_η^A may be approximated as the bracketed term in (3.3.4) where $\eta = \mu$.

The solutions to (3.2.10) in the harmonic approximation are well known —they are harmonic-oscillator functions and eigenvalues, with which we assume readers are familiar [Wilson, Decius, and Cross (1955)]. For a nonlinear molecule containing N atoms in the kth electronic state, the vibronic energy levels are

$$\mathcal{E}_{ki} = W_k(Q_0) + \sum_\eta^{3N-6} \left(v_\eta^i + \tfrac{1}{2}\right) h\nu_\eta \tag{3.5.2}$$

Here v_η^i and ν_η are respectively the quantum number and vibrational frequency of the ηth normal mode of electronic state k, and h is Planck's constant. The label i symbolizes the set of $3N - 6$ quantum numbers, $(v_1^i, v_2^i, v_3^i, \ldots, v_{3N-6}^i)$.

Excited-state potential minima do not necessarily coincide with that of the ground state. Consequently, excited-state potentials $W_k(Q)$ in the harmonic approximation do not in general have the simple form (3.5.1), and the eigenvalues do not necessarily have the simple form (3.5.2) *when ground-state normal coordinates are used.*

The harmonic approximation usually works well for low vibrational levels. Anharmonic terms in the potential (cubic, quartic, etc.) are in part responsible for the appearance of combination and overtone bands in the vibrational spectrum [Herzberg (1945, 1966); Wilson, Decius, and Cross (1955)].

3.6. The Transition-Moment Integral in the Born–Oppenheimer Approximation

Once we have solved the vibronic Hamiltonian for both ground and excited states, we are ready to determine selection rules for electronic transitions between them. This requires calculation of the transition-moment integral introduced in the previous chapter. Since we have eliminated rotational and translational degrees of freedom in Section 2.5, our electronic transition spans a manifold of vibronic transitions between BO states, that is, transitions from a specific vibrational level of electronic state A to a specific vibrational level of electronic state J. We label the BO states

$$\begin{aligned} |A\alpha g\rangle &\equiv \phi_{A\alpha}(q,Q)\chi_g(Q) \\ |J\lambda j\rangle &\equiv \phi_{J\lambda}(q,Q)\chi_j(Q) \end{aligned} \tag{3.6.1}$$

Here α and λ label components of electronic states A and J if they are degenerate, and g and j label vibrational states associated with A and J.

Let us now evaluate a typical vibronic matrix element of m_γ where γ labels the vector component of the electric dipole operator:

$$\langle A\alpha g|m_\gamma|J\lambda j\rangle = \langle A\alpha g|m_\gamma^e(q)|J\lambda j\rangle + \langle A\alpha g|m_\gamma^n(Q)|J\lambda j\rangle \tag{3.6.2}$$

The above decomposition of m_γ into an electronic, m_γ^e, and a nuclear, m_γ^n, part follows from the definition of m_γ in (2.6.8). Substituting (3.6.1) into (3.6.2) gives

$$\begin{aligned} \langle A\alpha g|m_\gamma|J\lambda j\rangle &= \big\langle \phi_{A\alpha}(q,Q)\chi_g(Q)\big|m_\gamma^e(q)\big|\phi_{J\lambda}(q,Q)\chi_j(Q)\big\rangle \\ &= \big\langle \chi_g(Q)\big[\langle\phi_{A\alpha}(q,Q)|m_\gamma^e(q)|\phi_{J\lambda}(q,Q)\rangle\big]\chi_j(Q)\big\rangle \end{aligned} \tag{3.6.3}$$

since the term in m_γ^n vanishes because of the orthogonality of the electronic functions. Hereafter we drop the superscript e on m_γ^e. Integration within the square brackets in (3.6.3) is over electronic coordinates only, the final integration over nuclear coordinates in the full bracket being performed subsequently.

The Franck–Condon Approximation

The relative intensities of vibronic transitions may be very usefully discussed using the FC principle. This principle, in its classical form (due to Franck), states that an electronic transition takes place so rapidly in comparison with the vibrational motion, that almost no changes in the nuclear positions or momenta occur in the process. Consequently, if we represent the transition by an arrow on a potential-energy diagram, *the arrow must be vertical* (no change in nuclear position) and must connect vibrational levels which have the same nuclear momentum at the nuclear configuration specified by the arrow. Thus in Figure 3.6.1, transitions 1 and 3 would have large transition moments, since (classically, and thus neglecting zero-point motion) the nuclear momentum is zero in both vibrational states, whereas transitions 2 and 4 would have much smaller values, since

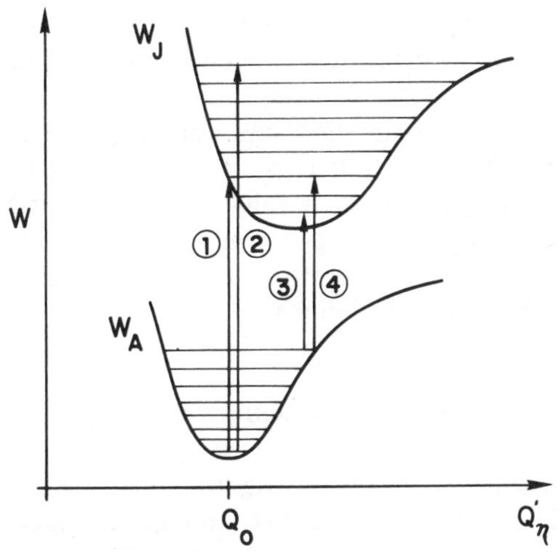

Figure 3.6.1. Potential-energy curves for the ground state A and excited state J as a function of one of the normal coordinates Q_η. The vertical arrows illustrate various Franck–Condon transitions.

the nuclei are moving rapidly in the upper vibrational states. Note also that transitions are most likely from the most probable ground-state values of Q_η (the classical turning points), and the actual intensities are weighted by Boltzmann population factors.

The quantum-mechanical formulation of this principle (due to Condon) is based on the assumption that the electronic functions are sufficiently slowly varying functions of Q that the square bracket in (3.6.3) can be approximated by its equilibrium value (Q_0) in the ground electronic state. Thus

$$\begin{aligned} \langle A\alpha g|m_\gamma|J\lambda j\rangle &\approx \big\langle \phi_{A\alpha}(q, Q_0)\big|m_\gamma(q)\big|\phi_{J\lambda}(q, Q_0)\big\rangle \big\langle \chi_g(Q)\big|\chi_j(Q)\big\rangle \\ &\equiv \langle \phi^0_{A\alpha}|m_\gamma|\phi^0_{J\lambda}\rangle \big\langle \chi_g(Q)\big|\chi_j(Q)\big\rangle \\ &\equiv \langle \phi_{A\alpha}|m_\gamma|\phi_{J\lambda}\rangle^0 \langle \chi_g|\chi_j\rangle \end{aligned} \tag{3.6.4}$$

In the second step we introduce the notation $\phi_{A\alpha}(q, Q_0) \equiv \phi^0_{A\alpha}$, and in the final step further simplify the notation. Equation (3.6.4) is the *Franck–Condon (FC) approximation.* It predicts that relative vibronic intensities are governed entirely by the Franck–Condon vibrational overlap factors, $\langle \chi_g|\chi_j\rangle$. The relation of (3.6.4) to the classical statement of the Franck–Condon principle may be readily appreciated by making sketches of the harmonic-oscillator functions involved in the transitions in Figure 3.6.1.

Note that the FC approximation says nothing about the approximation used for the BO electronic functions. It says that *whatever* the form of $\phi_{A\alpha}(q, Q)$ and $\phi_{J\lambda}(q, Q)$, only their value at Q_0 is required to evaluate the transition-moment integral. The approximation may also be used for matrix elements of other electronic operators, for example, for magnetic-moment matrix elements in MCD calculations.

Deviations from the Franck–Condon Approximation

The FC approximation is much less rigorous than the BO approximation, and its breakdown may be generally anticipated. Deviations from the FC approximation may be treated by expanding the electronic part of the transition moment integral, $\langle \phi_{A\alpha}(q, Q)|m_\gamma|\phi_{J\lambda}(q, Q)\rangle$ of (3.6.3), in a Taylor's series about the ground-state equilibrium internuclear configuration Q_0:

$$\langle \phi_{A\alpha}(q,Q)|m_\gamma|\phi_{J\lambda}(q,Q)\rangle$$

$$= \langle \phi^0_{A\alpha}|m_\gamma|\phi^0_{J\lambda}\rangle + \sum_\eta \left(\frac{\partial}{\partial Q_\eta} \langle \phi_{A\alpha}(q,Q)|m_\gamma|\phi_{J\lambda}(q,Q)\rangle \right)_{Q=Q_0} Q_\eta + \cdots$$

$$\equiv \langle A\alpha|m_\gamma|J\lambda\rangle^0 + \sum_\eta \langle A\alpha|m_\gamma|J\lambda\rangle'_\eta Q_\eta + \cdots \tag{3.6.5}$$

Note that integration here is only over electronic coordinates. To obtain selection rules, this expression must be substituted into (3.6.3). The leading term gives the FC approximation. The role of higher-order terms is discussed in Section 3.7. The Q_η are a set of coordinates describing the nuclear motion of the system. For an isolated polyatomic molecule with N atoms they would generally be the familiar $3N - 6$ vibrational normal coordinates. In all cases, the Q_η are chosen to belong to irreducible representations of the point group of the molecule.

Allowed Electronic Transitions: Bandshape in the Franck–Condon Approximation

We assume that the right-hand side of (3.6.5) is a rapidly converging series. If the first term is nonzero, we say that the transition is an *allowed electronic transition* or that it is *an electric-dipole-allowed electronic transition*. If we then make the FC approximation and ignore all other terms, (3.6.5) reduces to (3.6.4). Here $\langle \chi_g|\chi_j\rangle$, the *FC (vibrational overlap) factor*, is an overlap integral between vibrational wave functions in *two different electronic states*. In the BO–FC approximation it governs the vibronic structure of the band and hence determines its overall shape.

Inspection of (3.6.4) shows that any particular vibronic transition is allowed to the extent that its FC factor is nonzero. We note on the basis of group-theoretical arguments (see Section 9.2 et seq.) that the FC factor (or any matrix element) may be different from zero only if its integrand contains the totally symmetric irreducible representation (designated A_1) of the point group of the absorbing center.

Selection Rules When A and J Have Identical Minima and Potential Surfaces: Why Only a Single Line Is Predicted

Let us first suppose that the system is at a sufficiently low temperature that only the ground vibrational level of the ground electronic state is occupied.

Let us also assume for the moment that the two electronic states have identical potential surfaces: that is, that both show identical variation of the potential energy surface $W_k(Q)$ as a function of all the normal coordinates Q_η and have the same potential minima. We may depict this potential energy "hypersurface" by taking "cuts" along various Q_η. Two such cuts applicable to this section are shown in Fig. 3.6.2(a) and (b): one along $Q_\eta = Q_\eta(A_1)$, and one along a non-totally-symmetric mode which we always designate $Q_\eta(B)$. *We can conclude that the entire electronic band consists of a single line*. This follows because if the potential-energy hypersurfaces for the two electronic states are identical, the vibrational sets $\chi_g(Q)$ and $\chi_j(Q)$ are one and the same. Consequently the FC overlap factor of (3.6.4) is a simple orthogonality relation

$$\langle \chi_g(Q)|\chi_j(Q)\rangle = \delta_{v_g, v_j} \tag{3.6.6}$$

where v_g and v_j symbolize the degrees of excitation (that is, the set of

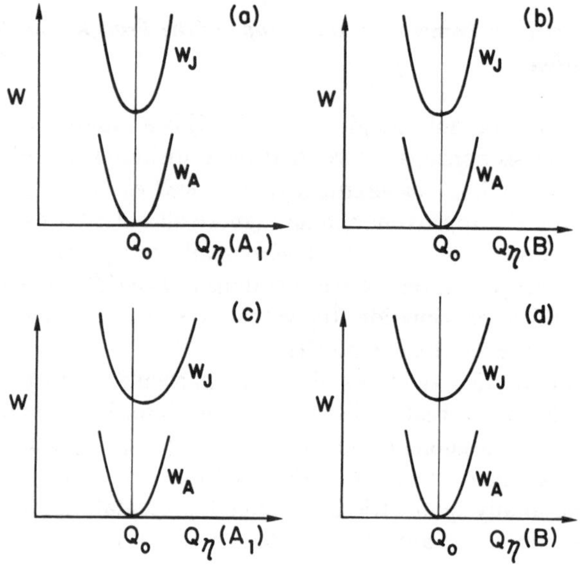

Figure 3.6.2. Potential-energy curves for the ground state A and an excited state J as a function of one of the normal coordinates. $Q_\eta(A_1)$ represents a totally symmetric mode while $Q_\eta(B)$ is non-totally-symmetric. Parts (a) and (b) show W_A and W_J with identical potential minima and force constants. In (c) W_A and W_J have different force constants and different minima; in (d), the same minimum but different force constants.

quantum numbers) of each of the vibrational modes in states A and J respectively (Section 3.5). [Here $v_g = v_j$ means $v_\eta^g = v_\eta^j$ for $\eta = 1$ to $3N - 6$ in (3.5.2).] Since by hypothesis we are in the ground vibrational state of electronic state A, we have $v_g = 0$, and from (3.6.6) we must have $v_j = v_g = 0$ for a nonzero result. And $v_g = 0$ means that *all* of the $3N - 6$ quantum numbers which define the vibrational state $\chi_g(Q)$ are zero. When $v_g = v_j = 0$, the line observed is called the 0–0 or *no-phonon line*, since it represents a vibronic transition between two electronic states in which no vibrational quanta (phonons) are excited in either state. It is sometimes referred to as the *origin* (line) of the electronic transition.

Selection Rules When A and J Have Different Minima: The Appearance of Totally Symmetric Progressions

Let us now imagine that the potential-energy hypersurfaces for A and J begin to differ. If it is assumed that the BO states A and J have the same symmetry (i.e. that the equilibrium configuration of the absorbing center belongs to the same point group in both electronic states), then (by definition) the potential minimum for each $Q_\eta(B)$ must coincide in the two states. This is illustrated in Fig. 3.6.2(d). If this were not the case, states A and J could not have the same symmetry. [If states A and J have different equilibrium symmetries, a new point group is defined by the symmetry elements which the two states have in common. Our entire subsequent discussion remains applicable if it is understood that all symmetry arguments refer to this new (lower-symmetry) point group.]

For $Q_\eta(A_1)$ there is no restriction on the relative position of the potential minima in A and J, since $\Delta Q_\eta(A_1)$ is symmetry-preserving [see Fig. 3.6.2(c)]. Potential minima at different values of $Q_\eta(A_1)$ for A and J mean different equilibrium internuclear distances and/or angles in the two states, *but the same symmetry*. Assuming such changes occur between A and J, the low-temperature spectrum, to a first (usually good) approximation, shows vibronic progressions only in totally symmetric modes. (A progression in a mode is the series of vibronic lines which arises when each successive vibrational quantum in that mode is excited. A simple example would be $0 \rightarrow 0, 0 \rightarrow 1, 0 \rightarrow 2, \ldots$, where the first number is v_η^g and the second is v_η^j. Figure 3.6.1 illustrates a progression if the five transitions between 1 and 2 are filled in. In this case, by the FC principle, transition 1 is most intense and successive lines on either side diminish in intensity. The $0 \rightarrow 0$ line will probably not be seen at all.) If we assume that the total vibrational function is a product of independent functions in each of the $3N - 6$ Q_η, then we may understand the lack of progressions in the $Q_\eta(B)$ in the following way. Since the potential minimum for each $Q_\eta(B)$ has the same value in both electronic states, orthogonality would apply as before if the potential surfaces were identical. Fortunately, this result continues to hold to a good approximation as the two surfaces begin to differ in shape. This fact and the orthogonality just referred to can be explicitly demonstrated in the harmonic approximation. In that case the vibrational wave functions can be written in the product form

$$\begin{aligned} |\chi_g(Q)\rangle &= \prod_{\eta=1}^{3N-6} |\chi_{v_\eta^g}(Q_\eta)\rangle \\ |\chi_j(Q)\rangle &= \prod_{\eta=1}^{3N-6} |\chi_{v_\eta^j}(Q_\eta)\rangle \end{aligned} \tag{3.6.7}$$

where the $\chi(Q_\eta)$ are harmonic-oscillator functions and the value of each v_η^g and v_η^j must be specified to define χ_g and χ_j uniquely [see Wilson, Decius, and Cross (1955)]. The FC factor now takes the form

$$\langle \chi_g(Q)|\chi_j(Q)\rangle = \prod_\eta \langle \chi_{v_\eta^g}(Q_\eta)|\chi_{v_\eta^j}(Q_\eta)\rangle \tag{3.6.8}$$

We note that a vibronic transition has zero intensity if any one of the $3N - 6$ FC factors vanishes. Each potential surface in Figure 3.6.2 is parabolic and is defined entirely by its force constant, $(\partial^2 W_k/\partial Q_\eta^2)_0$. An explicit formula may be derived for the FC factor of an individual oscillator when the force constants in the two states differ [see Yeakel and Schatz (1974), Eq. (A11), and Herzberg (1966), Eq. (II.29)]. One finds that even if the oscillator frequency changes by a factor of two (a factor-of-four change in the force constant)—a difference much larger than one would expect to encounter in practice—the $v_\eta^g = v_\eta^j$ ($\Delta v_\eta = 0$) transitions contain about 94% of the intensity if the oscillator potential minima coincide [see Herzberg (1966), p. 152]. Thus *to a very good approximation, when the BO approximation is valid and anharmonicity is small, we may assume that FC overlap factors are zero for non-totally-symmetric vibrations if* $\Delta v_\eta \neq 0$.

Since the potential surfaces for $Q_\eta(A_1)$ may have their relative potential minima displaced substantially (Fig. 3.6.2), there is no expectation of approximate orthogonality, and a progression is observed. The displacement of intensity (from the 0–0 line) into higher members of the progression is a measure of the change in equilibrium configuration along $Q_\eta(A_1)$ between the two electronic states. Thus, *in the BO approximation, if a molecule possesses a single* A_1 *mode, an allowed transition at low temperature should, to a first approximation, exhibit a single progression in this mode built on the no-phonon line*. The energy separation between successive members of the progression measures the A_1 vibrational frequency in the *excited* electronic state. If the molecule possesses more than one A_1 mode, then any member of one of the A_1 progressions may serve as an origin on which a progression in another A_1 mode may be built.

There are no group-theoretic restrictions on the occurrence of a progression in an A_1 mode, since a vibrational excitation of any degree in an A_1 mode is also A_1 [Wilson, Decius, and Cross (1955)]. But it is important to distinguish carefully the model-dependent and group-theoretic results in our analysis. Within the BO, the FC, and the harmonic approximations, one expects to see a spectrum almost completely dominated by totally symmetric progressions. This is a model-dependent result based on the fact that (3.6.6) usually remains a good approximation for B-type modes. However, many transitions are *permitted* in addition to the A_1 progressions, since group theory forbids a vibronic transition only if the FC integrand does not contain A_1. This is discussed well in Herzberg (1966). As the BO, the FC, and the harmonic approximations fail, these transitions gain intensity. Breakdown of the FC and BO approximations is discussed in Sections 3.7 and 3.8 respectively.

Hot Bands

As the temperature is raised, excited vibrational levels of the ground electronic state are populated, and transitions can originate from these states. Thus vibronic lines begin to appear, usually to the red of the origin; these are referred to as *hot lines* or *hot bands*, and their occurrence is illustrated in Figure 23.8.1. Again to a good first approximation, $\Delta v_\eta = 0$ applies except for $Q_\eta(A_1)$, and the A_1 progressions in the "cold" spectrum should be reflected about the origin in the hot lines. To a first approximation, corresponding partners (e.g., $0 \rightarrow 1$ and $1 \rightarrow 0$), should have intensity ratios in proportion to the Boltzmann populations of the corresponding vibrational levels in the ground electronic state. Hot lines can also appear in the "cold" spectrum region.

In the BO and FC approximations, $\Delta v_\eta \approx 0$ still applies for B modes. However, these hot lines do not coincide exactly with the corresponding cold lines if the potential surfaces in the two electronic states differ. Thus, for example, $0 \rightarrow 0, 1 \rightarrow 1, 2 \rightarrow 2, \ldots$ vibronic lines in a given mode all differ slightly in energy. Furthermore, in a progression, anharmonicity produces the same effect even if the potential surfaces are identical; i.e., $0 \rightarrow 1, 1 \rightarrow 2, 2 \rightarrow 3, \ldots$ do not coincide exactly. Again, many additional hot lines forbidden in the BO and FC approximations are not group-theory-forbidden. An extensive group-theoretical discussion of both cold and hot vibronic transitions may be found in Herzberg (1966).

In solids, vibronic lines tend to broaden rapidly with increasing temperature, so one is not usually able to resolve extensive structure in hot bands.

If the equilibrium point groups of the two electronic states involved in the transition differ, one may perfectly well use the previous analysis provided it is based on the lower-symmetry point group that the two states have in common, as was pointed out previously.

3.7. Electronically Forbidden Transitions in the Born–Oppenheimer and Franck–Condon Approximations

If the first term on the right-hand side of (3.6.5) is zero, the transition is said to be FC-forbidden or electronically forbidden, and one must consider the higher Q-dependent terms in the equation. If at least one of these is nonzero, the transition is said to be *vibrationally induced*. Discussion is usually restricted to terms linear in Q_η. Physically these (and higher) terms take account of the fact that the nuclei are not really fixed at Q_0 during an electronic transition.

How to Treat the Breakdown of the Franck–Condon Approximation: The Herzberg–Teller Approach

Our aim in this section is to find selection rules for the transition moment integral (3.6.3) in the case where the Taylor expansion of (3.6.5) is not limited by the FC approximation to a single term. Consequently, we need a systematic means of evaluating the coefficients of the terms linear in Q_η. We use the well-known *Herzberg–Teller* (*HT*) approach, which approximates the Q dependence of the electronic part of the transition moment (3.6.3) through the variation of the functions $\phi_k(q, Q)$ with Q. The method is useful as long as electronic functions vary slowly with Q. Perturbation theory is used exactly as described in Section 3.3 except that our notation is elaborated in this section to accommodate group-theoretic labels. These labels are defined in Chapter 9 and are used extensively in Chapter 22. We assume that all states belong to the same point group.

In the FC approximation, electronic functions in the transition-moment integral are approximated as eigenfunctions of $\mathcal{H}_{el}(q, Q_0)$. These functions become our zero-order functions in the HT treatment. Thus for ground and excited vibronic states we have

$$\begin{aligned} &|A\alpha\rangle^0 |g\xi\rangle \\ &|J\lambda\rangle^0 |j\zeta\rangle \end{aligned} \tag{3.7.1}$$

where $|A\alpha\rangle^0$ and $|J\lambda\rangle^0$ are eigenfunctions of $\mathcal{H}_{el}(q, Q_0)$, and $|g\xi\rangle$ and $|j\zeta\rangle$ are products of harmonic-oscillator functions as in (3.6.7), since we make the harmonic approximation. Here the function labels $A\alpha$, $J\lambda$, $g\xi$, and $j\zeta$ also serve as group-theoretic irreducible representation labels; the first and second symbols respectively designate irreducible representation and component thereof. Thus states belonging to a given degenerate set (for example, the $|A\alpha\rangle$ set) have the same irreducible representation label (A), but differ in component label (α).

Our exact electronic Hamiltonian $\mathcal{H}_{\text{el}}(q, Q)$ is approximated using a Taylor expansion in the normal coordinates, Q_η, of the system, exactly as in (3.3.1)–(3.3.3). If the Q_η are classified according to representations of the point group of the molecule using the representation labels $\eta = h\theta$, we obtain

$$\mathcal{H}_{\text{el}}(q, Q) = \mathcal{H}_{\text{el}}(q, Q_0) + \sum_{h\theta} U_{h\theta}^{\dagger} Q_{h\theta} \tag{3.7.2}$$

where $U_{h\theta} = (\partial V/\partial Q_{h\theta})_{Q_0}$ and the sum is over all irreducible representations h and their components θ to which the $3N - 6$ Q_η belong. [Since the Hamiltonian must be totally symmetric, only linear terms of the form $U_\eta^{\dagger} Q_\eta$ contribute to (3.7.2) and (3.3.3); if Q_η transforms as $h\theta$, then U must transform as the adjoint (symbolized by the dagger) of $h\theta$.] Higher-order terms in Q_η have been dropped. First-order perturbed electronic wavefunctions [the functions of (3.3.7)] are now

$$\begin{aligned} |A\alpha\rangle &= |A\alpha\rangle^0 + \sum_{h\theta} \sum_{N\varepsilon \neq A\alpha} \frac{\langle N\varepsilon | U_{h\theta}^{\dagger} | A\alpha\rangle^0}{W_A^0 - W_N^0} |N\varepsilon\rangle^0 Q_{h\theta} \\ |J\lambda\rangle &= |J\lambda\rangle^0 + \sum_{h\theta} \sum_{N\varepsilon \neq J\lambda} \frac{\langle N\varepsilon | U_{h\theta}^{\dagger} | J\lambda\rangle^0}{W_J^0 - W_N^0} |N\varepsilon\rangle^0 Q_{h\theta} \end{aligned} \tag{3.7.3}$$

where $W_A^0 = W_A(Q_0)$ and so on. To obtain the transition moment for a vibronic line as a function of Q we use these functions in (3.6.3) to obtain

$$\begin{aligned} \langle A\alpha, g\xi | m_\gamma | J\lambda, j\zeta\rangle &= \langle g\xi | (\langle A\alpha | m_\gamma | J\lambda\rangle) | j\zeta\rangle \\ &= \langle A\alpha | m_\gamma | J\lambda\rangle^0 \langle g\xi | j\zeta\rangle \\ &\quad + \sum_{h\theta} \left[\sum_{N\varepsilon \neq A\alpha} \frac{\langle A\alpha | U_{h\theta}^{\dagger} | N\varepsilon\rangle^0}{W_A^0 - W_N^0} \langle N\varepsilon | m_\gamma | J\lambda\rangle^0 \right. \\ &\quad \left. + \sum_{N\varepsilon \neq J\lambda} \langle A\alpha | m_\gamma | N\varepsilon\rangle^0 \frac{\langle N\varepsilon | U_{h\theta}^{\dagger} | J\lambda\rangle^0}{W_J^0 - W_N^0} \right] \\ &\quad \times \langle g\xi | Q_{h\theta} | j\zeta\rangle \end{aligned} \tag{3.7.4}$$

The zero superscripts indicate the integrals are evaluated at $Q = Q_0$. Comparison of this result with (3.6.5) shows that the HT method approximates $\langle A\alpha|m_\gamma|J\lambda\rangle'_{h\theta}$ with the bracketed expression and drops higher-order terms. Thus in the HT approximation we write

$$\langle A\alpha, g\xi|m_\gamma|J\lambda, j\zeta\rangle = \langle A\alpha|m_\gamma|J\lambda\rangle^0\langle g\xi|j\zeta\rangle + \sum_{h\theta}\langle A\alpha|m_\gamma|J\lambda\rangle'_{h\theta}\langle g\xi|Q_{h\theta}|j\zeta\rangle \quad (3.7.5)$$

where

$$\langle A\alpha|m_\gamma|J\lambda\rangle'_{h\theta} = [\text{bracketed expression in (3.7.4)}] \quad (3.7.6)$$

The two summation terms (over $N\varepsilon$) in (3.7.4), which represent ground- and excited-state mixing respectively, do not necessarily have the same sign, so some cancellation is possible.

Selection Rules in the Herzberg–Teller Approximation

Selection rules in the HT approximation depend on the group-theoretical selection rules for the four types of matrix elements which appear in (3.7.5). Those for $\langle A\alpha|m_\gamma|J\lambda\rangle^0$ are based on the Wigner–Eckart theorem and depend on the irreducible representations to which $^0\langle A\alpha|$, m_γ, and $|J\lambda\rangle^0$ belong. They may be determined using the methods described in Chapter 9 and following.

When $\langle A\alpha|m_\gamma|J\lambda\rangle^0 = 0$ and thus $|A\alpha\rangle^0 \rightarrow |J\lambda\rangle^0$ is electronically forbidden, it may still be magnetic-dipole-allowed. Magnetic-dipole transitions are very weak compared to electric-dipole transitions (Section 2.10). Their selection rules are found by replacing m_γ with μ_γ, the magnetic-dipole operator, and determining the irreducible representation to which it belongs. In octahedral (group O_h) systems m_γ has odd (or u = ungerade) parity, while μ_γ has even (or g = gerade) parity, so parity-forbidden electronic transitions may be magnetic-dipole-allowed.

Selection rules for $\langle g\xi|j\zeta\rangle$ have already been discussed in Section 3.6.

The HT approximation is used most frequently when $\langle A\alpha|m_\gamma|J_\lambda\rangle^0 = 0$, that is, when the electric-dipole transition moment vanishes at $Q = Q_0$. An example is the d–d transitions in octahedral (group O_h) transition-metal complexes, which are parity-forbidden at $Q = Q_0$. However, nonzero $\langle A\alpha|m_\gamma|J\lambda\rangle'_{h\theta}$ contributions to their transition moments arise as we move away from Q_0 in odd-parity vibrational modes. We recall from Section 3.6 that states A and J must have the same potential minima in all non-totally-symmetric normal coordinates, since only that configuration maintains the point-group symmetry of the molecule. But as we move away from Q_0 in either the ground or excited state, the electronic symmetry decreases, and O_h electronic selection rules break down. Away from $Q = Q_0$ in the ungerade (u) modes, parity is lost; this is described in the HT treatment above by the mixing of odd-parity $|N\varepsilon\rangle^0$ electronic states into $|A\alpha\rangle^0$ and/or $|J\lambda\rangle^0$ via coupling with the odd-parity vibrational modes. In fact $\langle A\alpha|m_\gamma|J\lambda\rangle'_{h\theta}$ is usually nonzero for some Q_η ($\eta = h\theta$).

Selection rules for $\langle A\alpha|m_\gamma|J\lambda\rangle'_{h\theta}$ in the HT approximation are found by determining, for each vibrational symmetry $h\theta$ and set of mixing states $|N\varepsilon\rangle^0$, the selection rules for the matrix elements in its expansion (3.7.6), using the Wigner–Eckart theorem as described in Chapter 9.

Finally we must determine the selection rules for the $\langle g\xi|Q_{h\theta}|j\zeta\rangle$. When the harmonic approximation is used for $\langle g\xi|$ and $|j\zeta\rangle$, these matrix elements are approximately zero unless $v^j_{h\theta} = v^g_{h\theta} \pm 1$, whereas the selection rules for the remaining vibrational quanta are exactly as described in Section 3.6. Thus, for example, if no vibrational quanta are excited in the ground vibrational state $\langle g\xi|$, as is true for systems at low temperatures, transitions are allowed to vibrational states $|j\zeta\rangle$ with one quantum of $v^j_{h\theta}$ excited ($v^j_{h\theta} = 1$) and any number of quanta of the totally symmetric mode(s) excited ($v^j_{A_1} = 0, 1, 2, \dots$). For all other modes, the approximate selection rule is $v^j_\eta = v^g_\eta = 0$. As the temperature is raised, higher vibrational levels of $\langle g\xi|$ become populated and transitions from $v^g_{h\theta} = 1$ to $v^j_{h\theta} = 0$ or 2 become possible, as do transitions from $v^g_{A_1} = 1$ to $v^j_{A_1} = 0, 1, 2, \dots$. All other modes Q_η must to a good approximation obey $v^g_\eta = v^j_\eta$, although now we may have $v^g_\eta \neq 0$. Transitions which become allowed as the temperature is increased are termed hot bands (Section 3.6).

The integrated intensity for the entire band increases with temperature, since non-totally-symmetric ground-state vibrational levels with $v^g_\eta \equiv m > 0$ gain population, and the intensity of $m \to m - 1$ and $m \to m + 1$ transitions increases with m. It can be shown [see Ballhausen (1962), p. 187, and also Section 7.10 below] that if HT coupling with a vibrational mode of frequency ν is solely responsible for the intensity of an electronic band, via the second term in (3.7.5), then the integrated band intensity at a temperature T obeys the rule

$$\text{Intensity}\,(T) = \coth\left(\frac{h\nu}{2kT}\right) \times \text{Intensity}\,(0\text{ K}) \tag{3.7.7}$$

Since the first term in (3.7.5)—the FC-allowed term—gives a temperature-independent contribution to the band intensity, [see (7.4.3)], it is often possible to identify vibrationally induced transitions by measuring the absorption spectrum as a function of T.

It should be noted that allowed transitions may also gain intensity by HT coupling with modes of the appropriate symmetries. But in this case, all the vibronic structure discussed for allowed transitions (see Section 3.6 above) is permitted and generally dominates the additional structure made allowed by the breakdown of the FC approximation.

An Example: The Bandshape Predicted for the $A_{1g} \to T_{1g}$ and $A_{1g} \to T_{2g}$ Electronic Transitions in Octahedral MX_6^{n-} *Complexes*

$A_{1g} \to T_{1g}$ and $A_{1g} \to T_{2g}$ electronic transitions are electric-dipole-forbidden (parity-forbidden), and so in the HT model, contributions to their intensity must come either from magnetic-dipole transitions or from vibrationally induced transitions. The selection rules for vibrationally induced transitions between HT–BO states allow transitions involving the change in one quantum of an activating vibration, $Q_{h\theta}$, and totally symmetric progressions built on these vibronic origins.

The detailed selection rules for the vibrationally induced transitions are found as described earlier. In octahedral systems, m_γ belongs to the irreducible representation T_{1u}, which has odd parity. Thus when $|A\alpha\rangle^0 = |A_{1g}\alpha\rangle^0$ and $|J\lambda\rangle^0 = |T_{1g}\lambda\rangle^0$ or $|T_{2g}\lambda\rangle^0$, (3.7.6) indicates that an odd-parity (u) vibration is required in (3.7.5) by parity selection rules (see Section 11.2). In the octahedral MX_6^{n-} complexes, the only odd-parity vibrations are those with $h = T_{1u}$ or T_{2u} and are labeled as $\nu_6(t_{2u})$, $\nu_4(t_{1u})$, and $\nu_3(t_{1u})$ in order of increasing energy. Using the methods of Chapter 9, we find that $|N\varepsilon\rangle^0$ must belong to $N = T_{1u}$ or T_{2u}. Since the magnetic-dipole operator $\mu_{\phi'}^{f'}$ belongs to $f' = T_{1g}$ in O_h, magnetic-dipole transitions are allowed to T_{1g} electronic states.

Thus a well-resolved $A_{1g} \to T_{1g}$ *d–d* absorption band of an octahedral transition-metal complex at liquid helium temperature might contain a sharp, relatively weak, magnetic-dipole-allowed no-phonon line followed by three vibronic sidebands to higher energy arising from the $A_{1g} \to T_{1g} + \nu_6(t_{2u})$, $A_{1g} \to T_{1g} + \nu_4(t_{1u})$, and $A_{1g} \to T_{1g} + \nu_3(t_{1u})$ transitions. Built upon each of these lines would be a progression in $\nu_1(a_{1g})$, the single totally symmetric mode, with the intensity distribution determined by the a_{1g} FC factors. Thus the *relative* intensities of the ν_6, ν_4, ν_3 lines should be the same for each member of the progression. An $A_{1g} \to T_{2g}$ transition could have approximately the same structure, except that the $A_{1g} \to T_{2g}$ magnetic-dipole transition is forbidden by symmetry, so the no-phonon line would be absent.

As the temperature is increased, hot bands appear which arise from $A_{1g} + \nu_6(t_{2u}) \to T_{1g}$ and similar transitions.

3.8. The Breakdown of the Born–Oppenheimer Approximation: How the Jahn–Teller Effect Is Treated Using Degenerate Perturbation Theory

Nearly all our discussion so far has been based on the BO approximation which is valid only so long as the potential surfaces of the states of interest are well separated from or are essentially noninteracting with all others (see Section 3.2). Thus while the BO approximation is excellent in a system of widely separated nondegenerate electronic states, it is risky when electronic states either are separated by an energy difference of the order of only a few vibrational quanta or are degenerate. In the first case, the system is susceptible to the pseudo-Jahn–Teller (PJT) effect and in the second to the Jahn–Teller (JT) effect. In these cases, the BO approximation breaks down and the wavefunctions $\phi_k(q, Q)$, which are eigenfunctions of $\mathcal{H}_{el}(q, Q)$, may be strongly mixed by vibrational coupling in the JT-active modes Q_μ. A consequence of this is that eigenfunctions no longer may be found in the simple adiabatic form (3.2.7). We do not discuss the PJT or JT effects in great detail in this book, but refer interested readers to Englman (1972), Longuet-Higgins (1961), Herzberg (1966), and Stephens (1976), Appendix I. Our discussion, which follows the last of these references closely, is intended as an introduction.

We follow the notation and methods of Section 3.3. We assume that the *ground state* has a single potential minimum, as would generally be the case for example if the state were nondegenerate or a Kramers doublet. We choose our normal coordinates so that this minimum occurs at $Q_0 = 0$ [i.e. $(Q_\eta)_0 = 0$ for $\eta = 1$ to $3N - 6$]. In that case, the terms linear in Q_η vanish in (3.3.4). In contrast to the treatment in Section 3.3, however, we consider an excited state k which is *n-fold degenerate*. Again $\mathcal{H}_{el}(q, Q)$ is expanded in a Taylor's series in Q to give (3.3.1)–(3.3.3). Now however we must use *degenerate* perturbation theory, and our first-order perturbed wave functions are therefore written

$$\phi'_{k\lambda} = \sum_{\kappa=1}^{n} c_{\lambda\kappa}\phi_{k\kappa} \tag{3.8.1}$$

where the $\phi_{k\kappa}$ are still given by (3.3.7):

$$\phi_{k\kappa} = \phi_{k\kappa}^0 + \sum_{j\neq k}\left[\frac{\sum_\eta \langle\phi_j^0|\left(\dfrac{\partial V}{\partial Q_\eta}\right)_0|\phi_{k\kappa}^0\rangle Q_\eta}{W_k^0 - W_j^0}\right]\phi_j^0 \qquad (\kappa = 1,\ldots, n) \tag{3.8.2}$$

where $W_k^0 \equiv W_k(Q_0)$, and so on. Note particularly that the sum over j now excludes all n degenerate components of state k, so that the $\phi_{k\kappa}$ are slowly varying functions of all Q_η.

The $c_{\lambda\kappa}$ in (3.8.1) and the new eigenvalues (potential surfaces), $W_{k\lambda}$ ($\lambda = 1,\ldots, n$), are found by solving the secular equations

$$\sum_{\kappa'=1}^{n} c_{\lambda\kappa'}(H_{\kappa\kappa'} - W_{k\lambda}\delta_{\kappa\kappa'}) = 0 \qquad (\kappa = 1,\ldots, n) \tag{3.8.3}$$

where

$$H_{\kappa\kappa'} = W_k^0\delta_{\kappa\kappa'} + \sum_\eta V_\eta^{\kappa\kappa'}Q_\eta$$
$$+\frac{1}{2}\sum_{\eta\mu}\left[V_{\eta\mu}^{\kappa\kappa'} + 2\sum_{j\neq k}\frac{V_\eta^{\kappa j}V_\mu^{j\kappa'}}{W_k^0 - W_j^0}\right]Q_\eta Q_\mu \tag{3.8.4}$$

and

$$V_\eta^{\kappa\kappa'} \equiv \langle\phi_{k\kappa}^0|\left(\frac{\partial V}{\partial Q_\eta}\right)_0|\phi_{k\kappa'}^0\rangle$$

$$V_{\eta\mu}^{\kappa j} \equiv \langle\phi_{k\kappa}^0|\left(\frac{\partial^2 V}{\partial Q_\eta\,\partial Q_\mu}\right)_0|\phi_j^0\rangle \tag{3.8.5}$$

If $V_\eta^{\kappa\kappa'} \neq V_\eta^k\delta_{\kappa\kappa'}$ and is nonzero, we note from (3.8.4) that the electronic degeneracy splits when $Q_\eta \neq 0$, thus giving rise to a first-order JT effect. Q_η is then said to be a JT-active normal coordinate. The $W_{k\lambda}$ potential surfaces are not degenerate (or parallel) away from $Q_\eta = 0$. The JT theorem states that for an orbitally degenerate state of a nonlinear molecule, at least one JT-active mode—generally itself degenerate—always exists. Moreover, $Q_\eta = 0$ is *not* the potential-energy minimum in electronic states k, since (3.8.4) is linear in Q_η. (If $V_\eta^{\kappa\kappa'} = V_\eta^k\delta_{\kappa\kappa'} \neq 0$, the potential minimum is displaced from $Q_\eta = 0$, but the potential surfaces do *not* split to first order and *no* first order JT effect is present.) In a similar way, the terms containing $Q_\eta Q_\mu$ in (3.8.4) can give rise to second-order JT effects. Note also that because of the presence of Q_η in (3.8.4), the $c_{\lambda\kappa}$ and therefore the $\phi'_{k\lambda}$ are in general *strong functions* of the JT-active normal coordinates. Consequently, the commutation with $T_n(Q)$ assumed in (3.2.8) is no longer justified, and the BO approximation (3.2.7) fails.

Under these circumstances, we seek solutions of the vibronic equations (3.2.5) (i.e., the dynamical problem) in the more general (nonadiabatic) form ($i = 1$ to ∞)

$$\Psi'_{ki} = \sum_{\lambda=1}^{n} \phi'_{k\lambda}(q, Q)\chi'_{\lambda i}(Q) \tag{3.8.6}$$

Substituting (3.8.6) into (3.2.5), multiplying the result in turn by each $\phi'^{*}_{k\lambda}$, and integrating over electronic coordinates, we obtain

$$\sum_{\lambda'=1}^{n} \left(W_{k\lambda}\delta_{\lambda\lambda'} + \langle\phi'_{k\lambda}|T_n|\phi'_{k\lambda'}\rangle - \mathcal{E}_{ki}\delta_{\lambda\lambda'}\right)\chi'_{\lambda' i} = 0 \qquad (\lambda = 1,\ldots, n) \tag{3.8.7}$$

where

$$\langle\phi'_{k\lambda}|T_n|\phi'_{k\lambda'}\rangle\chi'_{\lambda' i} \equiv \int \phi'^{*}_{k\lambda}(q, Q)T_n(Q)\phi'_{k\lambda'}(q, Q)\chi'_{\lambda' i}(Q)\, dq \tag{3.8.8}$$

We see from (3.8.7) that the n vibrational functions $\chi'_{\lambda i}$ associated with each vibronic state Ψ_{ki} are determined by a set of coupled equations which involve all n potential surfaces $W_{k\lambda}$ ($\lambda = 1,\ldots, n$). Thus the nuclear motion cannot be confined to any one surface. Note that when T_n commutes with the $\phi'_{k\lambda}$ (not our case here), (3.8.7) uncouples, and we return to n independent equations in the BO form, (3.2.7) and (3.2.10).

It is possible to solve (3.8.7) directly [Longuet-Higgins (1961)]. However, (3.8.7) has the undesirable feature that the electronic functions $\phi'_{k\lambda}$ *do not* commute with T_n, and so this operator must be applied to $\phi'_{k\lambda'}\chi'_{\lambda' i}$ before the integration over electronic coordinates can be performed to evaluate (3.8.8). It is therefore very useful to use the original (unprimed) $\phi_{k\kappa}$ of (3.8.1)–(3.8.2), which *do* commute with T_n (and permit the FC approximation). Thus we write in place of (3.8.6)

$$\Psi_{ki} = \sum_{\kappa=1}^{n} \phi_{k\kappa}(q, Q)\chi_{\kappa i}(Q) \tag{3.8.9}$$

If (3.8.9) is substituted into (3.2.5) and the previous procedure is repeated, the following set of coupled equations results:

$$\sum_{\kappa'=1}^{n} \left(H_{\kappa\kappa'} + T_n\delta_{\kappa\kappa'} - \mathcal{E}_{ki}\delta_{\kappa\kappa'}\right)\chi_{\kappa' i} = 0 \qquad (\kappa = 1,\ldots, n) \tag{3.8.10}$$

where $H_{\kappa\kappa'}$ is given by (3.8.4). Unlike the $\phi'_{k\lambda}$, the $\phi_{k\kappa}$ are not diagonal in $\mathcal{H}_{el}$. Nevertheless, the simplified role of T_n can more than compensate for this extra complication. Note that if $H_{\kappa\kappa'} = \delta_{\kappa\kappa'}H_k$—that is, if there is no JT effect—the $\phi_{k\kappa}$ uncouple and (3.8.10) reduces to n independent equations in the BO form, (3.2.7) and (3.2.10); when this is true, T_n also commutes with the $\phi'_{k\lambda}$ in (3.8.7)–(3.8.8).

It should be emphasized that the vibrational functions $\chi_{\kappa i}$ (or $\chi'_{\lambda i}$) are in general *not* harmonic-oscillator functions—and certainly not harmonic oscillators centered about $Q_0 = 0$—since the nuclear motion takes place on *all* of the JT-distorted surfaces. The coupled equations (3.8.7) and (3.8.10) are generally solved by numerical techniques. A more complete exposition of the perturbational approach may be found in Stephens [(1976), Appendix I].

To gain further understanding of the eigenfunctions (3.8.9), let us consider one direct approach to the solution of (3.8.10). In the absence of the JT effect where the BO approximation applies, the vibrational solutions (3.2.10) are (in the harmonic approximation) simply products of harmonic-oscillator functions, $\chi_{v_\mu}(Q_\mu)$—see (3.6.7). Since these are complete sets, we can expand the $\chi_{\kappa i}$ of (3.8.9) in terms of them. Thus the JT vibronic solutions can be expressed as (possibly very complicated) linear combinations of BO functions, a point which assumes great importance when moment analysis is applied to vibronic bands (Section 7.3).

Since $\mathcal{H}_{JT} \sim \Sigma_\mu(\partial V/\partial Q_\mu)_0 Q_\mu$, it links together basis functions $\chi_{v_\mu}(Q_\mu)$ which differ in v_μ by ± 1, because $\langle\chi_{v_\mu}(Q_\mu)|Q_\mu|\chi_{v'_\mu}(Q_\mu)\rangle = 0$ unless $v'_\mu = v_\mu \pm 1$ (exactly as in Section 3.7). Thus the vibronic matrix obtained from (3.8.10) is of infinite dimension in the harmonic-oscillator basis, and the $\chi_{\kappa i}$ are necessarily linear combinations of the harmonic oscillator functions χ_{v_μ} with $v_\mu = 0, 1, 2, \ldots$. Because of this, $v_\mu = 0$ character is distributed among the set of vibronic eigenfunctions (3.8.9), and all of these have some "no-phonon" character when all other $v_\mu(\eta \neq \mu) = 0$. The significance of this is discussed further in the sections ahead with respect to vibronic selection rules. An example of the use of a complete harmonic oscillator set about $Q_0 = 0$ to solve a JT-like problem is contained in the model used for mixed-valence systems which is referenced at the end of Section 3.2.

A Simple Jahn–Teller Case: The $E \otimes e$ System

In the simple "textbook" JT examples, such as that of an electronic state of E symmetry coupled to a JT-active vibrational mode of e symmetry in an octahedral or tetrahedral molecule, it is possible to solve (3.8.3) explicitly to obtain electronic eigenfunctions ($\phi'_{k\lambda}$) and eigenvalues ($W_{k\lambda}$) which are analytical functions in the Q_μ, the JT-active normal coordinates. In these

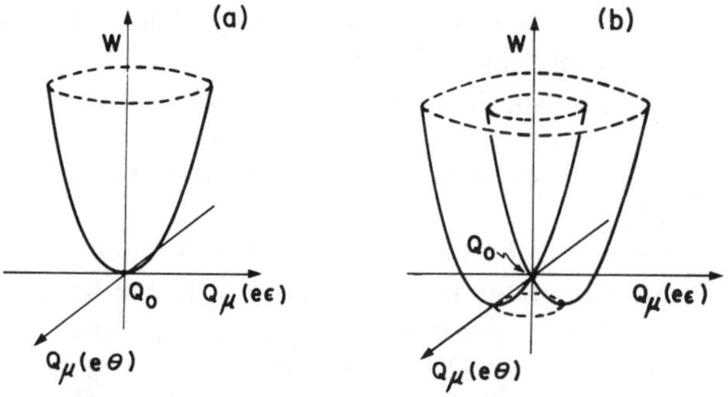

Figure 3.8.1. Potential-energy surfaces for a tetrahedral or octahedral molecule in a degenerate electronic state of symmetry E as a function of the two components of a degenerate normal coordinate of symmetry e. Part (a) is for zero JT coupling, so the surfaces for the E electronic state are everywhere degenerate. In (b) first-order (linear) JT coupling has been included (see Eq. (3.8.4)). The surfaces in (b), which are the $W_{k\lambda}$ of (3.8.3), remain degenerate at Q_0 but have minima away from Q_0. The solid curves show the surfaces in the $Q_\mu(e\epsilon)$ plane. The complete surface is obtained by rotating the solid curve about W, as indicated by the dashed lines.

simple examples, the new nonharmonic potential surfaces—the $W_{k\lambda}$—can then be plotted. These cases are helpful in giving a feel for the changes brought about by the JT effect. For example, in the octahedral or tetrahedral $E \otimes e$ case mentioned above, the JT potential surfaces form a Mexican hat as shown in Figure 3.8.1. In the absence of JT coupling, the potential surfaces of the degenerate E states are identical with minima at $Q_\mu = 0$. This case and others are discussed in some detail in Ballhausen (1962), Herzberg (1966), and Englman (1972).

Symmetry Classification of Jahn–Teller Vibronic Eigenfunctions

Note in the above example that the potential surface for the JT-active mode has minima away from Q_0. This is always the case because Q_μ occurs linearly in (3.8.4). At all these minima, the molecular system has a lower symmetry than that of the point group of the molecule, since any motion away from Q_0 in a non-totally-symmetric mode represents molecular distortion. Nonetheless, in the absence of other perturbations (such as strain in a crystal), as long as the high-symmetry point (i.e. Q_0) remains within the range of nuclear motion of the system, the vibronic *stationary-state wavefunctions of the molecule are usefully classified using the representations of the higher-symmetry point group*. This follows because $\mathcal{H}_{JT}$, like all terms in the molecular Hamiltonian, transforms as the totally symmetric (A_1) irreducible representation of the molecular point group defined at Q_0 (see discussion in Section 8.9). In the BO approximation, the symmetry of electronic and vibrational functions can be classified separately. In the JT case, the classification *must* be via vibronic functions. But in either case it follows that the potential surfaces must reflect the full symmetry of the molecular point group at Q_0.

Let us suppose when the system is cooled to low temperature that the zero-point vibrational energy is small compared to the barriers separating symmetrically equivalent potential minima. In this case, the symmetry classification with respect to Q_0 can still be retained, and indeed this must be done if one is concerned with tunneling phenomena. However, if one is concerned with a "local" effect, for example the ordinary vibrational or electronic spectrum, then one can perfectly well use the lower, local symmetry classification, since the presence of the other potential minima will be reflected only in unresolved degeneracies. We emphasize in this regard that JT effects by themselves can never split no-phonon lines, since the symmetry of a no-phonon state coincides with the electronic symmetry at Q_0.

JT systems are, however, notoriously unstable with respect to stress, and so often one of the minima becomes lower in energy than the others due to a small external, lower-symmetry perturbation [see Englman (1972)]. In such a case, the symmetry is truly lowered, and the system belongs to the lower-symmetry point group.

Static and Dynamic Jahn–Teller Effects

We simply note here that the terms "static" and "dynamic" are used in the literature in two different senses. One usage refers to a static JT effect when the system is strongly trapped in the various symmetrically equivalent potential minima. If this is not the case, then a dynamic JT effect is said to be operative. The other usage refers to the electronic problem, as for example in (3.8.3)–(3.8.5), as the static problem, since the nuclear kinetic-energy term T_n has not yet been considered. Solution of the full vibronic Schrödinger equation (3.2.5), resulting for example in (3.8.7) or (3.8.10), is then referred to as the dynamic problem.

Bandshapes and Vibronic Selection Rules in Jahn–Teller Systems

Let us assume for the sake of simplicity that the ground state is non-JT-susceptible and that we have solved the JT problem as discussed above to obtain vibronic stationary-state wavefunctions for a degenerate excited state. Our ground-state functions are thus BO products, while our excited-state functions are in the form (3.8.6) or (3.8.9).

We now consider selection rules for transitions to these JT vibronic eigenstates and compare them with those obtained for the simple adiabatic BO product states (3.2.7). We find that if we express our JT vibronic eigenfunctions in the form of (3.8.9), nearly all the arguments used in earlier sections carry over. The $\phi_{k\kappa}(q, Q)$ of (3.8.1)–(3.8.2) vary slowly with Q, as in Section 3.6, since they involve mixing only with electronic states *outside* the degenerate $k\kappa$ manifold. We may thus still make the FC approximation by replacing the $\phi_{k\kappa}(q, Q)$ with $\phi_{k\kappa}(q, Q_0) \equiv \phi^0_{k\kappa}$ in the transition-moment integral. Now, however, our vibronic eigenfunctions (3.8.9) are *nonadiabatic*, since they involve a linear combination of the $\phi^0_{k\kappa}$ functions. In addition, the vibrational functions χ_{ki} in (3.8.9) can be expressed as linear combinations of unperturbed (by $\mathcal{H}_{\mathrm{JT}}$) harmonic-oscillator functions about Q_0. This means that in the FC approximation, the "no-phonon" or 0–0 intensity is distributed over a series of vibronic lines, each of which has an intensity proportional to the coefficients which weight the no-phonon contribution to the line. Recall from Section 3.6 that in the non-JT case, the no-phonon intensity was limited to a single, purely electronic line.

Thus in the JT case the FC approximation predicts a series of lines with no-phonon intensity, instead of just one. The lowest of these lines has the same degeneracy as the electronic state in the absence of the JT effect as long as no other perturbations which split that degeneracy are present in addition to $\mathcal{H}_{\mathrm{JT}}$. When the JT coupling is small, this series of lines with no-phonon intensity forms an approximate progression in the JT-active mode. A large JT effect, however, can lead to substantial modification of the unperturbed line energies.

When the discussion of Section 3.6 is applied to the JT case, we see that each line with no-phonon character can serve as an origin upon which a totally symmetric progression may be built. Thus the dispersion of the spectrum may be quite complex. Nonetheless, the relative intensity distribution within each totally symmetric progression is determined by the same FC factors (of the totally symmetric vibration).

An Example of the Dispersion of a Jahn–Teller System: The $^2T_{2u}$ Band in $IrCl_6^{2-}$

$IrCl_6^{2-}$ is an octahedral d^5 system and thus its ground state configuration is t_{2g}^5. This configuration gives rise to a $^2T_{2g}$ state which is split into $E_g''(^2T_{2g})$ and $U_g'(^2T_{2g})$ states by spin–orbit coupling, with E_g'' lying lowest. The absorption spectrum is dominated in the visible and near uv regions by ligand-to-metal charge-transfer bands arising from $t_{1u}(\pi + \sigma) \rightarrow t_{2g}$ and $t_{2u}(\pi) \rightarrow t_{2g}$ orbital excitations. These excitations fill the t_{2g}^5 hole and leave a single hole in the t_{1u} or t_{2u} ligand orbital sets.

In this example (which is taken up in more detail in Chapter 23) we concentrate on the $t_{2u}(\pi)$ excitation which gives rise to a $^2T_{2u}$ excited state. Transitions to it are electronically allowed from the $E''_g(^2T_{2g})$ ground state. The $^2T_{2u}$ state is, however, susceptible to both spin–orbit and JT coupling, and so the dispersion of the vibronic lines of the band, as shown in Figure 23.3.1(d), is complex. The ground state is Kramers-degenerate and so is not JT-susceptible. The $^2T_{2u}$ state is split into $E''_u(^2T_{2u})$ and $U'_u(^2T_{2u})$ states by spin–orbit coupling. Thus at liquid-helium temperatures, in the absence of an excited-state JT effect, one would expect two overlapping totally symmetric progressions arising from the $E''_g \rightarrow E''_u(^2T_{2u})$ and $E''_g \rightarrow U'_u(^2T_{2u})$ bands respectively, for the reasons given in Section 3.6. The $E''_g \rightarrow U'_u(^2T_{2u})$ band origin would be predicted to be $\approx \frac{3}{4}\zeta_{\text{Cl}^-} \approx 440\ \text{cm}^{-1}$ to the blue of the $E''_g \rightarrow E''_u(^2T_{2u})$ origin, where ζ_{Cl^-} is the one-electron spin–orbit coupling constant for Cl^-. Instead of this, totally symmetric progressions built on a much more complex series of initial vibronic lines are observed.

Yeakel and Schatz (1974) explain the pattern by demonstrating that strong JT coupling in $\nu_5(t_{2g})$ and $\nu_3(e_g)$ modes redistributes the E''_u and U'_u no-phonon character throughout a number of vibronic E''_u and U'_u excited states and each of these becomes an allowed vibronic origin upon which a totally symmetric progression is built. Vibronic transitions to these states become electric-dipole-allowed from the ground level of the $E''_g(^2T_{2g})$ state in proportion to their no-phonon $E''_u(^2T_{2u})$ or $U'_u(^2T_{2u})$ character.

The Ham Effect

The lowest-lying E''_u and U'_u lines in the $^2T_{2u}$ band above are separated by only 5–6 cm^{-1} instead of the $\frac{3}{4}\zeta_{\text{Cl}^-} \approx 440\ \text{cm}^{-1}$ predicted in the absence of a JT effect. Thus it appears as if the spin–orbit splitting of the $^2T_{2u}$ no-phonon line had been quenched by JT coupling. This apparent quenching of electronic operators by JT coupling is known as the *Ham effect* after Ham (1965). Ham effects, which are usually detected by studying no-phonon lines, result because the JT effect leads to vibronic eigenfunctions in the form of (3.8.6) or (3.8.9). Off-diagonal matrix elements of electronic operators in the $\phi_{k\kappa}(q, Q)\chi_{\kappa i}$ basis of (3.8.9) are reduced by the poor overlap of the vibrational functions associated with $\phi_{k\kappa}$ and $\phi_{k\kappa'}$. For example, the quenching of spin–orbit coupling is a consequence of a reduction of the electronic orbital angular momentum by vibrational overlap factors of the form $\langle\chi_{\kappa i}|\chi_{\kappa' i}\rangle$. In the absence of a JT effect, the vibrational overlap factor is unity, since all components of a degenerate electronic state have identical potential surfaces, $W_{k\lambda}(Q) = W_{k\lambda'}(Q)$.

The Breakdown of the Franck–Condon Approximation in Jahn–Teller Systems

The breakdown of the FC approximation may be handled here in a similar way to that discussed in Section 3.7. HT coupling again may be used to allow mixing with additional electronic states. When transitions to the JT vibronic eigenstates are forbidden in the FC approximation (as, for example, is the case for all *d–d* transitions in octahedral systems), once again HT coupling with odd-parity modes allows transitions to gain intensity; here, however, HT vibronic lines may be associated with all JT vibronic eigenstates which contain no-phonon character.

Nonadiabatic Coupling as an Intensity Mechanism: The Effect of Small Deviations from the Born–Oppenhiemer Approximation

Another way a transition forbidden in the BO–FC approximations gains intensity is through deviations from the BO approximation. If these deviations are large, we say the system experiences a JT or PJT effect and account for them using methods such as those discussed for the JT effect in Sections 3.2 and 3.8. But if they are small, they may be treated via perturbation theory using the BO states of (3.2.5) as basis functions. The perturbation then arises from the terms in (3.2.8) which prevent the exact commutation of the electronic functions and the nuclear kinetic-energy operator T_n. Thus the operator T_n may couple BO states and lead to nonadiabatic wave functions; coupling by T_n is often termed *nonadiabatic* (*NA*) *coupling*. The selection rules for the transition moment (3.6.3) in the presence of NA coupling are nearly identical to those for HT coupling. However, unless HT coupling is extremely weak because of cancellation effects in (3.7.6), NA coupling is usually small compared to HT coupling and may be neglected. Readers interested in NA coupling as an intensity mechanism should see Ballhausen and Hansen (1972) and references therein, as well as Orlandi and Siebrand (1973).

4 *Magnetic Circular Dichroism (MCD) Spectroscopy*

MCD is a manifestation of the Faraday effect, which was studied in transparent regions of the spectrum for a wide variety of substances in the latter part of the nineteenth century.

4.1. The Faraday Effect

All substances exhibit the Faraday effect, which may be expressed at any nonabsorbing frequency ν by the equation

$$\phi = V(\nu)\int_0^l B(l)\,dl \tag{4.1.1}$$

where ϕ measures the rotation of the plane of polarization of a linearly polarized electromagnetic wave traveling through a homogeneous sample of thickness l parallel to an applied magnetic field $\mathbf{B}(l)$. V is a frequency-dependent proportionality constant (Verdet constant) which is characteristic of the material. If the magnetic field is constant along the entire path of the light beam through the substance (the usual experimental situation), the right-hand side of (4.1.1) becomes simply $V(\nu)Bl$.

Using results from Chapter 2 [(2.2.1), (2.2.2), (2.6.4), (2.6.9), Table 2.6.1, (2.11.3)], we may write explicit expressions for the electric and magnetic fields of a right (+) or left (−) circularly polarized wave (rcp or lcp wave) propagating in the $+Z$ direction. The electric field is

$$\begin{aligned}\mathbf{E}_{\pm} &= \Re e\,\frac{1}{\sqrt{2}}(\mathbf{e}_X \pm i\mathbf{e}_Y)E^0\exp\left[2\pi i\nu\left(t - \frac{n_{\pm}Z}{c}\right)\right]\\ &= \frac{E^0}{\sqrt{2}}\left[\mathbf{e}_X\cos 2\pi\nu(t - n_{\pm}Z/c)\right.\\ &\qquad \left.\mp\mathbf{e}_Y\sin 2\pi\nu(t - n_{\pm}Z/c)\right]\end{aligned} \tag{4.1.2}$$

where $n_{\pm}$ is the refractive index of the medium through which right $(+)$ or left $(-)$ circularly polarized light is propagating at frequency ν, and $\mathbf{e}_X, \mathbf{e}_Y, \mathbf{e}_Z$ are unit vectors in a right-handed laboratory-fixed Cartesian coordinate system. (We always omit the subscript "mac" on $\mathbf{E}$ and E^0, as discussed at the end of Section 2.11). As the terms rcp and lcp imply, the electric vectors in (4.1.2) execute circular motions while simultaneously advancing at velocity $c/n_{\pm}$; that is, the tips of the electric vectors move in helical paths. If lcp and rcp waves are combined with equal amplitudes but arbitrary relative phases [a permissible procedure, since Maxwell's equations (2.2.2) are linear], the resulting wave is linearly polarized provided $n_+ = n_-$. For example, for zero phase difference, from (4.1.2) we obtain

$$\mathbf{E} = \frac{1}{\sqrt{2}}(\mathbf{E}_+ + \mathbf{E}_-) = \mathbf{e}_X E^0 \cos 2\pi\nu\left(t - \frac{nZ}{c}\right) \tag{4.1.3}$$

which is clearly the electric vector for an X-polarized wave with $n \equiv (n_+ + n_-)/2$. Let us follow the same procedure but now assume $n_+ \neq n_-$. In this case, using standard trigonometric addition formulas, we find

$$\begin{aligned}\mathbf{E} &= \frac{1}{\sqrt{2}}(\mathbf{E}_+ + \mathbf{E}_-)\\ &= E^0\left[\mathbf{e}_X \cos\frac{\pi\nu Z\,\Delta n}{c}\cos 2\pi\nu\left(t - \frac{Zn}{c}\right)\right.\\ &\qquad \left. - \mathbf{e}_Y \sin\frac{\pi\nu Z\,\Delta n}{c}\cos 2\pi\nu\left(t - \frac{Zn}{c}\right)\right]\end{aligned} \tag{4.1.4}$$

where $\Delta n \equiv n_- - n_+$. We note that (4.1.4) still describes a linearly polarized wave, but the plane of polarization rotates as the wave advances. Following the sign convention which defines the rotation of the plane of polarization by

$$\phi = \arctan\left(-\frac{E_Y}{E_X}\right) \tag{4.1.5}$$

it follows from (4.1.4) that

$$\phi = \frac{\pi\nu l\,\Delta n}{c} \tag{4.1.6}$$

where ϕ is the rotation (in radians) of the plane of polarization when the wave travels a distance l through the medium.

Naturally optically active substances rotate linearly polarized light because an inherent net chirality is present which permits $n_+ \neq n_-$. The necessary and sufficient condition for this is that the molecular or crystal species possess no rotary reflection axis of any order. On the other hand, it can be shown that a longitudinal magnetic field causes $n_+ \neq n_-$ *in all substances*, and for this reason the Faraday effect is universal, the phenomenon sometimes being referred to as magnetically induced (or magnetic) optical activity. We emphasize at once the fundamentally different information content in natural and magnetically induced optical activity. Both produce the same physical effect—rotation of the plane of polarization of linearly polarized light, and circular dichroism (Section 4.2). But the former is sensitive to small inherent molecular dissymmetries, while the latter is a consequence of Zeeman interactions.

4.2. (Magnetic) Circular Dichroism

Our discussion so far has concerned only transparent spectral regions. We now generalize our results by permitting absorption to occur. It is clear from (2.11.4) that the form of the electromagnetic waves used thus far must be modified, since the intensity of the transmitted wave must be attenuated in an absorbing medium in accordance with the discussion of Section 2.1. We show now that this may be accomplished by the simple expedient of allowing the refractive index to be a complex quantity, which we designate as $\hat{n}$ (Born and Wolf (1959), Section 13.1]. Then by hypothesis

$$\hat{n} = n - ik \tag{4.2.1}$$

where n is the ordinary refractive index. We find just ahead that k is an absorption coefficient. Replacing n by $\hat{n}$ in (2.11.3), it is straightforward to recalculate $I(\nu)$ in (2.11.4) from (2.6.3) after applying (2.2.2) to (2.11.3). The result is, for linearly polarized light,

$$\begin{aligned} \mathbf{e}_Z I(\nu) &= \frac{\mathbf{e}_Z n \left(A^0_{\text{mac}}\right)^2 \pi\nu^2}{2c} \exp\left(-\frac{4\pi k\nu Z}{c}\right) \\ &= \mathbf{e}_Z I_0 \exp\left(-\frac{4\pi k\nu Z}{c}\right) \end{aligned} \tag{4.2.2}$$

where I_0 by comparison with (2.11.4) is identified as the intensity which would propagate through the medium in the absence of absorption. Equation (4.2.2) is clearly of the appropriate form, and comparison with (2.1.3) permits the identification

$$\kappa = \frac{4\pi k\nu}{c} \tag{4.2.3}$$

For rcp and lcp light

$$\hat{n}_{\pm} = n_{\pm} - ik_{\pm} \tag{4.2.4}$$

and one can confirm by the same method that (4.2.2) still applies with the appropriate subscript ($\pm$) appended to n and k.

Let us now suppose that a linearly polarized wave is incident on an optically active sample. The optical activity may be natural or magnetically induced. We suppose that the sample has thickness l, and we wish to examine the state of polarization of the wave which exits the sample (neglecting reflection at the surfaces). The incident wave is described by (4.1.3) with $n = 1$, even if a magnetic field is present, assuming the sample is in a vacuum (or to a good approximation, in air). Upon entering the medium at normal incidence,

$$\mathbf{E} = \frac{1}{\sqrt{2}}(\mathbf{E}_{+} + \mathbf{E}_{-}) \tag{4.2.5}$$

with

$$\mathbf{E}_{\pm} = \mathfrak{Re}\,\frac{1}{\sqrt{2}}(\mathbf{e}_X \pm i\mathbf{e}_Y)E^0 \exp\left[2\pi i\nu\left(t - \frac{\hat{n}_{\pm}Z}{c}\right)\right] \tag{4.2.6}$$

where $\hat{n}_{\pm}$ is defined by (4.2.4). As before, E_{+} and E_{-} propagate through the medium at different velocities ($n_{+} \neq n_{-}$), but both are now also attenuated, though to different extents ($k_{+} \neq k_{-}$). The amplitude and phase of $E_{\pm}$ upon leaving the medium may be ascertained by simply substituting $Z = l$ into (4.2.6), with the result

$$\mathbf{E}_{\pm} = \mathfrak{Re}\,\frac{1}{\sqrt{2}}(\mathbf{e}_X \pm i\mathbf{e}_Y)E^0_{\pm} e^{2\pi i\nu t} e^{-i\delta_{\pm}} \tag{4.2.7}$$

where

$$E^0_{\pm} \equiv E^0 e^{-2\pi\nu k_{\pm} l/c}, \qquad \delta_{\pm} \equiv \frac{2\pi\nu n_{\pm} l}{c}$$

Substituting (4.2.7) into (4.2.5), we may write down E_X and E_Y for the wave which reemerges from the sample into the vacuum:

$$E_X = a_1\cos(\tau + \delta_1)$$
$$E_Y = a_2\cos(\tau + \delta_2) \qquad (4.2.8)$$

where

$$\tau \equiv 2\pi\nu t, \qquad a_1 = \tfrac{1}{2}\left(W_+^2 + V_+^2\right)^{1/2}, \qquad a_2 = \tfrac{1}{2}\left(W_-^2 + V_-^2\right)^{1/2}$$

$$\tan\delta_1 = -\frac{V_+}{W_+}, \qquad \tan\delta_2 = \frac{W_-}{V_-} \qquad (4.2.9)$$

$$W_\pm = E_+^0\cos\delta_+ \pm E_-^0\cos\delta_-, \qquad V_\pm = E_+^0\sin\delta_+ \pm E_-^0\sin\delta_-$$

Equation (4.2.8) is the general form of an *elliptically polarized wave* propagating in the Z direction. That is, the tip of the electric (and magnetic) vector describes an ellipse if viewed in an XY plane as a function of time; it propagates in the Z direction in the form of a flattened helix.

A clear and detailed analysis of (4.2.8) is presented in Born and Wolf (1959), Section 1.4.2, and we simply summarize their results here with the

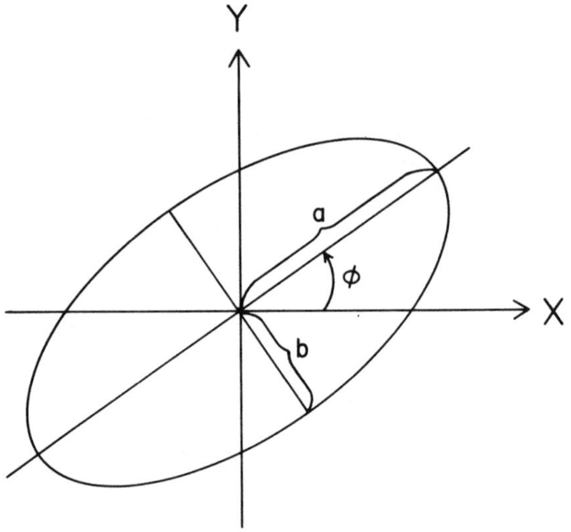

Figure 4.2.1. The path executed in a fixed XY plane by the electric vector of an elliptically polarized wave traveling in the Z direction.

aid of Figure 4.2.1, using our sign convention for ϕ [Eq. (4.1.5)]. The results are:

$$-\tan 2\phi = \tan 2\alpha \cos(\delta_2 - \delta_1)$$

$$|\tan \chi| = \frac{b}{a}, \qquad \sin 2\chi = \sin 2\alpha \sin(\delta_2 - \delta_1) \tag{4.2.10}$$

$$\tan \alpha = \frac{a_2}{a_1}$$

If one substitutes a_1, a_2, δ_1, and δ_2 from (4.2.9) into (4.2.10), one obtains after some algebraic and trigonometric reduction,

$$\phi = \frac{\pi \nu l \, \Delta n}{c}$$

$$\tan \chi = \tanh\left[\frac{\pi \nu l \, \Delta k}{c}\right] \tag{4.2.11}$$

where $\Delta k \equiv k_- - k_+$. The ratio $\tan \chi$ of minor and major axes of the ellipse (Fig. 4.2.1) is called the *ellipticity*, and tanh designates the hyperbolic tangent. We see from (4.2.11) that the expression for ϕ is the same as in the transparent case [Eq. (4.1.6)], but this angle now measures the rotation of the major axis of the vibrational ellipsoid of the electric vector (Fig. 4.2.1). In *most* practical situations, Δk is sufficiently small that the tanh in (4.2.11) may be approximated by its leading term, yielding

$$\tan \chi = \frac{\pi \nu l \, \Delta k}{c} \tag{4.2.12}$$

On the basis of (4.2.4), (4.2.11), and (4.2.12), we may *define* a complex rotation by

$$\hat{\phi} \equiv \phi - i\theta = \frac{\pi \nu l}{c}(\hat{n}_- - \hat{n}_+) \tag{4.2.13}$$

The imaginary part θ is proportional to Δk, and for this reason is called the *circular dichroism* (CD): if it arises through the application of a magnetic field, it is referred to as the *magnetic circular dichroism* (MCD). It is the quantity of fundamental interest to us both experimentally and theoretically, but is *linearly* related to the ellipticity only if the tanh in (4.2.11) can be approximated by its linear term. It should also be noted that due to the analytic properties of $\hat{\phi}$, the functions ϕ and θ are related by integral transforms (Kramers–Kronig relations), so that either can be calculated at any frequency from a knowledge of the other at all frequencies. These relations have been discussed in detail in the literature [Landau and Lifshitz (1960), Section 62]. Such relations are not presently of great practical significance, since θ is both the preferred theoretical quantity and the one measured (almost) exclusively in magnetic work.

Difficulties if a System is not Isotropic, Cubic, or Uniaxial

Inherent in the derivation of θ (or ϕ) is the assumption that circularly polarized waves continue to exist when the light propagates through the medium of interest, i.e., that (4.2.6) is a correct description. In fact, this in general is *not* the case in a biaxial system (see below). Thus if circularly polarized light is propagated in the Z direction through a medium in which the electrical properties vary with orientation of the electric vector in the XY plane, $n_{\pm}$ are no longer constants at a given frequency but vary with the orientation of the electric vector. Under such circumstances, an initially linearly polarized wave emerges elliptically polarized in a manner determined by the detailed optical properties of the medium, even though the medium shows no optical activity (natural or magnetic). Even if a "uniaxial" direction is found in such a system (always possible), this direction is a function of frequency ("dispersion of the axes"), particularly in regions of absorption. Under these circumstances, deducing the CD from the polarization properties of the emerging beam is a formidable problem in general. However, detailed theoretical analysis [Jensen, Schellman, and Troxell (1978)] shows that the CD can, in principle, always be deduced *if* the optical rotation (circular birefringence), linear dichroism, and linear birefringence are *all* also simultaneously measured. If the linear dichroism is sufficiently small, the theoretical analysis is substantially simplified, and it is relatively easy to measure the CD simply by correcting for the linear dichroism. CD measurements have been successfully made under such circumstances. If the linear dichroism is large, then severe problems may be anticipated in implementing the theoretical analysis in actual experimental situations.

These problems significantly limit or complicate CD (or MCD) measurements in solids. In this regard, systems can be classified as follows: isotropic, cubic, uniaxial, and biaxial. CD measurements may be made without the complications just discussed on all but the last of these classes. Isotropic system are generally solutions (or gases), and the same CD is obtained for all orientations. CD measurements can be made on cubic systems along any arbitrary direction, but the results are *not* always orientation-independent in the magnetic case (see Sections 4.6 and 17.5).

If two or more crystallographically equivalent directions may be chosen in one plane, and a third, perpendicular direction is crystallographically inequivalent, the crystal is said to be uniaxial. Straightforward CD measurements can again be made, but only along the unique axis. Uniaxial crystals can belong to the trigonal, tetragonal, or hexagonal systems. Crystals which do not contain two equivalent crystallographic directions are said to be biaxial; these crystals belong to the orthorhombic, monoclinic, or triclinic systems. For these, in general, and for uniaxial systems if the light is not propagated along the unique axis, measurement of the CD involves the additional complications outlined above.

Finally, it must be emphasized that samples may be unsatisfactory in practice because of the presence of strain birefringence or other imperfections which depolarize the light. Furthermore, samples which seem perfectly satisfactory at room temperature are sometimes quite unsatisfactory at low temperature because strain develops or a phase change occurs. The sharp lines which one often observes at low temperature tend to be much more sensitive to depolarization effects.

Fortunately, there is a simple method of testing for depolarization in MCD studies of substances possessing no natural optical activity. One measures the CD under run conditions (but at zero magnetic field), having first placed the sample in front of a naturally optically active substance. The resulting decrease in the known CD of the naturally optically active substance is a direct measure of the depolarization produced by the sample. If this depolarization is small, an approximate correction can be effected.

We do not discuss the experimental aspects of CD and MCD spectroscopy in this book. A good discussion with further references is given by Denning (1975). For earlier discussions see Ciardelli and Salvadori (1973). For a discussion of MCD and CD techniques in the vacuum ultraviolet and vibrational CD techniques, see Mason (1979).

4.3. The Relation of $\Delta k = k_- - k_+$ to Molecular Parameters

From our previous discussion, it is clear that the quantity of fundamental interest in MCD spectroscopy is $\Delta k \equiv k_- - k_+ \equiv k_L - k_R$, which we now relate to molecular parameters. Since explicit expressions for $\kappa_\pm$ have already been derived [use (2.11.5) with $\boldsymbol{\pi} = (\mathbf{e}_x \pm i\mathbf{e}_y)/\sqrt{2}$], and since k is related to κ via (4.2.3), one proceeds from this starting point. Such a procedure, however, requires a detailed justification. The reason is that we require Δk *in the presence* of a constant, longitudinal magnetic field B_z to discuss magnetically induced circular dichroism (MCD). When a constant magnetic field is applied to a system of point charges, the time-dependent Hamiltonian given by the sum of (2.2.4) and (2.9.1) must be modified. The vector potential must include the applied magnetic field (see also Section 2.9). This new vector potential designated by $\mathcal{Q}_j$, is

$$\mathcal{Q}_j = \mathbf{A}_j + \tfrac{1}{2}\mathbf{B} \times \mathbf{r}_j \tag{4.3.1}$$

where $\mathbf{A}_j$ as before may be obtained from (2.2.1), and the second term represents a constant (applied) field, as may be confirmed by applying the second equation of (2.2.2). In the present case, $\mathbf{B} = \mathbf{e}_z B$.

We proceed naively and assume that the only effect of the external magnetic field is to add the Zeeman term,

$$\mathcal{H}_Z = -\boldsymbol{\mu} \cdot \mathbf{B} = -\mu_z B \tag{4.3.2}$$

to $\mathcal{H}_T$, the time-independent vibronic Hamiltonian, where $\boldsymbol{\mu}$ is the magnetic-dipole operator defined in (2.9.2), and the field direction defines the z axis of a spaced-fixed coordinate system whose origin coincides with the molecular origin. The rigorous treatment of Stephens (1970) shows that to first order in $\mathcal{H}_Z$, $\boldsymbol{\mu}$ in (4.3.2) should contain the additional term $\Sigma_j(q_j/4m_i^2c^3)[\nabla_i V \cdot \mathbf{r}_i)\mathbf{s}_i - (\mathbf{s}_i \cdot \mathbf{r}_i)\nabla_i V]$, where $V = \Sigma_{i>j} q_i q_j / r_{ij}$ is the electrostatic term in $\mathcal{H}_T$. The net effect is to introduce a small field-dependent additional term into $\mu_\pm$ in (4.3.3) and (4.3.4), which we neglect. Thus we may use (2.11.5), recognizing that $|a\rangle$ and $|j\rangle$ are eigenfunctions of a field-dependent Hamiltonian, $\mathcal{H}_T + \mathcal{H}_Z$. Therefore, since $\boldsymbol{\pi} \equiv (\mathbf{e}_x \pm i\mathbf{e}_y)/\sqrt{2}$ for rcp (+) and lcp (−) light, (2.11.5) becomes

$$\kappa_\pm = \frac{8\pi^3\nu}{hc}(N_a - N_j)$$

$$\times \left| \frac{\alpha}{\sqrt{n}} \langle j|m_\mp|a\rangle \pm i\sqrt{n}\,\langle j|\mu_\mp|a\rangle \pm \sqrt{n}\,\frac{\pi\nu}{c}\langle j|Q_\pm|a\rangle \right|^2 \rho_{aj}(\nu) \tag{4.3.3}$$

where $m_\pm \equiv (m_x \pm im_y)/\sqrt{2}$, $\mu_\pm \equiv (\mu_x \pm i\mu_y)/\sqrt{2}$, $Q_\pm \equiv (Q_{yz} \pm iQ_{xz})/\sqrt{2}$, and corrections for medium effects have been included. Taking the absolute square in (4.3.3) and noting that $\langle j|O_\mp|a\rangle = \langle a|O_\pm|j\rangle^*$, where $O_{\pm}= m_{\pm}, \mu_{\pm}$, or $Q_{\pm}$, we obtain

$$\kappa_{\pm}=\frac{8\pi^3\nu}{hc}(N_a-N_j)\Big\{\alpha^2 n^{-1}|\langle a|m_{\pm}|j\rangle|^2+n\Big(|\langle a|Q_{\mp}|j\rangle|^2+|\langle a|\mu_{\pm}|j\rangle|^2\Big)$$

$$\mp 2\alpha\,\mathscr{I}\mathrm{m}\Big[\langle a|m_{\pm}|j\rangle\Big(\langle j|\mu_{\mp}|a\rangle-\frac{i\pi\nu}{c}\langle j|Q_{\pm}|a\rangle\Big)\Big]$$

$$+2n\,\mathscr{I}\mathrm{m}\,\frac{\pi\nu}{c}\langle a|\mu_{\pm}|j\rangle\langle j|Q_{\pm}|a\rangle\Big\}\rho_{aj}(\nu) \qquad (4.3.4)$$

where $\mathscr{I}\mathrm{m}$ designates imaginary part.

The CD is proportional to $\kappa_- - \kappa_+$. The first three terms in (4.3.4), that is, those containing the absolute squares of matrix elements, describe the absorption of circularly polarized light in electric-dipole, electric-quadrupole, and magnetic-dipole transitions respectively. These terms are shown later to give rise to zero CD in the absence of an applied magnetic field. The last three terms in (4.3.4) can give rise to CD in the absence of a magnetic field and thus are responsible for natural optical activity, the last of the three generally being much the smallest. We are not concerned with natural CD in this book and thus drop these latter terms. We also note that our derivation applies to a single transition, $a \to j$. In general, several transitions may contribute at a given energy— for example, because of degeneracy or because more than one state is populated. Consequently, we sum over all contributing transitions to obtain in the case of MCD

$$\Delta\kappa' \equiv \kappa'_- - \kappa'_+$$

$$=\frac{8\pi^3\nu}{hc}\sum_{aj}(N'_a-N'_j)$$

$$\times\Big[\alpha^2 n^{-1}\Big(|\langle a|m_-|j\rangle'|^2-|\langle a|m_+|j\rangle'|^2\Big)$$

$$+n\Big(|\langle a|Q_+|j\rangle'|^2-|\langle a|Q_-|j\rangle'|^2\Big)$$

$$+n\Big(|\langle a|\mu_-|j\rangle'|^2-|\langle a|\mu_+|j\rangle'|^2\Big)\Big]\rho'_{aj}(\nu) \qquad (4.3.5)$$

Hereafter primes designate magnetic-field-dependent quantities.

Since we are usually concerned with electric-dipole transitions, we omit the magnetic-dipole and electric-quadrupole terms. [Note, however, that aside from the difference in correction factor for medium effects, (4.3.5) has the same form for electric- and magnetic-dipole transitions. In general, in subsequent discussion, our formulas for electric-dipole transitions can be converted to those for magnetic-dipole transitions simply by everywhere replacing m_α with μ_α.] Thus for electric-dipole transitions, (4.3.5) becomes

$$\Delta\kappa' = \frac{8\pi^3\nu\alpha^2}{hcn}\sum_{aj}(N_a' - N_j')\left[|\langle a|m_-|j\rangle'|^2 - |\langle a|m_+|j\rangle'|^2\right]\rho_{aj}'(\nu) \tag{4.3.6}$$

and at *zero field*, neglecting natural CD,

$$\kappa = \frac{\kappa_- + \kappa_+}{2}$$
$$= \frac{4\pi^3\nu\alpha^2}{hcn}\sum_{aj}(N_a - N_j)\left[|\langle a|m_-|j\rangle|^2 + |\langle a|m_+|j\rangle|^2\right]\rho_{aj}(\nu) \tag{4.3.7}$$

The zero-field result follows immediately from the fact that a linearly polarized wave traveling through an optically *inactive* medium can be written as (4.2.5), where (4.2.6) applies with $n_- \equiv n_+$. Equation (4.3.7) also applies for unpolarized light [see (2.7.9)].

We may express (4.3.6) and (4.3.7) in conventional "chemical" units by noting from (2.1.3) and (2.1.6) that $A/cl = \epsilon = (\kappa/c)\log_{10}e$ and

$$\frac{N_a - N_j}{c} = \left(\frac{N_a - N_j}{N}\right)\frac{N}{c} = \frac{N_a - N_j}{N} \times N_0 \times 10^{-3},$$

where N_0 is Avogadro's number and N is the *total* number of absorbing molecules present per cm^3. Then, defining

$$\gamma = \frac{2N_0\pi^3\alpha^2 cl\log_{10}e}{250hcn} \tag{4.3.8}$$

we may finally write

$$\frac{\Delta A'}{\mathcal{E}} = \frac{\epsilon'_- - \epsilon'_+}{\mathcal{E}} cl$$

$$= \gamma \sum_{aj} \frac{N'_a - N'_j}{N} \left(|\langle a|m_-|j\rangle'|^2 - |\langle a|m_+|j\rangle'|^2 \right) \rho'_{aj}(\mathcal{E}) \quad (4.3.9)$$

$$\frac{A}{\mathcal{E}} = \frac{\epsilon_- + \epsilon_+}{2\mathcal{E}} cl$$

$$= \frac{\gamma}{2} \sum_{aj} \frac{(N_a - N_j)}{N} \left(|\langle a|m_-|j\rangle|^2 + |\langle a|m_+|j\rangle|^2 \right) \rho_{aj}(\mathcal{E}) \quad (4.3.10)$$

where A is the absorbance, ϵ the molar extinction coefficient, c the concentration in moles liter^{-1}, l the path length in cm (Section 2.1) and $\mathcal{E}$ ($= h\nu$) is the photon energy. For electronic transitions, $E_j - E_a$ is usually > 1000 cm^{-1}, so $N_j \approx 0$ at room temperature or below.

Equations (4.3.9) and (4.3.10) are the basic equations of MCD spectroscopy. The remainder of this chapter elaborates their detailed consequences.

It is important to keep in mind the distinction between field-dependent and zero-field quantities; we always designate the former with a prime. It is also important to recognize that the operators in (4.3.9) and (4.3.10) are referred to space-fixed axes, that is, axes fixed at the centers of gravity of the molecule but parallel to the laboratory axes (Section 2.5). Since we generally wish to use molecular wavefunctions, it will ultimately be necessary to relate space-fixed and molecule-fixed operators.

4.4. A Simple Example; $\mathcal{A}$, $\mathcal{B}$, and $\mathcal{C}$ Terms

Before discussing (4.3.9) and (4.3.10) in detail, we consider a simple example which illustrates their physical content. Suppose an atom with an s^2 ground configuration undergoes the excitation, $s^2 \to sp$. This can give rise (in the absence of spin–orbit coupling) to two spectroscopic transitions: $^1S \to {}^1P$ and $^1S \to {}^3P$. We assume here some basic familiarity with spectroscopic nomenclature and simply remind the reader that the left superscripts (1 and 3) designate spin singlet and triplet respectively, and S and P designate states with 0 and 1 unit of orbital angular momentum respectively. This subject is discussed at length in Chapter 11 et seq. Brief discussions may be found in most physical-chemistry texts. The $^1S \to {}^1P$ transition, both in the

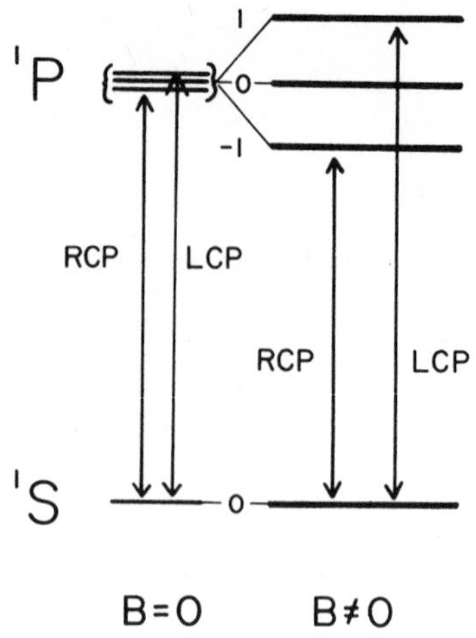

Figure 4.4.1. Zeeman splittings and selection rules for the transition $^1S \leftrightarrow {}^1P$ with light propagated parallel to **B**. The Zeeman subcomponents $^1P - 1$, $^1P0$, $^1P1$ are abbreviated -1, 0, 1 respectively, and they show respective Zeeman shifts of $-g\mu_B B$, 0, and $g\mu_B B$. In this example, $g = 1$.

absence and in the presence of a magnetic field, is illustrated in Fig. 4.4.1. The absorbances for left and right circularly polarized light, the total absorbance and the MCD deduced on the basis of Fig. 4.4.1 are shown in Fig. 4.4.2(a), assuming that linewidths are large compared to Zeeman splittings.

As indicated by the figures, the nonzero MCD, $\Delta A'$ (dashed curves), is entirely a consequence of the fact that lcp and rcp light are absorbed at different energies *in the presence of the field*. The characteristic derivative-shaped MCD is called an $\mathcal{A}$ *term*. [This nomenclature of $\mathcal{A}$, $\mathcal{B}$, $\mathcal{C}$ terms was introduced by Serber in 1932. For a review of early work, see Buckingham and Stephens (1966).] Note particularly that whereas the change in MCD is dramatic when the field is applied, the change in absorption is small or imperceptible. This is a consequence both of the difference-nature of the MCD and the sensitivity with which ΔA can be measured by phase-sensitive techniques. For example, if ΔA were $\sim 10^{-4}$ optical density units (ODU) in Fig. 4.4.2(a), it would be difficult experimentally to detect the change in A itself when the field is applied. On the other hand, a ΔA of 10^{-4} ODU can be measured with ease in MCD, since the state-of-the-art sensitivity is $\sim 1 \times 10^{-6}$ ODU. This immediately suggests one of the great powers of MCD spectroscopy: it can be used to study Zeeman splittings which cannot be resolved by conventional absorption techniques. Note also that $\Delta A'$ is a signed quantity; the $\mathcal{A}$ term in Fig. 4.4.2(a) would have the opposite sense if the Zeeman pattern were reversed —that is, if g were -1 instead of $+1$. The $\mathcal{A}$ term in Fig. 4.4.2(a) is termed positive when $\Delta A'$ is positive to high energy. (See Appendix A for a discussion of MCD sign conventions.)

Suppose now that we use Fig. 4.4.1 to describe the emission process ${}^1S \leftarrow {}^1P$. (We always observe the convention that the spectroscopic state on the left is the one at lower energy. The direction of the arrow distinguishes absorption and emission.) The previous selection rules for lcp and rcp light

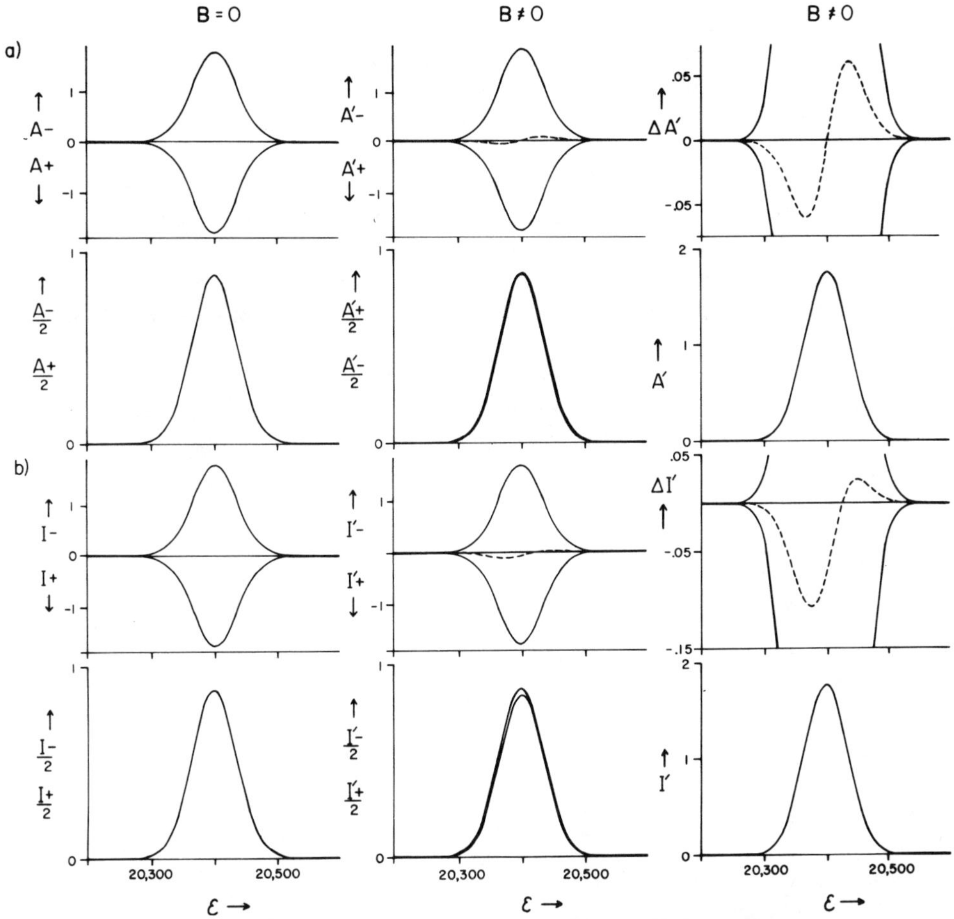

Figure 4.4.2. Absorption (A) and emission (I) patterns (solid curves) for left (−) and right (+) circularly polarized light for transition ${}^1S \leftrightarrow {}^1P$ (Fig. 4.4.1), and resulting MCD ($\Delta A'$) and MCPL ($\Delta I'$) dispersion (dashed curves). Parameters used with a Gaussian lineshape [Eq. (5.3.1)] are: $g\mu_B B = 1\ \text{cm}^{-1}$, $\Delta = 50\ \text{cm}^{-1}$, $kT = 50\ \text{cm}^{-1}$. Relative intensities are shown.

continue to apply as shown in the figure (see Section 4.7 for a proof and further discussion). We use the symbol I to designate emission, and define $\Delta I = (I_- - I_+)$ and $I = (I_- + I_+)/2$ in analogy to ΔA and A. We refer to ΔI as the (magnetic) circularly polarized luminescence, (MCPL), and to I as the total luminescence (or emission)—see Section 4.7. It is clear that an $\mathcal{A}$ term will be observed in emission, and its sense will be the same as in the absorption process. If it is assumed that a Boltzmann equilibrium is established among the Zeeman sublevels before emission, however, a population effect will also be observed. That is, the emission of rcp light will be more intense than that of lcp light, in accordance with the greater population of the $^1P-1$ level. The overall effect is the dashed, skewed curve shown in Fig. 4.4.2(b). Suppose this curve is decomposed into an $\mathcal{A}$ term and an absorptionlike curve in such a way that the two add to give the original dashed curve. Then the absorptionlike curve is referred to as a $\mathcal{C}$ *term*. It reflects the population difference in the two emitting levels and will show a strong temperature dependence. This apparently artificial decomposition into two distinct effects is justified in the next section. Like $\mathcal{A}$ terms, $\mathcal{C}$ terms occur with both positive and negative signs. The $\mathcal{C}$ term component in Fig. 4.4.2(b) is termed negative, because it corresponds to a negative MCD (see Appendix A).

There is one additional effect which is not apparent from Fig. 4.4.1. Application of a magnetic field mixes the zero-field eigenfunctions of any system. In the presence of the field, the ^1S and ^1P states contain a (usually) small amount of an infinite number of other zero-field states of the system,

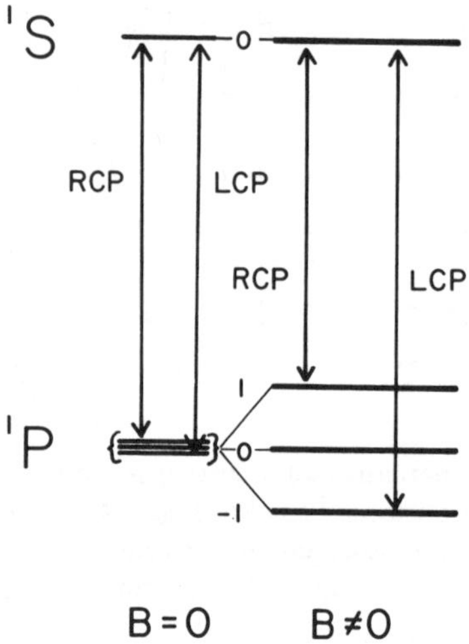

Figure 4.4.3. Zeeman splitting and selection rules for the transition $^1P \leftrightarrow {}^1S$ with light propagated parallel to **B**. Notation and Zeeman shifts are as in Fig. 4.4.1.

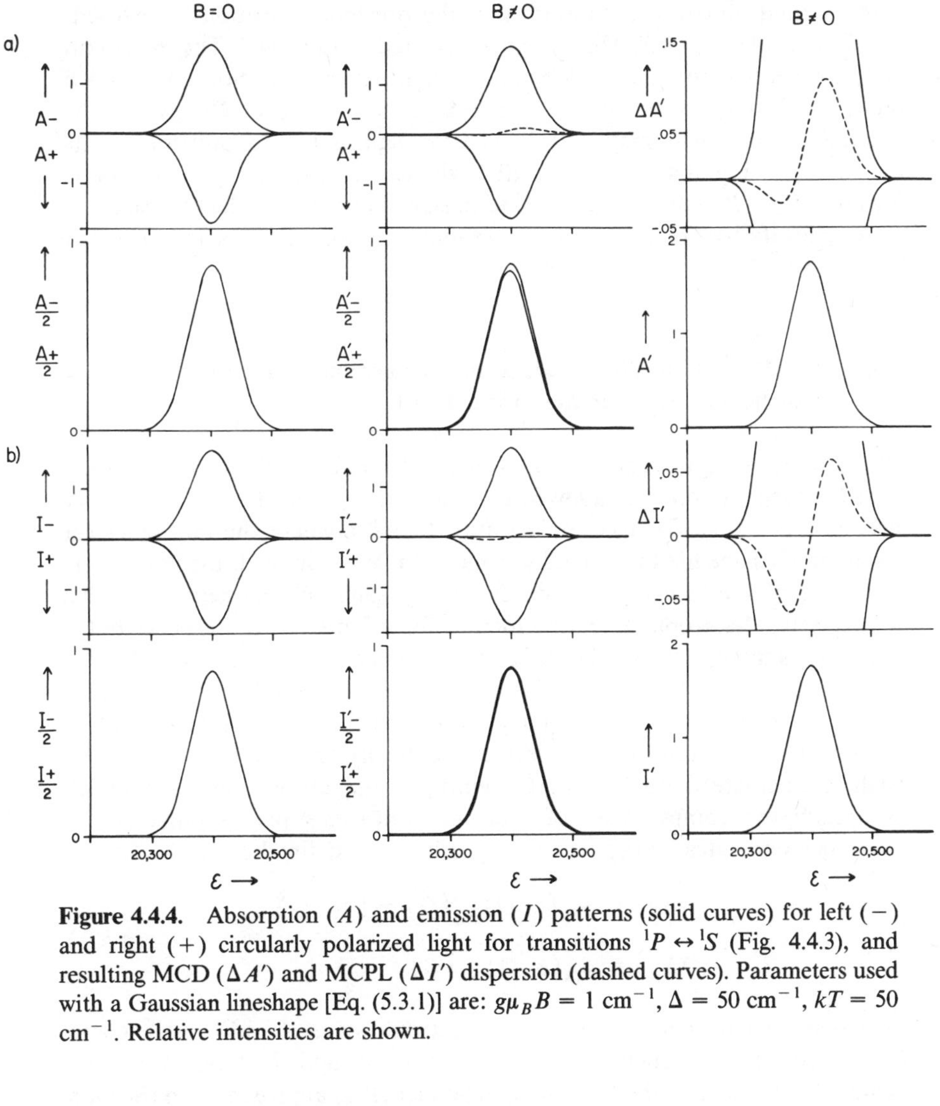

Figure 4.4.4. Absorption (A) and emission (I) patterns (solid curves) for left ($-$) and right ($+$) circularly polarized light for transitions $^1P \leftrightarrow {^1S}$ (Fig. 4.4.3), and resulting MCD ($\Delta A'$) and MCPL ($\Delta I'$) dispersion (dashed curves). Parameters used with a Gaussian lineshape [Eq. (5.3.1)] are: $g\mu_B B = 1\ \text{cm}^{-1}$, $\Delta = 50\ \text{cm}^{-1}$, $kT = 50\ \text{cm}^{-1}$. Relative intensities are shown.

the amount of each depending on the magnetic-moment matrix elements connecting the states and their energy separation. The net effect is that a temperature-independent absorptionlike curve (which can have either sign) arises, which is called a $\mathcal{B}$ *term*. Its exact form is derived in the next section. It is the $\mathcal{B}$ term which makes the Faraday effect (and hence MCD) a universal phenomenon. Clearly, the $\mathcal{A}$ and $\mathcal{C}$ terms vanish if no Zeeman splitting is possible.

As a final illustration, suppose that the previous transition is reversed; that is, consider ${}^1P \leftrightarrow {}^1S$. This case is illustrated in Fig. 4.4.3. The important difference is that the roles of lcp and rcp light reverse (see Section 2.8 for a proof) with respect to the corresponding ${}^1S \leftrightarrow {}^1P$ case. Otherwise, the qualitative arguments may be followed through as before, and the results are shown in Fig. 4.4.4. It is seen that whereas the $\mathcal{A}$-term sign is the same for both ${}^1S \leftrightarrow {}^1P$ and ${}^1P \leftrightarrow {}^1S$, the $\mathcal{C}$-term sign is opposite in the two cases.

A quantitative treatment of the example in Fig. 4.4.1 is presented in Section 5.1.

4.5. The MCD of an Allowed Electronic Transition between Born–Oppenheimer States in the Linear Limit

We now examine the consequences of (4.3.9) and (4.3.10) in the simplest possible case. Suppose an allowed electronic transition, $A \rightarrow J$, takes place between the molecular electronic states A and J where the molecules (or ions) are all imbedded in some inert medium. We start with the case where the molecules either have cubic symmetry (and are termed, somewhat imprecisely, "isotropic") or are identically oriented with respect to the applied magnetic field **B**, the direction of which defines the space-fixed z axis. This situation may be physically realized, for example, for "isotropic" molecules in solution, in a low-temperature matrix, or in a frozen solution or crystal, or for anisotropic molecules identically oriented in a crystal. We wish to calculate the MCD for the entire $A \rightarrow J$ absorption band, which comprises the complete set of vibronic transitions between the Born–Oppenheimer states $|A\alpha g\rangle$ and $|J\lambda j\rangle$, which are defined as in (3.6.1):

$$\begin{aligned} |A\alpha g\rangle &\equiv \phi_{A\alpha}(q, Q)\chi_g(Q) = |A\alpha\rangle|g\rangle \\ |J\lambda j\rangle &\equiv \phi_{J\lambda}(q, Q)\chi_j(Q) = |J\lambda\rangle|j\rangle \end{aligned} \tag{4.5.1}$$

Here α and λ label components of the electronic states A and J, while g and j label vibrational functions associated with A and J respectively. We assume the validity of the Born–Oppenheimer (BO) approximation (Section 3.2 et seq.), even though it is strictly applicable only for widely spaced nondegenerate or Kramers degenerate states. Using the notation above, (4.3.9) and (4.3.10) become

$$\frac{\Delta A'}{\mathcal{E}} = \gamma \sum_{\alpha\lambda gj} \frac{(N'_{A\alpha g} - N'_{J\lambda j})}{N} \left(|\langle A\alpha g|m_-|J\lambda j\rangle'|^2 - |\langle A\alpha g|m_+|J\lambda j\rangle'|^2\right) \times \rho'_{A\alpha g, J\lambda j}(\mathcal{E}) \tag{4.5.2}$$

and

$$\frac{A}{\mathcal{E}} = \frac{\gamma}{2} \sum_{\alpha\lambda gj} \left(\frac{N_{A\alpha g} - N_{J\lambda j}}{N} \right) \left(|\langle A\alpha g | m_- | J\lambda j \rangle|^2 + |\langle A\alpha g | m_+ | J\lambda j \rangle|^2 \right)$$
$$\times \rho_{A\alpha g, J\lambda j}(\mathcal{E}) \tag{4.5.3}$$

respectively, where in the first equation the primes designate magnetic-field-dependent quantities. We assume in the discussion which follows that the states A and J are well separated, so $N_{J\lambda j} = 0$. In addition, we make the FC approximation (Section 3.6) in the evaluation of all matrix elements of electronic operators. Thus, in the language of Chapters 3 and 7, we are working in the BO–FC approximation.

First consider the MCD equation (4.5.2). This expression has a deceptively simple appearance. The complications arise via the primes which designate field-dependent quantities, namely, the wave functions, the ground-state populations $N'_{A\alpha g}$ (if state A shows a Zeeman effect), and the position of an absorption line (if either A or J shows a Zeeman effect). In principle, one can proceed by brute force and determine all of these quantities by perturbation theory for any specific case, and thus evaluate (4.5.2) as a function of frequency. It is the purpose of this section to show that simple, general formulas for the MCD can be derived *if one assumes that Zeeman splittings are small compared both to kT and the bandwidth*, a circumstance we refer to as the *linear limit.*

If a field of several tesla (1 T = 10^4 gauss) is applied to an atomic or molecular system, very small changes in the electronic eigenfunctions and eigenvalues are anticipated. For example, for a free electron ($g \approx 2.0$) in an applied field of 5 T, the Zeeman splitting, $g\mu_B B$, is ≈ 4.7 cm^{-1}, where μ_B is the Bohr magneton. On the other hand, electronic excitations are typically of the order of 10^4 cm^{-1} with separations between electronic states of the order of 10^3 cm^{-1}. A perturbational approach clearly suggests itself, and we proceed on this basis. Basic elements of perturbation theory are summarized in most quantum-mechanics texts. Familiarity with the applications of this technique to degenerate systems, sometimes neglected in standard discussions, is particularly important for an understanding of material presented in this and several subsequent chapters.

It is convenient here to choose unperturbed electronic wavefunctions for A and J which are diagonal in the Zeeman perturbation, $-\mu_z B$, a procedure which is always possible. Thus *within* the A and J manifolds,

$$\begin{aligned} \langle A\alpha g| - \mu_z B | A\alpha' g' \rangle &= -\langle A\alpha | \mu_z | A\alpha \rangle^0 B \delta_{\alpha\alpha'} \delta_{gg'} \\ \langle J\lambda j| - \mu_z B | J\lambda' j' \rangle &= -\langle J\lambda | \mu_z | J\lambda \rangle^0 B \delta_{\lambda\lambda'} \delta_{jj'} \end{aligned} \tag{4.5.4}$$

The $\delta_{gg'}$ and $\delta_{jj'}$ factors follow from the FC approximation, and the zero-superscript notation is that of (3.6.4). Using first-order degenerate perturbation theory and the FC approximation, the perturbed electronic states (designated by primes on the kets) are

$$
\begin{aligned}
|A\alpha g\rangle' &= |A\alpha g\rangle - \sum_{\substack{K\kappa \\ (K\neq A)}} \frac{\langle K\kappa|\mu_z|A\alpha\rangle^0}{W_A^0 - W_K^0} B|K\kappa g\rangle \\
|J\lambda j\rangle' &= |J\lambda j\rangle - \sum_{\substack{K\kappa \\ (K\neq J)}} \frac{\langle K\kappa|\mu_z|J\lambda\rangle^0}{W_J^0 - W_K^0} B|K\kappa j\rangle
\end{aligned}
\tag{4.5.5}
$$

Here W_A^0, W_J^0, and W_K^0 are electronic state energies [Equation (3.2.6)] at $Q = Q_0$. We have made the FC approximation and have assumed the electronic states are well separated, so that energy denominators are not only large compared to Zeeman energies but can be assumed independent of vibrational state. The vibrational functions $|g\rangle$ and $|j\rangle$ appear in the extreme right-hand terms as a result of the relations $\Sigma_k|k\rangle\langle k|g\rangle = |g\rangle$, $\Sigma_k|k\rangle\langle k|j\rangle = |j\rangle$.

Making the same approximations, the electric-dipole matrix elements to first order in the magnetic field are

$$
\begin{aligned}
&\langle A\alpha g|m_\pm|J\lambda j\rangle' \\
&= \left\{\langle A\alpha|m_\pm|J\lambda\rangle^0 + \left[\sum_{\substack{K\kappa \\ (K\neq A)}} \frac{\langle A\alpha|\mu_z|K\kappa\rangle^0\langle K\kappa|m_\pm|J\lambda\rangle^0}{W_K^0 - W_A^0}\right.\right. \\
&\qquad \left.\left. + \sum_{\substack{K\kappa \\ (K\neq J)}} \frac{\langle A\lambda|m_\pm|K\kappa\rangle^0\langle K\kappa|\mu_z|J\lambda\rangle^0}{W_K^0 - W_J^0}\right] B\right\}\langle g|j\rangle
\end{aligned}
\tag{4.5.6}
$$

Again using perturbation theory, the vibronic energies [Equation (3.2.5)] to first order in B are

$$
\begin{aligned}
\mathcal{E}'_{A\alpha g} &= \mathcal{E}_{A\alpha g} - \langle A\alpha|\mu_z|A\alpha\rangle^0 B \\
\mathcal{E}'_{J\lambda j} &= \mathcal{E}_{J\lambda j} - \langle J\lambda|\mu_z|J\lambda\rangle^0 B
\end{aligned}
\tag{4.5.7}
$$

Finally we must evaluate the population factor in (4.5.2). Using (4.5.7) and the first-order expansion

$$
\exp\left\{\frac{\langle A\alpha|\mu_z|A\alpha\rangle^0 B}{kT}\right\} \approx 1 + \frac{\langle A\alpha|\mu_z|A\alpha\rangle^0 B}{kT} \tag{4.5.8}
$$

we obtain

$$\frac{N'_{A\alpha g}}{N} = \frac{N'_{A\alpha g}}{\sum_{\alpha g} N'_{A\alpha g}} = \frac{\exp(-\mathcal{E}'_{A\alpha g}/kT)}{\sum_{\alpha g} \exp(-\mathcal{E}'_{A\alpha g}/kT)}$$

$$= \frac{\exp(-\mathcal{E}_{A\alpha g}/kT)\exp(\langle A\alpha|\mu_z|A\alpha\rangle^0 B/kT)}{\sum_{\alpha g} \exp(-\mathcal{E}_{A\alpha g}/kT)\exp(\langle A\alpha|\mu_z|A\alpha\rangle^0 B/kT)}$$

$$\approx \frac{N_{A\alpha g}}{\sum_{\alpha g} N_{A\alpha g}}\left(1 + \frac{\langle A\alpha|\mu_z|A\alpha\rangle^0 B}{kT}\right)$$

$$= \frac{1}{|A|}\frac{N_g}{\sum_g N_g}\left(1 + \frac{\langle A\alpha|\mu_z|A\alpha\rangle^0 B}{kT}\right) \qquad (4.5.9)$$

The denominator in the second line simplifies because the first-order Zeeman splittings sum to zero: $\Sigma_\alpha \langle A\alpha|\mu_z|A\alpha\rangle^0 = 0$ (Section 14.11). In the last step we denote the electronic degeneracy of A by $|A|$, so that $\Sigma_{\alpha g} N_{A\alpha g} = |A|\Sigma_g N_{A\alpha g} \equiv |A|\Sigma_g N_g$. The expansion (4.5.8) clearly requires that the Zeeman splitting, $\langle A\alpha|\mu_z|A\alpha\rangle^0 B$, be small compared to kT. For a ground state which can be Zeeman-split, this approximation always fails if the temperature is lowered sufficiently. If this is the case, "saturation" effects are said to occur. This phenomenon is discussed in Chapter 6.

The Rigid-Shift Approximation

To complete the evaluation of (4.5.2) we must express $\rho'_{A\alpha g, J\lambda j}(\mathcal{E})$ in terms of the zero-field lineshape. To do this we make the so-called rigid-shift (RS) approximation: we assume that the magnetic field Zeeman-shifts each vibronic transition $|A\alpha g\rangle' \to |J\lambda j\rangle'$ contributing to the band, but does not change its shape, $\rho(\mathcal{E})$. Since a function $\rho(\mathcal{E})$ is translated $(+a)$ if $\mathcal{E}$ is replaced by $\mathcal{E} - a$, the rigid-shift approximation, strictly speaking, means

$$\rho'_{A\alpha g, J\lambda j}(\mathcal{E}) = \rho_{A\alpha g, J\lambda j}(\mathcal{E} - a_{\alpha\lambda}B) \qquad (4.5.10)$$

where $a_{\alpha\lambda}B$ is the Zeeman shift:

$$a_{\alpha\lambda}B \equiv (\mathcal{E}'_{J\lambda j} - \mathcal{E}'_{A\alpha g}) - (\mathcal{E}_{J\lambda j} - \mathcal{E}_{A\alpha g})$$

$$= -(\langle J\lambda|\mu_z|J\lambda\rangle^0 - \langle A\alpha|\mu_z|A\alpha\rangle^0)B \qquad (4.5.11)$$

As indicated ahead, we use the term "rigid-shift approximation" to include other approximations along with this one. The assumption (4.5.10) is eminently reasonable for single vibronic transitions and is exact in the delta-function limit where

$$\rho_{A\alpha g,J\lambda j}(\mathcal{E}) = \delta\left[\left(\mathcal{E}_{J\lambda_j} - \mathcal{E}_{A\alpha g}\right) - \mathcal{E}\right] \tag{4.5.12}$$

The right-hand side of (4.5.10) is next expanded in a Taylor series using the form $f(x+h) = f(x) + hf'(x) + (h^2/2)f''(x) + \cdots$. Derivatives are with respect to x, which is identified as $\mathcal{E}$ in (4.5.10). To first order in $a_{\alpha\lambda}B$ we obtain

$$\begin{aligned}\rho_{A\alpha g,J\lambda j}(\mathcal{E} - a_{\alpha\lambda}B) &= \rho_{A\alpha g,J\lambda j}(\mathcal{E}) + \left(\frac{\partial \rho_{A\alpha g,J\lambda j}(\mathcal{E})}{\partial \mathcal{E}}\right)(-a_{\alpha\lambda}B) + \cdots \\ &\approx \rho_{A\alpha g,J\lambda j}(\mathcal{E}) + \left(\langle J\lambda|\mu_z|J\lambda\rangle^0 - \langle A\alpha|\mu_z|A\alpha\rangle^0\right)B \\ &\quad \times \left(\frac{\partial \rho_{A\alpha g,J\lambda j}(\mathcal{E})}{\partial \mathcal{E}}\right)\end{aligned} \tag{4.5.13a}$$

The first-order approximation above is valid so long as the Zeeman shift, $a_{\alpha\lambda}B$, is small compared to the $|A\alpha g\rangle \to |J\lambda j\rangle$ linewidth Γ, since, as illustrated in Fig. 4.5.1, $\rho'_{A\alpha g,J\lambda j}(\mathcal{E})$ goes into $\rho_{A\alpha g,J\lambda j}(\mathcal{E})$ as $a_{\alpha\lambda}/\Gamma \to 0$. In fact, for a Gaussian lineshape in regions of appreciable absorption, the ratio of successive terms in the Taylor expansion (4.5.13a) goes as $a_{\alpha\lambda}/\Gamma$. Note well however (see definition of $\mathfrak{D}_0$ below) that if an unresolved band with bandshape $f(\mathcal{E})$ is considered, Γ is the width of the composite *band*.

We further assume that the zero-field lineshape function is independent of α and λ, and write

$$\rho_{A\alpha g,J\lambda j}(\mathcal{E}) = \rho_{gj}(\mathcal{E}) \tag{4.5.13b}$$

We use the term "rigid-shift approximation" in this book to mean that the approximations of (4.5.10) and (4.5.13) are *all* being made.

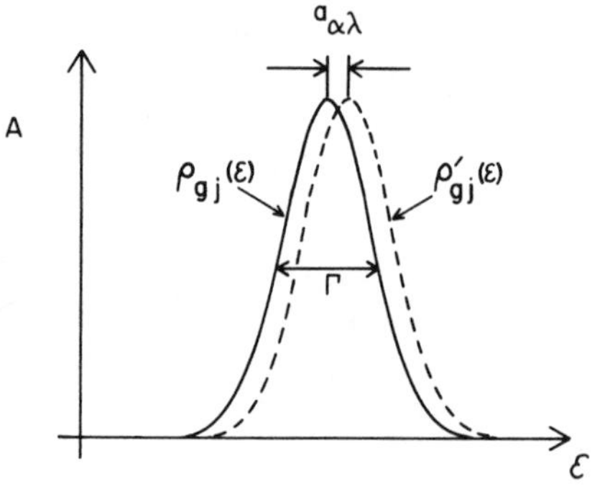

Figure 4.5.1. Zeeman shift per tesla ($a_{\alpha\lambda}$) compared with linewidth Γ.

Definition of $\mathscr{A}_1$, $\mathscr{B}_0$, and $\mathscr{C}_0$

We are now ready to express $\Delta A'/\mathcal{E}$ in terms of parameters which describe the $\mathscr{A}$, $\mathscr{B}$, and $\mathscr{C}$ terms. Substituting (4.5.6), (4.5.9), and (4.5.13) into (4.5.2) and discarding terms quadratic in the field, we obtain in the BO–FC–RS approximation

$$\frac{\Delta A'}{\mathcal{E}} = \gamma\mu_B B\left[\mathscr{A}_1\left(-\frac{\partial f(\mathcal{E})}{\partial \mathcal{E}}\right) + \left(\mathscr{B}_0 + \frac{\mathscr{C}_0}{kT}\right)f(\mathcal{E})\right] \tag{4.5.14}$$

We define $\mathscr{A}_1$, $\mathscr{B}_0$, and $\mathscr{C}_0$ using components of standard vector operators (see Section 9.8), so that [instead of $m_{\pm} = (1/\sqrt{2})(m_x \pm im_y)$]

$$m_{\pm 1} \equiv \mp\frac{1}{\sqrt{2}}(m_x \pm im_y) = \mp m_{\pm} \tag{4.5.15}$$

Basis functions for states A and J are chosen diagonal in μ_z. Thus the MCD parameters are found to be

$$\mathscr{A}_1 = \frac{1}{|A|}\sum_{\alpha\lambda}\left(\langle J\lambda|L_z + 2S_z|J\lambda\rangle^0 - \langle A\alpha|L_z + 2S_z|A\alpha\rangle^0\right)$$

$$\times\left(|\langle A\alpha|m_{-1}|J\lambda\rangle^0|^2 - |\langle A\alpha|m_{+1}|J\lambda\rangle^0|^2\right)$$

$$\mathscr{B}_0 = \frac{2}{|A|}\mathscr{Re}\sum_{\alpha\lambda}\left[\sum_{K\kappa(K\neq J)}\frac{\langle J\lambda|L_z + 2S_z|K\kappa\rangle^0}{W_K^0 - W_J^0}\right.$$

$$\times\left(\langle A\alpha|m_{-1}|J\lambda\rangle^0\langle K\kappa|m_{+1}|A\alpha\rangle^0 - \langle A\alpha|m_{+1}|J\lambda\rangle^0\langle K\kappa|m_{-1}|A\alpha\rangle^0\right)$$

$$+\sum_{\substack{K\kappa\\(K\neq A)}}\frac{\langle K\kappa|L_z + 2S_z|A\alpha\rangle^0}{(W_K^0 - W_A^0)}\left(\langle A\alpha|m_{-1}|J\lambda\rangle^0\langle J\lambda|m_{+1}|K\kappa\rangle^0\right.$$

$$\left.\left.-\langle A\alpha|m_{+1}|J\lambda\rangle^0\langle J\lambda|m_{-1}|K\kappa\rangle^0\right)\right]$$

$$\mathscr{C}_0 = -\frac{1}{|A|}\sum_{\alpha\lambda}\langle A\alpha|L_z + 2S_z|A\alpha\rangle^0$$

$$\times\left(|\langle A\alpha|m_{-1}|J\lambda\rangle^0|^2 - |\langle A\alpha|m_{+1}|J\lambda\rangle^0|^2\right) \tag{4.5.16}$$

In obtaining (4.5.16) we use the identity [see (14.10.5) and (14.10.11)]

$$\langle A\alpha|m_{\pm 1}|J\lambda\rangle^* = -\langle J\lambda|m_{\mp 1}|A\alpha\rangle \tag{4.5.17}$$

and make use of the fact that

$$\sum_{\alpha\lambda}\left(\left|\langle A\alpha|m_{-1}|J\lambda\rangle^0\right|^2 - \left|\langle A\alpha|m_{+1}|J\lambda\rangle^0\right|^2\right) = 0 \tag{4.5.18}$$

The latter is a symmetry property of any atomic or molecular system, which follows from Section 14.11. The left-hand side of (4.5.18) [times $f(\mathcal{E})$] is the only term obtained from (4.5.2) if $B = 0$.

The operator μ_z has been written

$$\mu_z = -\mu_B(L_z + 2S_z) \tag{4.5.19}$$

where μ_B is the Bohr magneton (which equals 9.274×10^{-24} J T^{-1} or 9.274×10^{-21} erg gauss^{-1} or 0.4669 cm^{-1} T^{-1}), and the g value of the electron ($\approx$ 2.00232) has been approximated as 2. Equation (4.5.19) requires that angular-momentum matrix elements be expressed in units of $h/2\pi$, since this factor is included in μ_B; i.e., *$h/2\pi$ is omitted in computing angular-momentum matrix elements.* We always follow this convention. We emphasize the distinction between the operators $m_{\pm 1}$ of (4.5.15) and $m_{\pm}$ used in all earlier sections of the book. The former, which follow the Condon–Shortley phase choice (Section 9.8), is used exclusively henceforth, for reasons which become clear when group-theoretic techniques are employed (Chaper 9 et seq.). We use the new MCD conventions of Stephens (1976), so $\mathcal{A}_1$, $\mathcal{B}_0$, $\mathcal{C}_0$, and $\mathcal{D}_0$ differ from earlier analogs which have been widely used in the literature. These matters are summarized in Appendix A.

Definition of the Dipole Strength $\mathcal{D}_0$

$\mathcal{D}_0$ is defined from (4.5.3). First we make use of the FC approximation to obtain

$$\langle A\alpha g|m_{\pm 1}|J\lambda j\rangle = \langle A\alpha|m_{\pm 1}|J\lambda\rangle^0\langle g|j\rangle \tag{4.5.20}$$

and then recall that

$$\frac{N_{A\alpha g}}{N} = \frac{N_{A\alpha g}}{\sum\limits_{\alpha g} N_{A\alpha g}} = \frac{N_g}{|A|N_G}$$

$$\rho_{A\alpha g, J\lambda j}(\mathcal{E}) \equiv \rho_{gj}(\mathcal{E}) \tag{4.5.21}$$

where $N_G \equiv \Sigma_g N_{A\alpha g} \equiv \Sigma_g N_g$. With these substitutions, (4.5.3) becomes

$$\frac{A}{\mathcal{E}} = \gamma \mathcal{D}_0 f(\mathcal{E}) \tag{4.5.22}$$

where

$$\mathcal{D}_0 = \frac{1}{|A|} \sum_{\alpha\lambda} |\langle A\alpha | m_{\pm 1} | J\lambda \rangle^0|^2$$

$$= \frac{1}{2|A|} \sum_{\alpha\lambda} \left(|\langle A\alpha | m_{-1} | J\lambda \rangle^0|^2 + |\langle A\alpha | m_{+1} | J\lambda \rangle^0|^2 \right) \tag{4.5.23}$$

We use the new conventions of Stephens (1976) in defining $\mathcal{D}_0$ (see Appendix A).

Definition of the Bandshape Function f(𝓔)

Finally, the bandshape function $f(\mathcal{E})$ is defined by

$$f(\mathcal{E}) = \frac{1}{N_G} \sum_g N_g \sum_j |\langle g | j \rangle|^2 \rho_{gj}(\mathcal{E}) \tag{4.5.24}$$

where

$$\int f(\mathcal{E})\, d\mathcal{E} = 1 \tag{4.5.25}$$

Equation (4.5.25) follows because from (2.3.2)

$$\int \rho_{gj}(\mathcal{E})\, d\mathcal{E} = 1$$

and

$$\frac{1}{N_G} \sum_g N_g \sum_j |\langle g | j \rangle|^2 = \frac{1}{N_G} \sum_g N_g = 1 \tag{4.5.26}$$

because the sums over g and j are over the complete sets of vibrational states associated with A and J. If $f(\mathcal{E})$ falls effectively to zero at both ends of the range of integration, then

$$\int \left(\frac{\partial f(\mathcal{E})}{\partial \mathcal{E}} \right) d\mathcal{E} = 0$$
$$\int \left(\frac{\partial f(\mathcal{E})}{\partial \mathcal{E}} \right) \mathcal{E}\, d\mathcal{E} = -1 \tag{4.5.27}$$

as may be confirmed by integrating by parts. We also define

$$\mathcal{E}^0 = \int \mathcal{E} f(\mathcal{E})\, d\mathcal{E} \tag{4.5.28}$$

We see from (4.5.24) that if the rigid shift approximation is applied to $f(\mathcal{E})$, Equations (4.5.10)–(4.5.13a) still apply if $\rho_{A\alpha g,\, J\lambda j}$ is replaced by $f(\mathcal{E})$. The approximation in (4.5.13a) however then requires only that the Zeeman shift be small compared to the width of the *band* contour $f(\mathcal{E})$.

How to Obtain $\mathcal{A}_1$, $\mathcal{B}_0$, $\mathcal{C}_0$, and $\mathcal{D}_0$ from Experimental Data

If (4.5.14) is valid, then application of (4.5.27) and (4.5.28) shows that $\mathcal{A}_1$ and $\mathcal{B}_0 + \mathcal{C}_0/kT$ can be obtained independently by numerically integrating the experimental $\Delta A'/\mathcal{E}$ and $\Delta A'$ over an *entire* MCD band and by similarly integrating $A/\mathcal{E}$ and A over the corresponding absorption band. Furthermore, if this is done as a function of temperature, $\mathcal{B}_0$ and $\mathcal{C}_0$ can be separated. If numerical integration is not feasible, for example because overlap with other bands occurs or because of baseline uncertainties, separation of the parameters may still be possible if the bandshape function is known *a priori*. (It is common to assume a Gaussian bandshape in such attempts—see Section 7.8.)

We note the distinction between the symbols $\mathcal{A}$, $\mathcal{B}$, $\mathcal{C}$ (and $\mathcal{D}$) used with and without subscripts. $\mathcal{A}$, $\mathcal{B}$, and $\mathcal{C}$ describe the dispersion forms (MCD "curves") discussed in Section 4.4. In contrast, $\mathcal{A}_1$, $\mathcal{B}_0$, $\mathcal{C}_0$ (and $\mathcal{D}_0$) are specific mathematical expressions associated with the "strengths" of these effects. In anticipation of our discussion of moments in Chapter 7, we have appended subscripts to the latter to denote a connection with zeroth and first moments.

Further discussion of the properties of $\mathcal{A}_1$, $\mathcal{B}_0$, and $\mathcal{C}_0$ is given in Chapter 5.

Limitations of the Model

Equations (4.5.14) and (4.5.22) are approximate, model-dependent equations. Both are limited to well-isolated electronic states, and both rest on the BO and FC approximations. The MCD equation, (4.5.14), requires further that Zeeman splittings be small compared to kT and the bandwidth, since terms quadratic or higher in the field are neglected.

In practice, the most drastic of the above approximations (besides the limitation to isolated electronic states) is the BO approximation, since MCD measurements are usually made on systems with degeneracy. These systems are often susceptible to vibronic interactions such as JT and PJT coupling which break down the BO approximation. Eigenfunctions in JT (or PJT) systems are not adiabatic, but are in the form of (3.8.6) or (3.8.9). Then vibronic eigenfunctions of the same electronic state are strongly mixed by the magnetic field, and (4.5.4), for example, becomes invalid. In such cases the MCD is no longer described by (4.5.14), and the method of moments, which is discussed in Chapter 7, becomes very useful. The calculation of MCD in systems with near-degenerate electronic states is also a subject of that chapter.

$\mathcal{A}_1$, $\mathcal{B}_0$, $\mathcal{C}_0$, and $\mathcal{D}_0$ are the most important parameters in MCD spectroscopy. Though derived for a very simple model, they continue to play a fundamental role when more sophisticated models are developed. Finally, it must be remembered that *their form as given by* (4.5.16) *and* (4.5.23) *applies only for oriented or isotropic molecules and requires in* (4.5.16) *that the eigenfunctions be diagonal in* μ_z. Generalizations removing these restrictions are developed in the next section.

4.6. Spatial Averaging and Basis Invariance

In carrying out MCD calculations, we usually use wavefunctions quantized with respect to molecular axes, since only then can the MCD be related to the polarization properties of the individual molecules. In order to adapt the $\mathcal{A}_1$, $\mathcal{B}_0$, $\mathcal{C}_0$, and $\mathcal{D}_0$ expressions of (4.5.16) and (4.5.23) to such calculations, we must recognize two distinct problems. First, the operators in these two equations are *space-fixed* (Section 2.5) and thus must be expressed in terms of their *molecule-fixed* counterparts. Second, and easily forgotten, (4.5.16) is valid only if (in degenerate cases) the zero-field wavefunctions ($|A\alpha\rangle$, $|A\alpha'\rangle,\ldots$, $|J\lambda\rangle$, $|J\lambda'\rangle,\ldots,$) are *diagonal in the space-fixed operator* μ_z. Thus if one has a distribution of molecular orientations with respect to the applied field, wavefunctions quantized with respect to the molecular axes cannot in general be used in (4.5.16), even after the transformation is made to molecule-fixed operators. Fortunately, there is a very simple and elegant solution to this problem, based on a theorem called the principle of spectroscopic stability, which we now state and prove.

Spectroscopic Stability

Given two arbitrary operators, $\mathcal{O}_1$ and $\mathcal{O}_2$, consider the expression

$$\sum_t \langle l|\mathcal{O}_1|t\rangle\langle t|\mathcal{O}_2|n\rangle \tag{4.6.1}$$

where $|l\rangle$ and $|n\rangle$ are arbitrary, and the summation is over any orthonormal set of functions, $|t\rangle$. Suppose the set $|t\rangle$ is related by a unitary transformation to another set $|\tau\rangle$ (which is thus also orthonormal). Then

$$|t\rangle = \sum_{\tau} |\tau\rangle\langle\tau|t\rangle$$
$$\langle t| = \sum_{\tau} \langle t|\tau\rangle\langle\tau| \tag{4.6.2}$$

for all $|t\rangle$ in the set. The principle of spectroscopic stability states:

$$\sum_{t} \langle l|\mathcal{O}_1|t\rangle\langle t|\mathcal{O}_2|n\rangle = \sum_{\tau} \langle l|\mathcal{O}_1|\tau\rangle\langle\tau|\mathcal{O}_2|n\rangle \tag{4.6.3}$$

This is proved by substituting (4.6.2) into the left-hand side of (4.6.3):

$$\sum_{t} \langle l|\mathcal{O}_1|t\rangle\langle t|\mathcal{O}_2|n\rangle = \sum_{t\tau\tau'} \langle l|\mathcal{O}_1|\tau\rangle\langle\tau'|\mathcal{O}_2|n\rangle\langle\tau|t\rangle\langle t|\tau'\rangle$$
$$= \sum_{\tau\tau'} \langle l|\mathcal{O}_1|\tau\rangle\langle\tau'|\mathcal{O}_2|n\rangle\delta_{\tau\tau'} = \sum_{\tau} \langle l|\mathcal{O}_1|\tau\rangle\langle\tau|\mathcal{O}_2|n\rangle$$

We have used the fact that the $\langle\tau|t\rangle$ matrix is defined to be unitary.

$\mathcal{A}_1$, $\mathcal{B}_0$, $\mathcal{C}_0$, $\mathcal{D}_0$ *in Basis-Invariant Form*

Let us now use this theorem to remove the restriction in (4.5.16) that (degenerate) zero-field functions must be diagonal in μ_z. We choose one of the terms in $\mathcal{C}_0$ to illustrate the argument. Consider

$$T \equiv \sum_{\alpha\lambda} |\langle A\alpha|m_{-1}|J\lambda\rangle|^2 \langle A\alpha|\mu_z|A\alpha\rangle \tag{4.6.4}$$

Since the $|A\alpha\rangle$ basis is diagonal in μ_z,

$$\langle A\alpha'|\mu_z|A\alpha\rangle = \delta_{\alpha\alpha'}\langle A\alpha|\mu_z|A\alpha\rangle \tag{4.6.5}$$

Therefore,

$$T = \sum_{\alpha\lambda} \langle A\alpha|m_{-1}|J\lambda\rangle\langle A\alpha|m_{-1}|J\lambda\rangle^* \langle A\alpha|\mu_z|A\alpha\rangle$$
$$= -\sum_{\alpha\alpha'\lambda} \langle A\alpha|m_{-1}|J\lambda\rangle\langle J\lambda|m_{+1}|A\alpha'\rangle\langle A\alpha'|\mu_z|A\alpha\rangle \tag{4.6.6}$$

Consider two sets of functions, $|A\bar{\alpha}\rangle$, $|A\bar{\alpha}'\rangle, \ldots$, and $|J\bar{\lambda}\rangle$, $|J\bar{\lambda}'\rangle, \ldots$, related by arbitrary unitary transformations respectively to the sets $|A\alpha\rangle$, $|A\alpha'\rangle, \ldots$, and $|J\lambda\rangle$, $|J\lambda'\rangle \ldots$. Applying (4.6.3) successively to $|J\lambda\rangle\langle J\lambda|$, $|A\alpha'\rangle\langle A\alpha'|$, and $|A\alpha\rangle\langle A\alpha|$ in (4.6.6) gives

$$T = -\sum_{\bar{\alpha}\bar{\alpha}'\bar{\lambda}} \langle A\bar{\alpha}|m_{-1}|J\bar{\lambda}\rangle\langle J\bar{\lambda}|m_{+1}|A\bar{\alpha}'\rangle\langle A\bar{\alpha}'|\mu_z|A\bar{\alpha}\rangle \tag{4.6.7}$$

Comparing (4.6.7) with (4.6.6), we conclude that (4.6.6) *is invariant to unitary transformations on the basis sets* $|A\alpha\rangle$ *and* $|J\lambda\rangle$. Consequently, if (4.6.4) is written in the form (4.6.6), the requirement that the $|A\alpha\rangle$ and $|J\lambda\rangle$ sets be diagonal in μ_z may be dropped, provided only that the actual bases used are related by unitary transformations to the diagonal bases. *A knowledge of the specific transformations is unnecessary.* We refer to (4.6.6) as a *basis-invariant form*. Using the type of argument just illustrated, we may write (4.5.16) and (4.5.23) in invariant form. The results, upon which the oriented or isotropic-case equations of Section 17.2 are based, are

$$\mathcal{A}_1 = -\frac{1}{|A|}\sum_{\substack{\alpha\alpha' \\ \lambda\lambda'}} \left(\langle J\lambda|L_z + 2S_z|J\lambda'\rangle^0\delta_{\alpha\alpha'} - \langle A\alpha'|L_z + 2S_z|A\alpha\rangle^0\delta_{\lambda\lambda'}\right)$$

$$\times \left(\langle A\alpha|m_{-1}|J\lambda\rangle^0\langle J\lambda'|m_{+1}|A\alpha'\rangle^0 - \langle A\alpha|m_{+1}|J\lambda\rangle^0\langle J\lambda'|m_{-1}|A\alpha'\rangle^0\right)$$

$$\mathcal{B}_0 = \text{same as in (4.5.16)}$$

$$\mathcal{C}_0 = \frac{1}{|A|}\sum_{\alpha\alpha'\lambda} \langle A\alpha'|L_z + 2S_z|A\alpha\rangle^0$$

$$\times \left(\langle A\alpha|m_{-1}|J\lambda\rangle^0\langle J\lambda|m_{+1}|A\alpha'\rangle^0 - \langle A\alpha|m_{+1}|J\lambda\rangle^0\langle J\lambda|m_{-1}|A\alpha'\rangle^0\right)$$

$$\mathcal{D}_0 = \text{same as (4.5.23)} \tag{4.6.8}$$

Useful identities if one wishes to use Cartesian components of the electric-dipole operator in (4.6.8) are

$$\sum_{\alpha\alpha'} \langle A\alpha'|L_z + 2S_z|A\alpha\rangle^0$$

$$\times \left(\langle A\alpha|m_{-1}|J\lambda\rangle^0\langle J\lambda|m_{+1}|A\alpha'\rangle^0 - \langle A\alpha|m_{+1}|J\lambda\rangle^0\langle J\lambda|m_{-1}|A\alpha'\rangle^0\right)$$

$$= 2\,\mathcal{I}\text{m}\sum_{\alpha\alpha'} \langle A\alpha'|L_z + 2S_z|A\alpha\rangle^0\langle A\alpha|m_x|J\lambda\rangle^0\langle J\lambda|m_y|A\alpha'\rangle^0$$

$$\sum_{\alpha\lambda}|\langle A\alpha|m_{+1}|J\lambda\rangle^0|^2 = \sum_{\alpha\lambda}|\langle A\alpha|m_{-1}|J\lambda\rangle^0|^2$$

$$= \frac{1}{2}\sum_{\alpha\lambda}\left(|\langle A\alpha|m_x|J\lambda\rangle^0|^2 + |\langle A\alpha|m_y|J\lambda\rangle^0|^2\right)$$

$$|\langle A\alpha|m_{+1}|J\lambda\rangle^0|^2 + |\langle A\alpha|m_{-1}|J\lambda\rangle^0|^2$$

$$= |\langle A\alpha|m_x|J\lambda\rangle^0|^2 + |\langle A\alpha|m_y|J\lambda\rangle^0|^2 \qquad (4.6.9)$$

where $\mathscr{I}\mathrm{m}$ designates the imaginary part of everything to its right: recall that $\mathscr{I}\mathrm{m}(a + ib) \equiv b$ for a, b real. These equations follow because m_x, m_y, and μ_z are Hermitian and wavefunctions can always be chosen real in the absence of a magnetic field. It is instructive to confirm (4.6.9)—the equations are not correct in the absence of the summations.

Molecule-Fixed Operators and Orientational Averaging; $\bar{\mathscr{A}}_1$, $\bar{\mathscr{B}}_0$, $\bar{\mathscr{C}}_0$, *and* $\bar{\mathscr{D}}_0$

Let us designate a set of molecule-fixed Cartesian axes (Section 2.5) by x', y', z'. These may be related to a space-fixed coordinate system (unprimed) with the same origin using the Eulerian angles defined in Goldstein (1950), Section 4-4. Specifically,

$$\begin{pmatrix} x \\ y \\ z \end{pmatrix} = (A)\begin{pmatrix} x' \\ y' \\ z' \end{pmatrix} \qquad (4.6.10a)$$

where the orthogonal matrix (A) has been derived by Goldstein (1950) and is as follows:

$$(A) =$$

$$\begin{pmatrix} \cos\psi\cos\phi - \cos\theta\sin\phi\sin\psi & \cos\psi\sin\phi + \cos\theta\cos\phi\sin\psi & \sin\psi\sin\theta \\ \sin\psi\cos\phi - \cos\theta\sin\phi\cos\psi & -\sin\psi\sin\phi + \cos\theta\cos\phi\cos\psi & \cos\psi\sin\theta \\ \sin\theta\sin\phi & -\sin\theta\cos\phi & \cos\theta \end{pmatrix}$$

$$(4.6.10b)$$

It follows immediately that

$$
\begin{aligned}
m_x &= A_{11}m_{x'} + A_{12}m_{y'} + A_{13}m_{z'} \\
m_y &= A_{21}m_{x'} + A_{22}m_{y'} + A_{23}m_{z'} \\
\mu_z &= A_{31}\mu_{x'} + A_{32}\mu_{y'} + A_{33}\mu_{z'}
\end{aligned}
\tag{4.6.11}
$$

If we are dealing with a single molecule held at some specified orientation with respect to the field, the A_{ij} are uniquely defined. Substitution of (4.6.11) into $\mathcal{A}_1$, $\mathcal{B}_0$, $\mathcal{C}_0$, and $\mathcal{D}_0$ of (4.6.8) yields the expressions we desire containing molecule-fixed operators everywhere. If the molecule possesses symmetry, further simplifications are generally possible, since some matrix elements may vanish and relationships among others may exist. Much of the material in later chapters is concerned with this most important matter.

More common is the situation in which a collection of molecules is randomly distributed. Then (4.6.8) must be averaged over orientation. To do this, we note that the orientational average of any function, say $F(\theta, \phi, \psi)$, over the Eulerian angles is given by

$$
\bar{F}(\theta, \phi, \psi) = \frac{\displaystyle\int_{\psi=0}^{2\pi}\int_{\phi=0}^{2\pi}\int_{\theta=0}^{\pi} F(\theta, \phi, \psi)\, d\tau}{\displaystyle\int_{\psi=0}^{2\pi}\int_{\phi=0}^{2\pi}\int_{\theta=0}^{\pi} d\tau}
\tag{4.6.12}
$$

where $d\tau = \sin\theta\, d\theta\, d\phi\, d\psi$. This equation may be understood by noting that an orientational average is accomplished by first rotating ψ from 0 to 2π about the molecule-fixed z' axis. The molecule-fixed z' axis is then swept over the unit sphere, precisely as is done with a radius vector in spherical polar coordinates (where the differential volume element of solid angle is $\sin\theta\, d\theta\, d\phi$). Thus $d\tau = \sin\theta\, d\theta\, d\phi\, d\psi$, and $\int d\tau = 8\pi^2$. The required averages may easily be performed in the present case, since [Andrews and Thirunamachandran (1977); Power and Thirunamachandran (1974), and references therein]

$$
\begin{aligned}
\frac{1}{8\pi^2}\int\!\!\int\!\!\int A_{i\lambda}A_{j\mu}\sin\theta\, d\theta\, d\phi\, d\psi &= \tfrac{1}{3}\delta_{ij}\delta_{\lambda\mu} \\
\frac{1}{8\pi^2}\int\!\!\int\!\!\int A_{i\lambda}A_{j\mu}A_{k\nu}\sin\theta\, d\theta\, d\phi\, d\psi &= \tfrac{1}{6}\epsilon_{ijk}\epsilon_{\lambda\mu\nu}
\end{aligned}
\tag{4.6.13}
$$

The *alternating tensor* ϵ_{ijk} has the following properties: $\epsilon_{ijk} = 0$ unless i, j, k

are all different; $\epsilon_{ijk} = 1$ for $i, j, k = 1,2,3$ or any even permutation of i, j, k, and $\epsilon_{ijk} = -1$ for any odd permutation. Equation (4.6.11) is substituted into the right-hand side of (4.6.9), which is then used in (4.6.8). Many terms result, but most are zero through (4.6.13), and the rest are $\pm\frac{1}{6}$ or $\frac{1}{3}$.

After some algebra we obtain for our orientationally averaged parameters

$$\bar{\mathcal{A}}_1 = \frac{i}{3|A|}\sum_{\substack{\alpha\alpha'\\ \lambda\lambda'}}\left(\langle J\lambda|(\mathbf{L}+2\mathbf{S})|J\lambda'\rangle^0\delta_{\alpha\alpha'} - \langle A\alpha'|(\mathbf{L}+2\mathbf{S})|A\alpha\rangle^0\delta_{\lambda\lambda'}\right)$$

$$\cdot\left(\langle A\alpha|\mathbf{m}|J\lambda\rangle^0 \times \langle J\lambda'|\mathbf{m}|A\alpha'\rangle^0\right)$$

$$\bar{\mathcal{B}}_0 = \frac{2\,\mathcal{I}m}{3|A|}\sum_{\alpha\lambda}\left[\sum_{\substack{K\kappa\\ (K\neq A)}}\frac{\langle K\kappa|(\mathbf{L}+2\mathbf{S})|A\alpha\rangle^0}{W_K^0 - W_A^0}\right.$$

$$\cdot\left(\langle A\alpha|\mathbf{m}|J\lambda\rangle^0 \times \langle J\lambda|\mathbf{m}|K\kappa\rangle^0\right)$$

$$+\sum_{\substack{K\kappa\\ (K\neq J)}}\frac{\langle J\lambda|(\mathbf{L}+2\mathbf{S})|K\kappa\rangle^0}{W_K^0 - W_J^0}$$

$$\left.\cdot\left(\langle A\alpha|\mathbf{m}|J\lambda\rangle^0 \times \langle K\kappa|\mathbf{m}|A\alpha\rangle^0\right)\right]$$

$$\bar{\mathcal{C}}_0 = -\frac{i}{3|A|}\sum_{\alpha\alpha'\lambda}\langle A\alpha'|(\mathbf{L}+2\mathbf{S})|A\alpha\rangle^0\cdot\left(\langle A\alpha|\mathbf{m}|J\lambda\rangle^0 \times \langle J\lambda|\mathbf{m}|A\alpha'\rangle^0\right)$$

$$\bar{\mathcal{D}}_0 = \frac{1}{3|A|}\sum_{\alpha\lambda}\left|\langle A\alpha|\mathbf{m}|J\lambda\rangle^0\right|^2 \tag{4.6.14}$$

where $\times$ and $\cdot$ designate the conventional vector and scalar products. Our space-averaged equations of Section 17.4 are based on (4.6.14). We have dropped the primes on the operators, but distinguish (4.6.14) from expressions containing space-fixed operators [(4.5.16), (4.5.23), (4.6.8)] by the bars over the Faraday parameters. *These always designate spatially averaged parameters expressed in terms of molecule-fixed operators*. Equation (4.6.14) also applies for freely rotating molecules, provided classical averaging is a satisfactory approximation. As before, we use the new MCD conventions (Appendix A), and $i \equiv \sqrt{-1}$.

The parameters in (4.6.14) may be written in an alternate form using the following identities [and the definitions (see Section 9.8) $m_{\pm 1} = \mp(1/\sqrt{2})(m_x \pm im_y)$, and $\mu_{\pm 1} = \mp(1/\sqrt{2})(\mu_x \pm i\mu_y)$]

$$i \sum_{\alpha\alpha'\lambda\lambda'} \langle A\alpha'|\boldsymbol{\mu}|A\alpha\rangle \cdot (\langle A\alpha|\mathbf{m}|J\lambda\rangle \times \langle J\lambda'|\mathbf{m}|A\alpha'\rangle)$$

$$= \mathcal{R}e \sum_{\alpha\alpha'\lambda\lambda'} [\langle A\alpha'|\mu_z|A\alpha\rangle$$

$$\times (\langle A\alpha|m_{-1}|J\lambda\rangle\langle A\alpha'|m_{-1}|J\lambda'\rangle^* - \langle A\alpha|m_{+1}|J\lambda\rangle\langle A\alpha'|m_{+1}|J\lambda'\rangle^*)$$

$$-2\langle A\alpha'|\mu_{-1}|A\alpha\rangle(\langle A\alpha|m_z|J\lambda\rangle\langle A\alpha'|m_{-1}|J\lambda'\rangle^*$$

$$+\langle A\alpha|m_{+1}|J\lambda\rangle\langle A\alpha'|m_z|J\lambda'\rangle^*)] \qquad (4.6.15a)$$

$$|\langle A\alpha|\mathbf{m}|J\lambda\rangle^0|^2 = \sum_{\gamma=x,y,z} |\langle A\alpha|m_\gamma|J\lambda\rangle^0|^2$$

$$= |\langle A\alpha|m_{+1}|J\lambda\rangle^0|^2 + |\langle A\alpha|m_{-1}|J\lambda\rangle^0|^2 + |\langle A\alpha|m_z|J\lambda\rangle^0|^2$$

$$(4.6.15b)$$

Note particularly that (4.6.15a) does *not* apply if the summations (over α, α', λ, and λ') are omitted. It is instructive to examine the reason for this.

Once again we emphasize that the presence of molecular symmetry permits important simplifications by application of group-theoretic techniques to be discussed in later chapters. Equation (4.6.15) is particularly useful in this regard.

Finally, it is physically obvious that orientational averaging is unnecessary for a system of isotropic absorbers, such as free atoms. In such a case, (4.6.14) and (4.6.8) are exactly equivalent; that is the equations of Section 17.2 for the oriented or isotropic case are equivalent to the equations of Section 17.4 for the space-averaged case. Furthermore, (4.5.16) can be used conveniently, since it is easy to choose basis functions diagonal in μ_z (molecule-fixed), and the "molecule-fixed" z axis of every atom can be chosen to coincide with the space-fixed axis. A more relevant question is whether these simplifications carry over to molecules despite the fact that they are never strictly isotropic. The answer is yes, in certain cases. If the x', y', and z' molecule-fixed axes can be chosen symmetrically equivalent (this is the case in the cubic point groups, e.g., O_h, O, T_d), then the equivalences described above always apply if the molecule (or ion) has an even number of electrons, and *usually* apply for an odd number. They always apply for the MCD in the linear limit (see Section 17.5). For example, given a randomly oriented collection of octahedral molecules, one can use (4.5.16) or (4.6.8), simply regarding the operators as molecule-fixed—provided, of course, in using the former that basis functions diagonal in $\mu_{z'}$ (molecule-fixed) are used. Thus in this case,

$$\begin{aligned} \bar{\mathcal{A}}_1 &= \mathcal{A}_1, & \bar{\mathcal{B}}_0 &= \mathcal{B}_0 \\ \bar{\mathcal{C}}_0 &= \mathcal{C}_0, & \bar{\mathcal{D}}_0 &= \mathcal{D}_0 \end{aligned} \qquad (4.6.16)$$

A formal proof of these statements requires group-theoretic methods which will be developed in later chapters. In fact, this "isotropy" of cubic systems is intuitively reasonable, and the surprising thing perhaps is that an exception exists in the Zeeman effect in odd-electron molecules.

4.7. Emission; Magnetic Circularly Polarized Luminescence (MCPL); Photoselection

Intimately connected with any transition in absorption, $A \to J$, is the corresponding emission, $A \leftarrow J$. (We reiterate the convention always followed in this book. In nomenclature of the form $A \to J$ and $A \leftarrow J$, the symbol on the left always designates the lower energy state, so that the direction of the arrow distinguishes absorption and emission.) It was first shown by Einstein that the emission process is the sum of an induced and a spontaneous part (see Section 2.4), the former being proportional to the radiation density at the resonance frequency; (2.4.3) gives the integrated relation between these two.

Since a longitudinal magnetic field induces a differential absorption of lcp and rcp light (MCD), we should expect to observe the same effect in emission. Since this effect is generally observed in spontaneous emission, we adopt the name *magnetic circularly polarized luminescence* (MCPL) to emphasize the spontaneous nature of the process.

Spontaneous Emission

With appropriate care, we can adapt the machinery previously developed for MCD (Section 4.5) to describe the MCPL process. Our fundamental guide is (2.4.2):

$$P_{a\to j} = P_{a\leftarrow j}$$

This equation states that for any arbitrary polarization of the radiation, the probability of inducing a transition $a \to j$ in absorption is exactly equal to the probability of *inducing* the reverse transition, $a \leftarrow j$, in emission. In particular, if a given transition, $a \to j$, happens to occur with absorption of a lcp (rcp) photon, the corresponding emission $a \leftarrow j$ will occur with the emission of a lcp (rcp) photon. Since spontaneous emission is related to induced emission by a simple factor, the same *selection rules* apply for spontaneous emission.

We derive a quantitative expression for spontaneous emission as follows. From the discussion leading to (2.4.1), we know that to relate $P_{a \to j}$ to $P_{j \leftarrow a}$ (for induced emission) we merely substitute $\mathbf{A}^0$ for $\mathbf{A}^{0*}$ [or $\boldsymbol{\pi}$ for $\boldsymbol{\pi}^*$ in view of (2.6.4)] and ρ_{ja} for ρ_{aj}. Let us for the moment confine ourselves to electric-dipole transitions. From (2.6.8) and the above discussion it follows that for induced emission, $P_{j \leftarrow a}$ is proportional to $|\langle j|\mathbf{m} \cdot \boldsymbol{\pi}|a\rangle|^2$. Equation (2.4.3) in turn suggests that the spontaneous emission is related to the induced emission by some scalar proportionality factor, though not the specific one in (2.4.3), which applies for emission summed over all orientations and polarizations. Thus we write for the spontaneous emission

$$P_{j \leftarrow a}(S) = K|\langle j|\mathbf{m} \cdot \boldsymbol{\pi}|a\rangle|^2 \tag{4.7.1}$$

In the spirit of our semiclassical treatment, we determine the proportionality constant K by requiring that $P_{j \leftarrow a}(S)$ be the Einstein coefficient of spontaneous emission when (4.7.1) is summed over both states of photon polarization and is integrated over all orientations. Thus

$$\sum_{\text{pol}} \int P_{j \leftarrow a}(S)\, d\Omega = \frac{64\pi^4}{3hc^3} \nu_{aj}^3 |\langle j|\mathbf{m}|a\rangle|^2 \tag{4.7.2}$$

where Ω represents solid angle and the right-hand side is the Einstein coefficient of spontaneous emission, which is obtained by the thermodynamic-type argument mentioned earlier [see (2.4.3)]. One finds [Griffith (1964), p. 53]

$$\sum_{\text{pol}} \int |\langle j|\mathbf{m} \cdot \boldsymbol{\pi}|a\rangle|^2\, d\Omega = \frac{8\pi}{3} |\langle j|\mathbf{m}|a\rangle|^2 \tag{4.7.3}$$

It follows immediately that $K = 8\pi^3 \nu_{aj}^3 / hc^3$, and thus (4.7.1) becomes

$$P_{j \leftarrow a}(S) = \frac{8\pi^3 \nu_{aj}^3}{hc^3} |\langle j|\mathbf{m} \cdot \boldsymbol{\pi}|a\rangle|^2 \tag{4.7.4}$$

Equation (4.7.4) is the basic emission equation; it gives the probability, per unit time per unit solid angle, of the spontaneous emission of a photon in a $\boldsymbol{\pi}$ state of polarization. The direction of propagation is perpendicular to $\boldsymbol{\pi}$. We emphasize again that if a fully quantum-mechanical treatment of the radiation field is carried out [Griffith (1964), Chapter 3; Loudon (1973)], (4.7.4) results directly without any need to appeal to the Einstein argument. Magnetic-dipole and electric-quadrupole terms can be added to (4.7.4) in analogy with (2.7.8) after dropping the stars on $\boldsymbol{\pi}$, π_1, and π_2 in the latter. To convert (4.7.4) to an intensity equation, we need only multiply by the concentration of molecules in the emitting state (N_a) and the energy of the emitted photon ($h\nu_{aj}$). To obtain a finite linewidth (or bandwidth) we append a lineshape function just as we did in the absorption case [Eq. (2.3.1)]. Thus the general equation for the spontaneous emission intensity is

$$I_{j\leftarrow a} = \frac{8\pi^3 N_a \nu^4}{c^3} |\langle j|\mathbf{m}\cdot\boldsymbol{\pi}|a\rangle|^2 \rho_{j\leftarrow a}(\nu) \tag{4.7.5}$$

[Hereafter, we write $\rho_{j\leftarrow a}(\nu)$ as $\rho_{ja}(\nu)$; see parenthetical note following (2.4.2).]

Magnetic Circularly Polarized Luminescence

For MCPL, by analogy to $A'_- - A'_+ = \Delta A'$ in MCD, we wish to calculate $I'_- - I'_+ \equiv \Delta I'$. From our previous discussion it follows that $I'_- - I'_+$ differs from $A'_- - A'_+$ by a proportionality factor *and in sign*. Specifically, for the emission $j' \leftarrow a'$, the equation analogous to (4.3.9) is

$$\frac{\Delta I'}{\mathcal{E}} = \frac{I'_- - I'_+}{\mathcal{E}}$$

$$= \frac{8\pi^3\nu^3 N}{c^3} \sum_{aj} \frac{N'_a}{N} \left[|\langle a|m_+|j\rangle'|^2 - |\langle a|m_-|j\rangle'|^2 \right] \rho'_{ja}(\mathcal{E}) \tag{4.7.6}$$

The change in sign between (4.7.6) and (4.3.9) is a consequence of the change ($\boldsymbol{\pi}^* \to \boldsymbol{\pi}$ and thus $m_\mp \to m_\pm$) in going from $a \to j$ (absorption) to $j \leftarrow a$ (emission).

Clearly, one can now apply the entire analysis of Sections 4.5–4.6 to (4.7.6), since it and (4.3.9) are identical in form. It is however necessary to consider two further factors. First, one must assume that (4.5.9) applies (where N, from now on written as N_e, designates the total number of molecules per cm^3 in the emitting manifold). This assumption, that the emitting Zeeman sublevels achieve a pseudo-Boltzmann distribution, must be justified experimentally on a case-by-case basis, since it clearly depends strongly on both the lifetime of the emitting state and the spin–lattice relaxation time. Second, we must consider the phenomenon of photoselection. This effect arises because, for a nonisotropic molecule, the probability of excitation is a function of the molecular orientation with respect to the direction of propagation and the state of polarization of the exciting light beam. (For example, suppose a linear molecule were excited via an electronic transition polarized along the molecular axis. Then clearly, molecules parallel to the light propagation direction would suffer no excitation, those perpendicular to it a maximum, and others an intermediate amount depending on their orientation.) Consequently, if the excited molecules do not randomize their orientation prior to emission, the subsequent emission properties will reflect the relative orientation (photoselection) imposed in the excitation process. We treat two limiting cases: complete randomization before emission, and none at all.

MCPL *with No Photoselection*

Let us first suppose that complete randomization occurs before emission. Then photoselection can be entirely ignored and the previous MCD analysis can be taken over in its entirety [assuming the validity of (4.5.9)] and applied to (4.7.6). The only change is that $A\alpha g$ and $J\lambda j$ must be interchanged in (4.5.11). This interchange is necessary because in the emission case, A is the higher-energy state. This has the effect of changing the sign of $\mathcal{A}_1$ relative to $\mathcal{B}_0$ and $\mathcal{C}_0$. The results for the allowed electronic transition $J \leftarrow A$ can be written, analogously, to (4.5.14),

$$\frac{\Delta I'}{\mathcal{E}} = -\frac{8\pi^3\nu^3 N_e \mu_B B}{c^3}\left[\bar{\mathcal{A}}_1\left(\frac{\partial f(\mathcal{E})}{\partial \mathcal{E}}\right) + \left(\bar{\mathcal{B}}_0 + \frac{\bar{\mathcal{C}}_0}{kT}\right) f(\mathcal{E})\right] \tag{4.7.7}$$

$\bar{\mathcal{B}}_0$, $\bar{\mathcal{C}}_0$, and $\bar{\mathcal{A}}_1$ are defined by (4.6.14) exactly as in the MCD case; that is, we define $\mathcal{A}_1(J \rightarrow A) \equiv \mathcal{A}_1(A \leftarrow J)$ and so on. [Strictly speaking, since ν^3 ($\approx \nu_{aj}^3$) in (4.7.6) is a function of orientation, it should be explicitly included and appropriately averaged in obtaining (4.7.7). For an electronic transition, this procedure has negligible consequences and can be ignored. Also, the right side of (4.7.7) should contain the approximate correction, α^2/n, for condensed-medium effects, as discussed earlier (Section 2.11). Since N_e is seldom known accurately (see later), we omit α^2/n.] Thus, *to convert the MCD expression* (4.5.14) *for the transition $A \rightarrow J$ to the MCPL expression for the transition $J \leftarrow A$, one simply multiplies $\mathcal{A}_1$ by -1 and replaces γ by*

$(-8\pi^3\nu^3 N_e)/c^3$. The Faraday parameters are then calculated, maintaining the position of states A and J in the (4.6.14) expressions. We emphasize that we directly relate $A \to J$ to $J \leftarrow A$ (rather than $A \to J$ to the corresponding emission, $A \leftarrow J$). However, having the former relation, the latter can be readily established.

The connection between the total emission and the dipole strength $\mathcal{D}_0$ may be immediately established. The result is

$$\frac{I}{\mathcal{E}} = \frac{8\pi^3\nu^3 N_e}{c^3} f(\mathcal{E})\overline{\mathcal{D}}_0 \tag{4.7.8}$$

and

$$\overline{\mathcal{D}}_0(A \to J) \equiv \overline{\mathcal{D}}_0(J \leftarrow A) \tag{4.7.9}$$

MCPL *and Total Emission with Photoselection*

We now consider the role of photoselection in MCPL, assuming that the sample molecules undergo no change in orientation during the period between excitation and emission. This limit is often realized in crystals, solid solutions, glasses, or matrices—particularly at low temperatures. To treat this case we simply multiply the emission probability by the absorption probability, before performing an average over orientation.

Let us suppose that the propagation direction of a linearly polarized excitation beam lies in the yz plane and is at an angle β with respect to the space-fixed z axis (the magnetic-field direction); the plane of polarization of the electric vector makes an angle α with respect to the space-fixed x axis (see Fig. 4.7.1). The unit polarization vector $\boldsymbol{\pi}_e$ for this excitation is

$$\boldsymbol{\pi}_e = \mathbf{e}_x \cos\alpha + \mathbf{e}_y \sin\alpha\cos\beta + \mathbf{e}_z \sin\alpha\sin\beta \tag{4.7.10}$$

where $\mathbf{e}_x$, $\mathbf{e}_y$, and $\mathbf{e}_z$ are unit vectors along x, y, and z. From (2.7.8) we know that the electric dipole absorption probability (P_{abs}) is proportional to $|\langle F|\boldsymbol{\pi}^* \cdot \mathbf{m}|I\rangle|^2$, where I and F represent the initial and final states in the excitation transition. I and F need have no relation to A and J.

When photoselection occurs, the MPCL expressions must be weighted by P_{abs} before an orientational average is performed. (If we are dealing with a collection of molecules all held at the same space-fixed orientation, photoselection is irrelevant.) In particular, we must multiply $\mathcal{A}_1$, $\mathcal{B}_0$, $\mathcal{C}_0$, and $\mathcal{D}_0$ [Eq. (4.6.8)] by P_{abs} before averaging over orientation, to obtain expressions

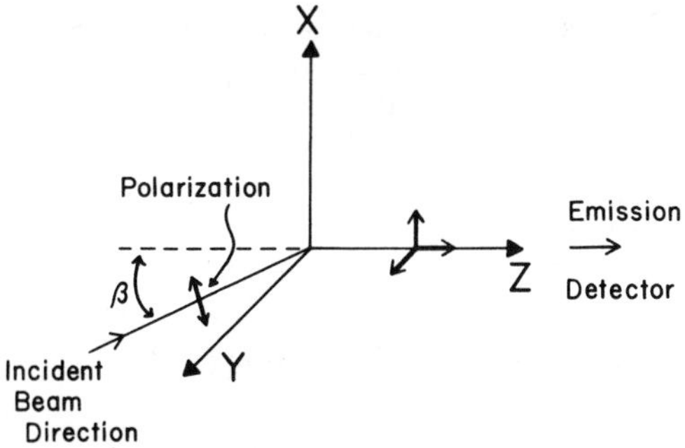

Figure 4.7.1. The polarization vector is at an angle α with respect to the X axis. The incident beam is in the YZ plane.

analogous to $\overline{\mathcal{A}}_1$, $\overline{\mathcal{B}}_0$, $\overline{\mathcal{C}}_0$, $\overline{\mathcal{D}}_0$ [Eq. (4.6.14)]. To illustrate this procedure, let us consider $\mathcal{C}_0$, assuming that the exciting radiation is head on [$\beta = 0°$ in (4.7.10)]. To average $\mathcal{C}_0$ over orientation we must write, using (4.6.12),

$$\overline{\mathcal{C}}_0(0°) = \frac{1}{8\pi^2}\int_0^{2\pi}\int_0^{2\pi}\int_0^{\pi} P_{\text{abs}}\mathcal{C}_0 \sin\theta\, d\theta\, d\phi\, d\psi \tag{4.7.11}$$

$\mathcal{C}_0$ in this integrand is obtained (as in Section 4.6) by substituting (4.6.11) into the right-hand side of (4.6.9), which is then used in (4.6.8). Then using (4.7.10) (with $\beta = 0°$),

$$P_{\text{abs}} = \mathfrak{N}\left|\overline{m}_x \cos\alpha + \overline{m}_y \sin\alpha\right|^2 \tag{4.7.12}$$

where $\overline{m}_x$ and $\overline{m}_y$ are the space-fixed transition dipole moments (numbers, not operators) for the transition producing the excitation, and $\mathfrak{N}$ is a proportionality constant. The space-fixed (unprimed) transition dipoles are related to the molecule-fixed (primed) transition dipoles according to (4.6.11):

$$\begin{aligned}\overline{m}_x &= A_{11}\overline{m}_{x'} + A_{12}\overline{m}_{y'} + A_{13}\overline{m}_{z'}\\ \overline{m}_y &= A_{21}\overline{m}_{x'} + A_{22}\overline{m}_{y'} + A_{23}\overline{m}_{z'}\end{aligned} \tag{4.7.13}$$

Next (4.7.13) is substituted into (4.7.12), and then the latter into (4.7.11). At this point the integrations may be performed if we have formulas analogous to (4.6.13) for products of four or five A_{ij}. Such formulas are available [Power and Thirunamachandran (1974), Eqs. (72) and (73)]. To simplify the resulting expressions, we chose the molecular z axis along the transition dipole so that $\overline{m}_{x'} = \overline{m}_{y'} = 0$ in (4.7.13). Doing the necessary bookkeeping, we obtain

$$\overline{\mathcal{C}}_0(0°) = -\frac{\mathcal{I}\mathfrak{m}}{15\mu_B|A|}$$

$$\times \sum_{\alpha\alpha'\lambda} \Big[(\cos^2\alpha + 3\sin^2\alpha)\big(\langle\mu_x\rangle\langle m_y\rangle\langle m_z\rangle - \langle\mu_y\rangle\langle m_x\rangle\langle m_z\rangle\big)$$

$$+ (3\cos^2\alpha + \sin^2\alpha)\big(\langle\mu_y\rangle\langle m_z\rangle\langle m_x\rangle - \langle\mu_x\rangle\langle m_z\rangle\langle m_y\rangle\big)$$

$$+\langle\mu_z\rangle\big(\langle m_x\rangle\langle m_y\rangle - \langle m_y\rangle\langle m_x\rangle\big)\Big] \tag{4.7.14}$$

where the proportionality factor $\mathcal{N}'(= \mathcal{N}\overline{m}_{z'}^2)$ has been incorporated into N_e in (4.7.7). The functions have been omitted from the bras and kets in (4.7.14), it being understood that they always occur in the order $A\alpha'$, $A\alpha$, $A\alpha$, $J\lambda$, $J\lambda$, $A\alpha'$. As before [$\overline{\mathcal{A}}_1 - \overline{\mathcal{D}}_0$, (4.6.14)], we drop the primes on the operators understanding that the bar on $\overline{\mathcal{C}}_0(0°)$ implies molecule-fixed operators. For unpolarized radiation we average over α, that is, $\cos^2\alpha$ and $\sin^2\alpha$ are replaced by $\frac{1}{2}$.

A numerical value of $\mathcal{N}'$ requires a knowledge of the absolute intensity of the exciting transition, but we seldom know N_e anyhow, because its value requires a knowledge of the quantum yield. The quantity of greatest utility in MCPL is the ratio $\Delta I'/I$, which is independent of such considerations and may be readily measured experimentally. If it is inconvenient to choose the molecular z axis along the transition dipole, (4.7.14) in general contains a variety of additional terms, but these may be worked out in a perfectly straightforward way using (4.7.13) and the formulas cited just below it.

By exactly the same methods, one can work out photoselected expressions for any of the Faraday parameters in any specified case. A detailed tabulation for unpolarized radiation has been given for 90° and 0° excitation, including in some cases magnetic-dipole and electric-quadrupole terms [Riehl and Richardson (1977)]. It should be emphasized again that the photoselection process is governed by the transition dipoles ($\overline{m}_x$, $\overline{m}_y$, $\overline{m}_z$) of the *excitation* process, which need have no relation whatever to the transition producing the emission.

Clearly, if one wishes to discuss the case in which some reorientation occurs between the time of excitation and emission, one must have a detailed knowledge of the dynamics of the reorientation process. Discussion of this problem may be found in the literature [Riehl and Richardson

(1977), Richardson (1982)]. It should be possible to obtain important information about the dynamics of relaxation and energy transfer in this intermediate case.

We may work out a photoselected value of $\mathfrak{D}_0$ in exactly the same manner, using the precise analog of (4.7.11). The result is

$$\overline{\mathfrak{D}}_0(0°) = \frac{1}{30|A|} \sum_{\alpha\lambda} \Big[3\Big(|\langle A\alpha|m_x|J\lambda\rangle^0|^2 + |\langle A\alpha|m_y|J\lambda\rangle^0|^2 \Big) + 4|\langle A\alpha|m_z|J\lambda\rangle^0|^2 \Big] \tag{4.7.15}$$

which is independent of the polarization angle α of the incident beam. The total emission in this case is given by (4.7.8) using (4.7.15); N_e is the same unknown quantity in $\Delta I'$ and I.

5 *Properties of the* MCD *Parameters*

In Chapter 4, using the rigid-shift (RS) model, we derived an explicit form for the MCD dispersion of an allowed electronic transition between Born–Oppenheimer (BO) states [Eq. (4.5.14)]. In this chapter we discuss the properties of this equation, which, despite its model-dependent nature, illustrates qualitatively (and often quantitatively) many of the important features of MCD spectroscopy. The parameters $\mathcal{A}_1$, $\mathcal{B}_0$, and $\mathcal{C}_0$ play a central role, and we begin with a discussion of their properties. We find later (Chapter 7) that these parameters have a generality which transcends (4.5.14).

5.1. Properties of $\mathcal{A}_1$, $\mathcal{B}_0$, and $\mathcal{C}_0$

In the linear limit, it is clear from (4.5.14) that the contributions of $\mathcal{A}_1$, $\mathcal{B}_0$, and $\mathcal{C}_0/kT$ are additive, the latter two having the same dispersion form as the absorption, and the former its derivative (in the RS model). In this section we discuss the properties of each contribution and their interrelations. Here, as elsewhere in the book, we use the new MCD conventions of Stephens (1976), so various expressions differ from those in the older literature, as discussed in Appendix A.

The Parameter $\mathcal{A}_1$

Referring to (4.5.16), we write ($m_1 \equiv m_{+1}$)

$$\mathcal{A}_1 = \frac{1}{\mu_B|A|}\sum_{\alpha\lambda} a_{\alpha\lambda}\left(|\langle A\alpha|m_{-1}|J\lambda\rangle^0|^2 - |\langle A\alpha|m_1|J\lambda\rangle^0|^2\right) \quad (5.1.1)$$

where

$$a_{\alpha\lambda} = -\left(\langle J\lambda|\mu_z|J\lambda\rangle^0 - \langle A\alpha|\mu_z|A\alpha\rangle^0\right) \quad (5.1.2)$$

$a_{\alpha\lambda}$ is the difference (per tesla) in Zeeman energies of the excited component $|J\lambda\rangle$ and the ground component $|A\alpha\rangle$. (Remember, $\mathcal{H}_z = -\mu_z B$.) Thus for the component transition $A\alpha \to J\lambda$ to contribute to $\mathcal{A}_1$, it must absorb lcp and rcp light to different extents, *and* the two states must Zeeman-shift by different amounts. $\mathcal{A}_1$ is obtained by summing over all possible transitions between Zeeman components of the two electronic states in accord with (5.1.1). Note that $\mathcal{A}_1$ must be zero for a transition between two nondegenerate states. This follows because $\langle K|\mu_z|K\rangle = 0$ if K is nondegenerate (Section 14.11).

As an illustration of (5.1.1), let us compute $\mathcal{A}_1$ for the example given in Fig. 4.4.1. By inspection, using the notation of (5.1.2),

$$a_{0,-1} = -g\mu_B$$

$$a_{0,0} = 0$$

$$a_{0,1} = g\mu_B \tag{5.1.3}$$

It is easy to show, either by direct computation [Slater (1960)] or by group-theoretic methods developed in Chapter 9, that

$$\begin{aligned} |\langle{}^1S|m_1|{}^1P-1\rangle| &= |\langle{}^1S|m_{-1}|{}^1P1\rangle| \neq 0 \\ |\langle{}^1S|m_{\pm1}|{}^1P0\rangle| &= |\langle{}^1S|m_{\pm1}|{}^1P\pm1\rangle| = 0 \end{aligned} \tag{5.1.4}$$

where either *both* upper or *both* lower signs apply in the second equation. Consequently, (5.1.1) gives

$$\mathcal{A}_1 = 2g|\langle{}^1S|m_{\pm1}|{}^1P\mp1\rangle|^2 \tag{5.1.5}$$

Using (4.5.23)

$$\mathcal{D}_0 = |\langle{}^1S|m_{\pm1}|{}^1P\mp1\rangle|^2 \tag{5.1.6}$$

and

$$\frac{\mathcal{A}_1}{\mathcal{D}_0} = 2g \tag{5.1.7}$$

In this example $g = 1$ (Chapter 13). The reader might confirm using (4.5.17) that the same calculation for Fig. 4.4.3 gives an $\mathcal{A}_1$ of the same sign and one-third the value of (5.1.5), since $|A| = 3$. Equation (5.1.7) continues to apply.

The dispersion form of the $\mathcal{A}$ terms is shown in Figs. 4.4.2 and 4.4.4. Remember, $\Delta A' \sim -\mathcal{A}_1(\partial f(\mathcal{E})/\partial\mathcal{E})$, so that a positive $\mathcal{A}_1$ value corresponds to positive MCD ($\Delta A'$) on the high-energy side of the band.

The present example is simple because transitions at different energies (in the field) are either pure left or pure right circularly polarized. In more complicated cases transitions at the same energy may be allowed in both circular polarizations (but to different extents if there is to be MCD). An example is $T_{1u} \leftrightarrow E_g$ in the octahedral case [Stephens (1976), Fig. 4(*b*)].

The Parameter $\mathcal{C}_0$

Referring to (4.5.16), we write ($m_1 \equiv m_{+1}$)

$$\frac{\mathcal{C}_0}{kT} = \frac{1}{\mu_B}\sum_{\alpha}\frac{\langle A\alpha|\mu_z|A\alpha\rangle^0}{|A|kT}\Delta\alpha \tag{5.1.8}$$

where

$$\Delta\alpha \equiv \sum_{\lambda}\left(|\langle A\alpha|m_{-1}|J\lambda\rangle^0|^2 - |\langle A\alpha|m_1|J\lambda\rangle^0|^2\right) \tag{5.1.9}$$

From (4.5.9) and (4.5.21)

$$\frac{N'_{A\alpha g} - N_{A\alpha g}}{N} = \frac{\Delta N'_{A\alpha g}}{N} = \left(\frac{\langle A\alpha|\mu_z|A\alpha\rangle^0 B}{|A|kT}\right)\frac{N_g}{N_G} \tag{5.1.10}$$

where $\Delta N'_{A\alpha g}/N$ is the fractional increase in the number of molecules in the substate $|A\alpha g\rangle'$ produced by the field B. If we take $N_g/N_G = 1$ for simplicity and drop the g label (as in the case of an atom), (5.1.8) becomes

$$\frac{\mathcal{C}_0}{kT} = \frac{1}{\mu_B}\sum_{\alpha}\left(\frac{\Delta N'_{A\alpha}}{NB}\right)\Delta\alpha \tag{5.1.11}$$

Thus for the component transition $A\alpha \to J\lambda$ to contribute to $\mathcal{C}_0/kT$, it must absorb lcp and rcp light to different extents (to obtain a nonzero $\Delta\alpha$), *and* $|A\alpha\rangle$ must undergo a population change when a field is applied (so that $\Delta N'_{A\alpha} \neq 0$).

Equation (5.1.11) can be further simplified using time-reversal symmetry arguments which are presented later (Section 14.11). We simply summarize the relevant conclusions here. For a system undergoing a first order Zeeman splitting:

1. The Zeeman sublevels may be divided into Kramers pairs, say $|A\alpha\rangle$ and $|A\alpha'\rangle$, which go into each other under time-reversal symmetry and split in opposite directions by equal amounts, that is,

$$\langle A\alpha|\mu_z|A\alpha\rangle = -\langle A\alpha'|\mu_z|A\alpha'\rangle \tag{5.1.12}$$

2. For every such pair, the following relation holds:

$$\Delta\alpha = -\Delta\alpha' \tag{5.1.13}$$

3. If the degeneracy of a level is odd (even-electron system), one of the sublevels is its own partner; this level will not change energy to first order in the field and satisfies

$$\Delta\alpha = 0 \tag{5.1.14}$$

Consequently (5.1.11) can be written

$$\frac{\mathcal{C}_0}{kT} = \frac{2}{\mu_B} \sum_{i\ \text{pairs}} \left(\frac{\Delta N'_{A\alpha_i}}{NB}\right) \Delta\alpha_i$$

$$= \frac{2}{\mu_B |A| kT} \sum_{i\ \text{pairs}} \langle A\alpha_i|\mu_z|A\alpha_i\rangle^0 \Delta\alpha_i \tag{5.1.15}$$

where (5.1.10), (5.1.12), and (5.1.13) have been used, and the sum includes *one* member, α_i or α'_i, of each Kramers pair.

Equation (5.1.15) displays clearly the dependence of $\mathcal{C}_0$ on both field-induced population shifts among Zeeman components of the ground state and differential absorption of lcp and rcp light. $\mathcal{C}_0$ is zero if the ground state is nondegenerate, since $\langle A|\mu_z|A\rangle = 0$, as discussed with respect to $\mathcal{A}_1$ above. If the excited state is nondegenerate, $\mathcal{C}_0 = \mathcal{A}_1$.

As an illustration, let us compute $\mathcal{C}_0$ for the example given in Fig. 4.4.3. By inspection we may confirm (5.1.12); Eqs. (5.1.13) and (5.1.14) follow from (5.1.4) and (4.5.17). Thus (5.1.15) becomes

$$\frac{\mathcal{C}_0}{kT} = \frac{2g}{3kT} |\langle {}^1P \pm 1|m_{\pm 1}|{}^1S\rangle|^2 \tag{5.1.16}$$

Using (4.5.23) we have

$$\mathcal{D}_0 = \tfrac{1}{3}|\langle {}^1P \pm 1|m_{\pm 1}|{}^1S\rangle|^2 \tag{5.1.17}$$

Finally

$$\frac{\mathcal{C}_0}{\mathcal{D}_0} = 2g \tag{5.1.18}$$

and again in this example $g = 1$.

The Parameter $\mathcal{B}_0$

The $\mathcal{B}$ term arises because the magnetic field mixes the (zero-field) wavefunctions of the system. This may be seen clearly by referring to (4.5.5). The sums on the right-hand side are the field-induced mixing terms. These couple $|A\alpha\rangle$ and $|J\lambda\rangle$ in $\langle A\alpha|m_{\pm 1}|J\lambda\rangle'$ [second and third terms in (4.5.6)], and the cross-products of these latter two terms with $\langle A\alpha|m_{\pm 1}|J\lambda\rangle^0$ when the square of the modulus of $\langle A\alpha|m_{\pm 1}|J\lambda\rangle'$ is substituted into (4.5.2) give rise to $\mathcal{B}_0$. The necessary condition for a mixing state $|K\kappa\rangle$ to contribute to the MCD of $A\alpha \rightarrow J\lambda$ is clear from (4.5.6). To mix with $|A\alpha\rangle$ or $|J\lambda\rangle$, a term of the form $\langle A\alpha|\mu_z|K\kappa\rangle\langle K\kappa|m_{\pm 1}|J\lambda\rangle$ or $\langle A\alpha|m_{\pm 1}|K\kappa\rangle\langle K\kappa|\mu_z|J\lambda\rangle$ respectively must be nonzero. That is, the mixing state component $K\kappa$ must simultaneously have a nonzero magnetic transition moment μ_z to the state with which it mixes *and* a nonzero electric-dipole transition moment $m_{\pm 1}$ to the other state involved in the transition.

As an example, let us consider the contribution to $\mathcal{B}_0$ which would arise if another ${}^1P$ state could be mixed by the field *with the excited* ${}^1P$ *state* in Fig. 4.4.1. (This in fact is not possible.) We use (4.5.16) and write

$$\mathcal{B}_0 = \frac{-2}{\mu_B|A|}\mathcal{R}e\left[\sum_{\alpha\lambda}\sum_{\kappa}\frac{\langle J\lambda|\mu_z|K\kappa\rangle^0}{W_K^0 - W_J^0}\left(\langle A\alpha|m_{-1}|J\lambda\rangle^0\langle K\kappa|m_1|A\alpha\rangle^0 - \langle A\alpha|m_1|J\lambda\rangle^0\langle K\kappa|m_{-1}|A\alpha\rangle^0\right)\right] \tag{5.1.19}$$

We have dropped the second term in $\mathcal{B}_0$ [Eq. (4.5.16)] because it involves mixing with the ground state, which we are not considering. Also, the sum over K has been omitted because we are considering a single mixing state K, which we designate ${}^1\bar{P}$ to distinguish it from $J = {}^1P$. It would be anticipated on the basis of the group-theoretic methods developed in later chapters that

$$\langle {}^1P1|\mu_z|{}^1\bar{P}1\rangle = -\langle {}^1P-1|\mu_z|{}^1\bar{P}-1\rangle \tag{5.1.20}$$

and that the other seven $\langle {}^1P\lambda|\mu_z|{}^1\bar{P}\kappa\rangle$ matrix elements would be zero. Using this, (4.5.17), and (5.1.4), it is straightforward algebra to evaluate (5.1.19). The result is

$$\mathfrak{B}_0 = \frac{4}{\mu_B}\frac{\langle {}^1P1|\mu_z|{}^1\bar{P}1\rangle}{W^0({}^1\bar{P}) - W^0({}^1P)}\langle {}^1S|m_{-1}|{}^1P1\rangle\langle {}^1S|m_{-1}|{}^1\bar{P}1\rangle^* \tag{5.1.21}$$

and one can show that the right-hand side of (5.1.21) is real.

Several points are noteworthy. First, $\mathfrak{B}_0$ is inversely proportional to the energy separation of the mixing states. Second, formation of the ratio $\mathfrak{B}_0/\mathfrak{D}_0$ does not cancel the transition dipole matrix elements (as is often the case for $\mathfrak{A}_1/\mathfrak{D}_0$ and $\mathfrak{C}_0/\mathfrak{D}_0$), because $\mathfrak{B}_0$ contains a transition dipole *to the mixing state* as well as the one connecting the states between which the transition occurs. The summation over K, dropped in (5.1.19), can be restored by simply summing (5.1.21) over all ${}^1\bar{P}$ states (${}^1\bar{P} \neq {}^1P$). In general, contributions from both ground- and excited-state mixing may be expected. Excited-state mixing is often more important, because the energy denominator is typically smaller. (An obvious exception occurs when very low-lying excited states can mix with the ground state.) Finally, we emphasize that our present example is numerically trivial because the matrix element of μ_z in (5.1.21) is zero. This is the case because the atomic functions ${}^1P1$ and ${}^1\bar{P}1$ are eigenfunctions of L_z (and S_z). Thus in fact there can be no $\mathfrak{B}$ terms *in atoms* for transitions between LS states.

In general it is much more difficult to calculate $\mathfrak{B}_0$ than $\mathfrak{A}_1$ or $\mathfrak{C}_0$. This is even more the case for the *ratio* $\mathfrak{B}_0/\mathfrak{D}_0$ vs. $\mathfrak{A}_1/\mathfrak{D}_0$ and/or $\mathfrak{C}_0/\mathfrak{D}_0$. In the ${}^1S \to {}^1P$ example, the ratios $\mathfrak{A}_1/\mathfrak{D}_0$ and $\mathfrak{C}_0/\mathfrak{D}_0$ involve only the magnetic moments of the ground and/or excited state, since the transition moment cancels. As discussed above, this does not happen in $\mathfrak{B}_0/\mathfrak{D}_0$. Moreover, the expression for $\mathfrak{B}_0$ involves sums over an infinity of mixing states and requires a knowledge of electric and magnetic transition dipoles involving these states. In general $\mathfrak{B}_0$ calculations are feasible only if very few mixing states need be considered. This is typically the case if a state is close in energy to either A or J and can therefore be assumed to dominate the mixing process because of the small energy denominator involved. An important instance of this occurs if a degenerate state is split by a small perturbation. If field-induced mixing can occur among the split components, the resulting $\mathcal{B}$ terms can dominate the MCD of transitions involving the state. We discuss this case next.

5.2. Overlapping Bands and Pseudo-$\mathcal{A}$ Terms

Let us consider the case in which two bands overlap because a near-degeneracy exists in either the excited or the ground state. We follow Stephens (1976), Section III.B. We make the rather drastic assumption that the pair of nearly degenerate states have parallel potential surfaces and thus identical vibrational functions. We continue to make the BO–FC–RS approximation. (Some of these restrictions are removed in Chapter 7.)

Excited-State Near-Degeneracy

Consider the two overlapping transitions, $A \to J_1$ and $A \to J_2$ where $\Delta W \equiv W_{J_2}^0 - W_{J_1}^0$. Let us assume that the Zeeman interactions between J_1 and J_2 are small compared to ΔW. That is for all λ, λ'

$$\frac{\langle J_1\lambda|\mu_z|J_2\lambda'\rangle^0 B}{\Delta W} \ll 1 \tag{5.2.1}$$

If the derivation in Section 4.5 is repeated, we obtain in place of (4.5.14)

$$\frac{\Delta A'}{\mathcal{E}} = \gamma\mu_B B \sum_{i=1}^{2}\left[\mathcal{A}_1(J_i)\left(-\frac{\partial f_i(\mathcal{E})}{\partial \mathcal{E}}\right) + \left(\mathcal{B}_0(J_i) + \frac{\mathcal{C}_0(J_i)}{kT}\right)f_i(\mathcal{E})\right] \tag{5.2.2}$$

where $\mathcal{A}_1(J_i) \equiv \mathcal{A}_1(A \to J_i)$, and so on, are defined as in (4.5.16), (4.6.8), or (4.6.14). Because of the assumption of identical potential surfaces, it also follows that

$$f_2(\mathcal{E}) = f_1(\mathcal{E} - \Delta W) \tag{5.2.3}$$

We see from (5.2.2) that the MCD is simply additive in the present case even if ΔW is small compared to the bandwidths [provided that (5.2.1) holds]. Similarly, for the zero-field absorption

$$\frac{A}{\mathcal{E}} = \gamma \sum_{i=1}^{2} \mathcal{D}_0(J_i) f_i(\mathcal{E}) \tag{5.2.4}$$

Let us first examine the $\mathcal{B}$ terms in (5.2.2) for small ΔW. We start by noting the following theorem. If field-induced mixing *only* between J_1 and J_2 is considered,

$$\mathcal{B}_0(J_1, J_2) = -\mathcal{B}_0(J_2, J_1) \tag{5.2.5}$$

where $\mathcal{B}_0(J_1, J_2)$ is the $\mathcal{B}_0$ term for $A \rightarrow J_1$ arising from the Zeeman interaction of J_1 with J_2, and so on. This result follows from the $\mathcal{B}_0$ expression, (4.5.16), and (4.5.17). Now consider once again the example in Fig. 4.4.1, but suppose that the ^1P-state degeneracy is completely split by some perturbation in the absence of the magnetic field. For example, suppose the atom is in a rhombohedral crystal field, so that the x, y, and z directions are all nonequivalent. This case is depicted in Fig. 5.2.1, where the order and magnitude of the splittings is chosen arbitrarily.

If the MCD of $^1S \rightarrow {}^1P$ is now measured, $\mathcal{A}_1(J_i) = 0$ for all i by our previous discussion (Section 5.1), because the transitions are all between nondegenerate states. Furthermore, this applies even if the perturbation is very small. At first glance this is a disturbing conclusion, since it seems to imply that an infinitesimal perturbation can cause the MCD in Fig. 4.4.2 to vanish. We now show that under such circumstances the MCD changes in a continuous manner from an $\mathcal{A}$ term to two $\mathcal{B}$ terms of opposite sign (a *pseudo-$\mathcal{A}$ term*).

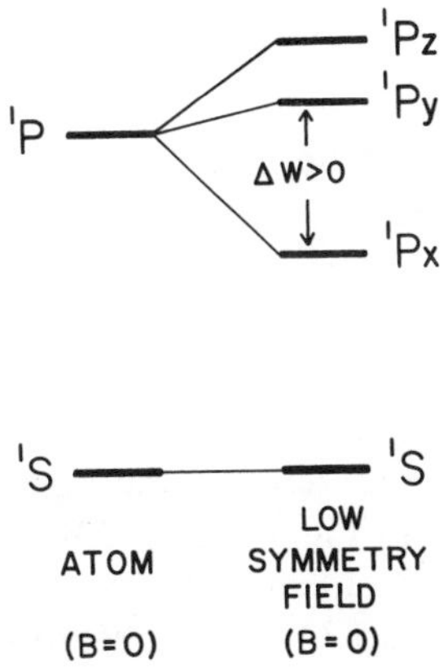

Figure 5.2.1. Effect of low-symmetry perturbation on the ^1P state of Fig. 4.4.1.

Let us calculate $\mathcal{B}_0$ for $^1S \to {}^1Px$, confining ourselves *only* to the states in Fig. 5.2.1. Using previous results [(5.1.4), (5.1.20)] and the definitions $^1P \pm 1 = \mp(1/\sqrt{2})(^1Px \pm i\,^1Py)$ and $^1P0 = {}^1Pz$, one finds (Chapter 11, 13)

$$\langle {}^1S|m_{\pm 1}|{}^1Px\rangle = \pm\frac{1}{\sqrt{2}}\langle {}^1S|m_{\pm 1}|{}^1P \mp 1\rangle$$

$$\langle {}^1S|m_{\pm 1}|{}^1Py\rangle = \frac{i}{\sqrt{2}}\langle {}^1S|m_{\pm 1}|{}^1P \mp 1\rangle$$

$$\begin{aligned}
\langle {}^1S|m_{\pm 1}|{}^1Pz\rangle &= 0 = \langle {}^1S|m_{\pm 1}|{}^1P \pm 1\rangle \\
\langle {}^1Px|L_z|{}^1Py\rangle &= -\langle {}^1Py|L_z|{}^1Px\rangle = -i \\
\langle {}^1Px|L_z|{}^1Px\rangle &= \langle {}^1Py|L_z|{}^1Py\rangle \\
&= \langle {}^1Px|L_z|{}^1Pz\rangle = \langle {}^1Py|L_z|{}^1Pz\rangle = 0
\end{aligned} \tag{5.2.6}$$

In expressions containing more than one $\pm$ sign, either *all* upper or *all* lower signs apply. Spin (the S_z operator) plays no role, since our basis states are singlets. Substituting (5.2.6) into $\mathcal{B}_0$ [Eq. (4.6.8)], one obtains

$$\begin{aligned}
\mathcal{B}_0(^1Px) &= -\frac{2}{\Delta W}|\langle {}^1S|m_{\pm 1}|{}^1P \pm 1\rangle|^2 \\
&= \mathcal{B}_0(^1Px, {}^1Py) \\
\mathcal{B}_0(^1Py) &= \mathcal{B}_0(^1Py, {}^1Px) \\
\mathcal{B}_0(^1Pz) &= 0
\end{aligned} \tag{5.2.7}$$

where $\Delta W = W^0(^1Py) - W^0(^1Px)$. Note that the only field-induced mixing is between ^1Px and ^1Py. From our earlier theorem [Eq. (5.2.5)],

$$\mathcal{B}_0(^1Py, {}^1Px) = -\mathcal{B}_0(^1Px, {}^1Py) \tag{5.2.8}$$

Thus the MCD consists of a negative $\mathcal{B}$ term to the low-energy side of the unperturbed zero-field transition and a positive $\mathcal{B}$ term to the high-energy side; that is, the MCD looks like the $\mathcal{A}$ term in Fig. 4.4.2a if ΔW is not too large. To get a quantitative result we use (5.2.2), identifying ^1Px and ^1Py with J_1 and J_2 respectively [noting that $\mathcal{A}_1(J_i) = \mathcal{C}_0(J_i) = 0$]. Substituting from (5.2.7) and (5.2.8), we obtain

$$\frac{\Delta A'}{\mathcal{E}} = \gamma\mu_B B\left[\mathcal{B}_0({}^1Px)f_1(\mathcal{E}) + \mathcal{B}_0({}^1Py)f_2(\mathcal{E})\right]$$

$$= \gamma\mu_B B\left[2|\langle {}^1S|m_{\pm 1}|{}^1P \mp 1\rangle|^2 \frac{f_2(\mathcal{E}) - f_1(\mathcal{E})}{\Delta W}\right] \quad (5.2.9)$$

We note from (5.2.3) that

$$\lim_{\Delta W \to 0} \frac{f_2(\mathcal{E}) - f_1(\mathcal{E})}{\Delta W} = -\frac{\partial f_1(\mathcal{E})}{\partial \mathcal{E}} \quad (5.2.10)$$

Consequently

$$\lim_{\Delta W \to 0} \frac{\Delta A'}{\mathcal{E}} = \gamma\mu_B B\left[-2|\langle {}^1S|m_{\pm 1}|{}^1P \mp 1\rangle|^2 \frac{\partial f_1(\mathcal{E})}{\partial \mathcal{E}}\right] \quad (5.2.11)$$

We conclude

$$\frac{\Delta A'}{\mathcal{E}} = \gamma\mu_B B\left[2|\langle {}^1S|m_{\pm 1}|{}^1P \mp 1\rangle|^2 \left(-\frac{\partial f_1(\mathcal{E})}{\partial \mathcal{E}}\right)\right]$$

$$= \gamma\mu_B B\left[\mathcal{A}_1'({}^1Px, {}^1Py)\left(-\frac{\partial f_1(\mathcal{E})}{\partial \mathcal{E}}\right)\right] \quad (5.2.12)$$

which is precisely the result obtained for $\Delta A'/\mathcal{E}$ from (4.5.14) in the absence of the rhombohedral perturbation [Eq. (5.1.5)]. Thus in the limit $\Delta W \to 0$, the pseudo-$\mathcal{A}$ term $\mathcal{A}_1'(J_1, J_2)$, obtained by the overlapping of $\mathcal{B}_0(J_1, J_2)$ with $\mathcal{B}_0(J_2, J_1)$, is identical to $\mathcal{A}_1$ for a state J with degenerate components J_1 and J_2:

$$\mathcal{A}_1'(J_1, J_2) = -\mathcal{B}_0(J_1)\,\Delta W$$

$$= \mathcal{A}_1(J) \quad (5.2.13)$$

Thus we have the satisfying result that the two $\mathcal{B}$ terms "collapse" to an $\mathcal{A}$ term as $\Delta W \to 0$.

It should also be noted that the same result would be obtained if the order of splitting of ${}^1Px$ and ${}^1Py$ were reversed. In that case, the signs of the two $\mathcal{B}_0$ terms interchange, as do their positions. Thus one can conclude nothing about the order of ${}^1Px$ and ${}^1Py$ from the $\mathcal{B}$-term pattern. It should also be noted that $^1P0$ ($= {}^1Pz$) plays no role in the MCD pattern [nor did it in the original unperturbed case (Fig. 4.4.1)]. Finally, additional contributions to $\mathcal{B}_0$ arise from states not included in Fig. 5.2.1 ("out-of-state" contributions). Such contributions are ordinarily much smaller, because very much larger energy denominators are involved.

Another important case occurs if we consider overlapping $\mathcal{C}$ terms in (5.2.2). For small ΔW it follows from (5.2.10) that (5.2.3) can be written

$$f_2(\mathcal{E}) \approx f_1(\mathcal{E}) - \Delta W \frac{\partial f_1(\mathcal{E})}{\partial \mathcal{E}} \tag{5.2.14}$$

For simplicity let us suppose that $\mathcal{A}_1(J_i) = \mathcal{B}_0(J_i) = 0$. Using (5.2.14), Eq. (5.2.2) becomes

$$\frac{\Delta A'}{\mathcal{E}} = \frac{\gamma \mu_B B}{kT} \left\{ [\mathcal{C}_0(J_1) + \mathcal{C}_0(J_2)] f_1(\mathcal{E}) - \mathcal{C}_0(J_2)\, \Delta W \frac{\partial f_1(\mathcal{E})}{\partial \mathcal{E}} \right\} \tag{5.2.15}$$

If $\mathcal{C}_0(J_1)$ and $\mathcal{C}_0(J_2)$ are of the same sign and of comparable magnitude, the term $[\mathcal{C}_0(J_1) + \mathcal{C}_0(J_2)] f_1(\mathcal{E})$ dominates for small ΔW. But if $\mathcal{C}_0(J_1)$ and $\mathcal{C}_0(J_2)$ are of opposite sign, the MCD can be very sensitive to ΔW. In the extreme case, $\mathcal{C}_0(J_1) = -\mathcal{C}_0(J_2)$, (5.2.15) becomes

$$\frac{\Delta A'}{\mathcal{E}} = \frac{\gamma \mu_B B}{kT} \mathcal{C}_0(J_2)\, \Delta W \left(- \frac{\partial f_1(\mathcal{E})}{\partial \mathcal{E}} \right) \tag{5.2.16}$$

Comparing with (4.5.14), we see that (5.2.16) has the dispersion form of an $\mathcal{A}$ term; it is sometimes referred to as a "temperature-dependent $\mathcal{A}$ term." The "pseudo-$\mathcal{A}$" $\mathcal{A}_1$ value is $\mathcal{C}_0(J_2)\,\Delta W/kT$. The MCD is directly proportional to ΔW and vanishes in the limit $\Delta W \to 0$.

Ground-State Near-Degeneracy

In this case we consider two transitions $A_1 \to J$ and $A_2 \to J$ where $\Delta W \equiv W^0_{A_2} - W^0_{A_1} > 0$. As before, we require parallel potential surfaces for our near-degenerate states—this time A_1 and A_2—and in analogy to (5.2.1) we require

$$\frac{\langle A_1 \alpha | \mu_z | A_2 \alpha' \rangle B}{\Delta W} \ll 1 \tag{5.2.17}$$

Again, if the derivation in Section 4.5 is repeated, we obtain for the MCD and zero-field absorption

$$\frac{\Delta A'}{\mathcal{E}} = \gamma\mu_B B \sum_{i=1}^{2} \delta_i \left[\mathcal{A}_1(A_i)\left(-\frac{\partial f_i(\mathcal{E})}{\partial \mathcal{E}}\right) + \left(\mathcal{B}_0(A_i) + \frac{\mathcal{C}_0(A_i)}{kT}\right) f_i(\mathcal{E})\right] \tag{5.2.18}$$

$$\frac{A}{\mathcal{E}} = \gamma \sum_{i=1}^{2} \delta_i \mathcal{D}_0(A_i) f_i(\mathcal{E}) \tag{5.2.19}$$

where $\mathcal{A}_1(A_i) \equiv \mathcal{A}_1(A_i \to J)$ and so on, and the states A_i have fractional populations δ_i:

$$\delta_1 = \frac{|A_1|}{|A_1| + |A_2| e^{-\Delta W/kT}}, \qquad \delta_2 = 1 - \delta_1 \tag{5.2.20}$$

Thus the MCD and absorption are additive functions of the individual transitions weighted by appropriate population factors.

The dependence of the MCD on ΔW arises through (5.2.3) and through an energy denominator if $\mathcal{B}$ terms occur between A_1 and A_2, *and* also through the population factors δ_1 and δ_2. A variety of behaviors are possible. One interesting case occurs if ΔW is much smaller than both the bandwidth and kT. A straightforward analysis assuming A_1, A_2, and J are nondegenerate [Stephens (1976), p. 223] gives

$$\frac{\Delta A'}{\mathcal{E}} = \frac{\gamma\mu_B B \Delta W}{2}\left[\mathcal{B}_0'(A_1)\left(-\frac{\partial f_1(\mathcal{E})}{\partial \mathcal{E}}\right) + \frac{\mathcal{B}_0'(A_1)}{kT} f_1(\mathcal{E})\right] \tag{5.2.21}$$

$$\frac{A}{\mathcal{E}} = \frac{\gamma}{2}\left[\mathcal{D}_0(A_1) + \mathcal{D}_0(A_2)\right] f_1(\mathcal{E}) \tag{5.2.22}$$

where $\mathcal{B}_0'(A_1)$ is the contribution to $\mathcal{B}_0(A_1)$ from field-induced mixing *only* between states A_1 and A_2. Note particularly that $\Delta W \mathcal{B}_0'(A_1)$ is *independent of* ΔW in this case, as therefore is the MCD in (5.2.21). We have here both pseudo-$\mathcal{A}$ and pseudo-$\mathcal{C}$ terms. An actual example of the case discussed here is given in Section 13.5.

Finally we comment on an apparent contradiction in our analysis of the overlapping-band cases. Our treatment requires the validity of (5.2.1) or (5.2.17), yet we have not hesitated to go to the limit $\Delta W \to 0$. In fact (5.2.1) or (5.2.17) can *always* be satisfied by appropriate choice of the sets of functions J_1 and J_2 or A_1 and A_2. This may perhaps be seen most clearly when $\Delta W = 0$. In that case we simply choose our degenerate sets (J_1 and J_2 or A_1 and A_2) diagonal in μ_z at the outset, and the left-hand side of (5.2.1) or (5.2.17) vanishes. In fact, starting with a diagonal basis and using (4.5.14), we see that Eq. (4.5.16) is a completely equivalent way of getting the limiting case $\Delta W = 0$.

Contours for several cases of overlapping bands are illustrated in the next section (Fig. 5.3.2). Moment analysis of overlapping bands is discussed in Section 7.6.

5.3. MCD Dispersion Forms; Quantitative Comparisons of $\mathcal{A}$, $\mathcal{B}$, and $\mathcal{C}$ Terms

The explicit form of the MCD for an allowed electronic transition between BO states follows directly from (4.5.14):

$$\frac{\Delta A'}{\mathcal{E}} = \gamma\mu_B B\left[\mathcal{A}_1\left(-\frac{\partial f(\mathcal{E})}{\partial \mathcal{E}}\right) + \left(\mathcal{B}_0 + \frac{\mathcal{C}_0}{kT}\right)f(\mathcal{E})\right]$$

where $\mathcal{A}_1$, $\mathcal{B}_0$, and $\mathcal{C}_0$ are given by (4.5.16) for oriented or isotropic molecules whose eigenfunctions are diagonal in (space-fixed) μ_z, by (4.6.8) with the latter restriction removed, or by (4.6.14) if the molecules are randomly distributed over all orientations.

We emphasize again the distinction between the symbols $\mathcal{A}$, $\mathcal{B}$, $\mathcal{C}$, and $\mathcal{D}$ used with and without subscripts. The former describe the dispersion forms (MCD and absorption "curves") which arise from the Zeeman shift, from field-induced mixing, from ground-state population differentials, and from the zero-field absorption, respectively. In contrast, $\mathcal{A}_1$, $\mathcal{B}_0$, $\mathcal{C}_0$ and $\mathcal{D}_0$ are specific mathematical expressions associated with the "strengths" of these effects. They appear in the expressions for the dispersion of the MCD and absorption, (4.5.14) and (4.5.22), and they are encountered along with their higher members (X_n, $n = 0, 1, 2, \ldots$, where $X = \mathcal{A}, \mathcal{B}, \mathcal{C}, \mathcal{D}$) when MCD and absorption moments are discussed (Chapter 7).

It is clear from (4.5.14) that $\mathcal{B}$ and $\mathcal{C}$ are associated with an absorptionlike dispersion, and $\mathcal{A}$ with its derivative. By inspection of (4.5.16) we see that *in order of magnitude* $\mathcal{A}_1 : \mathcal{B}_0 : \mathcal{C}_0 = 1 : (1/\Delta W) : 1$, where ΔW is of the order of the energy interval to the closest (field-induced) mixing state. We may also ascertain the relative contributions to the MCD if we assume a specific form for $f(\mathcal{E})$. For purposes of illustration consider the Gaussian,

$$f(\mathcal{E}) = \frac{1}{\Delta\sqrt{\pi}} \exp\left[-\frac{(\mathcal{E} - \mathcal{E}^0)^2}{\Delta^2} \right] \tag{5.3.1}$$

which satisfies $\int f(\mathcal{E})\, d\mathcal{E} = 1$. The function $f(\mathcal{E})$ and its first derivative are shown in Fig. 7.8.1. From this figure and (4.5.14), we see that the maximum contributions to $\Delta A'/\mathcal{E}$ for the three terms are $\mathcal{A}(\max) : \mathcal{B}(\max) : \mathcal{C}(\max) = (\mathcal{A}_1/1.17\Delta) : \mathcal{B}_0 : \mathcal{C}_0/kT$. Clearly, the three terms are favored respectively by a narrow bandwidth, nearby mixing states, and low temperature. In particular, the maximum amplitude of the $\mathcal{B}$ and $\mathcal{C}$ terms (and the absorption band) vary as the inverse first power of the bandwidth, whereas the maximum $\mathcal{A}$ term amplitude varies as the inverse second power. Thus while $\mathcal{C}$ terms typically dominate a broad band at room temperature, $\mathcal{A}$ terms can become very prominent at low temperature if individual, sharp lines are resolved, even though the $\mathcal{C}$-term amplitude itself increases dramatically because of its inverse temperature dependence.

We estimate the relative magnitudes of the MCD and absorption by examining $\Delta A'/A$, that is, Eq. (4.5.14) divided by (4.5.22). Thus for $\mathcal{A}$, $\mathcal{B}$, and $\mathcal{C}$ respectively, $(\Delta A'/A)_{\max}$ is roughly $g\mu_B B/\Delta$, $g\mu_B B/\Delta W$, and $g\mu_B B/kT$. For a typical case of a broad band at room temperature, we choose $g = 2$, $B = 1$ T, $\Delta = 10^3$ cm^{-1}, $\Delta W = 10^4$ cm^{-1}, and $kT = 200$ cm^{-1}. Then the $(\Delta A'/A)_{\max}$ values are respectively 1×10^{-3}, 1×10^{-4}, and 5×10^{-3}. For a sharp band ($\Delta = 10$ cm^{-1}) at 10 K ($kT = 7$ cm^{-1}) the approximate ratios are respectively 0.1, 1×10^{-4}, and 0.1. It is clear that in general, low-temperature MCD measurements are highly desirable. One often resolves fine structure as the temperature is lowered, as in absorption spectroscopy. This of course provides additional spectroscopic detail. The extra bonus for MCD, if this happens, is the potential orders-of-magnitude increase in the $\mathcal{A}$- and $\mathcal{C}$-term signals, since the experimental sensitivity is proportional to $\Delta A'/A$. Finally, one can in principle separate the contributions of $\mathcal{A}$, $\mathcal{B}$, and $\mathcal{C}$ by their differing dispersion and temperature behavior. In practice, this is best done by the method of moments (Chapter 7). In the rigid-shift approximation, the separation can then be made without the assumption of a specific bandshape [see also the discussion following (4.5.28)].

MCD and absorption plots are shown for a variety of cases in Figs. 5.3.1 and 5.3.2.

Whereas the assumption of a gaussian band shape (5.3.1) is strictly ad hoc, our general conclusions are independent of this assumption. (As a matter of experimental fact, a Gaussian is often a surprisingly good ap-

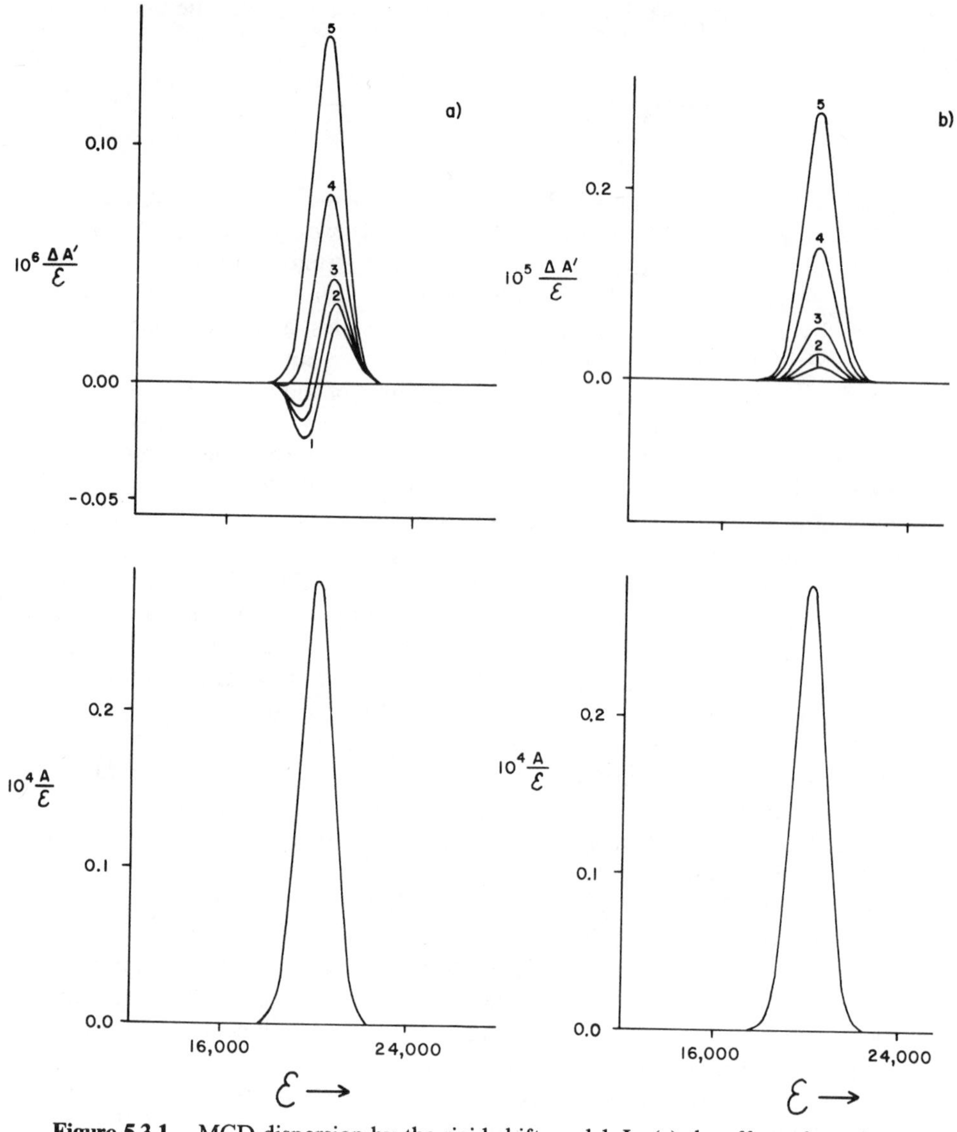

Figure 5.3.1. MCD dispersion by the rigid-shift model. In (a) the effect of varying the relative magnitudes of the $\mathscr{A}$ and $\mathscr{B} + \mathscr{C}/kT$ terms is shown. The parameters used are $\mathscr{D}_0 = 1$, $\gamma = 5 \times 10^{-2}$, $\mathscr{E}^0 = 20{,}000\ \text{cm}^{-1}$, $\Delta = 1000\ \text{cm}^{-1}$, $\mathscr{A}_1 = 1$, $\mu_B B = 1\ \text{cm}^{-1}$, and $\mathscr{B}_0 + \mathscr{C}_0/kT$ equal to: (1) 0, (2) 5×10^{-4}, (3) 10^{-3}, (4) 2.5×10^{-3}, (5) 5×10^{-3}—all based on (4.5.22), (4.5.14), and (5.3.1). In (b) the effect of varying the temperature is shown, using $\mathscr{A}_1 = \mathscr{B}_0 = 0$, $\mathscr{C}_0 = 1$, and kT equal to: (1) 200 cm^{-1}, (2) 100 cm^{-1}, (3) 50 cm^{-1}, (4) 20 cm^{-1}, (5) 10 cm^{-1}. [Reproduced from Figure 2, Stephens (1976) with permission.]

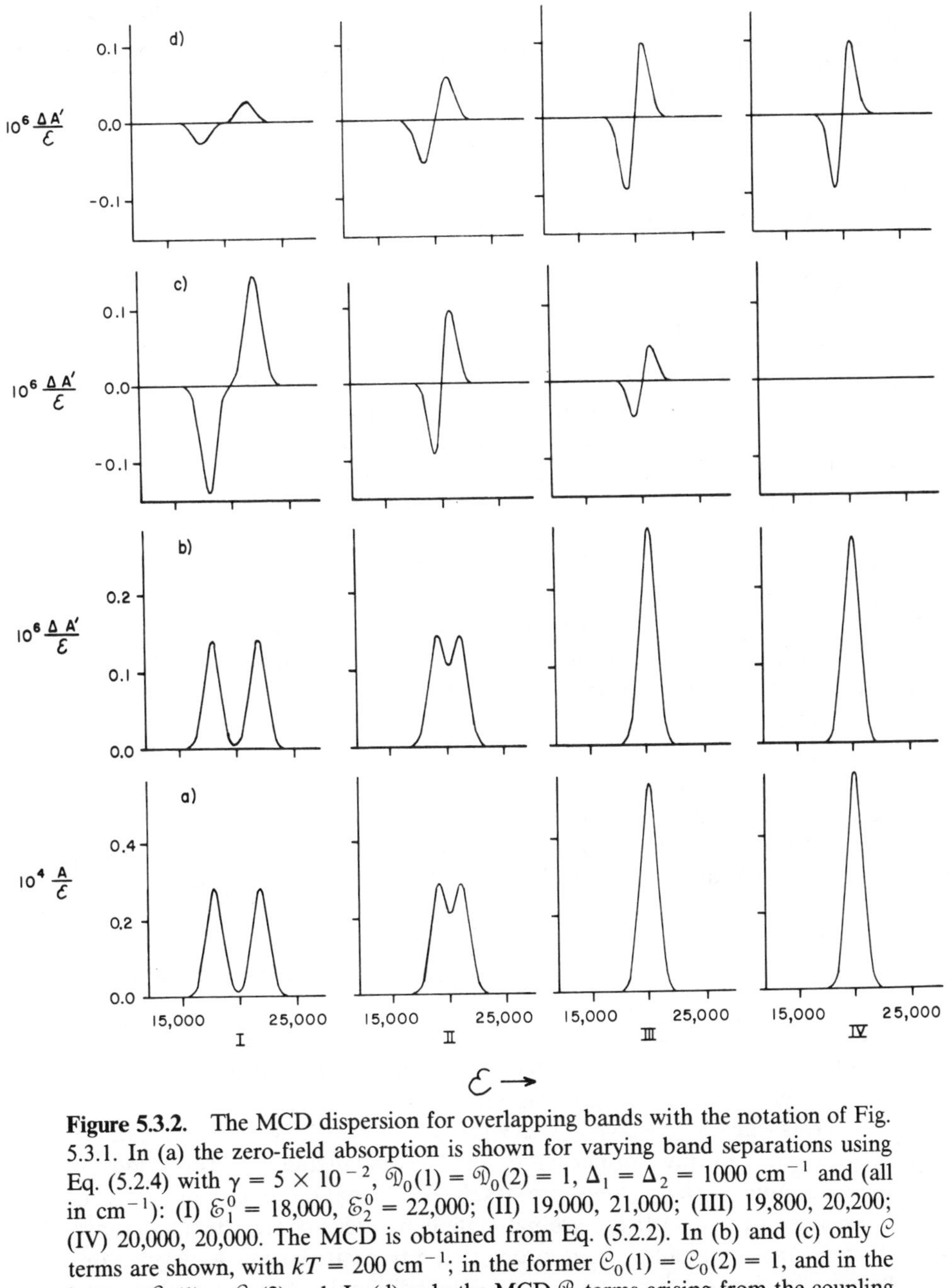

Figure 5.3.2. The MCD dispersion for overlapping bands with the notation of Fig. 5.3.1. In (a) the zero-field absorption is shown for varying band separations using Eq. (5.2.4) with $\gamma = 5 \times 10^{-2}$, $\mathcal{D}_0(1) = \mathcal{D}_0(2) = 1$, $\Delta_1 = \Delta_2 = 1000\ \text{cm}^{-1}$ and (all in cm^{-1}): (I) $\mathcal{E}_1^0 = 18{,}000$, $\mathcal{E}_2^0 = 22{,}000$; (II) 19,000, 21,000; (III) 19,800, 20,200; (IV) 20,000, 20,000. The MCD is obtained from Eq. (5.2.2). In (b) and (c) only $\mathcal{C}$ terms are shown, with $kT = 200\ \text{cm}^{-1}$; in the former $\mathcal{C}_0(1) = \mathcal{C}_0(2) = 1$, and in the latter $-\mathcal{C}_0(1) = \mathcal{C}_0(2) = 1$. In (d) only the MCD $\mathcal{B}$ terms arising from the coupling of the two excited states are shown, with $-\mathcal{B}_0(1) = \mathcal{B}_0(2) = 0.4/(\mathcal{E}_2^0 - \mathcal{E}_1^0)$. [Reproduced from Figure 5, Stephens (1976) with permission.]

proximation to the lineshapes and bandshapes observed in solutions and crystals.)

5.4 The Utility of MCD Spectroscopy

Having examined the properties of the MCD dispersion for a simple model, we are now in a position to consider, at least in a preliminary way, the utility of the technique.

Perhaps most obvious is the fact that MCD is a signed quantity which can come in two dispersion forms [$f(\mathcal{E})$ and $\partial f(\mathcal{E})/\partial \mathcal{E}$] with a possible strong temperature dependence ($\mathcal{C}$ term). Thus when taken in conjunction with the absorption spectrum (which comes in one sign and dispersion form), there is an obvious multiplication of available information.

We again emphasize the advantages over a conventional Zeeman experiment when the bandwidth is large compared to the Zeeman splitting (see Section 4.4). With an absorbance of ~ 1, $\Delta A \sim 10^{-6}$ is measurable with present instrumentation. The discussion of the previous section shows that $\mathcal{A}$ and $\mathcal{C}$ terms can be detected with orders of magnitude more sensitivity than their counterparts in a Zeeman measurement.

Considering individual terms, we note first that the appearance of a $\mathcal{C}$ term (often inferred simply by its large magnitude, but easily confirmed by its T^{-1} dependence) demonstrates the existence of a degenerate ground state. Furthermore, its sign conveys additional information which may allow one to distinguish among alternative assignments of a transition. Indeed, one of the first applications of MCD spectroscopy was the use of $\mathcal{C}$-term signs to assign charge-transfer transitions unambiguously in $Fe(CN)_6^{3-}$. (This application is discussed in Sections 11.2 and 13.5.) Furthermore, the quantity $\mathcal{C}_0/\mathcal{D}_0$ is usually a simple function of the magnetic moment of the ground state, which can thus be roughly determined experimentally. (As discussed in Section 6.3, this same information can be obtained without $\mathcal{D}_0$ if saturation effects can be observed—that is, if B/T can be made sufficiently large that Zeeman splittings are $\gtrsim kT$.)

$\mathcal{A}$ terms can be observed if the ground and/or excited state can be Zeeman-split. In the former case, the accompanying $\mathcal{C}$ terms tend to dominate unless sharp lines are resolved. (If only the ground state is degenerate, $\mathcal{A}_1 = \mathcal{C}_0$.) If the ground state is nondegenerate, the appearance of an $\mathcal{A}$ term is unambiguous proof that the excited state is degenerate. Again, the sign of the $\mathcal{A}$ term provides additional information that may distinguish among alternative assignments of transitions. If the ground state is nondegenerate, or if its magnetic moment is known, measurement of

$\mathcal{A}_1/\mathcal{D}_0$ usually yields an immediate value of the *excited-state* magnetic moment.

The $\mathcal{B}$-term contribution to the MCD dispersion is usually smaller than those for $\mathcal{A}$ and $\mathcal{C}$ terms (unless there are nearby mixing states). This is usually a happy circumstance, since the $\mathcal{A}$ and $\mathcal{C}$ terms are often of more theoretical interest and are much easier to calculate. In systems of low symmetry (no degeneracies), of course only $\mathcal{B}$ terms are present. Their use to effect comparisons with theory usually requires the assumption that mixing contributions from a relatively small number of states are dominant. A considerable amount of $\mathcal{B}$-term analysis has been done on organic systems.

Less obvious is the use of MCD to observe an MCD-"active" species in a mixture. For example, a paramagnetic species may be largely obscured in the absorption spectrum by a diamagnetic species present in greater concentration. Nevertheless, the former may entirely dominate the MCD, especially at low temperature or if sharp lines are present. Thus, one can often pick out the MCD spectrum clearly despite the fact that its absorption spectrum is largely obscured. A nice illustration of this may be found in Collingwood et al. (1974). Similarly, overlapping absorption bands (from the same or different species) may be much more clearly defined in the MCD if their dispersions are of opposite sign or of different form.

Applications of MCD appear throughout this book as illustrations, and Chapter 23 is devoted almost entirely to MCD examples.

Finally, we restate some obvious limitations of MCD. First and foremost, the experimental technique is straightforward only when applied to systems in which circularly polarized light retains its integrity, that is, in isotropic and cubic systems, and in uniaxial systems if light and field are along the unique axis. In other cases, the complications discussed in Section 4.2 must be considered, though successful *natural CD* measurements have been made. Furthermore, one must always guard against (or correct for) depolarization effects due to light scattering, sample imperfections, strain birefringence, and so on. Fortunately, it is almost always possible to assess the importance of these effects by measuring the known natural CD of an appropriately chosen substance after the light has traversed the proposed MCD sample (at zero field).

6 *Deviations from the* MCD *Linear Limit*

Our previous MCD discussion (Chapters 4 and 5) has used a dispersion form [Eq. (4.5.14)] which varies linearly with magnetic field; that is, we have worked in the linear limit. For this to be valid, three specific assumptions are required. In this chapter, we examine the consequences of dropping one or more of these.

6.1. Failure of First-Order Perturbation Theory

Referring back to Chapter 4, one finds that the entire MCD analysis hinges on the validity of (4.5.5) and (4.5.7), that is, on the validity of first-order perturbation theory. Since μ_z matrix elements are at most only a few Bohr magnetons ($\mu_B = 0.4669\ \mathrm{cm}^{-1}/\mathrm{T}$), the magnetic-field perturbation is inherently small (excluding colossal magnetic fields). Thus difficulty may be anticipated only if states mixed by the field are very close in energy, so that higher-order terms are required in (4.5.5) and (4.5.7). In such a case, a linear treatment can formally be restored by including all the offending terms in the zero-field manifolds of A and/or J. As before, these manifolds are first diagonalized with respect to μ_z. Terms with small energy denominators cannot then appear in (4.5.5), and the previous treatment can be carried through. The resulting MCD expression in general contains contributions from several (nearby) transitions, but each is in the linear form (4.5.14). Indeed, precisely this case was treated in Section 5.2. That analysis is applicable for states which are arbitrarily close in energy provided parallel potential surfaces are assumed (as well as the BO–FC–RS approximation).

Various of these restrictions are removed in Section 7.6, where the invariance of lower moments under unitary transformations on the basis set plays a fundamental role.

Thus granting the validity of the first-order perturbation expansion (4.5.5), we see that an MCD expression linear in the field [e.g., (4.5.14)] rests on two basic assumptions, namely that Zeeman splittings are small compared to kT and to bandwidth. These conditions permit truncation of the expansions in (4.5.8) and (4.5.13a) at terms linear in the field. In this and the next section we examine in turn the consequences of dropping each of these assumptions.

6.2. Saturation: Zeeman Splittings $\gtrsim kT$

From earlier discussion (Sections 4.4, 4.5, 5.1), we know that the MCD associated with a $\mathcal{C}$ term increases linearly with B/T. However, if B/T is increased indefinitely, a limit is always reached beyond which the MCD increases ever more slowly. Ultimately, further increases in B/T produce no further increase in this MCD contribution. As this deviation from linearity occurs, the system is said to be saturating, and when no further change occurs, the system is said to be saturated. This behavior can be understood qualitatively by recalling [Eq. (5.1.8) et seq.] that a $\mathcal{C}$ term arises because a magnetic field causes a population redistribution among Zeeman sublevels in accord with (5.1.11). It is clear that if B/T becomes sufficiently large, all molecules will have dropped into the lowest Zeeman sublevel, and further increases in B/T can produce no further population changes.

To treat this effect quantitatively, we return to (4.5.2). Again we make the BO–FC–RS approximation and follow the previous treatment [Eqs. (4.5.4)–(4.5.7), (4.5.10)–(4.5.13), (4.5.17), (4.5.24)]. Again we discard terms quadratic in B, *but we do not make the approximation* (4.5.8). The result is

$$\frac{\Delta A'}{\mathcal{E}} = \frac{\gamma\mu_B B}{\sum_{\alpha\lambda} \exp(\langle A\alpha|\mu_z|A\alpha\rangle^0 B/kT)}$$

$$\times \left\{ \sum_{\alpha\lambda} \exp\left(\frac{\langle A\alpha|\mu_z|A\alpha\rangle^0 B}{kT} \right) \right.$$

$$\times \left[\mathcal{A}_1(\alpha\lambda)\left(\frac{-\partial f(\mathcal{E})}{\partial \mathcal{E}} \right) + \mathcal{B}_0(\alpha\lambda) f(\mathcal{E}) \right.$$

$$\left.\left. + \left(|\langle A\alpha|m_{-1}|J\lambda\rangle^0|^2 - |\langle A\alpha|m_1|J\lambda\rangle^0|^2 \right) \frac{f(\mathcal{E})}{\mu_B B} \right] \right\} \qquad (6.2.1)$$

where $\mathcal{A}_1(\alpha\lambda)/|A|$ and $\mathcal{B}_0(\alpha\lambda)/|A|$ are defined respectively as $\mathcal{A}_1$ and $\mathcal{B}_0$

of (4.5.16) *with the sums over α and λ omitted*. That is,

$$\mathcal{A}_1 \equiv \frac{1}{|A|}\sum_{\alpha\lambda}\mathcal{A}_1(\alpha\lambda), \qquad \mathcal{B}_0 \equiv \frac{1}{|A|}\sum_{\alpha\lambda}\mathcal{B}_0(\alpha\lambda) \tag{6.2.2}$$

Equation (6.2.1) is the quantitative description of the saturation phenomenon. As with (4.5.16), this equation applies only for oriented or isotropic molecules and requires ground-state eigenfunctions diagonal in (space-fixed) μ_z. (See further discussion, Section 6.3.) The connection with the linear limit is immediately established if we substitute (4.5.8) into (6.2.1). Discarding terms quadratic in B and using (4.5.18), we obtain (4.5.14).

6.3. Saturation Analog of the $\mathcal{C}$ Term

Let us first consider the contribution which gives rise to the $\mathcal{C}$ term in the linear limit, that is, the last term in (6.2.1). Using (5.1.9), (5.1.12), and (5.1.13), we may write this term

$$\left(\frac{\Delta A'}{\mathcal{E}}\right)_{\mathcal{C}(\text{sat})} = \frac{\gamma \sum\limits_{i\text{ pairs}} \Delta\alpha_i \left[\exp\left(\frac{\langle A\alpha_i|\mu_z|A\alpha_i\rangle^0 B}{kT}\right) - \exp\left(-\frac{\langle A\alpha_i|\mu_z|A\alpha_i\rangle^0 B}{kT}\right)\right]}{\sum\limits_{i\text{ pairs}} \left[\exp\left(\frac{\langle A\alpha_i|\mu_z|A\alpha_i\rangle^0 B}{kT}\right) + \exp\left(-\frac{\langle A\alpha_i|\mu_z|A\alpha_i\rangle^0 B}{kT}\right)\right]} f(\mathcal{E}) \tag{6.3.1}$$

where the sum is over *either* member i of each Kramers pair (Section 5.1) and $\Delta\alpha_i$ is given by

$$\Delta\alpha_i \equiv \sum_{\lambda}\left(|\langle A\alpha_i|m_{-1}|J\lambda\rangle^0|^2 - |\langle A\alpha_i|m_1|J\lambda\rangle^0|^2\right) \tag{6.3.2}$$

If $|A\alpha_1\rangle$ is the lowest Zeeman sublevel, then at complete saturation

$$\left(\frac{\Delta A'}{\mathcal{E}}\right)_{\mathcal{C}(\text{sat})} = \gamma\Delta\alpha_1 f(\mathcal{E}) \tag{6.3.3}$$

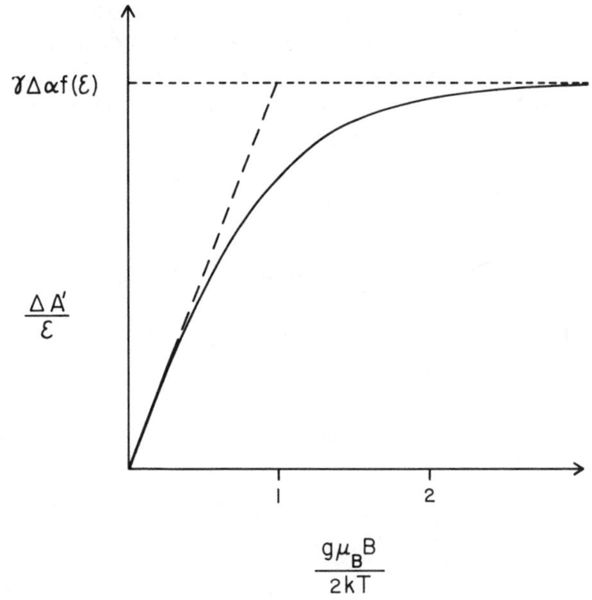

Figure 6.3.1. MCD saturation behavior for an isotropic or oriented-Kramers-doublet ground state [Eq. (6.3.4)]. The dashed line extrapolates normal $\mathcal{C}$-term behavior.

If the ground state consists of a single Kramers pair,

$$\left(\frac{\Delta A'}{\mathcal{E}}\right)_{\mathcal{C}(\text{sat})} = \gamma \Delta\alpha f(\mathcal{E}) \tanh\left(\frac{\langle A\alpha|\mu_z|A\alpha\rangle^0 B}{kT}\right)$$

$$\equiv \gamma \Delta\alpha\, f(\mathcal{E}) \tanh\left(\frac{g\mu_B B}{2kT}\right) \qquad (6.3.4)$$

A plot of (6.3.4) is shown in Fig. 6.3.1. As we show next, this simple tanh form does not apply if the (zero-field) ground state is not a single Kramers pair, nor does it in general apply even in that case if the molecules are randomly oriented.

Basis Invariance

In the linear limit, by using spectroscopic stability, it was a simple matter to remove the requirement that eigenfunctions be diagonal in μ_z (Section 4.6). When saturation occurs, this is still a trivial matter for the excited-state basis J, because these functions occur only in $\Delta\alpha_i$ of (6.3.1). Inspection of the definition of $\Delta\alpha_i$, (6.3.2), shows that the $|J\lambda\rangle$ indeed already occur in invariant form. Consequently, (6.3.1) applies for any excited-state basis related to the diagonal basis by a unitary transformation. On the other hand, this is *not* the case for the ground-state basis, because of the occurrence of the $|A\alpha\rangle$ in the exponentials in (6.3.1). This is a fundamental complication.

Molecule-Fixed Operators and Orientational Averaging

We may transform (6.3.1) to molecule-fixed operators by using (4.6.11). After some algebra, $\Delta\alpha_i$ takes the form

$$\Delta\alpha_i = \sum_\lambda \Big[\cos\theta\Big(\big|\langle A\alpha_i|m'_{-1}|J\lambda\rangle^0\big|^2 - \big|\langle A\alpha_i|m'_1|J\lambda\rangle^0\big|^2\Big) - \sqrt{2}\,\mathfrak{Re}\sin\theta e^{i\phi}$$

$$\times\big(\langle A\alpha_i|m'_1|J\lambda\rangle^0\langle A\alpha_i|m'_z|J\lambda\rangle^{0*}$$

$$-\langle A\alpha_i|m'_z|J\lambda\rangle^0\langle A\alpha_i|m'_{-1}|J\lambda\rangle^{0*}\big)\Big] \tag{6.3.5}$$

where $\mathfrak{Re}$ designates the real part of everything to its right. The trigonometric factors are a consequence of the relations $A_{11}A_{22} - A_{12}A_{21} = \cos\theta$, $A_{12}A_{23} - A_{22}A_{13} + i(A_{11}A_{23} - A_{21}A_{13}) = e^{i\phi}\sin\theta$; $m'_{\pm1}$ and m'_z are *molecule-fixed* operators. Similarly, μ_z in the exponentials may be replaced by its molecule-fixed equivalent, (4.6.11). If all the molecules have the same space-fixed orientation, (6.3.1) can then be expressed explicitly and completely in terms of molecule-fixed operators, remembering that all $|A\alpha_i\rangle$ must be diagonal in (space-fixed) μ_z.

On the other hand, if it is necessary to average (6.3.1) over orientation, a basic problem arises. The difficulty is that the $|A\alpha_i\rangle$, and therefore $\Delta\alpha_i$ and $\langle A\alpha_i|\mu_z|A\alpha_i\rangle$, are, in general, orientation-dependent. That is, molecules at different orientations with respect to the field have different Zeeman energies and eigenfunctions. [This of course was equally true in the linear limit, yet no difficulty was encountered in the orientational averaging. The reason was that the exponentials were expanded and only constant and linear terms were retained. It was then possible to apply spectroscopic stability to the *ground-state* eigenfunctions so that the *same* molecule-fixed set could be used for each molecule regardless of its orientation with respect to the field. Thus all orientation dependence appeared via (4.6.11), and the appropriate averages were performed once and for all to yield (4.6.14).

Because of the exponentials in (6.3.1), this procedure cannot be applied, and one must proceed on a case-by-case basis.]

Explicit, orientation-dependent ground-state eigenfunctions must be substituted into (6.3.1) before the averaging expressed by (4.6.12) can be performed. This process has been carried out for molecules in the point group $D_{\infty h}$ and many of its subgroups [Schatz et al. (1978)]. This extra, nontrivial complication emphasizes the great simplification effected by working in the linear limit.

The orientational averaging of (6.3.1) is simplified if the ground-state Zeeman splitting is isotropic (i.e., is independent of molecular orientation), since the exponentials are then orientation-independent. This is, for example, usually true for the cubic point groups. If in addition the ground state consists of a single Kramers pair, (6.3.1) becomes

$$\left(\frac{\Delta A}{\mathcal{E}}\right)_{\mathcal{C}(\mathrm{sat})} = \gamma\overline{\Delta\alpha}\tanh\left(\frac{\langle A\alpha|\mu_z|A\alpha\rangle^0 B}{kT}\right) f(\mathcal{E}) \qquad (6.3.6)$$

where the bar designates an orientational average, and the simple tanh form of Fig. 6.3.1 is again obtained.

In the case of a $\mathcal{C}$ term, it is usually possible to obtain the ground-state magnetic moment from the ratio $\mathcal{C}_0/\mathcal{D}_0$ (Section 5.3). The same information can be obtained from a saturation study *without a measurement of the absorption spectrum at all.* For example, if (6.3.6) applies, $\langle A\alpha|\mu_z|A\alpha\rangle^0$ can be estimated by finding that value which best fits the experimentally observed saturation behavior. Such measurements can be made, at least roughly, in a matter of minutes by sitting at the $\mathcal{C}$-term maximum and simply scanning the MCD deflection at constant temperature while running field in (or out) of the magnet at a constant rate. For a simple tanh form, a few additional minutes of hand calculation (or a good eye) will yield at least a rough value of the ground-state magnetic moment. Least-squares fitting procedures of course do better.

Saturation Analog of $\mathcal{A}$ and $\mathcal{B}$ Terms

In general the effect of saturation on $\mathcal{A}$ and $\mathcal{B}$ terms is of much less practical importance, since the $\mathcal{C}$ term, which is necessarily present, usually dominates the MCD. The first two terms in (6.2.1) show the saturation analogs of the $\mathcal{A}$ and $\mathcal{B}$ terms respectively. The same general complications arise as in the $\mathcal{C}$-term case, and they can be treated in an analogous way. Again simplifications are possible in special cases.

6.4. Zeeman Splittings Comparable to Bandwidth

Let us consider the case in which the Zeeman splitting is comparable to or greater than the bandwidth but is still small compared to kT. This occurs if sufficiently sharp bands persist appreciably above liquid-helium temperature.

We start again with (4.5.2), and use (4.5.9) but *not* (4.5.13). Thus

$$\frac{\Delta A'}{\mathcal{E}} = \frac{\gamma\mu_B B}{|A|}\sum_{\alpha\lambda} f'_{A\alpha,J\lambda}(\mathcal{E})\left[\mathcal{B}_0(\alpha\lambda) + \frac{\mathcal{C}_0(\alpha\lambda)}{kT} + \mathcal{R}_0(\alpha\lambda)\right] \quad (6.4.1)$$

where $\mathcal{C}_0(\alpha\lambda)$ is defined in exact analogy to $\mathcal{A}_1(\alpha\lambda)$ and $\mathcal{B}_0(\alpha\lambda)$ in (6.2.2), and $\mathcal{R}_0(\alpha\lambda)$ is defined by

$$\mu_B B\mathcal{R}_0(\alpha\lambda) \equiv \left(|\langle A\alpha|m_{-1}|J\lambda\rangle^0|^2 - |\langle A\alpha|m_1|J\lambda\rangle^0|^2\right) \quad (6.4.2)$$

In analogy to (4.5.24)

$$f'_{A\alpha,J\lambda}(\mathcal{E}) = \frac{1}{N_G}\sum_g N_g \sum_j |\langle g|j\rangle|^2 \rho'_{A\alpha g,J\lambda j}(\mathcal{E}) \quad (6.4.3)$$

The new feature in (6.4.1) is the term $\mu_B B\mathcal{R}_0(\alpha\lambda)$, which is independent of field; it reflects the fact that individual Zeeman transitions are circularly polarized. When a sum over all of these transitions is performed, such contributions vanish by (4.5.18). It should be emphasized that (6.4.1) has a complicated field and orientation dependence. The function $f'_{A\alpha,J\lambda}(\mathcal{E})$ moves linearly in the field according to (4.5.10), and $|A\alpha\rangle'$ and $|J\lambda\rangle'$ are in general functions of the orientation of the molecule with respect to the field and must be chosen diagonal in (space-fixed) μ_z.

6.5. Saturation and Large Zeeman Splitting

In this most general case, neither the expansion (4.5.8) nor (4.5.13a) can be truncated at terms linear in the field, and one must return to the general expression (4.5.2). Each Zeeman transition must be assigned its own field-dependent bandshape function (all identical and each moving rigidly with field in the rigid-shift model), and $N'_{A\alpha g}$ must be calculated by the first equation of (4.5.9). The MCD must be calculated at each frequency and orientation in accord with (4.5.2), and will in general be a complicated function of both field and temperature. The simple partition into separate contributions ($\mathcal{A}$, $\mathcal{B}$, $\mathcal{C}$) with easily understood physical origins is no longer possible. The computer comes to the fore.

Finally, it should be reiterated that (4.5.14) itself applies only for allowed electronic transition between BO states. When the BO approximation is dropped, one cannot hope to write a simple expression for the full MCD dispersion, and moment analysis is an appropriate approach. This topic is discussed in the next chapter.

7 Moments and Parameter Extraction

7.1. Relating the Experimental Lineshape to Theoretical Parameters

Once MCD and absorption spectra have been measured, the results for each band are compared with observables calculated using theoretical models for the system. If correlation between experiment and theory is good, and the model is unique in giving this close correlation, detailed information about molecular structure may be obtained. In this chapter we discuss two ways in which this comparison may be realized. The first, the method of moments (MM), is the more rigorous of the two. It is simple to apply to experimental data but requires discrete bands. The second method is based on the rigid-shift (RS) equations of Section 4.5 with a Gaussian lineshape function for $f(\mathscr{E})$. This procedure, which we term the RS–G method, requires more drastic theoretical approximations but has wider experimental applicability.

Our discussion in Sections 7.4–7.8 and in Section 7.10 applies to *entire bands*. We consider *vibronic lines* in Sections 7.9 and 7.11. Most of our discussion is limited to oriented systems. Generalization to solutions and randomly oriented systems is straightforward, as discussed in Section 4.6.

7.2. Introduction to the Method of Moments

The method of moments was first applied in electronic spectroscopy by Lax (1952). It may be used whenever some observable α has been measured as a function of the energy $\mathscr{E}$ to give a bandshape (or dispersion) $\alpha(\mathscr{E})$ such as that of Figure 7.2.1. The moments of a band with bandshape $\alpha(\mathscr{E})$ are defined by

$$\langle \alpha \rangle_n^{\mathscr{E}^0} = \int \alpha(\mathscr{E})(\mathscr{E} - \mathscr{E}^0)^n \, d\mathscr{E} \qquad (7.2.1)$$

where $\mathscr{E}^0$ is some chosen energy about which the moments of the band are taken. Integration is always between the ends of the band which must be points where $\alpha(\mathscr{E}) = 0$ ($\mathscr{E}_1$ and $\mathscr{E}_2$ in Fig. 7.2.1). Thus moments may only

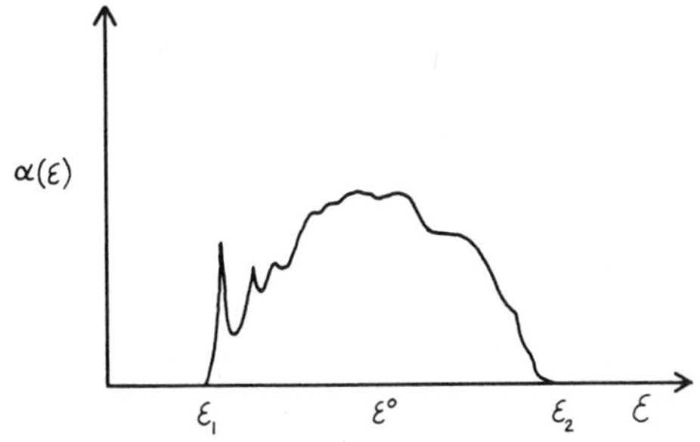

Figure 7.2.1. A typical bandshape of some observable α plotted as a function of the energy $\mathcal{E}$.

be evaluated for discrete bands, which in practice is the greatest limitation of the method. The zeroth moment

$$\langle\alpha\rangle_0 = \int \alpha(\mathcal{E})\, d\mathcal{E} \tag{7.2.2}$$

which is independent of $\mathcal{E}^0$, is simply the integrated area of $\alpha(\mathcal{E})$.

$\mathcal{E}^0$ is usually chosen to be the average energy $\overline{\mathcal{E}}$ of the band, defined so that $\langle\alpha\rangle_1^{\overline{\mathcal{E}}} = 0$. Thus

$$\overline{\mathcal{E}} = \frac{\int \alpha(\mathcal{E})\mathcal{E}\, d\mathcal{E}}{\int \alpha(\mathcal{E})\, d\mathcal{E}} \tag{7.2.3}$$

and

$$\langle\alpha\rangle_n^{\overline{\mathcal{E}}} = \int \alpha(\mathcal{E})(\mathcal{E} - \overline{\mathcal{E}})^n\, d\mathcal{E} \tag{7.2.4}$$

In absorption spectroscopy $\alpha(\mathcal{E}) = A_{\pm}/\mathcal{E}\ [= (\epsilon_{\pm} cl)/\mathcal{E}]$ of (4.3.10), and in MCD spectroscopy $\alpha(\mathcal{E}) = \Delta A'/\mathcal{E}$ of (4.3.9). $\overline{\mathcal{E}}$ is defined as in (7.2.3) for $\alpha(\mathcal{E}) = A_{\pm}/\mathcal{E}$ and is used for *both* absorption and MCD moments. Moment analysis was first applied to MCD by Henry, Schnatterly, and Slichter (1965) in their work on color centers. The theory was subsequently generalized by Stephens (1970, 1976) to apply to most systems of interest to chemists. In this chapter we draw continually from the work of Stephens, particularly from his review article (1976).

To apply the MM we integrate the experimental data as specified for the moment desired ($n = 0, 1, \ldots$) and compare the result with that obtained by integrating the theoretical expression for $\alpha(\mathcal{E})$ in the same moment equation. For low values of n, it is often possible to calculate $\langle\alpha\rangle_n^{\bar{\mathcal{E}}}$ fairly rigorously, while calculation of the full dispersion $\alpha(\mathcal{E})$ requires drastic approximations. The relative ease of moment calculations for bands (but not for lines) is due to the fact that $\langle\alpha\rangle_n^{\bar{\mathcal{E}}}$ may be evaluated in a convenient excited-state representation. This crucial point is demonstrated in Section 7.3. Thus, for example, while the full vibronic problem must be solved rigorously to obtain $\alpha(\mathcal{E})$ for absorption bands, $\langle\alpha\rangle_0$ and so on may be calculated in a Born–Oppenheimer (BO) basis. Of course, determination of $\langle\alpha\rangle_0$ or $\langle\alpha\rangle_1^{\bar{\mathcal{E}}}$ gives much less molecular information than the exact $\alpha(\mathcal{E})$. But a complete calculation of $\alpha(\mathcal{E})$ is a formidable problem. When the MM is not used, bandshapes in absorption and MCD spectroscopy are often assumed Gaussian, and the RS approximation is made. Experimental parameters are then determined from Gaussian fits. *The RS–G assumptions are not required in the MM.*

7.3. The Moment Equations in Absorption and MCD Spectroscopy

The nth moments of the absorption and MCD are obtained by substituting $A/\mathcal{E}$ of (4.3.10) and $\Delta A'/\mathcal{E}$ of (4.3.9) into the standard moment equation (7.2.4). Before substitution, however, we rewrite $A/\mathcal{E}$ and $\Delta A'/\mathcal{E}$ using $m_{\pm 1}$ of (4.5.15) [instead of $m_{\pm}$ of (2.6.11)] and use (4.5.17). For an absorption band arising from a set of vibronic transitions $|a\rangle \to |j\rangle$, where $|a\rangle$ and $|j\rangle$ are eigenstates of $\mathcal{H}_T$ [from (3.2.7) or (3.8.9)] and $N_j = 0$, we have

$$\left\langle \frac{A_{\pm}}{\mathcal{E}} \right\rangle_n^{\bar{\mathcal{E}}} = \int \left(\frac{A_{\pm}}{\mathcal{E}} \right) (\mathcal{E} - \bar{\mathcal{E}})^n \, d\mathcal{E}$$

$$= \gamma \left[-\sum_{aj} \left(\frac{N_a}{N} \right) \langle a|m_{\pm 1}|j\rangle\langle j|m_{\mp 1}|a\rangle \int \rho_{aj}(\mathcal{E})(\mathcal{E} - \bar{\mathcal{E}})^n \, d\mathcal{E} \right]$$

$$= \gamma \left[-\sum_{aj} \left(\frac{N_a}{N} \right) \langle a|m_{\pm 1}|j\rangle\langle j|m_{\mp 1}|a\rangle \right.$$

$$\left. \times \left(\langle j|\mathcal{H}_T|j\rangle - \langle a|\mathcal{H}_T|a\rangle - \bar{\mathcal{E}} \right)^n \right] \qquad (7.3.1)$$

where $A = A_+ = A_- = A_{\pm}$. In the last line we have made the assumption

$\rho_{aj}(\mathcal{E}) = \delta(\mathcal{E}_{ja} - \mathcal{E})$ (Section 2.3), since only then is it true *for all n* that

$$\int \rho_{aj}(\mathcal{E})(\mathcal{E} - \bar{\mathcal{E}})^n \, d\mathcal{E} = \left(\mathcal{E}_{ja} - \bar{\mathcal{E}}\right)^n \tag{7.3.2}$$

Equation (7.3.2) is implicit [via (7.3.1)] in all of our subsequent moment analysis. For $n = 0$ (7.3.2) applies for any $\rho_{aj}(\mathcal{E})$ satisfying $\int \rho_{aj}(\mathcal{E})\, d\mathcal{E} = 1$ [Eq. (2.3.2)], and for $n = 1$ it applies if $\rho_{aj}(\mathcal{E})$ is also an even function of $\mathcal{E} - \mathcal{E}_{ja}$—that is, it is symmetrical about $\mathcal{E}_{ja}$. However, in general (7.3.2) requires the assumption that the vibronic band contour is reasonably well described by a continuum of infinitely sharp lines [Kauzmann (1957), p. 580]. This is a plausible supposition in condensed phases.

The corresponding MCD band arises from a set of vibronic transitions $|a\rangle' \to |j\rangle'$ where $|a\rangle'$ and $|j\rangle'$ are eigenstates of $\mathcal{H}'_T = \mathcal{H}_T - \mu_z B$. As in Chapter 4, magnetic-field-dependent quantities are designated with primes, which in the case of functions are placed on the ket, not on the function in the ket. It follows from (7.2.4), (4.3.9), and (7.3.2) that the moments of the MCD are

$$\begin{aligned}
\left\langle \frac{\Delta A'}{\mathcal{E}} \right\rangle_n^{\bar{\mathcal{E}}} &= \int \frac{\Delta A'}{\mathcal{E}} (\mathcal{E} - \bar{\mathcal{E}})^n \, d\mathcal{E} \\
&= \gamma \left[-\sum_{aj} \frac{N'_a}{N} \left(\langle a|m_{-1}|j\rangle' \langle j|m_1|a\rangle' \right. \right. \\
&\qquad \left. \left. - \langle a|m_1|j\rangle' \langle j|m_{-1}|a\rangle' \right) \left(\mathcal{E}'_{ja} - \bar{\mathcal{E}}\right)^n \right] \\
&= \gamma \left[-\sum_{aj} \frac{N'_a}{N} \left(\langle a|m_{-1}|j\rangle' \langle j|m_1|a\rangle' - \langle a|m_1|j\rangle' \langle j|m_{-1}|a\rangle' \right) \right. \\
&\qquad \left. \times \left(\langle j|\mathcal{H}'_T|j\rangle' - \langle a|\mathcal{H}'_T|a\rangle' - \bar{\mathcal{E}} \right)^n \right]
\end{aligned} \tag{7.3.3}$$

In all of these equations $\bar{\mathcal{E}}$ is defined by $\langle A_{\pm}/\mathcal{E}\rangle_1^{\bar{\mathcal{E}}} = 0$, so

$$\bar{\mathcal{E}} = \frac{\int \left(\frac{A_{\pm}}{\mathcal{E}} \right) \mathcal{E} \, d\mathcal{E}}{\int \left(\frac{A_{\pm}}{\mathcal{E}} \right) d\mathcal{E}} \tag{7.3.4}$$

To evaluate these moments further we must define the basis states $|a\rangle$, $|j\rangle$, $|a\rangle'$, and $|j\rangle'$. Happily the equations can be brought into a form which is invariant under unitary transformations on the *excited-state* manifold, as we now explicitly demonstrate for the zeroth and first absorption and MCD moments. The invariance of the zeroth moment ($n = 0$) follows directly from the principle of spectroscopic stability, stated in (4.6.3). The principle, which is often abbreviated

$$\sum_t |t\rangle\langle t| = \sum_{\bar{t}} |\bar{t}\rangle\langle \bar{t}| \tag{7.3.5}$$

gives

$$\sum_j \langle a|m_{\mp 1}|j\rangle\langle j|m_{\pm 1}|a\rangle = \sum_{\bar{j}} \langle a|m_{\mp 1}|\bar{j}\rangle\langle \bar{j}|m_{\pm 1}|a\rangle \tag{7.3.6}$$

where $|\bar{j}\rangle$ represents any set of functions related to the $|j\rangle$ set by a unitary transformation. For the first moment ($n = 1$) consider the relevant portion of a typical term from (7.3.1) [or (7.3.3)]:

$$\sum_{aj} \left(-\frac{N_a}{N}\right)\langle a|m_{\mp 1}|j\rangle\langle j|m_{\pm 1}|a\rangle\left(\langle j|\mathcal{H}_T|j\rangle - \langle a|\mathcal{H}_T|a\rangle - \bar{\mathcal{E}}\right) \tag{7.3.7}$$

Since the $|j\rangle$ (and $|a\rangle$) are eigenstates of (and thus diagonal in) $\mathcal{H}_T$,

$$\langle j|\mathcal{H}_T|j\rangle - \langle a|\mathcal{H}_T|a\rangle = \sum_{j'}\left(\langle j|\mathcal{H}_T|j'\rangle - \langle a|\mathcal{H}_T|a\rangle\delta_{jj'}\right) \tag{7.3.8}$$

where j' runs over all members of the set $|j\rangle$. Substituting (7.3.8) into (7.3.7) and rearranging slightly gives

$$\sum_{ajj'} \left(-\frac{N_a}{N}\right)\langle a|m_{\mp 1}|j\rangle$$

$$\times \left(\langle j|\mathcal{H}_T|j'\rangle - \langle a|\mathcal{H}_T|a\rangle\delta_{jj'} - \bar{\mathcal{E}}\delta_{jj'}\right)\langle j'|m_{\pm 1}|a\rangle \tag{7.3.9}$$

since j' can be substituted for j in the last bra. It now follows immediately from the principle of spectroscopic stability that (7.3.9) remains unchanged (invariant) if the $|j\rangle$ set is replaced by any other set related to it by a unitary transformation (remember that $|j\rangle$ and $|j'\rangle$ are in the same set).

Consequently, we can choose any excited-state basis which is related to the $|j\rangle$ (or the field-dependent $|j\rangle'$) by a unitary transformation or, in fact, by a series of unitary transformations; that is, the basis need not be composed of the actual eigenfunctions of $\mathcal{H}_T$ (or $\mathcal{H}'_T$). Thus, a more general form of the moment equations may be written, which does not require $|j\rangle$ and $|j\rangle'$ to be diagonal in $\mathcal{H}_T$ and $\mathcal{H}'_T$ respectively. For $n = 0$ and 1

$$\left\langle \frac{A_\pm}{\mathcal{E}} \right\rangle_0 = \gamma \left[-\sum_{aj} \frac{N_a}{N} \langle a|m_{\pm1}|j\rangle\langle j|m_{\mp1}|a\rangle \right] \tag{7.3.10}$$

$$\left\langle \frac{A_\pm}{\mathcal{E}} \right\rangle_1 = \gamma \left[-\sum_{ajj'} \frac{N_a}{N} \langle a|m_{\pm1}|j\rangle\langle j'|m_{\mp1}|a\rangle \right.$$

$$\left. \times \left(\langle j|\mathcal{H}_T|j'\rangle - \langle a|\mathcal{H}_T|a\rangle\delta_{jj'} - \bar{\mathcal{E}}\delta_{jj'} \right) \right] \tag{7.3.11}$$

$$\left\langle \frac{\Delta A'}{\mathcal{E}} \right\rangle_0 = \gamma \left[-\sum_{aj} \frac{N'_a}{N} \left(\langle a|m_{-1}|j\rangle'\langle j|m_1|a\rangle' \right. \right.$$

$$\left. \left. - \langle a|m_1|j\rangle'\langle j|m_{-1}|a\rangle' \right) \right] \tag{7.3.12}$$

$$\left\langle \frac{\Delta A'}{\mathcal{E}} \right\rangle_1 = \gamma \left[-\sum_{ajj'} \frac{N'_a}{N} \left(\langle a|m_{-1}|j\rangle'\langle j'|m_1|a\rangle' \right. \right.$$

$$\left. - \langle a|m_1|j\rangle'\langle j'|m_{-1}|a\rangle' \right)$$

$$\left. \times \left(\langle j|\mathcal{H}'_T|j'\rangle' - \langle a|\mathcal{H}'_T|a\rangle'\delta_{jj'} - \bar{\mathcal{E}}\delta_{jj'} \right) \right] \tag{7.3.13}$$

Unfortunately, the invariant form of the moment equations does not hold true for the ground state, because of the population factors N_a/N. To make the treatment tractable, *we always assume the BO approximation holds for the ground state* and so in (7.3.10) and (7.3.11) the $|a\rangle$ are in the form

$$|a\rangle = |A\alpha\rangle|g\xi\rangle \tag{7.3.14}$$

where $|A\alpha\rangle$ and $|g\xi\rangle$ are eigenfunctions of (3.2.6) and (3.2.10) respectively. (Here, in anticipation of later group-theoretic applications, we designate electronic $|A\alpha\rangle$ and vibrational $|g\xi\rangle$ functions by two labels, the first being the irreducible representation and the second the component thereof. The indices g and ξ run over the complete set of ground-state vibrational functions.) Note that $N_a = N_{A\alpha,\, g\xi}$ has the same value for each electronic component α and vibrational component ξ. Thus we may write

$$\frac{N_a}{N} = \frac{N_a}{\sum_a N_a} = \frac{N_{A\alpha,\, g\xi}}{\sum_{\alpha g\xi} N_{A\alpha,\, g\xi}} = \frac{N_g}{|A|\sum_g |g| N_g}$$

$$\equiv \frac{1}{|A|}\,\frac{N_g}{N_G} \tag{7.3.15}$$

where $|A|$ is the degree of degeneracy of A, $|g|$ is the degree of degeneracy of g, and $N_a = N_{A\alpha,\, g\xi} \equiv N_g$. The last line defines N_G.

The great utility of the invariant form of the moment equations becomes apparent when we consider an excited state subject to the Jahn–Teller (JT) effect. A JT effect mixes BO states and, as discussed in Section 3.8, the determination of the eigenfunctions of $\mathcal{H}_T$ requires a laborious calculation —the solution of the dynamic JT problem. But this calculation is not necessary for the moment equations, since the JT vibronic eigenfunctions are related by a unitary transformation to the BO basis. Thus we can calculate moments using the BO basis, which we write $|J\lambda\rangle|j\zeta\rangle$. But in fact we can arrive at a *still* more useful basis in the following way. Suppose we expand the excited-state vibrational functions $|j\zeta\rangle$ in the complete orthonormal set of *ground-state* vibrational functions $|g\xi\rangle$:

$$|J\lambda\rangle|j\zeta\rangle = \sum_{g'\xi'} |J\lambda\rangle|g'\xi'\rangle\langle g'\xi'|j\zeta\rangle \tag{7.3.16}$$

It immediately follows that the $|J\lambda\rangle|g'\xi'\rangle$ are related by a unitary transformation to the $|J\lambda\rangle|j\zeta\rangle$ [simply by inverting (7.3.16)]. Consequently, because of the invariant form of the moment equations, we can use the $|J\lambda\rangle|g'\xi'\rangle$ functions as our basis:

$$|j\rangle = |J\lambda\rangle|g'\xi'\rangle \tag{7.3.17}$$

Note that we need never determine the transformation coefficients (the $\langle g'\xi'|j\zeta\rangle$), and the $|g'\xi'\rangle$ are *ground-state* vibrational functions. [Thus our final excited-state basis (7.3.17) is adiabatic (Section 3.2), but is not, strictly speaking, a BO basis.] The electronic functions $|J\lambda\rangle$ are in the form of the ϕ_k of Section 3.3 in the nondegenerate case. In degenerate (or near-degenerate) cases, the $|J\lambda\rangle$ are chosen in the form of the $\phi_{k\kappa}$ of Section 3.8, so that our electronic functions always commute with T_n to a good approximation.

7.4. Moments for Franck–Condon-Allowed Transitions

We now make the FC approximation (Section 3.6) for all purely electronic operators. In fact our $|j\rangle, |j'\rangle$ basis above was chosen to give FC matrix elements particularly simple forms. Thus

$$\langle A\alpha, g\xi|m_{\pm 1}|J\lambda, g'\xi'\rangle = \langle A\alpha|m_{\pm 1}|J\lambda\rangle^0 \delta_{gg'}\delta_{\xi\xi'} \tag{7.4.1}$$

(where $\delta_{gg'} = 0$ if g and g' designate different vibrational levels, even if both belong to the same irreducible representation).

With the above substitutions and those of Section 7.3, the zeroth absorption moment, (7.3.10), becomes

$$\left\langle \frac{A_\pm}{\mathcal{E}} \right\rangle_0 = \gamma \Bigg[-\sum_{\alpha\lambda} \frac{1}{|A|} \langle A\alpha|m_{\pm 1}|J\lambda\rangle^0 \langle J\lambda|m_{\mp 1}|A\alpha\rangle^0 \times \sum_{g\xi g'\xi'} \frac{N_g}{N_G} \langle g\xi|g'\xi'\rangle\langle g'\xi'|g\xi\rangle \Bigg] \tag{7.4.2}$$

The sums over vibrational functions give unity, so

$$\left\langle \frac{A_\pm}{\mathcal{E}} \right\rangle_0 = \gamma \mathcal{D}_0 \tag{7.4.3}$$

where $\mathcal{D}_0$ is given by (4.5.23). Thus the integrated band intensity of a FC-allowed transition is temperature-independent.

The first moment of absorption, (7.3.11), is far more complicated:

$$\left\langle \frac{A_\pm}{\mathcal{E}} \right\rangle_1^{\bar{\mathcal{E}}} = \gamma \Bigg\{ - \sum_{\alpha\lambda\lambda'} \frac{1}{|A|} \langle A\alpha | m_{\pm 1} | J\lambda \rangle^0 \langle J\lambda' | m_{\mp 1} | A\alpha \rangle^0$$

$$\times \Bigg[\sum_{g\xi} \frac{N_g}{N_G} \langle g\xi | \langle J\lambda | \mathcal{H}_{\text{el}} | J\lambda' \rangle + T_n \delta_{\lambda\lambda'}$$

$$- W_A \delta_{\lambda\lambda'} - T_n \delta_{\lambda\lambda'} | g\xi \rangle - \bar{\mathcal{E}} \delta_{\lambda\lambda'} \Bigg] \Bigg\}$$

$$\equiv \gamma \mathcal{D}_1 \qquad (7.4.4)$$

The T_n terms cancel. W_A is in the form (3.3.4), and $\mathcal{H}_{\text{el}}$ is approximated by (3.3.1)–(3.3.3). $\bar{\mathcal{E}}$, the barycenter of the absorption band, is defined by setting $\mathcal{D}_1 = 0$. While the theoretical expression for $\bar{\mathcal{E}}$ is rather intimidating, $\bar{\mathcal{E}}$ is easy to determine experimentally by integrating the data according to (7.3.4).

We next calculate the zeroth and first moments of the MCD in the low-field limit where the MCD is linear in the magnetic field B. Many of the approximations made in Section 4.5 are thus used again. The magnetic field perturbs electronic functions so that our $|a\rangle'$ and $|j\rangle'$ bases are

$$|A\alpha, g\xi\rangle' = |A\alpha, g\xi\rangle - \sum_{\substack{K\kappa \\ (K \neq A)}} \frac{\langle K\kappa | \mu_z | A\alpha \rangle^0}{W_A^0 - W_K^0} B |K\kappa, g\xi\rangle$$

$$|J\lambda, g'\xi'\rangle' = |J\lambda, g'\xi'\rangle - \sum_{\substack{K\kappa \\ (K \neq J)}} \frac{\langle K\kappa | \mu_z | J\lambda \rangle^0}{W_J^0 - W_K^0} B |K\kappa, g'\xi'\rangle \qquad (7.4.5)$$

As in Section 4.5, we have assumed $|A\alpha\rangle$ and $|J\lambda\rangle$ have been chosen diagonal in μ_z, and we have used the FC approximation. The excited-state basis is related to the $\mathcal{H}'_T$ eigenstates by a unitary transformation, as required. (Note that the $|J\lambda, g'\xi'\rangle$ basis cannot be used, since it is not related to $|J\lambda, g'\xi'\rangle'$ by a unitary transformation.) Since to first order in B

$$\mathcal{E}'_{A\alpha,\, g\xi} = \mathcal{E}_{A\alpha,\, g\xi} - \langle A\alpha|\mu_z|A\alpha\rangle^0 B \tag{7.4.6}$$

we have, using the same arguments that lead to (4.5.9),

$$\begin{aligned}\frac{N'_a}{N} &= \frac{N'_{A\alpha,\, g\xi}}{\sum\limits_{\alpha g\xi} N'_{A\alpha,\, g\xi}} \\ &\approx \frac{N_a}{N}\left(1 + \frac{\langle A\alpha|\mu_z|A\alpha\rangle^0 B}{kT}\right)\end{aligned} \tag{7.4.7}$$

where N_a/N is given by (7.3.15). As in Section 4.5, we have approximated the Zeeman exponentials by their first-order expansions in B: this is valid as long as $\langle A\alpha|\mu_z|A\alpha\rangle^0 B$ is small compared to kT.

The perturbed electric-dipole matrix elements in the FC approximation to first order in B are

$$\begin{aligned}\langle A\alpha, g\xi|m_{\pm 1}|J\lambda, g'\xi'\rangle' = \delta_{gg'}\delta_{\xi\xi'}\Bigg[&\langle A\alpha|m_{\pm 1}|J\lambda\rangle^0 \\ &+ \Bigg(\sum_{\substack{K\kappa \\ (K\neq J)}} \frac{\langle A\alpha|m_{\pm 1}|K\kappa\rangle^0\langle K\kappa|\mu_z|J\lambda\rangle^0}{W_K^0 - W_J^0} \\ &+ \sum_{\substack{K\kappa \\ (K\neq A)}} \frac{\langle K\kappa|m_{\pm 1}|J\lambda\rangle^0\langle A\alpha|\mu_z|K\kappa\rangle^0}{W_K^0 - W_A^0}\Bigg) B\Bigg] \\ \equiv \delta_{gg'}\delta_{\xi\xi'}\big[&\langle A\alpha|m_{\pm 1}|J\lambda\rangle^0 + \langle A\alpha|m_{\pm 1}|J\lambda\rangle' B\big]\end{aligned} \tag{7.4.8}$$

Inserting (7.4.7) and (7.4.8) into (7.3.12) and keeping terms to first order in

B, we obtain

$$\left\langle \frac{\Delta A'}{\mathcal{E}} \right\rangle_0 = \gamma \left\{ \frac{-1}{|A|} \sum_{\alpha\lambda} \frac{\langle A\alpha|\mu_z|A\alpha\rangle^0}{kT} \left(\langle A\alpha|m_{-1}|J\lambda\rangle^0 \langle J\lambda|m_1|A\alpha\rangle^0 \right. \right.$$

$$\left. - \langle A\alpha|m_1|J\lambda\rangle^0 \langle J\lambda|m_{-1}|A\alpha\rangle^0 \right)$$

$$-2\mathcal{R}e \frac{1}{|A|} \sum_{\alpha\lambda} \left[\sum_{\substack{K\kappa \\ (K\neq J)}} \frac{\langle J\lambda|\mu_z|K\kappa\rangle^0}{W_K^0 - W_J^0} \right.$$

$$\times \left(\langle A\alpha|m_{-1}|J\lambda\rangle^0 \langle K\kappa|m_1|A\alpha\rangle^0 - \langle A\alpha|m_1|J\lambda\rangle^0 \langle K\kappa|m_{-1}|A\alpha\rangle^0 \right)$$

$$+ \sum_{K\kappa(K\neq A)} \frac{\langle K\kappa|\mu_z|A\alpha\rangle^0}{W_K^0 - W_A^0}$$

$$\left. \left. \times \left(\langle A\alpha|m_{-1}|J\lambda\rangle^0 \langle J\lambda|m_1|K\kappa\rangle^0 - \langle A\alpha|m_1|J\lambda\rangle^0 \langle J\lambda|m_{-1}|K\kappa\rangle^0 \right) \right] \right\} B \tag{7.4.9}$$

Then, if we define $\mathcal{B}_0$ and $\mathcal{C}_0$ as in (4.5.16), this may be written

$$\left\langle \frac{\Delta A'}{\mathcal{E}} \right\rangle_0 = \gamma \left(\mathcal{B}_0 + \frac{\mathcal{C}_0}{kT} \right) \mu_B B \tag{7.4.10}$$

$\mathcal{B}_0$ and $\mathcal{C}_0$ may be obtained separately by determining the zeroth moment as a function of temperature.

We follow the same procedure for the first MCD moment, (7.3.13). To first order in B we have

$$\langle J\lambda, g\xi|\mathcal{H}_T'|J\lambda', g'\xi'\rangle' = \left\langle g\xi|\langle J\lambda|\mathcal{H}_{\mathrm{el}}|J\lambda'\rangle + T_n\delta_{\lambda\lambda'}|g'\xi'\right\rangle - \langle J\lambda|\mu_z|J\lambda\rangle^0 B\delta_{\lambda\lambda'}\delta_{gg'}\delta_{\xi\xi'} \qquad (7.4.11)$$

and

$$\langle A\alpha, g\xi|\mathcal{H}_T'|A\alpha, g\xi\rangle' = \langle g\xi|W_A + T_n|g\xi\rangle - \langle A\alpha|\mu_z|A\alpha\rangle^0 B \qquad (7.4.12)$$

since in our basis

$$\begin{aligned}\langle J\lambda|\mathcal{H}_{\mathrm{el}}|K\kappa\rangle &= \delta_{JK}\langle J\lambda|\mathcal{H}_{\mathrm{el}}|J\kappa\rangle \\ \langle J\lambda|T_n|K\kappa\rangle &= \delta_{JK}\delta_{\lambda\kappa}T_n\end{aligned} \qquad (7.4.13)$$

We have used the FC approximation in writing the μ_z matrix element. (The FC approximation cannot be used for $\mathcal{H}_{\mathrm{el}}$ or T_n matrix elements, since they are not purely electronic operators.) The relations (7.4.7)–(7.4.8) and (7.4.11)–(7.4.12) are inserted into (7.3.13) to give

$$\begin{aligned}\left\langle \frac{\Delta A'}{\mathcal{E}}\right\rangle_1^{\bar{\mathcal{E}}} = \gamma \sum_{\alpha\lambda\lambda'} -\frac{1}{|A|}\left(1 + \frac{\langle A\alpha|\mu_z|A\alpha\rangle^0 B}{kT}\right) \\ \times \Big[\big(\langle A\alpha|m_{-1}|J\lambda\rangle^0 + \langle A\alpha|m_{-1}|J\lambda\rangle' B\big) \\ \times \big(\langle J\lambda'|m_1|A\alpha\rangle^0 + \langle J\lambda'|m_1|A\alpha\rangle' B\big) \\ - \big(\langle A\alpha|m_1|J\lambda\rangle^0 + \langle A\alpha|m_1|J\lambda\rangle' B\big)\big(\langle J\lambda'|m_{-1}|A\alpha\rangle^0 \\ + \langle J\lambda'|m_{-1}|A\alpha\rangle' B\big)\Big] \\ \times \left[\sum_{g\xi}\frac{N_g}{N_G}\left\langle g\xi|\langle J\lambda|\mathcal{H}_{\mathrm{el}}|J\lambda'\rangle - W_A\delta_{\lambda\lambda'}|g\xi\right\rangle - \bar{\mathcal{E}}\delta_{\lambda\lambda'} \right. \\ \left. - \langle J\lambda|\mu_z|J\lambda\rangle^0 B\delta_{\lambda\lambda'} + \langle A\alpha|\mu_z|A\alpha\rangle^0 B\delta_{\lambda\lambda'}\right]\end{aligned} \qquad (7.4.14)$$

Thus to first order in B,

$$\left\langle \frac{\Delta A'}{\mathcal{E}} \right\rangle_1^{\bar{\mathcal{E}}} = \gamma\left(\mathcal{A}_1 + \mathcal{B}_1 + \frac{\mathcal{C}_1}{kT}\right)\mu_B B \tag{7.4.15}$$

where $\mathcal{A}_1$ is given in (4.5.16). $\mathcal{B}_1$ and $\mathcal{C}_1$, which are neglected in the model of Section 4.5, are defined by the equations

$$\begin{aligned}
\mathcal{B}_1 = \mathcal{Re}\Bigg\{ \frac{2}{|A|} \sum_{\alpha\lambda\lambda'} \Bigg[& \sum_{\substack{K\kappa \\ (K \neq J)}} \frac{\langle J\lambda'|L_z + 2S_z|K\kappa\rangle^0}{W_K^0 - W_J^0} \\
& \times \left(\langle A\alpha|m_{-1}|J\lambda\rangle^0 \langle K\kappa|m_1|A\alpha\rangle^0 - \langle A\alpha|m_1|J\lambda\rangle^0 \langle K\kappa|m_{-1}|A\alpha\rangle^0\right) \\
& + \sum_{\substack{K\kappa \\ (K \neq A)}} \frac{\langle K\kappa|L_z + 2S_z|A\alpha\rangle^0}{W_K^0 - W_A^0} \\
& \times \left(\langle A\alpha|m_{-1}|J\lambda\rangle^0 \langle J\lambda'|m_1|K\kappa\rangle^0 - \langle A\alpha|m_1|J\lambda\rangle^0 \langle J\lambda'|m_{-1}|K\kappa\rangle^0\right)\Bigg] \\
& \times \left[\sum_{g\xi} \frac{N_g}{N_G} \left\langle g\xi|\langle J\lambda|\mathcal{H}_{el}|J\lambda'\rangle - W_A \delta_{\lambda\lambda'}|g\xi\right\rangle - \bar{\mathcal{E}}\delta_{\lambda\lambda'}\right]\Bigg\}
\end{aligned} \tag{7.4.16}$$

$$\begin{aligned}
\mathcal{C}_1 = \frac{1}{|A|} \sum_{\alpha\lambda\lambda'} & \langle A\alpha|L_z + 2S_z|A\alpha\rangle^0 \\
& \times \left(\langle A\alpha|m_{-1}|J\lambda\rangle^0 \langle J\lambda'|m_1|A\alpha\rangle^0 - \langle A\alpha|m_1|J\lambda\rangle^0 \langle J\lambda'|m_{-1}|A\alpha\rangle^0\right) \\
& \times \left[\sum_{g\xi} \frac{N_g}{N_G} \left\langle g\xi|\langle J\lambda|\mathcal{H}_{el}|J\lambda'\rangle - W_A \delta_{\lambda\lambda'}|g\xi\right\rangle - \bar{\mathcal{E}}\delta_{\lambda\lambda'}\right]
\end{aligned} \tag{7.4.17}$$

We emphasize that the moment equations derived in this section are quite general. The *form* of the equations is independent of the excited-state

JT effect and of spin–orbit or symmetry-lowering interactions *as long as the moments are summed over the entire basis of interacting vibronic states.* Moreover, the *values* of all MCD parameters which do not involve $\mathcal{H}_{el}$ (i.e. $\mathcal{D}_0$, $\mathcal{A}_1$, $\mathcal{B}_0$, and $\mathcal{C}_0$) are independent of such interactions. Obviously the values of $\bar{\mathcal{E}}$, $\mathcal{B}_1$, and $\mathcal{C}_1$ are sensitive to such interactions through the $\langle J\lambda|\mathcal{H}_{el}|J\lambda'\rangle$ matrix, but the equations for these parameters are in invariant form, and an unperturbed BO basis may be used for their calculation.

We have, of course, assumed throughout that the BO approximation is valid in the ground state, and have calculated the MCD in the linear limit—that is, we assume that the MCD is linear in B and that $\langle A\alpha|\mu_z|A\alpha\rangle^0 B \ll kT$. JT interactions in the ground state invalidate our results, as do deviations from the FC approximation.

7.5. The Moment Equations in the Born–Oppenheimer Approximation

The equations of the last section simplify considerably in cases where

$$\langle J\lambda|\mathcal{H}_{el}|J\lambda'\rangle = W_J\delta_{\lambda\lambda'} \tag{7.5.1}$$

For $\mathcal{H}_{el}$ defined as in (3.3.1)–(3.3.3), Equation (7.5.1) is true only when the BO approximation is valid in the excited state and the $J\lambda$ basis is otherwise diagonal in $\mathcal{H}_{el}$; an obvious example is a system in which J is either nondegenerate or Kramers-degenerate. Equation (7.5.1) allows us to factor the purely electronic parameters, $\mathcal{D}_0$, $\mathcal{B}_0$, and $\mathcal{C}_0$, out of $\mathcal{D}_1$, $\mathcal{B}_1$, and $\mathcal{C}_1$. Moreover, we now show that the other factor is zero. If we define

$$\begin{aligned}\overline{W}_{JA} &= \sum_{g\xi} \frac{N_g}{N_G}\langle g\xi|W_J - W_A|g\xi\rangle \\ &= \overline{W}_J - \overline{W}_A \end{aligned} \tag{7.5.2}$$

(7.4.4) becomes

$$\begin{aligned}\left\langle \frac{A_\pm}{\mathcal{E}} \right\rangle_1^{\bar{\mathcal{E}}} &= \gamma\mathcal{D}_1 \\ &= \gamma\left[\mathcal{D}_0\left(\overline{W}_{JA} - \bar{\mathcal{E}}\right)\right] \end{aligned} \tag{7.5.3}$$

Since $\bar{\mathcal{E}}$ is defined by setting the first moment above equal to zero, we have

$$\bar{\mathcal{E}} = \overline{W}_{JA} \tag{7.5.4}$$

Similarly

$$\mathcal{B}_1 = \mathcal{B}_0(\overline{W}_{JA} - \overline{\mathcal{E}}) = 0 \tag{7.5.5}$$

$$\mathcal{C}_1 = \mathcal{C}_0(\overline{W}_{JA} - \overline{\mathcal{E}}) = 0 \tag{7.5.6}$$

so

$$\left\langle \frac{\Delta A'}{\mathcal{E}} \right\rangle_1^{\overline{\mathcal{E}}} = \gamma \mathcal{A}_1 \mu_B B \tag{7.5.7}$$

The zeroth moments are the same as in the previous section.

In many literature applications of moments to MCD, (7.5.7) has been used without justification. *Unless* it can be shown that $\mathcal{B}_1$ and $\mathcal{C}_1$ are negligible, this is usually equivalent to assuming that the BO approximation is valid in the excited state.

Other Situations in Which $\mathcal{B}_1$ and $\mathcal{C}_1$ are Zero

In some cases it is possible to show that $\mathcal{B}_1$ and $\mathcal{C}_1$ are zero *without* making the BO approximation, as illustrated for $\mathcal{B}$ terms in Stephens et al. (1971). One begins by writing $\mathcal{D}_0$, $\mathcal{D}_1$, $\mathcal{B}_0$, $\mathcal{B}_1$, $\mathcal{C}_0$, and $\mathcal{C}_1$ in the form

$$\mathcal{D}_0 = \sum_{\lambda\lambda'} d_{\lambda\lambda'} \delta_{\lambda\lambda'}$$

$$\mathcal{D}_1 = \sum_{\lambda\lambda'} d_{\lambda\lambda'}(x_{\lambda\lambda'} - \overline{\mathcal{E}}\delta_{\lambda\lambda'}) = \sum_{\lambda\lambda'} d_{\lambda\lambda'} x_{\lambda\lambda'} - \overline{\mathcal{E}}\mathcal{D}_0$$

$$\mathcal{B}_0 = \sum_{\lambda\lambda'} b_{\lambda\lambda'} \delta_{\lambda\lambda'}$$

$$\mathcal{B}_1 = \sum_{\lambda\lambda'} b_{\lambda\lambda'}(x_{\lambda\lambda'} - \overline{\mathcal{E}}\delta_{\lambda\lambda'}) = \sum_{\lambda\lambda'} b_{\lambda\lambda'} x_{\lambda\lambda'} - \overline{\mathcal{E}}\mathcal{B}_0$$

$$\mathcal{C}_0 = \sum_{\lambda\lambda'} c_{\lambda\lambda'} \delta_{\lambda\lambda'}$$

$$\mathcal{C}_1 = \sum_{\lambda\lambda'} c_{\lambda\lambda'}(x_{\lambda\lambda'} - \overline{\mathcal{E}}\delta_{\lambda\lambda'}) = \sum_{\lambda\lambda'} c_{\lambda\lambda'} x_{\lambda\lambda'} - \overline{\mathcal{E}}\mathcal{C}_0 \tag{7.5.8}$$

where

$$x_{\lambda\lambda'} = \sum_{g\xi} \frac{N_g}{N_G} \left\langle g\xi | \langle J\lambda | \mathcal{H}_{el} | J\lambda' \rangle - W_A \delta_{\lambda\lambda'} | g\xi \right\rangle$$

$$d_{\lambda\lambda'} = -\frac{1}{|A|} \sum_{\alpha} \langle A\alpha | m_{\pm 1} | J\lambda \rangle^0 \langle J\lambda' | m_{\mp 1} | A\alpha \rangle^0$$

$$b_{\lambda\lambda'} = \mathcal{R}e \left[\frac{2}{|A|} \sum_{\alpha} \text{[electronic part in square brackets in (7.4.16)]} \right]$$

$$c_{\lambda\lambda'} = \frac{1}{|A|} \sum_{\alpha} \langle A\alpha | L_z + 2S_z | A\alpha \rangle^0$$
$$\times \left(\langle A\alpha | m_{-1} | J\lambda \rangle^0 \langle J\lambda' | m_1 | A\alpha \rangle^0 - \langle A\alpha | m_1 | J\lambda \rangle^0 \langle J\lambda' | m_{-1} | A\alpha \rangle^0 \right) \tag{7.5.9}$$

Since $\overline{\mathcal{E}}$ is defined by setting $\mathcal{D}_1 = 0$, from (7.5.8) we have

$$\overline{\mathcal{E}} = \frac{\sum_{\lambda\lambda'} d_{\lambda\lambda'} x_{\lambda\lambda'}}{\mathcal{D}_0} \tag{7.5.10}$$

Let us now consider circumstances in which $\mathcal{B}_1$ and $\mathcal{C}_1$ are zero. First let us suppose that the $|J\lambda\rangle$ are exactly degenerate at $Q = Q_0$ and so are in every way like the $\phi_{K\kappa}(Q, q)$ of (3.8.2). Thus, although the $|J\lambda\rangle$ are not eigenfunctions of $\mathcal{H}_{el}$ and are mixed by $\mathcal{H}_{el}$, we have for the diagonal terms in the $\mathcal{H}_{el}$ matrix

$$\langle J\lambda | \mathcal{H}_{el} | J\lambda \rangle = W_J(Q) \qquad \text{for all } \lambda \tag{7.5.11}$$

Consequently when $\lambda = \lambda'$

$$x_{\lambda\lambda} = \overline{W}_J - \overline{W}_A = \overline{W}_{JA} \tag{7.5.12}$$

where $\overline{W}_{JA}$ is independent of λ. Next suppose that, by symmetry,

$$d_{\lambda\lambda'} = \delta_{\lambda\lambda'} d_{\lambda\lambda} \tag{7.5.13}$$

$$b_{\lambda\lambda'} = \delta_{\lambda\lambda'} b_{\lambda\lambda} \tag{7.5.14}$$

$$c_{\lambda\lambda'} = \delta_{\lambda\lambda'} c_{\lambda\lambda} \tag{7.5.15}$$

so that the only contributions to $\mathcal{D}_1$, $\mathcal{B}_1$, and $\mathcal{C}_1$ are for $\lambda = \lambda'$. These relations frequently hold when the ground state is A_1; for example, in the group O it is easy to show for $A_1 \to T_1$ using group theory [Eq. (10.2.2) and Section C.12] that

$$\langle A_1 a_1 | m^{t_1}_{\pm 1} | T_1 \lambda \rangle = \delta_{\lambda, \mp 1} \langle A_1 a_1 | m^{t_1}_{\pm 1} | T_1 \mp 1 \rangle$$

$$\langle T_1 \lambda' | m^{t_1}_{\mp 1} | A_1 a_1 \rangle = \delta_{\lambda', \mp 1} \langle T_1 \mp 1 | m^{t_1}_{\mp 1} | A_1 a_1 \rangle \qquad (7.5.16)$$

These immediately give (7.5.13) and (7.5.15). And in this case group theory also gives $b_{\lambda\lambda'} = \delta_{\lambda\lambda'} b_{\lambda\lambda}$.

When our suppositions of (7.5.11) and (7.5.13)–(7.5.15) are valid, it follows from (7.5.8) that

$$\begin{aligned} \mathcal{D}_1 &= \mathcal{D}_0 \left(\overline{W}_{JA} - \overline{\mathcal{E}} \right) \\ \overline{\mathcal{E}} &= \overline{W}_{JA} \\ \mathcal{B}_1 &= \mathcal{B}_0 \left(\overline{W}_{JA} - \overline{\mathcal{E}} \right) = 0 \\ \mathcal{C}_1 &= \mathcal{C}_0 \left(\overline{W}_{JA} - \overline{\mathcal{E}} \right) = 0 \end{aligned} \qquad (7.5.17)$$

Thus in our $A_1 \to T_1$ example, (7.5.7) gives the first moment *even in the presence of JT effects.*

Remember, however, that the above derivation depends on *both* (7.5.11) and (7.5.13)–(7.5.15), and if $\mathcal{H}_{el}$ includes perturbations (spin–orbit coupling, etc.) which split the $J\lambda$ degeneracy at $Q = Q_0$, then (7.5.11) is no longer valid. In such cases the approach above should be useful in pinpointing contributions to $\mathcal{D}_1$, $\mathcal{B}_1$, and $\mathcal{C}_1$.

More general conditions than (7.5.11) and (7.5.13)–(7.5.15) that require $\mathcal{B}_1$ and $\mathcal{C}_1$ to vanish are

$$\mathcal{B}_1 = \sum_{\lambda\lambda'} b_{\lambda\lambda'} x_{\lambda\lambda'} - \frac{\mathcal{B}_0}{\mathcal{D}_0} \sum_{\lambda\lambda'} d_{\lambda\lambda'} x_{\lambda\lambda'} = 0 \qquad (7.5.18)$$

or

$$b_{\lambda\lambda'} = \frac{\mathcal{B}_0}{\mathcal{D}_0} d_{\lambda\lambda'} \qquad \text{for all } \lambda, \lambda' \qquad (7.5.19)$$

and similarly, for $\mathcal{C}_1$, that

$$c_{\lambda\lambda'} = \frac{\mathcal{C}_0}{\mathcal{D}_0} d_{\lambda\lambda'} \qquad \text{for all } \lambda, \lambda' \qquad (7.5.20)$$

These equations are clearly valid if (7.5.1) is true.

Stephens et al. (1971) have estimated a rough upper limit to $\mathcal{B}_1$ of $\mathcal{B}_0 \times \Gamma$, where Γ is the bandwidth at half height.

7.6. Overlapping Bands

While it would make life easier for spectroscopists if every band in a spectrum were isolated and arose from a single transition, in practice overlapping bands are all too common. In this section we apply the MM to the two cases discussed earlier in Section 5.2. The first, near-degeneracy in the excited state, is more common than the second, near-degeneracy in the ground state. Our discussion is similar to that of Stephens (1976).

How to Deal with Excited-State Near-Degeneracy Using Moments

In the first case we have two nearly degenerate excited states, which we label J_1 and J_2. We do not require that these states have parallel potential surfaces (as in Sections 5.2 and 7.8) so $W_{J_2}(Q) - W_{J_1}(Q)$ is a function of Q, but we do assume no PJT mixing of J_1 and J_2 in calculating the $\mathcal{H}_T$ matrix. The moments of the absorption are found as in Section 7.4 above:

$$\left\langle \frac{A_\pm}{\mathcal{E}} \right\rangle_0 = \gamma[\mathcal{D}_0(J_1) + \mathcal{D}_0(J_2)]$$
$$\left\langle \frac{A_\pm}{\mathcal{E}} \right\rangle_1^{\bar{\mathcal{E}}} = \gamma[\mathcal{D}_1(J_1) + \mathcal{D}_1(J_2)] \tag{7.6.1}$$

We have used the relation

$$\langle J_i\lambda_i, g\xi|\mathcal{H}_T|J_{i'}\lambda'_{i'}, g'\xi'\rangle = \left\langle g\xi\middle|\langle J_i\lambda_i|\mathcal{H}_{\text{el}}|J_i\lambda'_i\rangle + \delta_{\lambda_i\lambda'_i}T_n\middle|g'\xi'\right\rangle\delta_{ii'} \tag{7.6.2}$$

[which is analogous to (7.4.13) for $J = J_i$ and $K = J_{i'}$] to obtain the first moment of absorption; the factor $\delta_{ii'}$ in (7.6.2) limits equations derived for $\mathcal{B}_1$, $\mathcal{C}_1$, and $\mathcal{D}_1$ to cases where there is no PJT effect.

The bases for the MCD moments are chosen as in Section 7.4 except that for J_1 and J_2 there is a slight modification. We choose

$$|J_i\lambda_i, g\xi\rangle' = |J_i\lambda_i, g\xi\rangle - \sum_{\substack{K\kappa \\ (K\neq J_1, J_2)}} \frac{\langle K\kappa|\mu_z|J_i\lambda_i\rangle^0 B}{W_{J_i}^0 - W_K^0}|K\kappa, g\xi\rangle \tag{7.6.3}$$

where the mixing is only with states *outside* the J_1 *and* J_2 manifolds. [Mixing of $|J_1\lambda_1, g\xi\rangle$ and $|J_2\lambda_2, g\xi\rangle$ by the field leads to a new basis which is related to this one by a unitary transformation. Thus we may use the simpler basis of (7.6.3) above in our moment equations.] Most matrix elements are analogous in form to those of Section 7.4. In addition we require $\mathcal{H}_T'$ matrix elements off-diagonal in $i = 1, 2$:

$$\langle J_i\lambda_i, g\xi|\mathcal{H}_T'|J_{i'}\lambda_{i'}', g'\xi'\rangle = \left\langle g\xi\middle|\langle J_i\lambda_i|\mathcal{H}_{\text{el}}|J_i\lambda_i'\rangle + T_n\delta_{\lambda_i\lambda_i'}\middle|g'\xi'\right\rangle\delta_{ii'}$$

$$-\langle J_i\lambda_i|\mu_z|J_{i'}\lambda_{i'}'\rangle^0 B\delta_{gg'}\delta_{\xi\xi'} \qquad (7.6.4)$$

Here $J_1\lambda_1$ and $J_2\lambda_2$ are assumed to have been chosen diagonal in μ_z. Note, however, that they are mixed by the field. The zeroth and first moments of the MCD are

$$\left\langle\frac{\Delta A'}{\mathcal{E}}\right\rangle_0 = \gamma\left[\mathcal{B}_0(J_1, K) + \frac{\mathcal{C}_0(J_1)}{kT} + \mathcal{B}_0(J_2, K) + \frac{\mathcal{C}_0(J_2)}{kT}\right]\mu_B B \qquad (7.6.5)$$

and

$$\left\langle\frac{\Delta A'}{\mathcal{E}}\right\rangle_1^{\bar{\mathcal{E}}} = \gamma\left[\mathcal{A}_1(J_1) + \mathcal{A}_2(J_2) + \mathcal{A}_1'(J_1, J_2)\right.$$

$$\left. + \mathcal{B}_1(J_1, K) + \frac{\mathcal{C}_1(J_1)}{kT} + \mathcal{B}_1(J_2, K) + \frac{\mathcal{C}_1(J_2)}{kT}\right]\mu_B B \qquad (7.6.6)$$

Here

$$\mathcal{A}_1'(J_1, J_2) = -\frac{1}{|A|}\sum_{\substack{ii'\\(i\neq i')}}\sum_{\alpha\lambda_i\lambda_{i'}}\langle J_i\lambda_i|L_z + 2S_z|J_{i'}\lambda_{i'}\rangle^0$$

$$\times\left(\langle A\alpha|m_{-1}|J_i\lambda_i\rangle^0\langle J_{i'}\lambda_{i'}|m_1|A\alpha\rangle^0\right.$$

$$\left.-\langle A\alpha|m_1|J_i\lambda_i\rangle^0\langle J_{i'}\lambda_{i'}|m_{-1}|A\alpha\rangle^0\right) \qquad (7.6.7)$$

is a "pseudo-$\mathcal{A}$" term (Section 5.2), and $\mathcal{B}_0(J_1, K)$ and $\mathcal{B}_0(J_2, K)$ are $\mathcal{B}$

terms which arise from the interaction of A and J with states K outside the J_1 *and* J_2 manifolds.

Applications of the Excited-State Near-Degeneracy Moment Equations

The expressions just derived come to life when we consider specific cases. First let us suppose that a degenerate electronic excited state J is split by a small perturbation into states J_1 and J_2 and that the splitting is small enough so that $A \to J_1$ and $A \to J_2$ are only partially resolved from one another in the spectrum. Suppose also that the ground state A is unsplit by the perturbation. For example, J could be an octahedral (group-O_h) T_{1u} state which is split by a small tetragonal crystal field into the group-D_{4h} states $J_1 = A_{2u}$ and $J_2 = E_u$. Alternatively J could be a $^2T_{1u}$ state in an octahedral system split by first-order spin–orbit coupling into the spin–orbit states $J_1 = E_u'$ and $J_2 = U_u'$ (Section 8.12 et seq.). For simplicity we also suppose that the ground state A is orbitally nondegenerate and that J_1 and J_2 are each diagonal in $\mathcal{H}_{el}$ for the perturbed system, so that (7.5.1) applies for $A \to J_1$ and $A \to J_2$ individually. For the combined band in the oriented case

$$\left\langle \frac{A_\pm}{\mathcal{E}} \right\rangle_0 = \gamma\left[\mathcal{D}_0(J_1) + \mathcal{D}_0(J_2)\right] \tag{7.6.8}$$

$$\bar{\mathcal{E}} = \frac{\mathcal{D}_0(J_1)\overline{W}_{J_1A} + \mathcal{D}_0(J_2)\overline{W}_{J_2A}}{\mathcal{D}_0(J_1) + \mathcal{D}_0(J_2)} \tag{7.6.9}$$

$$\left\langle \frac{\Delta A'}{\mathcal{E}} \right\rangle_0 = \gamma\left[\mathcal{B}_0(J_1, K) + \mathcal{B}_0(J_2, K)\right]\mu_B B \tag{7.6.10}$$

$$\left\langle \frac{\Delta A'}{\mathcal{E}} \right\rangle_1^{\bar{\mathcal{E}}} = \gamma\Bigg[\mathcal{A}_1(J_1) + \mathcal{A}_2(J_2) + \mathcal{A}_1'(J_1, J_2) + \mathcal{B}_0(J_1, K)\left(\overline{W}_{J_1A} - \bar{\mathcal{E}}\right) + \mathcal{B}_0(J_2, K)\left(\overline{W}_{J_2A} - \bar{\mathcal{E}}\right) + \frac{\mathcal{C}_0(J_1)}{kT}\overline{W}_{J_2J_1}\Bigg]\mu_B B \tag{7.6.11}$$

We have used the fact that here $\mathcal{C}_0(J_1) + \mathcal{C}_0(J_2) = 0$ since our ground state has no orbital degeneracy, and that $\overline{W}_{J_2A} - \overline{W}_{J_1A} = \overline{W}_{J_2J_1}$. Thus the zeroth moment of the MCD has no temperature dependence. $\mathcal{C}_0(J_1)$ and $\mathcal{C}_0(J_2)$

may, however, individually be nonzero. The $\mathcal{B}$ terms in (7.6.11) sum to zero only if $\overline{W}_{J_2A} = \overline{W}_{J_1A} = \overline{\mathcal{E}}$ (i.e. no splitting) or $\mathcal{D}_0(J_1)\mathcal{B}_0(J_2, K) = \mathcal{D}_0(J_2)\mathcal{B}_0(J_1, K)$, but in the discussion which follows we assume for simplicity that they are negligible.

Let us first consider the MCD predicted for the $J = T_{1u}(O_h)$, $J_1 = A_{2u}(D_{4h})$, and $J_2 = E_u(D_{4h})$ example above when the ground state is $A_{1g}(O_h) = A_{1g}(D_{4h})$. In this case $\mathcal{C}_0(J_1) = \mathcal{C}_0(J_2) = 0$ and we have

$$\left\langle \frac{A_\pm}{\mathcal{E}} \right\rangle_0 = \begin{cases} \gamma[\mathcal{D}_0(J_1) + \mathcal{D}_0(J_2)] & \text{in } D_{4h} \\ \gamma[\mathcal{D}_0(J)] & \text{in } O_h \end{cases} \tag{7.6.12}$$

and

$$\left\langle \frac{\Delta A'}{\mathcal{E}} \right\rangle_0 = \gamma[\mathcal{B}_0(J_1, K) + \mathcal{B}_0(J_2, K)]\mu_B B$$

$$\left\langle \frac{\Delta A'}{\mathcal{E}} \right\rangle_1^{\overline{\mathcal{E}}} \begin{cases} \approx \gamma[\mathcal{A}_1(J_1) + \mathcal{A}_1(J_2) + \mathcal{A}_1'(J_1, J_2)]\mu_B B & \text{in } D_{4h} \\ \approx \gamma[\mathcal{A}_1(J)]\mu_B B & \text{in } O_h \end{cases} \tag{7.6.13}$$

Thus we find that the zeroth and first moments are identical to those calculated for the $A_{1g}(O_h) \to T_{1u}(O_h)$ transition in the absence of the perturbation, and they can be calculated most easily by ignoring it and using our moment equations from Section 7.5. Actually this result was anticipated by our discussion at the close of Section 7.4. The calculation is agreeably simple, but uninformative in that we find out nothing about the perturbation. Higher moments are required for that.

It might seem that JT splittings of a state J could be handled by the formalism of this section. However, in general when J_1 and J_2 are eigenfunctions of $\mathcal{H}_{\text{el}}$, as are the $\phi'_{k\lambda}$ of (3.8.1), they do not commute with T_n as we require for our moments. The zeroth and first moments of $A \to J$ in the presence of the JT effect are calculated as described in Section 7.4. Little information about the JT perturbation is found from their measurement, since they are only altered from their BO values by the $\mathcal{B}_1$ and $\mathcal{C}_1$ contributions. Higher moments are very helpful in analyzing JT effects (Section 7.7).

If we choose a $^2A_{1g}$ ground state, the second example mentioned at the beginning of this subsection is the "classic" F-center case of Henry, Schnatterly, and Slichter (1965). Without spin–orbit coupling, $A = {}^2A_{1g}$ and $J = {}^2T_{1u}$. First-order spin–orbit coupling (Chapter 12) gives $A = E_g'({}^2A_{1g})$, $J_1 = E_u'({}^2T_{1u})$, and $J_2 = U_u'({}^2T_{1u})$. If we ignore the $\mathcal{B}_1(J_i, K)$ terms, we have for $A \to J_1 + J_2$

$$\left\langle \frac{A_\pm}{\mathcal{E}} \right\rangle_0 = \gamma[\mathcal{D}_0(J_1) + \mathcal{D}_0(J_2)]$$

$$\left\langle \frac{\Delta A'}{\mathcal{E}} \right\rangle_0 = \gamma[\mathcal{B}_0(J_1, K) + \mathcal{B}_0(J_2, K)]\mu_B B$$

$$\left\langle \frac{\Delta A'}{\mathcal{E}} \right\rangle_1^{\bar{\mathcal{E}}} \approx \gamma\left[\mathcal{A}_1(J_1) + \mathcal{A}_1(J_2) + \mathcal{A}_1'(J_1, J_2) + \frac{\mathcal{C}_0(J_1)\overline{W}_{J_2J_1}}{kT}\right]\mu_B B$$

$$= \gamma\left[\mathcal{A}_1(J) + \frac{\mathcal{C}_0(J_1)\overline{W}_{J_2J_1}}{kT}\right]\mu_B B \tag{7.6.14}$$

While the $\mathcal{A}_1$ part is conventional enough, the $\mathcal{C}$ term contribution to the first moment is extremely interesting, since it gives a temperature-dependent contribution which may dominate $\mathcal{A}_1$ at low temperatures, even when $\overline{W}_{J_2J_1}$, the "average" $J_2 - J_1$ splitting, is small—say $\sim 10\ \text{cm}^{-1}$. To the unsuspecting the band appears to be dominated by a temperature-dependent $\mathcal{A}$ term, which we see is actually $\mathcal{C}_0(J_1)\overline{W}_{J_2J_1}/kT$. This is a case where the first moment *is* strongly affected by the perturbation. In fact, in our approximation $\overline{W}_{J_2J_1}$ is obtained by measuring the first moment as a function of temperature.

Once again, these results were anticipated by our discussion at the end of Section 7.4. If the $^2T_{1u}$ BO basis were used and $\mathcal{H}_{SO}$ were included in $\mathcal{H}_{el}$, the same results would be obtained from (7.4.3), (7.4.10), and (7.4.15) if the same assumptions were made. The equations in Section 7.4 are, however, more general, and are valid for $\mathcal{B}_1$, $\mathcal{C}_1$, and $\mathcal{D}_1$ in the presence of strong vibronic effects which scramble the spin–orbit eigenfunctions and invalidate (7.6.2).

Accidental Degeneracy

Equations (7.6.1) and (7.6.6) also apply when two states J_1 and J_2 are accidentally very close in energy. Thus when $\mathcal{B}$ terms from mixing of states external to the band are neglected and $\mathcal{C}$ terms are zero,

$$\left\langle \frac{A_\pm}{\mathcal{E}} \right\rangle_0 = \gamma[\mathcal{D}_0(J_1) + \mathcal{D}_0(J_2)]$$

$$\left\langle \frac{\Delta A'}{\mathcal{E}} \right\rangle_1^{\bar{\mathcal{E}}} = \gamma[\mathcal{A}_1(J_1) + \mathcal{A}_1(J_2) + \mathcal{A}_1'(J_1, J_2)]\mu_B B \tag{7.6.15}$$

and the same results are obtained as if J_1 and J_2 were precisely degenerate.

When strong vibronic mixing of J_1 and J_2 is suspected, the equations in Section 7.4 should be used to obtain $\mathcal{B}_1$, $\mathcal{C}_1$, and $\mathcal{D}_1$.

Ground-State Near-Degeneracy

In this case two near-degenerate states, A_1 and A_2, are populated, and we consider transitions from them to J. As always, we must assume that the ground states obey the BO approximation. In addition we require that A_1 and A_2 have parallel potential surfaces, so that $\overline{W}_{A_2A_1}(Q) = \overline{W}_{A_2A_1}(Q_0)$. States A_1 and A_2 have fractional populations δ_1 and δ_2,

$$\delta_1 = \frac{|A_1|}{|A_1| + |A_2|\exp\left(-\overline{W}_{A_2A_1}/kT\right)}$$

$$\delta_2 = \frac{|A_2|\exp\left(-\overline{W}_{A_2A_1}/kT\right)}{|A_1| + |A_2|\exp\left(-\overline{W}_{A_2A_1}/kT\right)} \tag{7.6.16}$$

and the moments are a function of δ_1 and δ_2. If we assume that J also obeys the BO approximation, the moments are

$$\left\langle \frac{A_\pm}{\mathcal{E}} \right\rangle_0 = \gamma\left[\delta_1\mathcal{D}_0(A_1 \to J) + \delta_2\mathcal{D}_0(A_2 \to J)\right]$$

$$\left\langle \frac{A_\pm}{\mathcal{E}} \right\rangle_1^{\overline{\mathcal{E}}} = \gamma\left[\delta_1\mathcal{D}_0(A_1 \to J)\left(\overline{W}_{JA_1} - \overline{\mathcal{E}}\right) + \delta_2\mathcal{D}_0(A_2 \to J)\left(\overline{W}_{JA_2} - \overline{\mathcal{E}}\right)\right]$$

$$\left\langle \frac{\Delta A'}{\mathcal{E}} \right\rangle_0 = \gamma\left\{\delta_1\left[\mathcal{B}_0(A_1 \to J) + \frac{\mathcal{C}_0(A_1 \to J)}{kT}\right] + \delta_2\left[\mathcal{B}_0(A_2 \to J) + \frac{\mathcal{C}_0(A_2 \to J)}{kT}\right]\right\}\mu_B B \tag{7.6.17}$$

$$\left\langle \frac{\Delta A'}{\mathcal{E}} \right\rangle_1^{\overline{\mathcal{E}}} = \gamma\left\{\delta_1\mathcal{A}_1(A_1 \to J) + \delta_2\mathcal{A}_1(A_2 \to J) + \delta_1\left[\mathcal{B}_0(A_1 \to J) + \frac{\mathcal{C}_0(A_1 \to J)}{kT}\right]\left(\overline{W}_{JA_1} - \overline{\mathcal{E}}\right) + \delta_2\left[\mathcal{B}_0(A_2 \to J) + \frac{\mathcal{C}_0(A_2 \to J)}{kT}\right]\left(\overline{W}_{JA_2} - \overline{\mathcal{E}}\right)\right\}\mu_B B$$

In general *all* of these moments are temperature-dependent.

At high temperatures where $\overline{W}_{A_2A_1} \ll kT$, we have $\delta_1 \approx \delta_2$, and the MCD is equivalent to that when A_1 and A_2 are exactly degenerate. At temperatures where $\overline{W}_{A_2A_1}$ is comparable to kT, the MCD is very sensitive to temperature, and the form of the absorption and MCD are dependent on the relative magnitudes of the various $\mathcal{A}$, $\mathcal{B}$, $\mathcal{C}$, and $\mathcal{D}$ parameters for $A_1 \to J$ and $A_2 \to J$.

7.7. Higher Moments

Higher moments ($n > 1$) are increasingly sensitive to contributions from the wings of the band and hence may be measured accurately only for well-isolated bands with reliable baselines. Nevertheless, these moments provide valuable information about vibronic interactions and other perturbations.

Theoretical expressions for higher ($n > 1$) absorption and MCD moments may be written in a form analogous to (7.4.3), (7.4.4), (7.4.10), and (7.4.15):

$$\left\langle \frac{A_\pm}{\mathcal{E}} \right\rangle_n^{\bar{\mathcal{E}}} = \gamma \mathcal{D}_n$$
$$\left\langle \frac{\Delta A'}{\mathcal{E}} \right\rangle_n^{\bar{\mathcal{E}}} = \gamma \left(\mathcal{A}_n + \mathcal{B}_n + \frac{\mathcal{C}_n}{kT} \right) \mu_B B \tag{7.7.1}$$

General expressions for the parameters in (7.7.1) have been given by Stephens (1970) and Osborne and Stephens (1972) using the old conventions [see (A.5.7)]. The results for specific models become increasingly complex with increasing n, and we only quote one particularly interesting result as an example of the types of information obtainable. Consider a $^2A_{1g} \to {}^2T_{1u}$ transition in an octahedral system (the F-center case of Section 7.6). For sufficiently small spin–orbit coupling one finds

$$\left\langle \frac{A_\pm}{\mathcal{E}} \right\rangle_2^{\bar{\mathcal{E}}} \bigg/ \left\langle \frac{A_\pm}{\mathcal{E}} \right\rangle_0 = \Delta_{a_{1g}}^2 + \Delta_{t_{2g}}^2 + \Delta_{e_g}^2 \tag{7.7.2}$$

$$\left\langle \frac{\Delta A'}{\mathcal{E}} \right\rangle_3^{\bar{\mathcal{E}}} \bigg/ \left\langle \frac{\Delta A'}{\mathcal{E}} \right\rangle_1^{\bar{\mathcal{E}}} = 3\Delta_{a_{1g}}^2 + \tfrac{3}{2}\left(\Delta_{t_{2g}}^2 + \Delta_{e_g}^2 \right) \tag{7.7.3}$$

where Δ_Γ measures the relative contribution of all modes of symmetry Γ to the root-mean-square bandwidth. One can solve (7.7.2) and (7.7.3) simultaneously to obtain $\Delta_{a_{1g}}^2$ and $\Delta_{t_{2g}}^2 + \Delta_{e_g}^2$ *individually*. One thus has a powerful quantitative method for assessing the contribution of the JT modes (t_{2g}, e_g) to the bandwidth and hence the importance of the JT effect in the excited state. This application was first made by Henry et al. (1965) and is discussed by Stephens (1976) and Osborne and Stephens (1972). The latter reference provides an illuminating and detailed application of the method of moments to F centers.

7.8. Obtaining MCD Parameters from the Rigid-Shift Equations

Gaussian Fits of an Isolated Band

The use of Gaussian fitting to obtain MCD parameters is based on the RS–BO–FC equations of Section 4.5 with a Gaussian lineshape function, $f(\mathcal{E}) = f_G(\mathcal{E})$. Thus, repeating (4.5.22) and (4.5.14), we have

$$\frac{A_{\pm}}{\mathcal{E}} = \gamma \mathcal{D}_0 f(\mathcal{E}) \tag{7.8.1}$$

$$\frac{\Delta A'}{\mathcal{E}} = \gamma \left[\mathcal{A}_1 \left(\frac{-\partial f(\mathcal{E})}{\partial \mathcal{E}} \right) + \left(\mathcal{B}_0 + \frac{\mathcal{C}_0}{kT} \right) f(\mathcal{E}) \right] \mu_B B \tag{7.8.2}$$

where [as always; cf. (4.5.25) and (4.5.27)–(4.5.28)]

$$\begin{aligned} &\int f(\mathcal{E})\, d\mathcal{E} = 1, \qquad &&\int f(\mathcal{E}) \mathcal{E}\, d\mathcal{E} = \mathcal{E}^0 \\ &\int \frac{\partial f(\mathcal{E})}{\partial \mathcal{E}} d\mathcal{E} = 0, \qquad &&\int \frac{\partial f(\mathcal{E})}{\partial \mathcal{E}} \mathcal{E}\, d\mathcal{E} = -1 \end{aligned} \tag{7.8.3}$$

and

$$\begin{aligned} f(\mathcal{E}) &= f_G(\mathcal{E}) = \frac{1}{\Delta\sqrt{\pi}} e^{-(\mathcal{E}-\mathcal{E}^0)^2/\Delta^2} \\ \frac{\partial f(\mathcal{E})}{\partial \mathcal{E}} &= \frac{\partial f_G(\mathcal{E})}{\partial \mathcal{E}} = \frac{-2(\mathcal{E}-\mathcal{E}^0)}{\Delta^3\sqrt{\pi}} e^{-(\mathcal{E}-\mathcal{E}^0)^2/\Delta^2} \end{aligned} \tag{7.8.4}$$

As illustrated in Fig. 7.8.1, $\mathcal{E}^0$ is the energy of the Gaussian band maximum and Δ is a bandwidth parameter.

In Fig. 7.8.2 we show absorption and MCD data for an idealized transition with a truly Gaussian lineshape. The absorption is completely

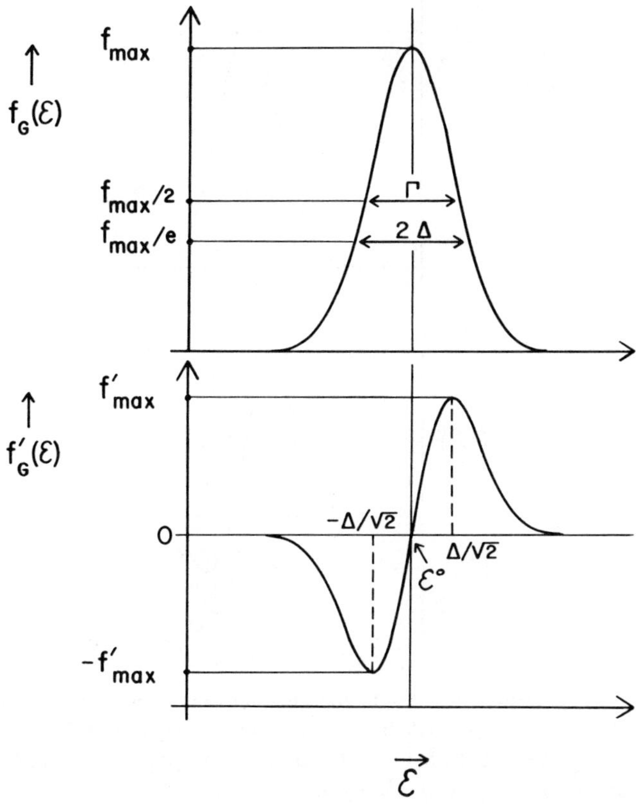

Figure 7.8.1. Plots of $f_G(\mathcal{E})$ and $f'_G(\mathcal{E})$ of (7.8.4) for a Gaussian band. For Gaussian curves $\Gamma = 2\Delta\sqrt{\ln 2}$, $f_{\max} = 1/\Delta\sqrt{\pi}$, $f'_{\max} = \sqrt{2e}/\Delta^2\sqrt{\pi}$, and $f_{\max}/f'_{\max} = 1.17\Delta$.

defined by $\mathcal{E}^0$, $(A_\pm/\mathcal{E})_{\mathcal{E}^0} \equiv (A_\pm/\mathcal{E})_0$, and the bandwidth parameter Δ. This follows because at $\mathcal{E} = \mathcal{E}^0$, $f(\mathcal{E}^0) = 1/\Delta\sqrt{\pi}$. Thus at $\mathcal{E} = \mathcal{E}^0$, (7.8.1) gives

$$\left(\frac{A_\pm}{\mathcal{E}}\right)_0 = \gamma\mathcal{D}_0\left(\frac{1}{\Delta\sqrt{\pi}}\right) \tag{7.8.5}$$

which substituted back into (7.8.1) leads to

$$\frac{A_\pm}{\mathcal{E}} = \left(\frac{A_\pm}{\mathcal{E}}\right)_0 e^{-(\mathcal{E}-\mathcal{E}^0)^2/\Delta^2} \tag{7.8.6}$$

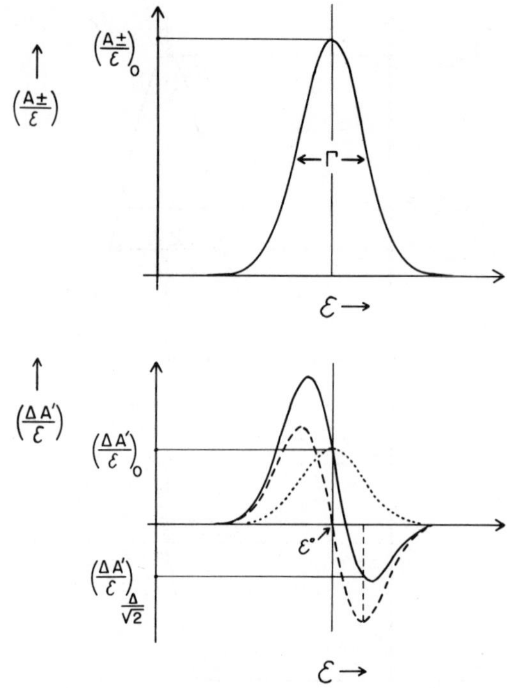

Figure 7.8.2. Absorption and MCD for an electronic transition with a Gaussian bandshape. Both the ground and the excited state are degenerate. In the lower plot, the solid line represents $\Delta A'/\mathcal{E}$ vs. $\mathcal{E}$, the dashed line $\gamma\mathcal{A}_1[-f'(\mathcal{E})]\mu_B B$ vs. $\mathcal{E}$, and the dotted line $\gamma(\mathcal{B}_0 + \mathcal{C}_0/kT)f(\mathcal{E})\mu_B B$.

Δ is often obtained from Γ, the experimental full bandwidth at half height —that is, at $f(\mathcal{E}) = 1/2f(\mathcal{E}^0)$—by using the relation $\Gamma = 2\Delta\sqrt{\ln 2}$, which is easily derived from (7.8.4). Solving (7.8.5) for $\mathcal{D}_0$ gives

$$\mathcal{D}_0 = \frac{\Delta\sqrt{\pi}}{\gamma}\left(\frac{A_\pm}{\mathcal{E}}\right)_0 \tag{7.8.7}$$

The MCD for a band with Gaussian lineshape is similarly defined by $\mathcal{E}^0$, Δ, $(\Delta A'/\mathcal{E})_0$, and $(\Delta A'/\mathcal{E})_{\mathcal{E}=\Delta/\sqrt{2}} \equiv (\Delta A'/\mathcal{E})_{\Delta/\sqrt{2}}$. At $\mathcal{E} = \mathcal{E}^0$, we have $\partial f(\mathcal{E})/\partial\mathcal{E} = 0$ and $f(\mathcal{E}) = 1/\Delta\sqrt{\pi}$. Thus

$$\left(\frac{\Delta A'}{\mathcal{E}}\right)_0 = \gamma\left(\mathcal{B}_0 + \frac{\mathcal{C}_0}{kT}\right)\left(\frac{1}{\Delta\sqrt{\pi}}\right)\mu_B B \tag{7.8.8}$$

and so

$$\left(\mathcal{B}_0 + \frac{\mathcal{C}_0}{kT}\right) = \frac{\Delta\sqrt{\pi}}{\gamma\mu_B B}\left(\frac{\Delta A'}{\mathcal{E}}\right)_0 \tag{7.8.9}$$

where Δ and $\mathcal{E}^0$ are taken from the absorption data. As always, $\Delta A'/\mathcal{E}$ must be measured as a function of T in order to determine $\mathcal{B}_0$ and $\mathcal{C}_0$ separately. As illustrated in Fig. 7.8.1, $\mathcal{E} = \Delta/\sqrt{2}$ corresponds to the maxima of $\partial f(\mathcal{E})/\partial\mathcal{E}$ with a value $\sqrt{2e}/\Delta^2\sqrt{\pi}$. Consequently at $\mathcal{E} = \Delta/\sqrt{2}$, (7.8.2) and (7.8.8) above give

$$\left(\frac{\Delta A'}{\mathcal{E}}\right)_{\Delta/\sqrt{2}} = \gamma\mathcal{A}_1\left(\frac{-\sqrt{2e}}{\Delta^2\sqrt{\pi}}\right)\mu_B B + \left(\frac{\Delta A'}{\mathcal{E}}\right)_0 e^{-(\Delta/\sqrt{2}-\mathcal{E}^0)^2/\Delta^2} \tag{7.8.10}$$

which is easily solved for $\mathcal{A}_1$.

Thus in our idealized case $\mathcal{A}_1$, $\mathcal{B}_0 + \mathcal{C}_0/kT$, and $\mathcal{D}_0$ are found in terms of the easily obtained experimental parameters $\mathcal{E}^0$, Δ, $(A_\pm/\mathcal{E})_0$, $(\Delta A'/\mathcal{E})_0$, and $(\Delta A'/\mathcal{E})_{\Delta/\sqrt{2}}$.

In the real world, the lineshape is never exactly Gaussian, and the MCD and absorption parameters above are generally found by a fitting procedure. One makes an initial guess of the relevant parameters: $\mathcal{D}_0$, Δ, and $\mathcal{E}^0$ for an absorption band, and $\mathcal{A}_1$, $\mathcal{B}_0 + \mathcal{C}_0/kT$, Δ, and $\mathcal{E}^0$ for an MCD band. This may be accomplished using (7.8.7) or (7.8.9) and (7.8.10) after Δ (or Γ) and $\mathcal{E}^0$ have been estimated from the experimental band (Fig. 7.8.1). One then may use a nonlinear least-squares fitting procedure which (if it works) refines the initial guesses in successive iterations to bring the root-mean-square deviation between the experimental and calculated absorption [Eq. (7.8.1)] or MCD [Eq. (7.8.2)] below some specific value. The success of such a procedure, and the quantitative significance of the parameters, depends upon how well the experimental bandshape approximates a Gaussian, especially if $\mathcal{B}_0 + \mathcal{C}_0/kT$ and $\mathcal{A}_1$ make comparable contributions to the MCD dispersion.

Moments in the RS–BO *Approximation*

Actually, when bands are isolated as assumed just above, it is much easier to determine MCD parameters using the method of moments, since no bandshape function $f(\mathcal{E})$ need be specified. Moreover, as we discussed in Section 7.2, using this method requires significantly less drastic approximations; it is not necessary to make the RS approximation or to assume the excited state obeys the BO approximation. However, if these assumptions are made, so that (7.8.1) and (7.8.2) are assumed to describe $A_\pm/\mathcal{E}$ and $\Delta A'/\mathcal{E}$, these equations may be used *directly* in our moment equations to obtain RS–BO–FC moments:

$$\left\langle \frac{A_\pm}{\mathcal{E}} \right\rangle_0 = \int \frac{A_\pm}{\mathcal{E}} d\mathcal{E} = \gamma \mathcal{D}_0$$

$$\left\langle \frac{A_\pm}{\mathcal{E}} \right\rangle_1^{\bar{\mathcal{E}}} = \int \frac{A_\pm}{\mathcal{E}} (\mathcal{E} - \bar{\mathcal{E}})\, d\mathcal{E} = \gamma \mathcal{D}_0 (\mathcal{E}^0 - \bar{\mathcal{E}}) = 0 \qquad (7.8.11)$$

$$\left\langle \frac{\Delta A'}{\mathcal{E}} \right\rangle_0 = \gamma \left(\mathcal{B}_0 + \frac{\mathcal{C}_0}{kT} \right) \mu_B B$$

$$\left\langle \frac{\Delta A'}{\mathcal{E}} \right\rangle_1^{\bar{\mathcal{E}}} = \gamma \mathcal{A}_1 \mu_B B$$

The second equation gives $\bar{\mathcal{E}} = \mathcal{E}^0$. The parameters $\mathcal{D}_0$, $\mathcal{B}_0 + \mathcal{C}_0/kT$, and $\mathcal{A}_1$ are found from these equations by simply integrating the experimental data in the prescribed manner—no iterations or initial estimates of Δ, $\mathcal{E}^0$, and so on are required. Furthermore, these RS–BO–FC moments do not require $f(\mathcal{E})$ to be Gaussian but only that the equations (7.8.3) be obeyed. These latter equations are satisfied by any normalized $f(\mathcal{E})$ that falls to zero at both ends of the range of integration.

A comparison of the moments above with those determined by the more rigorous methods described earlier is helpful in giving one a feel for what is left out or approximated when the RS–BO equations are used to describe the MCD. We see that Section 7.5 gives a theoretical description of $\bar{\mathcal{E}}$ as $\overline{W}_{JA}$. Both (7.5.3)–(7.5.7) and (7.8.1) require that (7.5.1) be valid [as does the use of (7.8.1) = (4.5.22) and (7.8.2) = (4.5.14) in Gaussian fitting].

Gaussian Fits of Overlapping Bands

Gaussian fits can be particularly useful when bands overlap. While moments give only combined parameters for a set of overlapping bands, individual parameters may be found by Gaussian fitting even when the wings of a band are obscured. The method works best when a plausible theoretical model can be proposed which accounts for the number of bands observed. The chances of a meaningful fit are much improved if it is known in each case that either $\mathcal{A}_1$ or $\mathcal{B}_0 + \mathcal{C}_0/kT$ dominates the MCD (or is zero).

Then the small (or zero) parameters may be held at zero throughout the fitting procedure. If the RS–BO–FC approximations are made and near-degenerate ground or excited states are assumed to have parallel potential surfaces, the absorption and MCD expressions are simply appropriate summations of (7.8.1) and (7.8.2) over all the bands involved.

A few warnings are in order. Calculation of the MCD of near-degenerate states is tricky, since interactions between BO basis states may dominate the MCD—these may be vibronic in origin, as in the JT or PJT effects, or may be magnetic-field-induced. Magnetic interactions may be large, even when the BO and parallel-potential-surface assumptions above are valid. For example, the transition $A \rightarrow J_1$ interacting with J_2 via the magnetic field gives rise to $\mathcal{B}_0(J_1, J_2)$ equal in magnitude but opposite in sign to $\mathcal{B}_0(J_2, J_1)$ for $A \rightarrow J_2$ interacting with J_1 [these together give rise to $\mathcal{A}'_1(J_1, J_2)$ in Section 7.6]. Thus when the energy separation $\overline{W}_{J_2J_1}$ between J_2 and J_1 becomes small compared to the bandwidth Γ of J_2 or J_1, a pseudo-$\mathcal{A}$ term results (Section 5.2). In the limit where $\overline{W}_{J_2J_1} \ll \Gamma$, the MCD is properly calculated by treating J_2 and J_1 as degenerate. When the near-degeneracy is accidental, the MCD can be particularly confusing, since the pseudo-$\mathcal{A}$ term may dominate $\mathcal{A}_1(J_1)$ and $\mathcal{A}_1(J_2)$, and the $A \rightarrow J_1$ and $A \rightarrow J_2$ absorption bands may be completely unresolved in the spectrum. Moment analysis in this and some other cases of near-degeneracy is discussed in Section 7.6 (see also Sections 5.2 and 13.5). Stephens (1976) shows in considerable detail how such situations are handled using the RS equations.

Unfortunately, while fitting procedures are extremely useful for plausibility arguments, they are rarely unique. It is often possible to propose numerous alternative sets of parameters which, in the absence of other corroborating evidence, reproduce the overall band contour about equally well. In some cases, no reliable MCD parameters are obtained. And at best, the RS–G method for bands is limited by the RS–BO–FC and parallel-potential-surface assumptions.

We emphasize that a multi-Gaussian fit of a complicated set of overlapping bands may look superb but may lead to less solid information than is obtainable with far less work via the MM. Unless a detailed theoretical model is available to explain the existence of the overlapping bands, trying to characterize them by Gaussian fits is apt to be a waste of time.

7.9. The MCD of Resolved Vibronic Lines in the Franck–Condon-Allowed Case

The spectra of crystals at low temperatures often contain resolved vibronic lines. While moments of entire bands may still be evaluated as discussed in

Section 7.4, analysis of the individual lines can yield more detailed information concerning excited-state vibronic interactions. This analysis can be carried out either by using the RS equations and fitting, or via moments with or without the RS assumption.

Use of the RS–BO *Model for Lines*

In this section we consider only FC-allowed transitions and assume the BO approximation is valid in both ground and excited states. Then, pursuing the RS approach and making the FC approximation as before, we find that (7.8.1) and (7.8.2) still hold except that $f(\mathcal{E})$ [Eq. (4.5.24)] is replaced by

$$f_l(\mathcal{E}) = \frac{1}{N_G}\sum_{g_l} N_{g_l}\sum_{j_l} |\langle g_l | j_l\rangle|^2 \rho_{g_l j_l}(\mathcal{E}) \qquad (7.9.1)$$

The summation is over those states $g = g_l$ and $j = j_l$ for which $g_l \to j_l$ falls within the frequency range defined by the particular "line" of interest. Thus, in contrast to the integral of $f(\mathcal{E})$ given in (7.8.3) and (4.5.25),

$$\begin{aligned} S_l &\equiv \int f_l(\mathcal{E})\, d\mathcal{E} \\ &= \sum_{g_l j_l} \frac{N_{g_l}}{N_G} |\langle g_l | j_l\rangle|^2 \neq 1 \end{aligned} \qquad (7.9.2)$$

since the sums over g and j are no longer complete.

However, using S_l we can write expressions for $A_{\pm}/\mathcal{E}$ and $\Delta A'/\mathcal{E}$ for individual absorption lines with a lineshape, $l(\mathcal{E}) = f_l(\mathcal{E})/S_l$, which does obey (7.8.3):

$$\begin{aligned} \left(\frac{A_{\pm}}{\mathcal{E}}\right)_l &= \gamma(\mathcal{D}_0 S_l)\, l(\mathcal{E}) \\ \left(\frac{\Delta A'}{\mathcal{E}}\right)_l &= \gamma\left[\mathcal{A}_1 S_l\left(-\frac{\partial l(\mathcal{E})}{\partial \mathcal{E}}\right) + \left(\mathcal{B}_0 S_l + \frac{\mathcal{C}_0 S_l}{kT}\right) l(\mathcal{E})\right]\mu_B B \end{aligned} \qquad (7.9.3)$$

These equations may be used in the same manner as our earlier RS equations for bands; Gaussian fits may be made, or RS–BO moments may be found by integrating (7.9.3) in standard moment equations. The parameters obtained are, however, weighted by S_l, a population-weighted sum of *FC* factors. Since the S_l factors cancel in MCD ratios, this model predicts that the same ratios $\mathcal{A}_1/\mathcal{D}_0$, $\mathcal{B}_0/\mathcal{D}_0$, and $\mathcal{C}_0/\mathcal{D}_0$ should be obtained for each resolved line in a progression, and that all should be equal to the corresponding MCD ratios for the entire band.

The MCD *of Vibronic Lines When the* BO *Approximation Breaks Down*

Systems with high symmetry are best suited for MCD spectroscopy, and they contain degenerate states. Since all orbitally degenerate states in nonlinear systems are subject to the JT effect, we must be ever watchful for vibronic interactions which break down the BO approximation and thus invalidate (7.9.3). As discussed in Section 3.8, JT systems have different vibronic selection rules, and so JT effects should be suspected whenever unusual vibronic patterns appear in the resolved or semiresolved lines of a band.

Here we consider how to analyze the MCD of such lines. The methods employed in Section 7.4 to obtain moments for a band cannot be used for the individual lines, since the invariance of (7.3.10)–(7.3.13) under unitary transformations on the excited-state manifold no longer applies. The analogous line equations are not invariant under such transformations. Thus we must solve the vibronic problem *before* proceeding further with either moments or the RS–G approach.

The full vibronic matrix relevant to the JT problem is described in Section 3.8. It is of infinite dimension and in general can only be solved numerically. A truncated vibrational basis may be used, as in Yeakel and Schatz (1974), to approximate the lower-energy vibronic lines of the band. More often, following Ham (1965), the analysis is restricted to a study of the changes in the no-phonon region brought about by the JT effect; these changes include the *Ham effect* (Section 3.8). Ham studied an orbital triplet (T_1 or T_2) in an octahedral or tetrahedral system. He showed how the JT effect quenches the spin–orbit and Zeeman interactions. These quenchings result because the T_1 or T_2 potential surfaces in the JT-active coordinates are not degenerate away from Q_0, and their associated JT electronic eigenfunctions [the $\phi'_{k\lambda}$ of (3.8.1)] have different minima. Thus matrix elements of electronic operators which are off-diagonal in the JT electronic eigenfunctions are reduced because of poor overlap between vibrational functions associated with the different minima. In favorable cases, the MCD of the no-phonon region can be calculated as a function of quenching factors.

We do not pursue this topic further, but refer interested readers to Ham (1965), Englman (1972), and Yeakel and Schatz (1974). It is tempting to extrapolate the quenching of spin–orbit and Zeeman interactions from the lowest-energy lines to the band as a whole. That, however, is incorrect. In Section 7.4 we show, for example, that $\mathcal{D}_0$, $\mathcal{A}_1$, $\mathcal{B}_0$, and $\mathcal{C}_0$ *for a band* are unchanged by JT effects. Thus in the non-BO case, MCD parameter ratios for lines may differ greatly from those for the band as a whole.

7.10. Franck–Condon-Forbidden Transitions in the Rigid-Shift Approximation

Here we consider the case where $\mathcal{D}_0$, $\mathcal{A}_1$, $\mathcal{B}_0$, and $\mathcal{C}_0$ are all zero in the FC approximation and all intensity is vibrationally induced, as in Section 3.7. We assume the BO approximation is valid in both ground and excited states, and make the harmonic approximation. Thus letting $|a\rangle = |A\alpha, g\xi\rangle$ and $|j\rangle = |J\lambda, j\zeta\rangle$ in (4.3.10) and using (7.3.15) and the HT approximation of (3.7.5), we have

$$\frac{A_\pm}{\mathcal{E}} = \gamma\left(-\sum_{\alpha\lambda}\sum_{g\xi j\zeta}\frac{N_{A\alpha,g\xi}}{N}\langle A\alpha, g\xi|m_{\pm1}|J\lambda, j\zeta\rangle \times\langle J\lambda, j\zeta|m_{\mp1}|A\alpha, g\xi\rangle\rho_{gj}(\mathcal{E})\right)$$

$$= \gamma\left[-\sum_{h\theta h'\theta'}\sum_{\alpha\lambda}\frac{1}{|A|}\langle A\alpha|m_{\pm1}|J\lambda\rangle'_{h\theta}\langle J\lambda|m_{\mp1}|A\alpha\rangle'_{(h'\theta')^\dagger} \times\frac{1}{N_G}\sum_{g\xi}N_g\sum_{j\zeta}\langle g\xi|Q^\dagger_{h\theta}|j\zeta\rangle\langle j\zeta|Q_{h'\theta'}|g\xi\rangle\rho_{gj}(\mathcal{E})\right] \quad (7.10.1)$$

In the harmonic approximation we must have $h' = h$ and $\theta' = \theta$ in order that the two vibrational integrals in any term in the sum above may be simultaneously nonzero. Thus

$$\frac{A_\pm}{\mathcal{E}} = \gamma\left[-\sum_{h\theta}\frac{1}{|A|}\sum_{\alpha\lambda}\langle A\alpha|m_{\pm1}|J\lambda\rangle'_{h\theta}\langle J\lambda|m_{\mp1}|A\alpha\rangle'_{(h\theta)^\dagger}f_{h\theta}(\mathcal{E})\right] \quad (7.10.2)$$

where

$$f_{h\theta}(\mathcal{E}) = \frac{1}{N_G}\sum_{g\xi}N_g\sum_{j\zeta}\langle g\xi|Q^\dagger_{h\theta}|j\zeta\rangle\langle j\zeta|Q_{h\theta}|g\xi\rangle\rho_{gj}(\mathcal{E}) \quad (7.10.3)$$

As always, we want to express $A_{\pm}/\mathcal{E}$ in terms of a bandshape function which integrates to one and otherwise obeys the equations for $f(\mathcal{E})$ of (7.8.3). Since the integral of $f_{h\theta}(\mathcal{E})$ is not equal to one, we define

$$\overline{Q_{h\theta}^2} \equiv \frac{1}{N_G}\sum_{g\xi} N_g \sum_{j\zeta} \langle g\xi | Q_{h\theta}^{\dagger} | j\zeta\rangle\langle j\zeta | Q_{h\theta} | g\xi\rangle$$

$$= \int f_{h\theta}(\mathcal{E})\, d\mathcal{E} \qquad (7.10.4)$$

and

$$g_{h\theta}(\mathcal{E}) = \frac{f_{h\theta}(\mathcal{E})}{\overline{Q_{h\theta}^2}} \qquad (7.10.5)$$

It follows that

$$\int g_{h\theta}(\mathcal{E})\, d\mathcal{E} = 1 \qquad (7.10.6)$$

Then defining

$$\mathcal{D}_0(h\theta) = \frac{1}{|A|}\sum_{\alpha\lambda} |\langle A\alpha | m_{\pm 1} | J\lambda\rangle'_{h\theta}|^2\, \overline{Q_{h\theta}^2} \qquad (7.10.7)$$

we can write $A_{\pm}/\mathcal{E}$ in the desired form:

$$\frac{A_{\pm}}{\mathcal{E}} = \sum_{h\theta}\left(\frac{A_{\pm}}{\mathcal{E}}\right)_{h\theta} = \sum_{h\theta} \gamma \mathcal{D}_0(h\theta)\, g_{h\theta}(\mathcal{E}) \qquad (7.10.8)$$

Similarly for the MCD, as shown explicitly in Stephens (1976),

$$\frac{\Delta A'}{\mathcal{E}} = \sum_{h\theta}\left(\frac{\Delta A'}{\mathcal{E}}\right)_{h\theta}$$

$$= \sum_{h\theta} \gamma\left[\mathcal{A}_1(h\theta)\left(\frac{-\partial g_{h\theta}(\mathcal{E})}{\partial \mathcal{E}}\right) + \left(\mathcal{B}_0(h\theta) + \frac{\mathcal{C}_0(h\theta)}{kT}\right) g_{h\theta}(\mathcal{E})\right]\mu_B B$$

$$(7.10.9)$$

where $\mathcal{A}_1(h\theta)$ and $\mathcal{C}_0(h\theta)$ are given in Section 22.3.

Since the sum in (7.10.4) is over the complete set $|j\zeta\rangle$,

$$\overline{Q_{h\theta}^2} = \frac{1}{N_G}\sum_{g\xi} N_g \sum_{j\zeta} \langle g\xi | Q_{h\theta}^\dagger | j\zeta\rangle\langle j\zeta | Q_{h\theta} | g\xi\rangle$$

$$= \frac{1}{N_G}\sum_{g\xi} N_g \langle g\xi | Q_{h\theta}^2 | g\xi\rangle \qquad (7.10.10)$$

Recall that in the harmonic approximation the $|g\xi\rangle$ have the form (3.6.7). Therefore, since the contributions to the Boltzmann factors from modes other than $h\theta$ cancel in the numerator and denominator, integration over all coordinates but $h\theta$ gives

$$\overline{Q_{h\theta}^2} = \frac{1}{\sum_{h\theta} N_{h\theta}} \sum_{h\theta} N_{h\theta}\langle h\theta | Q_{h\theta}^2 | h\theta\rangle \qquad (7.10.11)$$

Let n equal the number of quanta of $\nu_{h\theta}$ excited, and define $x = \exp(-h\nu_{h\theta}/kT)$, so that $N_{h\theta} = x^n$. Then in the harmonic approximation

$$\overline{Q_{h\theta}^2} = \frac{\hbar^2}{2\mathcal{E}(h\theta)} \frac{\sum_{n=0}^{\infty}(2n+1)x^n}{\sum_{n=0}^{\infty} x^n} = \frac{\hbar^2}{2\mathcal{E}(h\theta)}\left(\frac{1+x}{1-x}\right)$$

$$= \frac{\hbar^2}{2\mathcal{E}(h\theta)} \coth\left(\frac{\mathcal{E}(h\theta)}{2kT}\right) \qquad (7.10.12)$$

where $\mathcal{E}(h\theta) = h\nu_{h\theta}$. This result is used to obtain (3.7.7).

Equations (7.10.8) and (7.10.9) have the same RS form for each intensity-inducing vibration $h\theta$ as $A_{\pm}/\mathcal{E}$ and $\Delta A'/\mathcal{E}$ in (7.8.1) and (7.8.2). As long as (7.10.6) holds for each vibration, (7.10.8) and (7.10.9) may be used directly in the moment equations to obtain zeroth moments. Since in general $g_{h\theta}(\mathcal{E}) \neq g_{h'\theta'}(\mathcal{E})$, to obtain first moments we must assume

$$\int g_{h\theta}(\mathcal{E})\mathcal{E}\, d\mathcal{E} = \mathcal{E}_{h\theta}^0 = \mathcal{E}^0 \qquad (7.10.13)$$

for all $h\theta$. Equation (7.10.13) holds exactly if the only activating vibrations $(h\theta)$ are members of a degenerate set—say $\nu_4(t_{1u}x)$, $\nu_4(t_{1u}y)$, and $\nu_4(t_{1u}z)$ in an octahedral system. It should also hold approximately for broad,

Gaussian-shaped bands. Using (7.10.13) for the first moments, we obtain the RS–BO moments for FC-forbidden vibrationally induced MCD and absorption *bands*:

$$\left\langle \frac{A_{\pm}}{\mathcal{E}} \right\rangle_0 = \gamma \left[\sum_{h\theta} \mathcal{D}_0(h\theta) \right]$$

$$\left\langle \frac{A_{\pm}}{\mathcal{E}} \right\rangle_1^{\bar{\mathcal{E}}} = 0, \qquad \bar{\mathcal{E}} = \mathcal{E}^0$$

$$\left\langle \frac{\Delta A'}{\mathcal{E}} \right\rangle_0 = \gamma \left[\sum_{h\theta} \left(\mathcal{B}_0(h\theta) + \frac{\mathcal{C}_0(h\theta)}{kT} \right) \right] \mu_B B \tag{7.10.14}$$

$$\left\langle \frac{\Delta A'}{\mathcal{E}} \right\rangle_1^{\bar{\mathcal{E}}} = \gamma \left[\sum_{h\theta} \mathcal{A}_1(h\theta) \right] \mu_B B$$

Note that $\mathcal{D}_0(h\theta)$, $\mathcal{A}_1(h\theta)$, $\mathcal{B}_0(h\theta)$, and $\mathcal{C}_0(h\theta)$ are all temperature-dependent through their $\overline{Q_{h\theta}^2}$ factors. Thus when $\langle A_{\pm}/\mathcal{E} \rangle_0$ is temperature-dependent, vibration-induced transitions should be suspected, since in the FC-allowed case $\langle A_{\pm}/\mathcal{E} \rangle_0$ is temperature-independent (in the FC approximation).

7.11. Vibrationally Induced Lines in the Rigid-Shift Born–Oppenheimer Approximation

Equations similar to those of Section 7.10 may be derived for vibrationally induced lines (as opposed to bands). It is easy to show that (7.10.8) and (7.10.9) still hold if only $\overline{Q_{h\theta}^2}$ is replaced by $\overline{Q_{h\theta}(l)^2}$ in the expressions for $g_{h\theta}(\mathcal{E})$, $\mathcal{D}_0(h\theta)$, $\mathcal{A}_1(h\theta)$, $\mathcal{B}_0(h\theta)$, and $\mathcal{C}_0(h\theta)$, where

$$\begin{aligned} \overline{Q_{h\theta}(l)^2} &= \frac{1}{N_G} \int \sum_{g_l \xi_l} N_{g_l} \sum_{j_l \zeta_l} \left| \langle g_l \xi_l | Q_{h\theta}^{\dagger} | j_l \zeta_l \rangle \right|^2 \rho_{g_l j_l}(\mathcal{E})\, d\mathcal{E} \\ &= \frac{1}{N_G} \sum_{g_l \xi_l} N_{g_l} \sum_{j_l \zeta_l} \left| \langle g_l \xi_l | Q_{h\theta}^{\dagger} | j_l \zeta_l \rangle \right|^2 \end{aligned} \tag{7.11.1}$$

Of course, (7.10.12) does not hold for $\overline{Q_{h\theta}(l)^2}$, since the sums over $g_l \xi_l$ and $j_l \zeta_l$ are restricted to those $g\xi$ and $j\zeta$ which contribute to the line.

The moment equations (7.10.14) may likewise be used for lines if $\overline{Q_{h\theta}^2}$ is replaced everywhere by $\overline{Q_{h\theta}(l)}^2$. They are frequently more useful than the band equations, since a resolved vibronic line usually arises from activating vibrations of only one symmetry, h.

7.12. Hybrid Cases

We have limited our discussion in this chapter to (a) allowed transitions in the FC approximation and (b) transitions forbidden in the FC approximation for which *all* intensity is vibrationally induced. There are obviously in-between situations for which our equations are not applicable. For example, a FC-allowed transition may be weak, and vibronic transitions [as in (b)] may have comparable intensity. Usually, however, either mechanism (a) or (b) dominates.

8 Introduction to Group Theory

8.1. Introduction

In previous chapters, basic expressions were derived which connect observed MCD and absorption spectra with various matrix elements. Clearly, the evaluation of these matrix elements is the critical step in relating theory and experiment. It is the purpose of this and subsequent chapters to demonstrate the use of increasingly powerful group-theoretic techniques which allow us to reduce matrix elements quickly and easily to a minimum number whose ultimate evaluation requires the assumption of models.

The value of group theory in analyzing molecular systems is widely appreciated, and most chemists and physicists have had some exposure to actual applications, at least to the point of using character tables. However, until recently, the methods used by many have changed little since the early days of quantum mechanics, and as a consequence, all but the most mathematically adept are apt to be discouraged from applying group theory to a wide range of important, more complex problems. Over the last four decades, however, powerful techniques have emerged, primarily through the work of atomic spectroscopists, and these have been gradually extended to molecular systems by Griffith and others. Because of the inherent complexity of the derivations and resulting formulas, it is not fully appreciated that the actual applications of these methods are usually quite simple. Indeed, after the mastery of some notation, ease of application tends to be inversely proportional to complexity of basic formula.

In this chapter we review basic concepts of group theory. Then in Chapters 9–12 we develop the necessary machinery for the older methods where symmetry is used to construct multideterminantal wavefunctions, which in turn are used to evaluate matrix elements. In Chapter 13 we illustrate these methods with a variety of examples, and it becomes clear that more powerful techniques are a necessity in more complex applications. Such techniques are developed in Chapters 14–20, building on the concepts and nomenclature of Chapters 8–13. Matrix elements may then be evaluated without the explicit construction of multideterminantal wave functions. Symmetry is fully exploited, and a very clean and revealing delineation of the symmetry-determined and model-dependent aspects of problems emerges.

8.2. Vector Spaces and the Cartesian Representation

It is assumed that the reader has some previous familiarity with group theory, and in particular with the concept of group representations. We briefly review this and related concepts and introduce some of the mathematical machinery and nomenclature which is essential for what follows. An excellent discussion of basic concepts and mathematical techniques may be found in Bishop (1973).

Symmetry Elements and Groups

In this book we deal *exclusively* with point-group symmetry operations, that is, with rotations and rotary reflections that take a static object into itself. Thus if we consider a molecule with fixed nuclei, we can always associate a symmetry with it. This symmetry is uniquely described by specifying all of the distinct symmetry operations which can be performed upon the object. A *symmetry operation* is deemed to have been performed if, afterwards, it is impossible for a space-fixed observer to tell whether anything has been done —and indeed, doing nothing is the *identity operation*. The symmetry operations may be physically achievable (*proper*), as in the case of rotations, or may only be mathematically possible (*improper*), as for rotary reflections. The complete collection of such operations for any object always constitutes a group, which is called the *point group*, since at least one point in the object is invariant under all group operations. The operations can be put in one-to-one correspondence (i.e., are isomorphic) with the (abstract) elements of the group. The law of combination is "do the operation on the right followed by the operation to its left," and the operations then obey the group "multiplication" table.

Representations and Basis Vectors

A *representation* of a group is a set of square matrices, each of which can be associated with a different group element in such a way that the matrices obey the group multiplication table when the law of combination is matrix multiplication. The individual matrices are called *representatives*. Representations are the practical instrument for applying group theory, since they provide the mathematical means for expressing symmetry operations.

Of special interest to us is the Cartesian representation $\mathbf{R}$ obtained by applying symmetry operations R to the Cartesian basis vectors of the coordinate system used to describe the object (molecule). Let us choose an invariant point of an object as the origin for three mutually orthogonal unit vectors, which, for ease of later generalization, we designate $\mathbf{e}_1, \mathbf{e}_2, \mathbf{e}_3$ in preference to the more conventional $\mathbf{e}_x, \mathbf{e}_y, \mathbf{e}_z$ (or $\mathbf{i}, \mathbf{j}, \mathbf{k}$). We define three corresponding coordinate axes, X_1 (X), X_2 (Y), X_3 (Z), by requiring that X_i be the continuum of points x_i along the direction defined by the vector $\mathbf{r}_i = \mathbf{e}_i x_i$, $-\infty \leqslant x_i \leqslant \infty$. We use exclusively right-handed coordinate systems in this book, as illustrated in Fig. 8.2.1.

Any point in space (x_1, x_2, x_3) can be associated with a vector $\mathbf{r}$ in accordance with

$$\mathbf{r} = \mathbf{e}_1 x_1 + \mathbf{e}_2 x_2 + \mathbf{e}_3 x_3 \tag{8.2.1}$$

$\mathbf{e}_1, \mathbf{e}_2, \mathbf{e}_3$ are referred to as *basis vectors*, or collectively as *the basis* for the three-dimensional Cartesian space, since by (8.2.1), any vector can be expressed as a linear combination of these vectors. The numerical coefficients (x_1, x_2, x_3) are the components of the vector and are of course simply the Cartesian coordinates of the point to which the vector is drawn. Equation (8.2.1) may be written in matrix notation:

$$\mathbf{r} = (\mathbf{e}_1 \quad \mathbf{e}_2 \quad \mathbf{e}_3)\begin{pmatrix} x_1 \\ x_2 \\ x_3 \end{pmatrix} = \mathbf{ex} \tag{8.2.2}$$

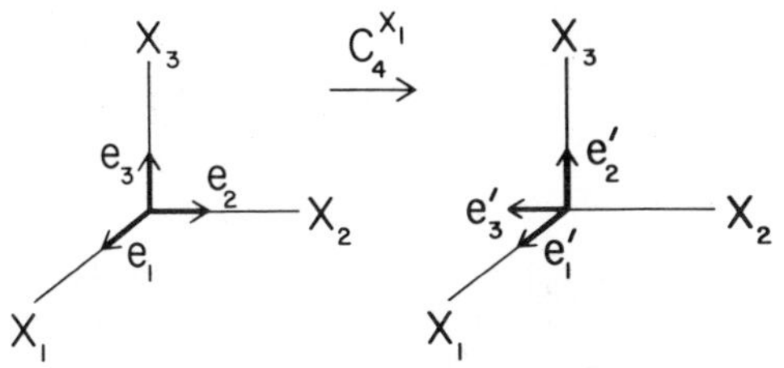

Figure 8.2.1. $\mathbf{e}_1, \mathbf{e}_2, \mathbf{e}_3$ are unit basis vectors for the right-handed Cartesian coordinate system X_1 (X), X_2 (Y), X_3 (Z). Note that the thumb of the right hand points along the positive direction of X_3 (Z) as the curl of the fingers carries X_1 into X_2. $\mathbf{e}_1', \mathbf{e}_2', \mathbf{e}_3'$ show how the basis vectors $\mathbf{e}_1, \mathbf{e}_2, \mathbf{e}_3$ are respectively transformed by the rotation operator $C_4^{X_1}$.

We designate a matrix by boldface type and understand that matrix multiplication applies when two matrices are placed side by side.

Since it is customary to write the vector components as a *column* matrix (often referred to, perhaps unfortunately, as a column vector), the basis vectors must appear as a *row* matrix *on the left*.

Let us now suppose that we "rotate"[‡] the basis vectors (and consequently, the attached coordinate system) by applying a point-group operation R, thus producing a new set of (mutually orthogonal) unit basis vectors, $\mathbf{e}' = (\mathbf{e}'_1 \ \mathbf{e}'_2 \ \mathbf{e}'_3)$. We then define R as the linear operator which performs the operation R on the basis vectors:

$$R\mathbf{e} = \mathbf{e}' = \mathbf{e}\mathbf{R} \tag{8.2.3}$$

The matrix $\mathbf{R}$ is the Cartesian representative for the operation R. For example, suppose $R = C_4^{X_1}$ is a positive (counterclockwise) rotation of 90° about the X_1 axis. Then $C_4^{X_1}$ rotates the basis vectors 90° counterclockwise about the X_1 axis as illustrated in Fig. 8.2.1. From the figure we see that

$$R(\mathbf{e}_1 \quad \mathbf{e}_2 \quad \mathbf{e}_3) = (\mathbf{e}'_1 \quad \mathbf{e}'_2 \quad \mathbf{e}'_3) = (\mathbf{e}_1 \quad \mathbf{e}_2 \quad \mathbf{e}_3)\begin{pmatrix} 1 & 0 & 0 \\ 0 & 0 & -1 \\ 0 & 1 & 0 \end{pmatrix} \tag{8.2.4}$$

so that here

$$\mathbf{R} = \mathbf{C}_4^{X_1} = \begin{pmatrix} 1 & 0 & 0 \\ 0 & 0 & -1 \\ 0 & 1 & 0 \end{pmatrix} \tag{8.2.5}$$

In more complicated cases, it is desirable to have a systematic procedure for generating Cartesian representatives. Referring to (8.2.3), we use vector addition (elementary trigonometry) to obtain

$$\mathbf{e}'_i = \mathbf{e}_1\cos[\mathbf{e}_1, \mathbf{e}'_i] + \mathbf{e}_2\cos[\mathbf{e}_2, \mathbf{e}'_i] + \mathbf{e}_3\cos[\mathbf{e}_3, \mathbf{e}'_i]$$

or

$$\mathbf{e}_i = \mathbf{e}'_1\cos[\mathbf{e}'_1, \mathbf{e}_i] + \mathbf{e}'_2\cos[\mathbf{e}'_2, \mathbf{e}_i] + \mathbf{e}'_3\cos[\mathbf{e}'_3, \mathbf{e}_i] \tag{8.2.6}$$

where $[\mathbf{e}_i, \mathbf{e}'_j]$ designates the angle between the vectors $\mathbf{e}_i$ and $\mathbf{e}'_j$. Equation (8.2.6) is simply the mathematical statement that an arbitrary (unit) vector

[‡]As a generic term, we refer to all point-group operations as "rotations." We distinguish between rotations and rotary reflections only when required in specific applications.

can be decomposed into components along three mutually orthogonal directions. Using (8.2.6), the matrix elements of **R** in (8.2.3) are given by

$$R_{ij} = \cos[\mathbf{e}_i, \mathbf{e}'_j] = \cos[\mathbf{e}'_j, \mathbf{e}_i] \tag{8.2.7}$$

We follow Griffith (1964) and Tinkham (1964) in defining the sign of a rotation by the right-hand rule; that is, a rotation is positive (negative) if the thumb of the right hand points along the positive (negative) direction of the rotation axis when the fingers curl along the sense of the rotation.

To illustrate (8.2.7), let $R = S_3^{X_3}$. The result of this operation is shown in Fig. 8.2.2. Inspection shows that $[\mathbf{e}_1, \mathbf{e}'_1] = [\mathbf{e}_2, \mathbf{e}'_2] = \frac{2}{3}\pi$; $[\mathbf{e}_3, \mathbf{e}'_3] = \pi$; $[\mathbf{e}_2, \mathbf{e}'_1] = \frac{1}{6}\pi$; $[\mathbf{e}_1, \mathbf{e}'_2] = \frac{7}{6}\pi$; $[\mathbf{e}_1, \mathbf{e}'_3] = [\mathbf{e}_2, \mathbf{e}'_3] = [\mathbf{e}_3, \mathbf{e}'_2] = [\mathbf{e}_3, \mathbf{e}'_1] = \frac{1}{2}\pi$. Thus

$$\mathbf{R} = \mathbf{S}_3^{X_3} = \begin{pmatrix} \cos\frac{2}{3}\pi & \cos\frac{7}{6}\pi & 0 \\ \cos\frac{1}{6}\pi & \cos\frac{2}{3}\pi & 0 \\ 0 & 0 & \cos\pi \end{pmatrix}$$

$$= \begin{pmatrix} -\frac{1}{2} & -\frac{1}{2}\sqrt{3} & 0 \\ \frac{1}{2}\sqrt{3} & -\frac{1}{2} & 0 \\ 0 & 0 & -1 \end{pmatrix} \tag{8.2.8}$$

We confirm that *the* **R** *matrices generated in this way are members of a representation*. If $T = PS$, then

$$PS\mathbf{e} = P\mathbf{e}\mathbf{S} = \mathbf{e}\mathbf{P}\mathbf{S} \tag{8.2.9}$$

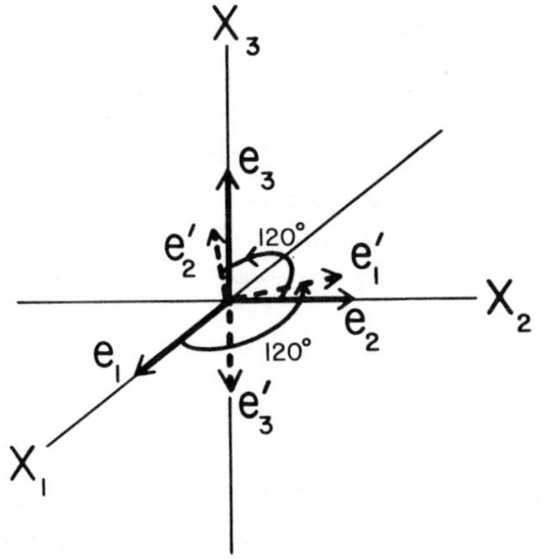

Figure 8.2.2. $\mathbf{e}'_1, \mathbf{e}'_2, \mathbf{e}'_3$ show how the basis vectors $\mathbf{e}_1, \mathbf{e}_2, \mathbf{e}_3$ are respectively transformed by the rotary reflection operator $S_3^{X_3}$.

where use is made of the fact that $P(\mathbf{eS}) = (P\mathbf{e})(\mathbf{S})$. [See Bishop (1973), Appendix A5-2. Keep in mind that all operations are defined with respect to the original (unprimed) basis.] Also

$$PS\mathbf{e} = T\mathbf{e} = \mathbf{eT} \tag{8.2.10}$$

whence

$$\mathbf{T} = \mathbf{PS} \tag{8.2.11}$$

These representatives are orthogonal (transpose $\tilde{\mathbf{R}}$ = inverse $\mathbf{R}^{-1}$), as may be confirmed using the elementary properties of the direction cosines defined by (8.2.6). We explicitly demonstrate this fact towards the end of this section adopting a slightly different point of view.

If we now operate with R on the arbitrary vector $\mathbf{r}$ defined in (8.2.2), it follows from (8.2.3) that

$$R\mathbf{r} \equiv \mathbf{r}' = R\mathbf{ex} = \mathbf{e}'\mathbf{x} \tag{8.2.12}$$

Thus R rotates an arbitrary vector in the same sense as it rotates the basis vectors, since by the last equality above, the vector components $\mathbf{x}$ remain the same but are expressed in terms of rotated basis vectors $\mathbf{e}'$. By the last equality in (8.2.3), we may also express $\mathbf{r}'$ in terms of the original basis vectors. Thus

$$\mathbf{r}' = R\mathbf{ex} = \mathbf{eRx} = \mathbf{ex}' \tag{8.2.13}$$

so that

$$\mathbf{x}' = \mathbf{Rx} \tag{8.2.14}$$

Equation (8.2.13) gives the components of the rotated vector in terms of the original basis vectors. Equation (8.2.14) expresses the relation between the components of the rotated vector and the components of the original vector, both expressed in the original coordinate system. The components $\mathbf{x}'$ of the rotated vector may be directly related to those of the original vector $\mathbf{x}$ using trigonometry [Bishop (1973), p. 76]. According to (8.2.14), $\mathbf{R}$ must be the same matrix obtained using (8.2.3) and (8.2.7). However, (8.2.14) must *not* be viewed in the operational sense, $R\mathbf{x} = \mathbf{x}'$. R operates *only* on basis vectors, not on scalar components.

The Active and Passive Conventions

The equation $\mathbf{x}' = \mathbf{Rx}$ can also be regarded as a relationship between the components of a *stationary vector viewed in two different coordinate systems.*

In fact, the relation between $\mathbf{r}$ and $\mathbf{r}'$ can be viewed in two entirely equivalent ways: either $\mathbf{r}'$ has been obtained by transforming $\mathbf{r}$ by R, or $\mathbf{r}$ has been held stationary and renamed $\mathbf{r}'$ in a new (primed) coordinate system transformed by R^{-1}. (This equivalence is of course destroyed in physical problems if a space-fixed field is present—e.g., a gravitational, magnetic, or electric field—since the isotropy of space is then removed.) The first view represents the *active convention*, and the second the *passive convention*.

It is clear that transforming the coordinate system by R produces the transformation matrix $\mathbf{R}^{-1}$, and so a potential ambiguity arises in the definition of symmetry operations. An *arbitrary* convention must be adopted to distinguish unambiguously between a symmetry operation and its inverse. Our choice is that of (8.2.3): *we use the active convention, in which symmetry operators rotate vectors (or later, functions), not coordinates*. Thus a series of transformations is expressed in terms of a single space-fixed coordinate system *to which the symmetry axes are fixed*. [This is convention (a), Tinkham (1964), Table 5-1, Section 5-7. Tinkham's table summarizes various equivalent points of view and alternative conventions.]

Dot and Scalar Products and Length-Preserving Transformations

The dot product of two vectors, $\mathbf{r}$ and $\mathbf{r}'$, is *defined* as the scalar quantity

$$\mathbf{r} \cdot \mathbf{r}' \equiv rr'\cos[\mathbf{r}, \mathbf{r}'] \tag{8.2.15}$$

where $r \equiv |\mathbf{r}|$. Since we use orthogonal coordinate systems exclusively, the (Cartesian) basis vectors satisfy

$$\mathbf{e}_i \cdot \mathbf{e}_j = \delta_{ij} \tag{8.2.16}$$

and therefore by (8.2.1)

$$\mathbf{r} \cdot \mathbf{r}' = \sum_{i=1}^{3} x_i x_i' \tag{8.2.17}$$

If we allow for complex vectors, we may generalize (8.2.17) and define a *scalar* product by

$$(\mathbf{r}, \mathbf{r}') \equiv \sum_{i=1}^{3} x_i^* x_i' \tag{8.2.18}$$

where the star, as always, designates complex conjugation.

We noted in connection with (8.2.11) that the Cartesian representatives generated by applying point-group operators to the arbitrary vector **r** are necessarily orthogonal in view of (8.2.7). We may explicitly demonstrate this from another point of view. Clearly, point-group operators preserve the length (magnitude) of **r** and thus may be referred to as length-preserving. Mathematically the length of a (real) vector is measured simply by its dot product with itself, and thus for such a length-preserving transformation

$$\mathbf{r}\cdot\mathbf{r} = \sum_{i=1}^{3} x_i^2 = \mathbf{r}'\cdot\mathbf{r}' = \sum_{i=1}^{3} x_i'^2 \tag{8.2.19}$$

or in matrix notation

$$\tilde{\mathbf{x}}\mathbf{x} = \tilde{\mathbf{x}}'\mathbf{x}' \tag{8.2.20}$$

Using (8.2.14), we have

$$\tilde{\mathbf{x}}\mathbf{x} = \tilde{\mathbf{x}}'\mathbf{x}' = (\widetilde{\mathbf{R}\mathbf{x}})(\mathbf{R}\mathbf{x}) = \tilde{\mathbf{x}}(\tilde{\mathbf{R}}\mathbf{R})\mathbf{x} \tag{8.2.21}$$

and thus

$$\tilde{\mathbf{R}}\mathbf{R} = \mathbf{E} \tag{8.2.22}$$

where **E** is the unit matrix. Since $\mathbf{R}^{-1}$ necessarily exists,

$$\tilde{\mathbf{R}} = \mathbf{R}^{-1} \tag{8.2.23}$$

Equation (8.2.23) is the definition of an *orthogonal matrix*. If the matrix elements in each row (or column) of **R** are regarded as the components of a vector, then the rows (or columns) define mutually orthonormal vectors, since (8.2.22) requires

$$\sum_{i=1}^{3} R_{ij}R_{ij'} = \delta_{jj'} = \sum_{i=1}^{3} R_{ji}R_{j'i} \tag{8.2.24}$$

We emphasize that the orthogonality of **R** results only if an orthogonal coordinate system is used. Conversely, one immediately notes from (8.2.24) that matrix multiplication of an orthonormal set of basis vectors by an orthogonal matrix necessarily produces another orthonormal set of basis vectors.

If we permit complex vector components, length is *defined* by the scalar product, and (8.2.19)–(8.2.24) are modified by appropriate complex conjugation. Thus

$$(\mathbf{r}, \mathbf{r}) = \tilde{\mathbf{x}}^* \mathbf{x} = (\mathbf{r}', \mathbf{r}') = \tilde{\mathbf{x}}'^* \mathbf{x}' \tag{8.2.25}$$

and therefore

$$\tilde{\mathbf{R}}^* \mathbf{R} = \mathbf{E}, \qquad \text{or} \quad \tilde{\mathbf{R}}^* = \mathbf{R}^{-1} \tag{8.2.26}$$

Equation (8.2.26) defines a *unitary* matrix, and (8.2.24) then reads

$$\sum_{i=1}^{3} R_{ij}^* R_{ij'} = \delta_{jj'} = \sum_{i=1}^{3} R_{ji}^* R_{j'i} \tag{8.2.27}$$

Unitary matrices are of great importance when many of the concepts developed for vector spaces are extended to function spaces in Section 8.6.

8.3. Change of Basis Vectors and Similarity Transformations

At the start of our discussion, we expressed an arbitrary vector $\mathbf{r}$ in terms of an arbitrary set of basis vectors, $\mathbf{e} = (\mathbf{e}_1 \ \mathbf{e}_2 \ \mathbf{e}_3)$. Applying a point-group operator R to this vector produced a new vector $\mathbf{r}'$ whose Cartesian components $\mathbf{x}'$, expressed in the same coordinate system, were related to those of the original vector by (8.2.14):

$$\mathbf{x}' = \mathbf{R}\mathbf{x}$$

It is of interest to determine the form (8.2.14) assumes if a different coordinate system (basis set) is used. Choosing the same origin, let us designate a new *orthonormal* basis set by $\bar{\mathbf{e}} = (\bar{\mathbf{e}}_1 \ \bar{\mathbf{e}}_2 \ \bar{\mathbf{e}}_3)$. The two bases must be related by an orthogonal transformation $\mathbf{Q}$:

$$\bar{\mathbf{e}} = \mathbf{e}\mathbf{Q} \tag{8.3.1}$$

In the original basis

$$\mathbf{r} = \mathbf{e}\mathbf{x}, \qquad R\mathbf{r} = \mathbf{r}' \equiv \mathbf{e}\mathbf{x}' \tag{8.3.2}$$

In the new basis we rename the vectors $\mathbf{r}, \mathbf{r}'$, as $\mathbf{s}, \mathbf{s}'$. Therefore

$$\mathbf{s} = \bar{\mathbf{e}}\mathbf{y}, \qquad R\mathbf{s} \equiv \mathbf{s}' \equiv \bar{\mathbf{e}}\mathbf{y}' \tag{8.3.3}$$

where $\mathbf{y}, \mathbf{y}'$ symbolize vector components in the new coordinate system.

Using (8.3.1), Eq. (8.3.3) becomes

$$\mathbf{s} = \mathbf{eQy}, \qquad \mathbf{s}' = \mathbf{eQy}' \tag{8.3.4}$$

But now, $\mathbf{s} \equiv \mathbf{r}$, $\mathbf{s}' \equiv \mathbf{r}'$. Therefore

$$\mathbf{x} = \mathbf{Qy}, \qquad \mathbf{x}' = \mathbf{Qy}' \tag{8.3.5}$$

Substituting this into (8.2.14) and premultiplying by $\mathbf{Q}^{-1}$ gives the desired relation,

$$\mathbf{y}' = \mathbf{Q}^{-1}\mathbf{RQy} \equiv \mathbf{R}'\mathbf{y} \tag{8.3.6}$$

Thus in the new coordinate system, the matrix $\mathbf{R}' = \mathbf{Q}^{-1}\mathbf{RQ}$ plays the same role as $\mathbf{R}$ did in the original coordinate system. $\mathbf{R}'$ and $\mathbf{R}$ are said to be related by a *similarity transformation*. If we use symmetry operators to generate representatives which in the original basis obey $\mathbf{T} = \mathbf{PS}$, then by (8.3.6), the corresponding result in the new basis must be

$$\mathbf{T}' = \mathbf{P}'\mathbf{S}' \tag{8.3.7}$$

where $\mathbf{T}' = \mathbf{Q}^{-1}\mathbf{TQ}$, $\mathbf{P}' = \mathbf{Q}^{-1}\mathbf{PQ}$, and $\mathbf{S}' = \mathbf{Q}^{-1}\mathbf{SQ}$. Two representations whose corresponding members are related by the *same* similarity transformation, as for example the primed and unprimed members here, are said to be *equivalent*. Because of the use of orthonormal bases, the similarity transformation matrix ($\mathbf{Q}$) is necessarily unitary. The physically significant results of any calculation must be independent of the coordinate system used. Consequently, the coordinate system should be chosen in any given problem to afford maximum simplification.

In Section 8.5, we extend these ideas to basis sets consisting of *functions*. The choice of basis set (basis functions) is then a matter of profound importance. Indeed, one of the vital roles of group theory is to tell us how to make this choice so as to maximize the simplifications which are possible when a system possesses symmetry.

8.4. *n*-Dimensional Vector Spaces

The extension of the previous discussion to an *n*-dimensional vector space is immediate. All previous equations are applicable with the understanding that

$$\mathbf{e} \equiv \begin{pmatrix} \mathbf{e}_1 & \mathbf{e}_2 & \cdots & \mathbf{e}_n \end{pmatrix} \tag{8.4.1}$$

so that vectors now have n components, and the corresponding Cartesian representation is n-dimensional (instead of three-dimensional); thus all summations that appear or are implied in the previous equations run from 1 to n instead of 1 to 3. The n-dimensional space is often referred to as *configuration space*, and specification of the values of all n Cartesian coordinates defines a *configuration*.

The n-dimensional case appears quite naturally in practice as soon as we deal with systems that require more than three independent coordinates, as for example in a many-particle system. Thus in a molecular system, an independent set of coordinates must be ascribed to each atom in order to describe the relative motions of the atoms in a vibrational problem; similarly, an independent set of electronic coordinates must be assigned to each atomic orbital centered on its atom in a molecular-orbital (MO) description.

As an example, let us consider the molecule BF_3, whose equilibrium nuclear configuration has point group symmetry D_{3h}. We define basis vectors on each atom as shown in Figure 8.4.1, where the relative orientations on different centers have been chosen for convenience. Let us now apply the symmetry operator, $S_3^{X_3}$, to this twelve-dimensional basis, $\mathbf{e} = (\mathbf{e}_1, \ldots, \mathbf{e}_{12})$. The corresponding twelve-dimensional representative is given by

$$S_3^{X_3}\mathbf{e} = \mathbf{e}\mathbf{S}_3^{X_3} \tag{8.4.2}$$

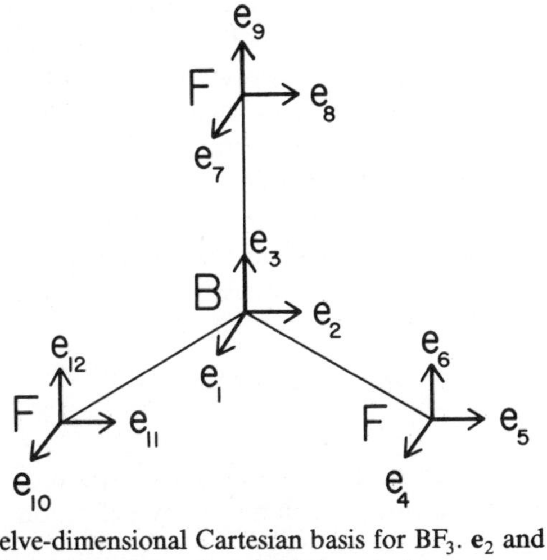

Figure 8.4.1. Twelve-dimensional Cartesian basis for BF_3. $\mathbf{e}_2$ and $\mathbf{e}_3$ define the BF_3 plane. For convenience, members of each of the sets $(\mathbf{e}_1, \mathbf{e}_4, \mathbf{e}_7, \mathbf{e}_{10})$, $(\mathbf{e}_2, \mathbf{e}_5, \mathbf{e}_8, \mathbf{e}_{11})$, and $(\mathbf{e}_3, \mathbf{e}_6, \mathbf{e}_9, \mathbf{e}_{12})$ have been chosen parallel.

where $\mathbf{S}_3^{X_3}$ is the supermatrix

$$\mathbf{S}_3^{X_3} = \begin{pmatrix} \mathbf{R} & \mathbf{0} & \mathbf{0} & \mathbf{0} \\ \mathbf{0} & \mathbf{0} & \mathbf{0} & \mathbf{R} \\ \mathbf{0} & \mathbf{R} & \mathbf{0} & \mathbf{0} \\ \mathbf{0} & \mathbf{0} & \mathbf{R} & \mathbf{0} \end{pmatrix} \tag{8.4.3}$$

R is given by (8.2.8), and **0** is the 3 × 3 null matrix. The three basis vectors associated with each atom are linearly independent of all others and hence may be regarded as orthogonal to all basis vectors on the other atoms. The only new feature arises from the fact that a symmetry operation may carry an atom into one of its symmetrically equivalent partners; in that event, the new (primed) basis vectors must be expressed in terms of the original set into which the atom was carried. This produces an *off-diagonal* submatrix in the representative. The previous three-dimensional Cartesian representation (Section 8.2) arises in the special case of a single center (atom), which goes into itself under all operations.

8.5. Function Space

In analogy with the discussion in Section 8.2, we may imagine a "space" in which the basis is a set of *functions* rather than a set of vectors. Then in analogy with a vector, an arbitrary *function* could be expressed as a linear combination of the members of the basis set. Such sets of basis functions lie at the heart of the application of group theory to quantum mechanics. The expansion of an arbitrary function in the complete set of eigenfunctions of a Hermitian operator illustrates precisely this concept. The function space of this complete set, or the isomorphous vector space defined by the expansion coefficients, is known as a Hilbert space.

In analogy with basis vectors (the $\mathbf{e}_i$ of Section 8.2), the basis of our function space is a linearly independent set of *basis functions* ψ_i. We *define* the scalar product of two arbitrary *functions* in the space by

$$\langle\phi|\phi'\rangle \equiv \int \phi^* \phi' \, d\tau \tag{8.5.1}$$

where integration is over all coordinates in the space. Our basis set can always be chosen orthonormal, which means by definition

$$\langle\psi_i|\psi_j\rangle = \delta_{ij} \tag{8.5.2}$$

This is clearly the function-space analog of a set of orthonormal unit vectors. Thus any arbitrary function ϕ in the space can be written

$$\phi = \sum_i a_i \psi_i \tag{8.5.3}$$

and the set of basis functions ψ_i are said to *span the space*. Taking the scalar product and using (8.5.2), we see that the expansion coefficients are given by

$$a_i = \langle \psi_i | \phi \rangle \tag{8.5.4}$$

"Completeness" guarantees the validity of such expansions. In group-theoretic applications, we are always concerned with subspaces of the (complete) Hilbert space. Thus let us suppose that an arbitrary subset of the complete set of ψ_i is chosen. This subset spans a subspace which contains all possible functions ϕ in (8.5.3) *where the sum runs only over the subset.*

Again in analogy with the vector case, we wish to define linear operators which perform point-group symmetry operations on arbitrary *functions.* We call these *transformation operators.*

8.6. The Transformation Operators $\mathcal{R}$

Suppose $f(\mathbf{x}) \equiv f(x_1, x_2, \ldots, x_n)$ is an arbitrary function of the independent variables $x_1, x_2, \ldots, x_n$. Then we *define* $\mathcal{R}$ to be an operator such that $\mathcal{R} f(\mathbf{x})$ is a function "rotated" by R with respect to $f(\mathbf{x})$. Thus, for example, if $R = C_4^Z$ [a positive (counterclockwise) rotation by 90° about the Z axis], $\mathcal{C}_4^Z f(\mathbf{x})$ is the new function which results when $f(\mathbf{x})$ is "rotated" positively by 90° about the Z axis.

To attach a precise meaning to such a transformation, we first note that (8.2.14) relates the components of a rotated vector (or function) to those of the original vector (or function):

$$\mathbf{x}' = \mathbf{R}\mathbf{x} \tag{8.6.1}$$

Thus the function $f(\mathbf{Rx})$ has the same value at point $\mathbf{x}$ that $f(\mathbf{x})$ has at point $\mathbf{x}'$. However, although $\mathbf{R}$ represents the rotation of an arbitrary vector by R in accord with (8.2.3), $f(\mathbf{Rx})$ is a function rotated by R^{-1} with respect to $f(\mathbf{x})$. It follows that since $\mathcal{R} f(\mathbf{x})$ is rotated by R with respect to $f(\mathbf{x})$, the rotated function must be $f(\mathbf{R}^{-1}\mathbf{x})$. Thus

$$\mathcal{R} f(\mathbf{x}) = f(\mathbf{R}^{-1}\mathbf{x}) \equiv f(\mathbf{x}'') \tag{8.6.2}$$

where

$$\mathbf{x}'' = \mathbf{R}^{-1}\mathbf{x} = \tilde{\mathbf{R}}\mathbf{x} \tag{8.6.3}$$

To illustrate (8.6.2) let us consider the function

$$f(x, y) = y - x^2 \tag{8.6.4}$$

If we plot the dependent variable $f(x, y)$ perpendicular to the xy plane [i.e., let $z = f(x, y)$], then the intersection of $f(x, y)$ with any plane parallel to xy, $f(x, y) = z$ = constant = C, is a parabola given by

$$y - x^2 = C \tag{8.6.5}$$

For convenience, let us consider the contour in the xy plane ($C = 0$) which is shown in Fig. 8.6.1. Choosing $R = C_4^Z$, we obtain $\mathbf{R}$ according to the methods of Section 8.2:

$$\mathbf{R} = \begin{pmatrix} 0 & -1 \\ 1 & 0 \end{pmatrix} \tag{8.6.6}$$

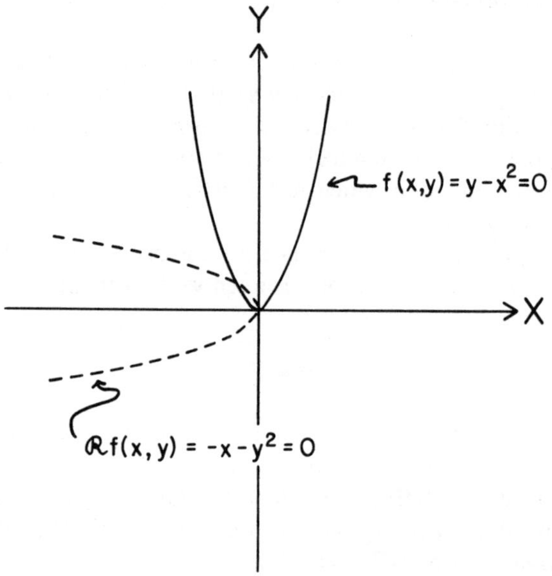

Figure 8.6.1. The solid line is a plot of $f(x, y) = y - x^2 = 0$, and the dashed line is a plot of $\mathfrak{R}f(x, y)$, where R is a 90° counterclockwise rotation in the XY plane.

Then according to (8.6.3) and (8.6.6)

$$\mathbf{x}'' = \begin{pmatrix} x'' \\ y'' \end{pmatrix} = \mathbf{R}^{-1}\mathbf{x} = \tilde{\mathbf{R}}\mathbf{x} = \begin{pmatrix} 0 & 1 \\ -1 & 0 \end{pmatrix}\begin{pmatrix} x \\ y \end{pmatrix} = \begin{pmatrix} y \\ -x \end{pmatrix} \tag{8.6.7}$$

Equation (8.6.2) then gives

$$\begin{aligned} \mathcal{C}_4^Z f(x, y) &= \mathcal{C}_4^Z(y - x^2) = f(x'', y'') = y'' - x''^2 \\ &= -x - y^2 \end{aligned} \tag{8.6.8}$$

We show $\mathcal{C}_4^Z f(x, y) = -x - y^2$ by the dashed curve in Fig. 8.6.1 and observe that contours of equal value of $f(x, y)$ are rotated 90° positively (counterclockwise) with respect to $f(x, y)$, as required.

Let us now examine the product of two transformation operators, say $\mathcal{P}$ and $\mathcal{S}$. If

$$\mathcal{S}f(\mathbf{x}) = f(\mathbf{S}^{-1}\mathbf{x}) \equiv g(\mathbf{x})$$

and

$$\mathcal{P}\mathcal{S}f(\mathbf{x}) \equiv \mathcal{P}[\mathcal{S}f(\mathbf{x})] = \mathcal{P}g(\mathbf{x}) = g(\mathbf{P}^{-1}\mathbf{x}) \tag{8.6.9}$$

Then by the definition of $g(\mathbf{x})$

$$\begin{aligned} g(\mathbf{P}^{-1}\mathbf{x}) &= f[\mathbf{S}^{-1}(\mathbf{P}^{-1}\mathbf{x})] = f[(\mathbf{S}^{-1}\mathbf{P}^{-1})\mathbf{x}] \\ &= f[(\mathbf{PS})^{-1}\mathbf{x}] \end{aligned} \tag{8.6.10}$$

Thus if $T = PS$, since $\mathcal{T}f(\mathbf{x}) = f(\mathbf{T}^{-1}\mathbf{x})$,

$$\mathcal{T} = \mathcal{P}\mathcal{S} \tag{8.6.11}$$

Comparing with the argument leading to (8.2.11), we see that the R, $\mathbf{R}$, and $\mathcal{R}$ multiply isomorphously. That is, if a group multiplication table reads $T = PS$, then

$$T = PS, \qquad \mathbf{T} = \mathbf{PS}, \qquad \mathcal{T} = \mathcal{P}\mathcal{S} \tag{8.6.12}$$

It is not difficult to prove that the $\mathcal{R}$ are linear [Bishop (1973), Section A.5-4].

8.7. Representations Generated by $\mathcal{R}$: The $\mathbf{D}(R)$

Let us consider a set of orthonormal functions, $\psi_1(\mathbf{x}), \psi_2(\mathbf{x}), \ldots, \psi_\nu(\mathbf{x})$. If a transformation operator $\mathcal{R}$ mixes members of this set only among themselves, the set is said to *form a basis for* $\mathcal{R}$. Then using (8.6.2), we have

$$\mathcal{R}\begin{pmatrix}\psi_1(\mathbf{x}) & \psi_2(\mathbf{x}) & \cdots & \psi_\nu(\mathbf{x})\end{pmatrix} \equiv \mathcal{R}\boldsymbol{\psi}$$

$$\equiv (\mathcal{R}\psi_1, \mathcal{R}\psi_2, \ldots, \mathcal{R}\psi_\nu)$$

$$= \begin{pmatrix}\psi_1(\mathbf{R}^{-1}\mathbf{x}) & \psi_2(\mathbf{R}^{-1}\mathbf{x}) & \cdots & \psi_\nu(\mathbf{R}^{-1}\mathbf{x})\end{pmatrix}$$

$$= \boldsymbol{\psi}\mathbf{D}(R)$$

or

$$\mathcal{R}\psi_\alpha = \sum_{\alpha'=1}^{\nu} \psi_{\alpha'} D(R)_{\alpha'\alpha}, \qquad \alpha = 1, 2, \ldots, \nu \tag{8.7.1}$$

which is analogous to $R\mathbf{e} = \mathbf{eR}$ of (8.2.3).

We may prove immediately that $\mathbf{D}(R)$ is unitary by substituting (8.7.1) into the left-hand side of (8.7.2):

$$\langle \mathcal{R}\psi_i | \mathcal{R}\psi_j \rangle = \langle \psi_i | \psi_j \rangle = \delta_{ij} \tag{8.7.2}$$

The first equality in (8.7.2) is most easily understood from the "rotate-the-coordinates" point of view. That is, $\mathcal{R}\psi_i$ and $\mathcal{R}\psi_j$ are simply the functions ψ_i and ψ_j expressed in a new coordinate system rotated by $\mathbf{R}^{-1}$. But the value of a definite integral over all space must be invariant to the choice of orthogonal coordinate system used in its evaluation.

Let us now suppose that our set, $\psi_1, \psi_2, \ldots, \psi_\nu$, form the basis for a *group* of $\mathcal{R}$. Then precisely in analogy with the arguments leading to (8.2.11), we see that the $\mathbf{D}(R)$ are members of a representation; that is, if

$$\mathcal{T} = \mathcal{P}\mathcal{S} \tag{8.7.3}$$

then

$$\mathbf{D}(T) = \mathbf{D}(P)\mathbf{D}(S) \tag{8.7.4}$$

The representation is said to be unitary, since the $\mathbf{D}(R)$ are unitary matrices; and the function ψ_α is said to belong to *row* α of the representation **D**, since, by (8.7.1), it is the matrix element in *row* α of **D** that determines the contribution (coefficient) of ψ_α in each of the $\mathcal{R}\psi_\lambda$ ($\lambda = 1, 2, \ldots, \nu$). (One could also argue that ψ_α belongs to *column* α, since it is that column which determines the contribution of each function to a given $\mathcal{R}\psi_\alpha$. We follow Wigner (1959) [and Tinkham (1964)] and use the term "row.")

Examples of the Use of the $\mathcal{R}$

Let us now do a standard example using an $\mathcal{R}$ to generate one of the $\mathbf{D}(R)$. Suppose that our basis functions are the set of p orbitals, p_x, p_y, p_z. The functional form of these orbitals may be expressed as

$$\begin{aligned} p_x &= xF(r) \\ p_y &= yF(r) \\ p_z &= zF(r) \end{aligned} \tag{8.7.5}$$

where $r = \sqrt{x^2 + y^2 + z^2}$, and $F(r)$ is a radial function containing an appropriate normalization constant. Suppose we wish to perform the rotary reflection S_4^X on this basis. The first step is to calculate the Cartesian representative in the relevant space, here ordinary three-space. Using (8.2.3) and (8.2.7),

$$S_4^X(\mathbf{e}_1 \quad \mathbf{e}_2 \quad \mathbf{e}_3) = \mathbf{eR} = (\mathbf{e}_1 \quad \mathbf{e}_2 \quad \mathbf{e}_3)\begin{pmatrix} -1 & 0 & 0 \\ 0 & 0 & -1 \\ 0 & 1 & 0 \end{pmatrix} \tag{8.7.6}$$

Then by (8.6.2) and (8.6.3)

$$\mathcal{S}_4^X\psi(x, y, z) \equiv \mathcal{S}_4^X\psi(\mathbf{x}) = \psi(\mathbf{x}'') = \psi(\mathbf{R}^{-1}\mathbf{x}) \tag{8.7.7}$$

where

$$\mathbf{x}'' = \begin{pmatrix} x'' \\ y'' \\ z'' \end{pmatrix} = \begin{pmatrix} -1 & 0 & 0 \\ 0 & 0 & 1 \\ 0 & -1 & 0 \end{pmatrix}\begin{pmatrix} x \\ y \\ z \end{pmatrix} = \begin{pmatrix} -x \\ z \\ -y \end{pmatrix} \tag{8.7.8}$$

Thus, for example,

$$\mathcal{S}_4^X p_y(x, y, z) = p_y(x'', y'', z'') = y''F(r'') \tag{8.7.9}$$

and from (8.7.8), $y'' = z$ (and $r'' = r$, as always). Therefore

$$\mathcal{S}_4^X p_y = zF(r) = p_z \tag{8.7.10}$$

as expected. Proceeding in the same way for p_x and p_z, one can, in analogy with $R\mathbf{e} = \mathbf{eR}$ of (8.2.3), summarize the results in the matrix equation

$$\mathcal{S}_4^X(p_x \quad p_y \quad p_z) = (p_x \quad p_y \quad p_z)\begin{pmatrix} -1 & 0 & 0 \\ 0 & 0 & -1 \\ 0 & 1 & 0 \end{pmatrix}$$

$$= (p_x \quad p_y \quad p_z)\mathbf{D}(S_4^X) \tag{8.7.11}$$

Note carefully that we use the active convention (Section 8.2).

The fact that $\mathbf{D}(R)$ in (8.7.11) happens to be $\mathbf{R}$ of (8.7.6) is an accident of this particular example. It is true here for all $\mathcal{R}$ because p_x, p_y, p_z, transform like x, y, z. It does *not* have any general significance, however. For example, let us apply $\mathcal{C}_4^Z$ to the basis set of three d orbitals (Section B.2) given by

$$d_{yz} = yzf(r)$$

$$d_{zx} = zxf(r) \tag{8.7.12}$$

$$d_{xy} = xyf(r)$$

where again $f(r)$ is an appropriate radial function. The analogs of (8.7.6) and (8.7.8) for C_4^Z give $x'' = y$, $y'' = -x$, $z'' = z$, with the result

$$\mathcal{C}_4^Z(d_{yz} \quad d_{zx} \quad d_{xy}) = (d_{yz} \quad d_{zx} \quad d_{xy})\begin{pmatrix} 0 & 1 & 0 \\ -1 & 0 & 0 \\ 0 & 0 & -1 \end{pmatrix} \tag{8.7.13}$$

whereas

$$\mathcal{C}_4^Z(p_x \quad p_y \quad p_z) = (p_x \quad p_y \quad p_z)\begin{pmatrix} 0 & -1 & 0 \\ 1 & 0 & 0 \\ 0 & 0 & 1 \end{pmatrix} \tag{8.7.14}$$

The $\mathbf{D}(R)$ of (8.7.14) is identical to $\mathbf{R} = \mathbf{C}_4^Z$ while that in (8.7.13) is clearly not.

The examples summarized above are in principle more complicated to depict than (8.6.8), since the p and d functions involve three independent variables, and contours of equal value are complicated surfaces in three-dimensional space. Nevertheless, sketches of these functions are very familiar, and one may obtain the $\mathbf{D}(R)$ by inspection simply by rotating the contours into each other, taking appropriate account of the signs of the various lobes. Such a procedure is illuminating and confirms the analytical procedure defined by (8.6.2), (8.6.3), and (8.7.1). However, this approach is useful only in certain special cases.

Consider the example where we apply the operation $\mathcal{C}_4^X$ to the pair of d functions

$$d_{z^2} = \frac{1}{2\sqrt{3}}(3z^2 - r^2)f(r)$$

$$d_{x^2-y^2} = \tfrac{1}{2}(x^2 - y^2)f(r) \tag{8.7.15}$$

Though this appears to be (and in fact is) a simple problem, inspection of sketches of these functions suggests that intuition is not very helpful. Assuming for the moment that $(d_{z^2}\ d_{x^2-y^2})$ indeed provides a basis for the operator $\mathcal{C}_4^X$—that is, that we can write

$$\mathcal{C}_4^X(d_{z^2}\ d_{x^2-y^2}) = (d_{z^2}\ d_{x^2-y^2})\begin{pmatrix} a_1 & a_3 \\ a_2 & a_4 \end{pmatrix} \tag{8.7.16}$$

—we seek to evaluate a_1, a_2, a_3, a_4. If we succeed, our initial assumption is justified. Working out $(\mathbf{C}_4^X)^{-1} = \tilde{\mathbf{C}}_4^X$, we have $x'' = x$, $y'' = z$, $z'' = -y$. Omitting $f(r)$ in (8.7.15) because it is invariant under the transformation, we write

$$\begin{aligned} \mathcal{C}_4^X d_{z^2} &= \frac{1}{2\sqrt{3}}(3z''^2 - r''^2) = \frac{1}{2\sqrt{3}}(3y^2 - r^2) \\ &= a_1 d_{z^2} + a_2 d_{x^2-y^2} \\ &= \frac{a_1}{2\sqrt{3}}(3z^2 - r^2) + \frac{a_2}{2}(x^2 - y^2) \end{aligned} \tag{8.7.17}$$

where we have substituted from (8.7.15) and (8.7.16). Since the quadratic quantities x^2, y^2, and z^2 are linearly independent, the coefficient of *each* of these quantities *must* be equated in (8.7.17), with the result

$$-\frac{1}{2} = -\frac{a_1}{2} + \frac{a_2\sqrt{3}}{2}$$

$$1 = -\frac{a_1}{2} - \frac{a_2\sqrt{3}}{2} \qquad (8.7.18)$$

$$-\frac{1}{2} = a_1$$

Solving this *overdetermined* set of equations yields the *unique* result

$$a_1 = -\frac{1}{2}, \qquad a_2 = -\frac{\sqrt{3}}{2} \qquad (8.7.19)$$

Proceeding in exactly the same way for $d_{x^2-y^2}$, we obtain

$$\mathcal{C}_4^X d_{x^2-y^2} = \tfrac{1}{2}(x''^2 - y''^2) = \tfrac{1}{2}(x^2 - z^2)$$

$$= \frac{a_3}{2\sqrt{3}}(3z^2 - r^2) + \frac{a_4}{2}(x^2 - y^2) \qquad (8.7.20)$$

with the *unique* result,

$$a_3 = -\frac{\sqrt{3}}{2}, \qquad a_4 = \frac{1}{2} \qquad (8.7.21)$$

The procedure just illustrated is perfectly general and should be used in all but the simplest cases. If one is unable to obtain a unique set of coefficients by this method, it means that the set of functions chosen is not a basis for the symmetry operation in question (see Section 8.9).

In anticipation of later discussion (Section 8.11) we observe that the transformation properties of spherical harmonics under arbitrary point-group operations can be deduced from general formulas [Tinkham (1964), Sections 5-6, 5-7], which can therefore always be used to form the point-group representatives of atomic orbitals. However, it is often much quicker to use the method just discussed.

8.8. Choice of Basis Functions

Since our basis set has been chosen orthonormal but is otherwise unspecified, we may consider some alternative orthonormal set in the same space. If we designate the new basis by ψ', then the two bases must be related by a unitary transformation to preserve orthonormality. Designating this unitary matrix by **U**,

$$\psi' = \psi \mathbf{U} \tag{8.8.1}$$

we obtain for all R

$$\mathbf{D}'(R) = \mathbf{U}^{-1}\mathbf{D}(R)\mathbf{U} \tag{8.8.2}$$

where $\mathbf{D}'(R)$ is the representative in the new basis equivalent to $\mathbf{D}(R)$ in the original one. The representations $\mathbf{D}'$ and $\mathbf{D}$ are said to be *equivalent*. Equation (8.8.2) is in the same form as (8.3.6), but **U** must in general be unitary (not simply orthogonal as for **Q** in Cartesian space), since the basis functions may be complex.

8.9. Group Theory and the Schrödinger Equation

The application of group theory to quantum mechanics depends upon the fact that the complete Hamiltonian of a system is unchanged by (i.e., is invariant under) various operations. These include translations, rotations, permutations, inversion of the coordinates of particles through the center of mass, and time reversal [Bunker (1979), Chapter 6]. It is important, however, to realize that the invariances which are actually useful depend crucially on the application at hand. For example, translational invariance plays a vital role in solid-state (crystallographic) applications, rotational invariance is critical in studying molecular rotations, and permutation and inversion symmetry can be used to classify the rovibronic states of nonrigid molecules.

However, the only type of system explicitly considered in this book is a fixed, isolated, nonrotating molecule (Section 2.5) in which the nuclear motion in any electronic state is confined to small vibrations about deep potential minima. Thus we are not concerned with translational or rotational invariances, nor with the permutation and inversion invariances which are useful in the discussion of nonrigid molecules. In our case, the only relevant symmetry operations are the point-group operations which take a static object into itself (Section 8.2). Time reversal symmetry is explicitly discussed later (Chapter 14).

We refer the interested reader to the treatise by Bunker (1979), which presents a detailed and illuminating discussion of all the molecular invariances.

The Molecular Point Group

Following Bunker (1979), we refer to the group of $\mathcal{R}$ which is isomorphic to a given point group [Equation (8.6.12)] as the *molecular point group*. [Tinkham (1964) and Wigner (1959) use the term "group of the Schrödinger equation."] The $\mathcal{R}$ perform the point-group operations on the vibronic variables, that is, on electronic *and* nuclear coordinates.

The fundamental theorem which permits our applications of group theory to quantum mechanics states that the vibronic Hamiltonian operator for any molecule commutes with all the $\mathcal{R}$ of the molecular point group. That is, for all R in the group

$$\mathcal{R}\mathcal{H}_T = \mathcal{H}_T\mathcal{R} \tag{8.9.1}$$

where $\mathcal{H}_T$ is defined in Section 3.2. This result is proved in Bishop (1973), Appendix A.8-1. Physically (8.9.1) simply states that the molecular vibronic energy cannot change if a point-group operation is performed. (Note that (8.9.1) does not apply exactly if rotational degrees of freedom are included in $\mathcal{H}_T$; in that case the molecular point group is referred to as a near symmetry group [Bunker (1979), Chapter 11].)

As a first step, we make the crude adiabatic approximation [Eq. (3.4.2)],

$$\begin{aligned}\Psi_{ki}(q, Q) &\approx \phi_k(q, Q_0)\chi_{ki}(Q)\\ &= \phi_k^0(q)\chi_{ki}(Q)\end{aligned} \tag{8.9.2}$$

That is, we assume that the ith vibronic wave function of the kth electronic state can be represented as a product of a vibrational and an electronic wave function, *the latter applying for nuclei held fixed at the equilibrium geometry of the electronic state*, $Q = Q_0$. We emphasize that our subsequent applications of group theory are *not* restricted to this assumption. However, if (8.9.2) is to be a useful *starting* point, it is necessary to assume that deviations from it due to electronic–vibrational (electron–phonon or "vibronic") coupling are sufficiently small that perturbation theory can be used to describe them, as for example in HT theory (Section 3.3) or in the JT case (Section 3.8). It is then useful to treat the electronic and vibrational problems independently as a first approximation (as we now do). Furthermore, the group-theoretic machinery thus developed can be carried over to treat vibronic effects, since these can be expressed in terms of the ϕ_k^0's and $\chi_{ki}(Q)$'s (Sections 3.7 and 3.8). Physically, we are making the familiar assumption that the nuclear motion in any electronic state is confined to small vibrations about equilibrium.

With the crude adiabatic approximation the electronic Schrödinger equation (3.2.6) becomes

$$\mathcal{H}_{el}(q, Q_0)\phi_k^0(q) \equiv \mathcal{H}_{el}^0\phi_k^0(q) = W_k^0\phi_k^0(q) \tag{8.9.3}$$

and the corresponding nuclear equation is given by (3.2.10). Note that $\mathcal{H}_{el}^0$ is a function only of electronic coordinates, since the nuclei are clamped at the equilibrium configuration ($Q = Q_0$), and the nuclear Hamiltonian $T_n(Q) + W_k(Q)$ is a function only of nuclear coordinates. When vibronic effects are included, so that $\mathcal{H}_{el} = \mathcal{H}_{el}(q, Q)$ [e.g., (3.3.1) and Section 3.8], the $\mathfrak{R}$ operate simultaneously on *both* electronic and vibrational coordinates. $\mathcal{H}_{el}(q, Q)$ is then invariant under (commutes with) the $\mathfrak{R}$ because the coordinates of *all* particles in the system are rotated.

Representations of the $\mathfrak{R}$

Let us now allow $\mathfrak{R}$ to operate on (8.9.3). [Since (3.2.10) is in exactly the same form, the following analysis applies equally well in that case.] Modifying our notation slightly ($\mathcal{H}_{el}^0 \equiv \mathcal{H}^0$, $\phi_k^0 \equiv \phi^k$), we have

$$\mathfrak{R}\left(\mathcal{H}_{el}^0\phi_k^0\right) \equiv \mathfrak{R}\left(\mathcal{H}^0\phi^k\right) = \mathfrak{R}\left(W_k^0\phi^k\right) \tag{8.9.4}$$

Then using (8.9.1), we obtain

$$\mathcal{H}^0\left(\mathfrak{R}\phi^k\right) = W_k^0\left(\mathfrak{R}\phi^k\right) \tag{8.9.5}$$

Equation (8.9.5) *and its nuclear analog are the basis for all our applications of group theory*. This equation states that if ϕ^k is an eigenfunction of $\mathcal{H}^0$, so also is $\mathfrak{R}\phi^k$ and with the same eigenvalue.

Let us suppose that W_k^0 is associated with an l_k-fold degenerate state whose eigenfunctions, chosen orthonormal, are designated ϕ_κ^k, $\kappa = 1, 2, \ldots, l_k$. Then as a necessary consequence of the linearity of the Schrödinger equation

$$\mathfrak{R}\begin{pmatrix}\phi_1^k & \phi_2^k & \cdots & \phi_{l_k}^k\end{pmatrix} \equiv \mathfrak{R}\boldsymbol{\phi}^k = \boldsymbol{\phi}^k\mathbf{D}^k(R) \tag{8.9.6}$$

Equation (8.9.6) is an example of (8.7.1) where $\nu = l_k$. $\mathbf{D}^k(R)$ constitutes a representative, since (8.9.6) clearly applies for all R in the group. The set of functions $\boldsymbol{\phi}^k$ are said to span the subspace of the eigenvalue W_k^0, since any function with this eigenvalue (and no other) can be expanded in the set.

The function ϕ_κ^k is said to belong to the *κth row* of the representation $\mathbf{D}^k$ [see discussion following (8.7.4)]. The set of functions ϕ_λ^k ($\lambda = 1, 2, \ldots, l_k$)

are often referred to as *partners*. Collectively, the basis set $\boldsymbol{\phi}^k$ is said to belong to the representation $\mathbf{D}^k$. In subsequent applications, *the same nomenclature is used for any set of functions which has the same transformation properties as the* $\boldsymbol{\phi}^k$, *that is, obeys* (8.9.6). In practice we seldom know the true eigenfunctions of $\mathcal{H}^0$ (the ϕ^k). Instead we work with approximate eigenfunctions which have the same transformation properties. Such functions of course do *not* belong to the subspace of eigenvalue W_k^0. Equations (8.7.11), (8.7.13), (8.7.14), and (8.7.16) show examples of such functions for the group O.

8.10. Irreducible Representations (Irreps)

If we choose a new orthonormal basis $\boldsymbol{\phi}'^k$, then by (8.8.1) and (8.8.2),

$$\boldsymbol{\phi}'^k = \boldsymbol{\phi}^k\mathbf{U} \tag{8.10.1}$$

and

$$\mathbf{D}'^k(R) = \mathbf{U}^{-1}\mathbf{D}^k(R)\mathbf{U} \tag{8.10.2}$$

A familiar simple example of (8.10.1) is the relation between the real and complex p orbitals, $\boldsymbol{\phi}^k = (p_x\ p_y\ p_z)$ and $\boldsymbol{\phi}'^k = (p_1\ p_0\ p_{-1})$, where Section B.2 gives

$$\mathbf{U} = \begin{pmatrix} -1/\sqrt{2} & 0 & 1/\sqrt{2} \\ -i/\sqrt{2} & 0 & -i/\sqrt{2} \\ 0 & 1 & 0 \end{pmatrix} \tag{8.10.3}$$

and by application of (8.10.2), Eq. (8.7.11) would assume the form

$$\mathcal{S}_4^X(p_1\quad p_0\quad p_{-1}) = (p_1\quad p_0\quad p_{-1})\begin{pmatrix} -\frac{1}{2} & -i/\sqrt{2} & \frac{1}{2} \\ -i/\sqrt{2} & 0 & -i/\sqrt{2} \\ \frac{1}{2} & -i/\sqrt{2} & -\frac{1}{2} \end{pmatrix} \tag{8.10.4}$$

If (8.10.2) applies for all R in the group, the representations $\mathbf{D}'^k$ and $\mathbf{D}^k$ are said to be *equivalent* (Sections 8.3, 8.8). If no similarity transformation (8.10.2) exists which *simultaneously* factors $\mathbf{D}^k(R)$ *for all* R into corresponding smaller blocks along the diagonal—a structure referred to as *reduced form* (Section 8.11)—the representation $\mathbf{D}^k$ is said to be *irreducible*. If such a factoring into reduced form is possible, the l_k-fold degeneracy assumed in (8.9.6) is said to be "accidental," since it apparently has no basis in symmetry. [A degeneracy is presumed to be "true" if it cannot be split by any possible symmetry-preserving perturbation. If this degeneracy is inconsistent with (is higher than) the assumed symmetry of the system, conventional wisdom holds that in fact some unrecognized higher symmetry is present, i.e., one is working in a subgroup of the true symmetry group. Putting it another way, there is no such thing as an "exact" accidental degeneracy. In our applications, the existence of accidental degeneracy is in general obvious; we may therefore determine a **U** matrix to effect the appropriate factoring into smaller (true) irreps, and we assume in subsequent discussion that this has been done.]

Thus we obtain the extraordinarily important result that the *exact* eigenfunctions of $\mathcal{H}_{\text{el}}^0$ always belong to *irreducible representations* (irreps) of the molecular point group. In particular, *the exact eigenfunctions of an n-fold degenerate state belong to an n-dimensional irrep* (and thus mix only among themselves under all $\mathcal{R}$). *The corresponding* $\mathbf{D}(R)$ *are said to describe the transformation properties of this set of functions.*

If we have several different sets of eigenfunctions, each one spanning a space defined by a *different* energy eigenvalue, the entire collection is a basis for $\mathcal{R}$. The corresponding representation is automatically in reduced form, in view of (8.9.6), since each set mixes only among its own members. The reduced form can in general be destroyed by an arbitrary similarity transformation as in (8.8.2), since new basis functions may be formed which can be linear combinations of members from all the sets. The reducibility of this new representation can be ascertained immediately by *character analysis* (Section 8.11), and the inverse similarity transformation will return the representation to reduced form. In practice, we are almost always concerned with the latter procedure; presented with a reducible representation in unreduced form, we must somehow find a similarity transformation which brings it to reduced form. The means by which this "unscrambling" may be accomplished is discussed in Chapter 9.

The Great Orthogonality Theorem

The remarkable properties of irreducible representations (irreps) are summarized in the Great Orthogonality Theorem, which we state here without proof in slightly restricted form. [Lucid proofs may be found in Wigner (1959), Tinkham (1964), and Bishop (1973).] For all the inequivalent, unitary irreps of a group

$$\sum_R D^i(R)^*_{\mu\nu} D^j(R)_{\alpha\beta} = \frac{h}{l_i}\delta_{ij}\delta_{\mu\alpha}\delta_{\nu\beta} \tag{8.10.5}$$

where $D^i(R)^*_{\mu\nu}$ is the complex conjugate of the $\mu\nu$th matrix element of the ith irreducible representative corresponding to the symmetry operation R, l_i is the dimension of the representation, and h is the order (number of elements) of the group. If the $D^i(R)_{\mu\nu}$ are regarded as the components of h-dimensional vectors, then (8.10.5) is a statement of the mutual *orthogonality* of all such vectors formed from inequivalent irreps. Since only h mutually orthogonal vectors can be constructed in an h-dimensional space, such an interpretation suggests that the total number of vectors is

$$\sum_i l_i^2 = h \tag{8.10.6}$$

where the sum is over all inequivalent irreps. Equation (8.10.6) may be readily proved [Tinkham (1964), Section 3-5; Bishop (1973), Section A.7-2].

The *character* of a representative (reducible or irreducible) is defined by

$$\chi^i(R) \equiv \sum_{\mu=1}^{l_i} D^i(R)_{\mu\mu} \tag{8.10.7}$$

That is, the character is the diagonal sum (or trace) of a representative. Letting $\mu = \nu$ and $\alpha = \beta$ in (8.10.5), and then summing this equation over both ν and β, one obtains

$$\sum_R \chi^i(R)^* \chi^j(R) = h\,\delta_{ij} \tag{8.10.8}$$

Equation (8.10.8) is often referred to as the *character orthogonality relation*. It states that if the $\chi^i(R)$ are regarded as vector components then the characters of inequivalent irreps of any group are orthogonal.

8.11. Character Analysis

Equation (8.10.8) is extraordinarily useful because it provides a very simple means for determining the structure of reducible representations, that is, the number of times each irrep occurs. Specifically, it tells one how a reducible representation blocks into reduced form without the need to actually carry out the reduction by determining the matrix **U**. Such information is remarkably far reaching in its implications and indeed is almost "the whole story" in simpler group-theoretic applications. For example, on the basis of character analysis and a knowledge of certain qualitative theorems about the symmetry properties of matrix elements (Chapter 9), one can, in a few minutes, construct qualitative MO energy-level diagrams once an atomic-orbital basis set is specified [Cotton (1971)]. Similarly, the symmetry classification of fundamental molecular vibrations (normal coordinates) and the deduction of the simpler spectroscopic selection rules for molecules is a trivial matter if a character table is available.

After introducing the concept of classes, we illustrate character analysis with several examples.

Classes

If two elements B and A in a group are connected by a third element X of the group according to

$$B = X^{-1}AX \tag{8.11.1}$$

then B and A are said to be conjugate. It is easy to show that two elements conjugate to a third are conjugate to each other. The collection of all elements from a group which are conjugates defines a *class*. It then follows that each element of a group belongs to one and only one class. If it is desired to find all the elements which belong to the same class as A, one uses (8.11.1) and allows X to run over all elements of the group.

When the group elements are symmetry operations, the existence of classes has a simple geometric interpretation; namely, two elements are in the same class if and only if one can be transformed into the other by applying a symmetry operation of the group to the coordinate system. For example, two C_3 operations about different axes are in the same class if some group symmetry operation will take the one C_3 axis into the other. (It should be emphasized that whereas two elements of different sorts, e.g., a rotation and a reflection, cannot possibly be in the same class, two elements of the same sort, e.g., two rotations by the same amount, are not necessarily in the same class.)

It is very easy to show that two (necessarily square) matrices related by a similarity transformation [e.g., (8.10.2)] have the same trace (diagonal sum). It thus follows from (8.11.1) that all representatives in the same class have the same character. Thus (8.10.8) can be rewritten

$$\sum_k N_k \chi^i(\mathcal{C}_k)^* \chi^j(\mathcal{C}_k) = h\,\delta_{ij} \tag{8.11.2}$$

where *the sum is now over all the classes*, and N_k is the number of elements in class $\mathcal{C}_k$. Equation (8.11.2) is in the form of an orthogonality relation in k space, *suggesting* the relation

$$\text{(number of inequivalent irreps)} = \text{(number of classes)} \qquad (8.11.3)$$

since the maximum number of mutually orthogonal vectors which can be constructed in a space is equal to its dimensionality. Equation (8.11.3) may be readily proved [Bishop, (1973), Section A.7-3].

Obtaining the Character-Analysis Equation

Let us suppose we have a reducible representation $\mathbf{D}^{\text{red}}$ of some group. (A representation whose dimensions are larger than any of the irreps of the group is always an example.) Since it is reducible, some similarity transformation exists which simultaneously brings all the representatives into the same block-diagonal form, so that for all R

$$\mathbf{U}^{-1}\mathbf{D}^{\text{red}}(R)\mathbf{U} = \begin{pmatrix} \mathbf{D}^1(R) & & & \\ & \mathbf{D}^2(R) & & \\ & & \ddots & \\ & & & \mathbf{D}^n(R) \end{pmatrix} \qquad (8.11.4)$$

where $\mathbf{D}^1(R)$, $\mathbf{D}^2(R),\ldots$ are all square matrices on the diagonal (some of which may be identical or equivalent to others). There are zeros everywhere else, and $\mathbf{D}^1$ has the same dimensions for all R, and likewise for $\mathbf{D}^2$, and so on. Since by hypothesis the $\mathbf{D}^{\text{red}}(R)$ (all R) constitute a representation, so also do the $\mathbf{D}^1$ (by themselves), the $\mathbf{D}^2$, and so on. This is an immediate consequence of the rule of matrix multiplication. The right side of (8.11.4) is referred to as a reduced form, provided $n \geqslant 2$. Now if *none* of these smaller representations, $\mathbf{D}^1, \mathbf{D}^2, \ldots, \mathbf{D}^n$ can themselves be further reduced (be brought into reduced form), then the representation $\mathbf{D}^{\text{red}}$ is said to have been completely reduced, and the $\mathbf{D}^i$ on the right-hand side of (8.11.4) are all irreps.

The goal of character analysis is to determine the number of times (n_j) each irrep occurs in the reduced form of $\mathbf{D}^{\text{red}}$, that is, the number of irreps of each symmetry which occur on the right-hand side of (8.11.4). To accomplish this goal we need only know the characters $\chi^{\text{red}}(R)$ of $\mathbf{D}^{\text{red}}$. It is not necessary to actually determine $\mathbf{U}$, since $\mathbf{D}^{\text{red}}(R)$ and its equivalent representative (8.11.4) have the same character, because they differ only by a similarity transformation. If irrep 1 occurs n_1 times, irrep 2 occurs n_2 times, and so on, then from the right-hand side of (8.11.4)

$$\chi^{\text{red}}(R) = n_1\chi^1(R) + n_2\chi^2(R) + \cdots$$
$$= \sum_i n_i \chi^i(R) \tag{8.11.5}$$

To determine the n_i we multiply both sides of (8.11.5) by $\chi^j(R)^*$ and sum over all operations of the group using (8.10.8) [or over classes using (8.11.2)]. The result is

$$n_j = \frac{1}{h}\sum_R \chi^j(R)^*\chi^{\text{red}}(R)$$
$$= \frac{1}{h}\sum_k N_k \chi^j(\mathcal{C}_k)^*\chi^{\text{red}}(\mathcal{C}_k) \tag{8.11.6}$$

Equation (8.11.6) is the basic equation for character analysis. The results of such an analysis are summarized in the form

$$\Gamma^{\text{red}} = n_1\Gamma_1 + n_2\Gamma_2 + \cdots$$
$$= \sum_i n_i \Gamma_i \tag{8.11.7}$$

which simply states that if the reduction described by (8.11.4) were performed, then on the right-hand side, irrep 1 would occur n_1 times, irrep 2 would occur n_2 times, and so on. It is to be emphasized that the + signs in (8.11.7) have a symbolic, *not* an operational significance; that is, they do not designate matrix addition. This is sometimes emphasized by using the symbol ⊕ rather than +. We do not follow this practice.

Significance of Character Analysis

The significance of character analysis becomes clearer when we associate basis functions with the representations as in (8.7.1). If a complete reduction is effected as in (8.11.4), then the new functions ψ' defined by (8.8.1) *belong to* irreps. That is, members of the first set, in number equal to l_1 [the dimension of $\mathbf{D}^1(R)$ of (8.11.4)], mix only among themselves under *all* $\mathcal{R}$ of the group, and so forth for each $\mathbf{D}^i(R)$. Thus in accord with the discussion in Section 8.10, the functions belonging, for example, to $\mathbf{D}^1$ have the same transformation properties (within a similarity transformation) as an l_1-fold degenerate set of exact eigenfunctions of $\mathcal{H}_{\text{el}}^0$ which belong to the same irrep.

We find in Chapter 9 that if basis functions are chosen to belong to irreps of the appropriate symmetry group, the evaluation of matrix elements is very much simplified through the use of the Wigner–Eckart theorem (Section 9.3). These simplifications are a consequence *only* of the transformation (symmetry) properties of the functions and thus are *entirely independent* of how well the functions approximate the true eigenfunctions. Putting it another way, the symmetry-determined aspects of problems in quantum mechanics can be completely evaluated using relatively simple algebraic methods which require no attempt whatever to solve the Schrödinger equation explicitly. It is this fact which makes group theory useful in quantum mechanics. It is also often the reason why absurdly simple quantum-mechanical models make qualitative or even quantitative sense.

In applying group theory to quantum mechanics and spectroscopy, the general approach is usually as follows. First, some finite (often small) basis set is chosen to approximate the true eigenfunctions, which are far too complicated to determine exactly. A character analysis is then performed to find out how many times each irrep is contained in the basis. This alone suffices to handle a variety of simple problems. To proceed further, one then actually determines the functions belonging to the irreps, that is, one determines the matrix **U** in (8.11.4). These "symmetry-adapted" functions are then the starting point (basis functions) in terms of which the problem of interest is formulated. Much of the remainder of this book concerns the uses of such functions and the far-reaching consequences of their symmetry properties. In any event, a character analysis is almost always the first step, and we consider several examples after a few remarks about character tables.

Character Tables

Performing a character analysis means determining the n_j in (8.11.6), and expressing the final result in the form (8.11.7). To perform such an analysis, one needs the characters of the irreps (the $\chi^j(\mathcal{C}_k)$) and the characters of the representation being reduced. The former are assembled once and for all for each group in a character table. (Some important tables are given in Sections C.1, D.1, and E.1.) The methods for constructing such tables have been discussed in many places [e.g., Tinkham (1964), Section 3-4; Bishop (1973), Section 7-7]. Character tables assume the form of a square matrix, in view of (8.11.3), since all the characters in a given class are the same. Each column is labeled by a class designation preceded by a coefficient which specifies the number of elements in that class. Each row is labeled by an irrep symbol which is simply an arbitrary name. The naming systems most commonly used are due respectively to Mulliken and Bethe. The alphabetic system of Mulliken [Bishop (1973), Section 7-8] is more informative than the Γ_n system of Bethe. A completely new numerical system which should gain popularity has been introduced by Butler (1981).

Some caution must be exercised in using character tables to insure that one is following the same choice of axes and symmetry designations that were used in constructing it. Conventions regarding these are often noted with the listing of the tables and are fairly standard. Some possible ambiguities are fairly trivial. For example, interchanging the definitions of C_2' and C_2'' in the group D_4 simply interchanges the definitions of irreps B_1 and B_2. On the other hand, interchanging the definitions of σ_d and σ_v in the group D_{6h} without a corresponding interchange of the definitions of C_2' and C_2'' yields nonsensical results in character analyses (i.e., noninteger n_j).

Most character tables list the transformation properties of some simple linear and quadratic functions of x, y, and z as well as those of the infinitesimal-rotation operators R_x, R_y, and R_z (which have the same transformation properties as the angular-momentum operators). That is, these functions are assigned to specific irreps. These assignments imply certain choices of the Cartesian axes, and one must be sure to make the same choices if such assignments are to be used without further examination. For example, in the group D_4 (Section D.1), assigning the function $x^2 - y^2$ to irrep B_1 and xy to B_2, rather than vice versa, shows how the x and y axes have been chosen with respect to the C_2' and C_2'' symmetry axes.

Reduction of the Generalized Cartesian Representation; Symmetry Coordinates and Normal Coordinates

With a character table in hand, reduction of the generalized Cartesian representation (GCR), in Section 8.4, is remarkably simple. All one need do is determine $\chi^{\text{red}}(R)$ for one R (the most convenient one) in each class and use (8.11.6). Let us do the BF_3 molecule (Fig. 8.4.1) as an example. The first point to note is that if an atom moves under a particular symmetry operation, its contribution to the character is zero. This follows because its three basis vectors move to a different symmetrically equivalent point in space; hence this part of the transformation is expressed in terms of a different set of basis vectors, producing a submatrix that is off diagonal in the representative [see (8.4.3)]. Thus we need fix our attention *only on unshifted* atoms. The next point to note is that for such atoms, the choice of orientation of the coordinate system (the three unit vectors) is arbitrary, since the infinity of possible orientations (all with the same origin) differ from each other only by similarity transformations, under which of course the character is invariant. Thus the choice is made on the basis of convenience, and a different choice can even be made for each different class. Finally, since the coordinate systems on different atoms can all be chosen parallel, one need carry out the analysis on only one set of unit vectors.

Thus the procedure is to choose a set of unit vectors $(\mathbf{e}_1\ \mathbf{e}_2\ \mathbf{e}_3) = \mathbf{e}$ at a symmetry-group-invariant point (regardless of whether the molecule under consideration has an atom at that point). Choosing the orientation of $\mathbf{e}$ for convenience, one computes the *diagonal* elements of the Cartesian representative (Section 8.2), sums them, and multiplies by the number of unshifted atoms (which may be zero). While it is possible to write a simple general formula for this procedure [see Bishop (1973), Eq. (9-6.1)], it is quicker to compute $\chi^{\text{GCR}}(R)$ than it is to look up the formula.

Applying this procedure to BF_3 (Fig. 8.4.1), we obtain $\chi(E) = 3 \times 4 = 12$ (the result for the identity operation is always three times the number of atoms), $\chi(C_3) = (1 + 2\cos 120°) \times 1$ (choosing the Z axis to coincide with the C_3 axis, and noting that there is only one unshifted atom), $\chi(C_2) = (-1) \times 2$ (since, e.g., $x \to -x$, $y \to -y$, $z \to z$ if the Z axis is chosen to coincide with C_2, and there are two unshifted atoms). With the same kind of reasoning, $\chi(\sigma_h) = 1 \times 4$, $\chi(S_3) = (-1 + 2\cos 120°) \times 1$, $\chi(\sigma_v) = 1 \times 2$. Using these characters in the second line of (8.11.6) along with the D_{3h} character table (Table 8.13.1 below) and remembering to include N_k, one obtains

$$\Gamma^{\text{GCR}}(BF_3) = A_1' + A_2' + 2A_2'' + 3E' + E'' \tag{8.11.8}$$

The meaning of (8.11.8), and the whole point of the character analysis, is the following. It is possible to find twelve linear combinations of the twelve Cartesian coordinates (defined with respect to the four atomic origins shown in Fig. 8.4.1) such that one linear combination transforms like A_1', one like A_2', three pairs (each pair independently) like E', two combinations (each independently) like A_2'', and one pair like E''. The exact form of these linear combinations depends on the specific orientations chosen for the coordinate systems on the four atoms. Once this choice is made, finding the linear combinations is exactly equivalent to determining the matrix $\mathbf{U}$ in (8.11.4). Methods for doing this are discussed in Chapter 9.

The utility of a character analysis of the GCR becomes immediately apparent if one wishes to apply group-theoretic methods to the vibrational Schrödinger equation, (3.2.10). [This topic is discussed in detail in Wilson, Decius, and Cross (1955) which we abbreviate, WDC.] In the harmonic approximation, the kinetic and potential energy are diagonal in appropriate linear combinations of the Cartesian coordinates called *normal coordinates*. Using these coordinates, the vibrational eigenfunctions can be written as products of harmonic-oscillator functions, each of which is a function of only one normal coordinate (Section 3.5). Furthermore, it is easy to show (WDC, Section 7-1) that a vibrational eigenfunction corresponding to the excitation of one vibrational quantum in one harmonic-oscillator function (a so-called fundamental vibration) has the same transformation properties as the corresponding normal coordinate. Since, by the discussion in Sections 8.9–8.10, vibrational eigenfunctions must belong to irreps of the molecular point group, the same applies for the normal coordinates.

Therefore (8.11.8) is a description of the transformation properties of the eigenfunctions of the fundamental vibrations of BF_3 *and* of the corresponding normal coordinates. It may be shown that six (five for linear molecules) of these normal coordinates are associated with zero-frequency vibrations, three corresponding to translations and three (or two) to rotations. Inspection of their explicit form (WDC, Section 2-5) shows that the three translational normal coordinates have the same transformation properties as the functions x, y, and z, and the three rotational coordinates have the same transformation properties as the infinitesimal-rotation operators (and angular-momentum operators) designated R_x, R_y, R_z [Tinkham (1964), Section 5.1]. Thus the irreps to which the translations and rotations belong may be ascertained by character analysis, or more usually simply by inspection, since the transformation properties of x, y, z, and R_x, R_y, R_z are included in most character tables. In the case of BF_3, $\Gamma^{\text{trans}} = A_2'' + E'$ and $\Gamma^{\text{rot}} = A_2' + E''$. Thus the "true" vibrational decomposition of BF_3 is

$$\Gamma^{\text{vib}}(BF_3) = A_1' + A_2'' + 2E' \tag{8.11.9}$$

Finally it should be emphasized that the actual normal coordinates with these symmetries cannot in general be found by group-theoretic methods alone. In Chapter 9 we show how to determine the matrix **U** of (8.11.4) group-theoretically. The linear combinations of Cartesian coordinates thus obtained are called symmetry coordinates (WDC).[‡] If only one set of normal coordinates belongs to a given irrep [i.e., if the coefficient of the irrep label in (8.11.9) is unity], the symmetry coordinate is in fact the true normal coordinate (aside from a normalization constant). However, in any other case, the normal coordinate is a linear combination of symmetry coordinates belonging to the same irrep, and a detailed model (a vibrational

[‡]In fact, it is often convenient to use internal rather than Cartesian coordinates. The former are linearly related to the latter (for infinitesimal displacements), are easy to visualize, and automatically exclude the translational and rotational degrees of freedom, thus leading directly to (8.11.9). Furthermore, a systematic method (the G-matrix method) is available to express the kinetic energy using such coordinates. See WDC.

potential function) must be assumed and the vibrational secular determinant solved to determine these linear combinations. If symmetry is properly used (Chapter 9), the dimensions of the secular determinant need only be equal to the coefficient of the irrep label in (8.11.9). Thus (8.11.9) states that BF_3 has single normal coordinates (fundamental vibrations) of symmetries A_1' and A_2'', and two doubly degenerate vibrations of symmetry E'.[‡] Each of the E' normal coordinates (four altogether) is a linear combination of *two* symmetry coordinates, and only one 2×2 secular determinant need be solved to determine the four normal coordinates.

Character Analysis of Atomic Basis Sets

Since atomic orbitals are almost always the basis for the description of the electronic properties of molecular systems, a character analysis of such sets is the first step in the application of group-theoretic methods to the electronic Schrödinger equation. Let us first consider atomic orbitals centered at an invariant point of the symmetry group.

By definition, an atomic orbital is the product of a radial function and a spherical harmonic (Section 9.4). We are concerned exclusively with the symmetry properties of the latter, since the radial functions are invariant under all $\mathfrak{R}$. Spherical harmonics are discussed in virtually all quantum-mechanics books, and we assume that the reader has some familiarity with these functions, particularly their role as angular-momentum eigenfunctions. We symbolize them by $|lm\rangle$ (or by $|jm\rangle$ as in Section 9.4).

Atoms belong to the group of all rotations, SO_3. From the point of view of transformation properties, the crucially important property of the spherical harmonics is that the $2l + 1$ functions of any given l value ($m = -l, -l + 1, \ldots, +l$) mix only among themselves under all possible point-group operations. Thus the $|lm\rangle$ form a basis for the irreps of SO_3, which are labeled by $l = 0, 1, \ldots,$ and have dimension $2l + 1$ [Tinkham (1964), Section 4.4]. The $2l + 1$ possible values of the quantum number m serve as partner labels for the lth irrep in the conventional SO_3 basis (this is called the $SO_3 \supset SO_2$ basis in the language of Section 15.2; $SO_2 \equiv C_\infty$).

In terms of atomic orbitals, $l = 0, 1, 2, 3, \ldots$ correspond to the familiar $s, p, d, f, \ldots$ sets of functions, and we now ascertain how each of these sets decomposes onto any point-group of lower symmetry than SO_3. To do this by character analysis, we first need the general formula which gives the character of any set of $|lm\rangle$ (all m) under an arbitrary rotation. This is derived in the following way. Since the choice of axes is arbitrary in a character analysis, we choose the Z axis of the atomic orbitals to coincide

[‡]If (8.11.9) read "$3E'$", that would mean three doubly degenerate vibrations, that is, three distinct frequencies.

with the rotation axis of the arbitrary symmetry operation. (This procedure is often expressed by the statement, "the functions are quantized with respect to the rotation axis.") Atomic orbitals are of the form

$$\begin{aligned}\psi_{nlm} &= R_{nl}(r)\Theta_{l,|m|}(\theta)\Phi_m(\phi) \\ &= R_{nl}(r)|lm\rangle \\ &\sim |lm\rangle \end{aligned} \tag{8.11.10}$$

where in the last line we have dropped $R(r)$ because it is invariant under all $\mathcal{R}$ and hence is irrelevant to the present discussion. To complete the argument, we need only consider the explicit form of Φ_m:

$$|lm\rangle = \Theta_{l,|m|}(\theta)\Phi_m(\phi) = \Theta_{l,|m|}(\theta)\frac{1}{\sqrt{2\pi}}e^{im\phi} \tag{8.11.11}$$

Consequently, a rotation of $|lm\rangle$ about the symmetry (Z) axis involves a change only in ϕ, and a rotation by ϕ' is given by

$$\begin{aligned}\mathcal{R}_{\phi'}|lm\rangle &= \mathcal{R}_{\phi'}\Theta_{l,|m|}(\theta)\Phi_m(\phi) = \Theta_{l,|m|}(\theta)\Phi(\phi-\phi') \\ &= \frac{1}{\sqrt{2\pi}}\Theta_{l,|m|}(\theta)e^{im(\phi-\phi')} = e^{-im\phi'}|lm\rangle \end{aligned} \tag{8.11.12}$$

Thus the character of any irreducible representative under a rotation by ϕ' (about any arbitrary rotation axis) is given by

$$\chi^l(C_{\phi'}) = \sum_{m=-l}^{+l} e^{-im\phi'} = e^{-il\phi'}\sum_{n=0}^{2l}(e^{i\phi'})^n \tag{8.11.13}$$

Summing the geometric series on the right-hand side and using the Euler relation, we have after a little rearranging

$$\chi^l(C_{\phi'}) = \frac{\sin(l+\frac{1}{2})\phi'}{\sin\frac{1}{2}\phi'} \tag{8.11.14}$$

To find the character for the corresponding rotary *inversion* $\chi^l(C_iC_{\phi'})$, we note that the inversion operation C_i changes ϕ to $\phi+\pi$ and θ to $\pi-\theta$. It is a well-known property of the spherical harmonics that

$$\Theta_{l,|m|}(\pi-\theta)\Phi_m(\phi+\pi) = (-1)^l\Theta_{l,|m|}(\theta)\Phi_m(\phi) \tag{8.11.15}$$

Thus, the character for a rotary inversion is

$$\chi^l(C_iC_{\phi'}) = (-1)^l\chi^l(C_{\phi'}) \tag{8.11.16}$$

Equation (8.11.15) is of course the basis for the parity classification of atomic orbitals, that is, whether a given orbital changes sign under inversion [and is thus (of) odd (parity) or ungerade (u)], or maintains the same sign [and is thus even or gerade (g)]. The full rotary-inversion (or rotary-reflection) group O_3 is the direct product (Section 8.13) of the group of all rotations SO_3 and the inversion group C_i. Thus it is sufficient to use simply the full rotation group and to append u or g subscripts when necessary.

It is customary to use rotary-*reflection* rather than rotary-inversion operations when discussing the finite point groups, and most character tables are labeled on that basis. It is easy to confirm that rotary reflection ($S_{\phi'}$) and rotary inversion ($C_iC_{\phi'}$) about the Z axis are related by $S_{\phi'} = C_iC_{\phi'-\pi}$. If this substitution is made in (8.11.16) using (8.11.14), after a little rearranging one obtains

$$\chi^l(S_{\phi'}) = \frac{\cos\left(l+\frac{1}{2}\right)\phi'}{\cos\frac{1}{2}\phi'} \tag{8.11.17}$$

Eqs. (8.11.14) and (8.11.17) permit a calculation of the character for *any* point-group operation. (In the event of a zero denominator, use L'Hôpital's rule.) We emphasize again that *these formulas are applicable for an axis of any arbitrary orientation*. To illustrate the use of these equations, let us reduce the d-orbitals ($l = 2$) on the point group D_{3h}: $\chi^{\text{red}}(E) = 5$, $\chi^{\text{red}}(C_3) = \chi^{\text{red}}(C_{120°}) = -1$, $\chi^{\text{red}}(C_2) = \chi^{\text{red}}(C_{180°}) = 1$, $\chi^{\text{red}}(\sigma_h) = \chi^{\text{red}}(S_1) = \chi^{\text{red}}(S_{360°}) = 1$, $\chi^{\text{red}}(S_3) = \chi^{\text{red}}(S_{120°}) = 1$, $\chi^{\text{red}}(\sigma_v) = \chi^{\text{red}}(S_1) = 1$. Using the character table for D_{3h} (Table 8.13.1) and (8.11.6), we obtain

$$\Gamma^{l=2}(D_{3h}) = A_1' + E' + E'' \tag{8.11.18}$$

Atomic Orbitals Not Centered at an Invariant Point

In this case, there are two possibilities. The first is that a symmetry operation may carry a particular atom into another identical atom. Atomic orbitals centered on such atoms contribute zero to the character for the operation in question, for the reason discussed earlier. The other possibility is that the atom does not move, in which case the character analysis proceeds just as if the atom were located at an invariant point.

It must of course always be understood that "symmetrically complete" basis sets are being used. For example, if a set of d orbitals are placed on an atom that is not at an invariant point, then an equivalent set must be placed on all other symmetrically equivalent atoms. Otherwise, the orbitals are not the basis for a representation, the basis set is physically nonsensical, and the application of character analysis (or group theory) is meaningless.

In simple cases (e.g., s and p orbitals) it is often easier to calculate χ^{red} by inspection than through the use of (8.11.14) and (8.11.17).

Symmetrically Equivalent Sets

Before performing a character analysis, it is often very useful to divide an atomic basis into subsets such that each *separately* forms the basis for a representation. If a subset cannot be further subdivided in this way, the members are said to be *symmetrically equivalent*, since they mix only among themselves under all group operations. Character analysis may then be performed separately on each symmetrically equivalent set, and one obtains extra information.

For example, suppose one wishes to reduce an atomic basis set consisting of the $2s$ and $2p$ orbitals on the boron and three fluorine atoms of BF_3 (point group D_{3h}). By inspection, the symmetrically equivalent sets are: (I) $2s(\text{B})$; (II) $2p_x(\text{B}), 2p_y(\text{B})$; (III) $2p_z(\text{B})$; (IV) $2p_z(\text{F}_1), 2p_z(\text{F}_2), 2p_z(\text{F}_3)$; (V) $2p_x(\text{F}_1), 2p_y(\text{F}_1), 2p_x(\text{F}_2), 2p_y(\text{F}_2), 2p_x(\text{F}_3), 2p_y(\text{F}_3)$. Reduction of each set separately then tells one what symmetry-adapted linear combinations can be formed from *each* set. However, using subsets, (8.11.14) and (8.11.17) are no longer applicable for the $2p$ orbitals on the boron (or on the F atoms), since they have been divided into two separate subsets. In this case the methods discussed earlier (Section 8.7) are very easily used to work out χ^{red}.

8.12. Spin Irreps and Double Groups

The double group concept arises when spin-orbit coupling is introduced (Chapter 12) because then the $\mathcal{R}$ commute with $\mathcal{H}$ only if they operate *simultaneously* in both orbital *and* spin space. Spin functions of half-integral quantum number, however, have the peculiar property of not being invariant under a rotation by 2π; rather, a rotation by 4π is required for invariance [Tinkham (1964), Chapter 5]. Thus a new group (the double group) can be defined in which it is supposed that the symmetry operation $C_1 \equiv R$ (a rotation by 2π about a symmetry axis) has the peculiar property $R \neq E$, but $R^2 = E$. The double group has twice as many elements as the original group, since for each element R' of the original group, we have elements R' and RR' ($= R'R$) in the double group. However, the double group does not necessarily have twice as many classes (irreps). One can proceed in a straightforward manner and derive the character table for this new group, and such tables are listed in Sections C.1, D.1, and E.1. Explicit and illuminating examples of the construction of double group character tables using this approach are given in Tinkham (1964), Section 4-7, and Ballhausen (1962), Section 3-g. The ultimate justification for such an approach rests on the transformation properties of functions of half-integral angular momentum (spin). Such functions are called *spinors*, and an excellent discussion of their properties is presented in Tinkham (1964), Chapter 5.

The double-group irreps are found to divide into two types. The first, the so-called *true irreps* (or *single-valued irreps*), come directly from the irreps of the original group, since for them, $\chi(RR') = \chi(R')$ for all R'. These irreps are given the same names they had in the original group, since they still have the same transformation properties. In fact, one can work with functions belonging to these irreps just as if one were working in the original group. The second type of irreps are called *spin irreps* (or *double-valued irreps*), because $\chi(RR') = -\chi(R')$ for all R'. They are given primed labels in Mulliken notation (e.g., E', E'', U') and half-integer labels in Butler (1981) notation (e.g. $\frac{1}{2}, \frac{\bar{1}}{2}, \frac{3}{2}$). Bethe's Γ_n notation does not differentiate between spin and true irreps. In addition, many authors label the double group corresponding to a group G as G^*. These points are exemplified by the character tables given in Sections C.1, D.1, and E.1.

Another way of treating the transformation properties of electron spin states is to weaken the representation condition to allow projective representations [see Butler (1981), Section 2.6]. This approach leads to the same irreps as the double-group approach, but avoids the necessity of defining ad hoc physical operations as described above to characterize the spin irreps. Authors who follow the projective-representation approach generally label *both* the group G and the double group G^* as G, a notation we adopt in this book.

In practice spin irreps become important in systems with half-integral spin, when spin–orbit coupling is large (Chapter 12). In our applications, the group theory for spin irreps is essentially the same as for true irreps. Readers interested in further discussion of the theory of spinors should see the references cited above.

8.13. Direct-Product Groups

It is often possible to divide a group into two subgroups in such a way that all the elements in one subgroup commute with all the elements in the other. In that event it is possible to make important conceptual and computational simplifications by introducing the concept of the *direct-product group*.

Let us first define the *direct product* of two matrices: a new matrix whose elements are the set of all possible pairs of products formed by multiplying together an element of each matrix. The direct-product matrix **C** of **A** and **B** is written **C** = **A** ⊗ **B** to distinguish it from ordinary matrix multiplication (**AB**). Explicitly, **C** is defined by

$$C_{ij;\,kl} = A_{ik}B_{jl} \tag{8.13.1}$$

The rows and columns of **C** are arranged in dictionary order, which means that the row labeled ij precedes $i'j'$ if $i < i'$, and if $i = i'$, then ij precedes $i'j'$ if $j < j'$. The same scheme applies for the columns. If **A** and **B** are square matrices of dimensions n and m respectively, it follows immediately that **C** is a square matrix of dimensions $n \times m$. Furthermore, it is easy to prove [Tinkham (1964), Section 3-9; Bishop (1973), Section 8-3] that

$$\chi(\mathbf{C}) = \chi^{A \otimes B} = \chi(\mathbf{A})\chi(\mathbf{B}) \tag{8.13.2}$$

Suppose now we have two groups of operators of orders h_1 and h_2, where all the operators of one group commute with all the operators of the other group. We also require that the two groups have only the identity element in common. Then the h_1h_2 unique elements obtained by taking all possible pairs of products between the two groups constitutes the *direct-product group*. It is now possible to prove the following theorems [Tinkham (1964), Section 3-9]: (1) the procedure just described indeed produces a group; (2) the direct product of any irrep of the one group with any irrep of the other group produces an irrep of the direct-product group; (3) all such possible products produce *all* the irreps of the direct-product group. With these propositions and (8.13.2), it follows that *the character table of the direct-product group is simply the direct product of the character tables, regarded as matrices, of the two groups.*

Example: Formation of the Character Table for $D_{3h} = D_3 \otimes C_s$

Consider the groups D_3 and C_s, for which the character tables (using only the single-valued irreps for simplicity) are as follows:

D_3	E	$2C_3$	$3C_2'$
A_1	1	1	1
A_2	1	1	−1
E	2	−1	0

C_s	E	σ_h
A'	1	1
A''	1	−1

We note that $C_3\sigma_h = \sigma_h C_3 = S_3$, $C_2'\sigma_h = \sigma_h C_2' = \sigma_v$, and we name the direct-product irreps as follows: $A_1' = A_1 \otimes A'$, $A_2' = A_2 \otimes A'$, $E' = E \otimes A'$, $A_1'' =$

$A_1 \otimes A''$, $A_2'' = A_2 \otimes A''$, $E'' = E \otimes A''$. Then taking the direct product of the two character tables (regarded as matrices), we obtain the character table of D_{3h} (Table 8.13.1). We include the parentheses in that table to emphasize that the D_{3h} character table can be regarded as four submatrices obtained by multiplying the D_3 character table by each of the four entries in the C_s character table.

How to Use the Direct-Product Group to Simplify Calculations

The concept of the direct-product group is useful in that, for example, we can initially use the group D_3 instead of D_{3h} to classify and couple (Chapter 9) functions. Then to realize the full information content of the group D_{3h}, it is only necessary at the end to note (usually by simple inspection) whether a function does not change sign (prime label) or does change sign (double-prime label) under the operation σ_h.

The other group frequently used in direct products is C_i (which contains the identity and inversion). The direct product with C_i illustrates the concept of *parity*, that is, whether a function does not (even, gerade, or g) or does (odd, ungerade, or u) change sign under inversion. A very familiar example is $O_h = O \otimes C_i$. In later chapters we use the group O extensively and then make subsequent parity classifications to deal with systems of O_h symmetry. Some other common examples of direct-product groups are $O_3 = SO_3 \otimes C_i$, $S_6 = C_3 \otimes C_i$, $D_{3d} = D_3 \otimes C_i$, $D_{2h} = D_2 \otimes C_i$, $D_{4h} = D_4 \otimes C_i$, $D_{5h} = D_5 \otimes C_s$, $D_{6h} = D_6 \otimes C_i$, and $D_{\infty h} = C_{\infty v} \otimes C_i$. The following rules should be self-evident: $(') \otimes (') = (')$, $(') \otimes ('') = ('')$, $('') \otimes ('') = (')$, and $g \otimes g = g$, $g \otimes u = u$, and $u \otimes u = g$.

Finally, we note that the term direct product is also used in a different but closely related sense when functions within a group are coupled. This gives rise to the extraordinarily important concept of the *direct product within a group* and is the subject of the next chapter.

Table 8.13.1. The D_{3h} character table formed from $D_3 \otimes C_s$

D_{3h}	E	$2C_3$	$3C_2$	σ_h	$2S_3$	$3\sigma_v$
A_1'	(1	1	1)	(1	1	1)
A_2'	1	1	−1	1	1	−1
E'	2	−1	0)	2	−1	0)
A_1''	(1	1	1)	(−1	−1	−1)
A_2''	1	1	−1	−1	−1	1
E''	2	−1	0)	−2	1	0)

9 *Vector Coupling Coefficients*

9.1. Introduction

The evaluation of matrix elements is the fundamental task in many spectroscopic and quantum mechanical calculations. In Chapters 3–6, for example, we show that absorption and MCD spectra are governed by electric-dipole and magnetic-dipole matrix elements. It is the purpose of much of the remainder of this book to show how the calculation of matrix elements can be simplified if a molecular system possesses symmetry. In this chapter we derive the fundamental equation which permits such simplifications, the Wigner–Eckart theorem, and we use it to develop the theory of vector coupling coefficients (v.c.c.). These coefficients suffice in many applications, and they serve as the basis for the development of more powerful techniques. These techniques require high-symmetry coefficients which are defined in terms of v.c.c. in Chapter 10.

We emphasize (as in Section 8.11) that our entire subsequent treatment depends on the transformation properties of functions and is entirely independent of their quality as approximate eigenfunctions. Any arbitrary function can be decomposed into a sum of functions, each of which belongs to a row of an irrep of the molecular point group [Tinkham (1964), Section 3.8]. Thus in actual applications we may always choose approximate functions in one-to-one correspondence with the true eigenfunctions and with the same transformation properties [Eq. (8.9.6)]. As is found in our subsequent discussion, such functions, regardless of their quality as approximate eigenfunctions, preserve the degeneracies of the true eigenfunctions. This is the basis for the statement that the indices designating irreps and their associated rows always remain "good quantum numbers," and it is for this reason that group theory is of practical value in quantum-mechanical calculations.

Before deriving the Wigner–Eckart theorem, we must first introduce the concept of direct-product representations and coupling.

9.2. Direct-Product Representations and Coupling

The direct product of two matrices ($\mathbf{A} \otimes \mathbf{B}$) was defined in Section 8.13, and reference was made to the fact that if **A** and **B** are square,

$$\chi^{A \otimes B} = \chi(\mathbf{A})\chi(\mathbf{B}) \tag{9.2.1}$$

Given four matrices $\mathbf{A}, \mathbf{A}', \mathbf{B}, \mathbf{B}'$, it can be shown [Wigner (1959), Chapter 2] that

$$(\mathbf{A} \otimes \mathbf{B})(\mathbf{A}' \otimes \mathbf{B}') = (\mathbf{AA}') \otimes (\mathbf{BB}'), \tag{9.2.2}$$

where it is understood that ($\mathbf{AA}'$) and ($\mathbf{BB}'$) are meaningful, that is, that these pairs of matrices are so dimensioned that the indicated (ordinary) matrix multiplication is defined.

Consider two sets of functions, ψ_n and ϕ_m, of dimensions n and m respectively. Suppose each set forms a basis for the same group of transformation operators, $\mathcal{R}_1, \mathcal{R}_2 \cdots \mathcal{R}_h$ (Section 8.7). By definition, for all R in the group,

$$\begin{aligned} \mathcal{R}\psi_n &= \psi_n \mathbf{D}^n(R) \\ \mathcal{R}\phi_m &= \phi_m \mathbf{D}^m(R) \end{aligned} \tag{9.2.3}$$

where $\mathbf{D}^n(R)$ and $\mathbf{D}^m(R)$ are representatives. Then

$$\begin{aligned} \mathcal{R}(\psi_n \otimes \phi_m) &= (\mathcal{R}\psi_n) \otimes (\mathcal{R}\psi_m) \\ &= [\psi_n \mathbf{D}^n(R)] \otimes [\phi_m \mathbf{D}^m(R)] \\ &= (\psi_n \otimes \phi_m)[\mathbf{D}^n(R) \otimes \mathbf{D}^m(R)] \end{aligned} \tag{9.2.4}$$

where (9.2.2) has been used in the last step. Thus, if the basis sets ψ_n and ϕ_m of dimensions n and m respectively belong to representations $\mathbf{D}^n$ and $\mathbf{D}^m$ of the same group, then the direct product $\psi_n \otimes \phi_m$ is also a basis set, has dimension $n \times m$, and belongs to the direct-product representation $\mathbf{D}^n \otimes \mathbf{D}^m$. Note that $\mathbf{D}^n \otimes \mathbf{D}^m$ is unitary if $\mathbf{D}^n$ and $\mathbf{D}^m$ are unitary.

Equation (9.2.4) is a result of fundamental importance because it provides the mathematical basis for using group theory to simplify the calculation of matrix elements. The process of forming direct-product basis functions is referred to as *coupling*. Since the coupled functions are the basis of a unitary representation (generally reducible), according to our previous discussion (Section 8.10) a unitary matrix **U** exists which relates this basis to another [Eq. (8.10.1)] in which the representation has been brought into reduced form [Eq. (8.11.4)] by the similarity transformation (8.10.2). The matrix elements of **U** via (8.8.1) thus constitute a prescription for forming linear combinations of coupled functions which belong to irreps of the group.

Let us illustrate these ideas with an example. Consider the octahedral point group O. Suppose that we have two sets of basis functions, $\boldsymbol{\psi}$ and $\boldsymbol{\phi}$, both belonging to the T_1 irrep. Specifically, let $\boldsymbol{\psi} = (p_x\ p_y\ p_z)$ and $\boldsymbol{\phi} = (x\ y\ z)$. (Note that these basis functions have *not* been chosen to transform as our *standard* T_1 bases, but their simple form clarifies the example. See Section 9.4 for a discussion of standard basis functions.) If we consider the operator $\mathbb{S}_4^X$, by (8.7.11)

$$\mathbb{S}_4^X \boldsymbol{\psi} = (p_x \quad p_y \quad p_z)\begin{pmatrix} -1 & 0 & 0 \\ 0 & 0 & -1 \\ 0 & 1 & 0 \end{pmatrix} = \boldsymbol{\psi}\mathbf{D}^{T_1}(S_4^X)$$

$$\mathbb{S}_4^X \boldsymbol{\phi} = (x \quad y \quad z)\begin{pmatrix} -1 & 0 & 0 \\ 0 & 0 & -1 \\ 0 & 1 & 0 \end{pmatrix} = \boldsymbol{\phi}\mathbf{D}^{T_1}(S_4^X) \tag{9.2.5}$$

By (9.2.4)

$$\mathbb{S}_4^X(\boldsymbol{\psi} \otimes \boldsymbol{\phi}) = (\boldsymbol{\psi} \otimes \boldsymbol{\phi})\left[\mathbf{D}^{T_1}(S_4^X) \otimes \mathbf{D}^{T_1}(S_4^X)\right] \tag{9.2.6}$$

where

$$(\boldsymbol{\psi} \otimes \boldsymbol{\phi}) = (p_x\boldsymbol{\phi} \quad p_y\boldsymbol{\phi} \quad p_z\boldsymbol{\phi}) \tag{9.2.7}$$

and

$$\mathbf{D}^{T_1}(S_4^X) \otimes \mathbf{D}^{T_1}(S_4^X) \equiv \mathbf{D}^{T_1 \otimes T_1}(S_4^X)$$

$$= \begin{pmatrix} -\mathbf{D}^{T_1}(S_4^X) & \mathbf{0} & \mathbf{0} \\ \mathbf{0} & \mathbf{0} & -\mathbf{D}^{T_1}(S_4^X) \\ \mathbf{0} & \mathbf{D}^{T_1}(S_4^X) & \mathbf{0} \end{pmatrix} \tag{9.2.8}$$

where $\mathbf{0}$ is the 3×3 null matrix. Clearly, a matrix analogous to (9.2.8) can be written for each symmetry element of the group. These 9×9 direct-product matrices must be members of a *reducible* representation, since the largest (single group) irreps in group O are 3×3. The irreducible structure of $\mathbf{D}^{T_1 \otimes T_1}$ may be immediately deduced by a character analysis (Section 8.11) in the following way. By (9.2.1), $\chi^{\text{red}}(R) = \chi^{T_1 \otimes T_1}(R) = \chi^{T_1}(R)\chi^{T_1}(R)$ and thus is immediately obtained by inspection of the group-O character table of Section C.1: $\chi^{\text{red}}(E) = 3 \times 3$, $\chi^{\text{red}}(C_3) = 0 \times 0$, $\chi^{\text{red}}(C_2) = (-1) \times (-1)$, and so on. Using (8.11.6), we find

$$\Gamma^{\text{red}} = T_1 \otimes T_1 = A_1 + E + T_1 + T_2 \tag{9.2.9}$$

This means that if the $\mathbf{D}^{T_1 \otimes T_1}$ representation is brought into reduced form (designated by a prime) it will have the following structure for each symmetry operation R of the group:

$$\mathbf{D}'(R)^{T_1 \otimes T_1} = \begin{pmatrix} \mathbf{D}^{A_1}(R) & \mathbf{0} & \mathbf{0} & \mathbf{0} \\ \mathbf{0} & \mathbf{D}^{E}(R) & \mathbf{0} & \mathbf{0} \\ \mathbf{0} & \mathbf{0} & \mathbf{D}^{T_1}(R) & \mathbf{0} \\ \mathbf{0} & \mathbf{0} & \mathbf{0} & \mathbf{D}^{T_2}(R) \end{pmatrix} \tag{9.2.10}$$

The $\mathbf{0}$ are null matrices of dimensions 1, 2, or 3.

Tabulations of all possible direct-product decompositions (i.e., direct-product tables) for some representative groups are given in Sections C.2, D.2, and E.2. By (8.10.1) and (8.10.2), the unitary matrix $\mathbf{U}$ which brings $\mathbf{D}^{T_1 \otimes T_1}$ into reduced form defines a new set of (nine) basis functions $\boldsymbol{\rho}$:

$$\boldsymbol{\rho} = (\boldsymbol{\psi} \otimes \boldsymbol{\phi})\mathbf{U} \tag{9.2.11}$$

The basis functions in $\boldsymbol{\rho}$ are linear combinations of the direct product basis functions. Furthermore, since $\boldsymbol{\rho}$ is the basis for the direct-product representation in reduced form [Eq. (9.2.10)], *each basis function in* $\boldsymbol{\rho}$ *belongs to a row of an irrep.*

The matrix elements of $\mathbf{U}$ in (9.2.11) are often referred to as *vector coupling* or *Clebsch–Gordan coefficients*. Their formation, properties, and great utility are the subject of the remainder of this chapter.

9.3. The Wigner–Eckart Theorem

Let us consider two functions, ϕ_α^a and ψ_β^b, which are in the same coordinate space and belong respectively to rows α and β of irreps a and b of a specified point group. They are otherwise arbitrary. The scalar product of these two functions is written

$$\int \phi_\alpha^{*a} \psi_\beta^b \, d\tau = \langle \phi_\alpha^a | \psi_\beta^b \rangle \equiv \langle a\alpha | b\beta \rangle \tag{9.3.1}$$

where we adopt the abbreviated notation $\langle a\alpha | b\beta \rangle$ when our discussion emphasizes the transformation properties of the functions. When this notation is employed, it must be remembered that *the sets of functions* $|a\alpha\rangle$ *and* $|b\beta\rangle$ *need not be identical* when $a = b$; the notation simply displays the transformation properties of two otherwise arbitrary sets of functions.

We may write

$$\begin{aligned} \langle a\alpha | b\beta \rangle &= \langle \mathfrak{R} a\alpha | \mathfrak{R} b\beta \rangle \\ &= \sum_{\alpha'\beta'} D^a(R)^*_{\alpha'\alpha} D^b(R)_{\beta'\beta} \langle a\alpha' | b\beta' \rangle \end{aligned} \tag{9.3.2}$$

where the first equality is simply a statement that a definite integral has the same value if evaluated in two coordinate systems which differ by an orthogonal transformation, and the second equality follows from (8.7.1). We assume always that the $|a\alpha\rangle$ and $|b\beta\rangle$ functions form orthonormal sets, so that $\mathbf{D}^a$ and $\mathbf{D}^b$ are unitary irreps. Summing both sides of (9.3.2) over all R using the Great Orthogonality Theorem (8.10.5), we have

$$\begin{aligned} \sum_R \langle \mathfrak{R} a\alpha | \mathfrak{R} b\beta \rangle &= h \langle a\alpha | b\beta \rangle \\ &= \sum_R \sum_{\alpha'\beta'} D^a(R)^*_{\alpha'\alpha} D^b(R)_{\beta'\beta} \langle a\alpha' | b\beta' \rangle \\ &= \sum_{\alpha'\beta'} \sum_R D^a(R)^*_{\alpha'\alpha} D^b(R)_{\beta'\beta} \langle a\alpha' | b\beta' \rangle \\ &= \sum_{\alpha'\beta'} \frac{h}{l_a} \delta_{ab} \, \delta_{\alpha'\beta'} \, \delta_{\alpha\beta} \langle a\alpha' | b\beta' \rangle \end{aligned} \tag{9.3.3}$$

where h is the order of the group and l_a is the dimension of irrep a. Summing over β' and dividing by h gives

$$\langle a\alpha | b\beta \rangle = \delta_{ab} \, \delta_{\alpha\beta} l_a^{-1} \sum_{\alpha'}^{l_a} \langle a\alpha' | a\alpha' \rangle \tag{9.3.4}$$

This extraordinarily important result states that functions belonging to different rows of any (nonequivalent) irreps are orthogonal. *If both functions belong to the same row* of the same irrep, *their scalar product is independent of the row.* (Recall that in general $\langle a\alpha|a\alpha\rangle \equiv \langle \phi_\alpha^a|\psi_\alpha^a\rangle$, and though $\langle \phi_\alpha^a|\phi_\beta^a\rangle = \langle \psi_\alpha^a|\psi_\beta^a\rangle = \delta_{\alpha\beta}$, $\langle a\alpha|a\alpha\rangle \neq 1$ unless $\phi \equiv \psi$.) This last statement is the basis of the Wigner–Eckart theorem and follows immediately because the sum in (9.3.4) is over *all rows* of the irrep.

Let us now consider a matrix element of the form

$$\langle \phi_\alpha^a|O_\phi^f|\psi_\beta^b\rangle \equiv \langle a\alpha|O_\phi^f|b\beta\rangle \tag{9.3.5}$$

where O_ϕ^f is a quantum-mechanical operator belonging to row ϕ of irrep f.[‡] We first consider the case in which O_ϕ^f belongs to the totally symmetric irrep (symbolized by A_1), i.e., $O_\phi^f = O_{a_1}^{A_1}$. If we form direct-product functions (Section 9.2) by coupling O_ϕ^f and ψ_β^b, then by the definition of A_1

$$O_{a_1}^{A_1}\psi_\beta^b = g_\beta^b \tag{9.3.6}$$

where g symbolizes the new, coupled (direct-product) functions. Using (9.3.6),

$$\langle \phi_\alpha^a|O_{a_1}^{A_1}|\psi_\beta^b\rangle = \langle \phi_\alpha^a|g_\beta^b\rangle \tag{9.3.7}$$

and (9.3.4) is immediately applicable. The matrix element is therefore zero unless $a = b$ and $\alpha = \beta$, in which case

$$\begin{aligned}\langle \phi_\alpha^a|O_{a_1}^{A_1}|\psi_\alpha^a\rangle &\equiv \langle a\alpha|O_{a_1}^{A_1}|a\alpha\rangle = \langle \phi_\alpha^a|g_\alpha^a\rangle \\ &\equiv |a|^{-1/2}\langle \phi^a\|O^{A_1}\|\psi^a\rangle \\ &\equiv |a|^{-1/2}\langle a\|O^{A_1}\|a\rangle \end{aligned} \tag{9.3.8}$$

Here $\langle \phi^a\|O^{A_1}\|\psi^a\rangle$ is by definition a *reduced matrix element* whose value is independent of α because of (9.3.4), and $|a|$ is the dimension of (or the degree of degeneracy of the functions belonging to) the irrep a (in this case

[‡]The expression $(\mathfrak{R}\mathcal{O})$ is undefined, since $\mathcal{O}$ must first operate on a function before the subsequent operation by $\mathfrak{R}$ can be performed. Formally, $\mathfrak{R}\mathcal{O} = \mathcal{O}' = \mathfrak{R}\mathcal{O}\mathfrak{R}^{-1}$ [Tinkham (1964), Section 5–11]. We note that if $\mathcal{O}$ is a function only of coordinates, then $\mathcal{O} = O$ and $\mathfrak{R}\mathcal{O}_\phi^f = \mathfrak{R}O_\phi^f = \Sigma_{\phi'}O_{\phi'}^f D^f(R)_{\phi'\phi}$ as in (8.7.1). If $\mathcal{O}$ contains momenta ($\not{p}$), then nevertheless, $\not{p}$ and $\mathbf{p}$ have the same transformation properties and the equation above still holds. Consequently, *as far as transformation properties are concerned*, one need make no special distinction between functions and operators.

$a = A_1$, so $|a| = 1$). Reduced matrix elements are symbolized by the *double* bar on either side of the operator. The normalization factor $|a|^{-1/2}$ is included to make the definition of the reduced matrix element in (9.3.8) consistent with that given in Section 10.2.

We now remove the restriction that $O = O_{a_1}^{A_1}$ and consider a general matrix element $\langle \phi_\alpha^a | O_\phi^f | \psi_\beta^b \rangle$. We recall from the example in Section 9.2, and particularly (9.2.9) and (9.2.10), that the direct product of irreps f and b is written $\mathbf{D}^f \otimes \mathbf{D}^b$ and must satisfy

$$\mathbf{D}^f \otimes \mathbf{D}^b = \sum_c n_c \mathbf{D}^c \tag{9.3.9}$$

where the n_c are integers and the summation runs over all the (nonequivalent) irreps of the group; (9.3.9) simply describes the reduction of the direct product on the group. We assume until noted otherwise that for all c, $n_c \leqslant 1$. Identifying g, O, and ψ respectively with ρ, ψ, and ϕ of (9.2.11), the generalization of (9.3.6) is

$$|g_\gamma^c\rangle = \sum_{\phi\beta} U_{f\phi b\beta}^{c\gamma} |O_\phi^f\rangle |\psi_\beta^b\rangle \tag{9.3.10}$$

This equation simply states that an appropriate linear combination of direct-product functions (symbolized by g) belongs to row γ of the irrep c. (If $f \otimes b$ does not contain c, i.e., $n_c = 0$, then all $U^{c\gamma} \equiv 0$.) We introduce Dirac notation for the expansion coefficients:

$$U_{f\phi b\beta}^{c\gamma} = \langle O_\phi^f \psi_\beta^b | g_\gamma^c \rangle \equiv (f\phi, b\beta | c\gamma) \tag{9.3.11}$$

These coefficients are usually referred to as *vector coupling coefficients* (v.c.c.). We follow the convention of using parentheses for coupling coefficients and Dirac brackets for matrix elements containing an operator.

The notation introduced in the last part of (9.3.11) emphasizes the symmetry-determined nature of the result. Thus $(f\phi, b\beta | c\gamma)$ is the coefficient which multiplies $|f\phi\rangle |b\beta\rangle$ when a function that transforms as $|c\gamma\rangle$ is formed from $\mathbf{O}^f \otimes \psi^b$, where $\mathbf{O}^f$ and ψ^b have the indicated transformation properties but are otherwise arbitrary. In this notation, (9.3.10) reads

$$|c\gamma\rangle = |(fb)c\gamma\rangle = \sum_{\phi\beta} (f\phi, b\beta | c\gamma) |f\phi\rangle |b\beta\rangle \tag{9.3.12}$$

Since the v.c.c. are the elements of a unitary matrix (Section 8.10), (9.3.12) can be immediately inverted, with the result

$$|f\phi\rangle|b\beta\rangle = \sum_{c\gamma}(f\phi, b\beta|c\gamma)^*|c\gamma\rangle$$

$$= \sum_{c\gamma}(c\gamma|f\phi, b\beta)|c\gamma\rangle \tag{9.3.13}$$

The second equation in (9.3.13) results from the general relation $(a|b) = (b|a)^*$. The v.c.c. are often real, in which case complex conjugation can be omitted. Then $(c\gamma|f\phi, b\beta) = (f\phi, b\beta|c\gamma)$.

Using (9.3.13), we can now write for a general matrix element

$$\langle\phi_\alpha^a|O_\phi^f|\psi_\beta^b\rangle \equiv \langle a\alpha|O_\phi^f|b\beta\rangle$$

$$= \sum_{c\gamma}(c\gamma|f\phi, b\beta)\langle\phi_\alpha^a|g_\gamma^c\rangle \tag{9.3.14}$$

In accord with (9.3.4), the matrix element is zero unless the sum over $c\gamma$ contains a function transforming as $|a\alpha\rangle$, in which case (9.3.14) becomes

$$\langle a\alpha|O_\phi^f|b\beta\rangle = (a\alpha|f\phi, b\beta)\langle\phi_\alpha^a|g_\alpha^a\rangle \tag{9.3.15}$$

As before, $\langle\phi_\alpha^a|g_\alpha^a\rangle \equiv |a|^{-1/2}\langle\phi^a\|O^f\|\psi^b\rangle = |a|^{-1/2}\langle a\|O^f\|b\rangle$ is a reduced matrix element. Thus finally

$$\langle a\alpha|O_\phi^f|b\beta\rangle = (a\alpha|f\phi, b\beta)|a|^{-1/2}\langle a\|O^f\|b\rangle$$

$$= (f\phi, b\beta|a\alpha)^*|a|^{-1/2}\langle a\|O^f\|b\rangle \tag{9.3.16}$$

This all-important equation is a restricted form of the Wigner–Eckart theorem. [The general form is given in (9.3.23).]

The derivation of (9.3.16) shows that a matrix element $\langle a\alpha|O_\phi^f|b\beta\rangle$ can be different from zero only if the direct product $f \otimes b$ contains the irrep a. But it is also seen that the direct-product function $|f\phi\rangle|b\beta\rangle$ must contain a contribution from a function *which belongs to row* α of the irrep a. Whether the first condition is met can be determined trivially by inspection of a direct-product table (e.g. Section C.2 for the group O) or by simple character analysis (Section 8.11). Note that (9.3.16) applies even if both conditions are not satisfied, since in that case the v.c.c. in (9.3.16) must be zero.

No irrep components are contained in the definition of the reduced matrix element. Thus (9.3.16) shows that an entire set of matrix elements of the form $\langle a\alpha|O_\phi^f|b\beta\rangle$ can be expressed in terms of a single quantity—the reduced matrix element—and a set of v.c.c., which are symmetry-determined and hence can be assembled once and for all in a (coupling) table. Furthermore, the reduced matrix element can be evaluated for that particular nonzero case for which the left-hand side of (9.3.16) is most easily evaluated. Thus (9.3.16) expresses matrix elements as the product of two factors, one completely symmetry-determined, and the other requiring a single, model-dependent calculation in a given problem. The enormous labor-saving potential of such a procedure becomes immediately apparent if one considers the case which often arises in cubic systems where ϕ^a, O^f, and ψ^b each are members of three-dimensional irreps. Without the use of (9.3.16), 27 different matrix elements arise, each of which must be independently evaluated to calculate the Faraday and absorption parameters discussed in Chapters 2–7. With the use of (9.3.16) and a coupling table, only one matrix element need be evaluated, since the other 26 are then symmetry-determined.

The Replacement Theorem

Suppose we have the two sets, $\boldsymbol{\phi}^a$, $\mathbf{O}^f$, $\boldsymbol{\psi}^b$, and $\bar{\boldsymbol{\phi}}^a$, $\bar{\mathbf{O}}^f$, $\bar{\boldsymbol{\psi}}^b$, which have the indicated transformation properties but need have no other particular relation to each other. If we write (9.3.16) for members of each of these sets and divide the two equations, the result is

$$\frac{\langle\phi_\alpha^a|O_\phi^f|\psi_\beta^b\rangle}{\langle\bar{\phi}_\alpha^a|\bar{O}_\phi^f|\bar{\psi}_\beta^b\rangle}=\frac{\langle\phi^a\|O^f\|\psi^b\rangle}{\langle\bar{\phi}^a\|\bar{O}^f\|\bar{\psi}^b\rangle} \tag{9.3.17}$$

assuming the denominator on the left side is nonzero. From the derivation of (9.3.16), we know that the right-hand side of (9.3.17) must be a number, say c, whose value is independent of α, ϕ, and β. Thus

$$\langle\phi_\alpha^a|O_\phi^f|\psi_\beta^b\rangle=c\langle\bar{\phi}_\alpha^a|\bar{O}_\phi^f|\bar{\psi}_\beta^b\rangle \tag{9.3.18}$$

This equation, which states that all corresponding matrix elements in the two sets are related by the same proportionality constant, is known as the *replacement theorem* [Griffith (1964)].

Repeated Representations

One further generalization of the Wigner–Eckart theorem given in (9.3.16) is required, since that was derived on the assumption that $n_c \leqslant 1$ in (9.3.9).

This is not always true. For example, when spin–orbit coupling is included in cubic systems, the four-dimensional U' irrep arises. The direct product of U' with either of the three-dimensional irreps, T_1 or T_2, gives n_c the value 2 for $c = U'$. Thus

$$T_1 \otimes U' = T_2 \otimes U' = E' + E'' + 2U' \tag{9.3.19}$$

Also

$$U' \otimes U' = A_1 + A_2 + E + 2T_1 + 2T_2 \tag{9.3.20}$$

The case $n_c > 1$ is called the *repeated representation* case. In the groups of usual interest to us, $n_c \leqslant 2$.

For $n_c = 2$, two different sets of $|c\gamma\rangle$ functions occur which must be distinguished. We do this with the label r. Thus in the general case, (9.3.12) and (9.3.13) may be written

$$|rc\gamma\rangle = |(ab)rc\gamma\rangle = \sum_{\alpha\beta}(a\alpha, b\beta|rc\gamma)|a\alpha\rangle|b\beta\rangle \tag{9.3.21}$$

$$\begin{aligned}|a\alpha\rangle|b\beta\rangle &= \sum_{rc\gamma}(rc\gamma|a\alpha, b\beta)|(ab)rc\gamma\rangle \\ &= \sum_{rc\gamma}(a\alpha, b\beta|rc\gamma)^*|(ab)rc\gamma\rangle\end{aligned} \tag{9.3.22}$$

with $r = 0$ and 1, and the Wigner–Eckart theorem (9.3.16) becomes

$$\begin{aligned}\langle a\alpha|O^f_\phi|b\beta\rangle &= \sum_r (ra\alpha|f\phi, b\beta)|a|^{-1/2}\langle a\|O^f\|b\rangle_r \\ &= \sum_r (f\phi, b\beta|ra\alpha)^*|a|^{-1/2}\langle a\|O^f\|b\rangle_r\end{aligned} \tag{9.3.23}$$

These equations may be used for all cases. When the extra label r is not needed, it may be dropped.

Thus *two* reduced matrix elements and *two* v.c.c. are required in the Wigner–Eckart theorem when $n_c = 2$. As before, the v.c.c. are symmetry-determined, but $\langle a\alpha|O^f_\phi|b\beta\rangle$ must be evaluated for two independent non-zero cases to permit evaluation of the two independent reduced matrix elements. Having done this, the remaining matrix elements are symmetry-determined.

To apply the Wigner–Eckart theorem, tabulations of v.c.c. are required. In Section 9.5 we show how such tables are assembled. Before proceeding with that illustration, standard basis functions must be defined; this is the topic of the next section.

9.4. Standard Basis Functions

Before proceeding further with the construction of v.c.c. tables one must define a choice of so-called *standard basis functions*. The need for this arises because any multidimensional unitary irrep $\mathbf{D}(R)$ can be transformed into a different (but equivalent) unitary irrep $\mathbf{D}'(R) = \mathbf{U}^{-1}\mathbf{D}(R)\mathbf{U}$ by an arbitrary unitary similarity transformation (Section 8.8). The basis functions ψ which belong to the rows of one particular irrep do not in general belong to the rows of any other equivalent irrep. Thus while it can make no difference which of the infinity of equivalent irreps is used in any particular problem, it is imperative that a single choice be used consistently throughout a calculation. Failure to observe this precaution will in general produce a disaster. Furthermore, in lengthy calculations where the values of different matrix elements are drawn from different sources, this problem can become quite subtle, and the greatest care must be exercised.

To minimize the possibility of errors of this sort, a choice of standard basis functions is made once and for all, and this choice is never changed. Any mutually orthonormal set of functions which spans the irrep in question is satisfactory. It is decreed that the components of these standard basis functions belong to the rows of an irrep. This defines from among the infinity of possibilities a *unique* irrep which is *always* used in *all* problems.

Even this procedure, however, is not unique. Suppose standard basis functions are chosen and the corresponding irrep is deduced. Then, nevertheless, there exists an infinity of related sets of basis functions which belong to the rows of this particular irrep. This is seen by noting in (8.7.1) or (8.9.6) that any set of basis functions obtained from the original ones by multiplying each and every function by the *same* arbitrary *phase factor*, $e^{i\alpha}$, also belongs to the rows of the same irrep. This is an inherent ambiguity which cannot be resolved. However, in calculations one is always concerned with matrix elements of the form $\langle \phi | \mathcal{O} | \psi \rangle$. If ϕ and ψ are members of the same symmetrically equivalent set, multiplying each member of the set by the *same* factor $e^{i\alpha}$ clearly leaves the matrix element unchanged. If ϕ and ψ are members of different sets, multiplying each member of one set by $e^{i\alpha_1}$ and each member of the other set by $e^{i\alpha_2}$ changes all the calculated matrix elements by the same modulus-one factor. However, *all calculated observables are invariant under such a change*.

We choose our standard basis functions for all groups following Butler (1981), Chapter 16. The Butler (1981) bases are actually defined and labeled in terms of group–subgroup chains (Section 15.2). We adopt an alternative, Griffith-type nomenclature in this book which simplifies the notation considerably for groups with numerous subgroups, and is more familiar to chemists. *Note carefully, however, that while the nomenclature is Griffith-like, our standard basis functions differ substantially both from those of Griffith* (1964) *and from those of Griffith* (1962).

The Standard Angular-Momentum Basis for SO_3

It is convenient to begin by defining the angular momentum, or $|jm\rangle$, basis functions of the full rotation group SO_3. Then all other point-group bases may be defined in terms of this standard SO_3 basis. The integer irreps of the $|jm\rangle$ basis (which in the language of Section 15.2 is the $SO_3 \supset SO_2$ basis) are spanned by the spherical harmonics for $j = 0, 1, 2, \ldots$. We define the spherical-harmonic (Y_m^j) basis to have Condon–Shortley phases and list the Y_m^j in Section B.1 for $j = 0$ through 5. The half-integer or spin irreps of SO_3 are spanned by *spinors*, which have no functional form in coordinate space and are defined in terms of their transformation properties and behavior under time reversal (Section 14.3). The transformation properties of our $\lfloor jm\rangle$ basis functions for both integer and half-integer j are those obtained via (8.9.6) using $\mathbf{D}^j(R)$ calculated according to Tinkham [(1964), Eq. (5-35)]. The behavior of the bases under time reversal is discussed in Section 14.3.

Standard Basis Functions for Subgroups of SO_3: *Group Generators*

Standard basis functions for a subgroup G of SO_3 may be expressed in terms of the $\lfloor jm\rangle$ discussed above. Then the $\mathbf{D}^k(R)$ for each irrep k of G follow directly from the standard $\mathbf{D}^j(R)$ defined for the $\lfloor jm\rangle$ basis, so no additional tabulation of the $\mathbf{D}^k(R)$ is required. It is very convenient, however, to have tables of the $\mathbf{D}^k(R)$ for the point groups, since they aid enormously in determining transformation properties of MOs. To define the basis for a group G in terms of the $\mathbf{D}^k(R)$ it is not necessary to specify the $\mathbf{D}^k(R)$ for all R. One can do with much less information by using the concept of group generators.

The multiplication table of a group defines a series of relationships among the elements, and thus among the $\mathbf{D}^k(R)$. *Group generators* are a minimum subset of elements from which the entire group can be generated using the multiplication table. Different choices are possible. This subset can be surprisingly small. For the group O (and the isomorphic group T_d) the size is 2 [Griffith (1964), p. 154]. The other 22 elements can be expressed as products involving only this pair via (8.7.4), though each member of the pair may occur several times in any particular product. The result therefore is that specification of the transformation properties of a set of functions under the group generators [i.e., specification of the $\mathbf{D}^k(R)$ of the group generators] is a *complete description* of the transformation properties of the set. Thus if one makes a choice of the $\mathbf{D}^k(R)$ for the group generators [or of the functions that produce them via (8.7.1) or (8.9.6)], the basis is completely specified. Any set of functions which transforms accordingly is a set of *standard basis functions*.

Choices of standard basis functions defined in terms of their behavior under the group generators are given in Sections C.5, D.4, and E.4. Then Sections C.6, D.5, and E.5 give the same standard basis functions expressed in terms of the $|jm\rangle$ of SO_3. A table of either the Section C.5 or the Section C.6 type is sufficient to define the basis.

We next illustrate by way of an example how tables of the Section C.5 type fix the basis.

Example: The Standard Basis for the T_1 Irrep in the Real Octahedral Basis

In Section C.5 we give two standard basis sets for the octahedral group, which are designated the real-O and the complex-O bases respectively. Here we illustrate the use of the tables in Section C.5 by considering how they define the standard real-O basis for the irrep T_1. Table C.5.1(b) gives the behavior of the three T_1 partners (T_1x, T_1y, T_1z) under the octahedral-group generators, C_4^Z, C_4^X, and C_3^{XYZ}. Since the table specifies that $\mathcal{C}_4^Z|T_1x\rangle = i|T_1y\rangle$, $\mathcal{C}_4^Z|T_1y\rangle = i|T_1x\rangle$, and $\mathcal{C}_4^Z|T_1z\rangle = |T_1z\rangle$, it follows from (8.7.1) or (8.9.6) that in the standard real-O basis

$$\mathcal{C}_4^Z\begin{pmatrix} T_1x & T_1y & T_1z \end{pmatrix} = \begin{pmatrix} T_1x & T_1y & T_1z \end{pmatrix}\begin{pmatrix} 0 & i & 0 \\ i & 0 & 0 \\ 0 & 0 & 1 \end{pmatrix}$$

$$= \begin{pmatrix} T_1x & T_1y & T_1z \end{pmatrix}\mathbf{D}^{T_1}(C_4^Z) \qquad (9.4.1)$$

Likewise $\mathbf{D}^{T_1}(C_4^X)$ and $\mathbf{D}^{T_1}(C_3^{XYZ})$ may be determined from the table. (Actually, specification of any two of these three is sufficient to define the standard real-O basis for T_1.) Any set of functions which give *identical* $\mathbf{D}^{T_1}(R)$ to those obtained from the table are defined to be standard real-O basis functions for T_1.

Examples of standard basis functions are given in the same Table C.5.1(b); we see that the set $(-x, iy, z)$ would be a satisfactory basis for (T_1x, T_1y, T_1z). (Thus the real-O Butler basis is not strictly speaking "real"; rather "real" means the basis is related to the $|jm\rangle$ by a real transformation.

For this reason the y component always appears as iy.) Likewise, from Section B.2 it follows that $(-p_x, ip_y, p_z)$ would also have the correct transformation properties. Note, however, that the T_1 basis sets (x, y, z) and (p_x, p_y, p_z) of our Section 9.2 example are *nonstandard*; they do *not* give the $\mathbf{D}^{T_1}(R)$ of Section C.5, and therefore are not satisfactory basis choices using our [Butler (1981)] standard basis definitions.

From the tables of Section C.5 and the atomic-orbital functional forms of Sections B.1 and B.2, standard basis functions for other irreps of the octahedral group may also be properly chosen with minimal effort. For example, standard d-orbital basis sets in the real-O basis are

$$\begin{aligned} |T_2\xi\rangle &= -id_{yz} \\ |T_2\eta\rangle &= d_{zx} \\ |T_2\zeta\rangle &= id_{xy} \\ |E\theta\rangle &= -d_{z^2} \\ |E\epsilon\rangle &= d_{x^2-y^2} \end{aligned} \tag{9.4.2}$$

where the atomic orbitals are defined in Section B.2. The basis functions are always chosen orthonormal. Section C.7 includes the inverse of the relations above—that is, $d_{yz} = i|T_2\xi\rangle$ and so on.

Further discussion of function and operator transformations is given in Section 9.8.

9.5. Construction of Coupling Tables

To apply the Wigner–Eckart theorem, it is clear from (9.3.16) and (9.3.23) that tabulations of v.c.c. are required. In this section we illustrate how such tabulations, generally referred to as *coupling tables*, are assembled. The cubic group O is used as an illustration, and spin–orbit coupling is neglected for the moment, so that (9.3.16) [or (9.3.23) with $r = 0$ only] applies.

Example: Construction of the $T_1 \otimes T_2$ Coupling Table in the Real Octahedral Basis

Coupling tables need be constructed only once for any point group. Many authors have constructed such tables using different standard basis functions. Thus many coupling tables already exist, and their use requires no knowledge whatever of the means by which they are constructed. Nevertheless, we now construct one such table. The procedure is completely straightforward, and the process illustrates some of the points discussed in the previous section. It also emphasizes the need for a choice of standard basis functions and shows what additional degrees of freedom arise.

Suppose we wish to construct the $T_1 \otimes T_2$ coupling table for the group O. First we note (Section C.2) that

$$T_1 \otimes T_2 = A_2 + E + T_1 + T_2 \tag{9.5.1}$$

—a decomposition which may be immediately deduced from the group-O character table by using character analysis (Section 8.11). Our problem is to determine the v.c.c. in (9.3.12) for $f = T_1$, $b = T_2$, when $c = A_2$, E, T_1, and T_2. Our procedure is to first find sets of functions belonging to the rows of the T_1 and T_2 irreps. We then form the nine-dimensional direct-product basis set $T_1 \otimes T_2$, and find nine appropriate linear combinations which belong to the rows of our standard A_2, E, T_1, and T_2 irreps. There is a standard group-theoretical procedure for carrying out the latter process, sometimes referred to as using the Van Vleck function-generating machine [Tinkham (1964), p. 41]. However, this is a laborious process which requires all the representatives of the group. If one instead uses the group generators, the desired results can be obtained with surprisingly little effort. Intuition is often the fastest way to proceed, since inspired guesses can always be checked and refined by reference to the group generator representatives (Section C.5). However, we proceed here in a systematic manner.

We use the real-O basis and first find the function $|A_2a_2\rangle$ obtained from the direct product $T_1 \otimes T_2$. To simplify the notation we abbreviate $|T_1x\rangle$, $|T_1y\rangle$, and $|T_1z\rangle$ as $|x\rangle$, $|y\rangle$, and $|z\rangle$, and $|T_2\xi\rangle$, $|T_2\eta\rangle$, and $|T_2\zeta\rangle$ as $|\xi\rangle$, $|\eta\rangle$, and $|\zeta\rangle$. (Note carefully that here by $|x\rangle$, $|y\rangle$, and $|z\rangle$ we do *not* mean the functions x, y, and z, since the functions x, y, and z do *not* transform as our standard T_1x, T_1y, and T_1z basis functions—see Section 9.4 above.) Our product is thus

$$\begin{aligned}
|A_2a_2\rangle &= [(|T_1x\rangle|T_1y\rangle|T_1z\rangle) \otimes (|T_2\xi\rangle|T_2\eta\rangle|T_2\zeta\rangle)]\mathbf{c} \\
&= [(|x\rangle|y\rangle|z\rangle) \otimes (|\xi\rangle|\eta\rangle|\zeta\rangle)]\mathbf{c} \\
&= c_1|x\rangle|\xi\rangle + c_2|x\rangle|\eta\rangle + c_3|x\rangle|\zeta\rangle + c_4|y\rangle|\xi\rangle + c_5|y\rangle|\eta\rangle \\
&\quad + c_6|y\rangle|\zeta\rangle + c_7|z\rangle|\xi\rangle + c_8|z\rangle|\eta\rangle + c_9|z\rangle|\zeta\rangle
\end{aligned} \tag{9.5.2}$$

since $|A_2a_2\rangle$ must be some linear combination of the nine $T_1 \otimes T_2$ direct-product basis functions, where **c** is the "A_2a_2" column of **U** in (9.2.11). [See also (9.3.10).] From Table C.5.1(b) we note that $\mathcal{C}_4^Z|A_2a_2\rangle = -|A_2a_2\rangle$. Furthermore, this table tells us how the direct-product basis functions transform; for example, with our abbreviations, $\mathcal{C}_4^Z|x\rangle|\xi\rangle = (\mathcal{C}_4^Z|x\rangle)(\mathcal{C}_4^Z|\xi\rangle) = (i|y\rangle)(i|\eta\rangle) = -|y\rangle|\eta\rangle$, and so on. Therefore

$$\begin{aligned}\mathcal{C}_4^Z|A_2a_2\rangle &= -c_1|y\rangle|\eta\rangle - c_2|y\rangle|\xi\rangle - c_3 i|y\rangle|\zeta\rangle - c_4|x\rangle|\eta\rangle \\ &\quad -c_5|x\rangle|\xi\rangle - c_6 i|x\rangle|\zeta\rangle + c_7 i|z\rangle|\eta\rangle + c_8 i|z\rangle|\xi\rangle - c_9|z\rangle|\zeta\rangle \\ &= -|A_2a_2\rangle \end{aligned} \tag{9.5.3}$$

Substituting from (9.5.2) for $-|A_2a_2\rangle$ and equating the coefficients of *each* direct-product basis function (since they are linearly independent) leads immediately to $c_5 = c_1$ and $c_4 = c_2$ with $c_3 = c_6 = c_7 = c_8 = 0$. Thus

$$|A_2a_2\rangle = c_1(|x\rangle|\xi\rangle + |y\rangle|\eta\rangle) + c_2(|x\rangle|\eta\rangle + |y\rangle|\xi\rangle) + c_9|z\rangle|\zeta\rangle \tag{9.5.4}$$

Following exactly the same procedure for the requirement in Table C.5.1(b), $\mathcal{C}_4^X|A_2a_2\rangle = -|A_2a_2\rangle$, leads to the further results $c_1 = c_9$, $c_2 = 0$. Thus

$$|A_2a_2\rangle = c_1(|x\rangle|\xi\rangle + |y\rangle|\eta\rangle + |z\rangle|\zeta\rangle) \tag{9.5.5}$$

This result can be double-checked against Table C.5.1(b) by confirming that $\mathcal{C}_3^{XYZ}|A_2a_2\rangle = |A_2a_2\rangle$. For reasons which will become clear shortly, we choose $c_1 = 1/\sqrt{3}$.

To work out the T_1 functions, one proceeds in a very similar manner. For example, $|T_1x\rangle$ is written in the form (9.5.2). Imposing the requirement $\mathcal{C}_4^X|T_1x\rangle = |T_1x\rangle$ from Table C.5.1(b) leads immediately to the result $c_5 = -c_9$, $c_6 = -c_8$, and the other coefficients are zero, so

$$|T_1x\rangle = c_5(|y\rangle|\eta\rangle - |z\rangle|\zeta\rangle) + c_6(|y\rangle|\zeta\rangle - |z\rangle|\eta\rangle) \tag{9.5.6}$$

Imposing the further requirements $\mathcal{C}_4^Z|T_1x\rangle = i|T_1y\rangle$ and $\mathcal{C}_4^Z|T_1y\rangle = i|T_1x\rangle$ yields $c_5 = 0$, so

$$\begin{aligned}|T_1x\rangle &= c_6(|y\rangle|\zeta\rangle - |z\rangle|\eta\rangle) \\ |T_1y\rangle &= -c_6(|x\rangle|\zeta\rangle + |z\rangle|\xi\rangle)\end{aligned} \tag{9.5.7}$$

and finally $\mathcal{C}_4^X|T_1y\rangle = i|T_1z\rangle$ gives

$$|T_1z\rangle = -c_6(|x\rangle|\eta\rangle - |y\rangle|\xi\rangle) \tag{9.5.8}$$

In this case we choose $c_6 = -1/\sqrt{2}$ as explained below.

By exactly the same kinds of arguments, one obtains the E functions

$$
\begin{aligned}
|E\theta\rangle &= \frac{-1}{\sqrt{2}}(|x\rangle|\xi\rangle - |y\rangle|\eta\rangle) \\
|E\epsilon\rangle &= \frac{1}{\sqrt{6}}(|x\rangle|\xi\rangle + |y\rangle|\eta\rangle - 2|z\rangle|\zeta\rangle)
\end{aligned}
\tag{9.5.9}
$$

and the T_2 functions,

$$
\begin{aligned}
|T_2\xi\rangle &= \frac{1}{\sqrt{2}}(|y\rangle|\zeta\rangle + |z\rangle|\eta\rangle) \\
|T_2\eta\rangle &= \frac{1}{\sqrt{2}}(-|x\rangle|\zeta\rangle + |z\rangle|\xi\rangle) \\
|T_2\zeta\rangle &= \frac{-1}{\sqrt{2}}(|x\rangle|\eta\rangle + |y\rangle|\xi\rangle)
\end{aligned}
\tag{9.5.10}
$$

where again in each case we have made a choice for the constant.

All of these results may be assembled in matrix form [see (9.2.11) or (9.3.10)]:

$$
(|A_2a_2\rangle, |E\theta\rangle, \ldots, |T_2\zeta\rangle) = (|x\rangle|\xi\rangle, |x\rangle|\eta\rangle, \ldots, |z\rangle|\zeta\rangle)\mathbf{U}
\tag{9.5.11}
$$

with the matrix elements of $\mathbf{U}$ determined by (9.5.5) and (9.5.7)–(9.5.10). Our representations are unitary, and therefore the constants in (9.5.5) and in (9.5.7)–(9.5.10) can be, and have been, chosen to make $\mathbf{U}$ unitary. Thus (9.5.11) may be immediately inverted to

$$
(|x\rangle|\xi\rangle, \ldots, |z\rangle|\zeta\rangle) = (|A_2a_2\rangle, \ldots, |T_2\zeta\rangle)\tilde{\mathbf{U}}^*
\tag{9.5.12}
$$

The matrix $\tilde{\mathbf{U}}^*$ is the $T_1 \otimes T_2$ coupling table. We tabulate it, together with the $T_1 \otimes T_1$ table, in Table 9.5.1. Reading across a v.c.c. table (and complex conjugating) gives the decomposition of any direct-product basis function into standard basis functions [Eq. (9.3.13) or (9.3.22)], and reading down gives just the inverse [Eq. (9.3.12) or (9.3.21)].

It is easy to confirm that the results in the table obey the relations in Table C.5.1(b). For example, from (9.3.12) or (9.3.21) and the $T_1 \otimes T_2$ v.c.c.

Table 9.5.1. Vector coupling coefficients for $T_1 \otimes T_2$ and $T_1 \otimes T_1$ in the real-O basis[a]

$T_1 \otimes T_2$		A_2	E		T_1			T_2		
		a_2	θ	ϵ	x	y	z	ξ	η	ζ
x	ξ	$1/\sqrt{3}$	$-1/\sqrt{2}$	$1/\sqrt{6}$	0	0	0	0	0	0
x	η	0	0	0	0	0	$1/\sqrt{2}$	0	0	$-1/\sqrt{2}$
x	ζ	0	0	0	0	$1/\sqrt{2}$	0	0	$-1/\sqrt{2}$	0
y	ξ	0	0	0	0	0	$-1/\sqrt{2}$	0	0	$-1/\sqrt{2}$
y	η	$1/\sqrt{3}$	$1/\sqrt{2}$	$1/\sqrt{6}$	0	0	0	0	0	0
y	ζ	0	0	0	$-1/\sqrt{2}$	0	0	$1/\sqrt{2}$	0	0
z	ξ	0	0	0	0	$1/\sqrt{2}$	0	0	$1/\sqrt{2}$	0
z	η	0	0	0	$1/\sqrt{2}$	0	0	$1/\sqrt{2}$	0	0
z	ζ	$1/\sqrt{3}$	0	$-\sqrt{2}/\sqrt{3}$	0	0	0	0	0	0

$T_1 \otimes T_1$		A_1	E		T_1			T_2		
		a_1	θ	ϵ	x	y	z	ξ	η	ζ
x	x	$-1/\sqrt{3}$	$1/\sqrt{6}$	$1/\sqrt{2}$	0	0	0	0	0	0
x	y	0	0	0	0	0	$-1/\sqrt{2}$	0	0	$-1/\sqrt{2}$
x	z	0	0	0	0	$-1/\sqrt{2}$	0	0	$-1/\sqrt{2}$	0
y	x	0	0	0	0	0	$1/\sqrt{2}$	0	0	$-1/\sqrt{2}$
y	y	$1/\sqrt{3}$	$-1/\sqrt{6}$	$1/\sqrt{2}$	0	0	0	0	0	0
y	z	0	0	0	$-1/\sqrt{2}$	0	0	$-1/\sqrt{2}$	0	0
z	x	0	0	0	0	$1/\sqrt{2}$	0	0	$-1/\sqrt{2}$	0
z	y	0	0	0	$1/\sqrt{2}$	0	0	$-1/\sqrt{2}$	0	0
z	z	$-1/\sqrt{3}$	$-\sqrt{2}/\sqrt{3}$	0	0	0	0	0	0	0

[a]The basis is that of Table C.5.1(b). The entries are the $(a\alpha, b\beta|rc\gamma)$. The $(rc\gamma|a\alpha, b\beta)$ are obtained by complex conjugation, since $(rc\gamma|a\alpha, b\beta) = (a\alpha, b\beta|rc\gamma)^*$. In this case $r = 0$ only and is thus dropped, and all v.c.c. are real.

in Table 9.5.1 we have

$$|T_1x\rangle = |(T_1T_2)T_1x\rangle = \sum_{\alpha\beta}(T_1\alpha, T_2\beta|T_1x)|T_1\alpha\rangle|T_2\beta\rangle$$

$$= \frac{-1}{\sqrt{2}}|T_1y\rangle|T_2\zeta\rangle + \frac{1}{\sqrt{2}}|T_1z\rangle|T_2\eta\rangle$$

$$|T_1y\rangle = |(T_1T_2)T_1y\rangle = \frac{1}{\sqrt{2}}|T_1x\rangle|T_2\zeta\rangle + \frac{1}{\sqrt{2}}|T_1z\rangle|T_2\xi\rangle \quad (9.5.13)$$

Table C.5.1(b) says that our standard basis functions must obey the relation $\mathcal{C}_4^Z|\mathrm{T}_1x\rangle = i|T_1y\rangle$. For the $|(T_1T_2)T_1x\rangle$ function above we have, using Table C.5.1(b),

$$\mathcal{C}_4^Z|(T_1T_2)T_1x\rangle = \mathcal{C}_4^Z\left(\frac{-1}{\sqrt{2}}|T_1y\rangle|T_2\zeta\rangle + \frac{1}{\sqrt{2}}|T_1z\rangle|T_2\eta\rangle\right)$$

$$= \frac{i}{\sqrt{2}}|T_1x\rangle|T_2\zeta\rangle + \frac{i}{\sqrt{2}}|T_1z\rangle|T_2\xi\rangle$$

$$= i|(T_1T_2)T_1y\rangle \quad (9.5.14)$$

Thus the $|(T_1T_2)T_1x\rangle$ function defined by the Table 9.5.1 v.c.c. obeys the Table C.5.1(b) relation for C_4^Z. In a similar manner it can be shown that all the entries in Table 9.5.1 define functions which obey our standard basis relations. The reader may wish to check the table as an exercise.

It should be clear from our $T_1 \otimes T_2$ example how to construct other coupling tables. In fact we have done the most difficult case. The next hardest one (also given in Table 9.5.1) is $T_1 \otimes T_1$.

We emphasize again (Section 9.4) that the normalization constant for any degenerate set of functions can be multiplied by an arbitrary phase factor, $e^{i\alpha}$. That is, c_1 in (9.5.5) could be multiplied by $e^{i\alpha_1}$, c_6 in (9.5.7) and (9.5.8) by $e^{i\alpha_2}$, and so forth in (9.5.9) and (9.5.10). This would have the effect of multiplying the corresponding sections of the coupling tables by these phase factors. However, it should be evident that the resulting coupling tables remain consistent with the standard basis relations. Thus for each section of the coupling table, an *arbitrary* phase choice must be made; the obvious simple alternative to our choices is $\alpha = \pi$. (If uniform behavior under time reversal is desired, the choice of phases is much more severely restricted, and in fact the only other choice is $\alpha = \pi$. Such restrictions are discussed in Chapter 14.) Our choices for group O are those of Butler (1981). Note that a different choice can be made in each problem provided it is used exclusively throughout. However, it clearly invites confusion to change choices from one problem to the next, and it results in no significant simplifications. Thus, choices are made once and for all and are *never* changed.

Since it proves convenient in most applications to use high-symmetry coefficients (the $3jm$ defined in the next chapter) rather than v.c.c., we tabulate very few v.c.c.. Any v.c.c. required are calculated from $3jm$ coefficients via Eq. (10.2.1). We emphasize again that v.c.c. found in this manner are dependent on the order of the coupling—that is, $(a\alpha, b\beta|rc\gamma)$ is not in general equal to $(b\beta, a\alpha|rc\gamma)$ when $a \neq b$, as it is in Griffith (1964), Table A16.

9.6. Construction of Coupling Tables in Another Basis

It is often convenient to define the so-called complex basis functions, which are more closely related to the angular-momentum basis functions (the $|jm\rangle$ of $SO_3 \supset SO_2$ or the $|j^{\pm}m\rangle$ of $O_3 \supset SO_3 \supset SO_2$ where $O_3 = SO_3 \otimes C_i$). In the Butler (1981) basis, the real-O and complex-O bases are related as specified in Table C.5.1(c). The complex-O basis functions are expressed in terms of the $|j^{+}m\rangle$ in Appendix C.6. Clearly, standard basis relations and coupling tables can be assembled for these functions from those for the real-O basis functions. The explicit matrix relation between the coupling tables in the two different bases may easily be established starting with (9.2.11). Let us suppose that the complex (primed) and real bases are related by

$$\begin{aligned} \boldsymbol{\rho}' &= \boldsymbol{\rho}\mathbf{V}_{\rho} \\ \boldsymbol{\psi}' &= \boldsymbol{\psi}\mathbf{V}_{\psi} \\ \boldsymbol{\phi}' &= \boldsymbol{\phi}\mathbf{V}_{\phi} \end{aligned} \tag{9.6.1}$$

where $\mathbf{V}_{\rho}$, $\mathbf{V}_{\psi}$, and $\mathbf{V}_{\phi}$ are unitary. Inverting (9.6.1), substituting the result into (9.2.11), using (9.2.2), and then multiplying on the right by $\mathbf{V}_{\rho}$ gives

$$\begin{aligned} \boldsymbol{\rho}' &= (\boldsymbol{\psi}' \otimes \boldsymbol{\phi}')\left(\mathbf{V}_{\psi}^{-1} \otimes \mathbf{V}_{\phi}^{-1}\right)\mathbf{U}\mathbf{V}_{\rho} \\ &= (\boldsymbol{\psi}' \otimes \boldsymbol{\phi}')\mathbf{U}' \end{aligned} \tag{9.6.2}$$

where

$$\mathbf{U}' = \left(\mathbf{V}_{\psi}^{-1} \otimes \mathbf{V}_{\phi}^{-1}\right)\mathbf{U}\mathbf{V}_{\rho} \tag{9.6.3}$$

Thus $\tilde{\mathbf{U}}'^{*}$ will be the coupling table in the new (complex-O) basis corresponding to $\tilde{\mathbf{U}}^*$ in the original (real-O) basis. Calculation of $\mathbf{U}'$ from $\mathbf{U}$ by (9.6.3) is completely straightforward.

One must obtain the same answer in any problem whether the real or the complex basis is used. Moreover, if our tables [or any others derived from Butler (1981)] are used, the reduced matrix elements are *identical* in the real and complex bases, since $\mathbf{U}'$ and $\mathbf{U}$ in (9.6.3) are phase-linked; that is, $\mathbf{U}'$ is completely defined by (9.6.3), not merely to within one or more phase factors.

9.7. Spin Irreps and Coupling

If Γ and Γ' symbolize respectively all functions belonging to single- and double-valued irreps (Section 8.12) of any group, the following general direct-product rules apply:

$$\begin{aligned} \Gamma \otimes \Gamma &= \Gamma \\ \Gamma \otimes \Gamma' &= \Gamma' \\ \Gamma' \otimes \Gamma' &= \Gamma \end{aligned} \tag{9.7.1}$$

These rules may be easily verified by character analysis.

V.c.c. involving spin irreps are found, for the most part, exactly like those involving only single-valued irreps. The only exceptional cases are those for direct products containing repeated representations (see Section 9.3). For example, in the group O,

$$T_1 \otimes U' = 2U' + E' + E'' \tag{9.7.2}$$

In this case, *two* mutually orthogonal four-dimensional sets of U' functions can be constructed from the 12-dimensional direct-product basis, and (9.3.21) contains the index r to distinguish them. The following problem arises in this (or any other) repeated-representation case. Given any pair of mutually orthonormal sets of U' functions (say i' and j') satisfying the standard basis relations, an infinity of other equally acceptable sets can be generated by the general orthogonal transformation,

$$(iU'\tau,\ jU'\tau) = (i'U'\tau,\ j'U'\tau)\begin{pmatrix} \cos\theta & \sin\theta \\ -\sin\theta & \cos\theta \end{pmatrix} \tag{9.7.3}$$

where $\tau = \kappa, \lambda, \mu$, or ν, and θ is any arbitrary angle of fixed value. Thus an arbitrary choice must be made. Griffith [(1964), Section 9.6.2] does this by using a proportionality that exists between $\mathcal{H}_{SO}$ matrix elements within an atomic ${}^{2S+1}P$ and an octahedral ${}^{2S+1}T_i$ term. The resulting U' sets can be classified by the atomic total-angular-momentum quantum number J, where $J = \frac{3}{2}$ or $\frac{5}{2}$. These sets have the further useful property that $\mathcal{H}_{SO}$ is completely diagonal in such a basis, i.e.,

$$\langle iU'\tau|\mathcal{H}_{SO}|jU'\tau'\rangle = \delta_{\tau\tau'}\delta_{ij}\langle iU'\tau|\mathcal{H}_{SO}|iU'\tau\rangle \tag{9.7.4}$$

The $\delta_{\tau\tau'}$ is simply a consequence of $\mathcal{H}_{SO}$ being totally symmetric and always applies, but *the* δ_{ij} *is a consequence of Griffith's particular choice of* iU' *and* jU' *sets. It is not true in general.* His U' sets are named by their associated J values. Thus for $T_i \otimes U'$, his two resulting U' sets are written $\frac{3}{2}U'$ and $\frac{5}{2}U'$.

We emphasize that Griffith's choice of repeated representations is only one arbitrary (though logical and useful) possibility. In fact, when the irreducible tensor method is applied to the group O in later chapters, it is necessary to make a different choice for U' states arising from $T_2 \otimes U'$ *but not* for $T_1 \otimes U'$ (Section 10.2). We follow Butler (1981) and make a different choice for *both* $T_2 \otimes U'$ and $T_1 \otimes U'$. This simplifies the tables somewhat, but represents a departure from Piepho (1979), who followed Dobosh (1972) and used Griffith's choice for $T_1 \otimes U'$ but not for $T_2 \otimes U'$.

9.8. Function and Operator Transformation Coefficients; Spherical Vectors

The transformation properties of any arbitrary function or operator are found using the standard basis information and the function behavior relations given in Sections B.2, C.5, D.4, and E.4, and in Chapter 16 of Butler (1981). We define function transformation coefficients $\langle a\alpha|\psi_i\rangle$ and operator transformation coefficients $\langle f\phi|\mathcal{O}_i\rangle$ by the equations

$$\psi_i = \sum_{a\alpha}\langle a\alpha|\psi_i\rangle|\psi_i a\alpha\rangle \tag{9.8.1}$$

$$\mathcal{O}_i = \sum_{f\phi}\langle f\phi|\mathcal{O}_i\rangle O^f_\phi \tag{9.8.2}$$

respectively. The function or operator on the left-hand side is arbitrary, whereas those on the right obey the standard basis relations. Whenever possible the basis is chosen so the sums are unnecessary.

The coefficients are found by first expressing functions or operators in terms of either rectangular coordinates or spherical harmonics. The electric-dipole operators m_x, m_y, and m_z transform respectively as x, y, and z, while the angular-momentum operators s_α, l_α, j_α, or μ_α transform as R_α ($\alpha = x, y$, or z). Thus in the real-O basis, Table C.5.1(b) gives

$$\begin{aligned} m_x &= \langle T_1 x | m_x \rangle m_x^{t_1} = -m_x^{t_1} \\ m_y &= \langle T_1 y | m_y \rangle m_y^{t_1} = -i m_y^{t_1} \\ m_z &= \langle T_1 z | m_z \rangle m_z^{t_1} = m_z^{t_1} \end{aligned} \tag{9.8.3}$$

so that $\langle T_1 x | m_x \rangle = -1$, $\langle T_1 y | m_y \rangle = -i$, and $\langle T_1 z | m_z \rangle = 1$. Note that since Table C.5.1(b) gives $iy \sim |T_1 y\rangle$ ("$\sim$" means "transforms as"), it follows that $y \sim -i|T_1 y\rangle$.

Similarly, in the real-O basis the atomic p functions, which also transform as x, y, and z, give

$$\begin{aligned} p_x &= \langle T_1 x | p_x \rangle | p_x\, t_1 x \rangle = -|p_x\, t_1 x\rangle = -|t_1 x\rangle \\ p_y &= \langle T_1 y | p_y \rangle | p_y\, t_1 y \rangle = -i|p_y\, t_1 y\rangle = -i|t_1 y\rangle \\ p_z &= \langle T_1 z | p_z \rangle | p_z\, t_1 z \rangle = |p_z\, t_1 z\rangle = |t_1 z\rangle \end{aligned} \tag{9.8.4}$$

We use lowercase irrep labels for operators and one-electron orbitals. In each case the final equality illustrates a common simplification in the notation.

The $\langle f\phi | \mathcal{O} \rangle$ and $\langle a\alpha | \psi \rangle$ coefficients for the group T_d differ from those for the group O when $\mathcal{O}$ or ψ are odd under inversion. This is because the odd-parity $\mathcal{O}$ and ψ belong to different irreps in O and T_d (see below). For example, x, y, and z belong to T_1 in O, but to T_2 in T_d. Thus in the real-T_d basis we find, using Table C.5.1(b), that

$$\begin{aligned} m_x &= -m_\xi^{t_2} = -m_x^{t_2} \\ m_y &= i m_\eta^{t_2} = i m_y^{t_2} \\ m_z &= m_\zeta^{t_2} = m_z^{t_2} \end{aligned} \tag{9.8.5}$$

where we have used two alternative labeling schemes for the T_2 irrep partners.

It is often convenient to express operators in terms of spherical vectors or spherical tensors rather than Cartesian coordinates. *Spherical tensors* (or irreducible tensors) are defined to have identical transformation properties to the $|jm\rangle$ of SO_3 in the angular momentum basis ($SO_3 \supset SO_2$). *Spherical vectors* are the special case of spherical tensors for which $j = 1$. Thus for a vector **V**,

$$\mathbf{V} = V_x\mathbf{e}_x + V_y\mathbf{e}_y + V_z\mathbf{e}_z$$

$$= -V_{-1}\mathbf{e}_1 + V_0\mathbf{e}_0 - V_1\mathbf{e}_{-1} \qquad (9.8.6)$$

where V_x, V_y, and V_z transform respectively as x, y, and z, and

$$V_{-1} \equiv \frac{1}{\sqrt{2}}(V_x - iV_y) \sim |1\ -1\rangle$$

$$V_0 \equiv V_z \sim |10\rangle \qquad (9.8.7)$$

$$V_1 \equiv \frac{-1}{\sqrt{2}}(V_x + iV_y) \sim |11\rangle$$

Note that $V_{\pm 1}$ operators differ from $V^{\pm}$ operators (often used as step-up and step-down operators):

$$V_{-1} = \frac{1}{\sqrt{2}}V^-$$

$$(9.8.8)$$

$$V_1 = \frac{-1}{\sqrt{2}}V^+$$

The **e**'s are unit vectors with $\mathbf{e}_{\pm 1} = \mp(1/\sqrt{2})(\mathbf{e}_x \pm i\mathbf{e}_y)$ and $\mathbf{e}_0 = \mathbf{e}_z$. They obey the orthonormality relations

$$\mathbf{e}_\mu \cdot \mathbf{e}_{\mu'} = \delta_{\mu,\,-\mu'}(-1)^\mu, \qquad \mu, \mu' = 0, 1, -1 \qquad (9.8.9)$$

so that the scalar (dot) product of two vectors becomes

$$\mathbf{V}\cdot\mathbf{W} = \sum_{\mu=0,1,-1} (-1)^\mu V_\mu W_{-\mu}$$

$$= -V_1W_{-1} + V_0W_0 - V_{-1}W_1$$

$$= V_xW_x + V_yW_y + V_zW_z \qquad (9.8.10)$$

Inverting (9.8.7) gives the useful relations

$$V_x = \frac{1}{\sqrt{2}}(V_{-1} - V_1) \sim \frac{1}{\sqrt{2}}(|1\,-1\rangle - |11\rangle)$$

$$V_y = \frac{i}{\sqrt{2}}(V_{-1} + V_1) \sim \frac{i}{\sqrt{2}}(|1\,-1\rangle + |11\rangle) \qquad (9.8.11)$$

$$V_z = V_0 \sim |10\rangle$$

We follow the convention of using **R** to denote a generalized vector operator which is even (superscript +) under inversion (such as **l**, **s**, **j**, or $\boldsymbol{\mu}$) and **V** to denote a vector operator which is odd (superscript −) under inversion (such as **m**). This notation is useful because in the full rotary-inversion group ($O_3 = SO_3 \otimes C_i$), in the angular-momentum basis, the components of **R** transform as the $|1^+m\rangle$ while those of **V** transform as the $|1^-m\rangle$ (see Section C.7). In subgroups of O_3 this difference may lead to different transformation properties for **R** and **V**. Thus in the group T_d, components of **R** transform as T_1 but those of **V** transform as T_2, while in the group O_h components of **V** transform as T_{1u} and those of **R** as T_{1g} (see Table C.5.1(b) and Section C.7).

Since the transformation properties of components of spherical vectors and tensors follow those of the $|jm\rangle$, it is easy to define the $\langle f\phi|\mathcal{O}\rangle$ for them using our standard basis definitions. In the complex-O basis, Section C.7 gives, for example, for **V** = **m**

$$m_{-1} = \langle T_1\,-1|m_{-1}\rangle m^{t_1}_{-1} = -m^{t_1}_{-1}$$

$$m_0 = \langle T_1 0|m_0\rangle m^{t_1}_0 = m^{t_1}_0 \qquad (9.8.12)$$

$$m_1 = \langle T_1 1|m_1\rangle m^{t_1}_1 = -m^{t_1}_1$$

Clearly, the $\langle T_1\phi|\mathcal{O}\rangle$ for analogous components of other vector operators in the same basis are identical. Values in the standard complex-T_d basis are given in Section C.7. We emphasize again that they differ from those in the complex-O basis.

Just as **R** and **V** may transform differently in O_3 subgroups because of their different parity, so may the $|j^+m\rangle$ and the $|j^-m\rangle$. The $|j^{\pm}m\rangle$ are the O_3 basis functions in the angular-momentum basis (this is the $O_3 \supset SO_3 \supset SO_2$ basis in chain notation—see Section 15.2). Spin functions have intrinsic even parity, so in the $O_3 \supset SO_3 \supset SO_2$ basis they transform as the $|j^+m\rangle$. On the other hand, if the angular-momentum states of interest are the spherical harmonics Y^j_m of Section B.1, the parity is + when j is an even integer and − when j is an odd integer. Thus Y^j_m transforms as $|j^+m\rangle$ for $j = 0, 2, 4, \ldots$, but as $|j^-m\rangle$ for $j = 1, 3, 5, \ldots$ in the $O_3 \supset SO_3 \supset SO_2$ basis. The transformation coefficients for the $|j^{\pm}m\rangle$ in a group (such as T_d or O_h) follow directly from the relations between the $O_3 \supset SO_3 \supset SO_2$ basis and the group basis of interest—see, for example, Section C.6.

Function and operator transformation coefficients are tabulated in Sections C.7, D.6, and E.6. It is important to realize that transformation coefficients are very much basis-dependent. In this regard, note that both the basis functions and the definitions of components of vector operators given in this book differ from the definitions used previously by the authors. Our current notation conforms to that of Butler and that used in all standard angular-momentum texts.

10 High-Symmetry Coupling Coefficients: The 3jm

In this chapter we introduce the high-symmetry coupling coefficients known as $3jm$. They replace, in nearly all applications in the remainder of the book, the vector coupling coefficients $(a\alpha, b\beta|rc\gamma)$ defined in Chapter 9. As demonstrated by Griffith (1962), the use of high-symmetry coefficients allows the very powerful methods of Racah algebra, originally developed for the group SO_3 of all rotations, to be applied to molecular point groups. Using these methods, calculations may be performed within the latter groups almost as easily as within SO_3 itself. Such calculations are faster and less prone to error than the older techniques [e.g. those of Griffith (1964)] and are also more straightforward than operator-equivalent methods and the like.

Examples of the use of the $3jm$ in relatively simple calculations are given in Chapters 11–13. Then in Chapters 17–23 we demonstrate more sophisticated uses of the $3jm$, and of the related $6j$ and $9j$ defined in Chapter 16. These chapters make clear the enormous advantage of high-symmetry coefficients in complicated situations.

10.1. Introduction

At this initial stage we introduce the $3jm$ and equations using them with little or no justification; that comes in Chapter 14. We have chosen this approach because it is far easier to use these results than to derive them, and once the notation is familiar it is easier to understand the derivations. The equations of this chapter are shown in Chapter 14 to apply to all common molecular point groups.

Advantages of the 3jm

Some readers may already be familiar with high-symmetry coefficients. Besides the $3jm$, examples include $\overline{V}$ and $3j$ coefficients used for atoms [see

Fano and Racah (1959), Edmonds (1963), Brink and Satchler (1968), and Rotenberg et al. (1959)], and V coefficients used for molecular point groups (see Griffith (1962), Dobosh (1972), and Piepho (1979)). The $3jm$ which are described by Butler (1981), have the same symmetry properties as V coefficients for molecular point groups and as the $3j$ for the group of all rotations, SO_3. Butler has, however, chosen defining equations for the $3jm$ that differ from those of Griffith (1962) and Dobosh (1972) for V coefficients. Therefore, reduced matrix elements in the $3jm$ formulation differ in phase from those in Griffith (1962), and the relation between the $3jm$ and v.c.c. differs from that between V coefficients and v.c.c.. Consequently, the general equations in our book and in Butler (1981) differ somewhat from those in Griffith (1962) and Piepho (1979).

We follow the $3jm$ formulation of Butler (1981) for several reasons. Perhaps most important is that Butler's book contains a complete tabulation of $3jm$ [in the form of $3jm$ factors (Section 15.3)] for the molecular point groups. Secondly, we strongly believe that standardization of notation and defining equations is essential. Other important advantages of the Butler formulation are described in Chapter 15.

10.2. Definition of the 3*jm* and Related Quantities

In this section we rewrite the Wigner–Eckart theorem (9.3.23) and the $|rc\gamma\rangle$ expansion formula (9.3.21) in terms of $3jm$, and show how $3jm$ and v.c.c. are related. To do this we must first introduce notation for conjugate irreps and irrep components (partners). This need arises because the transformation properties of a ket $|a\alpha\rangle$ and its conjugate $|a\alpha\rangle^\dagger \equiv \langle a\alpha|$ may differ. As discussed in Chapter 14, this difference requires the definition of a^* irreps and α^* components. It also leads to the appearance of $2jm$ phase factors $\begin{pmatrix} a \\ \alpha \end{pmatrix}$ in many equations. Conjugate irrep notation is not needed in v.c.c., since the bra-and-ket notation for the v.c.c. accomplishes the same purpose.

The Wigner-Eckart Theorem Using the 3jm

Following Butler (1981), we define the $3jm$

$$\begin{pmatrix} a & b & c^* \\ \alpha & \beta & \gamma^* \end{pmatrix}^{r}$$

by the equations

$$(rc\gamma|a\alpha, b\beta) \equiv |c|^{1/2}\begin{pmatrix} c \\ \gamma \end{pmatrix}\begin{pmatrix} a & b & c^* \\ \alpha & \beta & \gamma^* \end{pmatrix}^r$$

$$(a\alpha, b\beta|rc\gamma) \equiv (rc\gamma|a\alpha, b\beta)^*$$

$$= |c|^{1/2}\begin{pmatrix} c \\ \gamma \end{pmatrix}\begin{pmatrix} a & b & c^* \\ \alpha & \beta & \gamma^* \end{pmatrix}^{*\,r} \tag{10.2.1}$$

Thus (9.3.23) and (9.3.21) become

$$\langle a\alpha|O^f_\phi|b\beta\rangle = \begin{pmatrix} a \\ \alpha \end{pmatrix}\sum_r \begin{pmatrix} a^* & f & b \\ \alpha^* & \phi & \beta \end{pmatrix}^r \langle a\|O^f\|b\rangle_r \tag{10.2.2}$$

and

$$|(ab)rc\gamma\rangle = |c|^{1/2}\begin{pmatrix} c \\ \gamma \end{pmatrix}\sum_{\alpha\beta}\begin{pmatrix} a & b & c^* \\ \alpha & \beta & \gamma^* \end{pmatrix}^{*\,r} |a\alpha\rangle|b\beta\rangle \tag{10.2.3}$$

We tabulate the $3jm$ rather than the v.c.c. in our appendixes and calculate v.c.c. where required via (10.2.1).

An additional phase factor $H(abc)$ is inserted on the right side of (10.2.1) and (10.2.3) for the SO_3 or O_3 groups (see Section B.3). This factor is needed to obtain the "historical" Condon–Shortley phases for the SO_3 and O_3 v.c.c. in the angular-momentum basis [see Butler (1981)]. For SO_3 and O_3, $H(abc) = H(j_1 j_2 j_3) = (-1)^{-j_1+j_2-j_3}$.

For many groups the $3jm$ are real numbers, so that $3jm = 3jm^*$. The index r and the sum over r are required only when repeated irreps occur and may otherwise be dropped (Section 9.3).

To use (10.2.2) and (10.2.3) one simply looks up the appropriate $2jm$ phase factor (abbreviated $2jm$) and $3jm$ in a table. For the groups O and T_d they are given in Sections C.11 and C.12 for several common bases. $2jm$ and $3jm$ for several other groups are given in Sections B.3, D.8–D.9 and E.8–E.9. The $3j$ symbols of Rotenberg et al. (1959) are our $3jm$ for $SO_3 \supset SO_2$—see Section B.3. Those for all remaining point groups may be found in Butler (1981) in factored form (Section 15.3). Not all required $3jm$ are given in the tables, since

$$\begin{pmatrix} a & b & c \\ \alpha & \beta & \gamma \end{pmatrix}^r = \begin{pmatrix} b & c & a \\ \beta & \gamma & \alpha \end{pmatrix}^r = \begin{pmatrix} c & a & b \\ \gamma & \alpha & \beta \end{pmatrix}^r$$

$$= \{abcr\}\begin{pmatrix} b & a & c \\ \beta & \alpha & \gamma \end{pmatrix}^r = \{abcr\}\begin{pmatrix} c & b & a \\ \gamma & \beta & \alpha \end{pmatrix}^r$$

$$= \{abcr\}\begin{pmatrix} a & c & b \\ \alpha & \gamma & \beta \end{pmatrix}^r \tag{10.2.4}$$

Thus $3jm$ are defined so that cyclic (even) interchanges of columns leave the value unchanged and noncyclic (odd) interchanges multiply the coefficient by the $3j$ *phase* $\{abcr\}$; it is for this reason that they are called high-symmetry coefficients. Only one of the six $3jm$ in (10.2.4) is tabulated; but once the $3j$ phases $\{abcr\}$ are defined, all others may be found using (10.2.4). $3j$ phases for the groups O and T_d are given in Section C.11.

The 3j Phase $\{abcr\}$

The selection of the $\{abcr\}$ phases (which are analogous to the $(-1)^{j_1+j_2+j_3}$ phase factors of SO_3) involves more arbitrary choice than one might expect. The only limitation on the choice of phase arises when we couple two functions together which belong to the same irrep. As is discussed further below, when c occurs in the *symmetric product* (written $[a]^2$) of $a \otimes a$, we must have $\{aacr\} = +1$, and when c occurs in the *antisymmetric product* (written $(a)^2$), we must have $\{aacr\} = -1$. All other $\{abcr\}$ phases may be chosen arbitrarily. We always choose the $\{abcr\}$ phases to have values ± 1 and to be invariant to permutation of the irrep order abc. Thus $\{abcr\} = \{bacr\} = \{cabr\}$, and so on. Also $\{abcr\} = \{a^*b^*c^*r\}$.

For groups which are termed simply reducible, $3j$ phases may be chosen to fit a general formula. This formula, which expresses a $3j$ phase as a product of irrep phases, is extremely convenient in that it leads to simplification of later equations. *Simply reducible* groups are *both* multiplicity-free (i.e. contain no repeated representations in irrep direct products, so $r = 0$ only), and ambivalent. *Ambivalent* groups have all real characters; in such groups all irreps are self-conjugate—or, symbolically, $a = a^*$ for each irrep a. *Nonambivalent* groups have some irreps with complex characters; for these irreps, $a^* = b \neq a$.

For the simply reducible groups we therefore choose a phase $(-1)^a$ for each irrep such that all $3j$ phases obey

$$\{abcr\} = (-1)^a(-1)^b(-1)^c \tag{10.2.5}$$

Moreover, we choose $(-1)^a$ such that (10.2.5) is true for *non*-simply-reducible groups as far as possible. Thus, for example, $(-1)^a$ may be selected so that (10.2.5) is obeyed by *most* $\{abcr\}$ for O, T_d, and D_3, and by *all* $\{abcr\}$ of D_4 and SO_3. We term $3j$ phases that do not obey (10.2.5) *law-breaking factors*. In the case of O and T_d, if we assign $(-1)^a$ such that

$$\begin{gathered}(-1)^{A_1}=(-1)^{E}=(-1)^{T_2} = 1\\ (-1)^{A_2}=(-1)^{T_1} = -1\\ (-1)^{E'} = (-1)^{E''} = i, \qquad (-1)^{U'} = -i\end{gathered} \tag{10.2.6}$$

Equation (10.2.5) holds *except for* $\{U'U'T_2r\}$ for $r = 1$. $\{U'U'T_21\}$ must be given the value $+1$ (see below), whereas by (10.2.6) above it would equal -1.

The existence of law-breaking factors is one of the major differences between SO_3 in the angular-momentum basis and the molecular point groups. In $SO_3 \supset SO_2$, irreps and irrep components can be assigned numerical values: specifically the j and m quantum numbers for spin, orbital, or total angular momentum. Thus the equivalents of our $\{abcr\}$ and $\begin{pmatrix} a \\ \alpha \end{pmatrix}$ phases—namely $(-1)^{j_1+j_2+j_3}$ and $(-1)^{j_1-m_1}$—may be treated algebraically, since j_1, j_2, j_3, and m_1 have numerical values. Section B.3 gives the algebraic formulas for the various SO_3 phase factors.

Limitations on the Choice of $\{abcr\}$ Phases

The advantage of the $3jm$ over v.c.c. depends to a large extent on $\{abcr\}$ phases having the simplest form possible, and we therefore allow values ± 1 only. These values are always *required* when one of the irreps appears more than once in the $3jm$—i.e. say $a = b$—and no repeated irreps occur in $a \otimes a$. Then from (10.2.3) and (10.2.4),

$$|(aa)rc\gamma\rangle = |c|^{1/2}\begin{pmatrix} c \\ \gamma \end{pmatrix}\sum_{\alpha\beta}\frac{1}{2}\left[\begin{pmatrix} a & a & c^* \\ \alpha & \beta & \gamma^* \end{pmatrix}^{*\,r}|a\alpha\rangle|a\beta\rangle\right.$$
$$\left.+\begin{pmatrix} a & a & c^* \\ \beta & \alpha & \gamma^* \end{pmatrix}^{*\,r}|a\beta\rangle|a\alpha\rangle\right]$$
$$= |c|^{1/2}\begin{pmatrix} c \\ \gamma \end{pmatrix}\sum_{\alpha\beta}\begin{pmatrix} a & a & c^* \\ \alpha & \beta & \gamma^* \end{pmatrix}^{*\,r}$$
$$\times\tfrac{1}{2}\left[|a\alpha\rangle|a\beta\rangle + \{aac^*r\}|a\beta\rangle|a\alpha\rangle\right] \qquad (10.2.7)$$

When no repeated irreps occur in $a \otimes a$, a function $|(aa)c\gamma\rangle$ which obeys standard basis relations is either symmetric or antisymmetric with respect to the interchange of α and β, depending on whether c occurs in the symmetric (written $[a]^2$) or antisymmetric product (written $(a)^2$) of $a \otimes a$ (see Sections C.3, D.3, and E.3). Clearly for the symmetric functions above, $\{aac^*r\} = +1$, and for antisymmetric functions, $\{aac^*r\} = -1$. Thus these factors necessarily have our specified form.

When repeated irreps occur in $a \otimes b$, two linearly independent $|(ab)rc\gamma\rangle$ functions (which differ in the label r) may be constructed which obey standard basis relations. But any unitary transformation of these two functions obeys the same standard basis relations, and so the particular definition of the two "standard" linearly independent functions is arbitrary (see Section 9.7). For the group O when $|(aa)rc\gamma\rangle = |(U'U')rT_1\gamma\rangle$, any choice gives symmetric functions $|[U']^2rT_1\gamma\rangle$, since T_1 occurs twice in $[U']^2$ and not at all in $(U')^2$. Thus for all choices $\{U'U'T_1r\} = 1$. The situation is quite different for $|(U'U')rT_2\gamma\rangle$ functions, since T_2 occurs once in $[U']^2$ and once in $(U')^2$. While we can (and do) choose our two linearly independent functions to be the antisymmetric $|(U')^2 0T_2\gamma\rangle$ and the symmetric $|[U']^2 1T_2\gamma\rangle$ functions so that $\{U'U'T_2 0\} = -1$ and $\{U'U'T_2 1\} = +1$, other choices of standard $|(U'U')rT_2\rangle$ functions in general are *neither* symmetric *nor* antisymmetric with respect to interchange of α and β. They are not, therefore, consistent with our requirement that $\{abcr\}$ equal ± 1.

In summary, when using the irreducible tensor method we always require $|(aa)rc\gamma\rangle$ functions to be either symmetric or antisymmetric; thus

$$\{aac^*r\} = \begin{cases} 1 & \text{when} \quad rc \in [a]^2 \\ -1 & \text{when} \quad rc \in (a)^2 \end{cases} \tag{10.2.8}$$

$\{abcr\}$ in which no two irreps are identical may for convenience be defined as either ± 1. Our choices are made so that (10.2.5) is obeyed as far as possible. In all cases our $3j$ phases are chosen identical to those of Butler (1981).

The 2j Phase $\{a\}$

While the $2j$ phase $\{a\}$ is not used until later, we define it now, since it is closely related to the $3j$ phase:

$$\begin{aligned} \{a\} &\equiv \{aa^*A_1\} \\ &= (-1)^a(-1)^{a^*} \end{aligned} \tag{10.2.9}$$

Here A_1 is the totally symmetric irrep, and r is dropped because the A_1 irrep never appears as a repeated representation. For single-valued irreps $\{a\} = 1$, and for double-valued irreps $\{a\} = -1$. Thus for the group O, $\{a\} = 1$ for A_1, A_2, E, T_1, and T_2, and -1 for E', E'', and U'.

For any set of irreps appearing in a nonzero $3jm$,

$$\{a\}\{b\}\{c\} = 1 \tag{10.2.10}$$

This follows from the direct-product rules of (9.7.1); in any nonzero $3jm$, either two or none of the factors in (10.2.10) are -1. In the group O, for example, either two or none of the irreps are from the set E', E'', and U'.

The 2jm Phase $\begin{pmatrix} a \\ \alpha \end{pmatrix}$

The $2jm$ phase (or simply $2jm$) is used to relate the transformation properties of $|a\alpha\rangle^{\dagger} \equiv \langle a\alpha|$ to those of $|a\alpha\rangle$. We define it to be ± 1. The $2jm$ phase is simply related to the $3jm$ for aa^*A_1:

$$\begin{pmatrix} a \\ \alpha \end{pmatrix} = |a|^{1/2} \begin{pmatrix} a & a^* & A_1 \\ \alpha & \alpha^* & a_1 \end{pmatrix} \tag{10.2.11}$$

Thus separate tabulations of the $2jm$ are really unnecessary. We give $2jm$ for representative groups in Sections B.3, C.11, D.8, and E.8. Note that for SO_3 in the $\lfloor jm\rangle$ basis,

$$\begin{pmatrix} j \\ m \end{pmatrix} = (-1)^{j-m}$$

From (10.2.11), (10.2.4), and (10.2.9),

$$\begin{pmatrix} a^* \\ \alpha^* \end{pmatrix} = \{a\} \begin{pmatrix} a \\ \alpha \end{pmatrix} \tag{10.2.12}$$

Definition of a* and α*

The irrep a^* and irrep component α^* are defined using the time-reversal operator T (Chapter 14). The resulting definitions are summarized in $3jm$ tabulations, since $3jm$ containing the A_1 irrep contain $a\alpha$ and $a^*\alpha^*$ as column labels. All irreps with real characters have $a^* = a$. For SO_3 in the $\lfloor jm\rangle$ basis the algebraic rules $j^* = j$ and $m^* = -m$ apply. We list a^* and α^* in Sections C.11, D.8, and E.8 for the relevant groups.

10.3. Useful Equations Involving the 3jm

Relation of 3jm Containing an A_1 Representation to the 2jm

We show in Section 14.6 that $3jm$ containing A_1 have the value

$$\begin{pmatrix} a & b & A_1 \\ \alpha & \beta & a_1 \end{pmatrix} = \delta_{ba^*}\delta_{\beta\alpha^*}|a|^{-1/2}\begin{pmatrix} a \\ \alpha \end{pmatrix} \tag{10.3.1}$$

*Relation between 3jm and 3jm**

In Section 14.8 we show that our conventions lead to the definition of $3jm^*$:

$$\begin{pmatrix} a & b & c \\ \alpha & \beta & \gamma \end{pmatrix}^{* r} = \begin{pmatrix} a \\ \alpha \end{pmatrix}\begin{pmatrix} b \\ \beta \end{pmatrix}\begin{pmatrix} c \\ \gamma \end{pmatrix}\begin{pmatrix} a^* & b^* & c^* \\ \alpha^* & \beta^* & \gamma^* \end{pmatrix}^{r} \tag{10.3.2}$$

This equation together with (10.2.1) imposes a further phase standardization on v.c.c. beyond that of (10.2.4). It allows us to express $3jm^*$ in terms of $3jm$. This is essential for the derivation of general equations in Chapters 16–22.

The Coupling of Conjugate States

Conjugate states are coupled according to the equation

$$\begin{aligned} \langle (ab)rc\gamma| &= \sum_{\alpha\beta} (rc\gamma|a\alpha, b\beta)\langle a\alpha|\langle b\beta| \\ &= |c|^{1/2}\begin{pmatrix} c \\ \gamma \end{pmatrix}\sum_{\alpha\beta}\begin{pmatrix} a & b & c^* \\ \alpha & \beta & \gamma^* \end{pmatrix}^{r} \langle a\alpha|\langle b\beta| \end{aligned} \tag{10.3.3}$$

which follows from (10.2.1).

Orthonormality Equations for the 3jm

Orthonormality equations for the $3jm$ follow directly from the unitary properties of the v.c.c.. Since the $3jm$ are related to v.c.c. via (10.2.1), we have

$$|c|\sum_{\alpha\beta}\begin{pmatrix} a & b & c \\ \alpha & \beta & \gamma \end{pmatrix}^{r}\begin{pmatrix} a & b & c' \\ \alpha & \beta & \gamma' \end{pmatrix}^{* r'} = \delta_{rr'}\delta_{cc'}\delta_{\gamma\gamma'}\delta(abcr) \tag{10.3.4}$$

$$\sum_{rc\gamma}|c|\begin{pmatrix} a & b & c \\ \alpha & \beta & \gamma \end{pmatrix}^{r}\begin{pmatrix} a & b & c \\ \alpha' & \beta' & \gamma \end{pmatrix}^{* r} = \delta_{\alpha\alpha'}\delta_{\beta\beta'}\delta(abcr) \tag{10.3.5}$$

and

$$\sum_{\alpha\beta\gamma}\begin{pmatrix} a & b & c \\ \alpha & \beta & \gamma \end{pmatrix}^{r}\begin{pmatrix} a & b & c \\ \alpha & \beta & \gamma \end{pmatrix}^{* r'} = \delta_{rr'}\delta(abcr) \tag{10.3.6}$$

where $\delta(abcr) = 1$ if rc^* is contained in $a \otimes b$ and is zero otherwise. The first two equations are equivalent to the column and row orthonormality of a unitary matrix [Equation (8.2.27)].

These sum rules, together with (10.2.1), (10.2.4), (10.2.10), (10.2.12), and (10.3.1), prove very useful in deriving general formulas involving higher invariants—the $6j$ and $9j$ coefficients—in later chapters.

Inversion of the Reduced-Matrix-Element Equation

It is most convenient to derive the matrix-element equations in reduced form, since we can then relate one reduced matrix element to another, avoiding all appearance of components in our equations. Multiplying both sides of (10.2.2) by

$$\begin{pmatrix} a \\ \alpha \end{pmatrix} \begin{pmatrix} a^* & f & b \\ \alpha^* & \phi & \beta \end{pmatrix}^{*\,r'}$$

and summing over α, ϕ, and β using (10.3.6), we obtain an equation useful for such derivations:

$$\langle a \| O^f \| b \rangle_r = \sum_{\alpha\phi\beta} \begin{pmatrix} a \\ \alpha \end{pmatrix} \begin{pmatrix} a^* & f & b \\ \alpha^* & \phi & \beta \end{pmatrix}^{*\,r} \langle a\alpha | O^f_\phi | b\beta \rangle \qquad (10.3.7)$$

10.4. Example of Use of the Wigner–Eckart Theorem: Demonstration that T_1 and T_2 States in the Complex-*O* Basis are Diagonal in μ_z

In this section we illustrate the use of the Wigner–Eckart theorem in a practical problem. When MCD parameters are calculated via (4.5.16), a basis diagonal in μ_z is required. We show here with the help of the Wigner–Eckart theorem that the T_1 and T_2 partners in the complex-O basis of Section C.5 satisfy this requirement.

The transformation properties of operators are discussed in Section 9.8 and summarized in Section C.7 for the complex-O basis. We see from Section C.7 that μ_z transforms as T_10 in the complex-O basis with a transformation coefficient of 1: thus we write $\mu_z = \mu^{t_1}_0$. Our problem is therefore to show that

$$\langle T_i\alpha | \mu^{t_1}_0 | T_i\beta \rangle = \delta_{\alpha\beta} \langle T_i\beta | \mu^{t_1}_0 | T_i\beta \rangle \qquad (10.4.1)$$

for $i = 1$ and 2 and $\alpha, \beta = 0, 1, -1$. Since by (10.2.2) the Wigner–Eckart theorem reads

$$\langle T_i\alpha|\mu_0^{t_1}|T_i\beta\rangle = \begin{pmatrix} T_i \\ \alpha \end{pmatrix}\begin{pmatrix} T_i^* & T_1 & T_i \\ \alpha^* & 0 & \beta \end{pmatrix}\langle T_i\|\mu^{t_1}\|T_i\rangle \tag{10.4.2}$$

the T_1 and T_2 partners are diagonal in μ_z if

$$\begin{pmatrix} T_i^* & T_1 & T_i \\ \alpha^* & 0 & \beta \end{pmatrix} = \delta_{\alpha\beta}\begin{pmatrix} T_i^* & T_1 & T_i \\ \beta^* & 0 & \beta \end{pmatrix} \tag{10.4.3}$$

The $3jm$ table of Section C.12 allows us to demonstrate that the above relation holds in the complex-O basis. When $i = 2$, for example, (10.2.4) and Section C.11 give

$$\begin{aligned}\begin{pmatrix} T_2^* & T_1 & T_2 \\ \alpha^* & 0 & \beta \end{pmatrix} &= \{T_2T_1T_2\}\begin{pmatrix} T_2^* & T_2 & T_1 \\ \alpha^* & \beta & 0 \end{pmatrix} \\ &= -1\begin{pmatrix} T_2 & T_2 & T_1 \\ \alpha^* & \beta & 0 \end{pmatrix}\end{aligned} \tag{10.4.4}$$

Then, examining Section C.12, we find the table

T_2	T_2	T_1	$3jm^\dagger$
-1	0	-1	$1/\sqrt{6}$
-1	1	0	$1/\sqrt{6}$
0	1	1	$1/\sqrt{6}$

which means, for example, using the second row, that

$$\begin{pmatrix} T_2 & T_2 & T_1 \\ -1 & 1 & 0 \end{pmatrix} = \frac{1}{\sqrt{6}} = \begin{pmatrix} T_2 & T_2 & T_1 \\ \alpha^* & \beta & 0 \end{pmatrix} \tag{10.4.6}$$

Thus the only nonzero values for the $3jm$ occur when $\alpha^* = -1$ and $\beta = 1$ or, using (10.2.4), when $\alpha^* = 1$ and $\beta^* = -1$. All $3jm$ which are neither found in the table, nor are obtainable from those in the table via (10.2.4), are zero. Since by Section C.11 we have for the T_2 components that $(-1)^* = 1$ and $1^* = -1$, the nonzero values are those for which $\alpha = \beta$ in (10.4.4). Filling in the explicit $2jm$ and $3jm$ values from Sections C.11 and

C.12 [and using (10.2.4)], we find for the two nonzero cases that

$$\begin{aligned}\langle T_2 1|\mu_0^{t_1}|T_2 1\rangle &= \frac{1}{\sqrt{6}}\langle T_2\|\mu^{t_1}\|T_2\rangle \\ \langle T_2 -1|\mu_0^{t_1}|T_2 -1\rangle &= \frac{-1}{\sqrt{6}}\langle T_2\|\mu^{t_1}\|T_2\rangle\end{aligned} \tag{10.4.7}$$

Thus the T_2 states are indeed diagonal in μ_z in the complex-O basis. Similar arguments show the same is true for T_1 states, and in addition, that neither T_1 nor T_2 states are diagonal in μ_z in the real-O basis.

11 Elementary Uses of Coupling Tables

11.1. Introduction

In this chapter we present introductory applications of coupling. A classic MCD example is chosen to illustrate the use of coupling coefficients in matrix-element reduction via the Wigner–Eckart theorem (10.2.2), while the construction of *LS*-type wavefunctions provides examples which employ the coupling equation (10.2.3). We limit our group theory at present to equations from Chapter 10. Such methods are practical only for fairly simple cases. In more complicated situations, the irreducible-tensor method (Chapters 14–22) is essential because it completely obviates the need to construct explicit wavefunctions. Nevertheless, it is important to understand the relatively clumsy methods of this and the next few chapters. First they serve as a crucial introduction to the much more powerful and elegant methods which follow. Second, they can be the easiest way to attack simple problems. Lastly, appreciation of the limitations of these methods, particularly the staggering complexity which develops when systems with several open shells are considered, provides a strong impetus to confront the additional nomenclature, conventions, and more abstract formalism of the irreducible-tensor method.

11.2. Use of the Wigner–Eckart Theorem in a Simple Example: The MCD of $Fe(CN)_6^{3-}$ Neglecting Spin–Orbit Coupling

In this section we illustrate the utility of the Wigner–Eckart theorem by way of a classic MCD example, the calculation of the MCD of $Fe(CN)_6^{3-}$ [see Stephens (1965a) and Schatz et al. (1966)]. To keep the example simple, we neglect spin–orbit coupling; later (Section 13.5) we show that when we repair this omission, the numerical results change significantly, but the qualitative conclusions are unaffected. $Fe(CN)_6^{3-}$ is an octahedral (O_h) ion with a strong-field $d^5(t_{2g}^5)$ ground configuration. The room-temperature electronic absorption and MCD spectra in solution are shown in Fig. 11.2.1. The three moderately strong bands arise from allowed ligand-to-metal charge-transfer transitions. Since the bands are reasonably well separated, the zeroth and first MCD and absorption moments (Chapter 7) should in principle permit a rough experimental determination of the Faraday parameters $\mathcal{A}_1$, $\mathcal{B}_0 + \mathcal{C}_0/kT$, and $\mathcal{D}_0$ for each band. These parameters contain the molecular information and have been discussed in detail in Chapters 4 and 5.

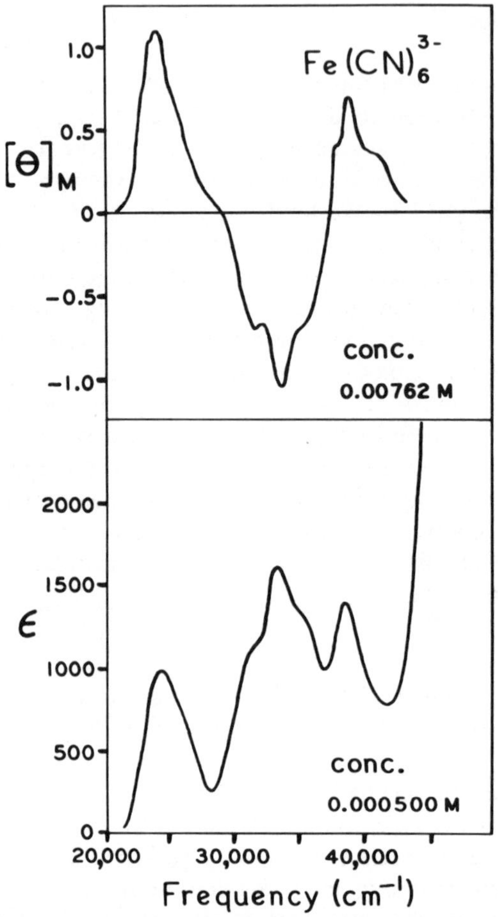

Figure 11.2.1. Room-temperature electronic absorption and MCD spectra of $Fe(CN)_6^{3-}$ in solution. $[\theta]_M$ is the MCD; the units are molar ellipticity per unit magnetic field (Section A.5). ϵ is the molar extinction coefficient.

As always, we wish to distinguish clearly what is symmetry-determined from what is model-dependent. Important conclusions can often be drawn from the former alone, and in any case, such an analysis is an essential first step, since it determines the minimum number of independent matrix elements required to connect theory and experiment; it also displays this relationship explicitly.

$\mathcal{C}$-Term Calculation

We first calculate the $Fe(CN)_6^{3-}$ $\mathcal{C}$ terms. Since we are dealing with the solution spectrum of an octahedral ion, (4.6.16) applies and we may use the $\mathcal{C}_0$ expression of (4.5.16), namely

$$\mathcal{C}_0 = \frac{1}{\mu_B |A|} \sum_{\alpha\lambda} \langle A\alpha|\mu_z|A\alpha\rangle \times \left(|\langle A\alpha|m_{-1}|J\lambda\rangle|^2 - |\langle A\alpha|m_1|J\lambda\rangle|^2 \right) \tag{11.2.1}$$

Note that this form requires basis functions $A\alpha$ (but *not* $J\lambda$) diagonal in μ_z. The operators μ_z and m_γ ($\gamma = 1, -1$) are respectively components of the magnetic- and electric-dipole operators, and $|A|$ is the degree of degeneracy of the ground state A. (To simplify the notation, we have dropped the superscript zeros on matrix elements.)

Figure 11.2.2 shows a schematic molecular-orbital diagram for $Fe(CN)_6^{3-}$, suggesting a plausible ordering for the allowed charge-transfer excitations. Since the lowest-energy octahedral configuration is t_{2g}^5, the single-hole ground state must be ${}^2T_{2g}$. Similarly, the single-hole charge-transfer excited states are ${}^2T_{1u}$ or ${}^2T_{2u}$. [Our states are labeled as ${}^{2\mathbb{S}+1}h$, where $\mathbb{S}$ is the total spin quantum number, $2\mathbb{S} + 1$ is the spin multiplicity, and h denotes the irrep to which the orbital state belongs in the point group of interest—here

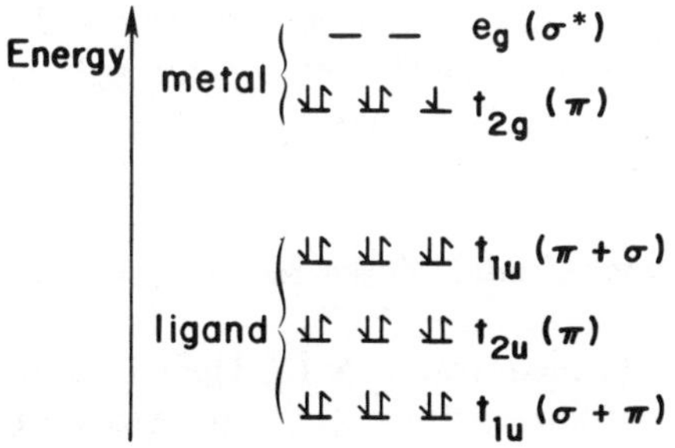

Figure 11.2.2. Orbital energy diagram for $Fe(CN)_6^{3-}$. Arrows denote electrons in the ground state configuration.

O_h. The construction of $|\mathcal{S}h\mathfrak{M}\theta\rangle$ wavefunctions (also called $|^{2\mathcal{S}+1}h\mathfrak{M}\theta\rangle$ wavefunctions) for these states is discussed in Sections 11.4–11.5.] In this simple example, each configuration corresponds to a single orbital state. Consequently, if spin–orbit coupling is neglected, A in (11.2.1) can be identified with $^2T_{2g}$, and J with $^2T_{1u}$ or $^2T_{2u}$.

To simplify (11.2.1) using the Wigner–Eckart theorem, we must first determine the transformation properties of the operators for the group O_h. This is accomplished as discussed in Section 9.8. We choose the complex-O (or, more exactly, the complex-O_h) basis for the problem, since (as demonstrated in Section 10.4) the T_1 and T_2 states are diagonal in μ_z in that basis. The transformation properties of the operators μ_z, m_{-1}, and m_1 in the complex-O_h basis are given in Section C.7:

$$\begin{aligned} \mu_z &= \mu_0 = \langle T_{1g}0|\mu_0\rangle \mu_0{}^{t_{1g}} = \mu_0{}^{t_{1g}} \\ &= (-\mu_B)\left(L_0^{t_{1g}} + 2S_0^{t_{1g}}\right) \\ m_1 &= \langle T_{1u}1|m_1\rangle m_1^{t_{1u}} = -m_1^{t_{1u}} \\ m_{-1} &= \langle T_{1u}-1|m_{-1}\rangle m_{-1}^{t_{1u}} = -m_{-1}^{t_{1u}} \end{aligned} \tag{11.2.2}$$

We show in Section 11.3 that when spin–orbit coupling is neglected and the $|\mathcal{S}h\mathfrak{M}\theta\rangle$ basis is used (as in this example), the S_z part of μ_z makes no contribution. We therefore drop this term, and we omit the explicit $\mathfrak{M}$ value from the wave functions. Substituting these results into (11.2.1) and noting that $|A| = |T_{2g}| = 3$, we obtain

$$\mathcal{C}_0 = -\frac{1}{3} \sum_{\substack{\alpha=1,0,-1 \\ \lambda=1,0,-1}} \langle {}^2T_{2g}\alpha|L_0^{t_{1g}}|{}^2T_{2g}\alpha\rangle \times \left(\left|\langle {}^2T_{2g}\alpha|m_{-1}^{t_{1u}}|{}^2T_{iu}\lambda\rangle\right|^2 - \left|\langle {}^2T_{2g}\alpha|m_1^{t_{1u}}|{}^2T_{iu}\lambda\rangle\right|^2 \right) \tag{11.2.3}$$

For the two transitions $^2T_{2g} \to {}^2T_{iu}$ $(i = 1,2)$, (11.2.3) requires the values of 42 matrix elements. However, many of these are zero, and the remainder can all be related to three independent reduced matrix elements through the Wigner–Eckart theorem (10.2.2). To use (10.2.2) we must first relate the O_h $3jm$ to those for the group O, since only the latter are tabulated.

Parity Labels and the Relation of a Direct-Product Group G_i to a Group G

The group O_h is a direct-product group formed from $O \otimes C_i$, where C_i is the inversion group (Section 8.13). The coupling coefficients for the group O_h

are not tabulated, but are found from those for the subgroup O, which has the same irrep labels without the parity (u, g) designations. Analogous statements apply to other groups G_i formed from $G \otimes C_i$.

The first step in using group-G $3jm$ (or v.c.c.) in a group-G_i problem is to check the parity (which are the C_i) rules:

$$\begin{aligned} g \otimes g &= g \\ u \otimes u &= g \\ g \otimes u = u \otimes g &= u \end{aligned} \tag{11.2.4}$$

It follows from these rules that nonzero v.c.c. or $3jm$ must contain either two u (odd-parity) irreps or none. If these rules are obeyed, the group-G_i $3jm$ (and other coefficients) are identical to those for the group G. As a consequence, in a group of type G_i (such as O_h), an odd-parity operator can only connect a bra and ket of opposite parities. And if the operator has even parity, it can only connect states of like parity. Otherwise the matrix element is zero.

Wigner–Eckart-Theorem Reductions

Following the above discussion, we note that all matrix elements in the $\mathcal{C}$-term expression of (11.2.3) obey the parity selection rules. Thus they may be reduced using (10.2.2), (10.2.4), and the group-O coefficients of Sections C.11–C.12. We find, for example, that

$$\begin{aligned} \langle {}^2T_{2g}1|m_1^{t_{1u}}|{}^2T_{1u}0\rangle &= \begin{pmatrix} T_2 \\ 1 \end{pmatrix}\begin{pmatrix} T_2^* & T_1 & T_1 \\ 1^* & 1 & 0 \end{pmatrix}\langle {}^2T_{2g}\|m^{t_{1u}}\|{}^2T_{1u}\rangle \\ &= (1)\begin{pmatrix} T_2 & T_1 & T_1 \\ -1 & 1 & 0 \end{pmatrix}\langle {}^2T_{2g}\|m^{t_{1u}}\|{}^2T_{1u}\rangle \\ &= \frac{1}{\sqrt{6}}\langle {}^2T_{2g}\|m^{t_{1u}}\|{}^2T_{1u}\rangle \end{aligned} \tag{11.2.5}$$

Similarly,

$$\begin{aligned} \langle {}^2T_{2g} \pm 1|L_0^{t_{1g}}|{}^2T_{2g} \pm 1\rangle &= \mp\frac{1}{\sqrt{6}}\langle {}^2T_{2g}\|L^{t_{1g}}\|{}^2T_{2g}\rangle \\ \langle {}^2T_{2g} \pm 1|m_{\pm 1}^{t_{1u}}|{}^2T_{1u}0\rangle &= \pm\frac{1}{\sqrt{6}}\langle {}^2T_{2g}\|m^{t_{1u}}\|{}^2T_{1u}\rangle \\ \langle {}^2T_{2g}0|m_{\pm 1}^{t_{1u}}|{}^2T_{1u} \pm 1\rangle &= \pm\frac{1}{\sqrt{6}}\langle {}^2T_{2g}\|m^{t_{1u}}\|{}^2T_{1u}\rangle \end{aligned} \tag{11.2.6}$$

where either all upper or all lower signs apply. All other combinations of α, λ in (11.2.3) give zero, because the $3jm$ required are zero [$3jm$ not listed in Section C.12 and not obtainable from those listed via (10.2.4) are zero].

Substituting (11.2.6) into (11.2.3), we obtain

$$\mathcal{C}_0\left({}^2T_{2g} \to {}^2T_{1u}\right) = \frac{-1}{9\sqrt{6}} \langle {}^2T_{2g} \| L^{t_{1g}} \| {}^2T_{2g} \rangle \times \left| \langle {}^2T_{2g} \| m^{t_{1u}} \| {}^2T_{1u} \rangle \right|^2 \qquad (11.2.7)$$

By a similar process, it is found that

$$\mathcal{C}_0\left({}^2T_{2g} \to {}^2T_{2u}\right) = \frac{1}{9\sqrt{6}} \langle {}^2T_{2g} \| L^{t_{1g}} \| {}^2T_{2g} \rangle \times \left| \langle {}^2T_{2g} \| m^{t_{1u}} \| {}^2T_{2u} \rangle \right|^2 \qquad (11.2.8)$$

Throughout, the new MCD sign conventions of Appendix A are used.

We emphasize that the results above are quite general and apply to any ${}^2T_{2g} \to {}^2T_{iu}$ ($i = 1, 2$) transition (with the assumption of zero spin–orbit coupling). The operators are all spin-independent, so it is only to identify the state involved that the doublet superscripts are retained in the reduced matrix elements. Indeed, the results really apply to *any* ${}^{2\mathbb{S}+1}T_{2g} \to {}^{2\mathbb{S}+1}T_{iu}$ transition for the group O_h. The numerical values of the reduced matrix elements differ in each specific case, but the form of the equations remains the same. Evaluation of reduced matrix elements is first discussed at length in Chapter 13.

Interpretation of the $Fe(CN)_6^{3-}$ MCD *Spectrum*

We can draw some interesting conclusions about the $Fe(CN)_6^{3-}$ MCD spectrum in Fig. 11.2.1 from the $\mathcal{C}$-term results above, since it is an experimental fact [Stephens (1965a); Schatz et al. (1966)] that this spectrum is dominated by $\mathcal{C}$ terms, as would be anticipated on the basis of the qualitative arguments presented in Section 5.3. We observe first from (11.2.7) and (11.2.8) that transitions to the ${}^2T_{1u}$ and ${}^2T_{2u}$ states *must* have opposite-signed $\mathcal{C}$ terms. Furthermore, if the three bands observed are associated with the three excitations in Fig. 11.2.2, the choice which places the ${}^2T_{2u}$ state between the two ${}^2T_{1u}$ states must be correct, since, experimentally, the first and third bands have $\mathcal{C}$ terms of the same sign.

We cannot however determine whether the signs of the $\mathcal{C}$ terms agree with experiment on the basis of symmetry arguments alone, because the sign of $\mathcal{C}_0$ is determined by the sign of the reduced matrix element $\langle {}^2T_{2g} \| L^{t_{1g}} \| {}^2T_{2g} \rangle$, which is a measure of the ground-state orbital angular momentum. The calculation of angular-momentum matrix elements is discussed in detail in Sections 13.2 and 13.4, where it is shown that

$$\langle {}^2T_{2g} \| L^{t_{1g}} \| {}^2T_{2g} \rangle = -\sqrt{6} \qquad (11.2.9)$$

if we represent the ground state of $Fe(CN)_6^{3-}$ as ${}^2T_{2g}(t_{2g}^5)$ with t_{2g} a pure metal d orbital. Thus

$$\begin{aligned} \mathcal{C}_0\left({}^2T_{2g} \to {}^2T_{1u}\right) &= \tfrac{1}{9}\left|\langle {}^2T_{2g} \| m^{t_{1u}} \| {}^2T_{1u} \rangle\right|^2 \\ \mathcal{C}_0\left({}^2T_{2g} \to {}^2T_{2u}\right) &= -\tfrac{1}{9}\left|\langle {}^2T_{2g} \| m^{t_{1u}} \| {}^2T_{2u} \rangle\right|^2 \end{aligned} \qquad (11.2.10)$$

The signs in (11.2.10) do indeed agree with the previous argument. While the numerical values are model-dependent, there can be no doubt at all that they are correct in sign and in rough magnitude.

The electric-dipole matrix elements in (11.2.10) are difficult to calculate reliably, and Stephens (1965a, b) first emphasized the virtue of eliminating them by calculating the ratio $\mathcal{C}_0/\mathcal{D}_0$. This procedure has the further advantage of eliminating the troublesome problem of condensed-medium effects (Section 2.11), since the correction factor should be the same for both $\mathcal{C}_0$ (and $\mathcal{A}_1$ or $\mathcal{B}_0$) and $\mathcal{D}_0$ and hence should cancel in the ratio. From (4.5.23) and (4.6.16), for a cubic system

$$\mathcal{D}_0 = \frac{1}{2|A|} \sum_{\alpha\lambda} \left(|\langle A\alpha | m_{-1} | J\lambda \rangle|^2 + |\langle A\alpha | m_1 | J\lambda \rangle|^2 \right) \qquad (11.2.11)$$

Proceeding exactly as before, one obtains

$$\begin{aligned} \mathcal{D}_0\left({}^2T_{2g} \to {}^2T_{1u}\right) &= \tfrac{1}{9}\left|\langle {}^2T_{2g} \| m^{t_{1u}} \| {}^2T_{1u} \rangle\right|^2 \\ \mathcal{D}_0\left({}^2T_{2g} \to {}^2T_{2u}\right) &= \tfrac{1}{9}\left|\langle {}^2T_{2g} \| m^{t_{1u}} \| {}^2T_{2u} \rangle\right|^2 \end{aligned} \qquad (11.2.12)$$

so that

$$\begin{aligned} \frac{\mathcal{C}_0}{\mathcal{D}_0}\left({}^2T_{2g} \to {}^2T_{1u}\right) &= 1.0 \\ \frac{\mathcal{C}_0}{\mathcal{D}_0}\left({}^2T_{2g} \to {}^2T_{2u}\right) &= -1.0 \end{aligned} \qquad (11.2.13)$$

Thus the predicted $\mathcal{C}_0/\mathcal{D}_0$ ratios for bands 1, 2, and 3 of Fig. 11.2.1 are 1.0, -1.0, and 1.0 (new convention—see Appendix A). The best experimental results are 1.2, -0.6, and 0.6; there is large experimental uncertainty in the values for bands 2 and 3. This theoretical analysis of the MCD spectrum of $Fe(CN)_6^{3-}$ is due to Stephens (1965a), and it marked the first instance in which the MCD technique was used to distinguish unambiguously between two proposed spectroscopic assignments in a molecular system. In fact, the generally accepted assignment at that time was the incorrect one which placed the ${}^2T_{2g} \to {}^2T_{2u}$ transition at lowest energy. What makes the analysis particularly attractive is the fact that, aside from the precise numerical value assigned to $\langle {}^2T_{2g}\|L^{t_{1g}}\|{}^2T_{2g}\rangle$, the argument is completely group-theoretical.

We return to this example in Section 17.3 and show how much simpler the calculation is using advanced group-theoretical methods. Equation (17.3.2) together with (11.2.9) gives our (11.2.13) results directly.

Calculation of $\mathcal{A}$ Terms for $Fe(CN)_6^{3-}$

The calculation of the corresponding $\mathcal{A}_1$ terms is now perfectly straightforward. In fact, most of the work is already done, since from (4.5.16) and (4.6.16), we have

$$\begin{aligned}\mathcal{A}_1 &= \frac{-1}{\mu_B|A|}\sum_{\alpha\lambda}\left(\langle J\lambda|\mu_z|J\lambda\rangle - \langle A\alpha|\mu_z|A\alpha\rangle\right)\\ &\quad\times\left(|\langle A\alpha|m_{-1}|J\lambda\rangle|^2 - |\langle A\alpha|m_1|J\lambda\rangle|^2\right)\\ &= \frac{-1}{\mu_B|A|}\sum_{\alpha\lambda}\langle J\lambda|\mu_z|J\lambda\rangle\\ &\quad\times\left(|\langle A\alpha|m_{-1}|J\lambda\rangle|^2 - |\langle A\alpha|m_1|J\lambda\rangle|^2\right) + \mathcal{C}_0 \end{aligned} \tag{11.2.14}$$

Again, S_z does not contribute when spin–orbit coupling is neglected—see Section 11.3. With the same techniques used previously, one obtains

$$\begin{aligned}\frac{\mathcal{A}_1}{\mathcal{D}_0}\left({}^2T_{2g}\to{}^2T_{1u}\right) &= \frac{1}{\sqrt{6}}\langle {}^2T_{1u}\|L^{t_{1g}}\|{}^2T_{1u}\rangle + \frac{\mathcal{C}_0}{\mathcal{D}_0}\left({}^2T_{2g}\to{}^2T_{1u}\right)\\ \frac{\mathcal{A}_1}{\mathcal{D}_0}\left({}^2T_{2g}\to{}^2T_{2u}\right) &= \frac{1}{\sqrt{6}}\langle {}^2T_{2u}\|L^{t_{1g}}\|{}^2T_{2u}\rangle + \frac{\mathcal{C}_0}{\mathcal{D}_0}\left({}^2T_{2g}\to{}^2T_{2u}\right)\end{aligned} \tag{11.2.15}$$

The method of calculating the "ligand" orbital-angular-momentum matrix elements which appear in (11.2.15) is discussed in Sections 13.2 and 13.4. These matrix elements are sensitive to the details of the excited-state wave functions and hence can provide valuable information. Unfortunately, the $\mathcal{C}$ terms completely dominate the MCD spectrum in Fig. 11.2.1, and so no information at all is obtainable about the magnitude or sign of the $\mathcal{A}$ terms. This is very often the case in broad-band spectra when $\mathcal{C}$ terms are present, that is, when the ground state can be Zeeman-split.

In the absence of spin–orbit coupling, $\mathcal{B}$ terms are not expected to be important in the $Fe(CN)_6^{3-}$ spectrum of Fig. 11.2.1, in view of the discussion in Section 5.3. This circumstance changes markedly when spin–orbit coupling is included (Section 13.5).

Selection Rules for Circularly Polarized Transitions in $Fe(CN)_6^{3-}$

It is illuminating to compare the selection rules for rcp and lcp light in $Fe(CN)_6^{3-}$ with the $^1S \rightarrow {}^1P$ atomic case of Figs. 4.4.1 and 4.4.3. The relevant $Fe(CN)_6^{3-}$ diagrams are shown in Fig. 11.2.3.

Ignoring spin for the moment, we note that four distinct transitions occur, two (of opposite polarization) terminating on the $^2T_{iu}0$ level of the excited state and two (of opposite polarization) originating on the $^2T_{2g}0$ level of the ground state. The contributions to $\mathcal{C}_0$ of these latter two cancel. The $\mathcal{C}$-term *sign* is determined by the polarization $(m_{\pm 1})$ of the transition moment from the lowest Zeeman sublevel, $^2T_{2g} - 1$, since that sublevel is the most highly populated. [$\mathcal{C}_0$ is directly proportional to the field-induced population shifts. See (5.1.11) et seq. for a quantitative discussion.] Since $\Delta A' = A'_L - A'_R = A'_- - A'_+$, it is clear that the $\mathcal{C}$ term for $^2T_{2g} \rightarrow {}^2T_{2u}$ corresponds to a negative MCD ($\mathcal{C}_0 < 0$) and that the opposite is the case for $^2T_{2g} \rightarrow {}^2T_{1u}$. Note particularly that *the sign of the $\mathcal{C}$ term is entirely independent of the excited-state Zeeman pattern*, that is, it is independent of the magnitude and sign of the excited-state g value. On the other hand, this is most certainly not the case for the $\mathcal{A}$ term. The reader might find it worthwhile to deduce the $\mathcal{A}$-term patterns corresponding to the transitions in Fig. 11.2.3. All four transitions will be found to contribute. (See also Section 5.1.) In the case of a $T_{1u} \rightarrow E_g$ transition [Stephens (1976), Fig. 4], one finds only two distinct circularly polarized transitions among Zeeman sublevels, each allowed (but to different extents) in *both* circular polarizations.

11.3. The Role of Spin in the Absence of Spin–Orbit Coupling

Though we defer a detailed discussion of spin and spin–orbit coupling until Chapter 12, it is shown in this section that spin can effectively be ignored in

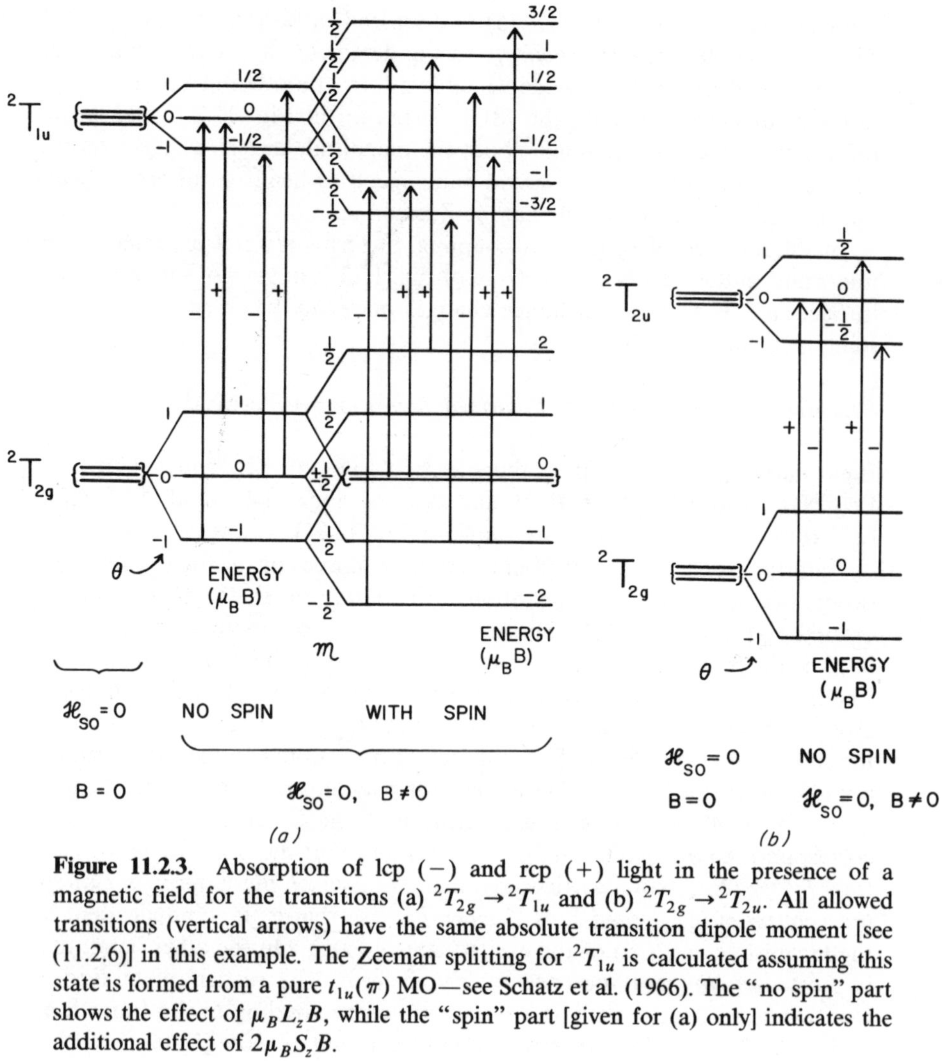

Figure 11.2.3. Absorption of lcp $(-)$ and rcp $(+)$ light in the presence of a magnetic field for the transitions (a) $^2T_{2g} \rightarrow {}^2T_{1u}$ and (b) $^2T_{2g} \rightarrow {}^2T_{2u}$. All allowed transitions (vertical arrows) have the same absolute transition dipole moment [see (11.2.6)] in this example. The Zeeman splitting for $^2T_{1u}$ is calculated assuming this state is formed from a pure $t_{1u}(\pi)$ MO—see Schatz et al. (1966). The "no spin" part shows the effect of $\mu_B L_z B$, while the "spin" part [given for (a) only] indicates the additional effect of $2\mu_B S_z B$.

the calculation of Faraday and absorption parameters if spin–orbit coupling is neglected. We recall from basic principles that any quantum-mechanical angular-momentum operator J can be associated with a complete set of functions, say $|jm\rangle$, such that

$$J^2|jm\rangle = j(j+1)|jm\rangle$$
$$J_z|jm\rangle = m|jm\rangle \tag{11.3.1}$$

where j is integer or half-integer only, and m ranges from $-j$ to $+j$ in integer steps. j and m are respectively the total and z-component angular-momentum quantum members. (As always, we express angular-momentum quantities in units of $\hbar$. Thus $\hbar^2$ and $\hbar$ are respectively omitted from the right-hand sides of the two equations just above.) For orbital angular momentum, j is integer only, and in the coordinate representation the $|jm\rangle$ are the familiar spherical harmonics (Sections 9.4 and B.1–B.2). For spin angular momentum, j may also be half-integer, and such $|jm\rangle$ are called spinors (Section 9.4).

For our purposes it is very useful to view angular-momentum functions group-theoretically (Section 8.11). We recall that j and m respectively denote an irrep and irrep partner, and in particular, that *every function* $|jm\rangle$ *transforms as a partner of an irrep of* SO_3, the infinite group of all rotations which take a sphere into itself. In the atomic case, *in the absence of spin–orbit coupling*, the operators L^2, L_z, S^2, and S_z all commute with each other and with the Hamiltonian operator $\mathcal{H}$, where $\mathbf{L}$ and $\mathbf{S}$ are respectively the total orbital and spin angular-momentum operators. The commutation of L^2 and L_z (or S^2 and S_z) with $\mathcal{H}$, in turn, is an immediate consequence of the invariance of the latter to all rotations in orbital (and spin) space (Landau and Lifshitz (1958), Tinkham (1964)). From basic principles of quantum mechanics it follows that a complete set of functions must exist which are simultaneous eigenfunctions of all of the commuting operators. We write such functions $|\alpha'\mathcal{S}\mathcal{L}\mathcal{M}_\mathcal{S}\mathcal{M}_\mathcal{L}\rangle$, and it follows from (11.3.1) that

$$\begin{aligned}
S^2|\alpha'\mathcal{S}\mathcal{L}\mathcal{M}_\mathcal{S}\mathcal{M}_\mathcal{L}\rangle &= \mathcal{S}(\mathcal{S}+1)|\alpha'\mathcal{S}\mathcal{L}\mathcal{M}_\mathcal{S}\mathcal{M}_\mathcal{L}\rangle \\
L^2|\alpha'\mathcal{S}\mathcal{L}\mathcal{M}_\mathcal{S}\mathcal{M}_\mathcal{L}\rangle &= \mathcal{L}(\mathcal{L}+1)|\alpha'\mathcal{S}\mathcal{L}\mathcal{M}_\mathcal{S}\mathcal{M}_\mathcal{L}\rangle \\
S_z|\alpha'\mathcal{S}\mathcal{L}\mathcal{M}_\mathcal{S}\mathcal{M}_\mathcal{L}\rangle &= \mathcal{M}_\mathcal{S}|\alpha'\mathcal{S}\mathcal{L}\mathcal{M}_\mathcal{S}\mathcal{M}_\mathcal{L}\rangle \\
L_z|\alpha'\mathcal{S}\mathcal{L}\mathcal{M}_\mathcal{S}\mathcal{M}_\mathcal{L}\rangle &= \mathcal{M}_\mathcal{L}|\alpha'\mathcal{S}\mathcal{L}\mathcal{M}_\mathcal{S}\mathcal{M}_\mathcal{L}\rangle
\end{aligned} \tag{11.3.2}$$

We use the script letters ($\mathcal{S}$, $\mathcal{L}$, etc.) to label irreps of SO_3 to avoid confusion with irrep labels for the molecular point groups. α' is a symbolic quantum number which represents the total energy and possibly other observables relevant to a given problem. (For example, if we were concerned with the hydrogen atom, α' would stand for the principal quantum number n. We in general omit α' from kets unless such observables require consideration in a particular case.)

The eigenkets $|\mathcal{S}\mathcal{L}\mathcal{M}_\mathcal{S}\mathcal{M}_\mathcal{L}\rangle$ are commonly referred to as Russell–Saunders or *LS* functions. The process whereby such functions are assembled is referred to as Russell–Saunders or *LS* coupling. We hereafter call it $\mathcal{S}\mathcal{L}$ coupling to indicate the order in which the coupled $\mathcal{S}$ and $\mathcal{L}$ angular-momentum functions are written. The quantum numbers $\mathcal{S}$, $\mathcal{L}$, $\mathcal{M}_\mathcal{S}$, $\mathcal{M}_\mathcal{L}$, are referred to as "good quantum numbers," because (11.3.2) is satisfied exactly by the true eigenfunctions of $\mathcal{H}$ (in the absence of spin–orbit coupling). That is, these quantum numbers classify irreps of the group of the Schrödinger equation.

A single set of functions of given α', $\mathcal{L}$, and $\mathcal{S}$ value, totaling $(2\mathcal{S}+1)\times(2\mathcal{L}+1)$ functions, is referred to as a *term* and is represented by the term symbol ${}^{2\mathcal{S}+1}\mathcal{L}$. We can regard $\mathcal{L}\mathcal{M}_{\mathcal{L}}$ as the group-SO_3 symmetry classification in orbital space and $\mathcal{S}\mathcal{M}_{\mathcal{S}}$ as the group-SO_3 symmetry classification in spin space, both in the angular momentum basis. Such a distinction is possible because $\mathcal{H}$ is invariant to any rotation in either space. No symmetry operation can convert an orbital function into a spin function or vice versa.

Let us now consider a molecule. The symmetry of such a system is always lower than that for an atom, and in general L^2 and L_z do not commute with $\mathcal{H}$, so that $\mathcal{L}$ and $\mathcal{M}_{\mathcal{L}}$ are no longer good quantum numbers. In such cases, as we already know, the transformation operators $\mathcal{R}$ of the group of the Schrödinger equation (the molecular point group) commute with $\mathcal{H}$ (though not necessarily with each other), and we can classify a function by the irreps of the point group to which the molecule belongs. However, if $\mathcal{H}$ contains no spin-dependent terms, S^2 and S_z clearly still commute with it, and thus $\mathcal{S}\mathcal{L}$-like functions can still be constructed. Following Griffith (1962), we designate these as $|\mathcal{S}h\mathcal{M}\theta\rangle$, where h and θ denote respectively the point-group irrep and partner, and spin functions continue to be classified in SO_3 in the angular momentum basis. We have abbreviated $\mathcal{M}_{\mathcal{S}}$ as $\mathcal{M}$. In analogy with the atomic case, these are referred to as $\mathcal{S}\mathcal{L}$ functions. For such functions, the first and third equations of (11.3.2) still apply, and the analogy to the second and fourth equations is (8.9.6) (with $l_k = |h|$) for all $\mathcal{R}$. Again in analogy to the atomic case, h and θ are sometimes referred to as "good quantum numbers."

In our applications, the only important spin-dependent term in $\mathcal{H}$ is the spin–orbit coupling operator $\mathcal{H}_{SO}$, which is discussed in Chapter 12. If we neglect this term, the $\mathcal{S}\mathcal{L}$ functions $|\mathcal{S}h\mathcal{M}\theta\rangle$ are appropriate basis functions for calculating Faraday parameters; that is, they are fully classified with respect to the symmetry of the zero-field Hamiltonian.

Calculation of $\mathcal{C}_0$ in the Absence of Spin–Orbit Coupling

Let us calculate a $\mathcal{C}_0$ term assuming that the spin–orbit coupling is zero. For simplicity, though this has no effect on the generality of the argument, we assume a cubic system as in Section 11.2. Thus

$$\mathcal{C}_0 = \frac{1}{\mu_B|A|}\sum_{\alpha\lambda}\langle A\alpha|\mu_z|A\alpha\rangle\left(|\langle A\alpha|m_{-1}|J\lambda\rangle|^2 - |\langle A\alpha|m_1|J\lambda\rangle|^2\right) \tag{11.3.3}$$

where the sum over α and λ is over $\mathcal{M}_{\mathcal{S}}$ ($\mathcal{M}$) as well as θ values. We choose functions diagonal in μ_z, and designate the ground-state functions $|\mathcal{S}h\mathcal{M}\theta\rangle$ and the excited-state functions $|\mathcal{S}'h'\mathcal{M}'\theta'\rangle$. Thus (11.3.3) becomes

$$\mathcal{C}_0 = \frac{1}{\mu_B|A|} \sum_{\mathcal{M}\theta\mathcal{M}'\theta'} \langle \mathcal{S}h\mathcal{M}\theta|\mu_z|\mathcal{S}h\mathcal{M}\theta\rangle \times \left(|\langle \mathcal{S}h\mathcal{M}\theta|m_{-1}|\mathcal{S}'h'\mathcal{M}'\theta'\rangle|^2 - |\langle \mathcal{S}h\mathcal{M}\theta|m_1|\mathcal{S}'h'\mathcal{M}'\theta'\rangle|^2\right) \tag{11.3.4}$$

We next note that

$$\langle \mathcal{S}h\mathcal{M}\theta|m_{\pm 1}|\mathcal{S}'h'\mathcal{M}'\theta'\rangle = \delta_{\mathcal{S}\mathcal{S}'}\delta_{\mathcal{M}\mathcal{M}'}\langle \mathcal{S}h\mathcal{M}\theta|m_{\pm 1}|\mathcal{S}h'\mathcal{M}\theta'\rangle \tag{11.3.5}$$

Equation (11.3.5) states that the electric dipole transition moment is zero unless ground- and excited-state functions have the same $\mathcal{M}$ and $\mathcal{S}$ values. This result may simply be regarded as an application of the Wigner–Eckart theorem in spin space. Since $m_{\pm 1}$ is spin-independent, it belongs to the identity (totally symmetric) irrep in spin space, and (11.3.5) follows immediately from (10.2.2). Furthermore, if $\mathcal{S}' = \mathcal{S}$ and $\mathcal{M}' = \mathcal{M}$, the matrix element in (11.3.5) is the same for all $\mathcal{M}$ values, again by the Wigner–Eckart theorem. The magnetic-dipole term for $\mathcal{S}' = \mathcal{S}$ and $\mathcal{M}' = \mathcal{M}$ then becomes

$$\left\langle \mathcal{S}h\mathcal{M}\theta|(-\mu_B)(L_z + 2S_z)|\mathcal{S}h\mathcal{M}\theta\right\rangle = \langle \mathcal{S}h\mathcal{M}\theta| - \mu_B L_z|\mathcal{S}h\mathcal{M}\theta\rangle - 2\mu_B\mathcal{M} \tag{11.3.6}$$

where the third equation in (11.3.2) has been used. The L_z matrix element is also the same for all $\mathcal{M}$ values by the Wigner–Eckart theorem. Substituting (11.3.5) and (11.3.6) into (11.3.4), we obtain

$$\mathcal{C}_0 = \frac{1}{\mu_B(2\mathcal{S}+1)|h|} \sum_{\theta\theta'\mathcal{M}} (\langle \mathcal{S}h\mathcal{M}\theta| - \mu_B L_z|\mathcal{S}h\mathcal{M}\theta\rangle - 2\mu_B\mathcal{M}) \cdot \times \left(|\langle \mathcal{S}h\mathcal{M}\theta|m_{-1}|\mathcal{S}h'\mathcal{M}\theta'\rangle|^2 - |\langle \mathcal{S}h\mathcal{M}\theta|m_1|\mathcal{S}h'\mathcal{M}\theta'\rangle|^2\right) \tag{11.3.7}$$

where $|A|$ in (11.3.4) is the product of the spin degeneracy $|\mathcal{S}| = 2\mathcal{S} + 1$, and $|h|$ the dimension of the irrep h of the ground state. The sum over $\mathcal{M}$ can be performed noting that $\Sigma_{\mathcal{M}=-\mathcal{S}}^{\mathcal{S}} \mathcal{M} = 0$. Since all matrix elements in (11.3.7) are independent of $\mathcal{M}$ value, the same expression is obtained $2\mathcal{S} + 1$ times. Equation (11.3.7) therefore becomes

$$\mathcal{C}_0 = \frac{1}{\mu_B|h|} \sum_{\theta\theta'} \langle \mathcal{S}h\mathcal{M}\theta| - \mu_B L_z|\mathcal{S}h\mathcal{M}\theta\rangle$$
$$\times \left(|\langle \mathcal{S}h\mathcal{M}\theta|m_{-1}|\mathcal{S}h'\mathcal{M}\theta'\rangle|^2 - |\langle \mathcal{S}h\mathcal{M}\theta|m_1|\mathcal{S}h'\mathcal{M}\theta'\rangle|^2\right) \tag{11.3.8}$$

This is in fact precisely the formula which was used for the $Fe(CN)_6^{3-}$ example in Section 11.2. In that case, the ground-state manifold consisted of the six $|^{2\mathcal{S}+1}h\mathcal{M}\theta\rangle$ functions $|^2T_{2g} \pm \frac{1}{2}0\rangle$, $|^2T_{2g} \pm \frac{1}{2} \pm 1\rangle$—that is, in the $|\mathcal{S}h\mathcal{M}\theta\rangle$ notation, $\mathcal{S} = \frac{1}{2}$, $h = T_{2g}$, $\mathcal{M} = \pm\frac{1}{2}$, and $\theta = \pm 1, 0$. In doing the calculation, the explicit $\mathcal{M}$ values were omitted, since the matrix elements in (11.3.8) are independent of $\mathcal{M}$.

It is left to the reader to prove that results exactly analogous to (11.3.8) apply not only for $\mathcal{C}_0$, but also for $\mathcal{A}_1$, $\mathcal{B}_0$, and $\mathcal{D}_0$ *in all cases* in which spin–orbit coupling is neglected. The physical reason for this result can be understood by examining the right-hand side of Fig. 11.2.3(a). When spin is included in that example, each level splits into two components at energies $\pm\mu_B B$ with respect to the no-spin case. But the transition energies and polarizations are seen to be unchanged, and since population changes go linearly with Zeeman splittings [Eq. (5.1.10)] in the linear limit (the case where $\mathcal{A}$, $\mathcal{B}$, $\mathcal{C}$ language has meaning), the spin invariance is an immediate consequence. When spin–orbit coupling is included (Chapter 12), the Zeeman sublevels are mixed and the neglect of spin is no longer valid.

The general prescription for calculating any of the Faraday parameters (or $\mathcal{D}_0$) when spin–orbit coupling is neglected is as follows. Set the operator $\boldsymbol{\mu} = -\mu_B\mathbf{L}$ and $|A| = |h|$, choose the appropriate sets of orbital functions $|h\theta\rangle$ and $|h'\theta'\rangle$, and carry out the calculation after constructing $|\mathcal{S}h\mathcal{M}\theta\rangle$ and $|\mathcal{S}h'\mathcal{M}\theta'\rangle$ for the single most convenient value of $\mathcal{S}$ and $\mathcal{M}$. The method of constructing $\mathcal{S}\mathcal{L}$ functions is discussed in Sections 11.4 and 11.5.

11.4. The First Step in Matrix-Element Evaluation: The Construction of Wavefunctions

The example in Section 11.2 should make it clear that we can in general divide calculations into two stages. The first is based entirely on symmetry (group-theoretic) arguments and is completely independent of specific models. Results are expressed in terms of a minimum number of reduced matrix elements, whose evaluation is the model-dependent second step. This latter step requires approximations, since accurate molecular wavefunctions are seldom available. Even in the simple example in Section 11.2, it was necessary to calculate (or measure), even if only roughly, the ground-state angular momentum to insure that the basic theoretical framework assumed at the start (Fig. 11.2.2) was sensible. However, it should be strongly emphasized that once a model is chosen, there is again great scope for group-theoretic techniques in dissecting out those aspects of the model which are symmetry determined, as for example in simplifying many-electron matrix elements.

Our working model is usually some form of LCAO–MO theory using a highly restricted basis set with crystal-field (or ligand-field) theory as a limiting case. This formalism permits us to write down specific wavefunctions which may be used to evaluate matrix elements. Since we are interested in angular-momentum and transition-moment matrix elements, we are dealing with one-electron operators. Group-theoretic techniques are of very great importance in reducing the many-electron reduced matrix elements of such operators to one-electron form.

We note at the outset that the more general methods discussed in Chapters 16–22 in most applications eliminate entirely the need to construct many-electron wave functions. Nevertheless, it is very important to be able to construct such functions. First, the use of explicit functions often proves the easiest way to do simple problems and may give one a "feel" for what is going on. Second, an understanding of the procedures involved allows one to judge when the more sophisticated methods are preferable. Finally, an appreciation of the immense labor that can be involved in the construction of wavefunctions in complex cases (plus the great difficulty in using such functions to produce answers which one believes), can provide considerable motivation to learn the methodology of the irreducible tensor method.

Our discussion in the remainder of this chapter is limited to the construction of $|\mathbb{S}h\mathfrak{M}\theta\rangle$ ($\mathbb{S}\mathfrak{L}$) functions. These are always a suitable basis, since appropriate linear combinations of them are also diagonal in $\mathcal{H}_{\mathrm{SO}}$ (as in the atomic case, where such linear combinations are designated $|(\mathbb{S}\mathfrak{L})\mathfrak{J}\mathfrak{M}\rangle$). The construction in the molecular case of such functions, which we designate $|(\mathbb{S}S, h)rt\tau\rangle$, is discussed in Chapter 12.

In the molecular case, $\mathbb{S}\mathfrak{L}$ functions are written $|\alpha'\mathbb{S}h\mathfrak{M}\theta\rangle$ (Section 11.3), and our task is to assemble an appropriate function α' which transforms simultaneously as the partner θ of the irrep h of the group G in orbital space, and as the partner $\mathfrak{M}$ of the irrep $\mathbb{S}$ of the group SO_3 in spin space. In all cases $\mathbb{S}$, $\mathfrak{M}$, h, and θ are defined according to standard basis relations. Such relations have been discussed in Section 9.4 and are tabulated in Sections C.5, D.4, and E.4 for some representative point groups.

How to Form $|\mathcal{S}h\mathcal{M}\theta\rangle$ Functions for a Configuration a^n

There are well-known procedures for constructing $|\alpha'\mathcal{S}h\mathcal{M}\theta\rangle$ functions. One starts with a well-defined basis set and couples separately in orbital and spin space using standard coupling tables. In the more complicated cases fractional-parentage coefficients are useful.

In this section we consider the group O as an example. Standard basis functions have been discussed in Section 9.4 and are tabulated for the group O in Section C.5. Our basis functions are always one-electron orbitals (MOs), and all wavefunctions are products (determinants) of such functions. The one-electron functions (always written in lowercase) are classified by the irreps of the group O: a_1, a_2, e, t_1, and t_2. Let us assume that we deal with sets of n equivalent electrons—that is, with functions of a single-open-shell configuration a^n. In view of the Pauli principle, the occupancy n of any one set of orbitals, a, is limited to $2|a|$, where $|a|$ is the dimension of the irrep a. In particular, $n \leqslant 2$ for a_1 and a_2 electrons, $n \leqslant 4$ for e electrons, and $n \leqslant 6$ for t_1 and t_2 electrons. Thus once and for all, we can construct all possible wavefunctions covering all degrees of occupancy of each of the above.

As a first example, let us consider e^n, where n covers the values 1 to 4. For the case $n = 1$, it is clear that

$$\begin{aligned}
|\tfrac{1}{2}E\tfrac{1}{2}\theta\rangle &\equiv |^2E\tfrac{1}{2}\theta\rangle = |\theta^+\rangle \\
|\tfrac{1}{2}E-\tfrac{1}{2}\theta\rangle &\equiv |^2E-\tfrac{1}{2}\theta\rangle = |\theta^-\rangle \\
|\tfrac{1}{2}E\tfrac{1}{2}\epsilon\rangle &\equiv |^2E\tfrac{1}{2}\epsilon\rangle = |\epsilon^+\rangle \\
|\tfrac{1}{2}E-\tfrac{1}{2}\epsilon\rangle &\equiv |^2E-\tfrac{1}{2}\epsilon\rangle = |\epsilon^-\rangle
\end{aligned} \tag{11.4.1}$$

where θ is an MO transforming as $|E\theta\rangle$, and the + or − superscript designates an $\mathcal{M} = \frac{1}{2}$ or $-\frac{1}{2}$ spin function, respectively, associated with the space orbital.

Next, for the configuration e^2 we proceed systematically as follows. First we couple orbital functions using (10.2.3). Thus since $e \otimes e = A_1 + A_2 + E$, using the real-$O$ $2jm$ and $3jm$ of Sections C.11–C.12,

$$\begin{aligned}
|(EE)A_1a_1\rangle &= |A_1|^{1/2}\begin{pmatrix} A_1 \\ a_1 \end{pmatrix}\sum_{\alpha\beta}\begin{pmatrix} E & E & A_1^* \\ \alpha & \beta & a_1^* \end{pmatrix}|E\alpha\rangle|E\beta\rangle \\
&= \frac{1}{\sqrt{2}}(|\theta\rangle|\theta\rangle + |\epsilon\rangle|\epsilon\rangle)
\end{aligned}$$

$$|(EE)A_2a_2\rangle = \frac{-1}{\sqrt{2}}(|\theta\rangle|\epsilon\rangle - |\epsilon\rangle|\theta\rangle) \tag{11.4.2}$$

$$|(EE)E\theta\rangle = \frac{-1}{\sqrt{2}}(|\theta\rangle|\theta\rangle - |\epsilon\rangle|\epsilon\rangle)$$

$$|(EE)E\epsilon\rangle = \frac{1}{\sqrt{2}}(|\theta\rangle|\epsilon\rangle + |\epsilon\rangle|\theta\rangle)$$

In writing products of one-electron space (or spin) functions, such as $|\theta\rangle|\theta\rangle$ or $|\theta\rangle|\epsilon\rangle$, it is always understood that electron 1 is associated with the first product function, electron 2 with the second, and so forth. *No antisymmetrization is implied.*

Now we must associate spins with the e^2 space functions (orbitals) in (11.4.2) so as to satisfy the Pauli exclusion principle: the wavefunction as a whole must be antisymmetric with respect to the interchange of (the coordinates of) the two electrons. The possible spin states are found by coupling the $|\frac{1}{2}\frac{1}{2}\rangle$ and $|\frac{1}{2}-\frac{1}{2}\rangle$ spin states using the form of (10.2.3) which includes the $H(abc)$ factor [Eq. (B.3.3) of Section B.3] and the $3jm$ of $SO_3 \supset SO_2$ [which are the $3j$ of Rotenberg et al. (1959)] or Wigner coefficients. The $3jm$ of $SO_3 \supset SO_2$ required for our examples are given in Section B.3. We obtain

$$\begin{aligned}|(\tfrac{1}{2}\tfrac{1}{2})00\rangle &= (-1)^{1/2-1/2+0}(1)(-1)^{0-0} \\ &\quad\times \sum_{m_1m_2}\begin{pmatrix} \frac{1}{2} & \frac{1}{2} & 0 \\ m_1 & m_2 & 0\end{pmatrix}|\tfrac{1}{2}m_1\rangle|\tfrac{1}{2}m_2\rangle \\ &= \frac{1}{\sqrt{2}}(|\tfrac{1}{2}\tfrac{1}{2}\rangle|\tfrac{1}{2}-\tfrac{1}{2}\rangle - |\tfrac{1}{2}-\tfrac{1}{2}\rangle|\tfrac{1}{2}\tfrac{1}{2}\rangle)\end{aligned} \tag{11.4.3}$$

$$|(\tfrac{1}{2}\tfrac{1}{2})11\rangle = |\tfrac{1}{2}\tfrac{1}{2}\rangle|\tfrac{1}{2}\tfrac{1}{2}\rangle$$

$$|(\tfrac{1}{2}\tfrac{1}{2})10\rangle = \frac{1}{\sqrt{2}}(|\tfrac{1}{2}\tfrac{1}{2}\rangle|\tfrac{1}{2}-\tfrac{1}{2}\rangle + |\tfrac{1}{2}-\tfrac{1}{2}\rangle|\tfrac{1}{2}\tfrac{1}{2}\rangle)$$

$$|(\tfrac{1}{2}\tfrac{1}{2})1\ -1\rangle = |\tfrac{1}{2}-\tfrac{1}{2}\rangle|\tfrac{1}{2}-\tfrac{1}{2}\rangle$$

Since the Pauli principle must be obeyed, the antisymmetric $\mathbb{S} = 0$ spin state (the singlet) must go with the symmetric orbital combinations of (11.4.2) while the symmetric $\mathbb{S} = 1$ spin state (the triplet) goes with the antisymmetric orbital combination. Thus the allowed $|e^2\ {}^1A_10a_1\rangle$ state is, for example,

$$|e^2\,0A_10A_1\rangle \equiv |e^2\,{}^1A_10A_1\rangle$$

$$= \frac{1}{\sqrt{2}}\left[\frac{1}{\sqrt{2}}(|\theta^+\rangle|\theta^-\rangle - |\theta^-\rangle|\theta^+\rangle) + \frac{1}{\sqrt{2}}(|\epsilon^+\rangle|\epsilon^-\rangle - |\epsilon^-\rangle|\epsilon^+\rangle)\right]$$

$$= \frac{1}{\sqrt{2}}(|\theta^+\theta^-\rangle + |\epsilon^+\epsilon^-\rangle) \tag{11.4.4}$$

In the last step we adopt the convention, always used hereafter, that kets of *products* of space–spin functions (spin–orbitals) are *normalized determinants* [see (19.1.2) for an explicit definition]. Thus $|\theta^+\theta^-\rangle$ and $|\epsilon^+\epsilon^-\rangle$ are 2×2 normalized determinantal functions. Using the notation, $|\theta^+\theta^-\rangle \equiv |\theta^2\rangle$, the final result is

$$|e^2\,{}^1A_10a_1\rangle = \frac{1}{\sqrt{2}}(|\theta^2\rangle + |\epsilon^2\rangle) \tag{11.4.5}$$

Precisely the same methods can be used for the remaining e^2 functions, with the results

$$\begin{aligned}
|e^2\,{}^1E0\theta\rangle &= \frac{-1}{\sqrt{2}}(|\theta^+\theta^-\rangle - |\epsilon^+\epsilon^-\rangle) \\
&= \frac{-1}{\sqrt{2}}(|\theta^2\rangle - |\epsilon^2\rangle) \\
|e^2\,{}^1E0\epsilon\rangle &= \frac{1}{\sqrt{2}}(|\theta^+\epsilon^-\rangle + |\epsilon^+\theta^-\rangle) \\
&= \frac{1}{\sqrt{2}}(|\theta^+\epsilon^-\rangle - |\theta^-\epsilon^+\rangle) \\
|e^2\,{}^3A_21a_2\rangle &= -|\theta^+\epsilon^+\rangle \\
|e^2\,{}^3A_20a_2\rangle &= \frac{-1}{\sqrt{2}}(|\theta^+\epsilon^-\rangle + |\theta^-\epsilon^+\rangle) \\
|e^2\,{}^3A_2-1a_2\rangle &= -|\theta^-\epsilon^-\rangle
\end{aligned} \tag{11.4.6}$$

The e^2 states defined above are seen to be properly antisymmetrized states, since interchanging two rows in a determinant ket changes its sign (Section 19.1).

The prescription for coupling the spin and space functions (without any antisymmetrization between the a^{n-1} and a^1 kets) may be described by the equation

$$\left|\left(a^{n-1}(\mathcal{S}_1 h_1), a\right)\mathcal{S}h\mathcal{M}\theta\right\rangle = \sum_{\mathcal{M}_1 m} (\mathcal{S}_1\mathcal{M}_1, \tfrac{1}{2}m|\mathcal{S}\mathcal{M})$$

$$\times \sum_{\theta_1\alpha} (h_1\theta_1, a\alpha|h\theta)|a^{n-1}\mathcal{S}_1 h_1 \mathcal{M}_1\theta_1\rangle|a\, m\alpha\rangle$$

$$= (-1)^{\mathcal{S}_1 - 1/2 + \mathcal{S}}|\mathcal{S}|^{1/2}(-1)^{\mathcal{S}-\mathcal{M}} \sum_{\mathcal{M}_1 m} \begin{pmatrix} \mathcal{S}_1 & \frac{1}{2} & \mathcal{S} \\ \mathcal{M}_1 & m & -\mathcal{M} \end{pmatrix}$$

$$\times |h|^{1/2}\binom{h}{\theta}\sum_{\theta_1\alpha}\begin{pmatrix} h_1 & a & h^* \\ \theta_1 & \alpha & \theta^* \end{pmatrix}^* |a^{n-1}\, \mathcal{S}_1 h_1 \mathcal{M}_1 \theta_1\rangle|a\, m\alpha\rangle \qquad (11.4.7)$$

which follows directly from (B.3.3) and (10.2.3). In the two-electron case, use of this equation by itself results in properly antisymmetrized states as long as symmetric spin states $|\mathcal{S}\mathcal{M}\rangle$ are combined with antisymmetric orbital states $|h\theta\rangle$ and vice versa (as was done in our e^2 example above). We shall see, however, in the general case that antisymmetrization often does not occur automatically and must be deliberately included if proper quantum-mechanical wavefunctions are to be obtained.

Proceeding next to the e^3 case, we can use the fact that since $2|e| = 4$, e^3 and e^1 are hole–particle equivalents [see Griffith (1964), Section 9.7]. Thus, as for e^1, only $|^2E \pm \frac{1}{2}\theta\rangle$ and $|^2E \pm \frac{1}{2}\epsilon\rangle$ states are possible. We consider first the construction of the $|e^3\ \frac{1}{2}E\frac{1}{2}\theta\rangle$ state as an example. Using (11.4.7) above, Section B.3, and Sections C.11–C.12 in the real-O basis, three different $|^2E\frac{1}{2}\theta\rangle$ states may be formed by coupling the e^2 and e^1 states:

$$\psi_1 = \left|\left(e^2(0A_1), e\right)\tfrac{1}{2}E\tfrac{1}{2}\theta\right\rangle$$

$$= |e^2\, 0A_1 0a_1\rangle|e\ \tfrac{1}{2}\theta\rangle$$

$$= \frac{1}{\sqrt{2}}(|\theta^+\theta^-\rangle|\theta^+\rangle + |\epsilon^+\epsilon^-\rangle|\theta^+\rangle)$$

$$\psi_2 = \left|\left(e^2(0E), e\right)\tfrac{1}{2}E\tfrac{1}{2}\theta\right\rangle$$

$$= \frac{-1}{\sqrt{2}}\left(|e^2\,0E0\theta\rangle|e\,\tfrac{1}{2}\theta\rangle - |e^2\,0E0\epsilon\rangle|e\,\tfrac{1}{2}\epsilon\rangle\right)$$

$$= \tfrac{1}{2}\left(|\theta^+\theta^-\rangle|\theta^+\rangle - |\epsilon^+\epsilon^-\rangle|\theta^+\rangle + |\theta^+\epsilon^-\rangle|\epsilon^+\rangle - |\theta^-\epsilon^+\rangle|\epsilon^+\rangle\right)$$

$$(11.4.8)$$

$$\psi_3 = \left|\left(e^2(1A_2), e\right)\tfrac{1}{2}E\tfrac{1}{2}\theta\right\rangle$$

$$= \frac{\sqrt{2}}{\sqrt{3}}|e^2\,1A_2 1a_2\rangle|e\,-\tfrac{1}{2}\epsilon\rangle - \frac{1}{\sqrt{3}}|e^2\,1A_2 0a_2\rangle|e\,\tfrac{1}{2}\epsilon\rangle$$

$$= -\frac{\sqrt{2}}{\sqrt{3}}|\theta^+\epsilon^+\rangle|\epsilon^-\rangle + \frac{1}{\sqrt{6}}|\theta^+\epsilon^-\rangle|\epsilon^+\rangle + \frac{1}{\sqrt{6}}|\theta^-\epsilon^+\rangle|\epsilon^+\rangle$$

None of these states, however, are linear combinations of fully antisymmetrized determinantal wavefunctions, and therefore none of them are satisfactory e^3 states. Note, however, that the linear combination

$$|e^3\,\tfrac{1}{2}E\tfrac{1}{2}\theta\rangle = \frac{1}{\sqrt{6}}\psi_1 - \frac{1}{\sqrt{3}}\psi_2 - \frac{1}{\sqrt{2}}\psi_3$$

$$= \frac{1}{\sqrt{3}}|\epsilon^+\epsilon^-\rangle|\theta^+\rangle - \frac{1}{\sqrt{3}}|\theta^+\epsilon^-\rangle|\epsilon^+\rangle + \frac{1}{\sqrt{3}}|\theta^+\epsilon^+\rangle|\epsilon^-\rangle$$

$$= |\theta^+\epsilon^+\epsilon^-\rangle \qquad (11.4.9)$$

is properly antisymmetrized. The coefficients in the first line of (11.4.9) are referred to as *coefficients of fractional parentage* (cfp). The derivation, properties and use of cfp are discussed in Section 19.3. We show in Chapter 19 that they are of immense value in reducing many-electron matrix elements to one-electron form. They also may be used, as in this example, to construct multielectron wavefunctions, but very often simpler means are available.

In this case, the easiest way to obtain the e^3 functions is by guessing. Since e^3 is the hole equivalent of e^1, one suspects that the required results, apart possibly from a phase factor, will simply be the hole equivalents, that is, $\theta \rightarrow \theta\epsilon^2$, $\epsilon \rightarrow \theta^2\epsilon$. So one guesses the answers

$$|e^3\, \tfrac{1}{2}E\tfrac{1}{2}\theta\rangle = |\theta^+\epsilon^2\rangle, \qquad |e^3\, \tfrac{1}{2}E - \tfrac{1}{2}\theta\rangle = |\theta^-\epsilon^2\rangle$$

$$|e^3\, \tfrac{1}{2}E\tfrac{1}{2}\epsilon\rangle = |\theta^2\epsilon^+\rangle, \qquad |e^3\, \tfrac{1}{2}E - \tfrac{1}{2}\epsilon\rangle = |\theta^2\epsilon^-\rangle \qquad (11.4.10)$$

One must then confirm that these satisfy the standard basis relations for space and spin and the designated hole–particle convention (see below). (For complex functions, hole-particle equivalents are not as intuitively obvious.) Similarly, for e^4 the single result is obviously

$$|e^4\, 0A_1 0a_1\rangle = |\theta^2\epsilon^2\rangle \qquad (11.4.11)$$

Hole–Particle Convention

Since standard basis relations and antisymmetrization criteria do not insure a unique phase choice for the hole-equivalent states, it is useful to adopt a hole–particle convention to make the choice unique. We follow the convention of Griffith [(1964), Section 9.7] and choose the phase of a state $|a^{2|a|-n}\, \mathbb{S}h\mathfrak{M}\theta\rangle$ for $n < |a|$ so that

$$\begin{aligned} |a^{2|a|}\, 0A_1 0a_1\rangle &= \sum_{\mathbb{S}_1 h_1} |\mathbb{S}_1|^{1/2}|h_1|^{1/2}\left[\frac{(2|a|)!}{n!\,(2|a|-n)!}\right]^{-1/2} \\ &\quad \times \left|\left(a^n(\mathbb{S}_1 h_1),\, a^{2|a|-n}(\mathbb{S}_1 h_1)\right)0A_1 0a_1\right\rangle \\ &\equiv \sum_{\mathbb{S}_1 h_1} c(a^n\, \mathbb{S}_1 h_1)\left|\left(a^n(\mathbb{S}_1 h_1),\, a^{2|a|-n}(\mathbb{S}_1 h_1)\right)0A_1 0a_1\right\rangle \end{aligned} \qquad (11.4.12)$$

The coupling is accomplished using an equation of the type (11.4.7) in which the ket for a^n is coupled to one for $a^{2|a|-n}$. When $n = 1$, $\mathbb{S}_1 h_1 = \tfrac{1}{2}a$ only and $c(a^n\, \mathbb{S}_1 h_1) = c(a\tfrac{1}{2}a) = 1$. Thus in the e^3 case, we want

$$|e^4\, 0A_1 0a_1\rangle = |\theta^+\theta^-\epsilon^+\epsilon^-\rangle = \left|(e, e^3)0A_1 0a_1\right\rangle \qquad (11.4.13)$$

where

$$|(e, e^3)0A_10a_1\rangle = \frac{1}{\sqrt{2}}\Bigg[\frac{1}{\sqrt{2}}\Big(|e\,\tfrac{1}{2}E\tfrac{1}{2}\theta\rangle|e^3\,\tfrac{1}{2}E - \tfrac{1}{2}\theta\rangle$$

$$- |e\,\tfrac{1}{2}E - \tfrac{1}{2}\theta\rangle|e^3\,\tfrac{1}{2}E\tfrac{1}{2}\theta\rangle\Big)$$

$$+ \frac{1}{\sqrt{2}}\Big(|e\,\tfrac{1}{2}E\tfrac{1}{2}\epsilon\rangle|e^3\,\tfrac{1}{2}E - \tfrac{1}{2}\epsilon\rangle - |e\,\tfrac{1}{2}E - \tfrac{1}{2}\epsilon\rangle|e^3\,\tfrac{1}{2}E\tfrac{1}{2}\epsilon\rangle\Big)\Bigg]$$

(11.4.14)

Substitution of the e^1 and e^3 kets of (11.4.1) and (11.4.10) into the equation above confirms that the hole–particle convention of (11.4.13) is obeyed for our e^3 states.

Fractional-Parentage Equation and the Formation of $|\mathbb{S}h\mathfrak{M}\theta\rangle$ States for t_1^n and t_2^n Configurations

The formal equation which describes the formation of properly antisymmetrized a^n states from kets of the type $|(a^{n-1}(\mathbb{S}_1h_1), a)\mathbb{S}h\mathfrak{M}\theta\rangle$, formed via (11.4.7), is

$$|a^n\,\mathbb{S}h\mathfrak{M}\theta\rangle = \sum_{\mathbb{S}_1h_1}\big(a^{n-1}\mathbb{S}_1h_1, a|a^n\,\mathbb{S}h\big)\big|\big(a^{n-1}(\mathbb{S}_1h_1), a\big)\mathbb{S}h\mathfrak{M}\theta\rangle$$

(11.4.15)

where $(a^{n-1}\mathbb{S}_1h_1, a|a^n\,\mathbb{S}h)$ is called a coefficient of fractional parentage (cfp). [We follow Butler (1981) and designate the cfp as above instead of as $(a^{n-1}\mathbb{S}_1h_1, a|\}a^n\,\mathbb{S}h)$; this new notation emphasizes the similarity between cfp and v.c.c.—see Section 19.3.] Equation (11.4.9) is simply this cfp equation applied to the e^3 case. We may also apply (11.4.15) to the e^2 and e^4 cases. In fact, our phase choices in the e^2 examples of (11.4.6) serve to define

$$\big(e, e|e^2\,\mathbb{S}h\big) = 1 \qquad (11.4.16)$$

for all the allowed terms of the e^2 configuration.

For t_1^2 and t_2^2 configurations it is not always convenient to define the analogous $n = 2$ cfp to be unity—see Section 19.3. Otherwise, the t_1^n and t_2^n

functions are found in the same manner as those for e^n. The $n = 1$ functions are obvious. Those for $n = 2$ and $n = 3$ are found using (11.4.7) together with (11.4.15) and the cfp of Section C.15 (see Section 19.3). The $n = 4$ functions may be determined in the same manner or by reasoning from the $n = 2$ functions and then checking that the hole–particle convention of (11.4.12) is observed. The $n = 5$ functions are obvious except for the phase factors which are required to satisfy the hole–particle convention of (11.4.12).

We list t_2^n and e^n functions obtained as described above in Section C.8. These functions are our analogs of the strong-field functions given by Griffith [(1964), Table A24]. The real-O basis has been used to couple the orbitals, and the standard angular-momentum basis for SO_3 ($SO_3 \supset SO_2$) with Condon–Shortley phases has been used to couple the spins. The functions in Section C.8 are given only for purposes of illustration and for simple applications, since we show in Chapter 19 that if the irreducible tensor method is used, explicit wavefunctions are not required—only the cfp.

11.5. Formation of $|\mathcal{S}h\mathfrak{M}\theta\rangle$ Functions for Configurations with More Than One Open Shell

Our final task is to form $|\mathcal{S}h\mathfrak{M}\theta\rangle$ functions for configurations $a^n b^m c^l \cdots$ made up of any arbitrary number of subconfigurations. We illustrate the process by forming $\mathcal{S}\mathcal{L}$ functions for the group-O configuration $t_2^m e^n$. The functions are formed by coupling together the t_2^m and e^n states according to the equation

$$|(a^m(\mathcal{S}_1h_1), b^n(\mathcal{S}_2h_2))\mathcal{S}h\mathfrak{M}\theta\rangle = \sum_{\mathfrak{M}_1\mathfrak{M}_2} (\mathcal{S}_1\mathfrak{M}_1, \mathcal{S}_2\mathfrak{M}_2|\mathcal{S}\mathfrak{M})$$

$$\times \sum_{\theta_1\theta_2} (h_1\theta_1, h_2\theta_2|h\theta)|a^m\, \mathcal{S}_1h_1\mathfrak{M}_1\theta_1\rangle|b^n\, \mathcal{S}_2h_2\mathfrak{M}_2\theta_2\rangle \tag{11.5.1}$$

and then antisymmetrizing between the a^m and b^n kets. The v.c.c. are evaluated using (B.3.2) for spin couplings [as in (11.4.3)] and (10.2.1) for orbital couplings.

Suppose first that the $|{}^3T_1 1x\rangle$ function from $t_2^5 e$ is required. This function must clearly be formed by coupling ${}^2T_2$ and ${}^2E$. As in the formation of the functions in Section C.8, we use the real-O basis for the orbitals. We always follow the convention of putting the t_2 functions before the e functions. A consistent convention must be adopted, since in some cases answers differing in sign (phase) result if the other order is followed. In this case, using (10.2.1) and Sections C.11–C.12 to evaluate the orbital v.c.c. in (11.5.1), we find for the orbital coupling

$$|(h_1, h_2)h\theta\rangle = |(T_2, E)T_1 x\rangle$$
$$= \frac{\sqrt{3}}{2}|T_2\xi\rangle|E\theta\rangle - \frac{1}{2}|T_2\xi\rangle|E\epsilon\rangle \qquad (11.5.2)$$

The spin coupling needed here is given in (11.4.3):

$$|(\mathcal{S}_1, \mathcal{S}_2)\mathcal{S}\mathcal{M}\rangle = |(\tfrac{1}{2}, \tfrac{1}{2})11\rangle = |\tfrac{1}{2}\tfrac{1}{2}\rangle|\tfrac{1}{2}\tfrac{1}{2}\rangle \qquad (11.5.3)$$

Substituting these results into (11.5.1) and antisymmetrizing, we have

$$|\mathcal{Q}(t_2^5(\tfrac{1}{2}T_2), e(\tfrac{1}{2}E))1T_1\, 1x\rangle \equiv |(t_2+e)^{5+1}\, 1T_1\, 1x\rangle$$
$$= \mathcal{Q}\left(\frac{\sqrt{3}}{2}|t_2^5\, \tfrac{1}{2}T_2\tfrac{1}{2}\xi\rangle|e\, \tfrac{1}{2}E\tfrac{1}{2}\theta\rangle - \tfrac{1}{2}|t_2^5\, \tfrac{1}{2}T_2\tfrac{1}{2}\xi\rangle|e\, \tfrac{1}{2}E\tfrac{1}{2}\epsilon\rangle\right)$$
$$(11.5.4)$$

where $\mathcal{Q}$ antisymmetrizes between the t_2^5 and e kets. If we now substitute from Section C.8 for $|t_2^5\, \frac{1}{2}T_2\frac{1}{2}\xi\rangle$, and so on, *and antisymmetrize*, the final result is

$$|1T_1\, 1x\rangle = \frac{-\sqrt{3}}{2}|\xi^+\eta^2\zeta^2\theta^+\rangle + \tfrac{1}{2}|\xi^+\eta^2\zeta^2\epsilon^+\rangle \qquad (11.5.5)$$

The explicit antisymmetrization procedure is given in (19.7.1).

Suppose now that $|^1T_1 0x\rangle$ from $t_2^5 e$ is required. In this case, the orbital coupling is identical to that for the $|^3T_1 1x\rangle$ state above, but the spins must be coupled to form the singlet state. The difference between the two states mirrors directly the difference between the $|(\frac{1}{2}\frac{1}{2})00\rangle$ and $|(\frac{1}{2}\frac{1}{2})11\rangle$ spin functions of (11.4.3). Thus the normalized final result is

$$|0T_1 0x\rangle = \frac{-\sqrt{3}}{2\sqrt{2}}(|\xi^+\eta^2\zeta^2\theta^-\rangle - |\xi^-\eta^2\zeta^2\theta^+\rangle)$$
$$+ \frac{1}{2\sqrt{2}}(|\xi^+\eta^2\zeta^2\epsilon^-\rangle - |\xi^-\eta^2\zeta^2\epsilon^+\rangle) \qquad (11.5.6)$$

A greater challenge is posed by the construction of a $|^2T_1\frac{1}{2}1\rangle$ function from the configuration t_2^2e. First we note that the problem itself is ambiguous, since two such functions can be formed, one from $^1T_2(t_2^2)\otimes{}^2E(e)$ and the other from $^3T_1(t_2^2)\otimes{}^2E(e)$. We do the second case. In this example, orbital states are coupled in the complex-O basis. Applying (11.5.1) and antisymmetrizing gives

$$\begin{aligned}
\left|\mathcal{Q}\left(t_2^2(1T_1), e\left(\tfrac{1}{2}E\right)\right)\tfrac{1}{2}T_1\tfrac{1}{2}1\right\rangle &= \left|(t_2+e)^{2+1}\,\tfrac{1}{2}T_1\tfrac{1}{2}1\right\rangle \\
&= \mathcal{Q}\left[\sqrt{\frac{2}{3}}\left(\frac{-\sqrt{3}}{2}|t_2^2\,1T_1 1-1\rangle|e\,\tfrac{1}{2}E-\tfrac{1}{2}\epsilon'\rangle\right.\right. \\
&\qquad\left. -\tfrac{1}{2}|t_2^2\,1T_1\,11\rangle|e\,\tfrac{1}{2}E-\tfrac{1}{2}\theta\rangle\right) \\
&\qquad \frac{-1}{\sqrt{3}}\left(\frac{-\sqrt{3}}{2}|t_2^2\,1T_1\,0-1\rangle|e\,\tfrac{1}{2}E\tfrac{1}{2}\epsilon'\rangle\right. \\
&\qquad\left.\left. -\tfrac{1}{2}|t_2^2\,1T_1\,01\rangle|e\,\tfrac{1}{2}E\tfrac{1}{2}\theta\rangle\right)\right]
\end{aligned}\tag{11.5.7}$$

The final step is to substitute explicit one-electron functions into (11.5.7) and antisymmetrize between t_2^2 and e.

We use Table C.5.1(c) to express the complex-O functions above in terms of real-O basis functions and then substitute in t_2^m and e^n functions. Following this procedure and using the functions in Section C.8, we obtain

$$\begin{aligned}
|t_2^2\,1T_1\,11\rangle &= \frac{1}{\sqrt{2}}(-|t_2^2\,1T_1\,1x\rangle + |t_2^2\,1T_1\,1y\rangle) \\
&= \frac{1}{\sqrt{2}}(-|\eta^+\zeta^+\rangle - |\xi^+\zeta^+\rangle) \\
|t_2^2\,1T_1\,1-1\rangle &= \frac{1}{\sqrt{2}}(|\eta^+\zeta^+\rangle - |\xi^+\zeta^+\rangle) \\
|e\,\tfrac{1}{2}E\tfrac{1}{2}\epsilon'\rangle &= -|e\,\tfrac{1}{2}E\tfrac{1}{2}\epsilon\rangle = -|\epsilon^+\rangle \\
|e\,\tfrac{1}{2}E-\tfrac{1}{2}\epsilon'\rangle &= -|e\,\tfrac{1}{2}E-\tfrac{1}{2}\epsilon\rangle = -|\epsilon^-\rangle
\end{aligned}\tag{11.5.8}$$

The $|e\ \frac{1}{2}E \pm \frac{1}{2}\theta\rangle$ functions are identical to those in the real-O basis given in (11.4.1). The t_2^2 spin-singlet functions, $|^3T_1 01\rangle$ and $|^3T_1 0-1\rangle$, are then generated from the corresponding functions of maximum $\mathfrak{M}$ value of (11.5.8). This is easily done, since the difference between the $\mathfrak{M}=1$, $\mathfrak{M}=0$, and $\mathfrak{M}=-1$ states is a direct reflection of the differences in the two-electron spin–spin couplings given in (11.4.3). Thus

$$|t_2^2\,1T_1 01\rangle = \tfrac{1}{2}(-|\eta^+\zeta^-\rangle - |\eta^-\zeta^+\rangle - |\xi^+\zeta^-\rangle - |\xi^-\zeta^+\rangle)$$
$$|t_2^2\,1T_1 0-1\rangle = \tfrac{1}{2}(|\eta^+\zeta^-\rangle + |\eta^-\zeta^+\rangle - |\xi^+\zeta^-\rangle - |\xi^-\zeta^+\rangle) \qquad (11.5.9)$$

The final result, after some rearranging of determinants, is

$$|^2T_1\tfrac{1}{2}1\rangle = |\mathcal{Q}\big(t_2^2(1T_1), e(\tfrac{1}{2}E)\big)\tfrac{1}{2}T_1\tfrac{1}{2}1\rangle$$
$$= \tfrac{1}{2}(|\eta^+\zeta^+\epsilon^-\rangle - |\xi^+\zeta^+\epsilon^-\rangle) + \frac{1}{2\sqrt{3}}(|\eta^+\zeta^+\theta^-\rangle + |\xi^+\zeta^+\theta^-\rangle)$$
$$+\tfrac{1}{4}(-|\eta^+\zeta^-\epsilon^+\rangle - |\eta^-\zeta^+\epsilon^+\rangle + |\xi^-\zeta^+\epsilon^+\rangle + |\xi^+\zeta^-\epsilon^+\rangle)$$
$$-\frac{1}{4\sqrt{3}}(|\eta^+\zeta^-\theta^+\rangle + |\eta^-\zeta^+\theta^+\rangle$$
$$+|\xi^-\zeta^+\theta^+\rangle + |\xi^+\zeta^-\theta^+\rangle) \qquad (11.5.10)$$

These examples should suffice to demonstrate how to couple any two subconfigurations to form $|\mathbb{S}h\mathfrak{M}\theta\rangle$ functions. If there are more than two subconfigurations to be coupled, that is, if there are more than two open shells, one can clearly proceed in a stepwise fashion by first coupling any pair, then coupling this result with the third, and so on. However, though simple in principle, this procedure can become very complicated in practice because a large number of states may arise.

For example, suppose one wished to form quartet $\mathbb{S}\mathfrak{L}$ states from the configuration $t_{2g}^4e_g^2t_{1u}^5$ (an excited charge-transfer configuration of high-spin octahedral Fe^{3+}). Then since $t_{2g}^4 \otimes e_g^2 \otimes t_{1u}^5 = (^3T_{1g} + {}^1A_{1g} + {}^1E_g + {}^1T_{2g}) \otimes (^3A_{2g} + {}^1A_{1g} + {}^1E_g) \otimes {}^2T_{1u}$, one finds that five different triple direct products contribute states, namely, $^3T_{1g} \otimes {}^3A_{2g} \otimes {}^2T_{1u}$, $^3T_{1g} \otimes {}^1A_{1g} \otimes {}^2T_{1u}$, $^3T_{1g} \otimes {}^1E_g \otimes {}^2T_{1u}$, $^1E_g \otimes {}^3A_{2g} \otimes {}^2T_{1u}$, $^1T_{2g} \otimes {}^3A_{2g} \otimes {}^2T_{1u}$. And if one wished to construct doublet states, the number of triple direct products contributing would be 11.

In such instances, the process of constructing explicit wavefunctions is tedious in the extreme, and the prospect of using such functions to calculate matrix elements is horrifying. Even many instances of two open shells—$t_{2g}^4 t_{1u}^5$ for example—are very complicated. To treat such problems, one turns to the irreducible-tensor method discussed in later chapters. One then does not have to construct wavefunctions at all, and if one proceeds in a careful and systematic fashion, one can calculate matrix elements quite quickly and with a high degree of confidence.

12 Spin–Orbit Coupling and Double-Group Calculations

12.1. The Spin–Orbit Coupling Operator

It is an experimental fact that an electron has an internal spin magnetic moment $\boldsymbol{\mu}_S$ given by

$$\boldsymbol{\mu}_S = g\gamma_e \mathbf{s} \approx 2\gamma_e \mathbf{s} \qquad (12.1.1)$$

where γ_e, the magnetogyric ratio, is defined as $\gamma_e = -e/2m_e c$, and the g-factor, which has a value of $g \approx 2.0023$, has been approximated by $g = 2$. In these equations $-e$, m_e, and $\mathbf{s}$ are respectively the charge (in statcoulombs), mass, and spin angular momentum of an electron. If we were to think classically and position ourselves on an electron with nonzero orbital angular momentum, we would feel a magnetic field arising from the orbital motion of the positively charged nucleus. It is the interaction of this magnetic field $\mathbf{B}$ and the spin magnetic moment of the electron $\boldsymbol{\mu}_S$ which gives rise to spin–orbit coupling. Since the magnetic field arising from the orbital motion of the positively charged nucleus increases with increasing nuclear charge, spin–orbit coupling is most pronounced for heavy atoms (and molecules or ions containing them). The interaction energy between the field $\mathbf{B}$ and $\boldsymbol{\mu}_S$ is $-\boldsymbol{\mu}_S \cdot \boldsymbol{B}$, where $\boldsymbol{B}$ is proportional to $\mathbf{l}$, the orbital angular momentum of the electron. Thus we expect the spin–orbit coupling operator $\mathcal{H}_{SO}$ for an electron in a hydrogenlike atom to involve the dot product of $\mathbf{s}$ and $\mathbf{l}$.

It is necessary, however, to use Dirac's relativistic wave equation to properly derive both $\mathcal{H}_{SO}$ and $\boldsymbol{\mu}_S$—the g (= 2) factor enters only through relativistic considerations. We refer the interested reader to Griffith [(1964), Chapter 5], and simply give the results here. The Dirac equation yields for the spin–orbit coupling operator of a hydrogenlike atom the expression $\xi(r)\mathbf{l} \cdot \mathbf{s}$, where $\xi(r) \equiv (1/2m_e^2 c^2 r)\, \partial U(r)/\partial r$. $U(r)$ is the central-field potential energy ($-Ze^2/r$ for a hydrogen atom), and $\mathbf{l}$ and $\mathbf{s}$ are the electronic orbital and spin angular-momentum operators respectively.

For an n-electron atom, the spin–orbit coupling operator is generally written in the approximate form

$$\mathcal{H}_{\text{SO}} = \sum_{k=1}^{n} \xi(r_k)\mathbf{l}(k) \cdot \mathbf{s}(k) \tag{12.1.2}$$

where $\mathbf{l}(k)$ and $\mathbf{s}(k)$ are the orbital and spin angular-momentum operators respectively for electron k. Equation (12.1.2) is obtained by simply summing $\mathcal{H}_{\text{SO}}$ of the hydrogen atom over the n electrons present. Such a procedure is not strictly correct. Furthermore, (12.1.2) neglects, except indirectly through $U(r)$, related magnetic effects such as orbit–orbit, spin–spin, and spin–other-orbit interactions. These are assumed to be unimportant in all of our applications. In addition we assume that (12.1.2) is always applicable, that is, that it applies for systems of lower than spherical symmetry even though the electrons no longer move in a central field $U(r)$. We never attempt calculations involving the radial part of $\mathcal{H}_{\text{SO}}$; rather we regard the radial integral over $\xi(r_k)$ as an adjustable parameter ζ—*the spin–orbit coupling constant*—whose value for each atom or ion is determined empirically. Thus one hopes that any inadequacies in the application of (12.1.2) to molecules is compensated in this way.

12.2. Spin–Orbit Basis Functions

The introduction of spin–orbit coupling has profound consequences because it adds a term to the zero-field Hamiltonian operator $\mathcal{H}$ which depends on *both* the spin and orbital angular momentum. The result, in the atomic case, is that the operators S^2, S_z, L^2, and L_z no longer commute with $\mathcal{H}$, and hence functions of the form $|\alpha'\mathcal{S}\mathcal{L}\mathcal{M}_\mathcal{S}\mathcal{M}_\mathcal{L}\rangle$ can no longer describe exact eigenfunctions of $\mathcal{H}$. Conversely, the exact eigenfunctions of $\mathcal{H}$ cannot satisfy (11.3.2). Thus $\mathcal{S}$, $\mathcal{L}$, $\mathcal{M}_\mathcal{S}$, and $\mathcal{M}_\mathcal{L}$ are no longer good quantum numbers; in the language of group theory this means that they no longer designate irreps of the group of the Schrödinger equation and so can no longer be used to classify the eigenfunctions of atoms or free ions. However, it is found [Griffith (1964), Chapter 2] that the total angular-momentum operator, $\mathbf{J} = \mathbf{L} + \mathbf{S}$, has the property that J^2, J_z, and $\mathcal{H}$ all commute (so that $\mathcal{H}$ is invariant to all rotations in spin-orbit (J) space). It therefore follows that a complete set of functions must exist which are eigenfunctions of these operators. We write these new functions as $|\alpha'(\mathcal{S}\mathcal{L})\mathcal{J}\mathcal{M}_\mathcal{J}\rangle$ and note that (in units of $\hbar$)

$$J^2|\alpha'(\mathbb{S}\mathcal{L})\mathcal{J}\mathcal{M}_{\mathcal{J}}\rangle = \mathcal{J}(\mathcal{J}+1)|\alpha'(\mathbb{S}\mathcal{L})\mathcal{J}\mathcal{M}_{\mathcal{J}}\rangle$$
$$J_z|\alpha'(\mathbb{S}\mathcal{L})\mathcal{J}\mathcal{M}_{\mathcal{J}}\rangle = \mathcal{M}_{\mathcal{J}}|\alpha'(\mathbb{S}\mathcal{L})\mathcal{J}\mathcal{M}_{\mathcal{J}}\rangle \qquad (12.2.1)$$

Thus in the presence of spin–orbit coupling, $\mathcal{J}$ and $\mathcal{M}_{\mathcal{J}}$ classify the eigenfunctions of atoms.

If spin–orbit coupling is small, that is, if the splitting of different terms by interelectronic repulsion is large compared to spin–orbit splittings, the $|\alpha'(\mathbb{S}\mathcal{L})\mathcal{J}\mathcal{M}_{\mathcal{J}}\rangle$ can be well approximated by linear combinations of $|\alpha'\mathbb{S}\mathcal{L}\mathcal{M}_{\mathbb{S}}\mathcal{M}_{\mathcal{L}}\rangle$ from a single term. In other words, $\mathcal{H}_{SO}$ can be treated satisfactorily by first-order perturbation theory. As $\mathcal{H}_{SO}$ gets larger, this is no longer the case, and functions of given $\mathcal{J}$ and $\mathcal{M}_{\mathcal{J}}$ from different terms mix significantly. Regardless of how large $\mathcal{H}_{SO}$ becomes, the $|\mathbb{S}\mathcal{L}\mathcal{M}_{\mathbb{S}}\mathcal{M}_{\mathcal{L}}\rangle$ are satisfactory basis functions *if enough* $\mathbb{S}\mathcal{L}$ *terms are used*, since the full set is complete. Although it is customary to speak of both the $|\mathbb{S}\mathcal{L}\mathcal{M}_{\mathbb{S}}\mathcal{M}_{\mathcal{L}}\rangle$ and $|(\mathbb{S}\mathcal{L})\mathcal{J}\mathcal{M}_{\mathcal{J}}\rangle$ as *LS* (Russell–Saunders) states (or functions), we usually refer to the latter (and their molecular analogs) as *spin–orbit states*.

In the case of molecules, an analogous situation arises. When $\mathcal{H}_{SO}$ is included, the Hamiltonian no longer commutes with S^2 and S_z; thus $\mathbb{S}$ and $\mathcal{M}_{\mathbb{S}}$ are no longer good quantum numbers. Likewise, the previous $h\theta$ classification is no longer applicable. It is now the double-group operators (Section, 8.12) which commute with $\mathcal{H}$. In exact analogy to the atomic case, spin and space are again coupled through a set of symmetry-determined coefficients. For this purpose, however, an SO_3 spin basis adapted to a lower symmetry group must be used, rather than the angular-momentum ($SO_3 \supset SO_2$) spin basis $|\mathbb{S}\mathcal{M}\rangle$. We label spin states in this symmetry-adapted SO_3 basis as $|\mathbb{S}SM\rangle$ when they transform as the irrep $\mathbb{S}$ in SO_3, and as the partner M of the irrep S in the lower symmetry group G to which the molecule belongs. (Section C.6, for example, may be used to relate such a group-O $|\mathbb{S}SM\rangle$ basis to the $|\mathbb{S}\mathcal{M}\rangle$ basis of SO_3.) Then the spin S is coupled to the orbital state h *in group* G to give states $|(\mathbb{S}S, h)rt\tau\rangle$ which are linear combinations of the $|\mathbb{S}ShM\theta\rangle$. The $|(\mathbb{S}S, h)rt\tau\rangle$ belong to irreps of the Hamiltonian containing spin–orbit coupling. Specifically, (10.2.3) gives

$$|(\mathbb{S}S, h)rt\tau\rangle = \sum_{M\theta}(SM, h\theta|rt\tau)|\mathbb{S}ShM\theta\rangle$$
$$= |t|^{1/2}\begin{pmatrix} t \\ \tau \end{pmatrix}\sum_{M\theta}\begin{pmatrix} S & h & t^* \\ M & \theta & \tau^* \end{pmatrix}^{*\,r}|\mathbb{S}ShM\theta\rangle \qquad (12.2.2)$$

where t and τ (but not r) are now the "good quantum numbers," in analogy respectively to $\mathcal{J}$ and $\mathcal{M}_{\mathcal{J}}$ in the atomic case. The spin–orbit state $|(\mathbb{S}S, h)rt\tau\rangle$ transforms as the partner τ of the irrep t in the group G. The extra index r (which equals 0 or 1) is required if the direct product of spin and space contains a repeated representation [as in the group O when $S \otimes h = U' \otimes T_1$ or $U' \otimes T_2$—see discussion leading to (9.3.22)]. The use of this index will become clear in later discussions and examples.

We shall see in Chapter 18 that the irreducible-tensor method eliminates the need to construct $|(\mathbb{S}S, h)rt\tau\rangle$ wavefunctions explicitly. Nevertheless, we give some examples of the procedure in Sections 12.3 and 12.4.

12.3. Construction of Spin–Orbit States $|(\mathbb{S}S, h)rt\tau\rangle$ for Integral $\mathbb{S}$

We now construct several spin–orbit functions to show that the process is completely straightforward. We start with the case where only single-valued irreps are involved; that is, $\mathbb{S}$ is an integer. Suppose one wishes to construct, for the group O, spin–orbit functions arising from the triplet terms, ${}^3A_1$, ${}^3A_2$, ${}^3E$, ${}^3T_1$, and ${}^3T_2$. Our problem is precisely the one previously considered (Section 9.5), namely, how to construct a linear combination of direct-product functions which transforms as a specified partner of a specific irrep. In the present case, each direct-product basis function $|\mathbb{S}ShM\theta\rangle$ is the product of a spin function $|\mathbb{S}SM\rangle$ and a space function $|h\theta\rangle$.

Spin functions of integer spin $\mathbb{S}$, just like orbital angular-momentum functions, transform *under rotations* in spin space exactly like the spherical harmonics (Sections 9.4 and B.1); thus both transform as the $|jm\rangle$ of $SO_3 \supset SO_2$ (see Section 9.8). However, it is very important to recognize that spin functions are *always* assigned gerade symmetry (even or + parity) *under inversion*, while the parity of the Y_m^j is + for even j and − for odd j (Section 9.8). Thus in general, *odd*-integer spinors transform *differently* than their spherical harmonic counterparts (which have odd or − parity) when improper transformation operators $S_\phi = C_i C_{\phi-\pi}$ are applied.

Recall that the use of (12.2.2) to construct $|(\mathbb{S}S, h)rt\tau\rangle$ states *requires* that spin states, as well as orbital states, be classified as discussed above according to the irreps of the molecular symmetry group; S and M are irrep labels of this symmetry group. As illustrated for the group O in Section C.6, quite often there is a one-to-one correspondence between the $|\mathbb{S}\mathcal{M}\rangle$ and the $|\mathbb{S}SM\rangle$. In all cases the relation of the $|\mathbb{S}\mathcal{M}\rangle$ to the $|\mathbb{S}SM\rangle$ is easily determined from tables such as those of Sections C.6, D.5, and E.5. Thus, for example, from Section C.6 we see that $\mathbb{S} = 2$ transforms as $S = E$ and

T_2 in O, and that

$$|\mathbb{S}SM\rangle = |2E\epsilon'\rangle = \frac{-1}{\sqrt{2}}\left(|2\,-2\rangle^{SO_3} + |22\rangle^{SO_3}\right) \tag{12.3.1}$$

and so on.

Returning to the problem of constructing spin–orbit functions from the triplet terms in group O, we consider the spin–orbit states $|(\mathbb{S}S, h)t\tau\rangle$ of $^3T_1$ as an example. No index r is needed, since no repeated representations occur here in $S \otimes h$. Section C.6 tells us that $\mathbb{S} = 1$ belongs to T_1 in the group O. Our direct-product (spin–orbit) basis is thus $T_1(\text{spin}, S) \otimes T_1(\text{space}, h) = (A_1 + E + T_1 + T_2)(\text{spin-orbit state}, t)$. That is, from the $^{2\mathbb{S}+1}h = {}^3T_1$ term, we can construct spin–orbit (t) functions transforming as A_1, E, T_1, and T_2. We follow the convention [opposite to that of Griffith (1964)] of always writing the spin part first. Thus, using (12.2.2) and the $3jm$ of Section C.12 in the complex-O basis, the $A_1(^3T_1)$ spin–orbit function is

$$\begin{aligned}|(1T_1, T_1)A_1a_1\rangle &= \frac{1}{\sqrt{3}}\left(|1T_1T_1\,-11\rangle - |1T_1T_100\rangle + |1T_1T_11\,-1\rangle\right)\\ &= \frac{1}{\sqrt{3}}\left(-|^3T_1\,-11\rangle - |^3T_100\rangle - |^3T_11\,-1\rangle\right)\end{aligned} \tag{12.3.2}$$

In the last line we use Section C.6 to express the $|\mathbb{S}ShM\theta\rangle = |1T_1T_1M\theta\rangle$ states in terms of the $|^{2\mathbb{S}+1}h\mathfrak{M}\theta\rangle$—that is, we convert the group-O-adapted $|\mathbb{S}SM\rangle$ to the $|\mathbb{S}\mathfrak{M}\rangle$ of $SO_3 \supset SO_2$. In exactly the same way, the $E(^3T_1)$, $T_1(^3T_1)$, and $T_2(^3T_1)$ functions are found to be

$$\begin{aligned}|(1T_1, T_1)E\theta\rangle &= \frac{-1}{\sqrt{6}}|1T_1T_1\,-11\rangle - \frac{\sqrt{2}}{\sqrt{3}}|1T_1T_100\rangle - \frac{1}{\sqrt{6}}|1T_1T_11\,-1\rangle\\ &= \frac{1}{\sqrt{6}}|^3T_1\,-11\rangle - \frac{\sqrt{2}}{\sqrt{3}}|^3T_100\rangle + \frac{1}{\sqrt{6}}|^3T_11\,-1\rangle\\ |(1T_1, T_1)T_11\rangle &= \frac{-1}{\sqrt{2}}|1T_1T_101\rangle + \frac{1}{\sqrt{2}}|1T_1T_110\rangle\\ &= \frac{-1}{\sqrt{2}}|^3T_101\rangle - \frac{1}{\sqrt{2}}|^3T_110\rangle\\ |(1T_1, T_1)T_21\rangle &= \frac{1}{\sqrt{2}}|1T_1T_101\rangle + \frac{1}{\sqrt{2}}|1T_1T_110\rangle\\ &= \frac{1}{\sqrt{2}}|^3T_101\rangle - \frac{1}{\sqrt{2}}|^3T_110\rangle\end{aligned} \tag{12.3.3}$$

where we list only one partner of E, T_1, and T_2.

Finally, note that we have already discussed the construction of $|\mathbb{S}h\mathfrak{M}\theta\rangle = |^{2\mathbb{S}+1}h\mathfrak{M}\theta\rangle$ functions in Sections 11.4 and 11.5. They may be inserted in the equations above to obtain the spin–orbit functions in terms of one-electron functions.

12.4 Construction of Spin–Orbit States When $\mathbb{S}$ Is Half-Integer

The construction of $|(\mathbb{S}S, h)rt\tau\rangle$ states when $\mathbb{S}$ is half-integer proceeds analogously to that when $\mathbb{S}$ is integer, except that extra care must be taken when repeated representations occur (Sections 9.3 and 9.7). Suppose we wish to form some of the spin–orbit functions arising from a $^4T_1$ term in the group O. From Section C.6 we see that $\mathbb{S} = \frac{3}{2}$ transforms as U', so that $U'(\text{spin}, S) \otimes T_1(\text{space}, h) = (E' + E'' + 2U')(\text{spin–orbit state}, t)$. The two U' states are distinguished by the repeated representation label r.

As in Section 12.3 above, we use (12.2.2), the complex-O $3jm$ of Section C.12, and Section C.6 to construct the $rt\tau(^4T_1)$ states. Examples formed in this manner are

$$\begin{aligned}
\left|\left(\tfrac{3}{2}U', T_1\right)E''\beta''\right\rangle &= \frac{1}{\sqrt{3}}|\tfrac{3}{2}U'T_1\nu 0\rangle + \frac{1}{\sqrt{2}}|\tfrac{3}{2}U'T_1\mu -1\rangle + \frac{1}{\sqrt{6}}|\tfrac{3}{2}U'T_1\kappa 1\rangle \\
&= \frac{1}{\sqrt{3}}|^4T_1 -\tfrac{3}{2}0\rangle - \frac{1}{\sqrt{2}}|^4T_1 -\tfrac{1}{2}-1\rangle + \frac{1}{\sqrt{6}}|^4T_1\tfrac{3}{2}1\rangle \\
\left|\left(\tfrac{3}{2}U', T_1\right)0U'\nu\right\rangle &= \frac{-1}{\sqrt{3}}|\tfrac{3}{2}U'T_1\nu 0\rangle + \frac{\sqrt{2}}{\sqrt{3}}|\tfrac{3}{2}U'T_1\kappa 1\rangle \\
&= \frac{-1}{\sqrt{3}}|^4T_1 -\tfrac{3}{2}0\rangle + \frac{\sqrt{2}}{\sqrt{3}}|^4T_1\tfrac{3}{2}1\rangle \qquad (12.4.1) \\
\left|\left(\tfrac{3}{2}U', T_1\right)1U'\nu\right\rangle &= \frac{-1}{\sqrt{3}}|\tfrac{3}{2}U'T_1\nu 0\rangle + \frac{1}{\sqrt{2}}|\tfrac{3}{2}U'T_1\mu -1\rangle - \frac{1}{\sqrt{6}}|\tfrac{3}{2}U'T_1\kappa 1\rangle \\
&= \frac{-1}{\sqrt{3}}|^4T_1 -\tfrac{3}{2}0\rangle - \frac{1}{\sqrt{2}}|^4T_1 -\tfrac{1}{2}-1\rangle - \frac{1}{\sqrt{6}}|^4T_1\tfrac{3}{2}1\rangle
\end{aligned}$$

The sets of $|(\mathbb{S}S, h)rt\tau\rangle$ states, $|(\tfrac{3}{2}U', T_1)0U'\tau\rangle$ and $|(\tfrac{3}{2}U', T_1)1U'\tau\rangle$, are mutually orthogonal, but in our basis are not diagonal in $\mathcal{H}_{SO}$—see Sections 12.5–12.7.

12.5. Behavior of $|\mathbb{S}h\mathfrak{M}\theta\rangle$ and $|(\mathbb{S}S, h)rt\tau\rangle$ Functions under $\mathcal{R}$ and $\mathcal{H}_{SO}$

We may explicitly illustrate the meaning of working in double group space by applying a transformation operator $\mathcal{R}$ to an $|\mathbb{S}ShM\theta\rangle$ function. Symbolically

$$\mathcal{R}|\mathbb{S}ShM\theta\rangle = (\mathcal{R}|\mathbb{S}SM\rangle)(\mathcal{R}|h\theta\rangle) \tag{12.5.1}$$

Consider the group O example in which $\mathcal{R} = \mathcal{C}_4^X$ and $|\mathbb{S}ShM\theta\rangle$ is the group-O state $|1T_1T_1 - 11\rangle$. Using Tables C.5.1(b) and C.5.1(c), we obtain for any group-O state transforming as T_11 or $T_1 - 1$,

$$\mathcal{C}_4^X|T_1 \pm 1\rangle = \tfrac{1}{2}(|T_1 \pm 1\rangle - |T_1 \mp 1\rangle) + \frac{i}{\sqrt{2}}|T_10\rangle \tag{12.5.2}$$

Using (12.5.1), we obtain in our example

$$\begin{aligned}\mathcal{C}_4^X|1T_1T_1 - 11\rangle &= \big(\mathcal{C}_4^X|1T_1 - 1\rangle\big)\big(\mathcal{C}_4^X|T_11\rangle\big)\\ &= \left[\tfrac{1}{2}(|1T_1 - 1\rangle - |1T_11\rangle) + \frac{i}{\sqrt{2}}|1T_10\rangle\right]\\ &\quad\times\left[\tfrac{1}{2}(|T_11\rangle - |T_1 - 1\rangle) + \frac{i}{\sqrt{2}}|T_10\rangle\right]\end{aligned} \tag{12.5.3}$$

where the first bracketed term is the transformed spin state $\mathcal{R}|\mathbb{S}SM\rangle$ and the second bracketed term is the transformed orbital state $\mathcal{R}|h\theta\rangle$. If we now "multiply" these two terms together and reassemble the product as a sum of $|\mathbb{S}ShM\theta\rangle$ states, we obtain

$$\begin{aligned}\mathcal{C}_4^X|1T_1T_1 - 11\rangle &= \tfrac{1}{4}(|1T_1T_1 - 11\rangle - |1T_1T_1 - 1 - 1\rangle\\ &\quad - |1T_1T_111\rangle + |1T_1T_11 - 1\rangle)\\ &\quad + \frac{i}{2\sqrt{2}}(|1T_1T_1 - 10\rangle - |1T_1T_110\rangle + |1T_1T_101\rangle\\ &\quad - |1T_1T_10 - 1\rangle) - \tfrac{1}{2}|1T_1T_100\rangle\end{aligned} \tag{12.5.4}$$

So for example to confirm explicitly that the spin-orbit function $|(1T_1, T_1)A_1a_1\rangle$ in (12.3.2) is correct, we would apply the group generators to the right hand side of that equation (as in the example above) to see if the standard basis relations are indeed obeyed. Clearly, the $\mathcal{R}$ may be applied to the $|\mathbb{S}h\mathfrak{M}\theta\rangle$ functions in an analogous way to (12.5.1).

$\mathcal{H}_{SO}$ transforms as the totally symmetric irrep of the double group. Thus, for example, $\mathcal{H}_{SO}$ splits $^3T_1$ into four levels, A_1, E, T_1, and T_2, of degeneracies one, two, three, and three respectively, while $^4T_1$ splits into four levels, E', E'', and $2U'$, where E' and E'' are twofold degenerate and the U' states are fourfold degenerate. In most cases, just as in the atomic case, the $|(\mathbb{S}S, h)rt\tau\rangle$-type functions for a term $^{2\mathbb{S}+1}h$ are diagonal in $\mathcal{H}_{SO}$. Thus they are good first-order spin–orbit eigenfunctions. [The important exceptions to this rule for $\mathbb{S} \leqslant \frac{3}{2}$ are the functions for $rU'(^4T_1)$, $r = 0, 1$, and $rU'(^4T_2)$, $r = 0, 1$, which in our Butler (1981) basis are not diagonal in $\mathcal{H}_{SO}$ for $r = 0, 1$—see Sections 12.6 and 18.2. While t and τ are good quantum numbers, r is not.] As first-order spin–orbit splittings become appreciable compared to the separation of $\mathbb{S}\mathbb{L}$ terms, functions from different terms belonging to the same row (τ) of the same irrep (t) of the double group begin to mix. In this manner functions of different spin multiplicity (e.g., triplet–singlet, doublet–quartet) are mixed under the influence of spin–orbit coupling. In Section 13.3 we first show how to calculate the matrix elements of $\mathcal{H}_{SO}$.

12.6. Repeated-Representation Considerations

In this section we extend the discussion of repeated representations (Sections 9.3, 9.7, and 10.2) and consider practical problems involving them.

Choice of the $U'U'T_1$ and $U'U'T_2$ Multiplicity Separations

As discussed in Section 9.7, when repeated representations occur in $a \otimes b$, there exists an infinity of different pairs of mutually orthogonal functions, $|(ab)ic\gamma\rangle$ and $|(ab)jc\gamma\rangle$, which satisfy standard basis relations of the type given in Section C.5. The multiplicity separation into states $r = i$ and $r = j$ is specified only when coupling coefficients are chosen. When the coefficients used are v.c.c., any of these possible multiplicity separations are acceptable. Some choices, however, are particularly convenient. Thus Griffith (1964) chose his $U' \otimes T_1$ and $U' \otimes T_2$ v.c.c. for the groups O and T_d so that his $rU'(^4T_1)$ and $rU'(^4T_2)$ spin–orbit functions are diagonal to first order in $\mathcal{H}_{SO}$. The r labels used in the Griffith (1964) multiplicity separation are $r = \frac{3}{2}$ and $\frac{5}{2}$.

If high-symmetry coefficients such as our $3jm$ are used, much less choice is available. As discussed in Section 10.2, the Griffith choice is unacceptable for $U' \otimes T_2$, since $|(U'T_2)rU'\tau\rangle$ states must be formed using a

$$\begin{pmatrix} U' & U' & T_2 \\ \alpha & \beta & \gamma \end{pmatrix}^r$$

table which obeys (10.2.4). While we could have used Griffith's choice of $U' \otimes T_1$, $r = \frac{3}{2}, \frac{5}{2}$, for our $U'U'T_1$ $3jm$, we follow Butler (1981), and choose the $U'U'T_1$ multiplicity separation to mimic the *required* $U'U'T_2$ separation, and label the states $r = 0$ and $r = 1$.

A consequence of our $r = 0, 1$ multiplicity separation is that the $rU'(^4T_1)$ and $rU'(^4T_2)$ states formed via (12.2.2) as in Section 12.4 above are *not* diagonal in $\mathcal{H}_{SO}$ in the index r. Thus to obtain first-order spin–orbit eigenfunctions for $U'(^4T_1)$ or $U'(^4T_2)$, the $\mathcal{H}_{SO}$ matrix for $|(\mathbb{S}S, h)rt\tau\rangle = |(\frac{3}{2}U', T_1)rU'\tau\rangle$, $r = 0, 1$, or $|(\frac{3}{2}U', T_2)rU'\tau\rangle$, $r = 0, 1$, must be diagonalized.

Determination of First-Order Spin–Orbit $U'(^4T_1)$ and $U'(^4T_2)$ Functions Using 3jm

While t and τ are always good quantum numbers, r is not. As discussed above, the $|(\frac{3}{2}U', T_1)rU'\tau\rangle$ or $|(\frac{3}{2}U', T_2)rU'\tau\rangle$ $\mathcal{H}_{SO}$ matrices for $r = 0, 1$ must be diagonalized to obtain first-order $U'(^4T_1)$ or $U'(^4T_2)$ eigenfunctions. This can be done once and for all, and we give the results below. Equation (18.4.6) is used to calculate the matrices in terms of the constants

$$c(^4T_i) \equiv \langle \tfrac{3}{2}T_i \| \sum_k s^1 u^{t_1}(k) \| \tfrac{3}{2}T_i \rangle^{SO_3, G} \qquad (12.6.1)$$

where the group G is O or T_d, and $i = 1$ for $^4T_1$ and 2 for $^4T_2$; the spin–orbit operator notation is that of Section 13.3. In units of $c(^4T_1)$ and $c(^4T_2)$ respectively, the matrices are

$\mathcal{H}_{SO}$	$0U'(^4T_1)$	$1U'(^4T_1)$
$0U'(^4T_1)$	$-1/3\sqrt{10}$	$1/3\sqrt{10}$
$1U'(^4T_1)$	$1/3\sqrt{10}$	$1/6\sqrt{10}$

(12.6.2)

$\mathcal{H}_{SO}$	$0U'(^4T_2)$	$1U'(^4T_2)$
$0U'(^4T_2)$	$1/3\sqrt{10}$	$1/3\sqrt{10}$
$1U'(^4T_2)$	$1/3\sqrt{10}$	$-1/6\sqrt{10}$

(12.6.3)

where $0U'$ and $1U'$ are the U' functions for $r = 0$ and 1. We have for the first-order eigenvalues and eigenvectors of (12.6.2) and (12.6.3) respectively,

$U'(^4T_1)$:

$$E[\alpha U'(^4T_1)] = \frac{1}{3\sqrt{2}}c(^4T_1), \qquad E[\beta U'(^4T_1)] = -\frac{1}{2\sqrt{2}}c(^4T_1)$$

$$|(\tfrac{3}{2}U', T_1)\alpha U'\tau\rangle = \frac{1}{\sqrt{5}}|(\tfrac{3}{2}U', T_1)0U'\tau\rangle + \frac{2}{\sqrt{5}}|(\tfrac{3}{2}U', T_1)1U'\tau\rangle$$

$$|(\tfrac{3}{2}U', T_1)\beta U'\tau\rangle = \frac{-2}{\sqrt{5}}|(\tfrac{3}{2}U', T_1)0U'\tau\rangle + \frac{1}{\sqrt{5}}|(\tfrac{3}{2}U', T_1)1U'\tau\rangle$$

(12.6.4a)

$U'(^4T_2)$:

$$E[\alpha U'(^4T_2)] = \frac{-1}{3\sqrt{2}}c(^4T_2), \qquad E[\beta U'(^4T_2)] = \frac{1}{2\sqrt{2}}c(^4T_2)$$

$$|(\tfrac{3}{2}U', T_2)\alpha U'\tau\rangle = \frac{-1}{\sqrt{5}}|(\tfrac{3}{2}U', T_2)0U'\tau\rangle + \frac{2}{\sqrt{5}}|(\tfrac{3}{2}U', T_2)1U'\tau\rangle$$

$$|(\tfrac{3}{2}U', T_2)\beta U'\tau\rangle = \frac{2}{\sqrt{5}}|(\tfrac{3}{2}U', T_2)0U'\tau\rangle + \frac{1}{\sqrt{5}}|(\tfrac{3}{2}U', T_2)1U'\tau\rangle$$

(12.6.4b)

The $\alpha U'$ and $\beta U'$ eigenfunctions differ in phase in a complicated way from the $\frac{3}{2}U'$ and $\frac{5}{2}U'$ functions in Griffith (1964), due to differences in the order of coupling of spin and space, and the like, but as is always the case, all observables are independent of phase differences provided the same phases are used consistently throughout a calculation. Thus the first-order eigenvalues labeled $\alpha U'$ and $\beta U'$ have the same spin–orbit energies respectively as the $\frac{3}{2}U'$ and $\frac{5}{2}U'$ states in Griffith (1964) for $^4T_1$ and $^4T_2$.

The results of (12.6.2)–(12.6.4) do *not* apply for $Sh = U'T_i$ states derived from spin sextets ($\mathbb{S} = \frac{5}{2}$). Once again, however, (18.4.6) may be used to determine the first-order matrices, which may then be diagonalized to obtain first-order eigenvalues and eigenvectors.

The Significance of Repeated-Representation Labels in Matrix-Element Reduction

It is essential that the significance of the various repeated-representation labels that may appear in matrix-element reduction equations be understood. For example, suppose we wish to reduce a matrix element of a T_1 operator involving the U' states $jU'\tau_1$ and $kU'\tau_2$,, where j and k are repeated-representation labels for the states and τ_1 and τ_2 are partner labels. Equation (10.2.2) is used to give

$$\langle jU'\tau_1|O_{\phi}^{t_1}|kU'\tau_2\rangle = \begin{pmatrix} U' \\ \tau_1 \end{pmatrix}\left[\begin{pmatrix} U' & T_1 & U' \\ \tau_1^* & \phi & \tau_2 \end{pmatrix}^0 \langle jU'\|O^{t_1}\|kU'\rangle_0 + \begin{pmatrix} U' & T_1 & U' \\ \tau_1^* & \phi & \tau_2 \end{pmatrix}^1 \langle jU'\|O^{t_1}\|kU'\rangle_1\right] \quad (12.6.5)$$

The repeated-representation labels $r = 0$ and 1 on the right-hand side above serve to label the two *independent* reduced matrix elements needed to express the $\langle jU'\tau_1|O_{\phi}^{t_1}|kU'\tau_2\rangle$ matrix. Two nonzero matrix elements with different sets of components must be evaluated to calculate $\langle jU'\|O^{t_1}\|kU'\rangle_0$ and $\langle jU'\|O^{t_1}\|kU'\rangle_1$. Once they are determined, (12.6.5) above may be used to evaluate matrix elements for all remaining sets.

Note carefully that the $r = 0, 1$ labels used on the right-hand side of (12.6.5) derive from (10.2.2); they give absolutely no information about the parentage of $|jU'\tau_1\rangle$ and $|kU'\tau_2\rangle$ ($j = 0$ or 1, $k = 0$ or 1) which can be constructed using (12.2.2). Conversely, the repeated representation labels j and k have no relation to the $r = 0$ or $r = 1$ labels on the right-hand side of (12.6.5), but serve only to define the coupling which produced the $|jU'\tau_1\rangle$ and $|kU'\tau_2\rangle$.

In all cases a repeated representation label r (or j, k, etc.) refers to a *combination of three irreps*; it really relates to the label on the $3jm$ used in (10.2.2) or in a coupling of the type (10.2.3).

13 Matrix-Element Evaluation: Elementary Concepts

13.1. Introduction

In our applications, we are interested almost exclusively in the matrix elements of one-electron operators, of which $\mathcal{H}_{SO}$, $\boldsymbol{\mu} \equiv (-\mu_B)(\mathbf{L} + 2\mathbf{S})$, and $\mathbf{m} \equiv \Sigma_i e_i \mathbf{r}_i$ are the most important. The evaluation of such matrix elements can be divided into three steps. First, a model must be chosen which defines wavefunctions for the states of interest. Our states are always described by products of one-electron functions in determinantal form, generally arising from an LCAO–MO model. Second, no matter how complicated these wavefunctions are, matrix elements of one-electron operators may ultimately be expressed as sums of one-electron matrix elements. Symmetry determines the number of independent one-electron reduced matrix elements in a given case. Furthermore, group-theoretic techniques play a powerful role in carrying out the reduction to one-electron form. Third, the one-electron reduced matrix elements must be evaluated. In general, the explicit form of the one-electron functions, as opposed to their transformation properties, has to be specified only in this final step.

The most obvious—but not the easiest—procedure is to adopt a model, write down explicit wavefunctions, and evaluate the necessary matrix elements. Several simple illustrations are offered in this chapter. We note, however, that while such a process is possible in principle no matter how complex the system, the limitations in practice are severe. The reason has been indicated in Section 11.5; namely, even for relatively simple systems, when wavefunctions must be constructed from more than one open shell (unfilled configuration), the process can be hopelessly complicated. In such circumstances, the use of the irreducible-tensor method discussed in Chapters 14–22 becomes essential, since it avoids entirely the need to construct many-electron wavefunctions. It is a formalism designed to extract all symmetry-determined information with maximum economy.

Following many of our examples, we refer to the equation or equations which would be used to accomplish the identical calculation using the irreducible-tensor method. The reader should appreciate that even for these simple examples, use of the referenced equations in Chapters 15 and 18–20 results in very significant time savings. Nevertheless, working through the examples in this chapter "the long way" should give a feeling for the steps involved in matrix-element calculations.

13.2. Reduction of Spin-Independent Matrix Elements to One-Electron Form: The Example of $Fe(CN)_6^{3-}$

Let us start with a simple example, the reduction to one-electron form of the spin-independent group-O reduced matrix elements in the $Fe(CN)_6^{3-}$ problem of Section 11.2. The relevant many-electron matrix elements are $\langle {}^2T_{2g}\|L^{t_{1g}}\|{}^2T_{2g}\rangle$, $\langle {}^2T_{2u}\|L^{t_{1g}}\|{}^2T_{2u}\rangle$, $\langle {}^2T_{1u}\|L^{t_{1g}}\|{}^2T_{1u}\rangle$, $\langle {}^2T_{2g}\|m^{t_{1u}}\|{}^2T_{1u}\rangle$, and $\langle {}^2T_{2g}\|m^{t_{1u}}\|{}^2T_{2u}\rangle$.

First we adopt a model. We use LCAO–MO theory, and assume that the ground state arises from the single strong-field configuration t_{2g}^5 and that the charge-transfer excited states arise from the single configurations t_{1u}^5 and t_{2u}^5. (See also the discussion in Section 11.2.) In writing configurations, if no confusion arises, we omit all filled subshells, since they are totally symmetric. Thus, $(t_{1u}^{(1)})^6 t_{2u}^6 (t_{1u}^{(2)})^6 t_{2g}^5 \equiv t_{2g}^5$, $(t_{1u}^{(1)})^6 t_{2u}^6 (t_{1u}^{(2)})^5 t_{2g}^6 \equiv (t_{1u}^{(2)})^5$, and so on—see Fig. 11.2.2. The t_{2g}, t_{1u}, and t_{2u} orbitals are LCAO–MOs whose exact form need not be specified until one-electron matrix elements are explicitly evaluated in Section 13.4. Using the method discussed in Section 11.4, it is very simple to write down the wavefunctions for the ground and excited states. However, since we wish to evaluate reduced matrix elements, we need not write these down for all the components of the wavefunctions; furthermore, we can often reduce the labor involved by making a careful choice of components.

Let us begin with $\langle {}^2T_{2g}\|L^{t_{1g}}\|{}^2T_{2g}\rangle$, the reduced matrix element for the ground-state orbital angular momentum. In Section 11.2 this was obtained from the reduction

$$\langle {}^2T_{2g}1|L_0^{t_{1g}}|{}^2T_{2g}1\rangle = \frac{-1}{\sqrt{6}}\langle {}^2T_{2g}\|L^{t_{1g}}\|{}^2T_{2g}\rangle \tag{13.2.1}$$

However, it is most convenient to work in the real-O basis, since wavefunctions for t_2^n in that basis are already listed in Section C.8. Whenever possible we work with the z component of the angular-momentum operator, since our one-electron atomic basis functions are often chosen to be eigenfunctions of this operator. Using (10.2.2), (10.2.4), and the real-O $2jm$ and $3jm$

of Sections C.11–C.12 to unreduce the matrix element, we find

$$\langle {}^2T_{2g}\|L^{t_{1g}}\|{}^2T_{2g}\rangle = \sqrt{6}\,\langle {}^2T_{2g}\eta|L_z^{t_{1g}}|{}^2T_{2g}\xi\rangle$$

$$\langle {}^2T_{2g}\|L^{t_{1g}}\|{}^2T_{2g}\rangle = \sqrt{6}\,\langle {}^2T_{2g}\xi|L_z^{t_{1g}}|{}^2T_{2g}\eta\rangle \tag{13.2.2}$$

We note that the same result may be obtained with more work from (13.2.1) by substituting $|T_21\rangle = (1/\sqrt{2})(-|T_2\xi\rangle + |T_2\eta\rangle)$ and $L_0^{t_{1g}} = L_z^{t_{1g}}$ from Table C.5.1(c). Arbitrarily choosing the first of (13.2.2) and spin component $\mathfrak{M} = \frac{1}{2}$, we can substitute from Section C.8 with the result

$$\langle {}^2T_{2g}\eta|L_z^{t_{1g}}|{}^2T_{2g}\xi\rangle = \langle t_{2g}^5\,\tfrac{1}{2}T_{2g}\tfrac{1}{2}\eta|L_z^{t_{1g}}|t_{2g}^5\,\tfrac{1}{2}T_{2g}\tfrac{1}{2}\xi\rangle$$

$$= -\langle \xi^2\eta^+\zeta^2|L_z^{t_{1g}}|\xi^+\eta^2\zeta^2\rangle \tag{13.2.3}$$

We now use some very important theorems about the matrix elements of one-electron operators proved in Section 19.1. Suppose we have two normalized, n-electron determinantal functions, $\psi_i = |\phi_1^i\phi_2^i \cdots \phi_n^i\rangle$ and $\psi_j = |\phi_1^j\phi_2^j \cdots \phi_n^j\rangle$, where the ϕ_k are spin–orbitals (one-electron space functions of specified spin) belonging to an orthonormal set. Let $V = \sum_{m=1}^n v(m)$ represent any quantum-mechanical one-electron operator. We first suppose that ψ_i and ψ_j contain precisely the same ϕ_k in identical order. Then $\psi_i = \psi_j$ and (19.1.11) gives

$$\langle\psi_i|V|\psi_i\rangle = \sum_{m=1}^{n} \left\langle \phi_m^i(m)|v(m)|\phi_m^i(m)\right\rangle \tag{13.2.4}$$

Suppose next that ψ_i and ψ_j differ in one spin–orbital, say ϕ_l^i and $\phi_{l'}^j$. Then (19.1.12) gives

$$\langle\psi_i|V|\psi_j\rangle = (-1)^p\langle\phi_l^i|v|\phi_{l'}^j\rangle \tag{13.2.5}$$

where p is the number of interchanges necessary to bring the identical ϕ_k in ψ_i and ψ_j into the same order, since the sign of a determinant changes if two columns are interchanged. Finally, if ψ_i and ψ_j differ in more than one ϕ_k,

$$\langle\psi_i|V|\psi_j\rangle = 0 \tag{13.2.6}$$

Returning to (13.2.3) and noting that $L_z^{t_{1g}} = \sum_{m=1}^5 l_z^{t_{1g}}(m)$, we apply (13.2.5) with the result

$$\langle \xi^2\eta^+\zeta^2|L_z^{t_{1g}}|\xi^+\eta^2\zeta^2\rangle = -\langle \xi^-|l_z^{t_{1g}}|\eta^-\rangle$$

$$= -\langle \xi|l_z^{t_{1g}}|\eta\rangle$$

$$= \frac{-1}{\sqrt{6}}\langle t_{2g}\|l^{t_{1g}}\|t_{2g}\rangle \tag{13.2.7}$$

We were able to integrate out spin in the second step of (13.2.7) because $l_z^{t_{1g}}$ is spin-independent. In the last step, (10.2.2) was used to obtain the one-electron reduced matrix element. Thus our final result from (13.2.2), (13.2.3), and (13.2.7) is

$$\langle {}^2T_{2g}\|L^{t_{1g}}\|{}^2T_{2g}\rangle = \langle t_{2g}\|l^{t_{1g}}\|t_{2g}\rangle \tag{13.2.8}$$

For the charge-transfer excited states, we proceed in much the same way. For the ${}^2T_{2u}$ state, the results are obtained from (13.2.8) by simply changing all g subscripts to u. For ${}^2T_{1u}$ the methods of Section 11.4 give

$$\begin{aligned} |t_{1u}^5 \tfrac{1}{2}T_{1u}\tfrac{1}{2}x\rangle &= -|x^+y^2z^2\rangle \\ |t_{1u}^5 \tfrac{1}{2}T_{1u}\tfrac{1}{2}y\rangle &= |x^2y^+z^2\rangle \end{aligned} \tag{13.2.9}$$

The results are

$$\begin{aligned} \langle {}^2T_{2u}\|L^{t_{1g}}\|{}^2T_{2u}\rangle &= \langle t_{2u}\|l^{t_{1g}}\|t_{2u}\rangle \\ \langle {}^2T_{1u}\|L^{t_{1g}}\|{}^2T_{1u}\rangle &= \langle t_{1u}\|l^{t_{1g}}\|t_{1u}\rangle \end{aligned} \tag{13.2.10}$$

Using the irreducible-tensor method, the results of (13.2.8) and (13.2.10) would be accomplished in only one step using (19.4.6) with the group-O $g(a^n, \mathfrak{S}hfh')$ coefficients of Section C.16. (See also Section B.4, which gives another approach useful for certain crystal-field calculations in the weak-field basis.)

The evaluation of the right-hand sides of (13.2.8) and (13.2.10) is discussed in Section 13.4.

Let us consider the electric-dipole reduced matrix element, $\langle {}^2T_{2g}\|m^{t_{1u}}\|{}^2T_{1u}\rangle$, which is off-diagonal in configuration. Proceeding initially as before, we write

$$\langle {}^2T_{2g}\|m^{t_{1u}}\|{}^2T_{1u}\rangle = -\sqrt{6}\,\langle {}^2T_{2g}\tfrac{1}{2}\xi|m_z^{t_{1u}}|{}^2T_{1u}\tfrac{1}{2}y\rangle \tag{13.2.11}$$

But now *both* the t_{1u} and t_{2g} shells must be explicitly included in the wavefunctions, with either placed first, but consistently so. The wavefunctions are formed by using (11.5.1) and then substituting in the a^m and the b^n kets and antisymmetrizing to give

$$\left|\mathcal{A}\left(t_{1u}^6(0A_{1g}), t_{2g}^5\left(\tfrac{1}{2}T_{2g}\right)\right)\tfrac{1}{2}T_{2g}\tfrac{1}{2}\xi\right\rangle = -|x^2y^2z^2\xi^+\eta^2\zeta^2\rangle \tag{13.2.12}$$

and

$$\left|\mathcal{A}\left(t_{1u}^5\left(\tfrac{1}{2}T_{1u}\right), t_{2g}^6(0A_{1g})\right)\tfrac{1}{2}T_{1u}\tfrac{1}{2}y\right\rangle = |x^2y^+z^2\xi^2\eta^2\zeta^2\rangle \tag{13.2.13}$$

Inserting these functions into the matrix element from (13.2.11) and applying (13.2.5) gives (since p is odd)

$$\langle {}^2T_{2g}\tfrac{1}{2}\xi|m_z^{t_{1u}}|{}^2T_{1u}\tfrac{1}{2}y\rangle = -\left(-\langle y^-|m^{t_{1u}}|\xi^-\rangle\right)$$

$$= \langle y|m_z^{t_{1u}}|\xi\rangle = \frac{1}{\sqrt{6}}\langle t_{1u}\|m^{t_{1u}}\|t_{2g}\rangle \quad (13.2.14)$$

Finally, combining (13.2.11) and (13.2.14), we have

$$\langle {}^2T_{2g}\|m^{t_{1u}}\|{}^2T_{1u}\rangle = -\langle t_{1u}\|m^{t_{1u}}\|t_{2g}\rangle \quad (13.2.15)$$

In exactly the same way,

$$\langle {}^2T_{2g}\|m^{t_{1u}}\|{}^2T_{2u}\rangle = \langle t_{2u}\|m^{t_{1u}}\|t_{2g}\rangle \quad (13.2.16)$$

In contrast with **L** (and **S**), we use the same symbol (**m**) for both the many- and the one-electron electric-dipole operators.

Our (13.2.15) and (13.2.16) results may be accomplished in one step using the irreducible-tensor method; this time the appropriate equation is (20.2.10).

13.3. Spin–Orbit-Coupling Matrix Elements

Like the operators of the previous section, the spin–orbit-coupling operator $\mathcal{H}_{SO}$ is a one-electron operator. In Section 12.1 we showed that it has the form

$$\mathcal{H}_{SO} = \sum_{k=1}^{n} \xi(r_k)\mathbf{l}(k)\cdot\mathbf{s}(k) \quad (13.3.1)$$

It proves convenient, however, to follow Griffith (1962) and write $\mathcal{H}_{SO}$ in a slightly different but equivalent form:

$$\mathcal{H}_{SO} = \sum_{k=1}^{n} \mathbf{s}(k)\cdot\mathbf{u}(k) \quad (13.3.2)$$

where

$$\mathbf{u}(k) \equiv \xi(r_k)\mathbf{l}(k) \quad (13.3.3)$$

Since $\mathbf{s}(k)\cdot\mathbf{u}(k)$ is the dot product of two vector operators, it may be expanded via (9.8.10) to give

$$\mathbf{s} \cdot \mathbf{u} = -s_1 u_{-1} + s_0 u_0 - s_{-1} u_1$$

$$= s_x u_x + s_y u_y + s_z u_z \qquad (13.3.4)$$

where the spherical vector components transform as the $|jm\rangle = |1m\rangle$ of the group SO_3 and as the $|j^+m\rangle = |1^+m\rangle$ of the group O_3 in the angular-momentum basis (Section 9.8). Since $s_0 = s_z$, $u_0 = u_z$, and $s_{\pm 1}$ and $u_{\pm 1}$ are related to the step-up and step-down operators $s^{\pm}$ and $u^{\pm}$ as given in (9.8.8), we may also write (13.3.4) in the form

$$\mathbf{s} \cdot \mathbf{u} = \tfrac{1}{2}s^+u^- + \tfrac{1}{2}s^-u^+ + s_z u_z \qquad (13.3.5)$$

Since $\mathcal{H}_{SO}$ is totally symmetric in the spin–orbit (molecular double-group) basis, the $|(\mathbb{S}S, h)rt\tau\rangle$ functions defined by (12.2.2) are the convenient choice for problems involving spin–orbit coupling. In this basis $\mathcal{H}_{SO} = (\mathcal{H}_{SO})^{A_1}_{a_1}$, and (10.2.2) and (10.3.1) give

$$\langle \alpha(\mathbb{S}S, h)rt\tau|\mathcal{H}_{SO}|\alpha'(\mathbb{S}'S', h')r't'\tau'\rangle$$

$$= \delta_{tt'}\delta_{\tau\tau'}|t|^{-1/2}\langle \alpha(\mathbb{S}S, h)rt\|(\mathcal{H}_{SO})^{A_1}\|\alpha'(\mathbb{S}'S', h')r't'\rangle \qquad (13.3.6)$$

Thus for a given configuration α and term $\mathbb{S}Sh$, the basis is diagonal in t and τ (but, unfortunately, not in r—see Sections 12.5 and 12.6).

We illustrate the reduction of spin–orbit matrix elements to one-electron form by way of an example.

Reduction of Spin–Orbit Matrix Elements to One-Electron Form: The Spin–Orbit Splittings of $Fe(CN)_6^{3-}$

As a specific illustration, we calculate the first-order spin–orbit splittings of the ground and excited states of $Fe(CN)_6^{3-}$ previously discussed in Sections 11.2 and 13.2. The ground state is ${}^2T_{2g}$, and the excited states are ${}^2T_{1u}$ and ${}^2T_{2u}$. $\mathbb{S} = \frac{1}{2}$ goes over to $S = E'$ in group O, so these states split under spin–orbit coupling according to the group-O direct products, $S \otimes h = E' \otimes T_1 = E' + U'$ for ${}^2T_1$ and $S \otimes h = E' \otimes T_2 = E'' + U'$ for ${}^2T_2$. We note that the group-theoretic aspect of the calculation is identical for ${}^2T_{2g}$ and ${}^2T_{2u}$ in this case. We need only add g and u subscripts respectively at the end.

Let us first consider ${}^2T_{1u}$, which is shown schematically in Fig. 13.3.1. The order of the splitting at this stage is arbitrary. Our spin–orbit basis states are $|(\mathbb{S}S, h)rt\tau\rangle = |(\frac{1}{2}E', T_1)E'\tau\rangle$, $\tau = \alpha'$ or β', and $|(\frac{1}{2}E', T_1)U'\tau\rangle$, $\tau = \kappa, \lambda, \mu$, or ν. In this case $S \otimes h$ contains no repeated representations, so $r = 0$ only; thus we drop the r label. For a given term, $\mathcal{H}_{SO}$ is diagonal in t and τ, so either of the E' partners and any of the four U' partners may be used to calculate the splitting. We calculate the first-order spin–orbit energies as

$$E_{SO}[E'({}^2T_1)] = \langle(\tfrac{1}{2}E', T_1)E'\alpha'|\mathcal{H}_{SO}|(\tfrac{1}{2}E', T_1)E'\alpha'\rangle$$

$$E_{SO}[U'({}^2T_1)] = \langle(\tfrac{1}{2}E', T_1)U'\kappa|\mathcal{H}_{SO}|(\tfrac{1}{2}E', T_1)U'\kappa\rangle \qquad (13.3.7)$$

To evaluate these matrix elements the $|(\mathbb{S}S, h)t\tau\rangle$ are expanded in terms of the $|\mathbb{S}ShM\theta\rangle$ using (12.2.2) and Sections C11–C12; then the $|\mathbb{S}ShM\theta\rangle$ are expressed in terms of the $|\mathbb{S}h\mathfrak{M}\theta\rangle = |{}^{2\mathbb{S}+1}h\mathfrak{M}\theta\rangle$ using the relations in Section C.6. This process gives

$$|(\tfrac{1}{2}E', T_1)E'\alpha'\rangle = \frac{\sqrt{2}}{\sqrt{3}}|\tfrac{1}{2}E'T_1\beta'1\rangle + \frac{1}{\sqrt{3}}|\tfrac{1}{2}E'T_1\alpha'0\rangle$$

$$= \frac{\sqrt{2}}{\sqrt{3}}|{}^2T_1 - \tfrac{1}{2}1\rangle + \frac{1}{\sqrt{3}}|{}^2T_1\tfrac{1}{2}0\rangle$$

$$|(\tfrac{1}{2}E', T_1)U'\kappa\rangle = |\tfrac{1}{2}E'T_1\alpha'1\rangle \qquad (13.3.8)$$

$$= |{}^2T_1\tfrac{1}{2}1\rangle$$

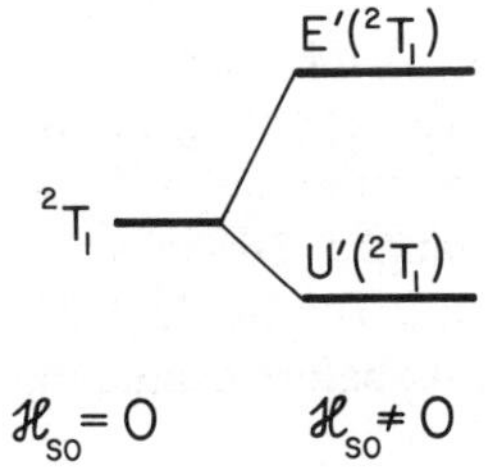

Figure 13.3.1. Splitting of ${}^2T_1$ under spin–orbit coupling.

Then using (13.3.2), (13.3.4), and Section C.7 [for $u(k)$ only], we have

$$\mathcal{H}_{\text{SO}} = \sum_k \left[-s(k)_1^1 u(k)_{-1}^1 + s(k)_0^1 u(k)_0^1 - s(k)_{-1}^1 u(k)_1^1 \right]$$

$$= \sum_k \left[s(k)_1^1 u(k)_{-1}^{t_1} + s(k)_0^1 u(k)_0^{t_1} + s(k)_{-1}^1 u(k)_1^{t_1} \right] \quad (13.3.9)$$

In the last line $s(k)$ is classified in $SO_3 \supset SO_2$, while $u(k)$ is classified in the complex-O basis. At this point we substitute (13.3.8) and (13.3.9) into (13.3.7) and reduce independently but simultaneously over spin coordinates (in the group SO_3) and over orbital coordinates (in the group O) using the Wigner–Eckart theorem (10.2.2). This is possible because in $\mathcal{H}_{\text{SO}}$, the $u(k) = \xi(r)l(k)$ operate only in orbital space $(h\theta)$ while the $s(k)$ operate only in spin space $(\mathcal{S}\mathcal{M})$. Thus if the Wigner–Eckart derivation of Section 9.3 is repeated using (12.5.1), we obtain

$$\langle \mathcal{S}h\mathcal{M}\theta | s(k)_m^1 \, u(k)_\phi^f | \mathcal{S}'h'\mathcal{M}'\theta' \rangle$$

$$= \begin{pmatrix} \mathcal{S} \\ \mathcal{M} \end{pmatrix} \begin{pmatrix} \mathcal{S} & 1 & \mathcal{S}' \\ -\mathcal{M} & m & \mathcal{M}' \end{pmatrix}^{SO_3} \begin{pmatrix} h \\ \theta \end{pmatrix} \begin{pmatrix} h^* & f & h' \\ \theta^* & \phi & \theta' \end{pmatrix}^{G}$$

$$\times \langle \mathcal{S}h \| s^1 u^f(k) \| \mathcal{S}'h' \rangle^{SO_3, G} \quad (13.3.10)$$

where in our example $G = O$. No repeated representations occur (except when $G = K$, T, or T_h, which are uncommon). For the group O, $f = t_1$.

We can anticipate in part the result of applying (13.3.10) to our problem, since $s_0^1 = s_z$, $s_1^1 = (-1/\sqrt{2})s^+$, and $s_{-1}^1 = (1/\sqrt{2})s^-$. Recall then that for an angular-momentum state $|jm\rangle$ (in units of $\hbar$),

$$J_z |jm\rangle = m|jm\rangle$$

$$J^{\pm}|jm\rangle = (j \mp m)^{1/2}(j \pm m + 1)^{1/2}|j\, m \pm 1\rangle \quad (13.3.11)$$

These relations apply here for $J = S$ and $|jm\rangle = |\mathcal{S}\mathcal{M}\rangle$. It follows that $s_z u_0^{t_1}$ connects only $|\mathcal{S}h\mathcal{M}\theta\rangle$ states of identical spin, while the $s_{\pm} u_{\mp 1}^{t_1}$ connect states which differ in $\mathcal{M}$ by ± 1. For spin–orbit-coupling matrix elements in the atomic $|\mathcal{S}\mathcal{L}\mathcal{M}_{\mathcal{S}}\mathcal{M}_{\mathcal{L}}\rangle$ basis the selection rules are $\Delta\mathcal{M}_{\mathcal{S}} = 0, \pm 1$ and $\Delta\mathcal{S} = 0, \pm 1$, and these carry over here unchanged. The selection rules are, of course, defined by the nonzero values of the $3jm$ in (13.3.10), but the above discussion should give a feel for the results before evaluating the $3jm$.

Substituting (13.3.8) and (13.3.9) into (13.3.7) and reducing the resulting matrix elements via (13.3.10) with the $2jm$ and $3jm$ of Sections B.3 and C.11–C.12 gives

$$\begin{aligned}
E_{\mathrm{SO}}[E'(^2T_1)] &= \left\langle \left(\tfrac{1}{2}E', T_1\right)E'\alpha' \middle| \mathcal{H}_{\mathrm{SO}} \middle| \left(\tfrac{1}{2}E', T_1\right)E'\alpha' \right\rangle \\
&= \tfrac{2}{3}\langle \tfrac{1}{2}T_1 -\tfrac{1}{2}1 | \sum_k s(k)^1_0 u(k)^{t_1}_0 | \tfrac{1}{2}T_1 -\tfrac{1}{2}1\rangle \\
&\quad + \tfrac{1}{3}\langle \tfrac{1}{2}T_1 \tfrac{1}{2}0 | \sum_k s(k)^1_0 u(k)^{t_1}_0 | \tfrac{1}{2}T_1 \tfrac{1}{2}0\rangle \\
&\quad + \frac{\sqrt{2}}{3}\langle \tfrac{1}{2}T_1 -\tfrac{1}{2}1 | \sum_k s(k)^1_{-1} u(k)^{t_1}_1 | \tfrac{1}{2}T_1 \tfrac{1}{2}0\rangle \\
&\quad + \frac{\sqrt{2}}{3}\langle \tfrac{1}{2}T_1 \tfrac{1}{2}0 | \sum_k s(k)^1_1 u(k)^{t_1}_{-1} | \tfrac{1}{2}T_1 -\tfrac{1}{2}1\rangle \\
&= \left[\left(\tfrac{2}{3}\right)\left(\tfrac{1}{6}\right) + \left(\tfrac{1}{3}\right)(0) + \left(\frac{\sqrt{2}}{3}\right)\left(\frac{1}{3\sqrt{2}}\right) + \left(\frac{\sqrt{2}}{3}\right)\left(\frac{1}{3\sqrt{2}}\right)\right] \\
&\quad \times \langle \tfrac{1}{2}T_1 \| \sum_k s^1 u^{t_1}(k) \| \tfrac{1}{2}T_1\rangle^{\mathrm{SO}_3,\, O} \\
&= \tfrac{1}{3}\langle \tfrac{1}{2}T_1 \| \sum_k s^1 u^{t_1}(k) \| \tfrac{1}{2}T_1\rangle^{\mathrm{SO}_3,\, O} \qquad (13.3.12)
\end{aligned}$$

$$\begin{aligned}
E_{\mathrm{SO}}[U'(^2T_1)] &= \left\langle \left(\tfrac{1}{2}E', T_1\right)U'\kappa \middle| \mathcal{H}_{\mathrm{SO}} \middle| \left(\tfrac{1}{2}E', T_1\right)U'\kappa \right\rangle \\
&= \langle \tfrac{1}{2}T_1 \tfrac{1}{2}1 | \sum_k s(k)^1_0 u(k)^{t_1}_0 | \tfrac{1}{2}T_1 \tfrac{1}{2}1\rangle \\
&= -\tfrac{1}{6}\langle \tfrac{1}{2}T_1 \| \sum_k s^1 u^{t_1}(k) \| \tfrac{1}{2}T_1\rangle^{\mathrm{SO}_3,\, O} \qquad (13.3.13)
\end{aligned}$$

To obtain the above results using the irreducible-tensor method one simply uses (18.4.6), which accomplishes the entire calculation in a single step.

Note that the results above give the first-order spin–orbit splitting of *any* $^2T_1$ state. The specific nature of the $^2T_1$ state is relevant only in the evaluation of the reduced matrix element common to both equations. This is our next concern. For this task we employ (13.3.10) in reverse to unreduce the reduced matrix element. But, since our t_{1u}^5 functions of (13.2.9) are given in a real-O orbital basis, we unreduce the orbital part of the reduced matrix element in the real-O basis, rather than in the complex-O basis. This is permissible because our reduced matrix elements are basis-independent. Using Section B.3 for SO_3, and Sections C.11 and C.12 for the group-O $2jm$ and $3jm$, we have

$$\langle \tfrac{1}{2}T_1 \| \sum_k s^1 u^{t_1}(k) \| \tfrac{1}{2}T_1 \rangle^{SO_3,O}$$

$$= \frac{\langle \tfrac{1}{2}T_1 \tfrac{1}{2}x | \sum_k s(k)_0^1 u(k)_z^{t_1} | \tfrac{1}{2}T_1 \tfrac{1}{2}y \rangle}{\begin{pmatrix} \frac{1}{2} \\ \frac{1}{2} \end{pmatrix}\begin{pmatrix} \frac{1}{2} & 1 & \frac{1}{2} \\ -\frac{1}{2} & 0 & \frac{1}{2} \end{pmatrix}\begin{pmatrix} T_1 \\ x \end{pmatrix}\begin{pmatrix} T_1 & T_1 & T_1 \\ x & z & y \end{pmatrix}}$$

$$= 6\langle \tfrac{1}{2}T_1 \tfrac{1}{2}x | \sum_k s(k)_0^1 u(k)_z^{t_1} | \tfrac{1}{2}T_1 \tfrac{1}{2}y \rangle \qquad (13.3.14)$$

Transforming the integral above to explicit functions via (13.2.9) and then applying (13.2.5) (since ψ_i and ψ_j differ in one spin–orbital), we have

$$\langle \tfrac{1}{2}T_1 \tfrac{1}{2}x | \sum_k s(k)_0^1 u(k)_z^{t_1} | \tfrac{1}{2}T_1 \tfrac{1}{2}y \rangle = -\langle x^+ y^2 z^2 | \sum_k s(k)_0^1 u(k)_z^{t_1} | x^2 y^+ z^2 \rangle$$

$$= \langle y^- | s_0^1 u_z^{t_1} | x^- \rangle \qquad (13.3.15)$$

Then since $s_0^1 = s_z$ and $s_z|x^-\rangle = -\tfrac{1}{2}|x^-\rangle$ in units of $\hbar$, (13.3.15) becomes

$$= -\tfrac{1}{2}\langle y | u_z^{t_1} | x \rangle = -\tfrac{1}{2}\langle t_1 y | u_z^{t_1} | t_1 x \rangle$$

$$= -\frac{1}{2}\begin{pmatrix} T_1 \\ y \end{pmatrix}\begin{pmatrix} T_1 & T_1 & T_1 \\ y & z & x \end{pmatrix}\langle t_1 \| u^{t_1} \| t_1 \rangle$$

$$= \frac{-1}{2\sqrt{6}}\langle t_1 \| u^{t_1} \| t_1 \rangle \qquad (13.3.16)$$

Substituting the above results into (13.3.14), we obtain

$$\langle \tfrac{1}{2}T_1 \| \sum_k s^1 u^{t_1}(k) \| \tfrac{1}{2}T_1 \rangle^{SO_3,O} = -\frac{\sqrt{6}}{2}\langle t_1 \| u^{t_1} \| t_1 \rangle \qquad (13.3.17)$$

Thus, (13.3.12) and (13.3.13) together with (13.3.17) give

$$E_{SO}[E_u'(^2T_{1u})] = \frac{-1}{\sqrt{6}}\langle t_{1u}\|u^{t_{1g}}\|t_{1u}\rangle$$

$$E_{SO}[U_u'(^2T_{1u})] = \frac{1}{2\sqrt{6}}\langle t_{1u}\|u^{t_{1g}}\|t_{1u}\rangle \tag{13.3.18}$$

In Chapter 19 we show that the calculation above becomes almost trivial using the irreducible-tensor method. One simply uses (19.5.5) with the group-O $G(a'', \mathcal{S}h\mathcal{S}'h')$ tabulated in Section C.17 and obtains (13.3.17) in one easy step. (See also Section B.4 for another approach useful for certain crystal-field calculations in the weak-field basis.)

Using the same methods as for the $^2T_2$ case, the reader may confirm that

$$E_{SO}[E''(^2T_2)] = \frac{1}{\sqrt{6}}\langle t_2\|u^{t_{1g}}\|t_2\rangle$$

$$E_{SO}[U'(^2T_2)] = \frac{-1}{2\sqrt{6}}\langle t_2\|u^{t_{1g}}\|t_2\rangle \tag{13.3.19}$$

For $^2T_{2g}$ and $^2T_{2u}$ one simply inserts g or u subscripts respectively because (and only because) both terms arise from t_2^5 configurations in this example.

Note that spin–orbit coupling matrix elements generally reduce to one-electron reduced matrix elements involving only $u = \xi(r)l$. The spin part of the operator can always be explicitly evaluated without detailed assumptions about the one-electron basis orbitals. Evaluation of the remaining one-electron angular-momentum matrix elements is discussed in the next section.

13.4. Evaluation of One-Electron Angular Momentum Reduced Matrix Elements

In Sections 13.2 and 13.3 we have seen that problems involving the operators $\boldsymbol{\mu}$ and $\mathcal{H}_{SO}$ ultimately reduce to the evaluation of one-electron matrix elements of l, the orbital-angular-momentum operator, or of $u = \xi(r)l$. In this section we evaluate such matrix elements. While previous sections required only that the transformation properties of one-electron functions be known, for this final step we must specify the explicit form of the one-electron functions. The calculations may be divided into two types:

those involving functions and operators on a single atomic center, and those in which the one-electron functions involve atoms displaced from the coordinate origin. In both cases the one-electron reduced matrix elements have the form $\langle a\|l^f\|b\rangle$ (or $\langle a\|u^f\|b\rangle$), where a and b are the irreps of the molecular point group to which the one-electron functions belong. We consider the single-center case first.

Single-Center One-Electron Calculations

When a and b involve functions and operators of a single atomic center, the $\langle a\|l^f\|b\rangle$ and $\langle a\|u^f\|b\rangle$ are very simple to evaluate. One simply expresses $|a\alpha\rangle$, $|b\beta\rangle$, and l^f_ϕ in terms of their equivalents in the $|jm\rangle$ basis of SO_3 and then evaluates the resulting atomic angular momentum integral(s).

Suppose, for example, we wish to evaluate the group-O_h reduced matrix element, $\langle t_{1u}\|l^{t_{1g}}\|t_{1u}\rangle$ within a basis of atomic p orbitals. We simply unreduce the matrix element in the complex-O basis with (10.2.2) and transform $|a\alpha\rangle$, $|b\beta\rangle$, and $l^{t_{1g}}_\phi$ to the $|jm\rangle$ basis, using the relations in Section C.6 (or C.7). The $|jm\rangle$ for integer j transform identically to the Condon–Shortley spherical harmonics (Section B.1), and the l^j_m are by definition standard angular-momentum operators (Section 9.8). Thus we have transformed back to a standard atomic angular-momentum integral which may be evaluated directly using (13.3.11) with $j = l$. Following the above procedure, we find (in units of $\hbar$)

$$\begin{aligned}\langle t_{1u}\|l^{t_{1g}}\|t_{1u}\rangle &= -\sqrt{6}\,\langle t_{1u}1\,|l^{t_{1g}}_0|\,t_{1u}1\rangle \\ &= -\sqrt{6}\,\langle 11|l^1_0|11\rangle \\ &= -\sqrt{6}\,\langle 11|l_z|11\rangle \\ &= -\sqrt{6} \qquad (13.4.1)\end{aligned}$$

In such calculations integration over the radial coordinate of the normalized wavefunctions is understood. Thus if one were instead evaluating $\langle t_{1u}\|\xi(r)l^{t_{1g}}\|t_{1u}\rangle$, which would arise in a problem involving $\mathcal{H}_{SO}$ (Section 13.3), the calculation would follow exactly as in (13.4.1), but the result would be written

$$\langle t_{1u}\|u^{t_{1g}}\|t_{1u}\rangle = \langle t_{1u}\|\xi(r)l^{t_{1g}}\|t_{1u}\rangle = -\sqrt{6}\,\zeta_p \qquad (13.4.2)$$

Here ζ_p is the spin–orbit coupling constant for the p orbital involved

($2p, 3p, \ldots$). It is simply the expectation value of $\xi(r)$ over the (unspecified) atomic radial functions, and is treated as a semiempirical parameter.

As a slightly more complicated group-O example let us consider $\langle t_{2g} \| l^{t_{1g}} \| t_{2g} \rangle$ arising from the ground state of $Fe(CN)_6^{3-}$, which was approximated by the single configuration, ${}^2T_{2g}(t_{2g}^5)$. Such an approximation is reasonable since the several ${}^2T_{2g}$ contributions from other metal configurations within the $d^5(Fe^{3+})$ manifold (i.e., from $t_{2g}^4 e_g \cdots t_{2g} e_g^4$) should be small in view of the large energy separation expected in this strong-field complex ($10Dq \equiv \Delta \approx 25{,}000$ cm^{-1}). (We generally avoid the unpleasant subject of how well a simple LCAO–MO model can represent a complex system.) In this case, the t_{2g} MO is an unknown linear combination of metal and ligand t_{2g} orbitals. Since the ligand t_{2g} orbital lies at considerably lower (more negative) energy than the metal one ($\gtrsim 30{,}000$ cm^{-1}), we neglect the ligand contribution entirely. (The method of calculating ligand orbital-angular-momentum matrix elements is discussed shortly.) With this approximation, our model is conventional ligand-field theory. The t_{2g} orbitals are now the so-called strong-field d functions. Using the complex-O basis again and following the same procedure as in (13.4.1), we find (in units of $\hbar$)

$$\begin{aligned} \langle t_{2g} \| l^{t_{1g}} \| t_{2g} \rangle &= -\sqrt{6}\,\langle t_{2g} 1 | l_0^{t_{1g}} | t_{2g} 1 \rangle \\ &= -\sqrt{6}\,\langle 21 | l_0^1 | 21 \rangle \\ &= -\sqrt{6}\,\langle 21 | l_z | 21 \rangle \\ &= -\sqrt{6} \end{aligned} \tag{13.4.3}$$

This is the result [via (13.2.8)] quoted in obtaining (11.2.9).

The fact that the MOs are not really true d orbitals is often corrected for empirically by noting that they nevertheless must have the same transformation properties. Thus,

$$\langle a \| l^f \| b \rangle = \kappa \langle \bar{a} \| l^f \| \bar{b} \rangle \tag{13.4.4}$$

where $\bar{a}$, $\bar{b}$ are true d orbitals. This is a simple example of the replacement theorem (9.3.18); κ, which is referred to as the *orbital reduction factor*, is a function of the orbitals involved (e.g., t_{2g} or e_g) but of course is independent of components. In practice, κ is found to be $\leqslant 1$ and is often interpreted as a measure of covalent character. If this correction is used, the right-hand side of (13.4.3) is multiplied by κ.

As a final example, we calculate $\langle t_{2g}\|l^{t_{1g}}\|e_g\rangle$ for d orbitals. Using exactly the same procedure, we write

$$
\begin{aligned}
\langle t_{2g}\|l^{t_{1g}}\|e_g\rangle &= \sqrt{3}\,\langle t_{2g}0|l_0^{t_{1g}}|e_g\epsilon'\rangle \\
&= \sqrt{3}\left\langle \frac{1}{\sqrt{2}}(-\langle 2\,-2| + \langle 22|)\left|l_z\right| -\frac{1}{\sqrt{2}}(|2\,-2\rangle + |22\rangle)\right\rangle \\
&= \sqrt{3}\,(-\tfrac{1}{2})(2+2) \\
&= -2\sqrt{3}
\end{aligned}
\tag{13.4.5}
$$

If κ were used in this case, it would in principle have a different value from the corresponding κ for $\langle t_{2g}\|l^{t_{1g}}\|t_{2g}\rangle$. Note that orbital reduction factors do not apply to the spin operators.

If atomic angular-momentum matrix elements of l_1 or l_{-1} result after transformation to the $|jm\rangle$ basis, as in systems of lower symmetry, precisely the same methods can be used if l_1 and/or l_{-1} are written as appropriate combinations of $l^{\pm}$ [see (9.8.8)]. Then (13.3.11) is again used to evaluate the atomic angular-momentum integral. Atomic matrix elements of l_x or l_y are handled by expressing the operators in terms of $l_{\pm 1}$ using (9.8.11) and then proceeding as above.

The single-center calculations above simplify considerably with the use of group-chain methods as illustrated in Section 15.4—in particular, see (15.4.6).

Many-Center Calculations in the LCAO–MO *Approximation*

Since our one-electron wavefunctions are in general LCAO–MOs, angular-momentum calculations usually extend over atomic orbitals on several centers. In such circumstances, a molecular origin must be (arbitrarily) chosen; this is usually a point which reflects the full symmetry of the molecular point group. Suppose again we wish to evaluate $\langle a\|l^f\|b\rangle$. Now using the LCAO approximation, we write

$$
\begin{aligned}
|a\alpha\rangle &= \sum_{i,p} c_i^p\phi_i^p \\
|b\beta\rangle &= \sum_{j,q} c_j^q\phi_j^q
\end{aligned}
\tag{13.4.6}
$$

where the sums over i and j are over atoms, the sums over p and q are over the orbitals on each atomic center, and ϕ_i^p and ϕ_j^q are atomic orbitals (Section B.2). Sections C.10, D.7, and E.7 give examples of such LCAO–MOs for some common cases. We can always relate $\langle a\|l^f\|b\rangle$ to atomic matrix elements in the real basis, $\langle \phi_i^p|l_\gamma|\phi_j^q\rangle$, where $\gamma = x, y$, or z. Using (10.2.2) and (C.7.2) respectively, we obtain

$$\langle a\alpha|l_\phi^f|b\beta\rangle = \begin{pmatrix} a \\ \alpha \end{pmatrix}\begin{pmatrix} a^* & f & b \\ \alpha^* & \phi & \beta \end{pmatrix}\langle a\|l^f\|b\rangle \tag{13.4.7}$$

and

$$\langle a\alpha|l_\phi^f|b\beta\rangle = \sum_{\gamma=x,y,z} \langle f\phi|l_\gamma\rangle^*\langle a\alpha|l_\gamma|b\beta\rangle \tag{13.4.8}$$

Then, expanding $|a\alpha\rangle$ and $|b\beta\rangle$ via (13.4.6), we have for each l_γ matrix element in (13.4.8)

$$\langle a\alpha|l_\gamma|b\beta\rangle = \sum_{ijpq} (c_i^p)^* c_j^q \langle \phi_i^p|l_\gamma|\phi_j^q\rangle \tag{13.4.9}$$

If our choice of origin coincides with the origin of one of the ϕ_i, say ϕ_k, then the atomic angular-momentum matrix element $\langle \phi_k^p|l_\gamma|\phi_k^q\rangle$ is evaluated as just discussed for the single-center case. One need only transform the real basis functions of SO_3 to the $|jm\rangle$ basis using the relations in Section B.2 and (9.8.11). The resulting matrix element is then evaluated by way of (13.3.11) [and (9.8.8) if $l_{\pm 1}$ are involved].

If, however, ϕ_i^p and ϕ_j^q are on different atomic centers, then as a first approximation we assume that

$$\langle \phi_i^p|l_\gamma|\phi_j^q\rangle = 0 \qquad \text{for} \quad i \neq j \tag{13.4.10}$$

The rationale is that such terms are proportional to the overlap integral, $\langle \phi_i^p|\phi_j^q\rangle$, and thus should be small compared to nonzero terms on a single center. Assuming the validity of (13.4.10), we are still left with matrix elements of the form

$$\langle \phi_k^p|l_\gamma|\phi_k^q\rangle \tag{13.4.11}$$

where the origin for the ϕ_k is at a different point in space than that for l_γ. The obvious way to evaluate (13.4.11) is to express the operator l_γ in terms of the coordinate origin of the ϕ_k. We consider two cases by way of example.

Transforming l_γ: The Octahedral Case

First we must define the coordinate system for each atomic center (and the origin). For octahedral systems we use exclusively right-handed systems with all positive x axes parallel and likewise for all y and z axes. This differs from the choice often made in LCAO–MO treatments of octahedral transition-metal systems. Our LCAO symmetry-MOs use the basis functions of Table C.5.1(b), defined with respect to the right-handed system just described. (See Section C.10).

Let us first consider (13.4.11) for the most common case. We choose the z axis of our origin (O) to point at atom k, which is assumed to be a distance R away. The coordinate systems are as shown in Fig. 13.4.1. Clearly

$$x = x_k, \qquad y = y_k, \qquad z = z_k + R \tag{13.4.12}$$

where x, y, and z are the coordinates of a point measured from O, and x_k, y_k, and z_k are the coordinates of the same point measured from k. By definition

$$\begin{aligned} l_x &= -i\hbar\left(y\frac{\partial}{\partial z} - z\frac{\partial}{\partial y}\right) \\ l_y &= -i\hbar\left(z\frac{\partial}{\partial x} - x\frac{\partial}{\partial z}\right) \\ l_z &= -i\hbar\left(x\frac{\partial}{\partial y} - y\frac{\partial}{\partial x}\right) \end{aligned} \tag{13.4.13}$$

where $i = \sqrt{-1}$. From (13.4.12) it follows by the chain rule that

$$\frac{\partial}{\partial x} = \frac{\partial}{\partial x_k}, \qquad \frac{\partial}{\partial y} = \frac{\partial}{\partial y_k}, \qquad \frac{\partial}{\partial z} = \frac{\partial}{\partial z_k} \tag{13.4.14}$$

Noting that (13.4.13) also applies for the coordinate system centered on k if

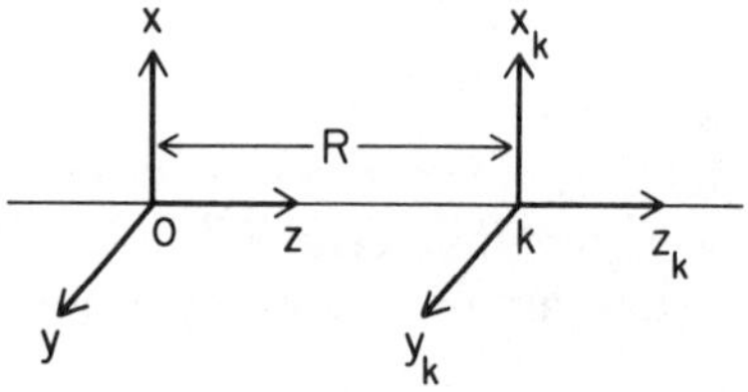

Figure 13.4.1. Coordinate system at origin and on atom k.

a subscript k is appended to x, y, and z everywhere, and using (13.4.14), we have our desired result:

$$l_x = l_{x_k} + i\hbar R\frac{\partial}{\partial y_k}$$

$$l_y = l_{y_k} - i\hbar R\frac{\partial}{\partial x_k} \qquad (13.4.15)$$

$$l_z = l_{z_k}$$

If we now wish to evaluate (13.4.11) for the case $\gamma = z$, we have

$$\langle \phi_k^p | l_z | \phi_k^q \rangle = \langle \phi_k^p | l_{z_k} | \phi_k^q \rangle \qquad (13.4.16)$$

Thus the problem is reduced precisely to the simple, single-center case previously discussed. However, if we choose $\gamma = x$ (or y) in (13.4.11), a new feature appears. For example, using (13.4.15),

$$\langle \phi_k^p | l_x | \phi_k^q \rangle = \langle \phi_k^p | l_{x_k} | \phi_k^q \rangle + i\hbar R \left\langle \phi_k^p \middle| \frac{\partial}{\partial y_k} \middle| \phi_k^q \right\rangle \qquad (13.4.17)$$

The first term on the right is very simple, as discussed before; however, the second term did not arise then. While it is not hard to evaluate in specific cases (see (13.4.27) et seq.), it has the unfortunate feature that a specific form for the radial wavefunction of ϕ_k^q must be assumed so that the indicated differentiation can be performed. It might be hoped that by clever choices of coordinate axes, changing each time (13.4.11) is evaluated for a different center (a perfectly legitimate procedure), terms of the form (13.4.17) could always be replaced by (13.4.16). Unfortunately, this is not the case. We also remark that if convenient, the x or y axis at the origin could be chosen to point at atom k. The changes this would produce in (13.4.15) are obvious.

Further octahedral examples are given in Section 19.6. The tetrahedral case is somewhat more complicated, so we treat it in detail below.

Transforming l_γ: The Tetrahedral Case

The coordinate system used in the tetrahedral ML_4^{n-} case is shown in Section C.10, where symmetry-adapted LCAO functions are also listed. Let us first rotate X, Y, Z respectively into X', Y', Z' chosen to be parallel respectively to X_1, Y_1, Z_1. Simple trigonometric considerations (Section 8.2) give

$$\begin{pmatrix} X' \\ Y' \\ Z' \end{pmatrix} = \begin{pmatrix} -1/\sqrt{6} & -1/\sqrt{6} & \sqrt{2}/\sqrt{3} \\ 1/\sqrt{2} & -1/\sqrt{2} & 0 \\ 1/\sqrt{3} & 1/\sqrt{3} & 1/\sqrt{3} \end{pmatrix} \begin{pmatrix} X \\ Y \\ Z \end{pmatrix} \qquad (13.4.18)$$

and the transformation matrix is orthogonal. In analogy to (13.4.12), we can write

$$x' = x_1, \qquad y' = y_1, \qquad z' = z_1 + R \qquad (13.4.19)$$

where (x', y', z') are the coordinates of a point measured from M, (x_1, y_1, z_1) are the coordinates of the same point measured from atom 1, and R (always > 0) is the ML_1 distance. Equations (13.4.13)–(13.4.15) are immediately applicable [now identifying (x, y, z) with (x', y', z') and (x_k, y_k, z_k) with (x_1, y_1, z_1)]. We next invert (13.4.18), note that (l_x, l_y, l_z) transform like (x, y, z) under a rotation (but *not* under a rotary reflection), and thus obtain

$$l_z = \frac{\sqrt{2}}{\sqrt{3}}\left(l_{x_1} + i\hbar R\frac{\partial}{\partial y_1}\right) + \frac{1}{\sqrt{3}} l_{z_1} \qquad (13.4.20)$$

This equation is appropriate for evaluating matrix elements of l_z (origin at M) in terms of functions on atom 1. Using the same method, the relevant equations for atoms 2–4 respectively are

$$l_z = \frac{\sqrt{2}}{\sqrt{3}}\left(l_{x_2} + i\hbar R\frac{\partial}{\partial y_2}\right) - \frac{1}{\sqrt{3}} l_{z_2}$$

$$l_z = \frac{\sqrt{2}}{\sqrt{3}}\left(l_{x_3} + i\hbar R\frac{\partial}{\partial y_3}\right) - \frac{1}{\sqrt{3}} l_{z_3} \qquad (13.4.21)$$

$$l_z = \frac{\sqrt{2}}{\sqrt{3}}\left(l_{x_4} + i\hbar R\frac{\partial}{\partial y_4}\right) + \frac{1}{\sqrt{3}} l_{z_4}$$

If instead we need l_x (or l_y) at origin M, we can appropriately "rotate" (13.4.20) or (13.4.21). For example, suppose we wish to find the l_x analog of (13.4.20). Applying a 90° rotation CCW about Z_1, we note that $Z \to X$,

$X_1 \to Y_1$, $Y_1 \to -X_1$, $Z_1 \to Z_1$. Therefore

$$l_x = \frac{\sqrt{2}}{\sqrt{3}}\left(l_{y_1} - i\hbar R\frac{\partial}{\partial x_1}\right) + \frac{1}{\sqrt{3}}l_{z_1} \tag{13.4.22}$$

Let us now apply these equations to specific cases. If both atomic orbitals are centered at the origin (atom M), the methods discussed earlier in this section for the single-center one-electron case apply. We therefore consider the many-center case. Suppose we wish to evaluate $\langle t_1 x|l_z|t_1 y\rangle$, where $|t_1 x\rangle$ and $|t_1 y\rangle$ are the ligand LCAOs of Table C.10.1(b). Then

$$\langle t_1 x|l_z|t_1 y\rangle = \frac{-i}{16}\langle\sqrt{3}\,(-x_1 - x_2 + x_3 + x_4) - y_1 - y_2 + y_3 + y_4|l_z$$
$$\times|\sqrt{3}\,(x_1 - x_2 + x_3 - x_4) - y_1 + y_2 - y_3 + y_4\rangle \tag{13.4.23}$$

where $x_1 \equiv p_{x_1}$, and so on. As before, we make the one-center approximation (13.4.10). We now express l_z by (13.4.20) and (13.4.21), matching the orbital and operator subscripts. A large number of matrix elements result. However, all are zero by symmetry (Wigner–Eckart theorem) except the following (remember $\partial/\partial y_j$ transforms like y_j):

$$\langle x_j|l_{z_j}|y_j\rangle = -\langle y_j|l_{z_j}|x_j\rangle = \langle x_k|l_{z_k}|y_k\rangle, \qquad j, k = 1,\ldots, 4 \tag{13.4.24}$$

Consequently (in units of $\hbar$)

$$\langle t_1 x|l_z|t_1 y\rangle = \frac{-i}{2}\langle x_1|l_{z_1}|y_1\rangle = -\frac{1}{2} \tag{13.4.25}$$

since by Section B.2 and (13.3.11), $\langle x_1|l_{z_1}|y_1\rangle = -i\hbar$.

As a final example, consider $\langle t_2\xi|l_z|t_2\eta\rangle$, where from Table C.10.1(b)

$$|t_2\xi\rangle = -a_1|x\rangle - ib_1|yz\rangle$$
$$-c_1|\tfrac{1}{2}(z_1 - z_2 + z_3 - z_4)\rangle - d_1|\tfrac{1}{2}(s_1 - s_2 + s_3 - s_4)\rangle$$
$$-e_1|\tfrac{1}{4}[x_1 + x_2 - x_3 - x_4 + \sqrt{3}\,(-y_1 - y_2 + y_3 + y_4)]\rangle \tag{13.4.26}$$

with the analogous expression for $|t_2\eta\rangle$. We proceed exactly as before. The only new feature is the occurrence of the nonzero element $\langle s_1|\partial/\partial y_1|y_1\rangle = -\langle y_1|\partial/\partial y_1|s_1\rangle$. To evaluate this integral, it is necessary to specify the radial parts of the atomic orbitals. Using Slater $2s$ and $2p$ functions, we have

$$|s\rangle = \left(\frac{k^5}{3\pi}\right)^{1/2} re^{-kr}, \qquad |z\rangle = \left(\frac{k^5}{\pi}\right)^{1/2} ze^{-kr} \tag{13.4.27}$$

where distances (including R) are in (atomic) units of a_0 (= 0.5292 Å). A straightforward calculation gives

$$\left\langle s_1\left|\frac{\partial}{\partial y_1}\right|y_1\right\rangle = \left\langle s_1\left|\frac{\partial}{\partial z_1}\right|z_1\right\rangle = \frac{k}{2\sqrt{3}} \tag{13.4.28}$$

(For the corresponding $3s$ and $3p$ orbitals, the result is $k/3\sqrt{3}$.) Using (13.4.28) and noting that $\langle yz|l_z|xz\rangle = i\hbar$, we obtain (in units of $\hbar$)

$$\langle t_2\xi|l_z|t_2\eta\rangle = \left(a^2 - b^2 + \frac{e^2}{2} - \frac{deRk}{\sqrt{6}} - ce\sqrt{2}\right) \tag{13.4.29}$$

Finally, we note that it is not difficult, and it is sometimes necessary, to calculate two-center contributions. For example, in the case of the ${}^1A_{1g} \to {}^1E_{1u}$ transition in benzene, one finds using $2p_\pi$ atomic orbitals that each one-center angular-momentum contribution is zero. It is then imperative to calculate the two-center contributions. An explicit expression for these is given by Snyder et al. [(1981), Eq. (5)].

13.5. The $\mathcal{C}$ Terms of $Fe(CN)_6^{3-}$ with Spin–Orbit Coupling

In Section 11.2 we calculated the $\mathcal{C}$ terms of $Fe(CN)_6^{3-}$ neglecting spin–orbit coupling. Thus we calculated $\mathcal{C}_0$ for ${}^2T_{2g} \to {}^2T_{2u}$ and ${}^2T_{2g} \to {}^2T_{1u}$. But in fact, as we showed explicitly in Section 13.3, all of these states are spin–orbit-split. Using (13.3.19) and (13.4.3), we note that the ground term (${}^2T_{2g}$) is split into U_g' and E_g'' states, with the latter lower in energy by approximately $\frac{3}{2}\zeta_{Fe^{3+}}$.

The excited-state splittings are given by (13.3.18) and (13.3.19), and the relevant reduced matrix elements are evaluated in Section 19.6 for the closely related case of $IrCl_6^{2-}$. The details are irrelevant here. We simply note that the excited-state splittings depend primarily on the *ligand* spin–orbit coupling constant (Table 19.6.1). For the CN^- ligand, this is a small number ($\lesssim 10^2$ cm^{-1}) compared to $Fe(CN)_6^{3-}$ solution bandwidths ($\sim 10^3$ cm^{-1}). Thus we may entirely neglect excited-state spin–orbit effects (see also the comments at the end of this section).

Our task therefore is to calculate the MCD due to transitions to $^2T_{1u}$ and $^2T_{2u}$ from the spin–orbit-split $^2T_{2g}$ ground state. We make no assumptions about the magnitude of the ground-state splitting except to place U_g' at higher energy. The MCD is expected to have a complex temperature dependence, since both the E_g'' and U_g' states can be populated. Furthermore, we find that an important $\mathcal{B}$ term arises from field-induced mixing between E_g'' and U_g' (Section 5.2).

We make the rigid-shift and Born–Oppenheimer approximations (Section 4.5), and using (5.2.18) and (5.2.19), we write for the MCD and zero-field absorption (with $f_i(\mathcal{E}) \equiv f_i$)

$$\frac{\Delta A'}{\mathcal{E}} = \gamma\mu_B B\left\{\delta_1\left[\mathcal{B}_0(1) + \frac{\mathcal{C}_0(1)}{kT}\right]f_1 + \delta_2\left[\mathcal{B}_0(2) + \frac{\mathcal{C}_0(2)}{kT}\right]f_2\right\} \tag{13.5.1}$$

$$\frac{A}{\mathcal{E}} = \gamma\left[\delta_1\mathcal{D}_0(1)f_1 + \delta_2\mathcal{D}_0(2)f_2\right] \tag{13.5.2}$$

where indices 1 and 2 designate the ground (E_g'') and low-lying excited (U_g') state respectively; δ_1 and δ_2 are the fractional populations of the two states [Eq. (5.2.20)]. $\mathcal{A}$ terms are not included in (13.5.1) because they are found experimentally to be swamped by the $\mathcal{C}$ terms, even at room temperature (Section 11.2).

To calculate $\mathcal{B}_0(i)$, $\mathcal{C}_0(i)$, and $\mathcal{D}_0(i)$, we choose ground-state functions diagonal in μ_z so that we can use (4.5.16)—the obvious choices are $(E_g''\alpha'', E_g''\beta'')$ and $(U_g'\kappa, U_g'\lambda, U_g'\mu, U_g'\nu)$ of Table C.5.1(d). Since the excited states are assumed to be unsplit, it is very convenient to express the ground-state functions in terms of their $|^{2S+1}h\mathfrak{M}\theta\rangle$ components right from the start. Thus for example, using the methods of Section 12.4,

$$\left|\left(\tfrac{1}{2}E_g', T_{2g}\right)E_g''\alpha''\right\rangle = \frac{-1}{\sqrt{3}}|^2T_{2g} - \tfrac{1}{2}0\rangle + \frac{\sqrt{2}}{\sqrt{3}}|^2T_{2g}\tfrac{1}{2}1\rangle$$

$$\left|\left(\tfrac{1}{2}E_g', T_{2g}\right)U_g'\kappa\right\rangle = \frac{-\sqrt{2}}{\sqrt{3}}|^2T_{2g} - \tfrac{1}{2}0\rangle - \frac{1}{\sqrt{3}}|^2T_{2g}\tfrac{1}{2}1\rangle \tag{13.5.3}$$

We now carry through the calculation as in Section 11.2. One new feature is the occurrence of matrix elements of the operator S_z. However, these are extremely simple to evaluate because the $|\mathcal{S}h\mathcal{M}\theta\rangle$ functions are eigenfunctions of S_z. The results are

$$\mathcal{C}_0\left(E_g'' \to {}^2T_{1u}\right) = \frac{1}{27}\left(1 - \frac{\sqrt{2}}{\sqrt{3}}\langle t_{2g}\|l^{t_{1g}}\|t_{2g}\rangle\right) \times \left|\langle {}^2T_{2g}\|m^{t_{1u}}\|{}^2T_{1u}\rangle\right|^2 \quad (13.5.4)$$

$$\mathcal{B}_0\left(E_g'', U_g' \to {}^2T_{1u}\right) = \frac{2}{27\Delta W}\left(2 - \frac{1}{\sqrt{6}}\langle t_{2g}\|l^{t_{1g}}\|t_{2g}\rangle\right) \times \left|\langle {}^2T_{2g}\|m^{t_{1u}}\|{}^2T_{1u}\rangle\right|^2 \quad (13.5.5)$$

$$\mathcal{D}_0\left(E_g'' \to {}^2T_{1u}\right) = \tfrac{1}{9}\left|\langle {}^2T_{2g}\|m^{t_{1u}}\|{}^2T_{1u}\rangle\right|^2 \quad (13.5.6)$$

$$\mathcal{C}_0\left(U_g' \to {}^2T_{1u}\right) = \frac{-5}{54}\left(1 + \frac{1}{\sqrt{6}}\langle t_{2g}\|l^{t_{1g}}\|t_{2g}\rangle\right) \times \left|\langle {}^2T_{2g}\|m^{t_{1u}}\|{}^2T_{2g}\rangle\right|^2 \quad (13.5.7)$$

$$4\mathcal{B}_0\left(U_g', E_g'' \to {}^2T_{1u}\right) = -2\mathcal{B}_0\left(E_g'', U_g' \to {}^2T_{1u}\right) \quad (13.5.8)$$

$$\mathcal{D}_0\left(U_g' \to {}^2T_{1u}\right) = \tfrac{1}{9}\left|\langle {}^2T_{2g}\|m^{t_{1u}}\|{}^2T_{1u}\rangle\right|^2 \quad (13.5.9)$$

(Alternatively, the same results may be obtained using the equations of Section 17.2 as in Section 23.2.) The same equations are found to apply for transitions to ${}^2T_{2u}$ provided the sign of each $\mathcal{B}_0$ and $\mathcal{C}_0$ value above is changed and $|\langle {}^2T_{2g}\|m^{t_{1u}}\|{}^2T_{1u}\rangle|^2$ is everywhere replaced by $|\langle {}^2T_{2g}\|m^{t_{1u}}\|{}^2T_{2u}\rangle|^2$. In these equations, the notation $\mathcal{B}_0(i, j \to {}^2T_{1u})$ means the $\mathcal{B}_0$ value for the transition $i \to {}^2T_{1u}$ considering mixing *only* between i and j. Then $\Delta W \equiv W_2^0 - W_1^0 = (W^0(U_g') - W^0(E_g'')) > 0$. [We note that (13.5.8) immediately generalizes for any system with multiple "ground" (i.e. populated) states:

$$|A|\mathcal{B}_0(A, B \to J) = -|B|\mathcal{B}_0(B, A \to J) \quad (13.5.10)$$

This result may be confirmed by inspecting $\mathcal{B}_0$ in (4.5.16). Note the contrast with the analogous case of excited-state mixing, (5.2.5), where the factors $|A|$ and $|B|$ are absent.] Finally, (13.4.3) and (13.4.4) give

$$\langle t_{2g} \| l^{t_{1g}} \| t_{2g} \rangle = -\sqrt{6}\,\kappa \tag{13.5.11}$$

where κ is the orbital reduction factor [see (13.4.4)].

Substituting (13.5.4)–(13.5.9) and (13.5.11) into (13.5.1) and (13.5.2), we obtain

$$\frac{\Delta \mathcal{A}'}{\mathcal{E}} = \frac{\gamma \mu_B B}{27} \left\{ \delta_1 \left[\frac{2(2+\kappa)}{\Delta W} + \frac{1+2\kappa}{kT} \right] f_1 + \delta_2 \left[-\frac{2+\kappa}{\Delta W} + \frac{5(-1+\kappa)}{2kT} \right] f_2 \right\} |m|^2 \tag{13.5.12}$$

$$\frac{\mathcal{A}}{\mathcal{E}} = \frac{\gamma}{9}(\delta_1 f_1 + \delta_2 f_2)|m|^2 \tag{13.5.13}$$

where $|m|^2 \equiv |\langle {}^2T_{2g} \| m^{t_{1u}} \| {}^2T_{1u} \rangle|^2$. [Recall that (13.5.12) and (13.5.13) apply for transitions to ${}^2T_{2u}$ if the right-hand side of (13.5.12) is multiplied by -1 and $|m|^2 \equiv |\langle {}^2T_{2g} \| m^{t_{1u}} \| {}^2T_{2u} \rangle|^2$ is used. Thus with these simple changes, all the rest of our analysis also applies to the ${}^2T_{2u}$ case.]

It is illuminating first to consider the case ΔW much smaller than kT and bandwidth. We can write $\delta_1 \approx (\frac{1}{3} + 2\,\Delta W/9kT)$, $\delta_2 \approx (\frac{2}{3} - 2\,\Delta W/9kT)$, and $f_2(\mathcal{E}) = f_1(\mathcal{E} + \Delta W) \approx f_1 + \Delta W(\partial f_1/\partial \mathcal{E})$, assuming that f_1 and f_2 have the same shape. Substituting these approximations into (13.5.12) and (13.5.13) and dropping small terms, we obtain

$$\frac{\Delta \mathcal{A}'}{\mathcal{E}} = \gamma \mu_B B \left[\frac{\kappa}{9kT} |m|^2 f_1 - \frac{2(2+\kappa)}{81} |m|^2 \left(\frac{\partial f_1}{\partial \mathcal{E}} \right) \right] \tag{13.5.14}$$

$$\frac{\mathcal{A}}{\mathcal{E}} = \frac{\gamma}{9} f_1 |m|^2 \tag{13.5.15}$$

This is an explicit example of the type of case described by (5.2.21) and (5.2.22) (but note that there all states are assumed nondegenerate). The term involving $\partial f_1/\partial \mathcal{E}$ is a pseudo-$\mathcal{A}$ term (see further discussion at end of section). We may now make a very important check of our calculation, since these equations must be consistent with the zero-spin–orbit case ($\Delta W = 0$) of (11.2.10) and (11.2.12). The coefficient of f_1/kT in (13.5.14) is $\mathcal{C}_0$ which is $\frac{1}{9}|m|^2$ (with $\kappa = 1$). Similarly, from (13.5.15), $\mathcal{D}_0 = \frac{1}{9}|m|^2$. These values indeed agree with the earlier calculation and hence give us considerable confidence in our general results, (13.5.12) and (13.5.13).

It is now interesting to compare these general results with the zero-spin–orbit treatment of Section 11.2. To do this we compare zeroth moments. We thereby eliminate $\mathcal{C}$-term contributions and any dependence of the results on the bandshape functions f_1 and f_2. Noting that $\int f_i \, d\mathcal{E} = 1$, $\int (\partial f_i/\partial \mathcal{E}) \, d\mathcal{E} = 0$ [see (4.5.25) and (4.5.27)], and $\delta_2 = 1 - \delta_1$, we obtain

$$\left\langle \frac{\Delta A'}{\mathcal{E}} \right\rangle_0 = \frac{\gamma \mu_B B}{27} \left[\frac{\delta_1(7-\kappa) + 5(\kappa - 1)}{2kT} + \frac{(3\delta_1 - 1)(2+\kappa)}{\Delta W} \right] |m|^2 \tag{13.5.16}$$

$$\left\langle \frac{A}{\mathcal{E}} \right\rangle_0 = \frac{\gamma}{9} |m|^2 \tag{13.5.17}$$

Equations (11.2.10) and (11.2.12) for $^2T_{1u}$ in zeroth-moment form read

$$\left\langle \frac{\Delta A'}{\mathcal{E}} \right\rangle_0 = \frac{\gamma \mu_B B}{9kT} |m|^2 \tag{13.5.18}$$

$$\left\langle \frac{A}{\mathcal{E}} \right\rangle_0 = \frac{\gamma}{9} |m|^2 \tag{13.5.19}$$

We define the fractional difference F as the right-hand side (RHS) of (13.5.16) minus the RHS of (13.5.18) divided by the RHS of (13.5.18). Then defining $V \equiv \Delta W/kT$, we obtain

$$F = \frac{1}{6} \left[\delta_1(7-\kappa) + (5\kappa - 11) + \frac{2(2+\kappa)(3\delta_1 - 1)}{V} \right] \tag{13.5.20}$$

with

$$\delta_1 = (1 + 2e^{-V})^{-1} \tag{13.5.21}$$

It is now straightforward to set $\partial F/\partial V = 0$ and numerically solve the resulting transcendental equation for specified values of κ to obtain the maximum fractional difference (F_{max}). Some numerical results are summarized in Table 13.5.1. For example, if $\kappa = 0.87$, we predict, at maximum, an MCD band area (zeroth moment) 37.5% larger than if spin–orbit coupling were neglected *and* κ were set equal to unity. The latter assumptions give our results (11.2.10), and are those of the literature [Stephens (1965a), Schatz et al. (1966)].

Table 13.5.1[a]

κ	V_{max}	δ_1 (%)	F_{max} (%)
1.0	2.643	87.6	49.1
0.87	2.722	88.4	37.5
0.8	2.766	88.8	31.3
0.7	2.831	89.5	22.5
0.6	2.898	90.1	13.6
0.5	2.969	90.7	4.9

[a] $V \equiv \Delta W/kT$. F_{max} is the maximum fractional difference, and δ_1 is the corresponding population of the E_g'' ground state.

We see that while the qualitative results of the 1965–1966 analysis are unaffected, the quantitative consequences of including spin–orbit coupling may be substantial, depending upon the κ value. Note also for a fixed value of κ that the temperature at which the error is a maximum (T_{max}) depends on ΔW. Thus for example if $\kappa = 0.87$ and $\Delta W = 420$ cm^{-1}, then $T_{max} = 222$ K, whereas for $\Delta W = 100$ cm^{-1}, $T_{max} = 52.9$ K.

On the basis of paramagnetic-resonance and magnetic-susceptibility data, Abragam and Bleaney (1970) suggest for $Fe(CN)_6^{3-}$ that $\kappa \approx 0.87$ (with 0.56 a less likely possibility) and $\zeta_{Fe^{3+}} \approx 280$ cm^{-1}. Using these values (with $\Delta W = 1.5\zeta_{Fe^{3+}}$), we find $F = 34.9\%$ at 298 K, fairly close to F_{max}, and $F = -4.9\%$ at 12 K (see below). For $Fe(CN)_6^{3-}$ in solution at room temperature, only the first band can be accurately moment-analyzed (see Fig. 11.2.1). The result [Schatz et al. (1966)], $\mathcal{C}_0/\mathcal{D}_0 = 1.2$, corresponds to $F = 20\%$, and thus deviates from the simple treatment in the predicted direction. Temperature-dependence studies of the MCD of $Fe(CN)_6^{3-}$ have been reported [Gale and McCaffery (1972), Kobayashi et al. (1970)]. The claim by Kobayashi et al. based on measurements in PVA (polyvinyl alcohol) films, that the MCD is dominated by $\mathcal{B}$ terms is clearly refuted by Gale and McCaffery's measurements at 290 and 12 K in two crystalline hosts. The latter work shows that $\mathcal{C}$ terms dominate, and the observed $\mathcal{C}_0/\mathcal{D}_0$ value for the first band (0.86 at 290 K and 0.66 at 12 K) indeed decreases by about 25% between room and low temperature, as compared to the 21% decrease (see above) predicted by our treatment. [However, note that Gale and McCaffery's $\mathcal{C}_0/\mathcal{D}_0$ value at room temperature (0.86) is substantially lower than the corresponding solution value (1.2).] It would clearly be interesting to make a new series of quantitative MCD measurements over a wide range of temperature. If the theoretical temperature dependence of (13.5.20) is confirmed, quantitative information about ΔW and/or κ will result.

The detailed theoretical behavior in any case can be ascertained from (13.5.20). If $\kappa = 1$, the treatment with spin–orbit coupling is seen to reduce to the simple result [Eq. (13.5.18)] in both limits: $\Delta W/kT \ll 1$ and $\Delta W/kT \gg 1$. This must be the case for the former, as discussed in connection with (13.5.14). The latter case is illustrated by $IrCl_6^{2-}$ and accounts for the similarity of the $Fe(CN)_6^{3-}$ and $IrCl_6^{2-}$ MCD room-temperature solution spectra (Section 23.2).

Finally, a similar analysis can be carried out for the $\mathcal{C}$ terms. If $\mathcal{C}_1$ is calculated for $E_g'' \rightarrow {}^2T_{1u}$ and for $U_g' \rightarrow {}^2T_{1u}$, the two contributions in the limit $\Delta W/kT \ll 1$ *plus* the $\mathcal{C}_1$ value from (13.5.14) $[+2(2+\kappa)|m|^2/81]$ should sum to the $\mathcal{C}_1$ value obtained for ${}^2T_{2g}$(unsplit) $\rightarrow {}^2T_{1u}$. (This exercise is left for the reader.)

If first-order excited state spin–orbit coupling is considered for ${}^2T_{1u}$ (or ${}^2T_{2u}$), the previous analysis continues to apply if moments include all contributions from both excited-state spin–orbit components. This is a consequence of the invariance of the zeroth moment under a unitary transformation of the excited-state eigenfunctions (Chapter 7). It is precisely for this reason that we can completely ignore excited-state spin–orbit coupling in $Fe(CN)_6^{3-}$. If individual transitions are considered, $\mathcal{B}$ terms due to excited-state mixing can contribute. Low-temperature crystal spectra of $IrCl_6^{2-}$ and $IrBr_6^{2-}$ (Chapter 23) are cases where the excited-state spin–orbit structure is clearly resolved, but of course only the E_g'' ground state is occupied because of the large $\zeta_{Ir^{4+}}$.

14 Theory of High-Symmetry Coefficients

In this chapter we outline the basic considerations which lead to the definition of the $3jm$ and related phase factors discussed in Chapter 10. Those readers more interested in applications may omit this chapter entirely.

14.1 Invariance and Contragredience

Of central importance to applications of group theory to quantum mechanics are quantities invariant under symmetry operations and under coordinate transformations. Examples are space-fixed vectors, scalar products, inner products of functions, and matrix elements. Any quantity, such as those above, invariant under coordinate transformations and under symmetry operations, has non-zero contributions only from sums of products of contragredient parts. Thus it is useful to understand what is meant by contragredience. Two sets whose unitary representations are complex conjugate to one another are termed *contragredient.*

A vector $\mathbf{r} = \mathbf{ex}$ is invariant with respect to transformations of the coordinate axes [Eq. (8.2.12)], since the unit vectors $\mathbf{e}$ and the coordinates $\mathbf{x}$ are contragredient to each other. Physically this can be understood in terms of the vector being a geometrical object existing independently of any system of coordinates. Of course, the components change as the transformation takes place, but the object, the vector, remains unchanged (this is the passive convention mentioned in Section 8.2.).

The sets of states $|\psi a\alpha\rangle$ and $\langle\phi a\alpha| \equiv |\phi a\alpha\rangle^{\dagger}$ are contragredient to each other. The *adjoint* (dagger, †) operation (conjugation) converts bras into kets (and vice versa), and numbers into their complex conjugates. Here a and α are irrep and irrep component labels, while ψ and ϕ label particular sets of functions with the indicated transformation properties. The sets transform (when both are written as row matrices) as

$$\mathcal{R}|\psi a\alpha\rangle = |\psi a\alpha\rangle' = \sum_{\kappa} |\psi a\kappa\rangle D^a(R)_{\kappa\alpha}$$
$$(\mathcal{R}|\phi a\alpha\rangle)^{\dagger} = '\langle\phi a\alpha| = \sum_{\kappa} \langle\phi a\kappa| D^a(R)^*_{\kappa\alpha} \tag{14.1.1}$$

and are thus contragredient to one another. $\mathcal{R}$ is a symmetry operator (Section 8.6). For simplicity we have assumed here that no repeated representations occur and have dropped the r label.

All coupling coefficients are defined for the coupling either of kets with kets, or of bras with bras. Thus the coefficients for the coupling of a function which transforms as $|\psi a\alpha\rangle$ with a function which transforms as $\langle\phi a\alpha|$ are undefined unless the transformation properties of the $\langle\phi a\alpha|$ are first related to those of the $|\phi a\alpha\rangle$; then the coefficients for the coupling of the $|\phi a\alpha\rangle$ to the $|\psi a\alpha\rangle$ may be used. If all irrep matrices are real, then $\mathbf{D}^a(R)^* = \mathbf{D}^a(R)$, and $\langle\phi a\alpha|$ and $|\phi a\alpha\rangle$ have identical transformation properties; thus the idea of contragredience need not be explicitly considered. This is the case, for example, for single-valued irreps in Griffith's (1962) real-O basis (but *not* in our real-O basis in Section C.5). Most often, however, some of the irrep matrices are complex, and the difference in the transformation properties of $|\phi a\alpha\rangle$ and $\langle\phi a\alpha|$ must be accounted for.

A fundamental way in which contragredience enters is in the formation of the invariant product of two sets. By *invariant product* we mean that component of the direct product which belongs to the totally symmetric (A_1) irrep (and so is invariant under symmetry operations of the group). From (9.3.21) we have

$$|(ab)A_1a_1\rangle = \sum_{\alpha\beta} (a\alpha, b\beta|A_1a_1)|\psi a\alpha\rangle|\phi b\beta\rangle \tag{14.1.2}$$

Since A_1 never occurs as a repeated representation, we have dropped the index r. In order to form an invariant product we need to couple $|\phi b\beta\rangle$ to some function $|\psi a\alpha\rangle$ that transforms contragrediently to $|\phi b\beta\rangle$—that is, that transforms as $|\psi b\beta\rangle^{\dagger}$—since the direct product must then contain A_1. In Section 14.6 we show that once the relation between the $\mathbf{D}^a(R)^*$ and the $\mathbf{D}^a(R)$ is established, a general expression for the $(a\alpha, b\beta|A_1a_1)$ above may be derived. Derivations of expressions for other invariants then make frequent use of this result. [In the group theory of SO_3 a simple algebraic formula for the analogous $(j_1m_1, j_2m_2|00)$ may be obtained directly from Wigner's general formula for the $(j_1m_1, j_2m_2|jm)$. Thus the problem is not approached in an analogous manner to that used here.]

14.2. The Relation between Contragredient Sets for Molecular Point Groups

Definition of Ambivalent and Nonambivalent Groups

In Sections 14.2–14.5 we determine the relation between the transformation properties of the contragredient sets $|a\alpha\rangle^\dagger$ and $|a\alpha\rangle$, where a and α are irrep labels for a molecular point group. To simplify the notation the label ϕ, which distinguishes between different functions with identical transformation properties, has been dropped. It is useful to classify molecular point groups into two general catagories—ambivalent (e.g. D_4, O, T_d, SO_3) and nonambivalent (e.g. C_3, C_4, D_3, T). Nonambivalent groups contain some irreps for which the conjugate kets $|a\alpha\rangle^\dagger$ have the transformation properties of another irrep, $b = a^*$, where $a^* \neq a$. Ambivalent groups have $a^* = a$ for all irreps.

Nonambivalent groups are easily recognized, since their character tables contain at least one pair of irreps (which we call P_1 and P_2) with complex characters. Corresponding characters of the pair are complex conjugate to each other: $\chi^{P_1}(R) = \chi^{P_2}(R)^*$ for all symmetry operations R. A typical nonambivalent group is D_3, which contains the nondegenerate irreps P_1 and P_2 (Section E.1) which fit our definitions of P_1 and P_2. (These are analogous to Griffith's [(1964), Table A 16] $E''\rho_1$ and $E''\rho_2$ entries for D_3.) Functions belonging to P_1- and P_2-type irreps go into themselves under all group symmetry operations (hence the separate character listings), but are transformed into one another under the time-reversal operation T (see below). Thus $P_1\rho_1$ and $P_2\rho_2$ are individually nondegenerate with respect to all point-group operations but $T|P_1\rho_1\rangle = c|P_2\rho_2\rangle$, where $c = +1$ or -1, depending on the basis choice made. Also $P_1^* = P_2$. Nonambivalent groups have special characteristics and thus are considered apart from the more common ambivalent groups, which we consider first.

The Relation between Contragredient Sets for Ambivalent Groups

In ambivalent groups the representatives $\mathbf{D}^a(R) = \mathbf{D}^{a^*}(R)$ and $\mathbf{D}^a(R)^*$ differ at most by a similarity transformation and are thus *equivalent* so that

$$\mathbf{D}^a(R) = \mathbf{U}^{-1}(a)\mathbf{D}^a(R)^*\mathbf{U}(a) \tag{14.2.1}$$

for all symmetry operations R [Eq. (8.8.2) with $\mathbf{D}'(R) = \mathbf{D}^a(R)$ and $\mathbf{D}(R) = \mathbf{D}^a(R)^*$]. It also follows that the transformation properties of functions $|a\alpha\rangle^\dagger$ belonging to $\mathbf{D}^a(R)^*$ are related to those of functions $|a\alpha\rangle$ belonging to $\mathbf{D}^a(R)$ by the unitary matrix above designated $\mathbf{U}(a)$:

$$|a\alpha\rangle \sim \sum_{\alpha'} |a\alpha'\rangle^{\dagger} U(a)_{\alpha'\alpha}$$

$$|a\alpha'\rangle^{\dagger} \sim \sum_{\alpha} |a\alpha\rangle U(a)^{-1}_{\alpha\alpha'} \tag{14.2.2}$$

This is the converse of the result given in Section 8.8. The " ~ " means "transforms as." $\mathbf{U}(a)$ is fixed by (14.2.1) *to within a phase factor*. [See Fano and Racah (1959) for further discussion of $\mathbf{U}(a)$.] Since there is a one-to-one correspondence between functions and their conjugates, we may expect that $\mathbf{U}(a)$ relates each irrep and irrep component $a\alpha$ to its conjugate $a^*\alpha^*$. Of course, $a = a^*$ for ambivalent groups. Thus, using the second of (14.2.2), let us suppose that

$$\begin{aligned} |a\alpha\rangle^{\dagger} &= |a\alpha^*\rangle \delta_{\alpha'\alpha^*} U(a)^{-1}_{\alpha^*\alpha} \\ &\equiv |a\alpha^*\rangle \begin{pmatrix} a \\ \alpha \end{pmatrix} \\ &= |a^*\alpha^*\rangle \begin{pmatrix} a \\ \alpha \end{pmatrix} \end{aligned} \tag{14.2.3}$$

The coefficients $\begin{pmatrix} a \\ \alpha \end{pmatrix}$ are known as $2jm$, since they relate one "jm" (here $a\alpha$) to another (here $a^*\alpha^*$).

To define $\mathbf{U}(a)$—and thus a^*, the α^*, and the $\begin{pmatrix} a \\ \alpha \end{pmatrix}$—we must first understand the relationship between the transformation properties of the $|a\alpha\rangle$ and the $|a\alpha\rangle^{\dagger}$. There are three cases to consider for each irrep a:

1. *The $|a\alpha\rangle$ are real functions, so $\phi^a_\alpha = (\phi^a_\alpha)^*$ for all α.* Since the $(\phi^a)^*$ belong to $(\mathbf{D}^a)^*$, $(\phi^a_\alpha)^*$ and $|a\alpha\rangle^{\dagger}$ differ by at most a phase factor. Thus $|a\alpha\rangle^{\dagger} \sim |a\alpha\rangle$ [recall that $\mathbf{U}(a)$ is fixed by (14.2.1) only to within a phase factor]. From (14.2.3) it follows that $a^* = a$, $\alpha^* = \alpha$, and we may choose

$$\begin{pmatrix} a \\ \alpha \end{pmatrix} = 1.$$

2. *The $|a\alpha\rangle$ are complex, but may be expressed in terms of real functions.* Symbolically,

$$\psi^a = \phi^a \mathbf{A} \tag{14.2.4}$$

where $\mathbf{A}$ is unitary and the ϕ^a are real. Thus the functions $(\psi^a)^*$ are well defined as $(\psi^a)^* = (\phi^a)^* A^*$ and the transformation properties of the $|a\alpha\rangle^\dagger$ are those of the $(\psi^a)^*$ though the two sets may differ by a common phase factor. For ambivalent groups all purely spatial functions fall into these first two categories, so $\mathbf{U}(a)$—and thus the α^* and $\begin{pmatrix} a \\ \alpha \end{pmatrix}$—are easy to define. Of course, $a = a^*$ by definition for ambivalent groups.

3. *The $|a\alpha\rangle$ are complex but are not expressible in terms of real functions.* This case occurs in systems with half-integral spin and includes states belonging to double-valued irreps (e.g. E', E'', and U'), and all states $|jm\rangle$ with half-integral j. The prototype example is that of the spin-one-half functions, $|\frac{1}{2}\frac{1}{2}\rangle$ and $|\frac{1}{2} - \frac{1}{2}\rangle$, of SO_3 in the $|jm\rangle$ basis. In this case what we mean by $|\frac{1}{2}\frac{1}{2}\rangle^\dagger$ and $|\frac{1}{2} - \frac{1}{2}\rangle^\dagger$ is not obvious, since $|\frac{1}{2}\frac{1}{2}\rangle$ and $|\frac{1}{2} - \frac{1}{2}\rangle$ cannot be expressed in terms of real functions. But once the transformation properties of these functions are defined, the transformation properties of any arbitrary function $|a\alpha\rangle^\dagger$ may be found, since all other functions may be expressed, using coupling tables, in terms of products of spin-one-half functions and functions of types 1 and 2 above. There are several ways to do this for the general case. In the next section we show how this may be done rather easily using the time-reversal operator T.

14.3. Use of the Time-Reversal Operation T

One way of defining $|\frac{1}{2}\frac{1}{2}\rangle^\dagger$ and $|\frac{1}{2} - \frac{1}{2}\rangle^\dagger$ of SO_3 in the angular-momentum basis—and thus, ultimately, all $|a\alpha\rangle^\dagger$—is by consideration of the behavior of the spin-one-half functions under the time-reversal operation T [see Tinkham (1964)] which takes the time t into $-t$. If no external magnetic field is present, time-independent Hamiltonians are invariant under time reversal, as well as under the group symmetry operations we have so far discussed. Thus if $|a\alpha\rangle$ is an eigenfunction of a zero-field time-invariant Hamiltonian with eigenvalue E, so is $T|a\alpha\rangle$.

The operation of T on a complex number is that of complex conjugation. Thus T is *not* a linear or a unitary operator, since

$$T(c\psi) = (Tc)(T\psi) = c^*(T\psi) \tag{14.3.1}$$

T is termed *antilinear* or *antiunitary*. The action of T on $|a\alpha\rangle$ may be defined by noting that if

$$\mathcal{R}|a\alpha\rangle = |a\alpha\rangle' = \sum_{\kappa} |a\kappa\rangle D^a(R)_{\kappa\alpha} \tag{14.3.2}$$

then, since T commutes with $\mathcal{R}$,

$$\mathcal{R}T|a\alpha\rangle = \sum_{\kappa}(T|a\kappa\rangle)(TD^a(R)_{\kappa\alpha}) = \sum_{\kappa}(T|a\kappa\rangle)D^a(R)^*_{\kappa\alpha} \tag{14.3.3}$$

We see that the functions $T|a\alpha\rangle$ have the same $\mathbf{D}^a(R)^*$ irrep matrices as the $|a\alpha\rangle^\dagger = \langle a\alpha|$, and so

$$T|a\alpha\rangle \sim |a\alpha\rangle^\dagger \tag{14.3.4}$$

to within a phase factor common to the entire $|a\alpha\rangle$ set.

Since for ambivalent groups, $\mathbf{D}^a(R)$ is equivalent to $\mathbf{D}^a(R)^*$, it follows that the $T|a\alpha\rangle$, like the $|a\alpha\rangle^\dagger$, are related to the $|a\alpha\rangle$ by a unitary transformation. Thus the transformation properties of the $|a\alpha\rangle^\dagger$ are (to within the phase factor mentioned above) those of $T|a\alpha\rangle$, and we may write

$$T|a\alpha\rangle = \sum_{\alpha'}|a\alpha'\rangle U(a)^{-1}_{\alpha'\alpha} \sim |a\alpha\rangle^\dagger \tag{14.3.5}$$

We emphasize that (14.3.4) and (14.3.5) are also true for $|a\alpha\rangle^\dagger$ multiplied by a phase factor; we use this phase freedom in (14.3.16) and (14.4.4). We must now define the $T|a\alpha\rangle$ to obtain $\mathbf{U}(a)^{-1}$ and hence our $2\,jm$. We do this first for the $|jm\rangle$ of $SO_3 \supset SO_2$.

The Behavior of the $|jm\rangle$ of SO_3 in the Angular-Momentum Basis under T

The behavior of the $|jm\rangle$ in the $SO_3 \supset SO_2$ (angular-momentum) basis under T depends on the spherical-harmonic phase convention used. If the spherical harmonics for integer j are chosen as

$$\mathcal{Y}^j_m = i^j Y^j_m \tag{14.3.6}$$

where the Y^j_m are the Condon–Shortley spherical harmonics (Section B.1), particularly simple results are obtained. This choice, which was made, for example, by Fano and Racah (1959), gives

$$\begin{aligned} T|jm\rangle &= |j-m\rangle(-1)^{j-m} \\ &= |jm^*\rangle\begin{pmatrix} j \\ m \end{pmatrix} \end{aligned} \tag{14.3.7}$$

for both integer and half-integer j. In the last line in this and subsequent equations we use the notation of Section 10.2; in the $SO_3 \supset SO_2$ basis, $j^* = j$ and $m^* = -m$.

When the $|jm\rangle$ for integer j are the more conventional Y^j_m,

$$T|jm\rangle = (-1)^p|j\,-m\rangle(-1)^{j-m} = (-1)^p|jm^*\rangle\begin{pmatrix} j \\ m \end{pmatrix} \tag{14.3.8}$$

for both integer and half-integer j. If $|jm\rangle = |lm\rangle = Y^l_m$, then $p = l$. More generally p is the sum of parities of the n orbitals which form the $|jm\rangle$ state [see Section 8.1 in Butler (1981)]:

$$p = \sum_{i=1}^{n} l_i \tag{14.3.9}$$

With appropriate conventions (14.3.7) and (14.3.8) are also obeyed by the coupled kets $|(j_1 j_2) j_3 m_3\rangle$ when the $3j$ coefficients of Rotenberg et al. (1959) for $SO_3 \supset SO_2$ (our $3jm$ for $SO_3 \supset SO_2$) are used for the coupling in the conventional manner. Thus either $T|jm\rangle$ equation may be used for coupled states of SO_3. Spin states have even parity and so do not contribute to p.

$\mathbf{U}(j)$ [and hence $\mathbf{U}(a)$ and the $2\,jm$] may be defined with either (14.3.7) or (14.3.8). If (14.3.8) is used rather than (14.3.7), the factor $(-1)^p$ is simply carried along in the equations, but the same $2\,jm$ factors are derived. We use (14.3.7) in our examples, but also give some results for (14.3.8). Combining (14.3.7) and (14.3.5) gives

$$T|jm\rangle = \sum_{m'}|jm'\rangle U(j)^{-1}_{m'm} = |j\,-m\rangle(-1)^{j-m} = |jm^*\rangle\begin{pmatrix} j \\ m \end{pmatrix} \tag{14.3.10}$$

so that

$$U(j)^{-1}_{m'm} = \delta_{m',-m}(-1)^{j-m} = \delta_{m',-m}\begin{pmatrix} j \\ m \end{pmatrix} \tag{14.3.11}$$

Thus we have

$$|jm\rangle^{\dagger} \sim T|jm\rangle = |jm^*\rangle\begin{pmatrix} j \\ m \end{pmatrix} \tag{14.3.12}$$

Then, since

$$(-1)^{j-m} = (-1)^{2j}(-1)^{j+m} \tag{14.3.13}$$

or, equivalently, in our notation of Section 10.2 [see Eq. (B.3.1)],

$$\begin{pmatrix} j \\ m \end{pmatrix} = \{j\}\begin{pmatrix} j^* \\ m^* \end{pmatrix} \tag{14.3.14}$$

it follows that

$$\begin{aligned} \tilde{\mathbf{U}}(j) &= (-1)^{2j}\mathbf{U}(j) \\ &= \{j\}\mathbf{U}(j) \end{aligned} \tag{14.3.15}$$

since $(-1)^{2j}$ is simply a $2j$ phase—see (10.2.9). We see that $\mathbf{U}(j)$ is symmetric for integer j and antisymmetric for half-integer j.

When the $T|jm\rangle$ are given instead by (14.3.8), Eq. (14.3.12) is replaced by

$$(-1)^{p}|jm\rangle^{\dagger} \sim T|jm\rangle = (-1)^{p}|jm^*\rangle\begin{pmatrix} j \\ m \end{pmatrix} \tag{14.3.16}$$

Here we have chosen the unspecified phase in (14.3.4) and (14.3.5) as $(-1)^{p}$, rather than as 1. Thus both (14.3.12) and (14.3.16) give

$$|jm\rangle^{\dagger} \sim |jm^*\rangle\begin{pmatrix} j \\ m \end{pmatrix} \tag{14.3.17}$$

These equations may be used to standardize the phases of the $|jm\rangle$ and the $\mathbf{U}(j)$ for $SO_3 \supset SO_2$. We are not interested in pursuing the topic further, but rather in showing how the $\mathbf{U}(a)$ of (14.2.3) and (14.3.5), and hence the $2jm$, may be found once the behavior of a $|jm\rangle$ basis under T is defined.

14.4. Obtaining U(a) and the 2jm for Ambivalent Groups Using the Time-Reversal Operator T

Let a basis $|a\alpha\rangle$ of a group be defined in terms of the $|jm\rangle$ of $SO_3 \supset SO_2$. Then $|a\alpha\rangle$ may be expressed in terms of the $|jm\rangle$ as

$$|a\alpha\rangle = \sum_{m} |jm\rangle\langle jm|a\alpha\rangle \tag{14.4.1}$$

or in matrix form as

$$|\mathbf{a}\rangle \equiv |\mathbf{j}\rangle\mathbf{B} \tag{14.4.2}$$

where $|\mathbf{a}\rangle$ symbolizes the row matrix of kets $|a\alpha_1\rangle, |a\alpha_2\rangle \cdots$. The $\langle jm|a\alpha\rangle$ are transformation coefficients, defined as in (9.8.1). Suppose then that the $|jm\rangle$ for integer j are the $\mathcal{Y}^j_m$ of (14.3.7) and we apply T to (14.4.1). Employing (14.3.1), (14.3.12), (14.3.4), and also (14.2.3), we obtain

$$\begin{aligned} T|a\alpha\rangle &= \sum_m (T|jm\rangle)(T\langle jm|a\alpha\rangle) \\ &= \sum_m |j-m\rangle(-1)^{j-m}\langle jm|a\alpha\rangle^* \\ &= \sum_m |jm^*\rangle\binom{j}{m}\langle jm|a\alpha\rangle^* \sim |a\alpha\rangle^\dagger \\ &= |a^*\alpha^*\rangle\binom{a}{\alpha} \end{aligned} \tag{14.4.3}$$

Of course, for ambivalent groups (our case here), $a = a^*$; we keep the a^* here and in (14.4.4) and (14.4.5) below for use later in Section 14.5 for the more general case.

If on the other hand the $|jm\rangle$ for integer j were the Y^j_m of (14.3.8), we would use (14.3.16) rather than (14.3.12), and then choose the unspecified phase in (14.3.4) as $(-1)^p$ rather than as 1 to obtain

$$\begin{aligned} T|a\alpha\rangle &= \sum_m (-1)^p|jm^*\rangle\binom{j}{m}\langle jm|a\alpha\rangle^* \\ &= (-1)^p|a^*\alpha^*\rangle\binom{a}{\alpha} \\ &\sim (-1)^p|a\alpha\rangle^\dagger \end{aligned} \tag{14.4.4}$$

In both (14.4.3) and (14.4.4) we have

$$|a\alpha\rangle^\dagger \sim |a^*\alpha^*\rangle\binom{a}{\alpha} \tag{14.4.5}$$

as in (14.2.3), and thus these equations all define the $2jm$ as

$$U(a)^{-1}_{\alpha^*\alpha} = \binom{a}{\alpha} \tag{14.4.6}$$

The importance of Eqs. (14.4.1)–(14.4.4) is that they give us a prescription for finding $2\,jm$ and for defining the a^* and α^*; we do this shortly in an example.

***Relation of* U(*a*) *and* Ũ(*a*)**

Rewriting (14.4.3) in matrix form gives

$$\begin{aligned} T|\mathbf{a}\rangle &= (T|\mathbf{j}\rangle)(T\mathbf{B}) \\ &= |\mathbf{j}\rangle \mathbf{U}(j)^{-1}\mathbf{B}^* \\ &= |\mathbf{a}\rangle \mathbf{B}^{-1}\mathbf{U}(j)^{-1}\mathbf{B}^* \\ &= |\mathbf{a}\rangle \mathbf{U}(a)^{-1} \end{aligned} \tag{14.4.7}$$

This result defines $\mathbf{U}(a)^{-1}$ as

$$\mathbf{U}(a)^{-1} = \mathbf{B}^{-1}\mathbf{U}(j)^{-1}\mathbf{B}^* \tag{14.4.8}$$

so that

$$\begin{aligned} \mathbf{U}(a) &= \left[\mathbf{B}^{-1}\mathbf{U}(j)^{-1}\mathbf{B}^*\right]^{-1} \\ &= \tilde{\mathbf{B}}\mathbf{U}(j)\mathbf{B} \end{aligned} \tag{14.4.9}$$

Then (14.4.9) and (14.3.15) give

$$\begin{aligned} \tilde{\mathbf{U}}(a) &= \widetilde{\tilde{\mathbf{B}}\mathbf{U}(j)\mathbf{B}} \\ &= \tilde{\mathbf{B}}\tilde{\mathbf{U}}(j)\mathbf{B} \\ &= (-1)^{2j}\mathbf{U}(a) \\ &= \{j\}\mathbf{U}(a) \end{aligned} \tag{14.4.10}$$

Thus $\mathbf{U}(a)$ is symmetric for integer j and antisymmetric for half-integer j. From (14.4.10) and our definition of the $2\,jm$, it follows that for ambivalent groups

$$\begin{pmatrix} a \\ \alpha^* \end{pmatrix} = \{a\}\begin{pmatrix} a \\ \alpha \end{pmatrix} \tag{14.4.11}$$

where $\{a\} = \{j\}$. Thus $\{a\} = 1$ for integer j and -1 for half-integer j.

Example: Definition of the 2jm and α^* for the E'' Irrep of the Group O

An example is helpful to demonstrate how the equations above may be used to define the $2\,jm$ and α^* for a set of $|a\alpha\rangle$ basis states expressed in terms of the $|jm\rangle$ of $SO_3 \supset SO_2$. As our example we consider the E'' irrep of the group O. From Section C.6,

$$\begin{aligned} |E''\alpha''\rangle &= \frac{1}{\sqrt{6}}\left(-|\tfrac{5}{2}-\tfrac{5}{2}\rangle + \sqrt{5}\,|\tfrac{5}{2}\tfrac{3}{2}\rangle\right) \\ |E''\beta''\rangle &= \frac{1}{\sqrt{6}}\left(-\sqrt{5}\,|\tfrac{5}{2}-\tfrac{3}{2}\rangle + |\tfrac{5}{2}\tfrac{5}{2}\rangle\right) \end{aligned} \tag{14.4.12}$$

where, for the purposes of defining the $2\,jm$, we assume the $|jm\rangle$ obey (14.3.7). Using (14.4.3) on $|E''\alpha''\rangle$, together with the definition of $|E''\beta''\rangle$ above, we have

$$\begin{aligned} T|E''\alpha''\rangle &= \frac{1}{\sqrt{6}}\left(|\tfrac{5}{2}\tfrac{5}{2}\rangle - \sqrt{5}\,|\tfrac{5}{2}-\tfrac{3}{2}\rangle\right) \\ &= |E''\beta''\rangle \\ &= |E''\alpha''^*\rangle\begin{pmatrix} E'' \\ \alpha'' \end{pmatrix} \end{aligned} \tag{14.4.13}$$

From the last two lines above it follows that

$$(\alpha'')^* = \beta'' \tag{14.4.14}$$

and

$$\begin{pmatrix} E'' \\ \alpha'' \end{pmatrix} = 1 \tag{14.4.15}$$

Since $j = \frac{5}{2}$ is half-integer, $\{E''\} = -1$. Thus from (14.4.11) we obtain

$$\begin{pmatrix} E'' \\ \beta'' \end{pmatrix} = \{E''\}\begin{pmatrix} E'' \\ \alpha'' \end{pmatrix} = -1 \tag{14.4.16}$$

These are the phases tabulated in Section C.11.

14.5. The Relation between Contragredient Sets for Nonambivalent Groups

Nonambivalent groups have some irreps with complex characters. For these irreps $a^* = b \neq a$, and $\mathbf{D}^a(R)$ and $\mathbf{D}^a(R)^*$ are *inequivalent*. However, in this case it can be shown that $\mathbf{D}^{a^*}(R)$ and $\mathbf{D}^a(R)^*$ are equivalent [Butler (1981), Section 2.5]. Thus

$$\mathbf{D}^{a^*}(R) = \mathbf{U}(a^*)^{-1}\mathbf{D}^a(R)^*\mathbf{U}(a^*) \tag{14.5.1}$$

and the analog of (14.2.2) is

$$\begin{aligned} |a^*\alpha\rangle &\sim \sum_{\alpha'} |a\alpha'\rangle^\dagger U(a^*)_{\alpha'\alpha} \\ |a\alpha'\rangle^\dagger &\sim \sum_{\alpha} |a^*\alpha\rangle U(a^*)^{-1}_{\alpha\alpha'} \end{aligned} \tag{14.5.2}$$

The logic of the last sections carries over if only we substitute a^* for a in the appropriate places. This has already been done in (14.4.3)–(14.4.5) for use here. Replacing (14.4.11) is

$$\begin{pmatrix} a^* \\ \alpha^* \end{pmatrix} = \{a\}\begin{pmatrix} a \\ \alpha \end{pmatrix} \tag{14.5.3}$$

With the substitution $a = a^*$, the more general nonambivalent-group equations reduce to those for ambivalent groups.

Example: Definition of $\begin{pmatrix} a \\ \alpha \end{pmatrix}$ *, a^*, and α^* for the Complex Irreps of D_3*

D_3 is a nonambivalent group with complex irreps P_1 and P_2. Section E.5 defines these irreps in terms of the $|jm\rangle$ of $SO_3 \supset SO_2$:

$$\begin{aligned} |P_1\rho_1\rangle &= \frac{1}{\sqrt{2}}\left(|\tfrac{3}{2}\tfrac{3}{2}\rangle - i|\tfrac{3}{2} - \tfrac{3}{2}\rangle\right) \\ |P_2\rho_2\rangle &= \frac{1}{\sqrt{2}}\left(i|\tfrac{3}{2}\tfrac{3}{2}\rangle - |\tfrac{3}{2} - \tfrac{3}{2}\rangle\right) \end{aligned} \tag{14.5.4}$$

As in the example in Section 14.4, we choose the $SO_3 \supset SO_2$ $|jm\rangle$ to obey (14.3.7). With the aid of (14.4.3) and (14.5.4) above, we obtain

$$\begin{aligned} T|P_1\rho_1\rangle &= -|P_2\rho_2\rangle = \begin{pmatrix} P_1 \\ \rho_1 \end{pmatrix}|P_1^*\rho_1^*\rangle \\ T|P_2\rho_2\rangle &= |P_1\rho_1\rangle = \begin{pmatrix} P_2 \\ \rho_2 \end{pmatrix}|P_2^*\rho_2^*\rangle \end{aligned} \tag{14.5.5}$$

It follows that

$$P_1^* = P_2, \qquad P_2^* = P_1$$

$$\rho_1^* = \rho_2, \qquad \rho_2^* = \rho_1$$

$$\begin{pmatrix} P_1 \\ \rho_1 \end{pmatrix} = -1 = \begin{pmatrix} P_2^* \\ \rho_2^* \end{pmatrix}$$

$$\begin{pmatrix} P_2 \\ \rho_2 \end{pmatrix} = 1 = \begin{pmatrix} P_1^* \\ \rho_1^* \end{pmatrix}$$

$$\{P_1\} = \{P_2\} = -1 \tag{14.5.6}$$

14.6. The Invariant Product of Two Sets

Now that the transformation properties of the $|a\alpha\rangle^\dagger$ have been related to those of the $|a\alpha\rangle$, we may proceed to obtain the coupling coefficients needed to form the invariant product of two sets. From (9.3.21) we have

$$|(ab)A_1a_1\rangle = \sum_{\alpha\beta} (a\alpha, b\beta|A_1a_1)|\psi a\alpha\rangle|\phi b\beta\rangle \tag{14.6.1}$$

Here ψ and ϕ label particular functions, while $a\alpha$ and $b\beta$ are irrep and irrep-partner labels. A_1 labels the totally symmetric (invariant) irrep for the group of interest. The scalar product of $a \otimes b$ is multiplicity-free, so the label r is not required. The only nonzero contributions to $|(ab)A_1a_1\rangle$ are sums of products of contragredient functions (see Section 14.1). Thus

$$|(ab)A_1a_1\rangle \sim N\sum_{\alpha} |\psi a\alpha\rangle|\phi a\alpha\rangle^\dagger \tag{14.6.2}$$

where N is a normalization constant. Then by (14.4.5)

$$|(ab)A_1a_1\rangle = N\sum_{\alpha} \begin{pmatrix} a \\ \alpha \end{pmatrix} |\psi a\alpha\rangle|\phi a^*\alpha^*\rangle \tag{14.6.3}$$

Comparing this with (14.6.1), we must have $b = a^*$ and $\beta = \alpha^*$; otherwise $|(ab)A_1a_1\rangle \equiv 0$. Hence

$$|(ab)A_1a_1\rangle = N\delta_{ba^*}\sum_{\alpha\beta} \delta_{\beta\alpha^*} \begin{pmatrix} a \\ \alpha \end{pmatrix} |\psi a\alpha\rangle|\phi b\beta\rangle \tag{14.6.4}$$

Consequently

$$(a\alpha, b\beta|A_1 a_1) = \delta_{ba^*}\delta_{\beta\alpha^*}|a|^{-1/2}\begin{pmatrix} a \\ \alpha \end{pmatrix} \tag{14.6.5}$$

where we have chosen $N = |a|^{-1/2}$ to make the v.c.c. matrix unitary. Then from (10.2.1) we obtain

$$\begin{pmatrix} a & b & A_1 \\ \alpha & \beta & a_1 \end{pmatrix}^* = \delta_{ba^*}\delta_{\beta\alpha^*}|a|^{-1/2}\begin{pmatrix} a \\ \alpha \end{pmatrix} \tag{14.6.6}$$

since

$$\begin{pmatrix} A_1 \\ a_1 \end{pmatrix} = 1 \quad \text{and} \quad |A_1| = 1$$

Complex conjugation of (14.6.6) gives (10.3.1), since the $2jm$ are real and $|a| = |a^*|$.

14.7. The Invariant Product of Three Sets: The $3jm$

The fundamental definition of the $3jm$ is not (10.2.1). Rather, they are more usefully defined in terms of the expansion coefficients for the invariant (or totally symmetric) part of the triple direct product of three irreps. The approach we take is similar to that of Fano and Racah (1959) but is adapted to the $3jm$. Thus we define the $3jm$ by

$$|((ab)c)rA_1a_1\rangle \equiv \sum_{\alpha\beta\gamma}\begin{pmatrix} a & b & c \\ \alpha & \beta & \gamma \end{pmatrix}^{*\,r} |a\alpha\rangle|b\beta\rangle|c\gamma\rangle \tag{14.7.1}$$

where, using (9.3.21) and (14.6.5),

$$\begin{aligned}
|((ab)c)rA_1a_1\rangle &\equiv |((ab)rd, c)A_1a_1\rangle \\
&= \sum_{\delta\gamma}(d\delta, c\gamma|A_1a_1)|(ab)rd\delta\rangle|c\gamma\rangle \\
&= \sum_{\gamma}|c|^{-1/2}\begin{pmatrix} c^* \\ \gamma^* \end{pmatrix}|(ab)rc^*\gamma^*\rangle|c\gamma\rangle \\
&= \sum_{\alpha\beta\gamma}|c|^{-1/2}\begin{pmatrix} c^* \\ \gamma^* \end{pmatrix}(a\alpha, b\beta|rc^*\gamma^*)|a\alpha\rangle|b\beta\rangle|c\gamma\rangle \\
&= \sum_{\alpha\beta\gamma}\begin{pmatrix} a & b & c \\ \alpha & \beta & \gamma \end{pmatrix}^{*\,r}|a\alpha\rangle|b\beta\rangle|c\gamma\rangle
\end{aligned} \tag{14.7.2}$$

Comparison of the last two lines above gives (10.2.1), since $|c| = |c^*|$ and the $2jm$ have values $+1$ or -1. It is possible to define the $3jm$ with many more symmetry properties than v.c.c., because the three irreps a, b, and c occur on a more equal footing in (14.7.2) than in (9.3.21) which defines the v.c.c.

We now consider the behavior of the scalars defined by (14.7.1) and (14.7.2) upon permutation of a, b, and c. If the invariant product of $a \otimes b \otimes c$ is multiplicity-free, only one linearly independent scalar exists, and permuting a, b, and c gives a function which differs from (14.7.2) by at most a phase factor. When multiplicity exists in the invariant product of $a \otimes b \otimes c$, there are two linearly independent scalars defined by (14.7.1) and (14.7.2) which differ in the index r. In this case permutation of a, b, and c may lead to functions which are linear combinations of the $r = 0$ and $r = 1$ functions of (14.7.2). It is always possible (for the point groups), however, to standardize phases so that permutation leads to functions which differ from (14.7.2) by at most a phase factor. In the next several sections we show how phases are standardized so that permuting the irreps a, b, and c in (14.7.1) leads to at most a phase change, and the phase change is that summarized by (10.2.4).

The Even-Permutation Rule for 3jm

The permutations $(ca)b$ and $(bc)a$ are cyclic (even) permutations of $(ab)c$, since they preserve the cyclic abc order and are achieved by an even number of interchanges of a, b, and c. From (14.7.1) and (14.7.2)

$$|((bc)a)rA_1a_1\rangle = \sum_{\alpha\beta\gamma} |a|^{-1/2}\begin{pmatrix} a^* \\ \alpha^* \end{pmatrix}(b\beta, c\gamma|ra^*\alpha^*)|a\alpha\rangle|b\beta\rangle|c\gamma\rangle$$

$$= \sum_{\alpha\beta\gamma} \begin{pmatrix} b & c & a \\ \beta & \gamma & \alpha \end{pmatrix}^{*\,r} |a\alpha\rangle|b\beta\rangle|c\gamma\rangle \qquad (14.7.3)$$

We wish to define our $3jm$ so that they are invariant under even permutations, as in (10.2.4). Thus we require

$$\begin{pmatrix} b & c & a \\ \beta & \gamma & \alpha \end{pmatrix}^{*\,r} = \begin{pmatrix} a & b & c \\ \alpha & \beta & \gamma \end{pmatrix}^{*\,r} \qquad (14.7.4)$$

This is true if v.c.c. phases are standardized so that

$$|a|^{-1/2}\begin{pmatrix} a^* \\ \alpha^* \end{pmatrix}(b\beta, c\gamma|ra^*\alpha^*) = |c|^{-1/2}\begin{pmatrix} c^* \\ \gamma^* \end{pmatrix}(a\alpha, b\beta|rc^*\gamma^*) \qquad (14.7.5)$$

Applying (14.7.5) to the $|((ca)b)A_1a_1\rangle$ case, it is easy to show that

$$\begin{pmatrix} c & a & b \\ \gamma & \alpha & \beta \end{pmatrix}^{*\ r} = \begin{pmatrix} a & b & c \\ \alpha & \beta & \gamma \end{pmatrix}^{*\ r} \tag{14.7.6}$$

The Odd-Permutation Rule for 3jm

The odd permutations of $(ab)c$ are $(ba)c$, $(cb)a$, and $(ac)b$. From (14.7.1) and (14.7.2)

$$|((ba)c)rA_1a_1\rangle = \sum_{\alpha\beta\gamma} |c|^{-1/2}\begin{pmatrix} c^* \\ \gamma^* \end{pmatrix}(b\beta,\, a\alpha|rc^*\gamma^*)|a\alpha\rangle|b\beta\rangle|c\gamma\rangle$$

$$= \sum_{\alpha\beta\gamma} \begin{pmatrix} b & a & c \\ \beta & \alpha & \gamma \end{pmatrix}^{*\ r} |a\alpha\rangle|b\beta\rangle|c\gamma\rangle \tag{14.7.7}$$

Comparison of (14.7.7) with (14.7.2) indicates that we obtain the (10.2.4) relationship

$$\begin{pmatrix} b & a & c \\ \beta & \alpha & \gamma \end{pmatrix}^{*\ r} = \{abcr\}\begin{pmatrix} a & b & c \\ \alpha & \beta & \gamma \end{pmatrix}^{*\ r} \tag{14.7.8}$$

if we standardize v.c.c. phases so that

$$(a\alpha,\, b\beta|rc^*\gamma^*) = \{abcr\}(b\beta,\, a\alpha|rc^*\gamma^*) \tag{14.7.9}$$

Here $\{abcr\}$ is a phase factor, which we designate the $3j$ phase. See Section 10.2 for a discussion of how $\{abcr\}$ phases are chosen.

Thus implicit in our $3jm$ symmetry rule of (10.2.4) are the v.c.c. phase standardizations of (14.7.5) and (14.7.9). In practice such standardization is easily achieved by defining

$$\begin{pmatrix} a & b & c \\ \alpha & \beta & \gamma \end{pmatrix}^{*\ r}$$

via (10.2.1) for one particular order—say, *abc*—for each, a, b, c set; then all other v.c.c. are found by expressing them in terms of that one, using (10.2.1) and (10.2.4).

In this way all v.c.c. used in the irreducible-tensor method for a given a, b, c set are linked through $3jm$ and are thus properly connected in phase.

Consequently, while all this sounds complicated, in practice one need not ever worry about phase standardization as long as one uses (10.2.1) and (10.2.4) to calculate all the v.c.c. needed from $3jm$.

The Coupling Is Not Associative

We have defined the $3jm$ via $|((ab)c)rA_1a_1)\rangle$ and have as yet not considered whether the coupling is associative. The answer is that it is *not*, since

$$
\begin{aligned}
|(a(bc))rA_1a_1\rangle &= \sum_{\alpha\beta\gamma} |a|^{-1/2}\begin{pmatrix} a \\ \alpha \end{pmatrix}(b\beta, c\gamma|ra^*\alpha^*)|a\alpha\rangle|b\beta\rangle|c\gamma\rangle \\
&= \{a\}\sum_{\alpha\beta\gamma}\begin{pmatrix} b & c & a \\ \beta & \gamma & \alpha \end{pmatrix}^{*\,r} |a\alpha\rangle|b\beta\rangle|c\gamma\rangle \\
&= \{a\}|((bc)a)rA_1a_1\rangle \\
&= \{a\}|((ab)c)rA_1a_1\rangle
\end{aligned}
\qquad (14.7.10)
$$

Here we have used (14.5.3) = (10.2.12), (14.7.3), and (14.7.4). The lack of associativity in the coupling is the result of (14.6.5); in (14.7.3) we are coupling a^* to a to give A_1, while in (14.7.10) we are coupling a to a^* to give A_1. Since (14.6.5) and (14.5.3) give

$$(a\alpha, a^*\alpha^*|A_1a_1) = \{a\}(a^*\alpha^*, a\alpha|A_1a_1) \qquad (14.7.11)$$

the $2j$ factor $\{a\}$ appears in (14.7.10).

It is possible (but not necessary) to define high-symmetry coefficients so that the coupling in the invariant triple direct product is associative. The form of the equation analogous to our (10.2.1) is, however, more involved, and includes an additional phase factor. The V coefficients of Dobosh (1972) and Piepho (1979), which are defined analogously to the $\overline{V}$ coefficients of Fano and Racah (1959), are of this type.

A Note on the Permutation Rules for Nonambivalent Groups

Of interest with respect to the permutation rules above for nonambivalent groups is that if the irreps P_1 and P_2 are treated together as components of a single (degenerate) representation of the group including time reversal, it is not possible to obtain $3jm$ involving such representations which obey (10.2.4) and have the $3j$ phase equal to $+1$ or -1. The P_1 and P_2 irreps are thus always treated by us as two distinct irreps so that high-symmetry $3jm$ may be obtained.

14.8. The Relation between 3*jm* and 3*jm**

The relation between $3jm$ and $3jm^*$ is derived by considering the expression for direct-product functions of conjugate states:

$$\langle (ab)rc\gamma| = \sum_{\alpha\beta} (rc\gamma|a\alpha, b\beta)\langle a\alpha|\langle b\beta|$$

$$= \sum_{\alpha\beta} (a\alpha, b\beta|rc\gamma)^* \langle a\alpha|\langle b\beta|$$

$$= |c|^{1/2} \begin{pmatrix} c \\ \gamma \end{pmatrix} \sum_{\alpha\beta} \begin{pmatrix} a & b & c^* \\ \alpha & \beta & \gamma^* \end{pmatrix}^r \langle a\alpha|\langle b\beta| \qquad (14.8.1)$$

We then use (14.4.5) to obtain the equation

$$\begin{pmatrix} c \\ \gamma \end{pmatrix} |(a^*b^*)rc^*\gamma^*\rangle = |c|^{1/2} \begin{pmatrix} c \\ \gamma \end{pmatrix} \sum_{\alpha\beta} \begin{pmatrix} a & b & c^* \\ \alpha & \beta & \gamma^* \end{pmatrix}^r \begin{pmatrix} a \\ \alpha \end{pmatrix} |a^*\alpha^*\rangle \begin{pmatrix} b \\ \beta \end{pmatrix} |b^*\beta^*\rangle \qquad (14.8.2)$$

If we multiply both sides by $\begin{pmatrix} c \\ \gamma \end{pmatrix}$ and use (10.2.12), we have

$$|(a^*b^*)rc^*\gamma^*\rangle = |c|^{1/2} \begin{pmatrix} c^* \\ \gamma^* \end{pmatrix} \sum_{\alpha\beta} \left[\begin{pmatrix} a & b & c^* \\ \alpha & \beta & \gamma^* \end{pmatrix}^r \begin{pmatrix} a \\ \alpha \end{pmatrix} \begin{pmatrix} b \\ \beta \end{pmatrix} \begin{pmatrix} c^* \\ \gamma^* \end{pmatrix} \right] |a^*\alpha^*\rangle |b^*\beta^*\rangle \qquad (14.8.3)$$

This equation is in every way equivalent to (10.2.3) when written for the same irreps in the same notation. Comparing (10.2.3) with the above, we see that we must have

$$\begin{pmatrix} a^* & b^* & c \\ \alpha^* & \beta^* & \gamma \end{pmatrix}^{*r} = \begin{pmatrix} a \\ \gamma \end{pmatrix} \begin{pmatrix} b \\ \beta \end{pmatrix} \begin{pmatrix} c^* \\ \gamma^* \end{pmatrix} \begin{pmatrix} a & b & c^* \\ \alpha & \beta & \gamma^* \end{pmatrix}^r \qquad (14.8.4)$$

and so the relation between $3jm$ and $3jm^*$ is

$$\begin{pmatrix} a & b & c \\ \alpha & \beta & \gamma \end{pmatrix}^{*r} = \begin{pmatrix} a^* \\ \alpha^* \end{pmatrix}\begin{pmatrix} b^* \\ \beta^* \end{pmatrix}\begin{pmatrix} c^* \\ \gamma^* \end{pmatrix}\begin{pmatrix} a^* & b^* & c^* \\ \alpha^* & \beta^* & \gamma^* \end{pmatrix}^r$$

$$= \begin{pmatrix} a \\ \alpha \end{pmatrix}\begin{pmatrix} b \\ \beta \end{pmatrix}\begin{pmatrix} c \\ \gamma \end{pmatrix}\begin{pmatrix} a^* & b^* & c^* \\ \alpha^* & \beta^* & \gamma^* \end{pmatrix}^r \qquad (14.8.5)$$

We have used (10.2.10) and (10.2.12) and assumed throughout that the $2jm$ are all $+1$ or -1.

Equation (14.8.5) places additional phase restrictions on the $3jm$ and, indirectly through (10.2.1), on all v.c.c. used in the irreducible-tensor method; together these equations require the v.c.c standardization

$$(a\alpha, b\beta|rc\gamma)^* = \begin{pmatrix} a \\ \alpha \end{pmatrix}\begin{pmatrix} b \\ \beta \end{pmatrix}\begin{pmatrix} c \\ \gamma \end{pmatrix}(a^*\alpha^*, b^*\beta^*|rc^*\gamma^*) \qquad (14.8.6)$$

This is equivalent to the requirement that v.c.c. must be chosen so that the coupled functions $|(ab)rc\gamma\rangle$ obey

$$T|(ab)rc\gamma\rangle = |(a^*b^*)rc^*\gamma^*\rangle\begin{pmatrix} c \\ \gamma \end{pmatrix} \qquad (14.8.7)$$

[assuming the convention (14.4.3) is used]. The practical significance of this is demonstrated in Section 14.9.

14.9. How to Find a Set of $3jm$ for a Molecular Point Group

In the following section we show how to find a set of $3jm$ for a molecular point group, using D_3 as an example. While our $3jm$ were actually calculated by Butler using a very elegant but highly abstract method [Butler (1981), Section 3.4], the more accessible method described below also produces a satisfactory set of $3jm$ for a given point-group basis. For our illustration we choose the group D_3 because, as a nonambivalent group, it contains complex irreps. Our procedure, which we present as a multistep process, may be used for both ambivalent and nonambivalent groups. An outline of the steps follows:

1. Define standard basis relations for each irrep in the manner of Section C.5. Then define the basis in terms of the $|jm\rangle$ of $SO_3 \supset SO_2$. Determine the behavior of all basis functions under T from the definition of the basis in terms of the $|jm\rangle$. If the $|jm\rangle$ for integer j are the $\mathcal{Y}^j_m$, do this

using (14.4.3). The resulting equations,

$$T|a\alpha\rangle = |a^*\alpha^*\rangle\binom{a}{\alpha} \tag{14.9.1}$$

define a^*, α^*, $\binom{a}{\alpha}$, and $\{a\}$ for all irreps. [Correspondingly, if the $|jm\rangle$ are the Y^j_m, use (14.4.4) instead.] Examples of the procedure are found in Sections 14.4–14.5.

2. Construct a set of v.c.c., $(a\alpha, b\beta|rc\gamma)$, for the group as described in Section 9.5. However, be sure to choose all $(a\alpha, b\beta|A_1a_1)$ v.c.c. according to (14.6.5). Remember to treat the nonambivalent (complex) P_1- and P_2-type irreps as two distinct, and completely independent, irreps. Check carefully that all coupled functions transform as the standard basis functions defined in step 1. v.c.c. tables are *not* needed for all possible direct products—see step 4 below.

3. Using the definitions of step 1, check to see whether all the coupled functions $|(ab)rc\gamma\rangle$, obtained via (9.3.21) using the tables constructed in step 2, obey (14.8.7); this is equivalent to requiring that the v.c.c. obey (14.8.6). If an $a \otimes b = c$ v.c.c. table fails this test, multiply that table by the appropriate phase factor to form a new $a \otimes b = c$ table which passes the test. In some cases the phase factor necessarily is i or $-i$, and so imaginary v.c.c. result.

The v.c.c. of step 2 are perfectly good ones for all calculations not using $3jm$, $6j$ or $9j$ coefficients (or analogous high-symmetry coefficients), since T is not a point-group symmetry operation. In fact they are often preferable to those of step 3, since they may more easily be chosen real. The additional phase standardization of step 3 is necessary, however, to insure that the $3jm$ defined in step 4 below obey (14.8.5).

4. Choose the $3j$ phases $\{abcr\}$ for the group as discussed in Section 10.2. Use selected v.c.c. tables from step 3 as a basis for defining $3jm$ via the relation (10.2.1):

$$\begin{pmatrix} a & b & c^* \\ \alpha & \beta & \gamma^* \end{pmatrix}^r = \frac{(a\alpha, b\beta|rc\gamma)^*}{|c|^{1/2}\binom{c}{\gamma}} \tag{14.9.2}$$

For each set of irreps a, b, c, the $3jm$ are calculated using the v.c.c. of *one table* (say, $a \otimes b = c$) only. $3jm$ for other orders of the irreps a, b, c, are found via (10.2.4). Note that if two irreps of an a, b, c set are conjugate—say, $a = b^*$—and the third is the totally symmetric (A_1) irrep, the $3jm$ must be calculated using v.c.c. defined by (14.6.5) so that the $3jm$ will obey (10.3.1).

If an irrep c occurs twice in $a \otimes a$ (as do T_1 and T_2 in $U' \otimes U'$ in the group O), it may be necessary to first take linear combinations of the $r = 0$ and $r = 1$ v.c.c. of step 3 above. This is required when c occurs in both $[a^2]$ and (a^2), but not when c occurs twice in $[a^2]$ or twice in (a^2) (see Section 10.2).

An Illustration: Determination of 3jm for D_3

We now illustrate the steps of the last section for D_3 in the complex-D_3 ($D_3 \supset C_3$) basis.

1. Table E.4.1(c) gives standard basis relations for our complex-D_3 basis. In Section E.5 we define the basis in terms of the $|jm\rangle$. These relations enable us to define the $2jm$ and $2j$ phases given in Section E.8.

2–3. v.c.c. for D_3 which satisfy the requirements of steps 2 and 3 are given in Table 14.9.1. Note that P_1 and P_2 are treated as two nondegenerate irreps. We suggest that, as an exercise, the reader check that these v.c.c. obey the standard basis relations of Table E.4.1(c).

Step 3 links the phase of $(a\alpha, b\beta|rc\gamma)^*$ to that of $(a^*\alpha^*, b^*\beta^*|rc^*\gamma^*)$ by requiring that v.c.c. obey (14.8.6). This insures that the $|(ab)rc\gamma\rangle$ formed with the v.c.c. via (9.3.21) obey (14.8.7). For example, (14.8.6) requires in our complex-D_3 basis

$$(P_1\rho_1, P_1\rho_1|A_2a_2') = -(P_2\rho_2, P_2\rho_2|A_2a_2')^* \qquad (14.9.3)$$

Without this step-3 requirement, and without the use of (14.6.5) to define $(a\alpha, b\beta|A_1a_1)$ required in step 2, all v.c.c. coupling nondegenerate irreps (here A_1, A_2, P_1, and P_2) could equally well be chosen equal to 1—such v.c.c. would define $|(ab)rc\gamma\rangle$ kets which transform correctly as standard basis functions [Table E.4.1(c)], as do those defined with the v.c.c. of Table 14.9.1. Also, without the requirement (14.8.6), the phases, for example, of the $E' \otimes E = P_1$ v.c.c. would be independent of those of the $E' \otimes E = P_2$ v.c.c. The further v.c.c. phase standardizations required by steps 2 and 3 are necessary only when v.c.c. are to be used to define high-symmetry coefficients as in (14.9.2).

The v.c.c. phases in Table 14.9.1 have also been chosen so that (14.7.5) is obeyed—a step *not* required by our procedure. We add it to insure conformity [assuming (14.7.9) is also used] between these v.c.c. and the $3jm$ defined in step 4 below. Equation (14.7.5) requires, for example, that

$$(P_1\rho_1, A_2a_2'|P_2\rho_2) = -(P_1\rho_1, P_1\rho_1|A_2a_2') \qquad (14.9.4)$$

Table 14.9.1. Vector Coupling Coefficients $(a\alpha, b\beta|rc\gamma)$ in the Complex-D_3 Basis

$$(A_1a_1, \Gamma\gamma|\Gamma\gamma) = 1 \qquad \text{for all } \Gamma\gamma$$

$$(A_2a_2', A_2a_2'|A_1a_1) = -1$$

$E \otimes A_2$		$E-1$	$E1$
-1	a_2'	1	0
1	a_2'	0	-1

$E \otimes$	E	A_1a_1	A_2a_2'	$E-1$	$E1$
-1	-1	0	0	0	1
-1	1	$1/\sqrt{2}$	$1/\sqrt{2}$	0	0
1	-1	$1/\sqrt{2}$	$-1/\sqrt{2}$	0	0
1	1	0	0	1	0

$E' \otimes A_2$		$E'\beta'$	$E'\alpha'$
β'	a_2'	-1	0
α'	a_2'	0	1

$E' \otimes$	E	$E'\beta'$	$E'\alpha'$	$P_1\rho_1$	$P_2\rho_2$
β'	-1	0	0	$-i/\sqrt{2}$	$-1/\sqrt{2}$
β'	1	0	-1	0	0
α'	-1	1	0	0	0
α'	1	0	0	$1/\sqrt{2}$	$i/\sqrt{2}$

$E' \otimes E'$		A_1a_1	A_2a_2'	$E-1$	$E1$
β'	β'	0	0	1	0
β'	α'	$-1/\sqrt{2}$	$-1/\sqrt{2}$	0	0
α'	β'	$1/\sqrt{2}$	$-1/\sqrt{2}$	0	0
α'	α'	0	0	0	1

$$(P_1\rho_1, A_2a_2'|P_2\rho_2) = i$$

$P_1 \otimes E$		$E'\beta'$	$E'\alpha'$
ρ_1	-1	0	1
ρ_1	1	i	0

$P_1 \otimes E'$		$E-1$	$E1$
ρ_1	β'	0	-1
ρ_1	α'	i	0

$$(P_1\rho_1, P_1\rho_1|A_2a_2') = -i$$

$$(P_2\rho_2, A_2a_2'|P_1\rho_1) = -i$$

$P_2 \otimes E$		$E'\beta'$	$E'\alpha'$
ρ_2	-1	0	$-i$
ρ_2	1	-1	0

$P_2 \otimes E'$		$E-1$	$E1$
ρ_2	β'	0	i
ρ_2	α'	-1	0

$$(P_2\rho_2, P_1\rho_1|A_1a_1) = 1$$

$$(P_2\rho_2, P_2\rho_2|A_2a_2') = -i$$

4. We are now ready to calculate $3jm$ for D_3 corresponding to the v.c.c. given in Table 14.9.1. To achieve this we use (14.9.2) with selected v.c.c. from Table 14.9.1 as described in our step-4 directions. The resulting $3jm$ are given in Section E.9. The $3jm$ for other orders of the irreps a, b, c are found with (10.2.4).

The $\{abc\}$ phases of Section E.8 are chosen using (10.2.8) and Section E.3 when $a = b$. The remaining factors are chosen so that as many as possible obey (10.2.5).

Groups for Which Our Formalism is Applicable

$3jm$ which satisfy our $3jm$ equations may be found using the methods described in this section for all crystallographic point groups and all common molecular point groups. We must only be able to choose the phase factors $\begin{pmatrix} a \\ \alpha \end{pmatrix}$, $\{a\}$, and $\{abcr\}$ for all irreps a of the group so that our defining equations are satisfied and all factors have values of $+1$ or -1 only. This is possible for all point groups of interest to us.

While this procedure works well and is highly illustrative, we urge readers to use instead the Butler (1981) tabulation of $3jm$ factors to determine $3jm$'s not found in our Appendixes—this should produce fewer errors and lead to more standardization of the literature. The method is described in Section 15.3.

14.10. Adjoint Operators $A^\dagger$ and Time-Reversed Operators A^*

Our discussion in Sections 14.2–14.5 shows that we may choose our conjugate kets $|a\alpha\rangle^\dagger = \langle a\alpha|$ to have the same transformation properties as time-reversed kets. Thus for kets, (14.4.3) gives

$$|a\alpha\rangle^\dagger \sim T|a\alpha\rangle = |a^*\alpha^*\rangle\begin{pmatrix} a \\ \alpha \end{pmatrix} \tag{14.10.1}$$

However, the adjoint of an operator $A^\dagger$—or Hermitian conjugate [Messiah (1961), p. 255], or conjugate linear operator [Griffith (1964), p. 206]—which is defined by $\langle a\alpha|A^\dagger|b\beta\rangle = \langle b\beta|A|a\alpha\rangle^*$, is not always equivalent to the complex conjugate or time-reversed operator [see Messiah (1961), p. 671]. For the operators A of interest to us,

$$A^\dagger = \eta A^* \tag{14.10.2}$$

where $\eta = \pm 1$.

The adjoints of the components of a standard vector operator defined as in Section 9.8 are given by the equations

$$V_\mu^\dagger = (-1)^\mu V_{-\mu} \qquad (\mu = 0, 1, -1) \tag{14.10.3}$$

$$V_\mu^\dagger = V_\mu \qquad (\mu = x, y, z) \tag{14.10.4}$$

These equations hold for vector operators of both odd and even parity under inversion. Thus, for example,

$$\begin{aligned} m_0^\dagger &= m_0 \\ m^\dagger_{\pm 1} &= -m_{\mp 1} \\ L_0^\dagger &= L_0 \\ L^\dagger_{\pm 1} &= -L_{\mp 1} \end{aligned} \tag{14.10.5}$$

while

$$\left.\begin{aligned} m_\mu^\dagger &= m_\mu \\ L_\mu^\dagger &= L_\mu \end{aligned}\right\} \quad \text{for} \quad \mu = x, y, z \tag{14.10.6}$$

The complex conjugate (or the time-reversed form) of an operator is found by expressing the operator in Cartesian form and then taking the complex conjugate. It is straightforward to show, for example, that

$$L_\mu^* = -L_\mu \qquad (\mu = x, y, z) \tag{14.10.7}$$

while

$$m_\mu^* = m_\mu \qquad (\mu = x, y, z) \tag{14.10.8}$$

Comparison of the results of (14.10.5)–(14.10.8) together with (9.8.7) gives

$$\begin{aligned} m_\mu^\dagger &= m_\mu^* \\ L_\mu^\dagger &= -L_\mu^* \end{aligned} \tag{14.10.9}$$

for $\mu = 0, 1, -1$, *and* for $\mu = x, y, z$. The behavior of other angular-momentum operators is analogous to that of the L_μ; all have $\eta = -1$ in (14.10.2). The spin–orbit coupling operator $\mathcal{H}_{SO}$ and the electric-dipole operators m_μ have $\eta = 1$.

Hermitian operators are self-adjoint, so by definition

$$A^\dagger = A \tag{14.10.10}$$

From (14.10.5) and (14.10.6) we see that L_μ, S_μ, J_μ, and m_μ for $\mu = x, y, z$ are self-adjoint, but not $L_{\pm 1}$, $S_{\pm 1}$, $J_{\pm 1}$, or $m_{\pm 1}$. The adjoint of a matrix is defined by the equations

$$A^\dagger_{mn} = \langle m|A^\dagger|n\rangle = \langle n|A|m\rangle^* = A^*_{nm} \tag{14.10.11}$$

Thus for a Hermitian (self-adjoint) operator, it follows that

$$\langle m|A|n\rangle = \langle n|A|m\rangle^* \tag{14.10.12}$$

Finally, the adjoint of an operator A multiplied by a constant c is given by

$$(cA)^\dagger = c^* A^\dagger \tag{14.10.13}$$

and the adjoint of an operator product by

$$(AB)^\dagger = B^\dagger A^\dagger \tag{14.10.14}$$

14.11. Use of Time-Reversal Symmetry to Simplify the MCD Equations

Time-reversal symmetry is useful in simplifying MCD equations. To illustrate this we begin by defining the behavior of Kramers pair states and of MCD operators under the time-reversal operation T.

Behavior of Kramers Pair States and MCD *Operators under T*

The Zeeman levels of a state Γ may be divided into Kramers pairs, $|\Gamma\gamma\rangle$ and $|\Gamma\gamma'\rangle$. *Kramers pair states* are by definition diagonal in μ_z and go into one another under time reversal:

$$T|\Gamma\gamma\rangle = c_{\Gamma\gamma}|\Gamma\gamma'\rangle \tag{14.11.1}$$

Here $c_{\Gamma\gamma}$ is a phase factor. When the degeneracy of the state is odd (even-electron system), one of the sublevels is its own partner, so

$$T|\Gamma\gamma\rangle = c_{\Gamma\gamma}|\Gamma\gamma\rangle \tag{14.11.2}$$

If $|\Gamma\gamma\rangle$ transforms as the γ component of irrep Γ, we can write (14.11.1) more specifically as

$$T|\Gamma\gamma\rangle = \begin{pmatrix}\Gamma\\ \gamma\end{pmatrix}|\Gamma^*\gamma^*\rangle \tag{14.11.3}$$

using (14.4.3). In this special case $c_{\Gamma\gamma}$ is simply the $2\,jm$ $\begin{pmatrix}\Gamma\\ \gamma\end{pmatrix}$. However, for the general case we must use (14.11.1) and not assume $\Gamma\gamma$ is an irrep label. The reason for this is that in groups which contain complex irreps it is not always possible to choose the basis so that it is diagonal in μ_z. For example, in D_3, the irreps P_1 and P_2 are complex with $P_1^* = P_2$. Since in the complex-D_3 basis of Section E.4 μ_z transforms as A_2a_2', $\langle P_1\rho_1|\mu_z|P_2\rho_2\rangle$ is nonzero, while the μ_z matrix elements diagonal in $P_1\rho_1$ and $P_2\rho_2$ are zero. Therefore, the Kramers pair states, $|\Gamma\gamma\rangle$ and $|\Gamma\gamma'\rangle$, are *necessarily* linear combinations of $P_1\rho_1$ and $P_2\rho_2$.

Operators have well-defined behavior under the time-reversal operation. The discussion in Section 14.10 leads to

$$\begin{aligned} T\mu_z &= \mu_z^* = -\mu_z \\ Tm_{\pm 1} &= m_{\pm 1}^* = -m_{\mp 1} \end{aligned} \tag{14.11.4}$$

Derivation of Some Results in Section 5.1

We first use (14.11.1), (14.11.2), and (14.11.4) to derive three equations which lead to (5.1.15). Together they show that both members of a Kramers pair, $A\alpha$ and $A\alpha'$, contribute identically to $\mathcal{C}_0$, so $\mathcal{C}_0$ may be calculated with one member of each pair. Later we use similar logic to simplify our MCD equations in a different way for use in Chapter 17 (see below).

Since all quantum-mechanical observables are real numbers, if $|\Gamma\gamma\rangle$ is diagonal in μ_z,

$$T\langle\Gamma\gamma|\mu_z|\Gamma\gamma\rangle = \langle\Gamma\gamma|\mu_z|\Gamma\gamma\rangle \tag{14.11.5}$$

Alternatively, with (14.11.1) and (14.11.4) we have

$$\begin{aligned} T\langle\Gamma\gamma|\mu_z|\Gamma\gamma\rangle &= c_{\Gamma\gamma}^*(-1)c_{\Gamma\gamma}\langle\Gamma\gamma'|\mu_z|\Gamma\gamma'\rangle \\ &= -\langle\Gamma\gamma'|\mu_z|\Gamma\gamma'\rangle \end{aligned} \tag{14.11.6}$$

Together these equations give the well-known result of (5.1.12) that Kramers pairs have equal and opposite first-order Zeeman splittings:

$$\langle \Gamma\gamma|\mu_z|\Gamma\gamma\rangle = -\langle \Gamma\gamma'|\mu_z|\Gamma\gamma'\rangle \tag{14.11.7}$$

When a state is its own partner, so that it obeys (14.11.2),

$$\langle \Gamma\gamma|\mu_z|\Gamma\gamma\rangle = 0 \tag{14.11.8}$$

The remaining equations in Section 5.1 that we wish to derive concern the quantity $\Delta\alpha$ defined in (5.1.9):

$$\Delta\alpha = \sum_{\lambda}\left(\left|\langle A\alpha|m_{-1}|J\lambda\rangle^0\right|^2 - \left|\langle A\alpha|m_1|J\lambda\rangle^0\right|^2\right) \tag{14.11.9}$$

We want to demonstrate that if $A\alpha$ and $A\alpha'$ form a Kramers pair, then [Eq. (5.1.13)]

$$\Delta\alpha = -\Delta\alpha' \tag{14.11.10}$$

and that if $|A\alpha\rangle$ obeys (14.11.2), then [Eq. (5.1.14)]

$$\Delta\alpha = 0 \tag{14.11.11}$$

To do this we first note that a definite integral is simply a (possibly complex) number of magnitude k. Thus

$$\begin{aligned} Tk &= k^* \\ |Tk|^2 &= |k^*|^2 = |k|^2 \end{aligned} \tag{14.11.12}$$

Applying the above to the integrals of interest gives

$$\begin{aligned} \left|\langle A\alpha|m_{\pm 1}|J\lambda\rangle^0\right|^2 &= \left|T\langle A\alpha|m_{\pm 1}|J\lambda\rangle^0\right|^2 \\ &= \left|c_{A\alpha}^*(-1)c_{J\lambda}\langle A\alpha'|m_{\mp 1}|J\lambda'\rangle^0\right|^2 \\ &= \left|\langle A\alpha'|m_{\mp 1}|J\lambda'\rangle^0\right|^2 \end{aligned} \tag{14.11.13}$$

Here we require that $|A\alpha\rangle$ and $|J\lambda\rangle$ obey (14.11.1) (but they need not be diagonal in μ_z). Substituting (14.11.13) into (14.11.9), we obtain (14.11.10):

$$\Delta\alpha = \sum_{\lambda}\left(|\langle A\alpha'|m_1|J\lambda'\rangle^0|^2 - |\langle A\alpha'|m_{-1}|J\lambda'\rangle^0|^2\right)$$
$$= -\Delta\alpha' \tag{14.11.14}$$

The last equality holds because the sum over λ includes the λ′. Thus λ and λ′ are dummy labels in (14.11.9) and (14.11.14). Equation (14.11.11) follows from (14.11.14) when $|A\alpha\rangle$ is its own partner; then (14.11.14) gives $\Delta\alpha = -\Delta\alpha$, so clearly $\Delta\alpha = 0$. Equation (4.5.18) also follows from (14.11.14).

Simplification of MCD *Equations for Use in Chapter 17*

We next use time-reversal symmetry to obtain identities which simplify our Chapter 4 MCD equations for use in Chapter 17. We begin with (4.5.16) for $\mathcal{A}_1$, $\mathcal{B}_0$, and $\mathcal{C}_0$ and (4.5.23) for $\mathcal{D}_0$. We show that the terms in $\mathcal{A}_1$, $\mathcal{C}_0$, and $\mathcal{D}_0$ containing $|\langle A\alpha|m_{-1}|J\lambda\rangle|^2$ give a total contribution identical to that of the terms containing $|\langle A\alpha|m_1|J\lambda\rangle|^2$, so that only one set need actually be calculated. An equivalent statement is true for the analogous terms in $\mathcal{B}_0$. However, unlike the (5.1.15) result, here *both* members of each Kramers pair must be used.

The $\mathcal{D}_0$ simplification, which follows immediately from (14.11.13), has already been given as the first line of (4.5.23). The proof for $\mathcal{C}_0$ is straightforward. We begin with (4.5.16) and use (14.11.7) and (14.11.13) in the first term. We assume the $|A\alpha\rangle$ are diagonal in μ_z and belong to Kramers pairs $|A\alpha\rangle$, $|A\alpha'\rangle$. The $|J\lambda\rangle$ are also chosen to obey (14.11.1), but need not be diagonal in μ_z. Thus

$$\mathcal{C}_0 = \frac{1}{\mu_B|A|}\left(\sum_{\alpha\lambda}\langle A\alpha|\mu_z|A\alpha\rangle|\langle A\alpha|m_{-1}|J\lambda\rangle|^2 - \sum_{\alpha\lambda}\langle A\alpha|\mu_z|A\alpha\rangle|\langle A\alpha|m_1|J\lambda\rangle|^2\right)$$
$$= \frac{1}{\mu_B|A|}\left(\sum_{\alpha\lambda}(-1)\langle A\alpha'|\mu_z|A\alpha'\rangle|\langle A\alpha'|m_1|J\lambda'\rangle|^2 - \sum_{\alpha\lambda}\langle A\alpha|\mu_z|A\alpha\rangle|\langle A\alpha|m_1|J\lambda\rangle|^2\right)$$
$$= \frac{-2}{\mu_B|A|}\sum_{\alpha\lambda}\langle A\alpha|\mu_z|A\alpha\rangle|\langle A\alpha|m_1|J\lambda\rangle|^2 \tag{14.11.15}$$

Much as in (14.11.14), the last equality holds because the sum over α and λ includes α' and λ'. The proof for $\mathcal{A}_1$ is analogous but requires the $|J\lambda\rangle$ to be Kramers pair states. To obtain the result for $\mathcal{B}_0$, we first note that terms contributing to $\mathcal{B}_0$ are real numbers and thus are invariant under time reversal. We apply T to all $\mu_z m_{-1} m_1$-type terms and find their sum is identical to the $\mu_z m_1 m_{-1}$ sum. Thus $\mathcal{B}_0$ may be calculated as twice the sum of the $\mu_z m_1 m_{-1}$ terms.

While our arguments above are based on $\mathcal{C}_0$ and $\mathcal{A}_1$ from (4.5.16), the results may be expressed in basis-invariant form as in (4.6.8). Such a form is preferred for most applications, since it does not require the $|A\alpha\rangle$ or $|J\lambda\rangle$ to be diagonal in μ_z. For example, (14.11.15) may be written in basis-invariant form as

$$\mathcal{C}_0 = \frac{2}{\mu_B|A|} \sum_{\alpha\alpha''\lambda} \langle A\alpha|\mu_z|A\alpha''\rangle\langle A\alpha''|m_1|J\lambda\rangle\langle J\lambda|m_{-1}|A\alpha\rangle \tag{14.11.16}$$

In obtaining this result we have employed (14.10.11) and (14.10.5). Also

$$\mathcal{A}_1 = \frac{-2}{\mu_B|A|} \sum_{\alpha\alpha''\lambda\lambda''} (\langle J\lambda|\mu_z|J\lambda''\rangle\delta_{\alpha\alpha''} - \langle A\alpha|\mu_z|A\alpha''\rangle\delta_{\lambda\lambda''})$$
$$\times\langle A\alpha''|m_1|J\lambda\rangle\langle J\lambda''|m_{-1}|A\alpha\rangle \tag{14.11.17}$$

The labels α'' and λ'' are used [rather than α' and λ' as in (4.6.8)] to avoid confusion with the Kramers-pair notation above.

When real basis sets are used, it is convenient to express $\mathcal{A}_1$, $\mathcal{B}_0$, $\mathcal{C}_0$, and $\mathcal{D}_0$ in terms of real operators. Recall that $\mu_0 = \mu_z$ and $m_{\pm 1} = \mp(1/\sqrt{2})\times(m_x \pm im_y)$. Thus in terms of real operators (4.5.16) becomes

$$\mathcal{C}_0 = \frac{i}{\mu_B|A|}\sum_{\alpha\lambda}\langle A\alpha|\mu_z|A\alpha\rangle$$
$$\times\left(\langle A\alpha|m_x|J\lambda\rangle\langle J\lambda|m_y|A\alpha\rangle - \langle A\alpha|m_y|J\lambda\rangle\langle J\lambda|m_x|A\alpha\rangle\right) \tag{14.11.18}$$

Then since m_x and m_y are self-adjoint, (14.10.12) may be used to condense (14.11.18) to

$$\mathcal{C}_0 = \frac{-2}{\mu_B|A|}\sum_{\alpha}\langle A\alpha|\mu_z|A\alpha\rangle \mathfrak{Im} \sum_{\lambda}\langle A\alpha|m_x|J\lambda\rangle\langle J\lambda|m_y|A\alpha\rangle \tag{14.11.19}$$

This expression may be further simplified, since any nonimaginary contributions to the sum over λ sum to zero; as a result of (14.11.7), these contributions are equal and opposite for the λ sum for each Kramers pair. Thus

$$\mathcal{C}_0 = \frac{2i}{\mu_B|A|}\sum_{\alpha\lambda}\langle A\alpha|\mu_z|A\alpha\rangle\langle A\alpha|m_x|J\lambda\rangle\langle J\lambda|m_y|A\alpha\rangle \quad (14.11.20)$$

This equation may also be written in invariant form.

Simplifified MCD equations for both real and complex operators derived as discussed above are given in (17.2.2)–(17.2.7) in basis-invariant form. These, of course, apply only for the oriented or isotropic case.

15 The Chain-of-Groups Approach to Symmetry Calculations

15.1. Introduction

In this chapter we describe some features of the Butler (1981) approach to group symmetry calculations. Our purpose is twofold. First of all we wish to make Dr. Butler's excellent book, *Point Group Symmetry Applications, Methods and Tables*, a useful companion to our text. Butler takes an elegant and unconventional approach to the calculation of symmetry coefficients and assumes a high level of sophistication on the part of his audience. Thus we feel an introduction to his methods and language will make his book more accessible. Secondly, we want to introduce readers to $2jm$ and $3jm$ factors, since they prove very useful in our calculations.

Butler uses a chain-of-groups approach to symmetry calculations. In practical applications this approach yields the greatest dividends when a system is slightly distorted so that it belongs to a subgroup (e.g. group O) of a higher group (e.g. the group of all rotations, SO_3). The Butler method allows direct calculation of the subgroup (O) matrix elements in terms of the higher-group (SO_3) matrix elements (Section 15.4). Thus the method is particularly useful for weak-field calculations and for all calculations involving spin-only and spin–orbit coupling operators.

Butler also describes briefly more advanced applications of his method. Readers interested in group theory should see in particular Chapter 7 of Butler (1981), which discusses the calculation of coefficients of fractional parentage using the chain-of-groups approach. We use this method in Section 19.9 to calculate cfp for t_1^n and t_2^n for the cubic groups.

In the calculation of group symmetry coefficients, Butler's approach is very different from that of Sections 9.5 and 14.9. He begins by defining the basis in terms of a group–subgroup chain (Section 15.2). The $6j$ coefficients are calculated, and then the $3jm$ factors (Section 15.3) which link the group to its first subgroup. $3jm$ coefficients are not tabulated at all (except when they are themselves simply $3jm$ factors), but are found by taking products of the $3jm$ factors appropriate for the group–subgroup scheme (Section 15.3).

15.2. Group–Subgroup Chains

The concept of a group–subgroup chain is central to symmetry calculations following Butler (1981). A group G_1 is a subgroup of a higher group G if all symmetry elements of the lower group G_1 also occur in G. A given group G may have more than one possible subgroup, and each subgroup in turn may have several different subgroups of its own. Each different choice of a group–subgroup chain, $G \supset G_1 \supset G_2 \supset \cdots \supset C_1$, corresponds to a different basis choice for the group G. Specification of a group–subgroup chain for G, together with the character tables for each group in the chain, defines the transformation properties of each component (partner) of each irrep of G for the basis. Thus specification of the chain gives nearly as much information about the basis as the transformation of the basis under group generators. The chain by itself does not completely define the basis functions, since the precise relation of the irrep partners to the x, y, and z coordinates (or to the $SO_3 \supset SO_2 \mid jm\rangle$) is not fixed by specifying the chain. Using the Butler (1981) approach, the final steps in the definition of basis functions comes *last* (Section 15.5).

The bases we give in Sections C.5, D.4, and E.4 have been defined both in terms of transformations under group generators (except for the O and T_d trigonal bases of Section E.4) and in terms of group–subgroup chains. Thus, for example, our real-tetragonal–octahedral (real-O) basis of Section C.5 is specified by the chain $O \supset D_4 \supset D_2 \supset C_2 \supset C_1$, and our complex-tetragonal–octahedral (complex-O) basis by the chain $O \supset D_4 \supset C_4 \supset C_1$. Then the real- and complex-trigonal–octahedral bases of Section E.4 are specified by the chains $O \supset D_3 \supset C_2 \supset C_1$ and $O \supset D_3 \supset C_3 \supset C_1$ respectively. In practice, the group chains contain no new information after the first C_n or S_n group is reached, since such groups have all one-dimensional irreps.‡ Thus we may truncate the chains after the first C_n or S_n (or $SO_2 \equiv C_\infty$) group is reached and label the first two basis chains above, for example, as $O \supset D_4 \supset D_2 \supset C_2$ and $O \supset D_4 \supset C_4$. In later discussion we regard the last step in the truncated chain as the last step in the chain. Chapter 12 of Butler (1981) shows diagramatically the group chains for the pure rotation and the crystallographic point groups.

How Basis States are Labeled by Specifying Irreps along the Chain

Butler labels basis states $|aa_1a_2 \cdots\rangle$ by specifying the irrep to which a state belongs for each group along the group–subgroup chain. As discussed above, the chain may be truncated at the first cyclic group. Thus a function belonging to the chain $O \supset D_4 \supset C_4$ would be labeled as $|aa_1a_2\rangle = |a(O)a_1(D_4)a_2(C_4)\rangle$ by determining the decomposition of the function along the group chain. Butler gives the decompositions in his branching-rule tables [Butler (1981), Chapter 12] using his own numerical irrep-labeling system. For the $O \supset D_4 \supset C_4$ chain these labels relate to the alphabetic Mulliken labels as follows:

$$
\begin{array}{ll}
O & \left\{\begin{array}{cccccccc} A_1 & A_2 & E & T_1 & T_2 & E' & E'' & U' \\ 0 & \tilde{0} & 2 & 1 & \tilde{1} & \frac{1}{2} & \tilde{\frac{1}{2}} & \frac{3}{2} \end{array}\right. \\[2ex]
D_4 & \left\{\begin{array}{ccccccc} A_1 & A_2 & B_1 & B_2 & E & E' & E'' \\ 0 & \tilde{0} & 2 & \tilde{2} & 1 & \frac{1}{2} & \frac{3}{2} \end{array}\right. \\[2ex]
C_4 & \left\{\begin{array}{cccccccc} & & \multicolumn{2}{c}{E} & \multicolumn{2}{c}{E'} & \multicolumn{2}{c}{E''} \\ A & B & P_1 & P_2 & P_1' & P_2' & P_1'' & P_2'' \\ 0 & 2 & 1 & -1 & -\frac{1}{2} & \frac{1}{2} & -\frac{3}{2} & \frac{3}{2} \end{array}\right.
\end{array}
\qquad (15.2.1)
$$

Note that Butler (correctly) gives separate irrep labels to the C_4 paired irreps. The P_1, P_2 notation is ours (Section 14.2).

For the T_1 irrep in the $O \supset D_4 \supset C_4$ chain the decomposition is

$$
\begin{array}{llcl}
O & & 1(T_1) & \\
D_4 & \tilde{0}(A_2) & + & 1(E) \\
C_4 & 0(A_1) + & 1(P_1) + & -1(P_2)
\end{array}
\qquad (15.2.2)
$$

Thus the three T_1 basis functions are labeled $|1\tilde{0}0\rangle$, $|111\rangle$, and $|11-1\rangle$. These functions are defined in Table C.5.1(c) (and also in Section C.6) to

‡Many authors show two-dimensional representations for the nonambivalent cyclic groups in their tables, but these are actually all reducible into sets of paired irreps. These reducible representations should properly be broken up into pairs of separate, nondegenerate irreps, since the partners *do not* mix under *any* group operation, but only under time reversal. We always do this, and so does Butler.

be, respectively, our $|T_1 0\rangle$, $|T_1 1\rangle$, and $|T_1\ -1\rangle$ complex-O basis functions. Each one of our bases is defined to be one of the Butler (1981) basis chains.

As another example, consider how a spin basis may be defined for distorted atoms or molecules when spin–orbit coupling is important. If spin–orbit coupling is neglected, the $SO_3 \supset SO_2$ ($|\mathbb{S}\mathfrak{M}\rangle$) spin basis is appropriate, but when spin–orbit interactions must be calculated, a symmetry-group-adapted SO_3 spin basis must be used (Section 12.2). In Section 12.2 we call this the $|\mathbb{S}SM\rangle$ basis where $\mathbb{S}$ is again an SO_3 label, and S and M are irrep and irrep-partner labels in a basis of the symmetry group G of the system; thus in this notation the three $\mathbb{S} = 1$ triplet spin functions in the complex-O-adapted SO_3 basis are labeled $|\mathbb{S}SM\rangle = |1\ T_1 0\rangle$, $|1\ T_1 1\rangle$, and $|1\ T_1 -1\rangle$. Using the Butler chain notation and (15.2.2) above, we would label these same spin states as the $SO_3 \supset O \supset D_4 \supset C_4$ kets $|\mathbb{S}SS_1S_2\rangle = |1\,1\,\tilde{0}\,0\rangle$, $|1\,1\,1\,1\rangle$, and $|1\,1\,1\,-1\rangle$, respectively (Section C.6 essentially gives this spin basis in both notations).

When the group decomposition of an irrep gives rise to a *branching multiplicity* (as in the decomposition of the $J = 5$ irrep of SO_3 to group O, which yields $E + 2T_1 + T_2$, and so has two T_1 branches), a further label is needed to distinguish between the branches. Since branching multiplicities do not occur in the examples of interest to us, we do not generalize our formulas to include this label.

15.3. The Racah Factorization Lemma and the Factoring of Coupling Coefficients

Since in the Butler chain notation a basis state $|a\alpha\rangle$ of the chain $G \supset G_1 \supset C_n$ is written as $|aa_1a_2\rangle = |a(G)a_1(G_1)a_2(C_n)\rangle$, in chain notation (10.2.1) becomes

$$(aa_1a_2, bb_1b_2|rcc_1c_2)^{G\supset G_1\supset C_n}$$

$$= H(abc)|c|^{1/2}\begin{pmatrix} c \\ c_1 \\ c_2 \end{pmatrix}\begin{matrix} G \\ G_1 \\ C_n \end{matrix}\begin{pmatrix} a & b & c^* \\ a_1 & b_1 & c_1^* \\ a_2 & b_2 & c_2^* \end{pmatrix}^{*}\begin{matrix} rG \\ G_1 \\ C_n \end{matrix} \qquad (15.3.1)$$

Here we have explicitly included the historical phase $H(abc)$. When $G = SO_3$ or O_3, $H(abc) = H(j_1j_2j_3) = (-1)^{-j_1+j_2-j_3}$; otherwise we define $H(abc) = 1$. Equation (15.3.1) is generalized for a $G \supset G_1 \supset \cdots \supset C_n$ basis by simply inserting further irrep labels between the G_1 and C_n labels.

The *Racah factorization lemma* allows us to factor the $G \supset G_1 \supset C_n$ v.c.c. in (15.3.1). The lemma says

$$(aa_1a_2, bb_1b_2|rcc_1c_2)^{G\supset G_1\supset C_n}$$
$$= \sum_{r_1}(aa_1, bb_1|rcc_1)_{r_1}^{G\supset G_1}(a_1a_2, b_1b_2|r_1c_1c_2)^{G_1\supset C_n} \qquad (15.3.2)$$

A proof is given in Butler (1981), Section 2.3. The v.c.c. with $G \supset G_1 \supset C_n$ and $G_1 \supset C_n$ superscripts are, respectively, v.c.c. for the $G \supset G_1 \supset C_n$ and the $G_1 \supset C_n$ bases. The *vector coupling factor* (or *isoscalar factor*) labeled $G \supset G_1$ is defined by this equation—note that it is independent of group-G_1 irrep-partner labels (a_2, b_2, c_2). Equation (10.2.1) is then generalized to define $G \supset G_1$ *3jm factors* in terms of these $G \supset G_1$ coupling factors:

$$(aa_1, bb_1|rcc_1)_{r_1}^{G\supset G_1} = H(abc)\frac{|c|^{1/2}}{|c_1|^{1/2}}\begin{pmatrix} c \\ c_1\end{pmatrix}\begin{matrix} G \\ G_1\end{matrix}\begin{pmatrix} a & b & c^* \\ a_1 & b_1 & c_1^*\end{pmatrix}^{*}\begin{matrix} rG \\ r_1G_1\end{matrix} \qquad (15.3.3)$$

When the group basis chain contains only one step (as in $D_4 \supset C_4$ or $SO_3 \supset SO_2$), (15.3.3) is identical to (10.2.1), since then $|c_1| = 1$. Thus *for the last step in a chain* (i.e., the step arriving at the first C_n or S_n group), *the* $3jm(2jm)$ *factor is a* $3jm(2jm)$ *and vice versa.* So for one-step bases, the $3jm$ *are* $3jm$ factors. Recall that $H(abc)$ must also be included in (10.2.1) when $G = SO_3$ or O_3 (Section 10.2). [For the case where $G \supset G_1$ is $O_3 \supset SO_3$, we define $H(abc) = 1$ in (15.3.3); it occurs only for $SO_3 \supset SO_2$ in the $O_3 \supset SO_3 \supset SO_2$ chain.]

If we use (15.3.1) and (15.3.3) to write (15.3.2) in terms of $3jm$ and $3jm$ factors, we obtain

$$\begin{pmatrix} c \\ c_1 \\ c_2\end{pmatrix}\begin{matrix} G \\ G_1 \\ C_n\end{matrix}\begin{pmatrix} a & b & c^* \\ a_1 & b_1 & c_1^* \\ a_2 & b_2 & c_2^*\end{pmatrix}^{*}\begin{matrix} rG \\ G_1 \\ C_n\end{matrix}$$
$$= \begin{pmatrix} c \\ c_1\end{pmatrix}\begin{matrix} G \\ G_1\end{matrix}\begin{pmatrix} c_1 \\ c_2\end{pmatrix}\begin{matrix} G_1 \\ C_n\end{matrix}\sum_{r_1}\begin{pmatrix} a & b & c^* \\ a_1 & b_1 & c_1^*\end{pmatrix}^{*}\begin{matrix} rG \\ r_1G_1\end{matrix}\begin{pmatrix} a_1 & b_1 & c_1^* \\ a_2 & b_2 & c_2^*\end{pmatrix}^{*}\begin{matrix} r_1G_1 \\ C_n\end{matrix} \qquad (15.3.4)$$

From (15.3.4) it follows that we may write 2 *jm* and 3*jm* in factored form:

$$\underbrace{\begin{pmatrix} c \\ \gamma \end{pmatrix} \equiv \begin{pmatrix} c \\ c_1 \\ c_2 \end{pmatrix}\begin{matrix} G \\ G_1 \\ C_n \end{matrix}}_{\substack{\text{group-}G \\ 2\,jm}} = \underbrace{\begin{pmatrix} c \\ c_1 \end{pmatrix}\begin{matrix} G \\ G_1 \end{matrix}}_{\substack{G\supset G_1 \\ 2\,jm\text{ factor}}} \underbrace{\begin{pmatrix} c_1 \\ c_2 \end{pmatrix}\begin{matrix} G_1 \\ C_n \end{matrix}}_{\substack{\text{group-}G_1 \\ 2\,jm}} = \begin{pmatrix} c_1 \\ c_2 \end{pmatrix}\begin{matrix} G_1 \\ C_n \end{matrix} \tag{15.3.5}$$

$$\underbrace{\begin{pmatrix} a & b & c \\ \alpha & \beta & \gamma \end{pmatrix}^{r} \equiv \begin{pmatrix} a & b & c \\ a_1 & b_1 & c_1 \\ a_2 & b_2 & c_2 \end{pmatrix}\begin{matrix} rG \\ G_1 \\ C_n \end{matrix}}_{\substack{\text{group-}G \\ 3jm}}$$

$$= \underbrace{\sum_{r_1} \begin{pmatrix} a & b & c \\ a_1 & b_1 & c_1 \end{pmatrix}\begin{matrix} rG \\ r_1G_1 \end{matrix}}_{\substack{G\supset G_1 \\ 3jm\text{ factor}}} \underbrace{\begin{pmatrix} a_1 & b_1 & c_1 \\ a_2 & b_2 & c_2 \end{pmatrix}\begin{matrix} r_1G_1 \\ C_n \end{matrix}}_{\substack{\text{group-}G_1 \\ 3jm}} \tag{15.3.6}$$

These are Butler (1981), Eqs. (3.1.8) and (3.1.9), but with branching-multiplicity labels omitted. The last equality in (15.3.5) occurs because *the* 2 *jm of Butler* (1981) *are defined so that for any group chain, all but the last* 2 *jm are unity*; thus the 2 *jm* phase, which is always ± 1, is governed completely by the final factor. Equations in the form of (15.3.5)–(15.3.6) may also be used to factor 2 *jm* and 3*jm factors*, such as the $G \supset G_1 \supset G_2$ 3*jm* factor in a $G \supset G_1 \supset G_2 \supset C_n$ chain—see (15.3.8).

The permutation rule for 3*jm* factors follows directly from (15.3.6) and (10.2.4):

$$\begin{pmatrix} a & b & c \\ a_1 & b_1 & c_1 \end{pmatrix}\begin{matrix} rG \\ r_1G_1 \end{matrix} = \{abcr\}\{a_1b_1c_1r_1\}\begin{pmatrix} b & a & c \\ b_1 & a_1 & c_1 \end{pmatrix}\begin{matrix} rG \\ r_1G_1 \end{matrix} \tag{15.3.7}$$

Equations (15.3.5)–(15.3.7) generalize in a straightforward manner for longer chains $G \supset G_1 \supset G_2 \supset \cdots \supset C_n$. Thus for the real-$O$ basis chain $O \supset D_4 \supset D_2 \supset C_2$ (since in D_4, D_2, and C_2 we are restricted to $r = 0$), we have

$$\begin{pmatrix} a & b & c \\ a_1 & b_1 & c_1 \\ a_2 & b_2 & c_2 \\ a_3 & b_3 & c_3 \end{pmatrix}\begin{matrix} rO \\ D_4 \\ D_2 \\ C_2 \end{matrix} = \begin{pmatrix} a & b & c \\ a_1 & b_1 & c_1 \end{pmatrix}\begin{matrix} rO \\ D_4 \end{matrix}\begin{pmatrix} a_1 & b_1 & c_1 \\ a_2 & b_2 & c_2 \end{pmatrix}\begin{matrix} D_4 \\ D_2 \end{matrix}$$

$$\times \begin{pmatrix} a_2 & b_2 & c_2 \\ a_3 & b_3 & c_3 \end{pmatrix}\begin{matrix} D_2 \\ C_2 \end{matrix} \qquad (15.3.8)$$

Some chains containing T_d and its subgroups end with S_n or C_s, rather than with C_n—see Sections C.5, D.4, and E.4—but the analogous equations apply.

Calculation of 2jm and 3jm Coefficients as a Product of 2jm and 3jm Factors

Because $2jm$ and $3jm$ may be factored as illustrated by (15.3.5)–(15.3.6), it is unnecessary to list the $2jm$ and $3jm$ for a given chain directly; rather only the $2jm$ and $3jm$ factors need be tabulated, and Butler (1981) gives only these factors. We have used his factors to obtain all our $3jm$. Thus our group-O and group-T_d $3jm$, given in Section C.12, were calculated from these Butler $3jm$ factors. For example, in the complex-O basis Section C.12 gives

$$\begin{pmatrix} T_2 & T_1 & E \\ -1 & 1 & \theta \end{pmatrix} \equiv \begin{pmatrix} \tilde{1} & 1 & 2 \\ 1 & 1 & 0 \\ -1 & 1 & 0 \end{pmatrix}\begin{matrix} O \\ D_4 \\ C_4 \end{matrix} = -\frac{1}{2} \qquad (15.3.9)$$

where we have used Table C.5.1(c) (or Section C.6) to write the irreps in chain notation. The value of the $3jm$ was calculated as follows, using (15.3.6) and the Butler $O \supset D_4$ and $D_4 \supset C_4$ $3jm$ factors:

$$\begin{pmatrix} \tilde{1} & 1 & 2 \\ 1 & 1 & 0 \\ -1 & 1 & 0 \end{pmatrix}\begin{matrix} O \\ D_4 \\ C_4 \end{matrix} = \begin{pmatrix} \tilde{1} & 1 & 2 \\ 1 & 1 & 0 \end{pmatrix}\begin{matrix} O \\ D_4 \end{matrix}\begin{pmatrix} 1 & 1 & 0 \\ -1 & 1 & 0 \end{pmatrix}\begin{matrix} D_4 \\ C_4 \end{matrix}$$

$$= \left(\frac{-1}{\sqrt{2}}\right)\left(\frac{1}{\sqrt{2}}\right) = -\frac{1}{2} \qquad (15.3.10)$$

In this case the required $3jm$ factors are also given in our appendixes.

However, Section C.6 [or (15.2.1)] must first be used to translate the Butler chain notation to that of Sections D.12 and D.9; thus, for example, $|\tilde{1}\, 1\, -1\rangle = |T_2\, E\, -1\rangle$, so (15.3.10) becomes

$$\begin{pmatrix} T_2 & T_1 & E \\ E & E & A_1 \\ -1 & 1 & a_1 \end{pmatrix} \begin{matrix} O \\ D_4 \\ C_4 \end{matrix} = \begin{pmatrix} T_2 & T_1 & E \\ E & E & A_1 \end{pmatrix} \begin{matrix} O \\ D_4 \end{matrix} \begin{pmatrix} E & E & A_1 \\ -1 & 1 & a_1 \end{pmatrix} \begin{matrix} D_4 \\ C_4 \end{matrix} \tag{15.3.11}$$

Then the (15.3.10) result is obtained with the table of $O \supset D_4$ $3jm$ factors in Section D.12 and the $D_4 \supset C_4$ $3jm$ in Section D.9.

15.4. Evaluating Reduced Matrix Elements in a Group–Subgroup Scheme

If a group–subgroup basis is used as described in Section 15.2, the Wigner–Eckart theorem may be applied in either group. Using (10.2.2) and the notation of the previous sections, we obtain

$$\begin{aligned} \langle aa_1a_2|O^{ff_1}_{f_2}|bb_1b_2\rangle &= \begin{pmatrix} a_1 \\ a_2 \end{pmatrix} \begin{matrix} G_1 \\ C_n \end{matrix} \sum_{r_1} \begin{pmatrix} a_1^* & f_1 & b_1 \\ a_2^* & f_2 & b_2 \end{pmatrix} \begin{matrix} r_1G_1 \\ C_n \end{matrix} \langle aa_1\|O^{ff_1}\|bb_1\rangle^{G_1}_{r_1} \\ &= \langle aa_1a_2|O^f_{f_1f_2}|bb_1b_2\rangle \\ &= \begin{pmatrix} a \\ a_1 \\ a_2 \end{pmatrix} \begin{matrix} G \\ G_1 \\ C_n \end{matrix} \sum_r \begin{pmatrix} a^* & f & b \\ a_1^* & f_1 & b_1 \\ a_2^* & f_2 & b_2 \end{pmatrix} \begin{matrix} rG \\ G_1 \\ C_n \end{matrix} \langle a\|O^f\|b\rangle^G_r \end{aligned} \tag{15.4.1}$$

Here the operator $O^{ff_1}_{f_2}$ is *identical* to $O^f_{f_1f_2}$; the notation simply serves to separate irrep labels (superscripts) from partner labels (subscripts) *in the basis in which reduction is being performed*. The $2jm$ and $3jm$ of the last line may be factored according to (15.3.5) and (15.3.6) respectively, so that (15.4.1) becomes

$$\begin{aligned} &= \begin{pmatrix} a \\ a_1 \end{pmatrix} \begin{matrix} G \\ G_1 \end{matrix} \begin{pmatrix} a_1 \\ a_2 \end{pmatrix} \begin{matrix} G_1 \\ C_n \end{matrix} \sum_r \sum_{r_1} \begin{pmatrix} a^* & f & b \\ a_1^* & f_1 & b_1 \end{pmatrix} \begin{matrix} rG \\ r_1G_1 \end{matrix} \\ &\times \begin{pmatrix} a_1^* & f_1 & b_1 \\ a_2^* & f_2 & b_2 \end{pmatrix} \begin{matrix} r_1G_1 \\ C_n \end{matrix} \langle a\|O^f\|b\rangle^G_r \end{aligned} \tag{15.4.2}$$

From these equalities it follows that

$$\langle aa_1 \| O^{ff_1} \| bb_1 \rangle^{G_1}_{r_1}$$

$$= \begin{pmatrix} a \\ a_1 \end{pmatrix}^{G}_{G_1} \sum_r \begin{pmatrix} a^* & f & b \\ a_1^* & f_1 & b_1 \end{pmatrix}^{rG}_{r_1 G_1} \langle a \| O^f \| b \rangle^G_r \tag{15.4.3}$$

The $G \supset G_1$ $2jm$ factor is unity unless $G_1 = C_n$, S_n, or SO_2 [and then (15.4.3) is really inapplicable, since (10.2.2) applies instead]. For chains $G \supset \cdots \supset G_1 \supset C_n$, the $2jm$ and $3jm$ $G \supset \cdots \supset G_1$ factors may themselves be factored as illustrated in (15.3.8).

Example: Orbital-Angular-Momentum Matrix Elements

An important use of (15.4.3) is to evaluate matrix elements in a weak-field basis. For example, in cubic systems the orbital parts of weak-field $\mathcal{SL}$ basis states belong to basis chains $O_3 \supset O_h \supset O \supset \cdots$ or $O_3 \supset O_h \supset T_d \supset \cdots$, where the end of the chain depends on the particular basis scheme employed (Sections C.5 and E.4 give examples). We may write these $\mathcal{SL}$ basis states in chain notation as $|l^n\, \mathcal{S}^+(O_3)\mathcal{L}^\pm(O_3)h^\pm(O_h)h_1(G_1)\cdots\rangle$, where $G_1 = T_d$ or O, and where $+$ applies for even-parity (g) states and $-$ for odd-parity (u) states in O_3 and O_h. Thus in the weak-field basis for $G_1 = T_d$, (15.4.3) gives for the orbital-angular-momentum reduced matrix elements between states of configuration l^n of identical spin $\mathcal{S}^+$

$$\langle h_1 \| L^{t_1} \| h_1' \rangle^{T_d}$$

$$= \langle \mathcal{L}^\pm h^\pm h_1 \| L^{1^+ t_{1g} t_1} \| \mathcal{L}'^\pm h'^\pm h_1' \rangle^{T_d}$$

$$= \begin{pmatrix} \mathcal{L}^\pm & 1^+ & \mathcal{L}'^\pm \\ h^\pm & T_{1g} & h'^\pm \end{pmatrix}^{O_3}_{O_h} \begin{pmatrix} h^\pm & T_{1g} & h'^\pm \\ h_1 & T_1 & h_1' \end{pmatrix}^{O_h}_{T_d} \langle \mathcal{L}^\pm \| L^{1^+} \| \mathcal{L}'^\pm \rangle^{O_3}$$

$$= \delta_{\mathcal{L}\mathcal{L}'} \begin{pmatrix} \mathcal{L} & 1 & \mathcal{L} \\ h & T_1 & h' \end{pmatrix}^{SO_3}_{O} \begin{pmatrix} h^\pm & T_{1g} & h'^\pm \\ h_1 & T_1 & h_1' \end{pmatrix}^{O_h}_{T_d} \sqrt{\mathcal{L}(\mathcal{L}+1)(2\mathcal{L}+1)} \tag{15.4.4}$$

In the last line we use the isomorphism of $O_3 \supset O_h$ to $SO_3 \supset O$ ($O_3 \supset O_h$

becomes $SO_3 \supset O$ if parity labels are dropped) and evaluate the final reduced matrix element using Section B.4. Note that for $\mathfrak{L}^+$, $h_1 = h$, but for $\mathfrak{L}^-$, $h_1 \neq h$. For the group O_h we obtain, in a similar fashion, for l^n states of identical spin

$$\langle h^{\pm}\|L^{t_{1g}}\|h'^{\pm}\rangle^{O_h} = \delta_{\mathfrak{L}\mathfrak{L}'}\begin{pmatrix} \mathfrak{L} & 1 & \mathfrak{L} \\ h & T_1 & h' \end{pmatrix}^{SO_3}_{O} \sqrt{\mathfrak{L}(\mathfrak{L}+1)(2\mathfrak{L}+1)} \tag{15.4.5}$$

The $SO_3 \supset O$ (Section B.5) and $O_h \supset T_d$ $3jm$ factors are given in Chapter 13 of Butler (1981). Recall that in Butler irrep notation, $T_1 = 1$. It is easy to derive analogous equations for subgroups of O_h and T_d.

The general method described above allows the tables of Nielson and Koster (1963) to be used for subgroups of O_3 or SO_3 in a weak-field basis—see Section B.4.

Even if a strong-field basis is employed, a weak-field basis is still appropriate for single-center, one-electron orbital-angular-momentum matrix elements. Thus the single-center matrix elements of Section 13.4 could have been evaluated using the equations above. For example, using (15.4.5), we obtain our result (13.4.5) as follows:

$$\begin{aligned}\langle t_{2g}\|l^{t_{1g}}\|e_g\rangle^{O_h} &= \begin{pmatrix} 2 & 1 & 2 \\ T_2 & T_1 & E \end{pmatrix}^{SO_3}_{O} \sqrt{30} \\ &= \left(\frac{-\sqrt{2}}{\sqrt{5}}\right)\sqrt{30} \\ &= -2\sqrt{3}\end{aligned} \tag{15.4.6}$$

Example: Spin-Angular-Momentum Matrix Elements

Spin-only matrix elements which are reduced in SO_3 are given by the first equation of (B.4.4) in Section B.4, regardless of the SO_3 basis used:

$$\langle \mathbb{S}\|S^1\|\mathbb{S}'\rangle^{SO_3} = \delta_{\mathbb{S}\mathbb{S}'}\sqrt{\mathbb{S}(\mathbb{S}+1)(2\mathbb{S}+1)} \tag{15.4.7}$$

In problems with spin–orbit coupling we use a symmetry-group-adapted SO_3 spin basis of the general sort $|\mathbb{S}^+(O_3)S(G)S_1(G_1)\cdots\rangle$ or $|\mathbb{S}(SO_3)S(G)S_1(G_1)\cdots\rangle$. [Here we use the chain notation of Section 15.2; in the notation of Section 12.2 the ket $|\mathbb{S}(SO_3)S(G)S_1(G_1)\cdots\rangle$ would be labeled $|\mathbb{S}SM\rangle$ if the system belonged to the group G, but $|\mathbb{S}S_1M_1\rangle$ if it belonged to G_1.] If spin matrix elements in this type of basis are reduced in an SO_3 subgroup (i.e. G or G_1), the resulting reduced matrix elements may be evaluated using (15.4.3). Moreover, the method is applicable *regardless* of whether a strong- or weak-field orbital basis is used. Spin reduced matrix elements for the group of interest are expressed in terms of SO_3 spin reduced matrix elements via (15.4.3); then (15.4.7) above is used to evaluate the SO_3 spin reduced matrix elements.

For group-O_h spin states $|\mathbb{S}^+(O_3)S^+(O_h)\cdots\rangle$ [or T_d spin states $|\mathbb{S}^+(O_3)S^+(O_h)S(T_d)\cdots\rangle$] the procedure gives

$$\begin{aligned}
\langle \mathbb{S}^+S^+\|S^{1^+t_{1g}}\|\mathbb{S}'^+S'^+\rangle_q^{O_h} &= \begin{pmatrix} \mathbb{S}^+ & 1^+ & \mathbb{S}'^+ \\ S^+ & T_{1g} & S'^+ \end{pmatrix}\begin{matrix} O_3 \\ qO_h \end{matrix} \langle \mathbb{S}^+\|S^{1^+}\|\mathbb{S}'^+\rangle^{O_3} \\
&= \begin{pmatrix} \mathbb{S} & 1 & \mathbb{S}' \\ S & T_1 & S' \end{pmatrix}\begin{matrix} SO_3 \\ qO \end{matrix} \langle \mathbb{S}\|S^1\|\mathbb{S}'\rangle^{SO_3} \\
&= \delta_{\mathbb{S}\mathbb{S}'}\begin{pmatrix} \mathbb{S} & 1 & \mathbb{S} \\ S & T_1 & S' \end{pmatrix}\begin{matrix} SO_3 \\ qO \end{matrix} \sqrt{\mathbb{S}(\mathbb{S}+1)(2\mathbb{S}+1)} \\
&= \langle \mathbb{S}^+S^+S\|S^{1^+t_{1g}t_1}\|\mathbb{S}'^+S'^+S'\rangle_q^{T_d} \qquad (15.4.8)
\end{aligned}$$

Here we use the isomorphism of $O_3 \supset O_h$ to $SO_3 \supset O$, and in the final equality, the isomorphism of $O_3 \supset O_h \supset T_d$ to $O_3 \supset O_h \supset O$ for even-parity states of O_3 (recall that spin states have instrinsic even parity).

Below we use (15.4.8) for the half-integer values of $\mathbb{S}^+$ and $\mathbb{S}'^+ \leqslant \frac{5}{2}$ for the groups O, O_h, and T_d. The matrix elements are written first in the $|\mathbb{S}^+S^+S\cdots\rangle$ chain notation used above, and then are rewritten in the $|\mathbb{S}SM\rangle$ basis notation of Section 12.2. To use these results for O_h, drop the last set of labels in the chain-notation matrix elements, and add g subscripts to the second set of labels in the $|\mathbb{S}SM\rangle$ matrix elements:

$$\begin{aligned}
\langle \tfrac{1}{2}^+\tfrac{1}{2}^+\tfrac{1}{2}\|S^{1^+t_{1g}t_1}\|\tfrac{1}{2}^+\tfrac{1}{2}^+\tfrac{1}{2}\rangle &= \langle \tfrac{1}{2}^+E'\|S^{t_1}\|\tfrac{1}{2}^+E'\rangle \\
&= (1)\frac{\sqrt{6}}{2} = \frac{\sqrt{6}}{2}
\end{aligned}$$

$$\begin{aligned}
\langle \tfrac{3}{2}^+\tfrac{3}{2}^+\tfrac{3}{2}\|S^{1^+t_{1g}t_1}\|\tfrac{3}{2}^+\tfrac{3}{2}^+\tfrac{3}{2}\rangle_q &= \langle \tfrac{3}{2}^+U'\|S^{t_1}\|\tfrac{3}{2}^+U'\rangle_q \\
&= \begin{cases} \dfrac{1}{\sqrt{5}}\sqrt{15} = \sqrt{3} & \text{for} \quad q = 0 \\[2ex] \dfrac{2}{\sqrt{5}}\sqrt{15} = 2\sqrt{3} & \text{for} \quad q = 1 \end{cases}
\end{aligned}$$

$$\langle \tfrac{5}{2}^{+} \tfrac{3}{2}^{+} \tfrac{3}{2} \| S^{1^{+} t_{1g} t_1} \| \tfrac{5}{2}^{+} \tfrac{3}{2}^{+} \tfrac{3}{2} \rangle_q = \langle \tfrac{5}{2}^{+} U' \| S^{t_1} \| \tfrac{5}{2}^{+} U' \rangle_q$$

$$= \begin{cases} \dfrac{-\sqrt{2 \times 7}}{3\sqrt{5}} \dfrac{\sqrt{5 \times 7 \times 3}}{\sqrt{2}} = -\dfrac{7}{\sqrt{3}} & \text{for} \quad q = 0 \\ \dfrac{-4\sqrt{2}}{3\sqrt{5 \times 7}} \dfrac{\sqrt{5 \times 7 \times 3}}{\sqrt{2}} = \dfrac{-4}{\sqrt{3}} & \text{for} \quad q = 1 \end{cases}$$

$$\langle \tfrac{5}{2}^{+} \tilde{\tfrac{1}{2}}^{+} \tilde{\tfrac{1}{2}} \| S^{1^{+} t_{1g} t_1} \| \tfrac{5}{2}^{+} \tilde{\tfrac{1}{2}}^{+} \tilde{\tfrac{1}{2}} \rangle = \langle \tfrac{5}{2}^{+} E'' \| S^{t_1} \| \tfrac{5}{2}^{+} E'' \rangle$$

$$= \frac{\sqrt{5}}{3\sqrt{7}} \frac{\sqrt{5 \times 7 \times 3}}{\sqrt{2}} = \frac{5}{\sqrt{6}}$$

$$\langle \tfrac{5}{2}^{+} \tilde{\tfrac{1}{2}}^{+} \tilde{\tfrac{1}{2}} \| S^{1^{+} t_{1g} t_1} \| \tfrac{5}{2}^{+} \tfrac{3}{2}^{+} \tfrac{3}{2} \rangle = \langle \tfrac{5}{2}^{+} E'' \| S^{t_1} \| \tfrac{5}{2}^{+} U' \rangle$$

$$= \frac{2\sqrt{10}}{\sqrt{3}}$$

$$\langle \tfrac{5}{2}^{+} \tfrac{3}{2}^{+} \tfrac{3}{2} \| S^{1^{+} t_{1g} t_1} \| \tfrac{5}{2}^{+} \tilde{\tfrac{1}{2}}^{+} \tilde{\tfrac{1}{2}} \rangle = \langle \tfrac{5}{2}^{+} U' \| S^{t_1} \| \tfrac{5}{2}^{+} E'' \rangle$$

$$= \frac{-2\sqrt{10}}{\sqrt{3}} \qquad (15.4.9)$$

Another example of the use of (15.4.3) for spin-only matrix elements is given in (19.4.20).

15.5. Basis Functions

The transformation properties of basis functions are completely specified by designating the group chain as discussed in Section 15.2. However, for chemical or physical applications more information is usually needed about the basis. The basis states must be expressed in terms of the functions x, y, and z and/or the $|jm\rangle$ of $SO_3 \supset SO_2$. This information is needed to determine transformation properties of operators (Section 9.8) and for the final evaluation of reduced matrix elements whose values are not fixed by group theory alone (Sections 13.4 and 19.6). The basis functions we use throughout this book are those of Butler (1981).

Basis functions for some important group chains are given in terms of the $SO_3 \supset SO_2$ $|jm\rangle$ in Butler (1981), Chapter 16. These basis functions were determined using the branching rules relating a group G to SO_3, together with the appropriate $3jm$ factors for the chain $SO_3 \supset \cdots \supset G$. Examples of the procedure, which readers may use to obtain basis functions for additional group chains, are given in Butler (1981), Section 5.3. It is noteworthy that the basis functions are found *last*, rather than *first* as in the more conventional method. The SO_3 basis for integer j may be expressed in terms of x, y, and z using the definitions of the spherical harmonics (Section B.1); for this purpose we assume the $|jm\rangle$ for integer j are the Y^j_m.

When using Butler's tables and those in our appendixes, one must remember that the bases differ from those of most other writers (e.g. Griffith). Thus readers must beware of unconscious assumptions they may make about a basis, since these may be invalid for our bases. For example, Butler basis functions corresponding to the real functions x, y, and z are not in general real. Thus in our real-O basis (which is chosen to be Butler's $O \supset D_4 \supset D_2 \supset C_2$ basis), the T_1 basis partners *do not* transform as x, y, and z, but as $-x$, iy, and z [Table C.5.1(b)]. Also the relation of the functions x and y to the E irrep partners in D_4, D_3, and so on, differ from group to group, so that there is no standard relation between the bases and the transformation properties of common operators. Consequently the operator transformation coefficients (Section 9.8) differ from group to group.

Another consequence of Butler's "real" basis not being truly real is that the $2jm$ for these bases are not always unity for single-valued irreps. The equivalent phases in the Griffith (1962) bases *are* always unity for single-valued irreps in the real basis.

Butler's multiplicity separation (and hence ours) for $U'U'T_1$ for the group O is also unconventional (Section 12.6). While his $U'U'T_2$ separation is analogous to that of Dobosh (1972), his for $U'U'T_1$ differs. For a discussion of how the Butler multiplicity separation affects the first-order spin–orbit coupling matrix for $^4T_1$ and $^4T_2$ states, see Section 12.6.

16 6j and 9j Coefficients

16.1 6*j* Coefficients

The great advantage of 3*jm* over v.c.c. is that sums over irrep partners in certain products of 3*jm* give useful invariants. A very simple example of this is (10.3.6): the right-hand side is independent of α, β, and γ and is trivial to evaluate compared to summing the products of 3*jm* on the left. Similarly we can sum over all irrep partners in the product of four 3*jm* to produce the extremely useful invariant known as a 6*j* *coefficient* (or 6*j* *symbol*):

$$\begin{Bmatrix} a & b & c \\ d & e & f \end{Bmatrix}_{r_1 r_2 r_3 r_4} = \sum_{\alpha\beta\gamma\delta\varepsilon\phi} \begin{pmatrix} a \\ \alpha \end{pmatrix} \begin{pmatrix} b \\ \beta \end{pmatrix} \begin{pmatrix} c \\ \gamma \end{pmatrix} \begin{pmatrix} d \\ \delta \end{pmatrix} \begin{pmatrix} e \\ \varepsilon \end{pmatrix} \begin{pmatrix} f \\ \phi \end{pmatrix}$$

$$\times \begin{pmatrix} a & e^* & f \\ \alpha & \varepsilon^* & \phi \end{pmatrix}^{r_1} \begin{pmatrix} d & b & f^* \\ \delta & \beta & \phi^* \end{pmatrix}^{r_2}$$

$$\times \begin{pmatrix} d^* & e & c \\ \delta^* & \varepsilon & \gamma \end{pmatrix}^{r_3} \begin{pmatrix} a^* & b^* & c^* \\ \alpha^* & \beta^* & \gamma^* \end{pmatrix}^{r_4} \qquad (16.1.1)$$

Once again the left-hand side is simply a number, and once it is tabulated, we have the value of the messy sum of 3*jm* to the right.

In contrast to the 3*jm*, the 6*j* for a group are independent of the basis used to define the irrep partners. All basis-dependent phases and coefficients have names ending in *jm* (e.g. 2*jm* and 3*jm*), while names of all basis-independent quantities lack the *m* label (e.g. 2*j*, 3*j*, 6*j*, and 9*j*). The 6*j* for SO_3 are identical to 6*j* coefficients of Rotenberg et al. (1959). For molecular point groups they are closely analogous to *W* coefficients [Griffith (1962), Dobosh (1972)]. We tabulate 6*j* for representative groups in Sections B.3, C.13, D.10, and E.10.

Since the 6*j* are defined as products of four 3*jm*, they are zero whenever

$$\delta(ae^*fr_1)\delta(dbf^*r_2)\delta(d^*ecr_3)\delta(a^*b^*c^*r_4) = 0 \qquad (16.1.2)$$

[Recall that $\delta(abc^*r) \equiv 1$ if $a \otimes b$ contains rc, and $\equiv 0$ otherwise.] The triads in the $6j$ coefficient to which the $3jm$ correspond are

$$\begin{Bmatrix} a & \bullet & \bullet \\ & \diagdown & \\ \bullet & e^* - & f \end{Bmatrix} r_1 \bullet\bullet\bullet, \qquad \begin{Bmatrix} \bullet & b & \bullet \\ / & & \diagdown \\ d & \bullet & f^* \end{Bmatrix} \bullet r_2 \bullet\bullet,$$

$$\begin{Bmatrix} \bullet & \bullet & c \\ & / & \\ d^* - & e & \bullet \end{Bmatrix} \bullet\bullet r_3 \bullet, \qquad \begin{Bmatrix} a^* - & b^* - & c^* \\ \bullet & \bullet & \bullet \end{Bmatrix} \bullet\bullet\bullet r_4 \tag{16.1.3}$$

The importance of the $6j$ derives from the fact that the sum of $3jm$ in (16.1.1) (or one easily related to it via the $3jm$ symmetry rules) occurs frequently in quantum-mechanical problems. We give numerous examples using $6j$ coefficients in Chapters 17–23.

Symmetry Rules for 6j for Simply Reducible Groups

The $6j$, like the $3jm$, have extensive symmetry properties. These are particularly simple for simply reducible groups, which are defined (Section 10.2) as ambivalent groups which are multiplicity-free. For these groups phase factors $(-1)^a$ may always be chosen so that (10.2.5) holds. The $6j$ of simply reducible groups are invariant under

1. Any cyclic (even) permutation of their columns.
2. Any noncyclic (odd) permutation of their columns.
3. Turning any pair of columns upside down.

An illustration of (3) is

$$\begin{Bmatrix} a & b & c \\ d & e & f \end{Bmatrix} = \begin{Bmatrix} d & e & c \\ a & b & f \end{Bmatrix} \tag{16.1.4}$$

Multiplicity indices are not needed, since for such groups $r = 0$ only.

Symmetry Rules for 6j in the General Case

The symmetry rules for the $6j$ for the general case [Butler (1981), Section 3.3] are easily derived from (10.2.4), (10.2.12), and our definition (16.1.1). The $6j$ are invariant under cyclic (even) permutation of their columns. Cyclic permutation of the multiplicity indices r_1, r_2, and r_3 always accompanies cyclic permutation of columns 1, 2, and 3. Thus

$$\begin{Bmatrix} a & b & c \\ d & e & f \end{Bmatrix}_{r_1r_2r_3r_4} = \begin{Bmatrix} c & a & b \\ f & d & e \end{Bmatrix}_{r_3r_1r_2r_4} = \begin{Bmatrix} b & c & a \\ e & f & d \end{Bmatrix}_{r_2r_3r_1r_4} \tag{16.1.5}$$

Noncyclic (odd) permutation of columns multiplies the $6j$ by a phase factor. The indices r_1, r_2, and r_3 are permuted in the same manner as columns 1, 2, and 3. Conjugation of d, e, and f also occurs. Thus

$$\begin{aligned}\begin{Bmatrix} a & b & c \\ d & e & f \end{Bmatrix}_{r_1r_2r_3r_4} &= (-1)^n \begin{Bmatrix} b & a & c \\ e^* & d^* & f^* \end{Bmatrix}_{r_2r_1r_3r_4} \\ &= (-1)^n \begin{Bmatrix} a & c & b \\ d^* & f^* & e^* \end{Bmatrix}_{r_1r_3r_2r_4} \\ &= (-1)^n \begin{Bmatrix} c & b & a \\ f^* & e^* & d^* \end{Bmatrix}_{r_3r_2r_1r_4} \end{aligned} \tag{16.1.6}$$

where

$$(-1)^n = \{d\}\{e\}\{f\}\{ae^*fr_1\}\{dbf^*r_2\}\{d^*ecr_3\}\{a^*b^*c^*r_4\} \tag{16.1.7}$$

Here, as pointed out by Dobosh (1972), n in $(-1)^n$ is equal to the number of "law-breaking" phases in (16.1.7) above—that is, the number of $3j$ phases for which (10.2.5) fails and

$$\{abcr\} = -(-1)^a(-1)^b(-1)^c \tag{16.1.8}$$

For the groups O and T_d this happens only for $\{U'U'T_21\}$, so n is the number of $U'U'T_21$ triads in the $6j$ coefficient. For the groups D_3 and C_{3v}, $\{EEE\}$ is the law-breaking factor (Section E.8).

Any of the first three triads of (16.1.3) may be interchanged with the upper (fourth) triad. This is equivalent to turning a pair of columns upside down. No phase change occurs, but conjugation of representations and permutation of multiplicity indices results:

$$\begin{aligned}\begin{Bmatrix} a & b & c \\ d & e & f \end{Bmatrix}_{r_1r_2r_3r_4} &= \begin{Bmatrix} a^* & e & f^* \\ d^* & b & c^* \end{Bmatrix}_{r_4r_3r_2r_1} \\ &= \begin{Bmatrix} d^* & b^* & f \\ a^* & e^* & c \end{Bmatrix}_{r_3r_4r_1r_2} = \begin{Bmatrix} d & e^* & c^* \\ a & b^* & f^* \end{Bmatrix}_{r_2r_1r_4r_3} \end{aligned} \tag{16.1.9}$$

The column not starred follows cyclicly the column not flipped. The multiplicity index of the latter column is interchanged with r_4, and the other two multiplicity indices are interchanged with each other.

The behavior of the $6j$ under complex conjugation follows from (10.3.2):

$$\begin{Bmatrix} a & b & c \\ d & e & f \end{Bmatrix}^*_{r_1r_2r_3r_4} = \begin{Bmatrix} a^* & b^* & c^* \\ d^* & e^* & f^* \end{Bmatrix}_{r_1r_2r_3r_4} \tag{16.1.10}$$

This relation means that all $6j$ for nonambivalent groups are real numbers.

The symmetry rules outlined above reduce enormously the number of $6j$ which must be tabulated for any group. Only one of the set of $6j$ connected by the symmetry rules need be given.

Simple Formula for 6j Containing an A_1 Irrep

A simple equation for $6j$ containing a totally symmetric (A_1) irrep may be derived with the help of (10.3.1), (10.3.6), and other equations in Chapter 10:

$$\begin{Bmatrix} A_1 & b & c \\ d & e & f \end{Bmatrix}_{0rs0} = \delta_{bc^*}\delta_{ef}\delta_{rs}\,\delta(dbf^*r)|b|^{-1/2}|e|^{-1/2}\{dbf^*r\}$$

$$= \begin{Bmatrix} e^* & c^* & d \\ b^* & f^* & A_1 \end{Bmatrix}_{00rs} \tag{16.1.11}$$

16.2. Other Equations Relating 3*jm* Products to 6*j*

There are other relationships besides (16.1.1) between $3jm$ products and $6j$, some of which are extremely useful in derivations of general formulas. They may be obtained from a variation in the $6j$ defining equation given below which differs from (16.1.1) in that it contains no sum over γ:

$$\delta_{\gamma\gamma'}\delta_{cc'}|c|^{-1}\begin{Bmatrix} a & b & c \\ d & e & f \end{Bmatrix}_{r_1r_2r_3r_4}$$

$$= \sum_{\alpha\beta\delta\varepsilon\phi} \begin{pmatrix} a \\ \alpha \end{pmatrix}\begin{pmatrix} b \\ \beta \end{pmatrix}\begin{pmatrix} c \\ \gamma \end{pmatrix}\begin{pmatrix} d \\ \delta \end{pmatrix}\begin{pmatrix} e \\ \varepsilon \end{pmatrix}\begin{pmatrix} f \\ \phi \end{pmatrix}\begin{pmatrix} a & e^* & f \\ \alpha & \varepsilon^* & \phi \end{pmatrix}^{r_1}$$

$$\times \begin{pmatrix} d & b & f^* \\ \delta & \beta & \phi^* \end{pmatrix}^{r_2}\begin{pmatrix} d^* & e & c' \\ \delta^* & \varepsilon & \gamma' \end{pmatrix}^{r_3}\begin{pmatrix} a^* & b^* & c^* \\ \alpha^* & \beta^* & \gamma^* \end{pmatrix}^{r_4} \tag{16.2.1}$$

This form may easily be derived from the equations in Section 16.3 below. Equation (16.2.1) is converted to (16.1.1) by simply summing both sides over γ. Multiplying both sides of (16.2.1) by

$$|c|\begin{pmatrix} a & b & c \\ \alpha' & \beta' & \gamma \end{pmatrix}^{r_4}$$

and summing over r_4, c, and γ gives

$$\sum_{r_4 c} \delta_{cc'} \begin{pmatrix} a & b & c \\ \alpha' & \beta' & \gamma \end{pmatrix}^{r_4} \begin{Bmatrix} a & b & c \\ d & e & f \end{Bmatrix}_{r_1 r_2 r_3 r_4}$$

$$= \sum_{r_4 c \gamma} \sum_{\alpha\beta\delta\varepsilon\phi} |c| \begin{pmatrix} d \\ \delta \end{pmatrix}\begin{pmatrix} e \\ \varepsilon \end{pmatrix}\begin{pmatrix} f \\ \phi \end{pmatrix}\begin{pmatrix} a & b & c \\ \alpha' & \beta' & \gamma \end{pmatrix}^{r_4}\begin{pmatrix} a & b & c \\ \alpha & \beta & \gamma \end{pmatrix}^{*\,r_4}$$

$$\times \begin{pmatrix} a & e^* & f \\ \alpha & \varepsilon^* & \phi \end{pmatrix}^{r_1}\begin{pmatrix} d & b & f^* \\ \delta & \beta & \phi^* \end{pmatrix}^{r_2}\begin{pmatrix} d^* & e & c' \\ \delta^* & \varepsilon & \gamma' \end{pmatrix}^{r_3} \qquad (16.2.2)$$

Here we have used (10.3.2). We then simplify the above using (10.3.5) and drop primes to obtain an extremely useful equation:

$$\sum_{\delta\varepsilon\phi} \begin{pmatrix} d \\ \delta \end{pmatrix}\begin{pmatrix} e \\ \varepsilon \end{pmatrix}\begin{pmatrix} f \\ \phi \end{pmatrix}\begin{pmatrix} a & e^* & f \\ \alpha & \varepsilon^* & \phi \end{pmatrix}^{r_1}\begin{pmatrix} d & b & f^* \\ \delta & \beta & \phi^* \end{pmatrix}^{r_2}\begin{pmatrix} d^* & e & c \\ \delta^* & \varepsilon & \gamma \end{pmatrix}^{r_3}$$

$$= \sum_{r_4} \begin{pmatrix} a & b & c \\ \alpha & \beta & \gamma \end{pmatrix}^{r_4} \begin{Bmatrix} a & b & c \\ d & e & f \end{Bmatrix}_{r_1 r_2 r_3 r_4} \qquad (16.2.3)$$

We use this equation repeatedly in Chapter 17. For simply reducible groups it may be simplified by dropping the sum over r_4 and all multiplicity indices.

Equation (16.2.3) is sometimes called the (3, 1) equation, since it has three $3jm$ on one side and one on the other. In a similar way (2, 2), (1, 3), and (0, 4) equations may be derived. We do not use them in this book, however.

16.3. Definition of $6j$ in Terms of Recoupling Coefficients

At this point it is of interest to consider a more fundamental definition of the $6j$ with respect to quantum mechanics. We show now that transformations between quantum-mechanical wavefunctions formed by coupling three kets in two different possible ways are related to $6j$ coefficients.

Generalizing the treatment by Griffith [(1962), Section 4.3], we may couple three irreps, e, f, and b, in two ways while still maintaining the order of irreps efb. Letting p, q, r, and s be multiplicity indices, we can couple e and f to form pa, and then couple a to b to form qc; alternatively we may couple f to b to form rd, and then couple e to d to form sc'. These two couplings are

$$|((ef)pa, b)qc\gamma\rangle = \sum_{\varepsilon\phi\alpha\beta} (e\varepsilon, f\phi|pa\alpha)(a\alpha, b\beta|qc\gamma) \times |e\varepsilon\rangle|f\phi\rangle|b\beta\rangle \tag{16.3.1}$$

$$|(e,(fb)rd)sc'\gamma'\rangle = \sum_{\phi\beta\varepsilon\delta} (f\phi, b\beta|rd\delta)(e\varepsilon, d\delta|sc'\gamma') \times |e\varepsilon\rangle|f\phi\rangle|b\beta\rangle \tag{16.3.2}$$

Here two multiplicity labels are needed for each ket, since c is not limited to A_1 as it was in Section 14.7.

The orthonormal functions formed in (16.3.1) and (16.3.2) above couple the same kets—$|e\varepsilon\rangle$, $|f\phi\rangle$, and $|b\beta\rangle$—in two different ways. Thus the $|(a, b)qc\gamma\rangle$ and the $|(e, d)sc'\gamma'\rangle$ kets are related to one another by a unitary matrix of so-called *recoupling coefficients*:

$$|((ef)pa, b)qc\gamma\rangle = \sum_{sc'\gamma'} (((ef)pa, b)qc\gamma|(e,(fb)rd)sc'\gamma') \times |(e,(fb)rd)sc'\gamma'\rangle \tag{16.3.3}$$

Clearly these coefficients are zero by group orthogonality relations unless $c = c'$ and $\gamma = \gamma'$. It is not, however, required that $q = s$. These coefficients are defined by the equation.

$$\delta_{cc'}\delta_{\gamma\gamma'}(((ef)pa, b)qc\gamma|(e,(fb)rd)sc'\gamma')$$

$$= \sum_{\alpha\beta\delta\varepsilon\phi} (qc\gamma|a\alpha, b\beta)(pa\alpha|e\varepsilon, f\phi)(f\phi, b\beta|rd\delta)(e\varepsilon, d\delta|sc'\gamma')$$

$$= \sum_{\alpha\beta\delta\varepsilon\phi} |c|^{1/2}\begin{pmatrix} c \\ \gamma \end{pmatrix}\begin{pmatrix} a & b & c^* \\ \alpha & \beta & \gamma^* \end{pmatrix} q \; |a|^{1/2}\begin{pmatrix} a \\ \alpha \end{pmatrix}\begin{pmatrix} e & f & a^* \\ \varepsilon & \phi & \alpha^* \end{pmatrix} p$$

$$\times |d|^{1/2}\begin{pmatrix} d \\ \delta \end{pmatrix}\begin{pmatrix} f & b & d^* \\ \phi & \beta & \delta^* \end{pmatrix}^* r \; |c'|^{1/2}\begin{pmatrix} c' \\ \gamma' \end{pmatrix}\begin{pmatrix} e & d & c'^* \\ \varepsilon & \delta & \gamma'^* \end{pmatrix}^* s$$

(16.3.4)

where we have used (10.2.1). Using (10.3.2), putting $c = c'$ and $\gamma = \gamma'$, and summing over γ (which gives $|c|$ identical terms on the left-hand side), we obtain

$$(((ef)pa, b)qc\gamma|(e,(fb)rd)sc\gamma)$$

$$= \sum_{\alpha\beta\gamma\delta\varepsilon\phi} |a|^{1/2}|d|^{1/2}\begin{pmatrix} a \\ \alpha \end{pmatrix}\begin{pmatrix} f \\ \phi \end{pmatrix}\begin{pmatrix} b \\ \beta \end{pmatrix}\begin{pmatrix} d^* \\ \delta^* \end{pmatrix}\begin{pmatrix} e \\ \varepsilon \end{pmatrix}\begin{pmatrix} c^* \\ \gamma^* \end{pmatrix}$$

$$\times \begin{pmatrix} e & f & a^* \\ \varepsilon & \phi & \alpha^* \end{pmatrix} p \begin{pmatrix} f^* & b^* & d \\ \phi^* & \beta^* & \delta \end{pmatrix} r$$

$$\times \begin{pmatrix} e^* & d^* & c \\ \varepsilon^* & \delta^* & \gamma \end{pmatrix} s \begin{pmatrix} a & b & c^* \\ \alpha & \beta & \gamma^* \end{pmatrix} q \qquad (16.3.5)$$

We next interchange columns f^* and b^* in the second $3jm$ and e^* and d^* in the third $3jm$, using (10.2.4), to obtain the same cyclic order of irreps as in (16.1.1). Then we use (10.2.12) to obtain all the $2jm$ in proper form, and (10.2.10) to simplify the $2j$ phase factors. With these operations we have

$$(((ef)pa, b)qc\gamma|(e,(fb)rd)sc\gamma)$$

$$= |a|^{1/2}|d|^{1/2}\{c\}\{f^*b^*dr\}\{e^*d^*cs\}\begin{Bmatrix} a^* & b^* & c \\ d & e^* & f \end{Bmatrix} prsq \qquad (16.3.6)$$

Steps similar to those outlined above are used in many of the derivations in Chapters 17–22. For this reason we have presented them in detail.

Note that recoupling coefficients, and hence the $6j$, are independent of irrep partners, since the only dependence on γ is that the coefficients must be diagonal in γ.

16.4. $9j$ Coefficients

$9j$ coefficients (or *$9j$ symbols*) are the next highest independent invariants after the $6j$. They are defined as

$$\begin{Bmatrix} a & b & c \\ d & e & f \\ g & h & i \end{Bmatrix}\begin{matrix} r_1 \\ r_2 \\ r_3 \end{matrix} \atop {\begin{matrix} s_1 & s_2 & s_3 \end{matrix}} = \sum_{\substack{\alpha\beta\gamma\delta\varepsilon \\ \phi\eta\theta\tau}} \begin{pmatrix} a & b & c \\ \alpha & \beta & \gamma \end{pmatrix}^{r_1} \begin{pmatrix} d & e & f \\ \delta & \varepsilon & \phi \end{pmatrix}^{r_2}$$

$$\times \begin{pmatrix} g & h & i \\ \eta & \theta & \tau \end{pmatrix}^{r_3} \begin{pmatrix} a & d & g \\ \alpha & \delta & \eta \end{pmatrix}^{* s_1}$$

$$\begin{pmatrix} b & e & h \\ \beta & \varepsilon & \theta \end{pmatrix}^{* s_2} \begin{pmatrix} c & f & i \\ \gamma & \phi & \tau \end{pmatrix}^{* s_3} \qquad (16.4.1)$$

where, as always, for simply reducible groups all multiplicity indices may be dropped. They are identical to the $9j$ for the group SO_3 (see Section B.3), and are directly analogous to the Griffith (1962) X coefficients.

Symmetry Rules for 9j for Simply Reducible Groups

The $9j$ have extensive symmetry properties. For simply reducible groups they may be summarized by the rules below:

1. A $9j$ coefficient is invariant under any even permutation of its rows or columns or both.
2. It is invariant under transposition; that is,

$$\begin{Bmatrix} a & b & c \\ d & e & f \\ g & h & i \end{Bmatrix} = \begin{Bmatrix} a & d & g \\ b & e & h \\ c & f & i \end{Bmatrix} \qquad (16.4.2)$$

3. Under odd permutations of its *rows*, a $9j$ coefficient is multiplied by the factor

$$\{adg\}\{beh\}\{cfi\} \tag{16.4.3}$$

and under odd permutations of its *columns* by

$$\{abc\}\{def\}\{ghi\} \tag{16.4.4}$$

For simply reducible groups these factors are identical, since all $3j$ phases obey (10.2.5).

Symmetry Rules for 9j in the General Case

Symmetry properties of the $9j$ in the general case are given in Butler (1981), Section 3.3. In all cases the multiplicity indices follow their respective triads. Interchange of two *row* triads gives

$$\begin{Bmatrix} a & b & c \\ g & h & i \\ d & e & f \end{Bmatrix}\begin{matrix} r_1 \\ r_3 \\ r_2 \end{matrix} \underset{s_1\ s_2\ s_3}{} = \{adgs_1\}\{behs_2\}\{cfis_3\}\begin{Bmatrix} a & b & c \\ d & e & f \\ g & h & i \end{Bmatrix}\begin{matrix} r_1 \\ r_2 \\ r_3 \end{matrix} \underset{s_1\ s_2\ s_3}{} \tag{16.4.5}$$

Interchange of two *column* triads gives

$$\begin{Bmatrix} a & c & b \\ d & f & e \\ g & i & h \end{Bmatrix}\begin{matrix} r_1 \\ r_2 \\ r_3 \end{matrix} \underset{s_1\ s_3\ s_2}{} = \{abcr_1\}\{defr_2\}\{ghir_3\}\begin{Bmatrix} a & b & c \\ d & e & f \\ g & h & i \end{Bmatrix}\begin{matrix} r_1 \\ r_2 \\ r_3 \end{matrix} \underset{s_1\ s_2\ s_3}{} \tag{16.4.6}$$

Behavior under transposition and complex conjugation follow from the relations

$$\begin{Bmatrix} a & d & g \\ b & e & h \\ c & f & i \end{Bmatrix}^{*}\begin{matrix} s_1 \\ s_2 \\ s_3 \end{matrix} \underset{r_1\ r_2\ r_3}{} = \begin{Bmatrix} a & b & c \\ d & e & f \\ g & h & i \end{Bmatrix}\begin{matrix} r_1 \\ r_2 \\ r_3 \end{matrix} \underset{s_1\ s_2\ s_3}{} = \begin{Bmatrix} a^* & b^* & c^* \\ d^* & e^* & f^* \\ g^* & h^* & i^* \end{Bmatrix}^{*}\begin{matrix} r_1 \\ r_2 \\ r_3 \end{matrix} \underset{s_1\ s_2\ s_3}{} \tag{16.4.7}$$

How to Reduce 9j Containing an A_1 Irrep to 6j

A $9j$ coefficient containing an A_1 irrep may be expressed in terms of a $6j$ coefficient. We simply use (10.3.1) in (16.4.1) to obtain for the general case

$$\begin{Bmatrix} A_1 & b & c \\ d & e & f \\ g & h & i \end{Bmatrix}\begin{matrix} 0 \\ r_2 \\ r_3 \end{matrix} \atop \begin{matrix} 0 & s_2 & s_3 \end{matrix} = \delta_{b^*c}\delta_{d^*g}|b|^{-1/2}|d|^{-1/2}\{i\}\{cfis_3\}\{ghir_3\}$$

$$\times \begin{Bmatrix} b & e & h \\ d & i & f^* \end{Bmatrix}_{s_3 r_2 r_3 s_2} \tag{16.4.8}$$

or equivalently,

$$\begin{Bmatrix} a & b & c \\ d & e & f \\ g & h & A_1 \end{Bmatrix}\begin{matrix} r_1 \\ r_2 \\ 0 \end{matrix} \atop \begin{matrix} s_1 & s_2 & 0 \end{matrix} = \delta_{gh^*}\delta_{cf^*}|c|^{-1/2}|g|^{-1/2}\{e\}\{behs_2\}\{defr_2\}$$

$$\times \begin{Bmatrix} a & d & g \\ e & b^* & c \end{Bmatrix}_{r_1 r_2 s_2 s_1} \tag{16.4.9}$$

These equations reduce to somewhat simpler forms for simply reducible groups, since (10.2.5) may be used to simplify the phase factors and all multiplicity indices may be dropped.

The Relation of the 9j to Recoupling Coefficients

Like $6j$, the $9j$ may be related to recoupling coefficients which link two different coupling schemes. Here each of the coupling schemes involves the coupling of four kets. The relation is given in (3.2.19) of Butler (1981).

17 MCD Equations Using 3jm and 6j: Allowed Transitions

17.1. Introduction

With this chapter we begin to illustrate more sophisticated applications of the $3jm$ and $6j$. Here the advantages of the irreducible-tensor method over the more conventional methods of Chapters 11–13 first become apparent. We show how the general MCD equations may be simplified enormously by using $3jm$ sum rules—in this case (16.2.3). We do this separately for the two types of systems which arise most commonly in MCD work: (1) oriented systems such as cubic or uniaxial crystals, and (2) solutions. Isotropic molecules in solution may also be classified under (1) for the purpose of calculations, while randomly oriented anisotropic guest molecules in an isotropic crystal, matrix, or glass are classified under (2). We refer to (1) as the *oriented* or *isotropic case* and to (2) as the *space-averaged* case.

In our examples in this chapter we assume transitions are allowed in the Franck–Condon (FC) approximation. The Herzberg–Teller-allowed case is considered in Chapter 22. The MCD equations used are derived in Chapter 4 and Section 14.11. They are those of Stephens (1976), which have recently been adopted by most workers in the field. The advantages of these new equations are that they give $\mathcal{B}_0$ and $\mathcal{C}_0$ the same signs as the $\mathcal{B}$- and $\mathcal{C}$-term MCD (rather than the opposite, which proved very confusing) and excess numerical factors have been omitted. They are related to the old A, B and C terms of Stephens et al. (1966) and Piepho et al. (1975) by (A.5.7) of

Appendix A:

$$\mathcal{A}_1(\text{new}) = \frac{2}{3\mu_B} A$$

$$\mathcal{B}_0(\text{new}) = \frac{-2}{3\mu_B} B$$

$$\mathcal{C}_0(\text{new}) = \frac{-2}{3\mu_B} C$$

$$\mathcal{D}_0(\text{new}) = \tfrac{1}{3} D \tag{17.1.1}$$

$$\mathcal{A}_1/\mathcal{D}_0(\text{new}) = \frac{2}{\mu_B}\frac{A}{D}$$

$$\mathcal{B}_0/\mathcal{D}_0(\text{new}) = \frac{-2}{\mu_B}\frac{B}{D}$$

$$\mathcal{C}_0/\mathcal{D}_0(\text{new}) = \frac{-2}{\mu_B}\frac{C}{D}$$

17.2. The MCD Equations for Ambivalent Groups: Oriented or Isotropic Case

It is straightforward to derive simplified expressions for the Faraday parameters of (4.6.8) using the irreducible-tensor method. In the MCD experiment the magnetic field is along the unique (z) axis of an oriented system; if the system is isotropic, linear-in-the-field MCD is independent of the orientation of the magnetic field relative to the molecular or crystal axes (Sections 17.5 and 4.6). The magnetic-dipole operator is symbolized by μ, and the electric-dipole operator by m. Operators are defined using the conventions of Section 9.8:

$$\mu_0 = \mu_z = -\mu_B(L_z + 2S_z)$$

$$\mu_{\pm 1} = \mp \frac{1}{\sqrt{2}}(\mu_x \pm i\mu_y)$$

$$= -\mu_B\left(\frac{\mp 1}{\sqrt{2}}\right)\left[L_x + 2S_x \pm i(L_y + 2S_y)\right] \tag{17.2.1}$$

$$m_0 = m_z$$

$$m_{\pm 1} = \mp \frac{1}{\sqrt{2}}(m_x \pm i m_y)$$

The MCD equations are derived in forms appropriate for use with both real and complex basis sets. We begin with the equations (4.6.8) but make the simplifications of Section 14.11, so that our initial MCD equations are

$$\mathcal{A}_1 = \frac{-2}{\mu_B|A|} \sum_{\alpha\alpha'\gamma\gamma'} (\langle J\gamma|\mu_0|J\gamma'\rangle\delta_{\alpha\alpha'} - \langle A\alpha|\mu_0|A\alpha'\rangle\delta_{\gamma\gamma'}) \times \langle A\alpha'|m_1|J\gamma\rangle\langle J\gamma'|m_{-1}|A\alpha\rangle \tag{17.2.2}$$

or

$$\mathcal{A}_1 = \frac{-2i}{\mu_B|A|} \sum_{\alpha\alpha'\gamma\gamma'} (\langle J\gamma|\mu_z|J\gamma'\rangle\delta_{\alpha\alpha'} - \langle A\alpha|\mu_z|A\alpha'\rangle\delta_{\gamma\gamma'}) \times \langle A\alpha'|m_x|J\gamma\rangle\langle J\gamma'|m_y|A\alpha\rangle \tag{17.2.3}$$

and

$$\mathcal{B}_0 = \mathcal{R}e\left\{\frac{4}{\mu_B|A|}\sum_{\alpha\gamma}\left[\sum_{K\neq A}\sum_{\kappa}\frac{\langle K\kappa|\mu_0|A\alpha\rangle\langle A\alpha|m_1|J\gamma\rangle\langle J\gamma|m_{-1}|K\kappa\rangle}{W_K - W_A} + \sum_{K\neq J}\sum_{\kappa}\frac{\langle J\gamma|\mu_0|K\kappa\rangle\langle A\alpha|m_1|J\gamma\rangle\langle K\kappa|m_{-1}|A\alpha\rangle}{W_K - W_J}\right]\right\} \tag{17.2.4}$$

or

$$\mathcal{B}_0 = \mathcal{I}m\left\{\frac{-2}{\mu_B|A|}\sum_{\alpha\gamma}\left[\sum_{K\neq A}\sum_{\kappa}\frac{\langle K\kappa|\mu_z|A\alpha\rangle}{W_K - W_A}\times(\langle A\alpha|m_x|J\gamma\rangle\langle J\gamma|m_y|K\kappa\rangle - \langle A\alpha|m_y|J\gamma\rangle\langle J\gamma|m_x|K\kappa\rangle) + \sum_{K\neq J}\sum_{\kappa}\frac{\langle J\gamma|\mu_z|K\kappa\rangle}{W_K - W_J}(\langle A\alpha|m_x|J\gamma\rangle\langle K\kappa|m_y|A\alpha\rangle - \langle A\alpha|m_y|J\gamma\rangle\langle K\kappa|m_x|A\alpha\rangle)\right]\right\} \tag{17.2.5}$$

(where $\mathcal{R}e$ means "take the real part," and $\mathcal{I}m$ "take the imaginary part"),

and

$$\mathcal{C}_0 = \frac{2}{\mu_B|A|} \sum_{\alpha\alpha'\gamma} \langle A\alpha|\mu_0|A\alpha'\rangle\langle A\alpha'|m_1|J\gamma\rangle\langle J\gamma|m_{-1}|A\alpha\rangle \quad (17.2.6)$$

or

$$\mathcal{C}_0 = \frac{2i}{\mu_B|A|} \sum_{\alpha\alpha'\gamma} \langle A\alpha|\mu_z|A\alpha'\rangle\langle A\alpha'|m_x|J\gamma\rangle\langle J\gamma|m_y|A\alpha\rangle \quad (17.2.7)$$

In these equations $|A|$ is the total degeneracy of the ground state A, and the sums over α, α', γ, γ', and κ are over all states degenerate with A, J, or K in the absence of the magnetic field. To reduce the number of equations we introduce the notation

$$\mathcal{A}_1 = \mathcal{A}_1(J) + \mathcal{A}_1(A) = \mathcal{A}_1(J) + \mathcal{C}_0 \quad (17.2.8)$$

When spin–orbit coupling is neglected and a $|\mathcal{S}h\mathcal{M}\theta\rangle$ or $|\mathcal{S}ShM\theta\rangle$ basis [rather than the $|(\mathcal{S}S, h)rt\tau\rangle$ basis of (12.2.2)] is used, spin terms do not contribute (Section 11.3). In such cases we may replace μ_0 by $(-\mu_B)L_0$ and μ_z by $(-\mu_B)L_z$ in these and subsequent equations, and define $A\alpha$ and $J\gamma$ to be purely orbital $(h\theta)$ states [rather than spin–orbit $(t\tau)$ states]; $|A|$ then represents the ground-state orbital degeneracy (rather than the degeneracy of a spin–orbit-coupled ground state).

Simplification of the $\mathcal{A}_1$, $\mathcal{B}_0$, and $\mathcal{C}_0$ Equations Using 3jm and 6j

To derive general MCD equations using (16.2.3), we must first express our operators in terms of group basis operators using operator transformation coefficients as in (9.8.2). We require that the group basis be chosen so that each operator $\mathcal{O}_i$ in the MCD equations used transforms as one of the partners of an irrep to within a phase factor; thus (9.8.2) reduces to

$$\begin{aligned} \mu_i &= \langle f'\phi_i'|\mu_i\rangle \mu_{\phi_i'}^{f'} \\ m_i &= \langle f\phi_i|m_i\rangle m_{\phi_i}^{f} \end{aligned} \quad (17.2.9)$$

where the operator transformation coefficients have absolute value one. Our discussion is limited to groups where the basis may be chosen so that m_x and m_y (and μ_x and μ_y) or m_1 and m_{-1} (and μ_1 and μ_{-1}) transform as partners of a degenerate irrep f (with $f = f^*$) to within the $\langle f\phi|\mathcal{O}_i\rangle$ phase factors of (17.2.9) above; $\mathcal{A}$ and $\mathcal{C}$ terms do not occur for lower symmetry groups, for reasons discussed in Section 5.1. Values of the operator transformation coefficients $\langle f\phi|\mathcal{O}_i\rangle$ are given in Sections C.7, D.6, and E.6 for representative groups.

We assume that the basis functions of the ground A and excited state J have been chosen to transform like the respective irrep partners α and γ of the molecular point group. The basis chosen for the functions must of course be the same one used for the operators.

The discussion in Sections 17.2–17.5 is limited to cases in which all irreps are real. Thus $f = f^*$, $A = A^*$, $J = J^*$, and $K = K^*$. Transitions between complex irreps are treated in Section 17.6. We keep the star on A, J, and K in the key equations anyway, for subsequent use in Section 17.6.

Once all states and operators have been classified by their behavior under group operations, general equations may be derived for $\mathcal{A}_1$, $\mathcal{B}_0$, and $\mathcal{C}_0$ in a manner similar to that of Dobosh (1974). The matrix elements are first reduced using (10.2.2). The resulting sum of $3jm$ is then manipulated into the form of (16.2.3) with the help of various equations in Chapter 10. We illustrate the method in detail for $\mathcal{C}_0$ using a complex basis. We begin with (17.2.6), and the initial steps give

$$\mathcal{C}_0 = \langle f'\phi_0'|\mu_0\rangle\langle f\phi_1|m_1\rangle\langle f\phi_{-1}|m_{-1}\rangle\frac{2}{\mu_B|A|}$$

$$\times\sum_{\alpha'\alpha\gamma}\langle A\alpha|\mu^{f'}_{\phi_0'}|A\alpha'\rangle\langle A\alpha'|m^f_{\phi_1}|J\gamma\rangle\langle J\gamma|m^f_{\phi_{-1}}|A\alpha\rangle$$

$$= \langle f'\phi_0'|\mu_0\rangle\langle f\phi_1|m_1\rangle\langle f\phi_{-1}|m_{-1}\rangle\frac{2}{\mu_B|A|}$$

$$\times\sum_{\alpha\alpha'\gamma}\sum_q\begin{pmatrix}A\\ \alpha\end{pmatrix}\begin{pmatrix}A^* & f' & A\\ \alpha^* & \phi_0' & \alpha'\end{pmatrix}^q\sum_r\begin{pmatrix}A\\ \alpha'\end{pmatrix}\begin{pmatrix}A^* & f & J\\ \alpha'^* & \phi_1 & \gamma\end{pmatrix}^r$$

$$\times\sum_{r'}\begin{pmatrix}J\\ \gamma\end{pmatrix}\begin{pmatrix}J^* & f & A\\ \gamma^* & \phi_{-1} & \alpha\end{pmatrix}^{r'}$$

$$\times\langle A\|\mu^{f'}\|A\rangle_q\langle A\|m^f\|J\rangle_r\langle J\|m^f\|A\rangle_{r'} \qquad (17.2.10)$$

Since the reduced matrix elements are independent of α, α', and γ, we may take them outside the sum over components. Then using (10.2.4), (10.2.12), and (16.2.3) we have

$$\mathcal{C}_0 = \langle f'\phi_0'|\mu_0\rangle\langle f\phi_1|m_1\rangle\langle f\phi_{-1}|m_{-1}\rangle\frac{2}{\mu_B|A|}$$

$$\times\sum_{pqrr'}\{J\}\begin{pmatrix}f' & f & f\\ \phi_0' & \phi_{-1} & \phi_1\end{pmatrix}^p\begin{Bmatrix}f' & f & f\\ J^* & A^* & A^*\end{Bmatrix}qr'rp$$

$$\times\langle A\|\mu^{f'}\|A\rangle_q\langle A\|m^f\|J\rangle_r\langle J\|m^f\|A\rangle_{r'} \qquad (17.2.11)$$

For all groups of interest $p = 0$. Thus, defining an "MCD factor" by the equation

$$(\text{MCD factor}) \equiv \langle f'\phi_0'|\mu_0\rangle\langle f\phi_1|m_1\rangle\langle f\phi_{-1}|m_{-1}\rangle\begin{pmatrix} f' & f & f \\ \phi_0' & \phi_{-1} & \phi_1 \end{pmatrix} \tag{17.2.12}$$

we have

$$\mathcal{C}_0 = (\text{MCD factor})\frac{2\{J\}}{\mu_B|A|}\sum_q \langle A\|\mu^{f'}\|A\rangle_q \times \sum_{rr'}\begin{Bmatrix} f' & f & f \\ J^* & A^* & A^* \end{Bmatrix}_{qr'r0}\langle A\|m^f\|J\rangle_r\langle J\|m^f\|A\rangle_{r'} \tag{17.2.13}$$

In cases where no repeated representations occur, the multiplicity labels and the sums over them may be dropped.

Equations for $\mathcal{A}_1$ and $\mathcal{B}_0$ using a complex basis are derived in much the same fashion from (17.2.2) and (17.2.4) respectively, with the results

$$\mathcal{A}_1 = \left[(\text{MCD factor})\frac{2\{A\}}{\mu_B|A|}\sum_q \langle J\|\mu^{f'}\|J\rangle_q \sum_{rr'}\begin{Bmatrix} f' & f & f \\ A^* & J^* & J^* \end{Bmatrix}_{qrr'0} \times \langle A\|m^f\|J\rangle_r\langle J\|m^f\|A\rangle_{r'}\right] + \mathcal{C}_0 \tag{17.2.14}$$

$$\mathcal{B}_0 = \mathcal{R}\text{e}\left\{(\text{MCD factor})\frac{4}{\mu_B|A|} \times \left[\left(\sum_{K\neq A}\frac{\{J\}}{W_K - W_A}\sum_q \langle K\|\mu^{f'}\|A\rangle_q \times \sum_{rr'}\begin{Bmatrix} f' & f & f \\ J^* & A^* & K^* \end{Bmatrix}_{qr'r0}\langle A\|m^f\|J\rangle_r\langle J\|m^f\|K\rangle_{r'}\right) + \left(\sum_{K\neq J}\frac{-\{A\}}{W_K - W_J}\sum_q \langle J\|\mu^{f'}\|K\rangle_q \sum_{rr'}\begin{Bmatrix} f' & f & f \\ A^* & K^* & J^* \end{Bmatrix}_{qrr'0} \times \langle A\|m^f\|J\rangle_r\langle K\|m^f\|A\rangle_{r'}\right)\right]\right\} \tag{17.2.15}$$

We have used the fact that $\{f'ff\} = -1$ for all groups of interest, and also that a system has either all even, or all odd, electron states; thus $\{J\} = \{A\} = \{K\}$.

Nearly identical equations are derived when real basis sets are used. $\mathcal{C}_0$, $\mathcal{A}_1$, and $\mathcal{B}_0$ are still given by (17.2.13), (17.2.14), and (17.2.15), respectively, if only the MCD factor of (17.2.12) is redefined as

$$\text{(MCD factor)} = i\langle f'\phi_z'|\mu_z\rangle\langle f\phi_x|m_x\rangle\langle f\phi_y|m_y\rangle\begin{pmatrix} f' & f & f \\ \phi_z' & \phi_y & \phi_x \end{pmatrix} \tag{17.2.16}$$

where $i = \sqrt{-1}$. In Table 17.2.1 we tabulate values of the MCD factor for

Table 17.2.1. Values of the MCD Factors Defined by (17.2.12) and (17.2.16)

Basis	f'	f	MCD factor
$O \supset D_4 \supset C_4$ $O \supset D_4 \supset D_2 \supset C_2$ $O \supset D_3 \supset C_3$ $O \supset D_3 \supset C_2$	T_1	T_1	$\frac{1}{\sqrt{6}}$
$T_d \supset D_{2d} \supset S_4$ $T_d \supset D_{2d} \supset D_2 \supset C_2$ $T_d \supset C_{3v} \supset C_3$ $T_d \supset C_{3v} \supset C_s$	T_1	T_2	$\frac{-1}{\sqrt{6}}$
$D_4 \supset C_4$ $D_4 \supset D_2 \supset C_2$	A_2	E	$\frac{-1}{\sqrt{2}}$
$D_{2d} \supset S_4$ $D_{2d} \supset D_2 \supset C_2$	A_2	E	$\frac{1}{\sqrt{2}}$
$D_3 \supset C_3$ $D_3 \supset C_2$	A_2	E	$\frac{-1}{\sqrt{2}}$
$C_{3v} \supset C_3$ $C_{3v} \supset C_s$	A_2	E	$\frac{-1}{\sqrt{2}}$

bases included in our appendixes. Note that our bases have been chosen so that for the different basis chains for a given group, the MCD factors of (17.2.12) and (17.2.16) are identical, as are the reduced matrix elements in (17.2.13)–(17.2.15).

Simplification of the $\mathfrak{D}_0$ Equation Using 3jm

The parameters of interest in MCD are usually the ratios $\mathcal{A}_1/\mathfrak{D}_0$, $\mathcal{B}_0/\mathfrak{D}_0$, and $\mathcal{C}_0/\mathfrak{D}_0$, where the dipole strength $\mathfrak{D}_0$ is given by (4.6.8). With a complex basis set, the simplifications in Section 14.11 together with (14.10.5) give

$$\begin{aligned}\mathfrak{D}_0 &= \frac{1}{2|A|}\sum_{\alpha\gamma}\left(|\langle A\alpha|m_1|J\gamma\rangle|^2 + |\langle A\alpha|m_{-1}|J\gamma\rangle|^2\right)\\ &= \frac{1}{|A|}\sum_{\alpha\gamma}|\langle A\alpha|m_1|J\gamma\rangle|^2\\ &= \frac{1}{|A|}\sum_{\alpha\gamma}\langle A\alpha|m_1|J\gamma\rangle\langle J\gamma|(m_1)^\dagger|A\alpha\rangle\\ &= \frac{-1}{|A|}\sum_{\alpha\gamma}\langle A\alpha|m_1|J\gamma\rangle\langle J\gamma|m_{-1}|A\alpha\rangle\end{aligned} \tag{17.2.17}$$

while with a real basis set,

$$\begin{aligned}\mathfrak{D}_0 &= \frac{1}{2|A|}\sum_{\alpha\gamma}\left(|\langle A\alpha|m_x|J\gamma\rangle|^2 + |\langle A\alpha|m_y|J\gamma\rangle|^2\right)\\ &= \frac{1}{2|A|}\sum_{\alpha\gamma}\big(\langle A\alpha|m_x|J\gamma\rangle\langle J\gamma|m_x|A\alpha\rangle\\ &\qquad + \langle A\alpha|m_y|J\gamma\rangle\langle J\gamma|m_y|A\alpha\rangle\big)\end{aligned} \tag{17.2.18}$$

since (14.10.6) applies.

Proceeding from the absolute-value-squared form of $\mathfrak{D}_0$ in either basis using (17.2.9), (10.2.2), and (10.3.4), we obtain

$$\mathfrak{D}_0 = \frac{1}{|f|\,|A|}\sum_r \delta(A^*fJr)\left|\langle A\|m^f\|J\rangle_r\right|^2 \tag{17.2.19}$$

Then our later discussion in Section 18.3, and in particular (18.3.12), allows us to rewrite this result in the alternate form (since here $f = f^*$)

$$\mathfrak{D}_0 = \frac{-\{A\}}{|f|\,|A|} \sum_r \delta(A^* f J r)\{A^* f J r\}\langle A\|m^f\|J\rangle_r \langle J\|m^f\|A\rangle_r \qquad (17.2.20)$$

This result is valid in either the real or the complex basis and is somewhat simpler, for example, than the equation

$$\mathfrak{D}_0 = -\langle f\phi_1|m_1\rangle\langle f\phi_{-1}|m_{-1}\rangle\begin{pmatrix} f \\ \phi_1 \end{pmatrix}$$
$$\times \frac{\{A\}}{|f|\,|A|} \sum_r \delta(A^* f J r)\{A^* f J r\}\langle A\|m^f\|J\rangle_r \langle J\|m^f\|A\rangle_r \qquad (17.2.21)$$

which is obtained from the last line of (17.2.17) using (17.2.9), (10.2.2), and —after rearranging of the $3jm$ into appropriate form—(10.3.4). Comparison of (17.2.20) and (17.2.21) gives

$$-\langle f\phi_1|m_1\rangle\langle f\phi_{-1}|m_{-1}\rangle\begin{pmatrix} f \\ \phi_1 \end{pmatrix} = -1 \qquad (17.2.22)$$

Similarly, comparison of the analog of (17.2.21) obtained from (17.2.18) with (17.2.20) gives the relation

$$\frac{1}{2}\left[\langle f\phi_x|m_x\rangle^2\begin{pmatrix} f \\ \phi_x \end{pmatrix} + \langle f\phi_y|m_y\rangle^2\begin{pmatrix} f \\ \phi_y \end{pmatrix}\right] = -1 \qquad (17.2.23)$$

Properly chosen operator transformation coefficients should obey these relations, as do all those given in our appendixes.

$\mathcal{A}_1 / \mathfrak{D}_0$ and $\mathcal{C}_0 / \mathfrak{D}_0$ for Simply Reducible Groups

A group which is simply reducible (Section 10.2) is both ambivalent (no complex irreps) and multiplicity-free. Examples include SO_3, D_4, D_6, and $D_{\infty h}$. While this category is a large one, it excludes the important groups O, T_d, and D_3—at least when A and/or J are certain double-valued irreps or are complex. Equations always simplify when a group is simply reducible, and the MCD ratios are a good example. In this case (17.2.14), (17.2.13), and (17.2.20) give

$$\frac{\mathcal{A}_1}{\mathcal{D}_0} = -\,(\text{MCD factor})\frac{2|f|}{\mu_B}\{AfJ\}$$

$$\times\left(\begin{Bmatrix} f' & f & f \\ A & J & J \end{Bmatrix}\langle J\|\mu^{f'}\|J\rangle + \begin{Bmatrix} f' & f & f \\ J & A & A \end{Bmatrix}\langle A\|\mu^{f'}\|A\rangle\right) \tag{17.2.24}$$

$$\frac{\mathcal{C}_0}{\mathcal{D}_0} = -\,(\text{MCD factor})\frac{2|f|}{\mu_B}\{AfJ\}\begin{Bmatrix} f' & f & f \\ J & A & A \end{Bmatrix}\langle A\|\mu^{f'}\|A\rangle \tag{17.2.25}$$

where the MCD factors are defined by (17.2.12) and (17.2.16), respectively, for complex and real bases; values for representative cases are given in Table 17.2.1. Recall that when spin–orbit coupling is neglected and a $|\mathbb{S}h\mathcal{M}\theta\rangle$ or $|\mathbb{S}ShM\theta\rangle$ basis is used, we may replace μ everywhere by $(-\mu_B)L$ [see paragraph following (17.2.8)].

17.3. The Oriented or Isotropic Case: Examples

$\mathcal{A}_1/\mathcal{D}_0$ and $\mathcal{C}_0/\mathcal{D}_0$ for the Groups O and T_d when A and J are Single-Valued Irreps

Equations (17.2.24) and (17.2.25) are applicable to the groups O and T_d when A and J are single-valued irreps. The MCD factors in Table 17.2.1 give

$$\frac{\mathcal{A}_1}{\mathcal{D}_0} = \mp\frac{\sqrt{6}}{\mu_B}\{AfJ\}$$

$$\times\left(\begin{Bmatrix} T_1 & f & f \\ A & J & J \end{Bmatrix}\langle J\|\mu^{t_1}\|J\rangle + \begin{Bmatrix} T_1 & f & f \\ J & A & A \end{Bmatrix}\langle A\|\mu^{t_1}\|A\rangle\right) \tag{17.3.1}$$

$$\frac{\mathcal{C}_0}{\mathcal{D}_0} = \mp\frac{\sqrt{6}}{\mu_B}\{AfJ\}\begin{Bmatrix} T_1 & f & f \\ J & A & A \end{Bmatrix}\langle A\|\mu^{t_1}\|A\rangle \tag{17.3.2}$$

where the − sign and $f = T_1$ applies for the group-O bases, and the + sign and $f = T_2$ for the group-T_d bases, of Table 17.2.1. The equations also hold for the group O_h if the appropriate labels u and g are affixed. Once again the remarks following (17.2.8) apply when spin–orbit coupling is neglected and a $|\mathbb{S}h\mathfrak{M}\theta\rangle$ or $|\mathbb{S}ShM\theta\rangle$ basis is used.

$Fe(CN)_6^{3-}$ *Revisited*

To demonstrate the power of the general equations just derived, we return to our old group-O_h $Fe(CN)_6^{3-}$ example of Section 11.2. Now instead of doing numerous matrix-element reductions and summations, we have immediately from (17.3.2) above and Sections C.11 and C.13 that for both the ${}^2T_{2g} \to {}^2T_{1u}$ transitions

$$\begin{aligned}\frac{\mathcal{C}_0}{\mathfrak{D}_0}\left({}^2T_{2g} \to {}^2T_{1u}\right) &= \frac{-\sqrt{6}}{\mu_B}\{T_{2g}T_{1u}T_{1u}\}\begin{Bmatrix} T_{1g} & T_{1u} & T_{1u} \\ T_{1u} & T_{2g} & T_{2g} \end{Bmatrix} \\ &\quad \times \left\langle {}^2T_{2g}\|(-\mu_B)L^{t_{1g}}\|{}^2T_{2g}\right\rangle \\ &= \sqrt{6}\,\{T_2T_1T_1\}\begin{Bmatrix} T_1 & T_1 & T_1 \\ T_1 & T_2 & T_2 \end{Bmatrix}\langle {}^2T_{2g}\|L^{t_{1g}}\|{}^2T_{2g}\rangle \\ &= \sqrt{6}\,(1)\left(-\tfrac{1}{6}\right)\langle {}^2T_{2g}\|L^{t_{1g}}\|{}^2T_{2g}\rangle \\ &= \frac{-1}{\sqrt{6}}\langle {}^2T_{2g}\|L^{t_{1g}}\|{}^2T_{2g}\rangle \end{aligned} \tag{17.3.3}$$

and for the ${}^2T_{2g} \to {}^2T_{2u}$ transition

$$\begin{aligned}\frac{\mathcal{C}_0}{\mathfrak{D}_0}\left({}^2T_{2g} \to {}^2T_{2u}\right) &= -\frac{\sqrt{6}}{\mu_B}(-1)\left(-\tfrac{1}{6}\right)\left\langle {}^2T_{2g}\|(-\mu_B)L^{t_{1g}}\|{}^2T_{2g}\right\rangle \\ &= \frac{1}{\sqrt{6}}\langle {}^2T_{2g}\|L^{t_{1g}}\|{}^2T_{2g}\rangle \end{aligned} \tag{17.3.4}$$

The tedious process of Section 11.2 has been replaced by looking up one $3j$ phase and one $6j$ coefficient. Here we have neglected spin–orbit coupling and have used a $|\mathbb{S}h\mathfrak{M}\theta\rangle$ basis, so μ has been replaced by $(-\mu_B)L$. If we substitute in the value, $-\sqrt{6}$, of the final orbital-angular-momentum reduced matrix element [from (13.2.8) and (13.4.3)], we obtain again our (11.2.13) results.

$\mathcal{A}_1 / \mathcal{D}_0$ and $\mathcal{C}_0 / \mathcal{D}_0$ for the Groups O and T_d in the More General Case

Unfortunately such simple expressions as those of (17.3.1)–(17.3.2) for $\mathcal{A}_1/\mathcal{D}_0$ and $\mathcal{C}_0/\mathcal{D}_0$ cannot be derived for the groups O and T_d in the general case, since in some cases the electric-dipole reduced matrix elements do not cancel, as a result of the differing r and r' dependence of the coefficients in the numerator ($\mathcal{A}_1$ and $\mathcal{C}_0$) and the denominator ($\mathcal{D}_0$). However, when A and J are *not both simultaneously* U' states, we can only have $r = 0$ in $\mathcal{D}_0$, and $r = r' = 0$ in $\mathcal{A}_1$ and $\mathcal{C}_0$. In this case the equations which apply are

$$\frac{\mathcal{A}_1}{\mathcal{D}_0} = \mp \frac{\sqrt{6}}{\mu_B}\{AfJ\}\left(\sum_q \begin{Bmatrix} T_1 & f & f \\ A & J & J \end{Bmatrix}_{q000} \langle J\|\mu^{t_1}\|J\rangle_q \right.$$

$$\left. + \sum_q \begin{Bmatrix} T_1 & f & f \\ J & A & A \end{Bmatrix}_{q000} \langle A\|\mu^{t_1}\|A\rangle_q \right) \qquad (17.3.5)$$

$$\frac{\mathcal{C}_0}{\mathcal{D}_0} = \mp \frac{\sqrt{6}}{\mu_B}\{AfJ\}\sum_q \begin{Bmatrix} T_1 & f & f \\ J & A & A \end{Bmatrix}_{q000} \langle A\|\mu^{t_1}\|A\rangle_q \qquad (17.3.6)$$

where—as in (17.3.1) and (17.3.2)—the − sign and $f = T_1$ apply for group-O bases, and the + sign and $f = T_2$ apply for group-T_d bases, of Table 17.2.1.

When *both* A and J are U', we must, for example, transform our $|(\mathbb{S}S, h)rt\tau\rangle$ reduced matrix elements into $|\mathbb{S}ShM\theta\rangle$ reduced matrix elements before cancellation of electric-dipole reduced matrix elements is possible. An example of this procedure is given as part of the $CoCl_4^{2-}$ calculation in Section 18.2.

17.4. The MCD Equations for Ambivalent Groups: Space-Averaged Case

In this section we use (16.2.3) to simplify the MCD equations for molecules or ions in solution, and for systems containing randomly oriented anisotropic guest molecules in an isotropic crystal, matrix, or glass. Here we must average over all orientations of the molecule-fixed axes with respect to the magnetic-field axis. Thus our MCD equations are those of (4.6.14) rather than those of (4.6.8). For an allowed electronic transition from state A to J

we now have

$$\bar{\mathcal{A}}_1 = \frac{-i}{3\mu_B|A|} \sum_{\alpha\alpha'\gamma\gamma'} (\langle J\gamma|\boldsymbol{\mu}|J\gamma'\rangle \delta_{\alpha\alpha'} - \langle A\alpha|\boldsymbol{\mu}|A\alpha'\rangle \delta_{\gamma\gamma'})$$

$$\cdot \langle A\alpha'|\mathbf{m}|J\gamma\rangle \times \langle J\gamma'|\mathbf{m}|A\alpha\rangle$$

$$\bar{\mathcal{B}}_0 = \mathcal{I}\mathrm{m}\left\{ \frac{-2}{3\mu_B|A|} \right.$$

$$\times \sum_{\alpha\gamma} \left[\left(\sum_{K\neq A} \sum_{\kappa} \frac{\langle K\kappa|\boldsymbol{\mu}|A\alpha\rangle}{W_K - W_A} \cdot \langle A\alpha|\mathbf{m}|J\gamma\rangle \times \langle J\gamma|\mathbf{m}|K\kappa\rangle \right) \right.$$

$$\left.\left. + \left(\sum_{K\neq J} \sum_{\kappa} \frac{\langle J\gamma|\boldsymbol{\mu}|K\kappa\rangle}{W_K - W_J} \cdot \langle A\alpha|\mathbf{m}|J\gamma\rangle \times \langle K\kappa|\mathbf{m}|A\alpha\rangle \right) \right] \right\}$$

(17.4.1)

$$\bar{\mathcal{C}}_0 = \frac{i}{3\mu_B|A|} \sum_{\alpha\alpha'\gamma} \langle A\alpha|\boldsymbol{\mu}|A\alpha'\rangle \cdot \langle A\alpha'|\mathbf{m}|J\gamma\rangle \times \langle J\gamma|\mathbf{m}|A\alpha\rangle$$

$$\bar{\mathcal{D}}_0 = \frac{1}{3|A|} \sum_{\alpha\gamma} |\langle A\alpha|\mathbf{m}|J\gamma\rangle|^2$$

$$= \frac{1}{3|A|} \sum_{i} \sum_{\alpha\gamma} \langle A\alpha|m_i|J\gamma\rangle \langle J\gamma|(m_i)^{\dagger}|A\alpha\rangle$$

where the operators are defined as in (17.2.1) and Section 9.8. The bar distinguishes these space-averaged parameters from those of (17.2.2)–(17.2.7).

Our procedure is first to expand the product $\boldsymbol{\mu} \cdot (\mathbf{m} \times \mathbf{m})$. Then, if the complex basis is used, operators are expressed in complex form using (17.2.1). In either the real or the complex basis the matrix-element product is then simplified much as in Section 17.2. We illustrate the steps for $\bar{\mathcal{A}}_1$. Expanding the product $\boldsymbol{\mu} \cdot (\mathbf{m} \times \mathbf{m})$, we obtain

$$\bar{\mathcal{A}}_1 = \frac{-i}{3\mu_B|A|} \sum_{\alpha\alpha'\gamma\gamma'} \sum_{\delta\varepsilon\kappa} \epsilon_{\delta\varepsilon\kappa} (\langle J\gamma|\mu_\delta|J\gamma'\rangle \delta_{\alpha\alpha'} - \langle A\alpha|\mu_\delta|A\alpha'\rangle \delta_{\gamma\gamma'})$$

$$\times \langle A\alpha'|m_\varepsilon|J\gamma\rangle \langle J\gamma'|m_\kappa|A\alpha\rangle \qquad (17.4.2)$$

where δ, ε, and κ are each x, y, or z, and $\epsilon_{\delta\varepsilon\kappa}$ is the alternating tensor defined so

$$\begin{aligned} \epsilon_{xyz} &= \epsilon_{yzx} = \epsilon_{zxy} = 1 \\ \epsilon_{yxz} &= \epsilon_{zyx} = \epsilon_{xzy} = -1 \end{aligned} \tag{17.4.3}$$

and all other $\epsilon_{\delta\varepsilon\kappa} = 0$. The relations (17.2.1) may be summarized by the matrix equation

$$\begin{pmatrix} g_x \\ g_y \\ g_z \end{pmatrix} = \begin{pmatrix} 1/\sqrt{2} & 0 & -1/\sqrt{2} \\ i/\sqrt{2} & 0 & i/\sqrt{2} \\ 0 & 1 & 0 \end{pmatrix} \begin{pmatrix} g_{-1} \\ g_0 \\ g_1 \end{pmatrix} = \mathbf{U} \begin{pmatrix} g_{-1} \\ g_0 \\ g_1 \end{pmatrix} \tag{17.4.4}$$

where $\mathbf{g} = \boldsymbol{\mu}$ or $\mathbf{m}$. Thus, with operators in complex form,

$$\begin{aligned} \bar{\mathcal{A}}_1 = \frac{-i}{3\mu_B|A|} \sum_{\alpha\alpha'\gamma\gamma'} \sum_{jkl} \sum_{\delta\varepsilon\kappa} \epsilon_{\delta\varepsilon\kappa} U_{\delta j} U_{\varepsilon k} U_{\kappa l} \big(\langle J\gamma|\mu_j|J\gamma'\rangle \delta_{\alpha\alpha'} - \langle A\alpha|\mu_j|A\alpha'\rangle \delta_{\gamma\gamma'} \big) \\ \times \langle A\alpha'|m_k|J\gamma\rangle \langle J\gamma'|m_l|A\alpha\rangle \end{aligned} \tag{17.4.5}$$

where j, k, and l each have the values -1, 0, or 1. Here, as the reader may verify,

$$\sum_{\delta\varepsilon\kappa} \epsilon_{\delta\varepsilon\kappa} U_{\delta j} U_{\varepsilon k} U_{\kappa l} = (\det \mathbf{U}) \epsilon_{jkl} = (-i) \epsilon_{jkl} \tag{17.4.6}$$

where $i = \sqrt{-1}$ and the alternating tensor ϵ_{jkl} is defined in the same manner as $\epsilon_{\delta\varepsilon\kappa}$ starting from $\epsilon_{-101} = 1$.

Thus, except for the additional factor $-i$, $\bar{\mathcal{A}}_1$ has the same form when a complex basis set is used as it does when the basis set is real:

$$\begin{aligned} \bar{\mathcal{A}}_1 = \frac{-1}{3\mu_B|A|} \sum_{\alpha\alpha'\gamma\gamma'} \sum_{jkl} \epsilon_{jkl} \big(\langle J\gamma|\mu_j|J\gamma'\rangle \delta_{\alpha\alpha'} - \langle A\alpha|\mu_j|A\alpha'\rangle \delta_{\gamma\gamma'} \big) \\ \times \langle A\alpha'|m_k|J\gamma\rangle \langle J\gamma'|m_l|A\alpha\rangle \end{aligned} \tag{17.4.7}$$

Equations similar in form to (17.4.2) and (17.4.7) may be written for $\bar{\mathcal{C}}_0$, $\bar{\mathcal{B}}_0$, and $\bar{\mathcal{D}}_0$. These equations are the most convenient starting point for deriving general space-averaged equations using the irreducible-tensor method. From them we proceed exactly as for the oriented case. We again use the notation (17.2.8):

$$\bar{\mathcal{A}}_1 = \bar{\mathcal{A}}_1(J) + \bar{\mathcal{C}}_0 \tag{17.4.8}$$

and make the assumptions in Section 17.2 about operator and wavefunction transformation properties. As in Section 17.2, our discussion is limited to cases involving no complex irreps. The notation to indicate the operator transformation properties of (17.2.9) must, however, be augmented to include a subscript on f and f'. We have $\mu_{\pm 1}$ (or μ_x and μ_y) and m_0 (or m_z) in addition to the operators in (17.2.2)–(17.2.7). And in most groups, $\mu_{\pm 1}$ belongs to a different irrep than μ_0, and so on. Thus the f and f' labels are insufficient, and we replace (17.2.9) with

$$\mu_i = \langle f_i'\phi_i' | \mu_i \rangle \mu_{\phi_i'}^{f_i'}$$

$$m_i = \langle f_i\phi_i | m_i \rangle m_{\phi_i}^{f_i} \tag{17.4.9}$$

where i is 0, 1, or -1 (complex bases) or x, y, or z (real bases). In the group D_4, for example, $f_0' = A_2$ for μ_0, while $f_1' = f_{-1}' = E$ for μ_1 and μ_{-1}. Defining for complex basis sets the new MCD factor [Table 17.2.1 applies only for (17.2.12) and (17.2.16)]

$$(\text{MCD}_{ijk}\ \text{factor}) = \langle f_i'\phi_i' | \mu_i \rangle \langle f_j\phi_j | m_j \rangle \langle f_k\phi_k | m_k \rangle \begin{pmatrix} f_i' & f_k & f_j \\ \phi_i' & \phi_k & \phi_j \end{pmatrix} \tag{17.4.10}$$

and using (17.4.7) and its $\bar{\mathcal{C}}_0$, $\bar{\mathcal{B}}_0$, and $\bar{\mathcal{D}}_0$ analogs, together with (17.4.9) and our methods of Section 17.2, we have

$$\bar{\mathcal{A}}_1 = \left[\sum_{ijk} \epsilon_{ijk} (\text{MCD}_{ijk}\ \text{factor}) \frac{\{A\}}{3\mu_B |A|} \sum_q \langle J \| \mu^{f_i'} \| J \rangle_q \right.$$

$$\left. \times \sum_{rr'} \begin{Bmatrix} f_i' & f_j & f_k \\ A^* & J^* & J^* \end{Bmatrix}_{qrr'0} \langle A \| m^{f_j} \| J \rangle_r \langle J \| m^{f_k} \| A \rangle_{r'} \right] + \bar{\mathcal{C}}_0 \tag{17.4.11}$$

$$\bar{\mathcal{C}}_0 = \sum_{ijk} \epsilon_{ijk} (\text{MCD}_{ijk} \text{ factor}) \frac{\{J\}}{3\mu_B |A|} \sum_q \langle A \| \mu^{f_i'} \| A \rangle_q$$

$$\times \sum_{rr'} \begin{Bmatrix} f_i' & f_k & f_j \\ J^* & A^* & A^* \end{Bmatrix}_{qr'r0} \langle A \| m^{f_j} \| J \rangle_r \langle J \| m^{f_k} \| A \rangle_{r'} \qquad (17.4.12)$$

$$\bar{\mathcal{B}}_0 = \mathcal{R}e \left\{ \sum_{ijk} \epsilon_{ijk} (\text{MCD}_{ijk} \text{ factor}) \frac{2}{3\mu_B |A|} \right.$$

$$\times \left[\left(\sum_{K \neq A} \frac{\{J\}}{W_K - W_A} \sum_q \langle K \| \mu^{f_i'} \| A \rangle_q \right. \right.$$

$$\left. \times \sum_{rr'} \begin{Bmatrix} f_i' & f_k & f_j \\ J^* & A^* & K^* \end{Bmatrix}_{qr'r0} \langle A \| m^{f_j} \| J \rangle_r \langle J \| m^{f_k} \| K \rangle_{r'} \right)$$

$$+ \left(\sum_{K \neq J} \frac{-\{A\}}{W_K - W_J} \sum_q \langle J \| \mu^{f_i'} \| K \rangle_q \right.$$

$$\left. \left. \left. \times \sum_{rr'} \begin{Bmatrix} f_i' & f_j & f_k \\ A^* & K^* & J^* \end{Bmatrix}_{qrr'0} \langle A \| m^{f_j} \| J \rangle_r \langle K \| m^{f_k} \| A \rangle_{r'} \right) \right] \right\}$$

$$(17.4.13)$$

$$\bar{\mathcal{D}}_0 = \frac{-\{A\}}{3|A|} \sum_j \frac{1}{|f_j|} \sum_r \delta(A^* f_j J r) \{A^* f_j J r\} \langle A \| m^{f_j} \| J \rangle_r \langle J \| m^{f_j} \| A \rangle_r$$

$$(17.4.14)$$

The ijk sums are over i, j, and k each equal to -1, 0, and 1. The equations are applicable for real basis sets if the sums are instead taken over i, j, and k each equal to x, y, and z, and the MCD factor of (17.4.10) is redefined as

$$(\text{MCD}_{ijk} \text{ factor}) = i[\text{Eq. } (17.4.10)] \qquad (17.4.15)$$

where $i = \sqrt{-1}$.

In all the above equations $\mu^{f_i'}$ may be replaced by $(-\mu_B)L^{f_i'}$ when spin–orbit coupling is neglected and a $|\mathbb{S}h\mathfrak{M}\theta\rangle$ or $|\mathbb{S}ShM\theta\rangle$ basis is used

—see the remarks following (17.2.8). The equations must be modified somewhat for nonambivalent groups.

Comparison of MCD Equations for the Space-Averaged and Oriented Cases

For the cubic groups O, O_h, and T_d, (17.4.11)–(17.4.14) reduce to the equations of Section 17.2 for the oriented or isotropic case, so

$$\begin{aligned} \overline{\mathcal{A}}_1 &= \mathcal{A}_1 \\ \overline{\mathcal{C}}_0 &= \mathcal{C}_0 \\ \overline{\mathcal{B}}_0 &= \mathcal{B}_0 \\ \overline{\mathcal{D}}_0 &= \mathcal{D}_0 \end{aligned} \tag{17.4.16}$$

This is not true, however, for anisotropic systems, and very different results may be obtained with (17.4.11)–(17.4.14) than with the equations of Section 17.2. This happens because μ_0 and (μ_1, μ_{-1}), and also m_0 and (m_1, m_{-1}), belong to different irreps.

Thus, as mentioned above, in the group D_4, when $i = 0$, $f_i' = A_2$, but when $i = 1$ or -1, $f_i' = E$. The same holds true for f_j and f_k. Consequently for D_4, (17.4.11) and Sections D.6 and D.9 give

$$\overline{\mathcal{A}}_1 = \langle A_2 a_2|\mu_0\rangle\langle E1|m_1\rangle\langle E-1|m_{-1}\rangle\begin{pmatrix} A_2 & E & E \\ a_2 & -1 & 1 \end{pmatrix}$$

$$\times \frac{2\{A\}}{3\mu_B|A|}\left[\left(\begin{Bmatrix} A_2 & E & E \\ A & J & J \end{Bmatrix}\langle J\|\mu^{A_2}\|J\rangle + \begin{Bmatrix} A_2 & E & E \\ J & A & A \end{Bmatrix}\langle A\|\mu^{A_2}\|A\rangle\right)\right.$$

$$\times\langle A\|m^E\|J\rangle\langle J\|m^E\|A\rangle$$

$$+\left(\begin{Bmatrix} E & E & A_2 \\ A & J & J \end{Bmatrix}\langle J\|\mu^E\|J\rangle + \begin{Bmatrix} E & A_2 & E \\ J & A & A \end{Bmatrix}\langle A\|\mu^E\|A\rangle\right)$$

$$\left.\times\left(\langle A\|m^E\|J\rangle\langle J\|m^{A_2}\|A\rangle + \langle A\|m^{A_2}\|J\rangle\langle J\|m^E\|A\rangle\right)\right] \tag{17.4.17}$$

Here we have used Section D.6 and the $2jm$ and $6j$ symmetry rules to simplify the expression. Then from (17.4.14) we obtain

$$\bar{\mathcal{D}}_0 = \frac{-\{A\}}{3|A|}\left(\delta(AEJ)\{AEJ\}\langle A\|m^E\|J\rangle\langle J\|m^E\|A\rangle\right.$$

$$\left.+\delta(AA_2J)\{AA_2J\}\langle A\|m^{A_2}\|J\rangle\langle J\|m^{A_2}\|A\rangle\right) \qquad (17.4.18)$$

The equations also hold for D_{4h} if appropriate labels u and g are affixed. It is easy to show that these results are not equivalent to (17.2.14) and (17.2.20) in most cases. For example, when $A \to J$ is $A_1 \to E$ in the group D_4, only the first $6j$ coefficient in $\bar{\mathcal{A}}_1$ and only the first δ coefficient in $\bar{\mathcal{D}}_0$ are nonzero. We then obtain

$$\bar{\mathcal{A}}_1 = \langle A_2 a_2|\mu_0\rangle\langle E1|m_1\rangle\langle E-1|m_{-1}\rangle\begin{pmatrix} A_2 & E & E \\ a_2 & -1 & 1\end{pmatrix}\frac{2}{3\mu_B}\begin{Bmatrix} A_2 & E & E \\ A_1 & E & E\end{Bmatrix}$$

$$\times\langle E\|\mu^{A_2}\|E\rangle\langle A_1\|m^E\|E\rangle\langle E\|m^E\|A_1\rangle \qquad (17.4.19)$$

$$\bar{\mathcal{D}}_0 = -\tfrac{1}{3}\langle A_1\|m^E\|E\rangle\langle E\|m^E\|A_1\rangle$$

These look very much like our equations of Section 17.2 for the same symmetries, but comparison shows $\bar{\mathcal{A}}_1$ and $\bar{\mathcal{D}}_0$ above have $\frac{1}{3}$ and $\frac{2}{3}$ the values obtained from (17.2.14) and (17.2.20) respectively. Thus $\bar{\mathcal{A}}_1/\bar{\mathcal{D}}_0$ has $\frac{1}{2}$ the value of $\mathcal{A}_1/\mathcal{D}_0$.

An even greater difference occurs between the oriented and space-averaged results for $E_g'' \to E_u''$ in the group D_{4h}, since in this case all $6j$ and δ factors in (17.4.17) and (17.4.18) are nonzero; the equations contain nonzero m^{A_2} electric-dipole reduced matrix elements which do not enter at all into (17.2.14) and (17.2.20). For the oriented case (17.2.24) gives

$$\frac{\mathcal{A}_1}{\mathcal{D}_0} = -\langle A_{2g}a_{2g}|\mu_0\rangle\langle E_u 1|m_1\rangle\langle E_u -1|m_{-1}\rangle\begin{pmatrix} A_{2g} & E_u & E_u \\ a_{2g} & -1 & 1\end{pmatrix}$$

$$\times\frac{4}{\mu_B}\left(\begin{Bmatrix} A_{2g} & E_u & E_u \\ E_g'' & E_u'' & E_u''\end{Bmatrix}\langle E_u''\|\mu^{A_{2g}}\|E_u''\rangle\right.$$

$$\left.+\begin{Bmatrix} A_{2g} & E_u & E_u \\ E_u'' & E_g'' & E_g''\end{Bmatrix}\langle E_g''\|\mu^{A_{2g}}\|E_g''\rangle\right) \qquad (17.4.20)$$

while for the space-averaged case no simple expression for $\bar{\mathcal{A}}_1/\bar{\mathcal{D}}_0$ may be written at this stage, since the electric-dipole reduced matrix elements in $\bar{\mathcal{A}}_1$ and $\bar{\mathcal{D}}_0$ of (17.4.17) and (17.4.18) do not cancel.

17.5. Orientation Dependence of the MCD in Cubic Crystals

As demonstrated by Satten et al. (1968), Satten (1971), and Yeakel (1977), the Zeeman splitting of U' electronic or vibronic states in octahedral (or tetrahedral) molecules, except in a few special cases, is anisotropic. We wish to show here that, in contrast to the Zeeman effect, MCD transitions to all states in cubic systems are isotropic *as long as the MCD is linear in the field*; and this is the ordinary MCD case to which all the MCD equations in this chapter pertain. Consequently MCD results are independent of the orientation of the magnetic field with respect to the cubic crystal axes. This has practical significance in that octahedral crystals used in spectroscopic studies are frequently grown from the melt and are most easily polished without regard to the orientation of the crystal axes.

We demonstrate the above point for $\mathcal{A}_1$ only, but analogous results are obtained for $\mathcal{B}_0$ and $\mathcal{C}_0$. For an oriented crystal in a magnetic field parallel to the crystalline z axis, $\mathcal{A}_1$ is given by (17.2.3):

$$\mathcal{A}_1(z) = \frac{-2i}{\mu_B|A|} \sum_{\alpha\alpha'\gamma\gamma'} \left(\langle J\gamma|\mu_z|J\gamma'\rangle\delta_{\alpha\alpha'} - \langle A\alpha|\mu_z|A\alpha'\rangle\delta_{\gamma\gamma'}\right) \times \langle A\alpha'|m_x|J\gamma\rangle\langle J\gamma'|m_y|A\alpha\rangle \tag{17.5.1}$$

When the field is along an arbitrary direction z' with respect to the principal crystal axis z, the same formula holds and the same functions may be used except that the operators $\mu_{z'}$, $m_{x'}$, and $m_{y'}$, replace μ_z, m_x, and m_y respectively. We demonstrate that the same value of $\mathcal{A}_1$ $[\mathcal{A}_1(z')]$ is calculated in this case.

The operators $\mu_{z'}$, $m_{x'}$, and $m_{y'}$, are first expressed in terms of μ_x, μ_y, μ_z and m_x, m_y, m_z using the rotation matrix we call $\mathbf{R}$. After taking the sums outside, we have

$$\mathcal{A}_1(z') = \frac{-2i}{\mu_B|A|} \sum_{\delta\varepsilon\kappa} R_{z'\delta}R_{x'\varepsilon}R_{y'\kappa} \times \sum_{\alpha\alpha'\gamma\gamma'} \left(\langle J\gamma|\mu_\delta|J\gamma'\rangle\delta_{\alpha\alpha'} - \langle A\alpha|\mu_\delta|A\alpha'\rangle\delta_{\gamma\gamma'}\right) \times \langle A\alpha'|m_\varepsilon|J\gamma\rangle\langle J\gamma'|m_\kappa|A\alpha\rangle \tag{17.5.2}$$

where the sum is over δ, ε, and κ each equal to x, y, and z. We next use (17.4.9) and the methods of Section 17.2 to simplify the equation and obtain

$$\mathscr{A}_1(z') = \sum_{\delta\varepsilon\kappa} R_{z'\delta} R_{x'\varepsilon} R_{y'\kappa} \langle f'\phi'_\delta|\mu_\delta\rangle\langle f\phi_\varepsilon|m_\varepsilon\rangle\langle f\phi_\kappa|m_\kappa\rangle \begin{pmatrix} f' & f & f \\ \phi'_\delta & \phi_\kappa & \phi_\varepsilon \end{pmatrix}$$

$$\times \frac{2i}{\mu_B|A|}\Bigg[\{A\}\sum_q \langle J\|\mu^{f'}\|J\rangle_q \sum_{rr'} \begin{Bmatrix} f' & f & f \\ A & J & J \end{Bmatrix}_{qrr'0}$$

$$\times \langle A\|m^f\|J\rangle_r \langle J\|m^f\|A\rangle_{r'} + \{J\}\sum_q \langle A\|\mu^{f'}\|A\rangle_q$$

$$\times \sum_{rr'} \begin{Bmatrix} f' & f & f \\ J & A & A \end{Bmatrix}_{qr'r0} \langle A\|m^f\|J\rangle_r \langle J\|m^f\|A\rangle_{r'}\Bigg] \quad (17.5.3)$$

Here $f' = T_1$; and $f = T_1$ for the group O, but $f = T_2$ for T_d. If we use any of our real-O or -T_d bases, we find

$$\langle f'\phi'_\delta|\mu_\delta\rangle\langle f\phi_\varepsilon|m_\varepsilon\rangle\langle f\phi_\kappa|m_\kappa\rangle \begin{pmatrix} f' & f & f \\ \phi'_\delta & \phi_\kappa & \phi_\varepsilon \end{pmatrix}$$

$$= \epsilon_{\delta\varepsilon\kappa} \langle f'\phi'_z|\mu_z\rangle\langle f\phi_x|m_x\rangle\langle f\phi_y|m_y\rangle \begin{pmatrix} f' & f & f \\ \phi'_z & \phi_y & \phi_x \end{pmatrix} \quad (17.5.4)$$

where $\epsilon_{\delta\varepsilon\kappa}$ is defined as in (17.4.3). Furthermore

$$\sum_{\delta\varepsilon\kappa} R_{z'\delta} R_{x'\varepsilon} R_{y'\kappa} \epsilon_{\delta\varepsilon\kappa} = \det \mathbf{R} = 1 \quad (17.5.5)$$

since the *rotation* matrix $\mathbf{R}$, which expresses x', y', and z' in terms of x, y, and z, is orthogonal and thus has a determinant of 1.

Making all of these substitutions,

$$\mathscr{A}_1(z') = i\langle f'\phi'_z|\mu_z\rangle\langle f\phi_x|m_x\rangle\langle f\phi_y|m_y\rangle \begin{pmatrix} f' & f & f \\ \phi'_z & \phi_y & \phi_x \end{pmatrix}$$

$$\times \frac{2}{\mu_B|A|}[\text{same as } (17.5.3)] \quad (17.5.6)$$

This is precisely our result in Section 17.2 for real basis sets. It is clear that the same expression would be obtained for $\mathcal{A}_1(z)$ from (17.5.1), and hence $\mathcal{A}_1(z) = \mathcal{A}_1(z')$. Moreover, our results hold for vibronic transitions as well.

17.6. The MCD Equations for Nonambivalent Groups

Nonambivalent groups contain one or more pairs of complex irreps (a, b) for which $a^* = b \neq a$. In D_3, for example, the pair is (P_1, P_2), where $P_1^* = P_2$ and $P_2^* = P_1$ (Section 14.2). These irreps have complex characters and so are easily recognized by a glance at the character table for a group. In the absence of a magnetic field or time-dependent Hamiltonians, states belonging to a set of paired irreps are always degenerate. Consequently, spectroscopists often treat P_1- and P_2-type irreps theoretically as the components of a degenerate representation. In D_3 this degenerate representation is commonly called E'' [e.g. Griffith (1964)].

No one has yet been able, however, to define high-symmetry $3jm$ for nonambivalent groups while treating these paired irreps as a degenerate representation. This is not surprising, since the (P_1, P_2) degeneracy is purely a consequence of time-reversal symmetry, but the time-reversal operation is not a point-group operation. Thus we must, for example, in D_3 treat the E'' states as a pair of states belonging to separate P_1 and P_2 irreps.

Here we focus on the practical problem of how to calculate the MCD of the transitions $A \to J$ of the sort $A \to (Q_1, Q_2)$, $(P_1, P_2) \to J$, and $(P_1, P_2) \to (Q_1, Q_2)$, where (P_1, P_2) and (Q_1, Q_2) are pairs of complex (nonambivalent) irreps. P_1, P_2, Q_1, and Q_2 themselves may or may not be degenerate. While our simplified MCD equations of Sections 17.2 and 17.4 are correct, for example, for $P_1 \to J$ or $P_2 \to J$ individually, they do not take into account the degeneracy of P_1 and P_2 and are thus no good for $(P_1, P_2) \to J$.

In this section we derive MCD equations using a complex basis set for the oriented or isotropic case only. Similar equations may be easily obtained for the space-averaged case. Analogous changes are simply made in (17.4.11)–(17.4.14).

To derive the $\mathcal{C}_0$ equation for the oriented or isotropic case for $(P_1, P_2) \to J$, we start with (17.2.6) except that $A\alpha$, $A\alpha'$ are replaced by $A\alpha$, $A'\alpha'$ and a summation over $A, A' = P_1, P_2$ is added to the equation. Also $|A|$ is replaced by

$$|P_1| + |P_2| = 2|P_1| \qquad (17.6.1)$$

Thus instead of (17.2.6) we begin with

$$\mathcal{C}_0[(P_1, P_2) \to J]$$

$$= \frac{2}{\mu_B 2|P_1|}\Bigg(\sum_{\rho_1}\langle P_1\rho_1|\mu_0|P_1\rho_1\rangle\sum_{\gamma}\langle P_1\rho_1|m_1|J\gamma\rangle\langle J\gamma|m_{-1}|P_1\rho_1\rangle$$

$$+\sum_{\rho_2}\langle P_2\rho_2|\mu_0|P_2\rho_2\rangle\sum_{\gamma}\langle P_2\rho_2|m_1|J\gamma\rangle\langle J\gamma|m_{-1}|P_2\rho_2\rangle$$

$$+\sum_{\rho_1\rho_2}\langle P_1\rho_1|\mu_0|P_2\rho_2\rangle\sum_{\gamma}\langle P_2\rho_2|m_1|J\gamma\rangle\langle J\gamma|m_{-1}|P_1\rho_1\rangle$$

$$+\sum_{\rho_1\rho_2}\langle P_2\rho_2|\mu_0|P_1\rho_1\rangle\sum_{\gamma}\langle P_1\rho_1|m_1|J\gamma\rangle\langle J\gamma|m_{-1}|P_2\rho_2\rangle\Bigg) \tag{17.6.2}$$

From here the derivation proceeds exactly as for (17.2.13), with the result

$$\mathcal{C}_0[(P_1, P_2) \to J] = \langle f'\phi_0'|\mu_0\rangle\langle f\phi_1|m_1\rangle\langle f^*\phi_{-1}|m_{-1}\rangle\begin{pmatrix} f' & f^* & f \\ \phi_0' & \phi_{-1} & \phi_1 \end{pmatrix}$$

$$\times\frac{\{J\}}{\mu_B|P_1|}\sum_{A, A'=P_1, P_2}\sum_{q}\langle A\|\mu^{f'}\|A'\rangle_q$$

$$\times\sum_{rr'}\begin{Bmatrix} f' & f^* & f \\ J^* & A'^* & A^* \end{Bmatrix}_{qr'r0}\langle A'\|m^f\|J\rangle_r\langle J\|m^{f^*}\|A\rangle_{r'} \tag{17.6.3}$$

The equation above may be simplified using the methods of 14.11 in some cases. This can best be seen by returning to (17.6.2). Since $\mathcal{C}_0$ is an observable, it is invariant under the time-reversal operation T. Thus only real numbers contribute, and these real parts are individually invariant under T. Since $\mu_0 = \mu_z$, (14.4.3) and (14.11.4) give for the second term of

(17.6.2)

$$T\langle P_2\rho_2|\mu_z|P_2\rho_2\rangle\sum_{\gamma}\langle P_2\rho_2|m_1|J\gamma\rangle\langle J\gamma|m_{-1}|P_2\rho_2\rangle$$

$$= -\langle P_1\rho_1|\mu_z|P_1\rho_1\rangle\sum_{\gamma}\langle P_1\rho_1|m_{-1}|J^*\gamma^*\rangle\langle J^*\gamma^*|m_1|P_1\rho_1\rangle$$

$$= -\langle P_1\rho_1|\mu_z|P_1\rho_1\rangle\sum_{\gamma}\langle P_1\rho_1|m_{-1}|J\gamma\rangle\langle J\gamma|m_1|P_1\rho_1\rangle$$

$$= -\left[\text{first term of (17.6.2) with } m_1 \text{ and } m_{-1} \text{ interchanged}\right] \qquad (17.6.4)$$

since here $J = J^*$ and the sum over γ includes γ^*. Similarly for the fourth term

$$T\langle P_2\rho_2|\mu_z|P_1\rho_1\rangle\sum_{\gamma}\langle P_1\rho_1|m_1|J\gamma\rangle\langle J\gamma|m_{-1}|P_2\rho_2\rangle$$

$$= -\langle P_1\rho_1|\mu_z|P_2\rho_2\rangle\sum_{\gamma}\langle P_2\rho_2|m_{-1}|J\gamma\rangle\langle J\gamma|m_1|P_1\rho_1\rangle$$

$$= -\left[\text{third term of (17.6.2) with } m_1 \text{ and } m_{-1} \text{ interchanged}\right] \qquad (17.6.5)$$

We then proceed again through the same process for each term as we did in the derivation of (17.2.13). We find that the first and second terms and the third and fourth terms differ only in that the second and fourth are multiplied by

$$(-1)\begin{pmatrix} f' & f & f^* \\ \phi_0' & \phi_1 & \phi_{-1} \end{pmatrix} \quad \text{rather than} \quad \begin{pmatrix} f' & f^* & f \\ \phi_0' & \phi_{-1} & \phi_1 \end{pmatrix}$$

and f and f^* are interchanged in the $6j$ coefficient and as operator labels in the reduced matrix elements. For reasons discussed in Section 4.2, we have $f = f^*$ for all groups of interest for MCD. In such cases $\{f'ff^*\} = -1$. With these assumptions, we obtain

$$\mathcal{C}_0[(P_1, P_2) \to J]$$

$$= \mathcal{R}e\Bigg[(\text{MCD factor})\frac{2\{J\}}{\mu_B|P_1|}\Bigg(\sum_q \langle P_1\|\mu^{f'}\|P_1\rangle_q$$

$$\times \sum_{rr'} \begin{Bmatrix} f' & f & f \\ J^* & P_1^* & P_1^* \end{Bmatrix}_{qr'r0} \langle P_1\|m^f\|J\rangle_r \langle J\|m^f\|P_1\rangle_{r'}$$

$$+ \sum_q \langle P_1\|\mu^{f'}\|P_2\rangle_q \sum_{rr'} \begin{Bmatrix} f' & f & f \\ J^* & P_2^* & P_1^* \end{Bmatrix} qr'r0 \,\langle P_2\|m^f\|J\rangle_r \langle J\|m^f\|P_1\rangle_{r'}\Bigg)\Bigg] \tag{17.6.6}$$

where the MCD factor is that of (17.2.12).

The equation above is useful, for example, for the group D_3 in the $D_3 \supset C_3$ basis. There $f' = A_2, f = f^* = E$, and $|P_1| = 1$. The $6j$ for the first term is zero by (16.1.2), since $\delta(f'P_1P_1^*) = \delta(A_2P_1P_2) = 0$, and the MCD factor is given in Table 17.2.1. Thus we have

$$\mathcal{C}_0[(P_1, P_2) \to J] = \mathcal{R}e\Bigg[\left(\frac{-1}{\sqrt{2}}\right)\frac{2\{J\}}{\mu_B}\langle P_1\|\mu^{A_2}\|P_2\rangle$$

$$\begin{Bmatrix} A_2 & E & E \\ J^* & P_1 & P_2 \end{Bmatrix}\langle P_2\|m^E\|J\rangle\langle J\|m^E\|P_1\rangle\Bigg] \tag{17.6.7}$$

Here we have used the relations $P_1^* = P_2$, $P_2^* = P_1$ in the $6j$. When this equation is appropriate, $J = J^*$, but we keep the star to make the derivation clear.

Little change in $\mathcal{C}_0$ is involved when J is nonambivalent. Thus for $(P_1, P_2) \to (Q_1, Q_2)$ we may use (17.6.3) or (17.6.6) if only we sum over $J = Q_1, Q_2$ in each equation. Similarly, for $A \to (Q_1, Q_2)$ we may use (17.2.13) if we sum the equation over $J = Q_1, Q_2$. Thus

$$\mathcal{C}_0[(P_1, P_2) \to (Q_1, Q_2)] = \sum_{J=Q_1, Q_2} [\text{Eq. (17.6.3)}] \tag{17.6.8}$$

$$\mathcal{C}_0[A \to (Q_1, Q_2)] = \sum_{J=Q_1, Q_2} [\text{Eq. (17.2.13)}] \tag{17.6.9}$$

The $A \to (Q_1, Q_2)$ equation simplifies, using methods of Section 14.11, under the same conditions as those which allow (17.6.3) to be reduced to (17.6.6). When these conditions are satisfied, (17.6.9) may be replaced by

$$\mathcal{C}_0[A \to (Q_1, Q_2)] = \mathcal{R}e\Bigg[(\text{MCD factor})\frac{4\{Q_1\}}{\mu_B|A|} \times \sum_q \langle A\|\mu^{f'}\|A\rangle_q \sum_{rr'} \begin{Bmatrix} f' & f & f \\ Q_2 & A^* & A^* \end{Bmatrix} qr'r0 \times \langle A\|m^f\|Q_1\rangle_r \langle Q_1\|m^f\|A\rangle_{r'}\Bigg] \tag{17.6.10}$$

In this case, of course, $A = A^*$ (or else we would be using one of the P_1, P_2 equations).

Since $\mathcal{A}_1(A) = \mathcal{C}_0$, we need now only evaluate $\mathcal{A}_1(J)$ in order to obtain $\mathcal{A}_1$ when A or J or both are nonambivalent. Using the same methods which led to our equations above we find for the oriented or isotropic case that

$$\mathcal{A}_1(J)[A \to (Q_1, Q_2)] = (\text{MCD factor})\frac{2\{A\}}{\mu_B|A|} \sum_{J, J' = Q_1, Q_2} \times \sum_q \langle J\|\mu^{f'}\|J'\rangle_q \sum_{rr'} \begin{Bmatrix} f' & f & f \\ A^* & J'^* & J^* \end{Bmatrix} qrr'0 \times \langle A\|m^f\|J\rangle_r \langle J'\|m^f\|A\rangle_{r'} \tag{17.6.11}$$

$$\mathcal{A}_1(J)[(P_1, P_2) \to J] = (\text{MCD factor})\frac{\{P_1\}}{\mu_B|P_1|} \times \sum_{A = P_1, P_2} \sum_q \langle J\|\mu^{f'}\|J\rangle_q \times \sum_{rr'} \begin{Bmatrix} f' & f & f \\ A^* & J^* & J^* \end{Bmatrix} qrr'0 \times \langle A\|m^f\|J\rangle_r \langle J\|m^f\|A\rangle_{r'} \tag{17.6.12}$$

$$\mathcal{A}_1(J)[(P_1, P_2) \to (Q_1, Q_2)] = \sum_{A = P_1, P_2} [\text{Eq. (17.6.11) with } |A| \text{ replaced by } 2|P_1|] \tag{17.6.13}$$

where in each case, as always,

$$\mathcal{A}_1 = \mathcal{A}_1(J) + \mathcal{C}_0 \tag{17.6.14}$$

We have assumed $f = f^*$, which is true for all groups of interest for MCD. The equations for $\mathcal{A}_1(J)$ may be simplified in some cases by methods analogous to those for $\mathcal{C}_0$.

For the dipole strength we obtain

$$\mathfrak{D}_0[(P_1, P_2) \to J] = \frac{-\{P_1\}}{2|f|\,|P_1|} \sum_{A=P_1, P_2} \sum_r \delta(A^*fJr) \times \{A^*fJr\}\langle A\|m^f\|J\rangle_r \langle J\|m^f\|A\rangle_r \tag{17.6.15}$$

$$\mathfrak{D}_0[A \to (Q_1, Q_2)] = \sum_{J=Q_1, Q_2} [\text{Eq. (17.2.20)}] \tag{17.6.16}$$

$$\mathfrak{D}_0[(P_1, P_2) \to (Q_1, Q_2)] = \sum_{J=Q_1, Q_2} [\text{Eq. (17.6.15)}] \tag{17.6.17}$$

where again we assume $f = f^*$. The equations may be simplified to give

$$\mathfrak{D}_0[(P_1, P_2) \to J] = \frac{-\{P_1\}}{|f|\,|P_1|} \sum_r \delta(P_2 fJr)\{P_2 fJr\} \times \langle P_1\|m^f\|J\rangle_r \langle J\|m^f\|P_1\rangle_r \tag{17.6.18}$$

$$\mathfrak{D}_0[A \to (Q_1, Q_2)] = \frac{-2\{A\}}{|f|\,|A|} \sum_r \delta(A^*fQ_1 r)\{A^*fQ_1 r\} \times \langle A\|m^f\|Q_1\rangle_r \langle Q_1\|m^f\|A\rangle_r \tag{17.6.19}$$

Then when all reduced matrix elements are multiplicity-free, (17.6.10) and (17.6.19) give

$$\frac{\mathcal{C}_0}{\mathfrak{D}_0}[A \to (Q_1, Q_2)] = \mathfrak{Re}\left[-(\text{MCD factor})\frac{2|f|}{\mu_B}\{A^*fQ_1\} \times \begin{Bmatrix} f' & f & f \\ Q_2 & A^* & A^* \end{Bmatrix} \langle A\|\mu^{f'}\|A\rangle\right] \tag{17.6.20}$$

Note how similar the equation is to (17.2.25).

18 *Matrix-Element Simplification Using the Irreducible-Tensor Method: Fundamental Equations*

The most elegant applications of the irreducible-tensor method lie in the area of matrix-element simplification. We begin this process in this chapter with the derivation of some extremely useful general equations which may be used whenever a one-electron operator acts on only one part of a coupled system. A few applications of these equations are given at this time, and the reader will find they are also used extensively in Chapters 19 and 20 in a number of rather different contexts. In this chapter they are used to express $|(\mathbb{S}S, h)rt\tau\rangle$ matrix elements in terms of $|\mathbb{S}SM\rangle$, $|h\theta\rangle$, or $|\mathbb{S}ShM\theta\rangle$ matrix elements, the choice depending on the nature of the operator.

The equations derived are applicable to both ambivalent and nonambivalent groups. However, to obtain the equations for $SO_3 \supset SO_2$ (or $O_3 \supset SO_3 \supset SO_2$) with Condon–Shortley phases, the historical phase $H(abc)$ must be included whenever (10.2.3), (10.3.3), or (10.2.1) is used in the derivations. Since we are interested primarily in molecular systems, we include the $H(abc)$ factors only when they are needed for molecular applications—for example, in Section 18.5, where we couple spins in SO_3.

18.1. Simplification of Matrix Elements of Operators Which Act on Only One Part of a Coupled System

Often in spectroscopic applications we are interested in matrix elements of spin-independent operators, such as L_x, L_y, and L_z and m_x, m_y, and m_z, or of space-independent operators such as S_z. If our basis is a spin–orbit-coupled $|(\mathbb{S}S, h)rt\tau\rangle$ basis, these operators operate only on the space functions $|h\theta\rangle$, or only on the spin functions $|\mathbb{S}SM\rangle$, of the $|(\mathbb{S}S, h)rt\tau\rangle$ coupled functions. In such a situation, matrix elements may be simplified enormously using irreducible-tensor techniques. And in Chapters 19 and 20 we show how we can break down most *one-electron* matrix elements in such a way that they may be expressed in terms of matrix elements of an operator which acts on only one part of a coupled system. Thus the equations we derive in this section have a very wide applicability indeed, and are not limited to the situation described above for spin–orbit-coupled functions.

Suppose that we are interested in evaluating matrix elements of a one-electron operator which belongs to an irrep d of the point group G of the system, and that this operator D^d *operates only on the first part* (a *and* a'), *and not on the second part* (b *and* b'), *of the coupled kets* $|(ab)rc\gamma\rangle$ *and* $|(a'b')r'c'\gamma'\rangle$. Using (10.3.7) and expanding out the coupled wavefunctions with (10.2.3) and (10.3.3), we obtain

$$\left\langle (ab)rc \middle\| D^d \middle\| (a'b')r'c' \right\rangle_p$$

$$= \sum_{\gamma\gamma'\delta} \begin{pmatrix} c \\ \gamma \end{pmatrix} \begin{pmatrix} c^* & d & c' \\ \gamma^* & \delta & \gamma' \end{pmatrix}^{*p} \left\langle (ab)rc\gamma \middle| D^d_\delta \middle| (a'b')r'c'\gamma' \right\rangle$$

$$= \sum_{\gamma\gamma'\delta} \begin{pmatrix} c \\ \gamma \end{pmatrix} \begin{pmatrix} c^* & d & c' \\ \gamma^* & \delta & \gamma' \end{pmatrix}^{*p} |c|^{1/2} \begin{pmatrix} c \\ \gamma \end{pmatrix} \sum_{\alpha\beta} \begin{pmatrix} a & b & c^* \\ \alpha & \beta & \gamma^* \end{pmatrix}^{r}$$

$$\times |c'|^{1/2} \begin{pmatrix} c' \\ \gamma' \end{pmatrix} \sum_{\alpha'\beta'} \begin{pmatrix} a' & b' & c'^* \\ \alpha' & \beta' & \gamma'^* \end{pmatrix}^{*r'} \langle a\alpha, b\beta | D^d_\delta | a'\alpha', b'\beta' \rangle$$

$$= \delta_{bb'} |c|^{1/2} |c'|^{1/2} \sum_{\alpha\alpha'\beta\gamma\gamma'\delta} \begin{pmatrix} c' \\ \gamma' \end{pmatrix} \begin{pmatrix} a \\ \alpha \end{pmatrix} \begin{pmatrix} c^* & d & c' \\ \gamma^* & \delta & \gamma' \end{pmatrix}^{*p} \begin{pmatrix} a & b & c^* \\ \alpha & \beta & \gamma^* \end{pmatrix}^{r}$$

$$\times \begin{pmatrix} a' & b & c'^* \\ \alpha' & \beta & \gamma'^* \end{pmatrix}^{*r'} \sum_q \begin{pmatrix} a^* & d & a' \\ \alpha^* & \delta & \alpha' \end{pmatrix}^{q} \langle a \| D^d \| a' \rangle_q \tag{18.1.1}$$

In the last step we use (10.2.2) and our assumption that D^d acts on a and a' only. The sum of a product of four $3jm$ reminds us immediately of our $6j$ equation (16.1.1). Appropriate rearrangements of the $3jm$ and manipulation of phase factors puts (18.1.1) into a form suitable for substitution with (16.1.1); the method is shown in detail in Section 16.3. We obtain

$$\left\langle (ab)rc \middle\| D^d \middle\| (a'b')r'c' \right\rangle_p = \delta_{bb'} |c|^{1/2} |c'|^{1/2} \{abc^*r\}\{a'\} \sum_q \{a^*da'q\}$$

$$\times \begin{Bmatrix} d & c' & c^* \\ b^* & a & a' \end{Bmatrix}_{qr'rp} \langle a \| D^d \| a' \rangle_q \tag{18.1.2}$$

This equation for $\langle (ab)rc \| D^d \| (a'b')r'c' \rangle$ is by no means unique; different rearrangements of the $3jm$ and different correspondences between the irreps here and those of (16.1.1) lead to equations equivalent to this one but in another form; the $6j$ coefficient occurs in another arrangement and phase factors may differ. For example, Butler (1981) gives the D^d equation in the form

$$\langle (ab)rc \| D^d \| (a'b')r'c' \rangle_p = \delta_{bb'} |c|^{1/2} |c'|^{1/2} \{abc^* r\} \{a'\} \sum_q \{a^* d a' q\} \times \begin{Bmatrix} c^* & d & c' \\ a' & b^* & a \end{Bmatrix}_{rqr'p} \langle a \| D^d \| a' \rangle_q \tag{18.1.3}$$

which the reader should confirm, using the $6j$ symmetry rules of Section 16.1, is equal to (18.1.2).

Similarly we may derive an equation for the case where an operator E^e *operates only on the second part of a coupled system*:

$$\langle (ab)rc \| E^e \| (a'b')r'c' \rangle_p = \delta_{aa'} |c|^{1/2} |c'|^{1/2} \{a'b'c'^* r'\} \{b'\} \sum_q \{b^* e b' q\} \times \begin{Bmatrix} e & c' & c^* \\ a^* & b & b' \end{Bmatrix}_{qr'rp} \langle b \| E^e \| b' \rangle_q \tag{18.1.4}$$

Again this may be written in numerous equivalent forms, such as that of Eq. (4.3.9) of Butler (1981).

All these equations may, of course, be simplified when multiplicity indices are unnecessary and all $3j$ phases for the group obey (10.2.5). Then they reduce to the forms

$$\langle (ab)c \| D^d \| (a'b')c' \rangle = \delta_{bb'} |c|^{1/2} |c'|^{1/2} (-1)^{a'+b+c^*+d} \begin{Bmatrix} d & c' & c^* \\ b^* & a & a' \end{Bmatrix} \langle a \| D^d \| a' \rangle \tag{18.1.5}$$

$$\langle (ab)c \| E^e \| (a'b')c' \rangle = \delta_{aa'} |c|^{1/2} |c'|^{1/2} (-1)^{a'^*+b^*+c'+e} \begin{Bmatrix} e & c' & c^* \\ a^* & b & b' \end{Bmatrix} \langle b \| E^e \| b' \rangle \tag{18.1.6}$$

which are similar to those of Griffith (1962). In obtaining the equations we have used (10.2.9), which gives $(-1)^a(-1)^{a^*} = \{a\}$. Then in (18.1.2) we use the fact that here $\{a\} = \{a'\}$, and in (18.1.4) that $\{a'b'c'^*r'\} = \{a'^*b'^*c'r'\}$.

Equations (18.1.2) and (18.1.4) also simplify considerably for totally symmetric operators. For example, when $e = A_1$, (18.1.4) reduces to

$$\langle (ab)rc \| E^{A_1} \| (a'b')r'c' \rangle$$
$$= \delta_{aa'}\delta_{bb'}\delta_{cc'}\delta_{rr'}\delta(a'b'c'^*r')|c|^{1/2}|b|^{-1/2}\langle b\|E^{A_1}\|b'\rangle \quad (18.1.7)$$

since the $6j$ contains an A_1 irrep and so may be simplified with (16.1.11); then the phase factors cancel. Here by $\delta_{bb'}$ (and $\delta_{cc'}$) we mean that b and b' (and c and c') must belong to the same irrep; they may be different functions.

Example: Simplification of Matrix Elements of Spin-Independent and Space-Independent Operators

We now show how the equations of the previous section may be used to simplify matrix elements of spin-independent or space-independent operators in the spin–orbit-coupled $|(\mathbb{S}S, h)rt\tau\rangle$ basis defined by (12.2.2). A spin-independent operator O^f is of the type E^e in this basis, while a space-independent operator is of the type D^d. Thus (18.1.4) and (18.1.2) give directly the useful equations

$$\langle (\mathbb{S}S, h)rt \| O^f \| (\mathbb{S}'S', h')r't' \rangle_p = \delta_{\mathbb{S}\mathbb{S}'}\delta_{SS'}|t|^{1/2}|t'|^{1/2}\{S'h't'^*r'\}\{h^*fh'\}$$
$$\times \begin{Bmatrix} f & t' & t^* \\ S^* & h & h' \end{Bmatrix}_{0r'rp} \langle h\|O^f\|h'\rangle \quad (18.1.8)$$

$$\langle (\mathbb{S}S, h)rt \| S^f \| (\mathbb{S}'S', h')r't' \rangle_p = \delta_{hh'}|t|^{1/2}|t'|^{1/2}\{Sht^*r\}\{S'\}\sum_q \{S^*fS'q\}$$
$$\times \begin{Bmatrix} f & t' & t^* \\ h^* & S & S' \end{Bmatrix}_{qr'rp} \langle \mathbb{S}S\|S^f\|\mathbb{S}'S'\rangle_q \quad (18.1.9)$$

Since h and h' are single-valued irreps, $\{h'\} = 1$. The sum over q is omitted in (18.1.8) because repeated representations hardly ever occur in $f \otimes h'$ (they do for the uncommon groups T, T_h, and K). Here we assume that spin functions have been classified as discussed in Section 12.2. All symmetry labels in (18.1.8) and (18.1.9) (except $\mathbb{S}$ and $\mathbb{S}'$) are those of the molecular point group G to which the system belongs.

More detailed notation is sometimes used for the reduced matrix elements on the right of (18.1.8) and (18.1.9). For example,

$$\begin{aligned}\langle h\|O^f\|h'\rangle &= \langle \mathbb{S}Sh\|O^f\|\mathbb{S}Sh'\rangle \\ &= \langle \mathbb{S}ShM\|O^f\|\mathbb{S}Sh'M\rangle \\ &= \langle \mathbb{S}h\|O^f\|\mathbb{S}h'\rangle = \langle {}^{2\mathbb{S}+1}h\|O^f\|{}^{2\mathbb{S}+1}h'\rangle \\ \langle \mathbb{S}S\|S^f\|\mathbb{S}'S'\rangle_q &= \langle \mathbb{S}Sh\theta\|S^f\|\mathbb{S}'S'h\theta\rangle_q\end{aligned} \quad (18.1.10)$$

and so on.

The purely spin reduced matrix elements above may be evaluated using (15.4.3)—see, for example, (15.4.8), (15.4.9), and (19.4.20); so can the purely orbital-angular-momentum matrix elements ($O^f = L^f$) in a weak-field basis. In a strong-field basis, orbital-angular-momentum matrix elements are usually evaluated as discussed in Chapters 19 and 20.

18.2. Uses of the O^f and S^f Equations in MCD Calculations

Example: The $E_g''(^2T_{2g}) \to E_u''(^2T_{2u})$ $\mathcal{C}$ Term in $IrBr_6^{2-}$

The equations above completely eliminate the need to explicitly expand $|(\mathbb{S}S, h)rt\tau\rangle$ functions in terms of $|\mathbb{S}ShM\theta\rangle$ functions. Thus, for example, if we wish to calculate the dipole strength $\mathcal{D}_0$ and $\mathcal{C}_0/\mathcal{D}_0$ for the $E_g''(^2T_{2g}) \to E_u''(^2T_{2u})$ charge-transfer transition in octahedral (group-O_h) $IrBr_6^{2-}$ (see Figure 23.8.2), we first use (17.2.19), (17.3.6), and Sections C.11 and C.13 to obtain

$$\mathcal{D}_0 = \tfrac{1}{6}\left|\langle E_g''\|m^{t_{1u}}\|E_u''\rangle\right|^2 \quad (18.2.1)$$

$$\begin{aligned}\frac{\mathcal{C}_0}{\mathcal{D}_0} &= \frac{-\sqrt{6}}{\mu_B}(1)(-\tfrac{1}{3})\langle E_g''\|\mu^{t_{1g}}\|E_g''\rangle \\ &= \frac{-\sqrt{6}}{3}\left(\langle E_g''\|L^{t_{1g}}\|E_g''\rangle + 2\langle E_g''\|S^{t_{1g}}\|E_g''\rangle\right)\end{aligned} \quad (18.2.2)$$

Spin-one-half functions belong to the E' irrep in the group O. Thus using

(18.1.8) and Sections C.11 and C.13, we find

$$\langle E_g''\|m^{t_{1u}}\|E_u''\rangle = \left\langle \left(\tfrac{1}{2}^+ E_g', T_{2g}\right) E_g'' \middle\| m^{t_{1u}} \middle\| \left(\tfrac{1}{2}^+ E_g', T_{2u}\right) E_u'' \right\rangle$$

$$= \sqrt{2}\sqrt{2}(-1)(-1)(-\tfrac{1}{3})\langle {}^2T_{2g}\|m^{t_{1u}}\|{}^2T_{2u}\rangle$$

$$= -\tfrac{2}{3}\langle {}^2T_{2g}\|m^{t_{1u}}\|{}^2T_{2u}\rangle \tag{18.2.3}$$

$$\langle E_g''\|L^{t_{1g}}\|E_g''\rangle = \left\langle \left(\tfrac{1}{2}^+ E_g', T_{2g}\right) E_g'' \middle\| L^{t_{1g}} \middle\| \left(\tfrac{1}{2}^+ E_g', T_{2g}\right) E_g'' \right\rangle$$

$$= -\tfrac{2}{3}\langle {}^2T_{2g}\|L^{t_{1g}}\|{}^2T_{2g}\rangle \tag{18.2.4}$$

Then (18.1.9) gives

$$\langle E_g''\|S^{t_{1g}}\|E_g''\rangle = \left\langle \left(\tfrac{1}{2}^+ E_g', T_{2g}\right) E_g'' \middle\| S^{t_{1g}} \middle\| \left(\tfrac{1}{2}^+ E_g', T_{2g}\right) E_g'' \right\rangle$$

$$= \tfrac{1}{3}\langle \tfrac{1}{2}^+ E_g'\|S^{t_{1g}}\|\tfrac{1}{2}^+ E_g'\rangle \tag{18.2.5}$$

At this stage we have

$$\mathcal{D}_0 = \tfrac{2}{27}\left|\langle {}^2T_{2g}\|m^{t_{1u}}\|{}^2T_{2u}\rangle\right|^2 \tag{18.2.6}$$

$$\frac{\mathcal{C}_0}{\mathcal{D}_0} = \frac{-2\sqrt{6}}{9}\left(-\langle {}^2T_{2g}\|L^{t_{1g}}\|{}^2T_{2g}\rangle + \langle \tfrac{1}{2}^+ E_g'\|S^{t_{1g}}\|\tfrac{1}{2}^+ E_g'\rangle\right) \tag{18.2.7}$$

The ground-state configuration in $IrBr_6^{2-}$ is t_{2g}^5, and if we approximate the t_{2g} molecular orbitals as pure d orbitals, we find using (13.2.8) [or (19.4.6)] and (13.4.3) [or (15.4.5)] that

$$\langle {}^2T_{2g}\|L^{t_{1g}}\|{}^2T_{2g}\rangle = \langle t_{2g}\|l^{t_{1g}}\|t_{2g}\rangle = -\sqrt{6} \tag{18.2.8}$$

Then (15.4.9) gives $\langle \tfrac{1}{2}^+E_g'\|S^{t_{1g}}\|\tfrac{1}{2}^+E_g'\rangle = \sqrt{6}/2$. Substituting these into (18.2.7) above, we obtain

$$\frac{\mathcal{C}_0}{\mathcal{D}_0} = -2 \tag{18.2.9}$$

Finally, if desired, the $\mathcal{D}_0$ reduced matrix element may be calculated in terms of a one-electron reduced matrix element using the methods of

Section 13.2, or much more directly, using (20.2.10). As a promise of things to come, we show here how (20.2.10) may be used to obtain our earlier result (13.2.16). The ground state arises from a $t_{2u}^6 t_{2g}^5$ configuration, and the excited state from a $t_{2u}^5 t_{2g}^6$ configuration. The coefficients of fractional parentage (Section C.15) are both equal to 1, and the group-SO_3 $6j$ and the group-O $9j$ coefficients are easily calculated using (16.1.11) and (16.4.8). Thus (20.2.10) gives

$$\begin{aligned}\langle {}^2T_{2g}\|m^{t_{1u}}\|{}^2T_{2u}\rangle &= \left\langle \mathcal{Q}\left(t_{2u}^6(0A_{1g}), t_{2g}^5\left(\tfrac{1}{2}T_{2g}\right)\right)\tfrac{1}{2}T_{2g}\right\| m^{t_{1u}} \\ &\quad \times \left\| \mathcal{Q}\left(t_{2u}^5\left(\tfrac{1}{2}T_{2u}\right), t_{2g}^6(0A_{1g})\right)\tfrac{1}{2}T_{2u}\right\rangle \\ &= 18\begin{Bmatrix} 0 & \frac{1}{2} & \frac{1}{2} \\ 0 & \frac{1}{2} & \frac{1}{2}\end{Bmatrix}\begin{Bmatrix} A_1 & T_2 & T_2 \\ T_2 & A_1 & T_2 \\ T_2 & T_2 & T_1\end{Bmatrix}\langle t_{2u}\|m^{t_{1u}}\|t_{2g}\rangle \\ &= 18\left(-\tfrac{1}{2}\right)\left(-\tfrac{1}{9}\right)\langle t_{2u}\|m^{t_{1u}}\|t_{2g}\rangle \\ &= \langle t_{2u}\|m^{t_{1u}}\|t_{2g}\rangle \end{aligned} \tag{18.2.10}$$

Note that in all of the above calculations we never once constructed multielectron wavefunctions. As we demonstrate in Chapters 19 and 20, this characteristic of irreducible-tensor calculations remains even in far more complex situations; we require only configuration and symmetry information about states (and, of course, some knowledge of the individual molecular orbitals) to calculate an observable such as $\mathcal{C}_0/\mathcal{D}_0$.

Example: The MCD *of* $CoCl_4^{2-}$

In odd-electron octahedral or tetrahedral systems with spin–orbit coupling, extra care must be taken, since repeated representation labels occur in matrix elements between U' states. A misunderstanding of these labels has led researchers to the wrong result. In this section we give a detailed example of such a calculation for a tetrahedral system. Before reading further it may be useful to review Section 12.6.

It is well known that the $CoCl_4^{2-}$ ion is tetrahedral with a ${}^4A_2$, d^7 ground state and that the lowest-energy spin-allowed d–d transitions, at $\approx$ 15,000 cm^{-1}, arise from ${}^4A_2 \to {}^4T_1$ transitions [Denning and Spencer (1969)]. The ${}^4A_2$ ground state is not split by spin–orbit coupling; in $|(\mathcal{S}S, h)t\tau\rangle$ notation it is labeled $|(\tfrac{3}{2}U', A_2)U'\tau\rangle$, and in $t({}^{2\mathcal{S}+1}h)$ notation as $U'({}^4A_2)$. The ${}^4T_1$

state is split by first-order spin–orbit coupling into E', E'', and two U' spin–orbit-coupled states; the U' states are designated $rt = \alpha U'$ and $\beta U'$ (Section 12.6). The first-order energies of these states may be easily calculated using (18.4.6) below in terms of a single parameter, $c(^4T_1) \equiv \langle \frac{3}{2} T_1 \| \Sigma_k su(k) \| \frac{3}{2} T_1 \rangle^{\mathrm{SO}_3, T_d}$. With the Butler (1981) U' multiplicity separation, (18.4.6) is diagonal in the index r for neither $rU'(^4T_1)$ nor $rU'(^4T_2)$ states (Section 12.6). The value of $c(^4T_1)$ depends, of course, on the precise nature of the $^4T_1$ state. From (12.6.4) we have for the first-order energies

$$E[\alpha U'(^4T_1)] = \frac{1}{3\sqrt{2}} c(^4T_1)$$
$$E[\beta U'(^4T_1)] = \frac{-1}{2\sqrt{2}} c(^4T_1) \qquad (18.2.11)$$

where $\alpha U'$ and $\beta U'$ are the linear combinations of $0U'$ and $1U'$ states specified in (12.6.4). Then (18.4.6) gives for the diagonal $E'(^4T_1)$ and $E''(^4T_1)$ states

$$E[E'(^4T_1)] = \frac{5}{6\sqrt{2}} c(^4T_1)$$
$$E[E''(^4T_1)] = \frac{-1}{2\sqrt{2}} c(^4T_1) \qquad (18.2.12)$$

For the purpose of our MCD calculation we do not need to know the exact magnitude of $c(^4T_1)$ and do not need to further define the $^4T_1$ state. Our results thus hold for *any* $^4T_1$ in $CoCl_4^{2-}$ for which first-order spin–orbit coupling is a good approximation; they are valid for weak-field, intermediate-field, or strong-field $^4T_1$ states. They also hold for $^4T_1$ charge-transfer states.

In this example we calculate $\mathcal{C}_0/\mathcal{D}_0$ for the $U'(^4A_2) \to E'(^4T_1)$, $E''(^4T_1)$, $\alpha U'(^4T_1)$, and $\beta U'(^4T_1)$ Franck–Condon-allowed transitions. While we see from (18.2.11) and (18.2.12) that the $E''(^4T_1)$ and $\beta U'(^4T_1)$ states are degenerate to first order in $\mathcal{H}_{\mathrm{SO}}$, we assume for simplicity in the remainder of this section that this degeneracy has been lifted by some (unspecified) perturbation. Thus we calculate $\mathcal{C}_0/\mathcal{D}_0$ for $U' \to E''(^4T_1)$ and $U' \to \beta U'(^4T_1)$ individually, rather than for the combined transition, $U' \to [E''(^4T_1) + \beta U'(^4T_1)]$. For the $U' \to E'$ and $U' \to E''$ calculations we may use (17.3.6). But for the $U' \to U'$ transitions we must instead use (17.2.13) and (17.2.20), and then the equations in Section 18.1 with $\alpha U'$ and $\beta U'$

defined by (12.6.4a), before cancellation of electric-dipole reduced matrix elements is possible. As made explicit by (17.4.16), these equations may be used for both the space-averaged and the oriented tetrahedral case. Since $CoCl_4^{2-}$ is tetrahedral, the electric-dipole operator belongs to the T_2 irrep. The complex-T_d basis of Section C.5 and the tables in Sections C.11–C.13 are employed throughout.

Proceeding as described above, we have for $U' \to E'$ and $U' \to E''$ respectively

$$\frac{\mathscr{C}_0}{\mathscr{D}_0}[U' \to E'] = \frac{\sqrt{6}}{\mu_B}(1)\left(\begin{Bmatrix} T_1 & T_2 & T_2 \\ E' & U' & U' \end{Bmatrix}_{0000} \langle U' \| \mu^{t_1} \| U' \rangle_0 + \begin{Bmatrix} T_1 & T_2 & T_2 \\ E' & U' & U' \end{Bmatrix}_{1000} \langle U' \| \mu^{t_1} \| U' \rangle_1\right)$$

$$= \frac{\sqrt{6}}{\mu_B}\left(-\frac{1}{6\sqrt{2}}\langle U' \| \mu^{t_1} \| U' \rangle_0 + \frac{1}{3\sqrt{2}}\langle U' \| \mu^{t_1} \| U' \rangle_1\right) \tag{18.2.13}$$

$$\frac{\mathscr{C}_0}{\mathscr{D}_0}[U' \to E''] = \frac{\sqrt{6}}{\mu_B}(1)\left(\begin{Bmatrix} T_1 & T_2 & T_2 \\ E'' & U' & U' \end{Bmatrix}_{0000} \langle U' \| \mu^{t_1} \| U' \rangle_0 + \begin{Bmatrix} T_1 & T_2 & T_2 \\ E'' & U' & U' \end{Bmatrix}_{1000} \langle U' \| \mu^{t_1} \| U' \rangle_1\right)$$

$$= \frac{\sqrt{6}}{\mu_B}\left(\frac{-1}{6\sqrt{2}}\langle U' \| \mu^{t_1} \| U' \rangle_0 - \frac{1}{3\sqrt{2}}\langle U' \| \mu^{t_1} \| U' \rangle_1\right) \tag{18.2.14}$$

Note that these ratios depend on the symmetry but are *independent of the precise nature of the E' and E'' excited-state wavefunctions*; these equations are quite general and are not limited to the ${}^4T_1$ case.

We now use the equations of Section 18.1 to evaluate $\langle U' \| \mu^{t_1} \| U' \rangle_p$ for $p = 0$ and 1. However, since ${}^4A_2$ states have no orbital angular momentum,

$$\left\langle \left(\tfrac{3}{2}U', A_2\right)U' \left\| L^{t_1} \right\| \left(\tfrac{3}{2}U', A_2\right)U' \right\rangle_p = 0, \qquad p = 0, 1 \tag{18.2.15}$$

This result may also be obtained with (18.1.8), since the $6j$ required are zero. Next, (18.1.9) and (15.4.9) give

$$\left\langle\left(\tfrac{3}{2}U', A_2\right)U'\middle\|S^{t_1}\middle\|\left(\tfrac{3}{2}U', A_2\right)U'\right\rangle_0 = \langle\tfrac{3}{2}U'\|S^{t_1}\|\tfrac{3}{2}U'\rangle_0 = \sqrt{3}$$

$$\left\langle\left(\tfrac{3}{2}U', A_2\right)U'\middle\|S^{t_1}\middle\|\left(\tfrac{3}{2}U', A_2\right)U'\right\rangle_1 = -\langle\tfrac{3}{2}U'\|S^{t_1}\|\tfrac{3}{2}U'\rangle_1 = -2\sqrt{3}$$

(18.2.16)

Thus

$$\left\langle\left(\tfrac{3}{2}U', A_2\right)U'\middle\|\mu^{t_1}\middle\|\left(\tfrac{3}{2}U', A_2\right)U'\right\rangle_p = \begin{cases} -\mu_B 2\sqrt{3} & \text{for} \quad p=0 \\ \mu_B 4\sqrt{3} & \text{for} \quad p=1 \end{cases}$$

(18.2.17)

Substituting these results into (18.2.13) and (18.2.14) above, we conclude

$$\frac{\mathcal{C}_0}{\mathcal{D}_0}\left[U'(^4A_2) \to E'\right] = 5$$

$$\frac{\mathcal{C}_0}{\mathcal{D}_0}\left[U'(^4A_2) \to E''\right] = -3 \qquad (18.2.18)$$

These equations indicate that as long as the ground state remains essentially $^4A_2$, $\mathcal{C}_0/\mathcal{D}_0$ is positive for all Franck–Condon-allowed $U' \to E'$ transitions and negative for all $U' \to E''$ transitions—regardless of the details of the excited states.

To obtain $\mathcal{C}_0/\mathcal{D}_0$ for the $U' \to U'$ transitions, a model for the U' excited states is required. This is because (17.2.13), (17.2.20), and Table 17.2.1 give for our starting equation

$$\frac{\mathcal{C}_0}{\mathcal{D}_0}\left[U'(^4A_2) \to \gamma U'\right]$$
$$= \frac{\sqrt{6}}{\mu_B}\left[\frac{\begin{array}{r}\displaystyle\sum_{qss'}\left\langle\left(\tfrac{3}{2}U', A_2\right)U'\middle\|\mu^{t_1}\middle\|\left(\tfrac{3}{2}U', A_2\right)U'\right\rangle_q\begin{Bmatrix} T_1 & T_2 & T_2 \\ U' & U' & U' \end{Bmatrix}_{qs's0} \\ \times\left\langle\left(\tfrac{3}{2}U', A_2\right)U'\middle\|m^{t_2}\middle\|\gamma U'\right\rangle_s\left\langle\gamma U'\middle\|m^{t_2}\middle\|\left(\tfrac{3}{2}U', A_2\right)U'\right\rangle_{s'}\end{array}}{\displaystyle\sum_s\{U'T_2U's\}\left\langle\left(\tfrac{3}{2}U', A_2\right)U'\middle\|m^{t_2}\middle\|\gamma U'\right\rangle_s\left\langle\gamma U'\middle\|m^{t_2}\middle\|\left(\tfrac{3}{2}U', A_2\right)U'\right\rangle_s}\right]$$

(18.2.19)

where the excited U' state is $\gamma U' = \alpha U'(^4T_1)$ or $\beta U'(^4T_1)$, which are defined by (12.6.4). Thus the electric-dipole matrix elements do not cancel at this stage, as they did for the $U' \to E'$ and $U' \to E''$ transitions. When the ground state is $U'(^4A_2)$, the only nonzero contributions to electric-dipole matrix elements come from U' states with $^4T_1$ character; transitions to states with $\mathbb{S} \neq \frac{3}{2}$ are spin-forbidden, and, since $A_2 \otimes f = A_2 \otimes T_2 = T_1$, transitions to states with $h \neq T_1$ are orbitally forbidden. For simplicity we assume here that our U' excited states are purely $U'(^4T_1)$ states and express all electric-dipole matrix elements in terms of

$$\begin{aligned} m &= \langle {}^4A_2 \| m^{t_2} \| {}^4T_1 \rangle \\ m' &= \langle {}^4T_1 \| m^{t_2} \| {}^4A_2 \rangle \end{aligned} \tag{18.2.20}$$

using (12.6.4) and (18.1.8).

We begin the evaluation of (18.2.19) by expanding the $\gamma U'(^4T_1)$ states ($\gamma = \alpha, \beta$) in the electric-dipole reduced matrix elements in terms of $0U'(^4T_1)$ and $1U'(^4T_1)$ states using (12.6.4); this gives for $s = 0, 1$

$$\begin{aligned} \left\langle \left(\tfrac{3}{2}U', A_2\right)U' \middle\| m^{t_2} \middle\| \alpha U' \right\rangle_s &= \frac{1}{\sqrt{5}} \left\langle \left(\tfrac{3}{2}U', A_2\right)U' \middle\| m^{t_2} \middle\| 0U' \right\rangle_s \\ &\quad + \frac{2}{\sqrt{5}} \left\langle \left(\tfrac{3}{2}U', A_2\right)U' \middle\| m^{t_2} \middle\| 1U' \right\rangle_s \\ \left\langle \left(\tfrac{3}{2}U', A_2\right)U' \middle\| m^{t_2} \middle\| \beta U' \right\rangle_s &= \frac{-2}{\sqrt{5}} \left\langle \left(\tfrac{3}{2}U', A_2\right)U' \middle\| m^{t_2} \middle\| 0U' \right\rangle_s \\ &\quad + \frac{1}{\sqrt{5}} \left\langle \left(\tfrac{3}{2}U', A_2\right)U' \middle\| m^{t_2} \middle\| 1U' \right\rangle_s \end{aligned} \tag{18.2.21}$$

Exactly analogous equations apply with the bras and kets reversed. Then with (18.1.8) we obtain as the only nonzero results

$$\begin{aligned} \left\langle \left(\tfrac{3}{2}U', A_2\right)U' \middle\| m^{t_2} \middle\| \left(\tfrac{3}{2}U', T_1\right)0U' \right\rangle_0 &= \frac{1}{2\sqrt{3}}m \\ \left\langle \left(\tfrac{3}{2}U', A_2\right)U' \middle\| m^{t_2} \middle\| \left(\tfrac{3}{2}U', T_1\right)1U' \right\rangle_1 &= \frac{1}{2\sqrt{3}}m \\ \left\langle \left(\tfrac{3}{2}U', T_1\right)0U' \middle\| m^{t_2} \middle\| \left(\tfrac{3}{2}U', A_2\right)U' \right\rangle_0 &= \frac{1}{2\sqrt{3}}m' \\ \left\langle \left(\tfrac{3}{2}U', T_1\right)1U' \middle\| m^{t_2} \middle\| \left(\tfrac{3}{2}U', A_2\right)U' \right\rangle_1 &= \frac{-1}{2\sqrt{3}}m' \end{aligned} \tag{18.2.22}$$

Substituting the (18.2.22) results into (18.2.21), and also into its analog with bras and kets reversed, is the next step. For the $\alpha U'$ case we find, for example,

$$\left\langle \left(\tfrac{3}{2}U', A_2\right)U' \| m^{t_2} \| \alpha U' \right\rangle_s = \begin{cases} \dfrac{1}{\sqrt{5}}\left(\dfrac{1}{2\sqrt{3}} m\right) & \text{for} \quad s = 0 \\ \dfrac{2}{\sqrt{5}}\left(\dfrac{1}{2\sqrt{3}} m\right) & \text{for} \quad s = 1 \end{cases}$$

$$\left\langle \alpha U' \| m^{t_2} \| \left(\tfrac{3}{2}U', A_2\right)U' \right\rangle_{s'} = \begin{cases} \dfrac{1}{\sqrt{5}}\left(\dfrac{1}{2\sqrt{3}} m'\right) & \text{for} \quad s' = 0 \\ \dfrac{2}{\sqrt{5}}\left(\dfrac{-1}{2\sqrt{3}} m'\right) & \text{for} \quad s' = 1 \end{cases} \tag{18.2.23}$$

Next these results (or their $\beta U'$ equivalents), together with our earlier results (18.2.17) and the appropriate $6j$ and $3j$, are substituted into (18.2.19). We find

$$\begin{aligned} \frac{\mathcal{C}_0}{\mathcal{D}_0}\left[U'(^4A_2) \to \alpha U'(^4T_1)\right] &= 2 \\ \frac{\mathcal{C}_0}{\mathcal{D}_0}\left[U'(^4A_2) \to \beta U'(^4T_1)\right] &= -3 \end{aligned} \tag{18.2.24}$$

The reduced-matrix-element product $mm' = \langle ^4A_2 \| m^{t_2} \| ^4T_1 \rangle \langle ^4T_1 \| m^{t_2} \| ^4A_2 \rangle$ now cancels, since it is common to all terms in the numerator and denominator.

Note that mixing states with doublet character into the U' excited states would not change these $\mathcal{C}_0/\mathcal{D}_0$ ratios, since it would multiply both $\mathcal{C}_0$ and $\mathcal{D}_0$ reduced matrix elements by a common factor. Mixing of $U'(^4T_2)$ character into the ground state, however, does change these ratios, since the ground-state magnetic moment is altered (the g value increases) and we are no longer restricted to $^4A_2 \to {}^4T_1$ allowed transitions, since transitions from $^4T_2$ are symmetry-allowed to $^4T_1$, $^4T_2$, ^4E and $^4A_1$ states.

18.3. Transformation Properties of the Dot Product of Tensor Operators

In this section we define the transformation properties of operators which are dot (scalar) products of spherical (irreducible) tensor operators. The most common operator of this type is the spin–orbit coupling operator $\mathcal{H}_{SO}$. The components T_q^k of a spherical tensor operator transform under rotations identically to the $|jm\rangle = |kq\rangle$ of $SO_3 \supset SO_2$. For j integer the tensor components T^j_m obey the adjoint equation (Section 14.10)

$$\left(T^j_m\right)^\dagger = (-1)^m T^j_{-m} \tag{18.3.1}$$

and are said to be Hermitian conjugate.‡ The dot (scalar) product of two such tensor operators is an immediate generalization of (9.8.10):

$$\begin{aligned}(\mathbf{T}^j \cdot \mathbf{U}^j) &= \sum_m T^j_m \left(U^j_m\right)^\dagger \\ &= \sum_m (-1)^m T^j_m U^j_{-m} \\ &= (-1)^j \sum_m \begin{pmatrix} j \\ m \end{pmatrix} T^j_m U^j_{-m} \\ &= (-1)^j |j|^{1/2} \sum_m (00|jm, j^*m^*) T^j_m U^{j^*}_{m^*} \\ &\equiv (-1)^j |j|^{1/2} \{T^j U^j\}^0_0 \end{aligned} \tag{18.3.2}$$

We have used the fact that the $2jm$ have values ± 1, (14.6.5), and—in the last line—(14.6.1). The $(-1)^j$ phase is "historical" and is not needed for invariance; thus it was not used in Section 14.6. For the scalar product of tensor operators it must, however, be carried along.

For subgroups of SO_3 or O_3 the transformation properties of the scalar product may be obtained from the last line, since $\{T^j U^j\}^0_0$ is basis-independent and may be expressed in any SO_3 (or O_3) subgroup basis. We do this using (14.6.1), (14.6.5), the Racah factorization lemma (15.3.2), and then Butler [(1981), Eq. (3.3.7)] and our (14.6.5). We have, for example, for the

‡They are also said to be Hermitian tensors. Both the terminology and the notation in the literature are confusing here, since clearly the *components* in (18.3.1) are not self-adjoint in the ordinary sense of (14.10.10). The *entire tensorial set*, however, is termed self-adjoint [Edmonds (1963), p. 77]; thus Edmonds writes $\mathbf{T}^\dagger = \mathbf{T}$, and his discussion (p. 77) leads to the relation $T^\dagger(kq) = T(kq)$, where his $T(kq)$ is our $T^k_q = T^j_m$. In such discussions, however, the position of the adjoint symbol has great significance; in particular note that Edmonds writes $T(kq)^\dagger = (-1)^q T(k-q)$—which is just (18.3.1) above—so clearly $T^\dagger(kq) \neq T(kq)^\dagger$. To simplify matters we avoid the $T^\dagger(kq)$ type of notation entirely, since all our tensor operators have $\mathbf{T}^\dagger = \mathbf{T}$ with the Edmonds definition.

group O, where f and ϕ designate group-O irrep and irrep partner,

$$\begin{aligned}(\mathbf{T}^j \cdot \mathbf{U}^j) &= (-1)^j |j|^{1/2} \{T^j U^j\}_0^0 \\ &= (-1)^j |j|^{1/2} \sum_{f\phi} (0A_1 a_1 | jf\phi, jf^*\phi^*) T^j_{f\phi} U^j_{f^*\phi^*} \\ &= (-1)^j |j|^{1/2} \sum_{f\phi} (0A_1 | jf, jf^*)^{\mathrm{SO}_3 \supset O} (A_1 a_1 | f\phi, f^*\phi^*)^O T^{jf}_{\phi} U^{jf^*}_{\phi^*} \\ &= (-1)^j |j|^{1/2} \sum_{f\phi} \frac{|f|^{1/2}}{|j|^{1/2}} \begin{pmatrix} j \\ f \end{pmatrix}^{\mathrm{SO}_3}_{O} |f|^{-1/2} \begin{pmatrix} f \\ \phi \end{pmatrix} T^{jf}_{\phi} U^{jf^*}_{\phi^*} \\ &= (-1)^j \sum_{f\phi} \begin{pmatrix} f \\ \phi \end{pmatrix} T^{jf}_{\phi} U^{jf^*}_{\phi^*} \\ &= (-1)^j \sum_{f} |f|^{1/2} \{T^{jf} U^{jf^*}\}^{0A_1}_{0A_1} \end{aligned} \quad (18.3.3)$$

For a subgroup G_1 of the group O the process is repeated in an iterative fashion. In the subgroup, tensor operators transform as the irreps $j(\mathrm{SO}_3)f(O)f_1(G_1)$. Thus

$$(\mathbf{T}^j \cdot \mathbf{U}^j) = (-1)^j \sum_{ff_1\phi_1} \begin{pmatrix} f_1 \\ \phi_1 \end{pmatrix} T^{jff_1}_{\phi_1} U^{jf^*f_1^*}_{\phi_1^*} \quad (18.3.4)$$

Equations (18.3.3) and (18.3.4) may be used with either real or complex basis sets.

The equations above may be applied directly to the spin–orbit coupling operator, which is defined in (13.3.1)–(13.3.3) as

$$\mathcal{H}_{\mathrm{SO}} = \sum_k \xi(r_k) \mathbf{l}(k) \cdot \mathbf{s}(k) = \sum_k \mathbf{s}(k) \cdot \mathbf{u}(k) \quad (18.3.5)$$

Both $\mathbf{s}$ and $\mathbf{u}$ transform as even-parity spherical vectors (Section 9.8), so $j = 1$ in (18.3.3)–(18.3.4). Thus we may write for the cubic groups O, T_d, and O_h (for O_h replace T_1 by T_{1g})

$$\mathcal{H}_{\mathrm{SO}} = (-1) \sum_{\phi} \begin{pmatrix} T_1 \\ \phi \end{pmatrix} \sum_k s(k)^{1t_1}_{\phi} u(k)^{1t_1}_{\phi^*} \quad (18.3.6)$$

For lower symmetry groups the double ff_1 irrep labeling in (18.3.4) is usually redundant for $j = 1$. For any group with a C_2 or C_3 symmetry axis, it is sufficient to drop the f labels in (18.3.4) and write

$$\mathcal{H}_{SO} = (-1)\sum_{f\phi} \begin{pmatrix} f \\ \phi \end{pmatrix} \sum_k s(k)^{1f}_{\phi} u(k)^{1f^*}_{\phi^*} \tag{18.3.7}$$

where f and ϕ are now irrep and irrep-partner labels in the subgroup of interest. In D_4, for example, $f = A_2$ and E. For cubic groups, $f = T_1$ only and (18.3.7) reduces to (18.3.6).

The operator products in (18.3.7) are examples of *double tensor* operators [Griffith (1962)], since they operate in two independent function spaces. A convenient abbreviation for such operators (which we used in (13.3.10)) is

$$s^{1f}_{\phi} u^{1f^*}_{\phi^*}(k) \equiv s(k)^{1f}_{\phi} u(k)^{1f^*}_{\phi^*} \tag{18.3.8}$$

Equation (18.3.3) or (18.3.4) may be used to define the adjoint of a symmetry-adapted spherical tensor operator. For example, if we write (18.3.3) as

$$(\mathbf{T}^j \cdot \mathbf{U}^j) = \sum_{f\phi} T^{jf}_{\phi} \left(U^{jf}_{\phi}\right)^{\dagger} \tag{18.3.9}$$

it follows that

$$\left(U^{jf}_{\phi}\right)^{\dagger} = (-1)^j \begin{pmatrix} f \\ \phi \end{pmatrix} U^{jf^*}_{\phi^*} \tag{18.3.10}$$

The result above may be used to relate reduced matrix elements of symmetry-adapted vector operators obtained from the matrix-element adjoint equation (14.10.11). Suppose $V^{1f} = V^f$ is a symmetry-adapted vector operator (spherical tensor operator with $j = 1$). Then it follows from (14.10.11), (10.2.2), and our discussion above that

$$\langle a\alpha | V^f_{\phi} | b\beta \rangle^* = \langle b\beta | \left(V^f_{\phi}\right)^{\dagger} | a\alpha \rangle = \begin{pmatrix} a \\ \alpha \end{pmatrix} \sum_r \begin{pmatrix} a^* & f & b \\ \alpha^* & \phi & \beta \end{pmatrix}^{*\,r} \langle a \| V^f \| b \rangle^*_r$$

$$= (-1) \begin{pmatrix} f \\ \phi \end{pmatrix} \begin{pmatrix} b \\ \beta \end{pmatrix} \sum_r \begin{pmatrix} b^* & f^* & a \\ \beta^* & \phi^* & \alpha \end{pmatrix}^{r} \langle b \| V^{f^*} \| a \rangle_r \tag{18.3.11}$$

Then using (14.8.5) and rearranging the $3jm$ to identical form gives

$$\{a\}\{a^* f b r\} \langle a \| V^f \| b \rangle^*_r = (-1) \langle b \| V^{f^*} \| a \rangle_r \tag{18.3.12}$$

since the reduced matrix elements are linearly independent with respect to r. We have used this result in the calculation of $\mathcal{D}_0$ in Chapter 17.

For the coupled scalar-product operator defined completely within a group G (with $G \neq SO_3$ or O_3), there is no reason to include a phase analogous to $(-1)^j$. Such scalar-product operators arise, for example, in (3.7.2) in the Taylor expansion of the electronic Hamiltonian in normal coordinates. For operators of this type

$$\left(O_\phi^f\right)^\dagger = \begin{pmatrix} f \\ \phi \end{pmatrix} O_{\phi^*}^{f^*} \tag{18.3.13}$$

applies instead of (18.3.10), and

$$\{a\}\{a^*fbr\}\langle a\|O^f\|b\rangle_r^* = \langle b\|O^{f^*}\|a\rangle_r \tag{18.3.14}$$

replaces (18.3.12).

18.4. Matrix Elements of Dot-Product Operators: Spin–Orbit-Coupling Matrix Elements

We are now ready to return to the calculation of spin–orbit-coupling matrix elements, a topic introduced in Section 13.3. We show how the irreducible-tensor method may be used to express spin–orbit matrix elements in the $|(\mathbb{S}S, h)rt\tau\rangle$ basis in terms of $\sum_k su(k)$ reduced matrix elements in the $|\mathbb{S}h\mathfrak{M}\theta\rangle$ basis. The lengthy method of Section 13.3 is here replaced by a single equation. The notation is that of Sections 12.2 and 13.3, and a review of those sections is recommended before proceeding.

To derive the result desired we simply expand our $|(\mathbb{S}S, h)rt\tau\rangle$ states via (12.2.2) [or, equivalently, (10.2.3)], expand $\mathcal{H}_{\text{SO}}$ in terms of symmetry-adapted operators according to (18.3.7), and reduce the resulting $|\mathbb{S}ShM\theta\rangle$ matrix element over spin and space simultaneously as in (13.3.10):

$$\begin{aligned}
\langle(\mathbb{S}S, h)rt\tau|\mathcal{H}_{\text{SO}}|(\mathbb{S}'S', h')r't'\tau'\rangle &= |t|^{1/2}\begin{pmatrix} t \\ \tau \end{pmatrix}\sum_{M\theta}\begin{pmatrix} S & h & t^* \\ M & \theta & \tau^* \end{pmatrix}^r \\
&\times\sum_{f\phi}(-1)\begin{pmatrix} f \\ \phi \end{pmatrix}|t'|^{1/2}\begin{pmatrix} t' \\ \tau' \end{pmatrix}\sum_{M'\theta'}\begin{pmatrix} S' & h' & t'^* \\ M' & \theta' & \tau'^* \end{pmatrix}^{*\,r'} \\
&\times\sum_y\begin{pmatrix} S \\ M \end{pmatrix}\begin{pmatrix} S^* & f & S' \\ M^* & \phi & M' \end{pmatrix}^y\sum_z\begin{pmatrix} h \\ \theta \end{pmatrix}\begin{pmatrix} h^* & f^* & h' \\ \theta^* & \phi^* & \theta' \end{pmatrix}^z \\
&\times\langle\mathbb{S}Sh\|\sum_k s^{1f}u^{1f^*}(k)\|\mathbb{S}'S'h'\rangle_{y,z}
\end{aligned} \tag{18.4.1}$$

Except for several uncommon groups (T, T_h, and K), $z = 0$ only, so we drop the multiplicity index z. We also drop the $j = 1$ SO_3 irrep label on the orbital part of the operator, $u(k)$, since at this stage it serves only to complicate the notation. Then (16.2.1) is used to simplify the $3jm$ product, with the result

$$\langle (\mathbb{S}S, h)rt\tau | \mathcal{H}_{\mathrm{SO}} | (\mathbb{S}'S', h')r't'\tau' \rangle$$
$$= \delta_{tt'}\delta_{\tau\tau'}(-1)\{Sht^*r\}\sum_f \{h^*f^*h'\}$$
$$\times \sum_y \begin{Bmatrix} S' & h' & t'^* \\ h^* & S & f \end{Bmatrix}_{y0rr'} \langle \mathbb{S}Sh \| \sum_k s^{1f}u^{f^*}(k) \| \mathbb{S}'S'h' \rangle_y \tag{18.4.2}$$

For the groups O, T_d, and O_h, the sum over f is unnecessary, since $f = T_1$ (or T_{1g} for O_h) only.

At this point (15.4.3) may be used to obtain the final $\mathbb{S}Sh$ matrix element (which is reduced over both spin and space in G, the point group of the molecule) in terms of an $\mathbb{S}h$ matrix element (which is reduced over spin in SO_3 and space in group G). Using our notation of Section 12.2, we have

$$\langle \mathbb{S}Sh \| \sum_k s^{1f}u^{f^*}(k) \| \mathbb{S}'S'h' \rangle_y^{G,G}$$
$$= \begin{pmatrix} \mathbb{S} & 1 & \mathbb{S}' \\ S^* & f & S' \end{pmatrix}_{yG}^{SO_3} \langle \mathbb{S}h \| \sum_k s^1 u^{f^*}(k) \| \mathbb{S}'h' \rangle^{SO_3, G} \tag{18.4.3}$$

Exactly analogous results pertain when G is a subgroup of O_3.

The equation above may be written more precisely using the Butler chain notation for spin functions and operators (Section 15.4), since when G is not a direct subgroup of SO_3, the $3jm$ factor in (18.4.3) above factors. For example, suppose we are calculating the $t = E$ spin–orbit matrix for the ^3E and ^1E terms of a group-D_4 configuration. The possible spin–orbit states for $t = E$ for these terms are $|(\mathbb{S}S, h)t\tau\rangle = |(1A_2, E)E\tau\rangle$ and $|(0A_1, E)E\tau\rangle$. Application of (18.4.2) gives

$$\langle (1A_2, E)E\tau | \mathcal{H}_{\mathrm{SO}} | (1A_2, E)E\tau \rangle = 0$$
$$\langle (0A_1, E)E\tau | \mathcal{H}_{\mathrm{SO}} | (0A_1, E)E\tau \rangle = 0$$
$$\langle (0A_1, E)E\tau | \mathcal{H}_{\mathrm{SO}} | (1A_2, E)E\tau \rangle = \frac{-1}{\sqrt{2}} \langle 0A_1E \| \sum_k s^{1a_2}u^{a_2}(k) \| 1A_2E \rangle^{D_4, D_4}$$
$$\langle (1A_2, E)E\tau | \mathcal{H}_{\mathrm{SO}} | (0A_1, E)E\tau \rangle = \frac{1}{\sqrt{2}} \langle 1A_2E \| \sum_k s^{1a_2}u^{a_2}(k) \| 0A_1E \rangle^{D_4, D_4} \tag{18.4.4}$$

Then (18.4.3) gives, for example,

$$\langle 0A_1E\|\sum_k s^{1a_2}u^{a_2}(k)\|1A_2E\rangle^{D_4,\,D_4}$$

$$\equiv \langle 0A_1A_1,\, E\|\sum_k s^{1t_1a_2}u^{a_2}(k)\|1T_1A_2,\, E\rangle^{SO_3\supset O\supset D_4,\,D_4}$$

$$= \begin{pmatrix} 0 & 1 & 1 \\ A_1 & T_1 & T_1 \\ A_1 & A_2 & A_2 \end{pmatrix}\begin{matrix} SO_3 \\ O \\ D_4 \end{matrix} \langle 0E\|\sum_k s^1u^{a_2}(k)\|1E\rangle^{SO_3,\,D_4}$$

$$= \begin{pmatrix} 0 & 1 & 1 \\ A_1 & T_1 & T_1 \end{pmatrix}\begin{matrix} SO_3 \\ O \end{matrix}\begin{pmatrix} A_1 & T_1 & T_1 \\ A_1 & A_2 & A_2 \end{pmatrix}\begin{matrix} O \\ D_4 \end{matrix}\langle \cdots \| \cdots \| \cdots \rangle$$

$$= \frac{1}{\sqrt{3}}\langle 0E\|\sum_k s^1u^{a_2}(k)\|1E\rangle^{SO_3,\,D_4} \qquad (18.4.5)$$

Here we use the Butler chain notation (Section 15.2) and Sections B.5 and D.12 to evaluate the $3jm$ factors.

For the group O we may combine (18.4.2) and (18.4.3) to obtain

$$\langle (\mathcal{S}S,\, h)rt\tau|\mathcal{H}_{SO}|(\mathcal{S}'S',\, h')r't'\tau'\rangle$$

$$= \delta_{tt'}\delta_{\tau\tau'}(-1)\{Shtr\}\{hT_1h'\}$$

$$\times\left[\sum_y \begin{Bmatrix} S' & h' & t' \\ h & S & T_1 \end{Bmatrix}_{y0rr'}\begin{pmatrix} \mathcal{S} & 1 & \mathcal{S}' \\ S & T_1 & S' \end{pmatrix}\begin{matrix} SO_3 \\ yO \end{matrix}\right]$$

$$\times\langle \mathcal{S}h\|\sum_k s^1u^{t_1}(k)\|\mathcal{S}'h'\rangle^{SO_3,\,O} \qquad (18.4.6)$$

The equation also applies for T_d—see the discussion following (15.4.8). The required $3j$ and $6j$ are given in Sections C.11 and C.13 respectively, while the necessary $SO_3 \supset O$ $3jm$ factors are given in Section B.5 (partial list) and in Chapter 13 of Butler (1981).

Equation (18.4.6) accomplishes in one easy step the calculations of (13.3.12)–(13.3.13). We suggest the reader check those results with (18.4.6). It becomes clear in Chapters 19 and 20 that (18.4.2)–(18.4.3) or (18.4.6) serves as a very useful starting point for the reduction of $\mathcal{H}_{SO}$ matrix elements from n-electron to one-electron form; there we replace the clumsy methods of Section 13.3 for evaluating the $\mathbb{S}h$ reduced matrix elements in (18.4.3) and (18.4.6) with much more elegant methods.

Other dot-product operators may be treated in a similar fashion to $\mathcal{H}_{SO}$ using the formalism of Section 18.3 and procedures analogous to the above.

18.5 The Double-Tensor Analogs of the D^d and E^e Equations of Section 18.1

Matrix elements of double tensor operators may be simplified using the methods of Section 18.1 above when the double tensor operator acts on only one part (either $|\mathbb{S}_1 h_1 \mathfrak{M}_1 \theta_1\rangle$ or $|\mathbb{S}_2 h_2 \mathfrak{M}_2 \theta_2\rangle$) of an $|(\mathbb{S}_1 h_1, \mathbb{S}_2 h_2)\mathbb{S}h\mathfrak{M}\theta\rangle$ coupled system. Just when an operator behaves in this way may not be apparent at this point, since we have yet to discuss the simplification of n-electron matrix elements, the topic which is the focus of Chapters 19 and 20.

The procedure for obtaining double-tensor analogs of the equations in Section 18.1 is straightforward, since we can proceed exactly as in that section except that we double the number of expansions and reductions, doing them once for all spin functions (in SO_3) as well as once for all space functions (in the point group G of the molecule). Let D^{jd} be a double tensor operator which transforms as irrep j of SO_3 in spin space and as irrep d of group G in orbital space, and suppose that D^{jd} operates only on the first part ($\mathbb{S}_1 a$ and $\mathbb{S}'_1 a'$) of a coupled system. We find then, analogous to (18.1.2),

$$\langle(\mathbb{S}_1 a, \mathbb{S}_2 b)\mathbb{S}rc\|D^{jd}\|(\mathbb{S}'_1 a', \mathbb{S}'_2 b')\mathbb{S}'r'c'\rangle_p^{SO_3, G}$$

$$= \delta_{\mathbb{S}_2\mathbb{S}'_2}\delta_{bb'} H(\mathbb{S}_1\mathbb{S}_2\mathbb{S}) H(\mathbb{S}'_1\mathbb{S}'_2\mathbb{S}')$$

$$\times (|\mathbb{S}||\mathbb{S}'|)^{1/2}\{\mathbb{S}_1\mathbb{S}_2\mathbb{S}\}\{\mathbb{S}'_1\}\{\mathbb{S}_1 j \mathbb{S}'_1\}\begin{Bmatrix} j & \mathbb{S}' & \mathbb{S} \\ \mathbb{S}_2 & \mathbb{S}_1 & \mathbb{S}'_1 \end{Bmatrix}$$

$$\times (|c||c'|)^{1/2}\{abc^*r\}\{a'\}\sum_q \{a^*da'q\}\begin{Bmatrix} d & c' & c^* \\ b^* & a & a' \end{Bmatrix}_{qr'rp}$$

$$\times \langle\mathbb{S}_1 a\|D^{jd}\|\mathbb{S}'_1 a'\rangle_q^{SO_3, G} \qquad (18.5.1)$$

No repeated representations occur in SO_3 direct products; thus r, r', p, and q are all group-G repeated representation labels. The historical phase factors, $H(abc)$, must be inserted whenever (10.2.3) is used in SO_3 in derivations—thus $H(\mathbb{S}_1\mathbb{S}_2\mathbb{S})$ and $H(\mathbb{S}_1'\mathbb{S}_2'\mathbb{S}')$ arise here from the spin–spin couplings. And when E^{je} is a double tensor operator which operates only on the second part ($\mathbb{S}_2 b$ and $\mathbb{S}_2' b'$) of a coupled system, we obtain, analogous to (18.1.4),

$$\begin{aligned}
&\big\langle(\mathbb{S}_1 a, \mathbb{S}_2 b)\mathbb{S} rc\big\|E^{je}\big\|(\mathbb{S}_1' a', \mathbb{S}_2' b')\mathbb{S}' r'c'\big\rangle_p^{SO_3, G} \\
&\quad = \delta_{\mathbb{S}_1\mathbb{S}_1'}\delta_{aa'} H(\mathbb{S}_1\mathbb{S}_2\mathbb{S}) H(\mathbb{S}_1'\mathbb{S}_2'\mathbb{S}') \\
&\quad \times (|\mathbb{S}||\mathbb{S}'|)^{1/2}\{\mathbb{S}_1'\mathbb{S}_2'\mathbb{S}'\}\{\mathbb{S}_2'\}\{\mathbb{S}_2 j \mathbb{S}_2'\}\begin{Bmatrix} j & \mathbb{S}' & \mathbb{S} \\ \mathbb{S}_1 & \mathbb{S}_2 & \mathbb{S}_2' \end{Bmatrix} \\
&\quad \times (|c||c'|)^{1/2}\{a'b'c'^{*}r'\}\{b'\}\sum_q \{b^{*}eb'q\}\begin{Bmatrix} e & c' & c^{*} \\ a^{*} & b & b' \end{Bmatrix}_{qr'rp} \\
&\quad \times \langle \mathbb{S}_2 b\| E^{je} \|\mathbb{S}_2' b'\rangle_q^{SO_3, G}
\end{aligned} \tag{18.5.2}$$

The definitions in Section B.3 may be used to simplify the $2j$, $3j$, and historical phases for SO_3.

19 Reduction of Multielectron Matrix Elements to One-Electron Form Using the Irreducible-Tensor Method

The power of the irreducible-tensor method is most apparent in the calculation of matrix elements of one-electron operators between states of multielectron configurations with two or more open shells. No wavefunctions need be constructed, and the matrix elements may be expressed, in a matter of minutes, as a sum of one-electron reduced matrix elements multiplied by symmetry-determined factors. Moreover, since the methods of reduction are systematic, the calculations may be computerized.

In this chapter we show in detail how very useful matrix-element formulas are derived using the results of the preceding chapters. The equations hold for a wide variety of molecular symmetry groups. We require only that the $3jm$ for the group be defined in the manner indicated in Chapters 10 and 14 and that repeated representation labels not be required to define the $|\mathbb{S}h\mathfrak{M}\theta\rangle$ states of a^n configurations; the latter condition is satisfied except for the uncommon groups T, T_h, and K. The equations apply to $SO_3 \supset SO_2$ (or $O_3 \supset SO_3 \supset SO_2$) with the same reservations as those given at the beginning of Chapter 18.

When a given a^n configuration gives rise to more than one state with the same $\mathbb{S}$ and h values, an additional label is needed to distinguish between the states. Thus, for example, $|a^n\, \mathbb{S}h\mathfrak{M}\theta\rangle$ is replaced by $|a^n\, \alpha\mathbb{S}h\mathfrak{M}\theta\rangle$ and so on. We usually omit these labels, but readers should supply them and carry them through derivations when necessary.

We make extensive use of the properties of determinantal wavefunctions in our derivations, and so we begin by reviewing some of those properties.

19.1 Permutation Properties of Determinantal Wavefunctions

No matter what basis we use, our wavefunctions $|\Gamma\gamma\rangle$ are always linear combinations of fully antisymmetrized determinantal wavefunctions $|\psi_j\rangle$:

$$|\Gamma\gamma\rangle = \sum_j c_{\gamma j}|\psi_j\rangle \tag{19.1.1}$$

where

$$\begin{aligned} |\psi_j\rangle &= |\phi_1^j\phi_2^j \cdots \phi_n^j\rangle \\ &= (n!)^{-1/2} \sum_{\mu=1}^{n!} (-1)^\mu P_\mu\{\phi_1^j(1)\phi_2^j(2) \cdots \phi_n^j(n)\} \end{aligned} \tag{19.1.2}$$

Here P_μ permutes the electrons, and the sum over μ includes the $n!$ possible permutations. The factor $(-1)^\mu$ is -1 when P_μ represents an odd number of individual electron exchanges, and $+1$ for an even number of exchanges. Thus, for example, when $n = 3$

$$\begin{aligned} |\psi_j\rangle = \frac{1}{\sqrt{6}}[&\phi_1^j(1)\phi_2^j(2)\phi_3^j(3) - \phi_1^j(1)\phi_2^j(3)\phi_3^j(2) \\ &+\phi_1^j(2)\phi_2^j(3)\phi_3^j(1) - \phi_1^j(2)\phi_2^j(1)\phi_3^j(3) \\ &+\phi_1^j(3)\phi_2^j(1)\phi_3^j(2) - \phi_1^j(3)\phi_2^j(2)\phi_3^j(1)] \end{aligned}$$

$$\equiv \frac{1}{\sqrt{6}} \begin{vmatrix} \phi_1^j(1) & \phi_2^j(1) & \phi_3^j(1) \\ \phi_1^j(2) & \phi_2^j(2) & \phi_3^j(2) \\ \phi_1^j(3) & \phi_2^j(3) & \phi_3^j(3) \end{vmatrix} \tag{19.1.3}$$

In this chapter the $\phi^j{}_k$ represent occupied spin–orbitals of the jth determinantal function; they may also represent "jj" orbitals (Chapter 21). For our $|\psi_j\rangle$ functions,

$$\sum_{\mu=1}^{n!} (-1)^\mu P_\mu|\psi_j\rangle = n!|\psi_j\rangle \tag{19.1.4}$$

and thus

$$\sum_{\mu=1}^{n!} (-1)^{\mu} P_{\mu}|\Gamma\gamma\rangle = n!|\Gamma\gamma\rangle \tag{19.1.5}$$

These equations hold because the effect of the operator $(-1)^{\mu}P_{\mu}$ on $|\psi_j\rangle$ is to produce the same terms as in the sum (19.1.2) but in a different order. Any particular permutation gives the result

$$P_{\mu}|\psi_j\rangle = (-1)^{\mu}|\psi_j\rangle \tag{19.1.6}$$

where μ is an even integer when P_{μ} represents an even number of permutations of the electrons, and an odd integer for an odd number of permutations.

We shall be calculating matrices of one-electron operators $V = \sum_{m=1}^{n} v(m)$. For these operators

$$\langle\psi_i|V|\psi_j\rangle = \sum_{m=1}^{n} \langle\psi_i|v(m)|\psi_j\rangle$$

$$= n\langle\psi_i|v(m)|\psi_j\rangle \tag{19.1.7}$$

and thus

$$\langle\Gamma\gamma|V|\Gamma'\gamma'\rangle = n\langle\Gamma\gamma|v(m)|\Gamma'\gamma'\rangle \tag{19.1.8}$$

since if P_{kl} is the permutation operator which permutes the kth and the lth electron,

$$\langle\psi_i|v(k)|\psi_j\rangle = P_{kl}\langle\psi_i|v(k)|\psi_j\rangle$$

$$= \langle P_{kl}\psi_i|P_{kl}v(k)|P_{kl}\psi_j\rangle$$

$$= \langle\psi_i|v(l)|\psi_j\rangle \tag{19.1.9}$$

The first equality holds because a matrix element is simply a number and its value cannot depend on the labeling of the arguments, and the final step follows from (19.1.6).

Another useful result for functions of the type (19.1.2) and one-electron operators is

$$\langle\psi_i|V|\psi_j\rangle = \frac{1}{\sqrt{n!}}\sum_{\mu=1}^{n!}(-1)^{\mu}P_{\mu}^{-1}\langle\psi_i|V|P_{\mu}\{\phi_1^j(1)\phi_2^j(2)\cdots\phi_n^j(n)\}\rangle$$

$$= \frac{1}{\sqrt{n!}}\sum_{\mu=1}^{n!}(-1)^{\mu}\langle P_{\mu}^{-1}\psi_i|P_{\mu}^{-1}V|P_{\mu}^{-1}P_{\mu}\{\cdots\}\rangle$$

$$= \sqrt{n!}\,\langle\psi_i|V|\{\phi_1^j(1)\phi_2^j(2)\cdots\phi_n^j(n)\}\rangle \tag{19.1.10}$$

In the first step we expand $|\psi_j\rangle$ with (19.1.2) and also permute the *entire integral* by P_{μ}^{-1}, which cannot change its value, since it simply relabels the arguments. P_{μ}^{-1} is the permutation operator which restores the normal order of the variables in $P_{\mu}\{\phi_1^j(1)\phi_2^j(2)\cdots\phi_n^j(n)\}$. Using this along with the relation $P_{\mu}^{-1}V = V$ and the result (19.1.4), we obtain the final equality. Thus all the $n!$ terms in the sum over μ give an identical contribution to the integral.

When $\psi_i = \psi_j$, (19.1.10) reduces to

$$\langle\psi_i|V|\psi_i\rangle = \sum_{m=1}^{n}\langle\phi_m^i(m)|v(m)|\phi_m^i(m)\rangle \tag{19.1.11}$$

On the other hand, when ψ_i and ψ_j differ only in one spin–orbital so that $\phi_l^i \neq \phi_{l'}^j$, (19.1.10) simplifies to

$$\langle\psi_i|V|\psi_j\rangle = (-1)^p\langle\phi_l^i|v|\phi_{l'}^j\rangle \tag{19.1.12}$$

where p is the number of interchanges necessary to bring the identical ϕ_k^i and ϕ_k^j into the same order. Finally, if ψ_i and ψ_j differ in more than one spin–orbital, (19.1.10) gives zero. These results, all of which follow from the orthonormality of the spin–orbitals, were used in Chapter 13.

19.2. Outline of the Method Used to Simplify Matrix Elements of a Configuration a^n

The simplest multielectron matrix elements are those of the type

$$\langle a^n\,\mathcal{S}h\mathfrak{M}\theta|V|a^n\,\mathcal{S}'h'\mathfrak{M}'\theta'\rangle \tag{19.2.1}$$

where $V = \Sigma_{m=1}^{n} v(m)$ is a one-electron operator and $|\Gamma\gamma\rangle = |a^n\, \mathbb{S}h\mathfrak{M}\theta\rangle$ is the $|\mathbb{S}h\mathfrak{M}\theta\rangle$ function for a state of the configuration a^n. Thus we begin by outlining the method used in evaluating these matrix elements. Very similar methods are employed, however, in more complicated cases.

We assume here and elsewhere in this chapter that no repeated representation labels are needed to uniquely define the $|\mathbb{S}h\mathfrak{M}\theta\rangle$ states of a configuration a^n. This is true for all crystallographic point groups except for the groups T and T_h, which are uncommon, and all chemically important molecular symmetry groups except for the icosahedral groups.

Our first step in evaluating (19.2.1) is to use (19.1.8) with $m = n$, so that (19.2.1) becomes

$$n\langle a^n\, \mathbb{S}h\mathfrak{M}\theta | v(n) | a^n\, \mathbb{S}'h'\mathfrak{M}'\theta'\rangle \tag{19.2.2}$$

where $v(n)$ operates only on the variables of the nth electron. We then decompose the completely antisymmetrized $|a^n\, \mathbb{S}h\mathfrak{M}\theta\rangle$ functions into linear combinations of functions of the type (11.4.7):

$$\left|\left(a^{n-1}(\mathbb{S}_1 h_1), a\right)\mathbb{S}h\mathfrak{M}\theta\right\rangle = \sum_{\mathfrak{M}_1 \mathfrak{m}} \left(\mathbb{S}_1\mathfrak{M}_1, \tfrac{1}{2}\mathfrak{m}|\mathbb{S}\mathfrak{M}\right) \times \sum_{\theta_1 \alpha} (h_1\theta_1, a\alpha|h\theta)\left|a^{n-1}\mathbb{S}_1 h_1 \mathfrak{M}_1\theta_1\right\rangle \left|a\, \mathfrak{m}\alpha\right\rangle \tag{19.2.3}$$

where the spin and space labels $(\frac{1}{2}a)$ have been dropped for $a^1 = a$, since they are redundant. The notation above, which was introduced in Section 11.4 has a significance which we emphasize here. These $|(a^{n-1}(\mathbb{S}_1 h_1), a)\mathbb{S}h\mathfrak{M}\theta\rangle$ functions are *not* of the type (19.1.1). They are not completely antisymmetrized (though $\mathbb{S}$, h, $\mathfrak{M}$, θ are still correct symmetry labels); the first $n-1$ electrons are restricted to $|a^{n-1}\mathbb{S}_1 h_1\mathfrak{M}_1\theta_1\rangle$, and the nth electron is in $|a\,\mathfrak{m}\alpha\rangle$. The a^{n-1} and a^1 functions [which *are* of the type (19.1.1)] occupy totally different spaces. They have no variables in common. Thus, since the operator $v(n)$ operates only on the variables of the nth electron, it operates only on $|a\,\mathfrak{m}\alpha\rangle$. Therefore, we have precisely the type of matrix element described in Chapter 18, in which the operator acts on only one part of a coupled system—in this case the second part. Consequently, the matrix element (19.2.2) may be expressed, using formulas such as (18.1.4) or (18.5.2), as a symmetry-determined factor times $\langle a\|v\|a\rangle$ or $\langle \frac{1}{2}a\|v\|\frac{1}{2}a\rangle$. The latter are reduced matrix elements for one-electron MO functions.

Our initial step is therefore to express our $|a^n\, \mathbb{S}h\mathfrak{M}\theta\rangle$ functions in terms of the functions defined in (19.2.3) above. This is accomplished using *coefficients of fractional parentage* (cfp) [Eq. (11.4.15)]:

$$|a^n \mathbb{S}h\mathfrak{M}\theta\rangle = \sum_{\mathbb{S}_1 h_1} \left(a^{n-1}\mathbb{S}_1 h_1, a | a^n \mathbb{S}h\right)\left|\left(a^{n-1}(\mathbb{S}_1 h_1), a\right)\mathbb{S}h\mathfrak{M}\theta\right\rangle \tag{19.2.4}$$

The coefficients are known as cfp because they specify the contribution of each "parent" $a^{n-1}(\mathbb{S}_1 h_1)$ state required to form a properly antisymmetrized $|a^n \mathbb{S}h\mathfrak{M}\theta\rangle$ wavefunction. Many authors denote the cfp above as $(a^{n-1}(\mathbb{S}_1 h_1), a|\}a^n\mathbb{S}h)$, but we prefer the simplified notation of Butler used in (19.2.4) above, since it emphasizes the similarity of the cfp to vector coupling factors (Section 15.3).

19.3. Coefficients of Fractional Parentage

Coefficients of fractional parentage (cfp) are extensively employed in the decomposition of a^n functions as in (19.2.4). They are essentially vector coupling factors (Section 15.3). As discussed in Chapter 7 of Butler (1981), the cfp for the configuration a^n are basically products of coupling factors for the group chain beginning with the group $U_{2|a|}$ of all unitary transformations in a space of dimension $2|a|$, and ending with the symmetry group to which a^n belongs. The cfp of any higher group in the chain may be related to those of a lower group by the vector coupling factor (or product of factors) which connects the groups. Thus the cfp for a^n where $|a| = 1, 3, 5$, or 7 are related respectively to those for s^n, p^n, d^n, and f^n given by Nielson and Koster (1963). Of practical significance is that our cfp for the octahedral or tetrahedral pseudo-p^n configurations, t_1^n and t_2^n, have been obtained in the above manner from the cfp for p^n using the Butler (1981) $SO_3 \supset O$ and $O_3 \supset O_h \supset T_d$ $3jm$ factors; we demonstrate the method in Section 19.9.

The cfp may also be found using less elegant methods. We illustrate one such procedure below, which involves the construction of a^n states with cfp. The method is somewhat lengthy but gives a good introduction to the cfp concept.

How to Calculate cfp *and Use Them to Construct Octahedral* $|t_2^3 \mathbb{S}h\mathfrak{M}\theta\rangle$ *States*

We now show how to construct t_2^3 states for the group O by successively coupling states of t_2 electrons. In the process the values of the

$(t_2^2\,\mathbb{S}_1h_1, t_2|t_2^3\,\mathbb{S}h)$ and $(t_2, t_2|t_2^2\,\mathbb{S}h)$ cfp used in the expansions are determined. The reader may find it useful at this point to review Section 11.4. We use the $SO_3 \supset SO_2$ $|\mathbb{S}\mathfrak{M}\rangle$ basis for spins and the real-O basis of Section C.5 for orbitals.

Functions for t_2^1 may simply be written down:

$$\begin{aligned}|t_2^1\tfrac{1}{2}T_2\tfrac{1}{2}\xi\rangle &\equiv |t_2\tfrac{1}{2}\xi\rangle \equiv |\xi^+\rangle\\ |t_2^1\tfrac{1}{2}T_2\tfrac{1}{2}\eta\rangle &\equiv |t_2\tfrac{1}{2}\eta\rangle \equiv |\eta^+\rangle\\ |t_2^1\tfrac{1}{2}T_2\tfrac{1}{2}\zeta\rangle &\equiv |t_2\tfrac{1}{2}\zeta\rangle \equiv |\zeta^+\rangle\end{aligned} \tag{19.3.1}$$

It follows that for the cfp in (19.2.4) in the one-electron case we need

$$\left(t_2^0, t_2|t_2\right) = 1 \tag{19.3.2}$$

which is the conventional choice. Here and elsewhere we simplify our notation to eliminate redundant labels.

Next t_2^2 states are formed from (19.2.4) and (19.2.3). For example, (19.2.4) gives for the $^3T_1 1x$ state

$$|t_2^2\,1T_1 1x\rangle = \left(t_2, t_2|t_2^2\,1T_1\right)|(t_2, t_2)1T_1 1x\rangle \tag{19.3.3}$$

Then (19.2.3), (B3.2), and (10.2.1) [which together give (11.4.7)] give

$$\begin{aligned}|(t_2, t_2)1T_1 1x\rangle &= \sum_{\mathfrak{M}_1 m} (-1)^{\frac{1}{2}-\frac{1}{2}+1}\sqrt{3}\,(-1)^{1-1}\begin{pmatrix}\frac{1}{2} & \frac{1}{2} & 1\\ \mathfrak{M}_1 & m & -1\end{pmatrix}\\ &\quad\times\sum_{\theta_1\alpha}|T_1|^{1/2}\begin{pmatrix}T_1\\ x\end{pmatrix}\begin{pmatrix}T_2 & T_2 & T_1\\ \theta_1 & \alpha & x\end{pmatrix}^* |t_2\mathfrak{M}_1\theta_1\rangle|t_2\, m\alpha\rangle\\ &= \frac{-1}{\sqrt{2}}\left(|\eta^+\rangle|\zeta^+\rangle - |\zeta^+\rangle|\eta^+\rangle\right)\\ &= -|\eta^+\zeta^+\rangle\end{aligned} \tag{19.3.4}$$

where we use the $SO_3 \supset SO_2$ $3jm$ of Section B.3 and the real-O $3jm$ of Section C.12. In this case the cfp in (19.3.3) is simply equal to a phase factor. We choose the phase of this cfp, and all other t_1^n and t_2^n cfp, to correspond with the phase calculated from the tables of Nielson and Koster (1963)—see Section 19.9. Thus $(t_2, t_2|t_2^2\,1T_1) = -1$ (Section C.15). With this choice

$$|t_2^2\,1T_1 1x\rangle = |\eta^+\zeta^+\rangle \tag{19.3.5}$$

Other t_2^2 functions found by the same method are

$$
\begin{aligned}
|t_2^2\,1T_1 1y\rangle &= -|\xi^+\zeta^+\rangle \\
|t_2^2\,1T_1 1z\rangle &= |\xi^+\eta^+\rangle \\
|t_2^2\,0A_1 0a_1\rangle &= \frac{1}{\sqrt{3}}(-|\xi^+\xi^-\rangle + |\eta^+\eta^-\rangle - |\zeta^+\zeta^-\rangle) \\
|t_2^2\,0E0\theta\rangle &= \frac{1}{\sqrt{6}}(|\xi^+\xi^-\rangle - |\eta^+\eta^-\rangle - 2|\zeta^+\zeta^-\rangle) \\
|t_2^2\,0E0\epsilon\rangle &= \frac{1}{\sqrt{2}}(|\xi^+\xi^-\rangle + |\eta^+\eta^-\rangle) \\
|t_2^2\,0T_2 0\xi\rangle &= \frac{1}{\sqrt{2}}(|\eta^+\zeta^-\rangle - |\eta^-\zeta^+\rangle) \\
|t_2^2\,0T_2 0\eta\rangle &= \frac{1}{\sqrt{2}}(-|\xi^+\zeta^-\rangle + |\xi^-\zeta^+\rangle) \\
|t_2^2\,0T_2 0\zeta\rangle &= \frac{1}{\sqrt{2}}(|\xi^+\eta^-\rangle - |\xi^-\eta^+\rangle)
\end{aligned} \tag{19.3.6}
$$

where we give only the $\mathfrak{M} = 1$ functions for the spin triplets.

Construction of t_2^3 wavefunctions is far more complex, and the determination of cfp consequently more complicated. We illustrate the process for $\left|t_2^3\,{}^2T_1\tfrac{1}{2}x\right\rangle$. Equation (19.2.4) gives

$$
\begin{aligned}
|t_2^3\,\tfrac{1}{2}T_1\tfrac{1}{2}x\rangle = {} & \left(t_2^2\,1T_1, t_2|t_2^3\,\tfrac{1}{2}T_1\right)\left|\left(t_2^2(1T_1), t_2\right)\tfrac{1}{2}T_1\tfrac{1}{2}x\right\rangle \\
& + \left(t_2^2\,0E, t_2|t_2^3\,\tfrac{1}{2}T_1\right)\left|\left(t_2^2(0E), t_2\right)\tfrac{1}{2}T_1\tfrac{1}{2}x\right\rangle \\
& + \left(t_2^2\,0T_2, t_2|t_2^3\,\tfrac{1}{2}T_1\right)\left|\left(t_2^2(0T_2), t_2\right)\tfrac{1}{2}T_1\tfrac{1}{2}x\right\rangle
\end{aligned} \tag{19.3.7}
$$

Here $t_2^2(\mathcal{S}_1 h_1) = t_2^2(0A_1)$ does not contribute, since $A_1 \otimes T_2 \neq T_1$. When the

$3jm$ of Sections B.3 and C.12 are used, (19.2.3) gives

$$\begin{aligned}
&\left|(t_2^2(1T_1), t_2)\tfrac{1}{2}T_1\tfrac{1}{2}x\right\rangle \\
&= \left(\frac{\sqrt{2}}{\sqrt{3}}\right)\left(\frac{-1}{\sqrt{2}}\right)\left(|t_2^2\,1T_1 1y\rangle|t_2 - \tfrac{1}{2}\zeta\rangle - |t_2^2\,1T_1 1z\rangle|t_2 - \tfrac{1}{2}\eta\rangle\right) \\
&+ \left(\frac{-1}{\sqrt{3}}\right)\left(\frac{-1}{\sqrt{2}}\right)\left(|t_2^2\,1T_1 0y\rangle|t_2\,\tfrac{1}{2}\zeta\rangle - |t_2^2\,1T_1 0z\rangle|t_2\,\tfrac{1}{2}\eta\rangle\right) \\
&= \frac{1}{\sqrt{3}}\left(|\xi^+\zeta^+\rangle|\zeta^-\rangle + |\xi^+\eta^+\rangle|\eta^-\rangle\right) \\
&+ \frac{1}{2\sqrt{3}}\left(-|\xi^+\zeta^-\rangle|\zeta^+\rangle - |\xi^-\zeta^+\rangle|\zeta^+\rangle - |\xi^+\eta^-\rangle|\eta^+\rangle\right. \\
&\qquad \left. - |\xi^-\eta^+\rangle|\eta^+\rangle\right)
\end{aligned} \tag{19.3.8}$$

$$\begin{aligned}
\left|(t_2^2(0E), t_2)\tfrac{1}{2}T_1\tfrac{1}{2}x\right\rangle &= \frac{-\sqrt{3}}{2}|t_2^2\,0E0\theta\rangle|t_2\,\tfrac{1}{2}\xi\rangle + \tfrac{1}{2}|t_2^2\,0E0\epsilon\rangle|t_2\,\tfrac{1}{2}\xi\rangle \\
&= \frac{1}{\sqrt{2}}\left(|\eta^+\eta^-\rangle|\xi^+\rangle + |\zeta^+\zeta^-\rangle|\xi^+\rangle\right)
\end{aligned}$$

$$\begin{aligned}
&\left|(t_2^2(0T_2), t_2)\tfrac{1}{2}T_1\tfrac{1}{2}x\right\rangle \\
&= \frac{-1}{\sqrt{2}}|t_2^2\,0T_2 0\eta\rangle|t_2\,\tfrac{1}{2}\zeta\rangle + \frac{1}{\sqrt{2}}|t_2^2\,0T_2 0\zeta\rangle|t_2\,\tfrac{1}{2}\eta\rangle \\
&= \tfrac{1}{2}\left(|\xi^+\zeta^-\rangle|\zeta^+\rangle - |\xi^-\zeta^+\rangle|\zeta^+\rangle + |\xi^+\eta^-\rangle|\eta^+\rangle - |\xi^-\eta^+\rangle|\eta^+\rangle\right)
\end{aligned}$$

Remember that kets of products, such as $|\xi^+\zeta^+\rangle$, always represent normalized determinantal functions. Recall, however, that the notation of the $|(t_2^2(\mathbb{S}_1 h_1), t_2)\frac{1}{2}T_1\frac{1}{2}x\rangle$ functions tells us that they are *not* in general by themselves properly antisymmetric wavefunctions. We must therefore choose our cfp in (19.3.7) so that the *total* $|t_2^3\,\frac{1}{2}T_1\frac{1}{2}x\rangle$ function *is* antisymmetric and obeys the Pauli principle. Clearly we must have

$$\left(t_2^2\,0T_2, t_2|t_2^3\,\tfrac{1}{2}T_1\right) = \frac{-1}{\sqrt{3}}\left(t_2^2\,1T_1, t_2|t_2^3\,\tfrac{1}{2}T_1\right) \tag{19.3.9}$$

if we are to obey the Pauli principle, since terms of the type $|\xi^-\zeta^+\rangle|\zeta^+\rangle$

must sum to zero. Then we need

$$\left(t_2^2 0E, t_2 | t_2^3 \tfrac{1}{2} T_1\right) = \frac{\sqrt{2}}{\sqrt{3}} \left(t_2^2 1T_1, t_2 | t_2^3 \tfrac{1}{2} T_1\right) \qquad (19.3.10)$$

in order to obtain $|\xi^+ \eta^+ \eta^-\rangle$ and $|\xi^+ \zeta^+ \zeta^-\rangle$ as properly antisymmetric kets of the type (19.1.2). Normalization fixes the sum of the squares of the cfp as 1, and these conditions fix the cfp values to within a phase factor p as

$$\left(t_2^2 1T_1, t_2 | t_2^3 \tfrac{1}{2} T_1\right) = \frac{1}{\sqrt{2}} p$$

$$\left(t_2^2 0E, t_2 | t_2^3 \tfrac{1}{2} T_1\right) = \frac{1}{\sqrt{3}} p \qquad (19.3.11)$$

$$\left(t_2^2 0T_2, t_2 | t_2^3 \tfrac{1}{2} T_1\right) = \frac{-1}{\sqrt{6}} p$$

Substitution of all these results back into (19.3.7) gives

$$|t_2^3 \tfrac{1}{2} T_1 \tfrac{1}{2} x\rangle = \frac{p}{\sqrt{2}} \left(|\xi^+ \eta^+ \eta^-\rangle + |\xi^+ \zeta^+ \zeta^-\rangle\right) \qquad (19.3.12)$$

We choose $p = 1$ to give the cfp the same phases as those derived from the cfp of Nielson and Koster (1963).

General Equations for cfp

Proper wavefunctions for all states of an a^n configuration may be formed using (19.2.3) and (19.2.4). While the method is systematic and straightforward (particularly if cfp tables are available), it is still cumbersome. The beauty of the irreducible-tensor method is that the actual wavefunctions are not required but *only the cfp*—in fact, only selected cfp are needed.

The general expression for cfp of which (19.2.4) is a special case is

$$|a^n \mathbb{S} h \mathfrak{M} \theta\rangle = \sum_{\mathbb{S}_1 h_1 \mathbb{S}_2 h_2} \left(a^p \mathbb{S}_1 h_1, a^q \mathbb{S}_2 h_2 | a^n \mathbb{S} h\right) \times \left|\left(a^p(\mathbb{S}_1 h_1), a^q(\mathbb{S}_2 h_2)\right) \mathbb{S} h \mathfrak{M} \theta\right\rangle \qquad (19.3.13)$$

with

$$|(a^p(\mathcal{S}_1h_1), a^q(\mathcal{S}_2h_2))\mathcal{S}h\mathcal{M}\theta\rangle = \sum_{\mathcal{M}_1\mathcal{M}_2} (\mathcal{S}_1\mathcal{M}_1, \mathcal{S}_2\mathcal{M}_2|\mathcal{S}\mathcal{M})$$

$$\times \sum_{\theta_1\theta_2} (h_1\theta_1, h_2\theta_2|h\theta)\, |a^p\, \mathcal{S}_1h_1\mathcal{M}_1\theta_1\rangle |a^q\, \mathcal{S}_2h_2\mathcal{M}_2\theta_2\rangle \tag{19.3.14}$$

We, however, only require coefficients for either $p = n - 1$, $q = 1$ [Eq. (19.2.4)] or for $p = 1$, $q = n - 1$. We tabulate cfp for the first case only, but the a, a^{n-1} coefficients may be calculated from these using the relation

$$\left(a, a^{n-1}\mathcal{S}'h'|a^n\,\mathcal{S}h\right) = (-1)^{n-1}\{h'ah^*\}(-1)^{\mathcal{S}'+\frac{1}{2}-\mathcal{S}}\left(a^{n-1}\mathcal{S}'h', a|a^n\,\mathcal{S}h\right) \tag{19.3.15}$$

which is derived by comparison of the $|a^n\,\mathcal{S}h\mathcal{M}\theta\rangle$ expansions for the two orders of coupling, a, a^{n-1} and a^{n-1}, a.

We give a complete tabulation of the a^{n-1}, a cfp for the groups O and T_d in Section C.15; they also apply for the analogous O_h configurations if u and g parity labels are affixed. Those for t_1^n and t_2^n were calculated from Nielson and Koster's (1963) p^n cfp as illustrated in Section 19.9.

The cfp we require for a_1^n and a_2^n for the cubic groups, and all cfp for their subgroups, may be calculated from general formulas which we now give. Unless stated otherwise, these also apply for the cfp of Section C.15. For $n = 1$ we always have

$$\left(a^0, a|a\right) = 1 \tag{19.3.16}$$

As required by (11.4.12) (for $|a| \neq 1$), we define

$$\left(a^{2|a|-1}, a|a^{2|a|}\right) = \left(a, a^{2|a|-1}|a^{2|a|}\right) = 1 \tag{19.3.17}$$

[But note when $|a| = 1$, (19.3.17) and (19.2.3)–(19.2.4) give

$$|a^2 0A_1\rangle = \begin{pmatrix} a \\ \alpha \end{pmatrix} |\alpha^+\alpha^-\rangle.]$$

At this point we have defined all cfp for configurations a^n for which $|a| = 1$. Since there is no atomic configuration l^n with $|l| = 2l + 1 = 2$ (or any other even number), cfp for a^n with $|a| = 2$, 4, and so on, cannot be derived from the tables of Nielson and Koster (1963) as they were for t_1^n and t_2^n (Section 19.9). Thus we define the a^2 cfp (which are at most a phase factor) *for even* $|a|$ to be unity:

$$(a, a|a^2 \mathbb{S}h) = 1 \tag{19.3.18}$$

This also holds for $|a| = 1$, since in that case (19.3.18) is identical to (19.3.17). Note, however, we do not use (19.3.18) for $|a| = 3$, but instead use the values in Section C.15.

Since we tabulate the cfp for e^n, t_1^n, and t_2^n for the groups O and T_d in Section C.15, it only remains to define the cfp for e^3 configurations. For groups O, T_d, and D_3 ($D_3 \supset C_2$ basis) with $p = 1$, and for groups D_4, D_3 ($D_3 \supset C_3$ basis), and D_∞ (Appendix F basis) with $p = -1$, we have

$$\begin{aligned}
(e^2 0A_1, e|e^3) &= \frac{p}{\sqrt{6}} \\
(e^2 0h', e|e^3) &= \frac{-p}{\sqrt{6}}|h'|^{1/2} \qquad (\text{for} \quad h' \neq A_1) \\
(e^2 \mathbb{S}'h', e|e^3) &= \frac{-p}{\sqrt{6}}(|\mathbb{S}'|\,|h'|)^{1/2} \qquad (\text{for } \mathbb{S}'h' \neq 0A_1)
\end{aligned} \tag{19.3.19}$$

Equation (19.3.19) is a special case of the first of two very useful cfp equations of Racah for $|a| \geqslant 2$. These were originally derived in Racah [(1943), Section 4], but were later shown in Racah (1949) to follow from the interchange symmetry of the cfp regarded as vector coupling factors. The equations are

$$\begin{aligned}
&(a^{2|a|-2}\, \mathbb{S}'h', a|a^{2|a|-1}) \\
&= \left[\frac{|\mathbb{S}'|\,|h'|}{|a|(2|a|-1)}\right]^{1/2} (-1)^{\mathbb{S}'} \{aah'^*\} \mu(\mathbb{S}'h')(a, a|a^2\, \mathbb{S}'h')
\end{aligned} \tag{19.3.20}$$

$$\begin{aligned}
&(a^3\, \mathbb{S}'h', a|a^4\, \mathbb{S}h) \\
&= (-1)^{\frac{1}{2}+a+\mathbb{S}'+h'+1} \left[\frac{3|\mathbb{S}'|\,|h'|}{4|\mathbb{S}|\,|h|}\right]^{1/2} \mu(\mathbb{S}'h')(a^2\, \mathbb{S}h, a|a^3\, \mathbb{S}'h')
\end{aligned} \tag{19.3.21}$$

The last equation applies only for $|a| = 3$.

The phase $\mu(\mathbb{S}'h')$ is defined for the cases of interest to us in Griffith [(1964), Section 9.7.1)] and [(1962), Section 7.2]. When $n \neq |a|$, $\mu(\mathbb{S}'h') \equiv 1$. However, when $\mathbb{S}'h'$ labels a term from a half-filled shell (i.e., $n = |a|$), $\mu(\mathbb{S}'h')$ may be ± 1. For O and T_d (but *not* most dihedral groups), when $|a| = n = 2$, we have $\mu(\mathbb{S}'h') = 1$ for ${}^1A_1$ and -1 otherwise. This is the $a = e$ case. The only other $\mu(\mathbb{S}'h')$ we commonly require are those for t_1^3 and t_2^3 configurations for O and T_d. For the t_2^3 configuration, $\mu(\mathbb{S}'h')$ is 1 for ${}^2T_2$, and -1 for ${}^4A_2$, ${}^2E$, and ${}^2T_1$. For the t_1^3 configuration, $\mu(\mathbb{S}'h')$ is 1 for ${}^2T_1$, and -1 for ${}^4A_1$, ${}^2E$, and ${}^2T_2$. The cfp of Section C.15 are consistent with these values. Our $\mu(\mathbb{S}'h')$ for dihedral groups differ from Griffith's.

In the derivation of (19.3.21) we have used (10.2.5) and have factored out the term $(-1)^{h+\mathbb{S}}$ for $\mathbb{S}h$ states of two-electron hole (a^4) or particle (a^2) configurations; for such states in SO_3, $(-1)^{h+\mathbb{S}} = 1$ as a result of the Pauli principle [Griffith (1962), Section 6.1], and the relation also applies for the $\mathbb{S}h$ states of e^2, t_1^2, and t_2^2 configurations for the cubic groups. In the e^3 case (19.3.20) gives

$$(e^2\,\mathbb{S}'h', e|e^3) = \frac{1}{\sqrt{6}}(|\mathbb{S}'|\,|h'|)^{1/2}(-1)^{\mathbb{S}'}\{aah'^*\}\mu(\mathbb{S}'h') \quad (19.3.22)$$

from which (19.3.19) follows for the dihedral and cubic groups.

19.4. One Open Shell: Matrix Elements of a^n

In this section we calculate the matrix elements of a^n for one-electron operators using the methods of Sections 19.2 and 15.4 for spin-independent and spin-only matrix elements respectively. If a spin–orbit basis is used, these matrix elements result after application of (18.1.8) or (18.1.9).

Spin-Independent Matrix Elements of a^n and the $g(a^n, \mathbb{S}hfh')$

We begin by calculating the matrix elements for a spin-independent one-electron operator O_ϕ^f, where f and ϕ are the group irrep and partner labels for the operator. For example, $\mathcal{O}$ could be the orbital-angular-momentum operator L or the electric dipole operator m. The spin independence of the operator, (19.2.2), and (10.2.2) give

$$\begin{aligned}\langle a^n\,\mathbb{S}h\mathfrak{M}\theta|O_\phi^f|a^n\,\mathbb{S}'h'\mathfrak{M}'\theta'\rangle &= \delta_{\mathbb{S}\mathbb{S}'}\delta_{\mathfrak{M}\mathfrak{M}'}\langle a^n\,\mathbb{S}h\mathfrak{M}\theta|O_\phi^f|a^n\,\mathbb{S}h'\mathfrak{M}\theta'\rangle \\ &= \delta_{\mathbb{S}\mathbb{S}'}\delta_{\mathfrak{M}\mathfrak{M}'}\,n\begin{pmatrix}h\\ \theta\end{pmatrix}\begin{pmatrix}h^* & f & h'\\ \theta^* & \phi & \theta'\end{pmatrix} \\ &\quad\times\langle a^n\,\mathbb{S}h\mathfrak{M}\|o(n)^f\|a^n\,\mathbb{S}h'\mathfrak{M}\rangle \quad (19.4.1)\end{aligned}$$

To evaluate these matrix elements we use the method outlined in Section 19.2 except that it is convenient to rearrange (19.2.4) and (19.2.3) respectively to

$$|a^n \mathcal{S}h\mathfrak{M}\theta\rangle = \sum_{\mathcal{S}_1 h_1} \left(a^{n-1}\mathcal{S}_1 h_1, a|a^n \mathcal{S}h\right) \times \sum_{\mathfrak{M}_1 m} \left(\mathcal{S}_1\mathfrak{M}_1, \tfrac{1}{2}m|\mathcal{S}\mathfrak{M}\right)\left|\left(a^{n-1}(\mathcal{S}_1 h_1\mathfrak{M}_1), a(m)\right)h\theta\right\rangle \tag{19.4.2}$$

$$\left|\left(a^{n-1}(\mathcal{S}_1 h_1\mathfrak{M}_1), a(m)\right)h\theta\right\rangle = \sum_{\theta_1\alpha}(h_1\theta_1, a\alpha|h\theta)\,|a^{n-1}\mathcal{S}_1 h_1\mathfrak{M}_1\theta_1\rangle|a\,m\alpha\rangle \tag{19.4.3}$$

which together accomplish the identical transformation. These are more efficient when the operator is spin-independent, since spin is uncoupled at an earlier stage. The notation has the usual meaning—the first $n-1$ electrons are always in the first ket of (19.4.3), and the second ket is a function of the coordinates of the nth electron only.

Expanding the functions of (19.4.1) using (19.4.2) gives

$$\begin{aligned}\delta_{\mathcal{S}\mathcal{S}'}\delta_{\mathfrak{M}\mathfrak{M}'}\,n \sum_{\mathcal{S}_1 h_1}\left(a^n \mathcal{S}h|a^{n-1}\mathcal{S}_1 h_1, a\right)\sum_{\mathfrak{M}_1 m}\left(\mathcal{S}\mathfrak{M}|\mathcal{S}_1\mathfrak{M}_1, \tfrac{1}{2}m\right)\\ \times \sum_{\mathcal{S}_1' h_1'}\left(a^{n-1}\mathcal{S}_1' h_1', a|a^n \mathcal{S}h'\right)\sum_{\mathfrak{M}_1' m'}\left(\mathcal{S}_1'\mathfrak{M}_1', \tfrac{1}{2}m'|\mathcal{S}\mathfrak{M}\right)\\ \times\Big\langle\left(a^{n-1}(\mathcal{S}_1 h_1\mathfrak{M}_1), a(m)\right)h\theta\Big|o(n)^f_\phi\Big|\\ \left(a^{n-1}(\mathcal{S}_1' h_1'\mathfrak{M}_1'), a(m')\right)h'\theta'\Big\rangle\end{aligned} \tag{19.4.4}$$

Since the operator $o(n)^f_\phi$ is spin-independent and operates only on functions of the nth electron, clearly in the above expression we must have $h_1 = h_1'$, $\mathcal{S}_1 = \mathcal{S}_1'$, $\mathfrak{M}_1 = \mathfrak{M}_1'$, and $m = m'$. And when these equalities are satisfied, the matrix element remaining in (19.4.4) is independent of spin quantum numbers. Then, since the spin v.c.c. above are the elements of a unitary matrix, we can sum over $\mathfrak{M}_1$ and m to obtain

$$\begin{aligned}&\langle a^n \mathcal{S}h\mathfrak{M}\theta|O^f_\phi|a^n \mathcal{S}'h'\mathfrak{M}'\theta'\rangle\\ &= \delta_{\mathcal{S}\mathcal{S}'}\delta_{\mathfrak{M}\mathfrak{M}'}\,n\sum_{\mathcal{S}_1 h_1}\left(a^n \mathcal{S}h|a^{n-1}\mathcal{S}_1 h_1, a\right)\left(a^{n-1}\mathcal{S}_1 h_1, a|a^n \mathcal{S}h'\right)\\ &\quad\times\Big\langle\left(a^{n-1}(\mathcal{S}_1 h_1\mathfrak{M}_1), a(m)\right)h\theta\Big|o(n)^f_\phi\Big|\\ &\quad\left(a^{n-1}(\mathcal{S}_1 h_1\mathfrak{M}_1), a(m)\right)h'\theta'\Big\rangle\end{aligned} \tag{19.4.5}$$

After reduction of both sides with respect to θ, θ', and ϕ, (18.1.4) [or (18.1.6) for simply reducible groups] may be applied directly, since we have $|(ab)c\gamma\rangle = |(a^{n-1}(\mathbb{S}_1 h_1 \mathfrak{M}_1), a(m))h\theta\rangle$ and our operator $o(n)^f_\phi$ operates only on the b part of the system. Multiplicity labels are unnecessary here for all cases of interest (including the groups O and T_d), and $\{a\} = 1$. Thus, analogously to Griffith [(1962), Eq. (7.14)], we write

$$\langle a^n\,\mathbb{S}h\mathfrak{M}\|O^f\|a^n\,\mathbb{S}'h'\mathfrak{M}'\rangle = \delta_{\mathbb{S}\mathbb{S}'}\delta_{\mathfrak{M}\mathfrak{M}'}g(a^n,\mathbb{S}hfh')\langle a\|o^f\|a\rangle \tag{19.4.6}$$

where

$$g(a^n,\mathbb{S}hfh') \equiv n\sum_{\mathbb{S}_1 h_1}\left(a^n\,\mathbb{S}h|a^{n-1}\,\mathbb{S}_1 h_1, a\right)\left(a^{n-1}\,\mathbb{S}_1 h_1, a|a^n\,\mathbb{S}h'\right)$$
$$\times |h|^{1/2}|h'|^{1/2}\{h_1 a h'^*\}\{a^* f a\}\begin{Bmatrix} f & h' & h^* \\ h_1^* & a & a \end{Bmatrix} \tag{19.4.7}$$

In practice, we frequently drop the labels $\mathfrak{M}$, $\mathfrak{M}'$ in (19.4.6); then $\mathfrak{M} = \mathfrak{M}'$ is implicit (as in some of the equations in Sections 18.1 and 19.9).

Spin-independent matrix elements are independent of the spin basis used. While we have derived our results for the $SO_3 \supset SO_2$ $|\mathbb{S}\mathfrak{M}\rangle$ basis, the results are equally valid in the $|\mathbb{S}SM\rangle$ basis if only $\delta_{\mathfrak{M}\mathfrak{M}'}$ is replaced by $\delta_{SS'}\delta_{MM'}$ in (19.4.6).

Note that the $g(a^n, \mathbb{S}hfh')$ are completely symmetry-determined numbers for a given group. For any group they may be calculated once and for all and tabulated for the possible states h, h' of a given configuration a^n for the various possible symmetries of one-electron operators. Moreover, *no wavefunctions need be constructed*. The evaluation of one-electron reduced matrix elements such as $\langle a\|o^f\|a\rangle$ is discussed in Sections 13.4 and 19.6.

In Section C.16 we tabulate the $g(a^n, \mathbb{S}hfh')$ for the groups O and T_d; the table may also be used for O_h if only the proper u and g labels are affixed. While the $g(a^n, \mathbb{S}hfh')$ may, of course, be calculated directly from (19.4.7), for the pseudo-p^n configurations t_1^n and t_2^n it is easier to obtain them via a method of Butler outlined in Section 19.9.

In many cases (19.4.7) simplifies. This is particularly true when $|a| \leqslant 2$, so we do not tabulate the $g(a^n, \mathbb{S}hfh')$ for subgroups of the cubic groups. We give simplified equations below which apply for all groups unless otherwise noted. All are from Griffith [(1962), Section 7.2]. When $f = A_1$

$$g(a^n,\mathbb{S}hA_1h') = \delta_{hh'}n|h|^{1/2}|a|^{-1/2} \tag{19.4.8}$$

Then

$$g\left(a^{1}, \tfrac{1}{2}hfh'\right) = \delta\left(a^{*}fa\right) \qquad (\text{for all } f)$$

$$g\left(a^{2|a|}, 0hfh'\right) = 0 \qquad (\text{for} \quad f \neq A_1) \tag{19.4.9}$$

These are the only possibilities when $|a| = 1$. For $|a| = 2$ and $n = 2$, we have $\mathbb{S}_1 = \frac{1}{2}$ and $h_1 = a = e$, so the required cfp are all unity by (19.3.18). Thus

$$g\left(e^{2}, \mathbb{S}hfh'\right) = 2|h|^{1/2}|h'|^{1/2}\{eeh'^{*}\}\{e^{*}fe\}\begin{Bmatrix} f & h' & h^{*} \\ e^{*} & e & e \end{Bmatrix} \tag{19.4.10}$$

When $|a| = 3$ and $n = 2$, we have $a^2 = t_1^2$ or t_2^2, so the $g(a^2, \mathbb{S}hfh')$ are given in Section C.16. Finally, when $n < |a|$ and $f \neq A_1$,

$$\begin{aligned} g\left(a^{2|a|-n}, \mathbb{S}hfh'\right) &= -\eta g\left(a^{n}, \mathbb{S}hfh'\right) \\ &= -\{a^{*}fa\}g\left(a^{n}, \mathbb{S}hfh'\right) \end{aligned} \tag{19.4.11}$$

Here η is defined by (14.10.2). When the $3j$ phase obeys (10.2.5) we obtain

$$g\left(a^{2|a|-n}, \mathbb{S}hfh'\right) = -(-1)^{f}g\left(a^{n}, \mathbb{S}hfh'\right) \tag{19.4.12}$$

The derivation of (19.4.11) is lengthy. The first equality follows from the relations between matrices of hole and particle states for one-electron operators [Griffith (1964), Section 9.7.2]. The second equality of (19.4.11) is derived essentially as in Griffith [(1962), Section 2.6]. It depends on the Section 14.10 relations

$$\langle a\alpha|O_\mu|b\beta\rangle^{*} = T\langle a\alpha|O_\mu|b\beta\rangle = \langle b\beta|\left(O_\mu\right)^{\dagger}|a\alpha\rangle = \eta\langle b\beta|\left(O_\mu\right)^{*}|a\alpha\rangle \tag{19.4.13}$$

where from (14.11.3) and (14.10.1) we have

$$T\langle a\alpha|O_\mu|b\beta\rangle = \begin{pmatrix} a \\ \alpha \end{pmatrix}\begin{pmatrix} b \\ \beta \end{pmatrix}\langle a^{*}\alpha^{*}|\left(O_\mu\right)^{*}|b^{*}\beta^{*}\rangle \tag{19.4.14}$$

We leave it to the reader to prove that if

$$O_\mu = \langle f\phi_\mu | O_\mu \rangle O^f_{\phi_\mu}$$

$$(O_\mu)^\dagger = (\text{any phase}) O_{-\mu} \tag{19.4.15}$$

$$O_{-\mu} = \langle f^*\phi^*_\mu | O_{-\mu} \rangle O^{f^*}_{\phi^*_\mu}$$

(see Sections 9.8 and 18.3) then

$$\eta \langle b \| O^{f^*} \| a \rangle_r = \{a\}\{a^* f b r\} \langle a^* \| O^{f^*} \| b^* \rangle_r \tag{19.4.16}$$

so that when $a = a^*$, $b = b^*$, $f = f^*$, $r = 0$, and $\{a\} = 1$ (the usual case),

$$\eta \langle a \| O^f \| a \rangle = \{afa\} \langle a \| O^f \| a \rangle \tag{19.4.17}$$

Thus $\eta = \{afa\}$ for nonzero $\langle a \| O^f \| a \rangle$.

A final useful equation relates the $g(a^n, \mathbb{S}hfh')$ to the $g(a^n, \mathbb{S}h'fh)$ when all nonspin irreps are real and obey (10.2.5):

$$g(a^n, \mathbb{S}h'fh) = (-1)^{h'+h} g(a^n, \mathbb{S}hfh') \tag{19.4.18}$$

Spin-Only Matrix Elements of a^n

Spin-only matrix elements are diagonal in configuration. They are also diagonal in all orbital irrep labels and all SO_3 spin irrep labels (i.e. $\mathbb{S}$ and $\mathbb{S}'$) for each open shell. Spin-only matrix elements *reduced in* SO_3 are given by (15.4.7). On the other hand, for spin-only matrix elements reduced in a subgroup of SO_3 [as in (18.1.9)] we have

$$\langle a^n \mathbb{S} S h \theta \| S^f \| a^n \mathbb{S}' S' h' \theta' \rangle_q = \delta_{hh'} \delta_{\theta\theta'} \delta_{\mathbb{S}\mathbb{S}'} \langle \mathbb{S} S \| S^f \| \mathbb{S} S' \rangle_q \tag{19.4.19}$$

where the spin-reduced matrix element on the right is evaluated as described in Section 15.4. Its value is independent of all orbital and configuration labels; in (15.4.9) we give examples for the cubic groups.

As a further example consider spin triplets ($\mathbb{S} = 1$) in the group D_4. Section C.6 tells us that in D_4 a spin triplet transforms as $A_2 + E$ (Mulliken labels) or $\tilde{0} + 1$ (Butler labels). The operator S also transforms as $A_2 + E$.

To evaluate the triplet spin-only matrix elements reduced in D_4 we use (15.4.3) for the chain $SO_3 \supset O \supset D_4$. For example, using chain notation, the $3jm$ factors of Section B.5 for $SO_3 \supset O$ and of Section D.12 for $O \supset D_4$, and (15.4.7), we obtain

$$\langle A_2\|S^e\|E\rangle^{D_4} \equiv \langle 1T_1A_2\|S^{1t_1e}\|1T_1E\rangle^{SO_3\supset O\supset D_4}$$

$$= \begin{pmatrix} 1 & 1 & 1 \\ T_1 & T_1 & T_1 \end{pmatrix}\begin{matrix} SO_3 \\ O \end{matrix}\begin{pmatrix} T_1 & T_1 & T_1 \\ A_2 & E & E \end{pmatrix}\begin{matrix} O \\ D_4 \end{matrix}\langle 1\|S^1\|1\rangle^{SO_3}$$

$$= (-1)\left(\frac{-1}{\sqrt{3}}\right)\sqrt{6} = \sqrt{2} \tag{19.4.20}$$

Example: Magnetic-Moment Matrix Elements for an Octahedral t_{2g}^4 Configuration

As an example we calculate $|a^n \mathbb{S}h\mathfrak{M}\theta\rangle$ matrix elements for the magnetic-moment operator $\mu_z = (-\mu_B)(L_z + 2S_z)$ within an octahedral t_{2g}^4 configuration. We use the $SO_3 \supset SO_2$ basis for spin and the complex-O basis of Section C.5 for orbital states. The $|\mathbb{S}\mathfrak{M}\rangle$ are eigenfunctions of S_z. Section C.7 gives $L_z = L_0 = L_0^{t_{1g}}$. Thus (10.2.2) and (19.4.6) give

$$\langle a^n \mathbb{S}h\mathfrak{M}\theta|\mu_z|a^n \mathbb{S}'h'\mathfrak{M}'\theta'\rangle$$

$$= \delta_{\mathbb{S}\mathbb{S}'}\delta_{\mathfrak{M}\mathfrak{M}'}\left[\begin{pmatrix} h \\ \theta \end{pmatrix}\begin{pmatrix} h & T_{1g} & h' \\ \theta^* & 0 & \theta' \end{pmatrix}(-\mu_B)g(a^n, \mathbb{S}hT_1h')\right.$$

$$\left.\times\langle a\|l^{t_{1g}}\|a\rangle + (-2\mu_B)\mathfrak{M}\right] \tag{19.4.21}$$

where the complex-O $2jm$ and $3jm$ are given in Sections C.11 and C.12, and the $g(a^n, \mathbb{S}hT_1h')$ in Section C.16.

The ease with which such general formulas may be derived is one of the great advantages of the irreducible-tensor method. With (19.4.21) it is straightforward to calculate the μ_z matrix for the $^3T_{1g}$, $^1A_{1g}$, ^1E_g, and $^1T_{2g}$ states of a t_{2g}^4 configuration. Thus, for example, we obtain

$$\langle t_{2g}^4\, 1T_1 1 1|\mu_z|t_{2g}^4\, 1T_1 1 1\rangle$$

$$= (1)\left(\frac{-1}{\sqrt{6}}\right)(-\mu_B)g(t_2^4, 1T_1T_1T_1)\langle t_{2g}\|l^{t_{1g}}\|t_{2g}\rangle - 2\mu_B$$

$$= -\frac{\mu_B}{\sqrt{6}}\langle t_{2g}\|l^{t_{1g}}\|t_{2g}\rangle - 2\mu_B \tag{19.4.22}$$

and

$$\langle t_{2g}^4 0T_2 00|\mu_z|t_{2g}^4 0E0\epsilon'\rangle$$

$$= (-1)\left(\frac{-1}{\sqrt{3}}\right)(-\mu_B)g\left(t_2^4, 0T_2T_1E\right)\langle t_{2g}\|l^{t_{1g}}\|t_{2g}\rangle + 0$$

$$= \frac{\sqrt{2}}{\sqrt{3}}\mu_B\langle t_{2g}\|l^{t_{1g}}\|t_{2g}\rangle \tag{19.4.23}$$

where we have dropped most of the parity labels to simplify the notation. If we assume the t_{2g} MOs are purely symmetry-adapted metal d orbitals, (15.4.5) gives

$$\langle t_{2g}\|l^{t_{1g}}\|t_{2g}\rangle = \begin{pmatrix} 2 & 1 & 2 \\ T_2 & T_1 & T_2 \end{pmatrix}^{SO_3}_{O} \sqrt{30} = \left(\frac{-1}{\sqrt{5}}\right)\sqrt{30} = -\sqrt{6} \tag{19.4.24}$$

where we use Table B.5.1 for the $SO_3 \supset O$ $3jm$ factors. This is precisely our result (13.4.3) obtained by the methods of Section 15.4.

19.5. One Open Shell: Spin–Orbit Matrix Elements for a^n and the $G(a^n, \mathbb{S}hf\mathbb{S}'h')$

In Section 18.4 we show that spin–orbit matrix elements in the $|(\mathbb{S}S, h)r t\tau\rangle$ basis may be expressed in terms of $\Sigma_k su(k)$ matrix elements in the $|\mathbb{S}h\mathfrak{M}\theta\rangle$ basis using (18.4.2)–(18.4.3) or (18.4.6). After applying (18.4.2)–(18.4.3) or (18.4.6) the problem of determining $\mathcal{H}_{SO}$ matrix elements reduces to that of calculating the $\langle a^n\,\mathbb{S}h\|\Sigma_k s^1 u^{f*}(k)\|a^n\,\mathbb{S}'h'\rangle^{SO_3,G}$. Using reasoning analogous to that for the O^f (spin-independent operator) case above, but with the expansion (19.2.4) rather than (19.4.2), we find

$$\langle a^n\,\mathbb{S}h\|\sum_k s^1 u^{f*}(k)\|a^n\,\mathbb{S}'h'\rangle^{SO_3,G}$$

$$= n\sum_{\mathbb{S}_1 h_1}\left(a^n\,\mathbb{S}h|a^{n-1}\,\mathbb{S}_1 h_1, a\right)\left(a^{n-1}\,\mathbb{S}_1 h_1, a|a^n\,\mathbb{S}'h'\right)$$

$$\times\left\langle\left(a^{n-1}(\mathbb{S}_1 h_1), a\right)\mathbb{S}h\left\|s^1 u^{f*}(n)\right\|\left(a^{n-1}(\mathbb{S}_1 h_1), a\right)\mathbb{S}'h'\right\rangle \tag{19.5.1}$$

In the final integral our functions are in the $|(\mathbb{S}_1 a, \mathbb{S}_2 b)\mathbb{S} c\rangle$ form, and thus we may use (18.5.2) to obtain the simple result

$$\langle a^n \mathbb{S}h \| \sum_k s^1 u^{f^*}(k) \| a^n \mathbb{S}'h'\rangle^{\mathrm{SO}_3, G}$$

$$= G(a^n, \mathbb{S}hf\mathbb{S}'h')\langle \tfrac{1}{2}a \| s^1 u^{f^*} \| \tfrac{1}{2}a\rangle^{\mathrm{SO}_3, G} \tag{19.5.2}$$

where by analogy to Griffith [(1962), Eq. (10.6)], we define

$$\begin{aligned} G(a^n, \mathbb{S}hf\mathbb{S}'h') = n \sum_{\mathbb{S}_1 h_1} & \left(a^n \mathbb{S}h | a^{n-1}\mathbb{S}_1 h_1, a\right)\left(a^{n-1}\mathbb{S}_1 h_1, a | a^n \mathbb{S}'h'\right) \\ & \times H(\mathbb{S}_1 \tfrac{1}{2}\mathbb{S}) H(\mathbb{S}_1 \tfrac{1}{2}\mathbb{S}')(|\mathbb{S}||\mathbb{S}'|)^{1/2} \\ & \times \{\mathbb{S}_1 \tfrac{1}{2}\mathbb{S}'\}\{\tfrac{1}{2}\}\{\tfrac{1}{2} 1 \tfrac{1}{2}\}\begin{Bmatrix} 1 & \mathbb{S}' & \mathbb{S} \\ \mathbb{S}_1 & \tfrac{1}{2} & \tfrac{1}{2} \end{Bmatrix} \\ & \times (|h||h'|)^{1/2}\{h_1 a h'^*\}\{a^* f^* a\}\begin{Bmatrix} f^* & h' & h^* \\ h_1^* & a & a \end{Bmatrix} \end{aligned} \tag{19.5.3}$$

No repeated representations occur in (19.5.3) for the cubic groups, so those labels have been dropped. Also $\{a\} = 1$. The spin phases and $6j$ coefficients are given in Section B.3.

The one-electron reduced matrix element in (19.5.2) factors into spin and orbital parts. The spin part is then given by (15.4.7) and u is defined by (13.3.3), so that

$$\begin{aligned} \langle \tfrac{1}{2}a \| s^1 u^{f^*} \| \tfrac{1}{2}a\rangle^{\mathrm{SO}_3, G} &= \langle \tfrac{1}{2} \| s^1 \| \tfrac{1}{2}\rangle^{\mathrm{SO}_3}\langle a \| u^{f^*} \| a\rangle^G \\ &= \frac{\sqrt{6}}{2}\langle a \| \xi l^{f^*} \| a\rangle^G \end{aligned} \tag{19.5.4}$$

Thus (19.5.2) may be rewritten as

$$\langle a^n \mathbb{S}h \| \sum_k s^1 u^{f^*}(k) \| a^n \mathbb{S}'h'\rangle^{\mathrm{SO}_3, G} = \frac{\sqrt{6}}{2} G(a^n, \mathbb{S}hf\mathbb{S}'h')\langle a \| \xi l^{f^*} \| a\rangle^G \tag{19.5.5}$$

where the G superscripts on the reduced matrix elements indicate that they have been reduced in orbital space in the point group G to which the molecule belongs. Evaluation of the $\langle a \| \xi l^{f^*} \| a \rangle$ is discussed in Sections 13.4 and 19.6.

The $G(a^n, \mathbb{S}hf\mathbb{S}'h')$ are completely symmetry-determined numbers which may be calculated and then tabulated, once and for all, for each set of irrep labels for each symmetry group G. For the cubic groups $f^* = f = T_1$, so we drop the $f = T_1$ label in $G(a^n, \mathbb{S}hf\mathbb{S}'h')$; also $\langle a \| \xi l^{t_1} \| a \rangle$ is nonzero only for $a = t_1$ or t_2; thus the $G(a^n, \mathbb{S}h\mathbb{S}'h')$ are nonzero only for $a^n = t_1^n$ and t_2^n. We tabulate these $G(a^n, \mathbb{S}h\mathbb{S}'h')$ for the groups O and T_d in Section C.17; the tables also apply for O_h if the appropriate parity labels are affixed. While the results in Section C.17 could be calculated directly from (19.5.3), for the pseudo-p^n configurations t_1^n and t_2^n it is easier to use the Butler method described in Section 19.9 below.

For other groups of interest to us (we exclude the uncommon groups T, T_h, and the icosohedral groups), simple enough equations for the $G(a^n, \mathbb{S}hf\mathbb{S}'h')$ may be derived from (19.5.3) that explicit tabulation is unnecessary. We find for all groups when $f^* = f$

$$G(a^1, \mathbb{S}hf\mathbb{S}'h') = \delta(a^* f^* a) \tag{19.5.6}$$

$$G(a^{2|a|}, \mathbb{S}hf\mathbb{S}'h') = 0 \tag{19.5.7}$$

These take care of all possibilities for $|a| = 1$, while Section C.17 takes care of the $|a| = 3$ case. When $|a| = 2$ we have $a = e$. Thus (19.3.18) applies and $E^* = E$, so

$$G(e^2, \mathbb{S}hf\mathbb{S}'h') = 2(|\mathbb{S}||\mathbb{S}'||h||h'|)^{1/2}(-1)^{\mathbb{S}+2\mathbb{S}'} \times \begin{Bmatrix} 1 & \mathbb{S}' & \mathbb{S} \\ \frac{1}{2} & \frac{1}{2} & \frac{1}{2} \end{Bmatrix} \{EEh'^*\}\{Ef^*E\} \begin{Bmatrix} f^* & h' & h^* \\ E & E & E \end{Bmatrix} \tag{19.5.8}$$

where we have used the definitions in Section B.3 to simplify the SO_3 phases. It remains only to define the $G(e^3, \mathbb{S}hf\mathbb{S}'h')$, which may be found using the general equation which holds in all cases when $n \neq |a|$:

$$\begin{aligned} G(a^{2|a|-n}, \mathbb{S}hf\mathbb{S}'h') &= -\eta\, G(a^n, \mathbb{S}hf\mathbb{S}'h') \\ &= -G(a^n, \mathbb{S}hf\mathbb{S}'h') \end{aligned} \tag{19.5.9}$$

The equation follows from the discussion in Griffith [(1964), Section 9.7.2], since $\eta = 1$ for $\mathcal{H}_{SO}$ (see our Section 14.10).

A final useful equation holds in all cases when all irreps are real and obey (10.2.5):

$$G(a^n, \mathcal{S}'h'f\mathcal{S}h) = \{\mathcal{S}'\}(-1)^{\mathcal{S}'+\mathcal{S}+h'+h}G(a^n, \mathcal{S}hf\mathcal{S}'h') \quad (19.5.10)$$

Example: Spin–Orbit Matrix Elements for an Octahedral t_{2g}^4 Configuration

As an example of an application of (18.4.6) and (19.5.5) we first calculate the spin–orbit matrix in the $|a^n(\mathcal{S}S, h)r t \tau\rangle$ basis for the octahedral $|t_{2g}^4(\mathcal{S}S, h)A_1a_1\rangle$ states. The two A_1 states for this configuration are $|t_{2g}^4(1T_1, T_1)A_1a_1\rangle$ and $|t_{2g}^4(0A_1, A_1)A_1a_1\rangle$. For simplicity we drop most of the g labels. We use Sections C.11, C.13, B.5, and C.17 to obtain

$$\left\langle t_2^4(1T_1, T_1)A_1a_1\middle|\mathcal{H}_{SO}\middle|t_2^4(1T_1, T_1)A_1a_1\right\rangle$$

$$= \tfrac{1}{3}\langle t_2^4\, 1T_1\|\sum_k s^1u^{t_1}(k)\|t_2^4\, 1T_1\rangle$$

$$= \left(\frac{1}{3}\right)\left(\frac{\sqrt{6}}{2}\right)(1)\langle t_{2g}\|\xi l^{t_{1g}}\|t_{2g}\rangle$$

$$= \frac{1}{\sqrt{6}}\langle t_{2g}\|\xi l^{t_{1g}}\|t_{2g}\rangle$$

$$\left\langle t_2^4(0A_1, A_1)A_1a_1\middle|\mathcal{H}_{SO}\middle|t_2^4(1T_1, T_1)A_1a_1\right\rangle$$

$$= \frac{-1}{\sqrt{3}}\langle t_2^4\, 0A_1\|\sum_k s^1u^{t_1}(k)\|t_2^4\, 1T_1\rangle \qquad (19.5.11)$$

$$= \left(\frac{-1}{\sqrt{3}}\right)\left(\frac{\sqrt{6}}{2}\right)\left(-\frac{\sqrt{2}}{\sqrt{3}}\right)\langle t_{2g}\|\xi l^{t_{1g}}\|t_{2g}\rangle$$

$$= \frac{1}{\sqrt{3}}\langle t_{2g}\|\xi l^{t_{1g}}\|t_{2g}\rangle$$

$$\left\langle t_2^4(1T_1, T_1)A_1a_1\middle|\mathcal{H}_{SO}\middle|t_2^4(0A_1, A_1)A_1a_1\right\rangle = \frac{1}{\sqrt{3}}\langle t_{2g}\|\xi l^{t_{1g}}\|t_{2g}\rangle$$

$$\left\langle t_2^4(0A_1, A_1)A_1a_1\middle|\mathcal{H}_{SO}\middle|t_2^4(0A_1, A_1)A_1a_1\right\rangle = 0$$

When the t_{2g} MO may be approximated as a purely metal $t_{2g}(d)$ atomic orbital (AO), the t_{2g} matrix element is a single-center integral. Thus from (19.4.24) and our discussion in Section 13.4,

$$\langle t_{2g} \| \xi l^{t_{1g}} \| t_{2g} \rangle = -\sqrt{6}\,\zeta_d \tag{19.5.12}$$

where ζ_d is the spin–orbit coupling constant for the d orbital involved ($3d$, $4d$, or $5d$). Thus our matrix is

$A_1 a_1$	$1T_1$	$0A_1$
$1T_1$	$-\zeta_d$	$-\sqrt{2}\,\zeta_d$
$0A_1$	$-\sqrt{2}\,\zeta_d$	0

(19.5.13)

We return to this example in Chapter 21 when we consider calculations for $OsCl_6^{2-}$ in the "jj"-coupling basis.

Example: Calculation of the Spin–Orbit Matrix for the $^2T_{2g}$, $^2T_{1u}$, and $^2T_{2u}$ States of $IrCl_6^{2-}$

$IrCl_6^{2-}$ is an octahedral complex ion with a $d^5(t_{2g}^5)$ ground state (Figure 23.2.2). The $^2T_{2g}$, $^2T_{1u}$, and $^2T_{2u}$ states of interest with respect to the visible-absorption and MCD spectra (Section 23.2) are all doublets, so $\mathcal{S} = \frac{1}{2}$ and $S = E'$. Since in the group O_h we have $S \otimes h = E' \otimes T_1 = U' + E'$ and $E' \otimes T_2 = E'' + U'$ (Section C.2), our spin–orbit matrix partitions into E_g'', U_g', U_u', E_u', and E_u'' blocks. We use the $|a^n(\mathcal{S}S, h)t\tau\rangle$ basis, together with (18.4.6), (19.5.5), and Sections C.11, C.13, B.5, and C.17, to calculate the spin–orbit matrix elements. Note that the entire calculation is independent of which group-O_h basis is used.

For the $E_g''(^2T_{2g})$ state (dropping most parity labels to simplify notation), we obtain

$$\left\langle t_{2g}^5\left(\tfrac{1}{2}E', T_2\right)E''\tau\middle|\mathcal{H}_{SO}\middle|t_{2g}^5\left(\tfrac{1}{2}E', T_2\right)E''\tau'\right\rangle$$

$$= \delta_{\tau\tau'}\left(-\tfrac{1}{3}\right)\langle t_{2g}^5\,\tfrac{1}{2}T_2 \| \sum_k s^1 u^{t_1}(k) \| t_{2g}^5\,\tfrac{1}{2}T_2\rangle$$

$$= \delta_{\tau\tau'}\left(-\frac{1}{3}\right)\left(\frac{\sqrt{6}}{2}\right)(-1)\langle t_{2g} \| \xi l^{t_{1g}} \| t_{2g}\rangle$$

$$= \delta_{\tau\tau'}\left(\frac{1}{\sqrt{6}}\right)\langle t_{2g} \| \xi l^{t_{1g}} \| t_{2g}\rangle \tag{19.5.14}$$

Table 19.5.1. Spin–Orbit Matrices for $IrCl_6^{2-}$ in Terms of One-Electron Reduced Matrix Elements of $u \equiv \xi l^{t_{1g}}$

E''_g	${}^2T_{2g}(d)$
${}^2T_{2g}(d)$	$(1/\sqrt{6})\langle t_{2g}\|u\|t_{2g}\rangle$

U'_g	${}^2T_{2g}(d)$
${}^2T_{2g}(d)$	$(-1/2\sqrt{6})\langle t_{2g}\|u\|t_{2g}\rangle$

E'_u	${}^2T_{1u}(\pi+\sigma)$
${}^2T_{1u}(\pi+\sigma)$	$(-1/\sqrt{6})\langle t_{1u}\|u\|t_{1u}\rangle$

E''_u	${}^2T_{2u}(\pi)$
${}^2T_{2u}(\pi)$	$(1/\sqrt{6})\langle t_{2u}\|u\|t_{2u}\rangle$

U'_u	${}^2T_{1u}(\pi+\sigma)$	${}^2T_{2u}(\pi)$
${}^2T_{1u}(\pi+\sigma)$	$(1/2\sqrt{6})\langle t_{1u}\|u\|t_{1u}\rangle$	$(1/2\sqrt{2})\langle t_{2u}\|u\|t_{1u}\rangle$
${}^2T_{2u}(\pi)$	$\dfrac{1}{2\sqrt{2}}\langle t_{2u}\|u\|t_{1u}\rangle^*$	$(-1/2\sqrt{6})\langle t_{2u}\|u\|t_{2u}\rangle$

and for $U'({}^2T_{2g})$,

$$\left\langle t_{2g}^5\left(\tfrac{1}{2}E', T_2\right)U'\tau\middle|\mathcal{H}_{SO}\middle|t_{2g}^5\left(\tfrac{1}{2}E', T_2\right)U'\tau'\right\rangle$$

$$= \delta_{\tau\tau'}\left(\tfrac{1}{6}\right)\langle t_{2g}^5\ \tfrac{1}{2}T_2\|\sum_k s^1u^{t_1}(k)\|t_{2g}^5\ \tfrac{1}{2}T_2\rangle$$

$$= \delta_{\tau\tau'}\left(\frac{-1}{2\sqrt{6}}\right)\langle t_{2g}\|\xi l^{t_{1g}}\|t_{2g}\rangle \qquad (19.5.15)$$

where for the ground state, (19.5.12) is a good approximation.

The calculation proceeds in much the same fashion for all other matrix elements diagonal in configuration. There are two configurations—$t_{1u}(\pi+\sigma)^5$ and $t_{1u}(\sigma+\pi)^5$—which give rise to ${}^2T_{1u}$ states; we, however, do not include the $t_{1u}(\sigma+\pi)^5$ configuration in our calculation. Only the $t_{2u}(\pi)^5$ configuration gives rise to a ${}^2T_{2u}$ state. Matrix elements off-diagonal in configuration must be calculated using (20.2.13) and (20.2.14) as illustrated in (20.2.17). Results are given in Table 19.5.1 in matrix form.

19.6. Calculation of One-Electron Matrix Elements: Multicenter Examples

We have come this far without saying very much about one-electron MOs, but they must be defined in order to calculate one-electron reduced matrix elements using our procedures in Section 13.4. Here we calculate the group-O_h one-electron reduced matrix elements of Table 19.5.1 as an example. The MOs are formed by taking linear combinations of the symmetry orbitals given in Section C.10. Since these are written in the real-O basis, it is most convenient here to work in that basis. We have from Section C.10

$$\begin{aligned}
|t_{2g}(d)\xi\rangle &= -i\left[a'|d_{yz}\rangle + b'\left|\tfrac{1}{2}(y_5 - y_6 + z_2 - z_4)\right\rangle\right] \\
&\approx -i|d_{yz}\rangle \\
|t_{1u}(\pi+\sigma)x\rangle &\approx -\left[a\left|\frac{1}{\sqrt{2}}(x_1 + x_3)\right\rangle + b\left|\tfrac{1}{2}(x_2 + x_4 + x_5 + x_6)\right\rangle\right] \\
|t_{1u}(\sigma+\pi)x\rangle &\approx -\left[b\left|\frac{1}{\sqrt{2}}(x_1 + x_3)\right\rangle - a\left|\tfrac{1}{2}(x_2 + x_4 + x_5 + x_6)\right\rangle\right] \\
|t_{2u}(\pi)\xi\rangle &= -i\left|\tfrac{1}{2}(x_2 + x_4 - x_5 - x_6)\right\rangle
\end{aligned} \tag{19.6.1}$$

and so on, where we have retained only the most important contributions to the molecular orbitals. One-electron reduced matrix elements for $t_{2g}(d)$ for l and $u = \xi l$ were already evaluated in (19.4.24) and (19.5.12) respectively.

One-electron matrix elements involving multicenter ligand functions require the more involved procedure described in Section 13.4. We make the one-center approximation and find, for example, that

$$\begin{aligned}
\langle t_{1u}\|\xi l^{t_{1g}}\|t_{1u}\rangle &= \sqrt{6}\,\langle t_{1u}(\pi+\sigma)x|\xi l_z^{t_{1g}}|t_{1u}(\pi+\sigma)y\rangle \\
&= \sqrt{6}\,(-i)\zeta_{\mathrm{Cl}^-}\left[a\left\langle\frac{1}{\sqrt{2}}(x_1 + x_3)\right|\right. \\
&\qquad \left. + b\left\langle\tfrac{1}{2}(x_2 + x_4 + x_5 + x_6)\right|\right] \\
&\qquad \times |l_z^{t_1}|\left[a\left|\frac{1}{\sqrt{2}}(y_2 + y_4)\right\rangle + b\left|\tfrac{1}{2}(y_5 + y_6 + y_1 + y_3)\right\rangle\right] \\
&= -i\sqrt{6}\,\zeta_{\mathrm{Cl}^-}\left[\frac{b^2}{4}\left(\langle x_5|l_z^{t_1}|y_5\rangle + \langle x_6|l_z^{t_1}|y_6\rangle\right)\right. \\
&\qquad + \frac{ab}{2\sqrt{2}}\left(\langle x_1|l_z^{t_1}|y_1\rangle + \langle x_3|l_z^{t_1}|y_3\rangle\right. \\
&\qquad \left.\left. + \langle x_2|l_z^{t_1}|y_2\rangle + \langle x_4|l_z^{t_1}|y_4\rangle\right)\right]
\end{aligned} \tag{19.6.2}$$

In our real-O basis Section C.7 gives $l_z^{t_1} = l_z$. Then, letting R be the metal-to-ligand bond length, we have from (13.4.15) that

$$l_z = l_{z_5} \tag{19.6.3}$$

The same kind of analysis (Section 13.4) for the other ligands in our coordinate system (Section C.10) gives

$$\begin{aligned} l_z &= l_{z_6} \\ l_z &= l_{z_1} - i\hbar R\frac{\partial}{\partial y_1} \\ l_z &= l_{z_2} + i\hbar R\frac{\partial}{\partial x_2} \\ l_z &= l_{z_3} + i\hbar R\frac{\partial}{\partial y_3} \\ l_z &= l_{z_4} - i\hbar R\frac{\partial}{\partial x_4} \end{aligned} \tag{19.6.4}$$

Parity considerations prevent the second terms of the ligand-transformed operators above from contributing to (19.6.2). Thus we have, for example, from the above discussion and Section B.2,

$$\begin{aligned} \langle x_1|l_z^{t_1}|y_1\rangle &= \langle x_1|l_z|y_1\rangle \\ &= \langle x_1|l_{z_1}|y_1\rangle - i\hbar R\langle x_1|\frac{\partial}{\partial y_1}|y_1\rangle \\ &= \langle p_x|l_z|p_y\rangle \\ &= \frac{1}{\sqrt{2}}(\langle 1\,-1| - \langle 11|)|l_z|\frac{i}{\sqrt{2}}(|1\,-1\rangle + |11\rangle) \\ &= \frac{i}{2}[(-1) - (1)] = -i \end{aligned} \tag{19.6.5}$$

The same result is obtained for the other integrals in (19.6.2). Thus

$$\langle t_{1u}\|\xi l^{t_{1g}}\|t_{1u}\rangle = \zeta_{\mathrm{Cl}^-}\frac{-\sqrt{6}}{2}(b^2 + 2\sqrt{2}\,ab) \tag{19.6.6}$$

here $a^2 + b^2 = 1$, and b varies from $\sqrt{2}/\sqrt{3}$ to 1. Both a and b must have the same sign for the $t_{1u}(\pi + \sigma)$ orbital, since it is the antibonding σ–π combination (to see this, sketch the MO).

The other one-electron matrix elements we require are found in a similar fashion and are tabulated in Table 19.6.1. Those involving the operator l rather than $u = \xi l$ are also given, since they are closely related in our approximation. Substituting these results into the spin–orbit matrices of Table 19.5.1 leads to the $\mathcal{H}_{SO}$ matrices of Table 19.6.2.

Table 19.6.1. One-Electron Matrix Elements for $IrCl_6^{2-}$

For Spin–Orbit Matrices[a]	For Magnetic-Moment Matrices[a]
$\langle t_{2g}(d)\Vert u\Vert t_{2g}(d)\rangle$ $= -\sqrt{6}\,\zeta_{Ir^{4+}}$	$\langle t_{2g}(d)\Vert l\Vert t_{2g}(d)\rangle$ $= -\sqrt{6}$
$\langle t_{1u}(\pi+\sigma)\Vert u\Vert t_{1u}(\pi+\sigma)\rangle$ $= -(\sqrt{6}/2)\zeta_{Cl^-}(b^2 + 2\sqrt{2}\,ab)$	$\langle t_{1u}(\pi+\sigma)\Vert l\Vert t_{1u}(\pi+\sigma)\rangle$ $= -(\sqrt{6}/2)(b^2 + 2\sqrt{2}\,ab)$
$\langle t_{2u}(\pi)\Vert u\Vert t_{2u}(\pi)\rangle$ $= -(\sqrt{6}/2)\zeta_{Cl^-}$	$\langle t_{2u}(\pi)\Vert l\Vert t_{2u}(\pi)\rangle$ $= -\sqrt{6}/2$
$\langle t_{2u}(\pi)\Vert u\Vert t_{1u}(\pi+\sigma)\rangle$ $= \langle t_{1u}(\pi+\sigma)\Vert u\Vert t_{2u}(\pi)\rangle$ $= i(\sqrt{6}/2)\zeta_{Cl^-}(b - \sqrt{2}\,a)$	$\langle t_{2u}(\pi)\Vert l\Vert t_{1u}(\pi+\sigma)\rangle$ $= \langle t_{1u}(\pi+\sigma)\Vert l\Vert t_{2u}(\pi)\rangle$ $= i(\sqrt{6}/2)(b - \sqrt{2}\,a)$

[a] $u \equiv \xi l^{t_{1g}}$ and $l = l^{t_{1g}}$.

Table 19.6.2. Spin–Orbit Matrices for $IrCl_6^{2-}$

E''_g	$^2T_{2g}(d)$
$^2T_{2g}(d)$	$-\zeta_{Ir^{4+}}$

U'_g	$^2T_{2g}(d)$
$^2T_{2g}(d)$	$\frac{1}{2}\zeta_{Ir^{4+}}$

E'_u	$^2T_{1u}(\pi+\sigma)$
$^2T_{1u}(\pi+\sigma)$	$\frac{1}{2}\zeta_{Cl^-}(b^2 + 2\sqrt{2}\,ab)$

E''_u	$^2T_{2u}(\pi)$
$^2T_{2u}(\pi)$	$-\frac{1}{2}\zeta_{Cl^-}$

U'_u	$^2T_{1u}(\pi+\sigma)$	$^2T_{2u}(\pi)$
$^2T_{1u}(\pi+\sigma)$	$-\frac{1}{4}\zeta_{Cl^-}(b^2 + 2\sqrt{2}\,ab)$	$(\sqrt{3}/4)\zeta_{Cl^-}(b - \sqrt{2}\,a)$
$^2T_{2u}(\pi)$	$(\sqrt{3}/4)\zeta_{Cl^-}(b - \sqrt{2}\,a)$	$\frac{1}{4}\zeta_{Cl^-}$

19.7. Two Open Shells: Expansion of Functions of an $a^m b^n$ Configuration

We now consider the expansion of functions of the type $|\mathcal{Q}(a^m(\mathcal{S}_1 h_1), b^n(\mathcal{S}_2 h_2))\mathcal{S}h\mathfrak{M}\theta\rangle$ formed from two-open-shell configurations. Here $\mathcal{Q}$ is the operator which antisymmetrizes the a^m and b^n electrons to form completely antisymmetrized $(a+b)^{m+n}$ states. Expansion of $(a+b)^{m+n}$ functions cannot be accomplished using equations of the type (19.3.13) and (19.3.14), since, for example, (19.3.13) gives properly antisymmetric functions only when all the electrons occupy the *same* open shell. Instead the correct formulas here are

$$|\mathcal{Q}(a^m(\mathcal{S}_1 h_1), b^n(\mathcal{S}_2 h_2))\mathcal{S}h\mathfrak{M}\theta\rangle$$
$$\equiv |(a+b)^{m+n}(a^m(\mathcal{S}_1 h_1), b^n(\mathcal{S}_2 h_2))\mathcal{S}h\mathfrak{M}\theta\rangle$$
$$= [m!n!(m+n)!]^{-1/2} \sum_{\nu=1}^{(m+n)!} (-1)^\nu P_\nu$$
$$\times |(a^m(\mathcal{S}_1 h_1), b^n(\mathcal{S}_2 h_2))\mathcal{S}h\mathfrak{M}\theta\rangle \qquad (19.7.1)$$

where P_ν permutes the electrons and the sum over ν includes the $(m+n)!$ possible permutations, and

$$|(a^m(\mathcal{S}_1 h_1), b^n(\mathcal{S}_2 h_2))\mathcal{S}h\mathfrak{M}\theta\rangle$$
$$= \sum_{\mathfrak{M}_1 \mathfrak{M}_2} (\mathcal{S}_1\mathfrak{M}_1, \mathcal{S}_2\mathfrak{M}_2|\mathcal{S}\mathfrak{M}) \sum_{\theta_1\theta_2} (h_1\theta_1, h_2\theta_2|h\theta)$$
$$\times |a^m\, \mathcal{S}_1 h_1 \mathfrak{M}_1 \theta_1\rangle |b^n\, \mathcal{S}_2 h_2 \mathfrak{M}_2 \theta_2\rangle \qquad (19.7.2)$$

which is (11.5.1). It is not difficult to demonstrate that (19.7.1) gives functions of the form we require for our $|\Gamma\gamma\rangle$. Substituting (19.7.2) into (19.7.1), we have

$$|\mathcal{Q}(a^m(\mathcal{S}_1 h_1), b^n(\mathcal{S}_2 h_2))\mathcal{S}h\mathfrak{M}\theta\rangle$$
$$= [m!n!(m+n)!]^{-1/2} \sum_{\mathfrak{M}_1\mathfrak{M}_2} (\mathcal{S}_1\mathfrak{M}_1, \mathcal{S}_2\mathfrak{M}_2|\mathcal{S}\mathfrak{M})$$
$$\times \sum_{\theta_1\theta_2} (h_1\theta_1, h_2\theta_2|h\theta)$$
$$\times \sum_{\nu=1}^{(m+n)!} (-1)^\nu P_\nu |a^m\, \mathcal{S}_1 h_1 \mathfrak{M}_1 \theta_1\rangle |b^n\, \mathcal{S}_2 h_2 \mathfrak{M}_2 \theta_2\rangle$$
$$(19.7.3)$$

The functions $|a^m\, \mathcal{S}_1 h_1 \mathfrak{M}_1 \theta_1\rangle$ and $|b^n\, \mathcal{S}_2 h_2 \mathfrak{M}_2 \theta_2\rangle$ are, of course, functions of the type (19.1.1):

$$|a^m \mathcal{S}_1 h_1 \mathfrak{M}_1 \theta_1\rangle = \sum_i c_{1i}|a_1^i a_2^i \cdots a_m^i\rangle$$
$$|b^n \mathcal{S}_2 h_2 \mathfrak{M}_2 \theta_2\rangle = \sum_j c_{2j}|b_{m+1}^j b_{m+2}^j \cdots b_{m+n}^j\rangle \tag{19.7.4}$$

After interchanging the order of summation over i and j with that over ν, (19.7.3) becomes

$$[m!n!(m+n)!]^{-1/2} \sum_{\mathfrak{M}_1\mathfrak{M}_2} (\mathcal{S}_1\mathfrak{M}_1, \mathcal{S}_2\mathfrak{M}_2|\mathcal{S}\mathfrak{M}) \sum_{\theta_1\theta_2} (h_1\theta_1, h_2\theta_2|h\theta)$$
$$\times \sum_i \sum_j c_{1i}c_{2j} \sum_{\nu=1}^{(m+n)!} (-1)^\nu P_\nu |a_1^i \cdots a_m^i\rangle |b_{m+1}^j \cdots b_{m+n}^j\rangle \tag{19.7.5}$$

The $(m+n)!$ permutations in the sum over ν includes the $m!$ possible permutations of the a^m electrons in the first ket and the $n!$ possible permutations of the b^n electrons in the second ket. Thus, using (19.1.5) and remembering normalization factors, we may write (19.7.5) as

$$|\mathcal{A}(a^m(\mathcal{S}_1h_1), b^n(\mathcal{S}_2h_2))\mathcal{S}h\mathfrak{M}\theta\rangle$$
$$= [m!n!(m+n)!]^{-1/2}$$
$$+ \sum_{\mathfrak{M}_1\mathfrak{M}_2} (\mathcal{S}_1\mathfrak{M}_1, \mathcal{S}_2\mathfrak{M}_2|\mathcal{S}\mathfrak{M}) \sum_{\theta_1\theta_2} (h_1\theta_1, h_2\theta_2|h\theta) \sum_i \sum_j c_{1i}c_{2j}$$
$$\times \frac{n!m!}{\sqrt{m!n!}} \sum_{\nu=1}^{(m+n)!} (-1)^\nu P_\nu \{a_1^i(1) \cdots a_m^i(m) b_{m+1}^j$$
$$\times (m+1) \cdots b_{m+n}^j(m+n)\}$$
$$= \sum_{\mathfrak{M}_1\mathfrak{M}_2} (\mathcal{S}_1\mathfrak{M}_1, \mathcal{S}_2\mathfrak{M}_2|\mathcal{S}\mathfrak{M}) \sum_{\theta_1\theta_2} (h_1\theta_1, h_2\theta_2|h\theta)$$
$$\times \sum_i \sum_j c_{1i}c_{2j}|a_1^i a_2^i \cdots a_m^i b_{m+1}^j b_{m+2}^j \cdots b_{m+n}^j\rangle \tag{19.7.6}$$

Thus, (19.7.1) gives us the $|\mathcal{A}(a^m(\mathcal{S}_1h_1), b^n(\mathcal{S}_2h_2))\mathcal{S}h\mathfrak{M}\theta\rangle$ functions we require since the equivalent functions of (19.7.6) are clearly antisymmetrized and normalized and are formed by coupling together the specified a^m and b^n states.

For spin-independent operators it is once again convenient to uncouple spins at an earlier stage. Thus, analogous to (19.4.2) and (19.4.3), we have the equations

$$|\mathcal{A}(a^m(\mathbb{S}_1h_1), b^n(\mathbb{S}_2h_2))\mathbb{S}h\mathfrak{M}\theta\rangle$$
$$= [m!n!(m+n)!]^{-1/2} \sum_{\mathfrak{M}_1\mathfrak{M}_2} (\mathbb{S}_1\mathfrak{M}_1, \mathbb{S}_2\mathfrak{M}_2|\mathbb{S}\mathfrak{M})$$
$$\times \sum_\nu (-1)^\nu P_\nu |(a^m(\mathbb{S}_1h_1\mathfrak{M}_1), b^n(\mathbb{S}_2h_2\mathfrak{M}_2))h\theta\rangle \tag{19.7.7}$$

$$|(a^m(\mathbb{S}_1h_1\mathfrak{M}_1), b^n(\mathbb{S}_2h_2\mathfrak{M}_2))h\theta\rangle$$
$$= \sum_{\theta_1\theta_2} (h_1\theta_1, h_2\theta_2|h\theta)|a^m\, \mathbb{S}_1h_1\mathfrak{M}_1\theta_1\rangle|b^n\, \mathbb{S}_2h_2\mathfrak{M}_2\theta_2\rangle \tag{19.7.8}$$

which together accomplish the same transformation as (19.7.1) and (19.7.2).

Example: Construction of a $^3T_2$ Wavefunction for a t_2e Configuration

We illustrate how (19.7.1) and (19.7.2) may be used to construct the function $|\mathcal{A}(a^m(\mathbb{S}_1h_1), b^n(\mathbb{S}_2h_2))\mathbb{S}h\mathfrak{M}\theta\rangle = |\mathcal{A}(t_2(\frac{1}{2}T_2), e(\frac{1}{2}E))1T_21\eta\rangle$. Employing these equations and the $3jm$ of Sections B.3 and C.12, we find

$$|\mathcal{A}(t_2(\tfrac{1}{2}T_2), e(\tfrac{1}{2}E))1T_21\eta\rangle$$
$$= \frac{1}{\sqrt{2}} \sum_\nu (-1)^\nu P_\nu |(t_2, e)1T_21\eta\rangle$$
$$= \frac{1}{\sqrt{2}} \sum_\nu (-1)^\nu P_\nu \left\{ -\tfrac{1}{2}|\eta^+\rangle|\theta^+\rangle + \frac{\sqrt{3}}{2}|\eta^+\rangle|\epsilon^+\rangle \right\}$$
$$= \frac{1}{\sqrt{2}} \left\{ -\tfrac{1}{2}\left[|\eta^+(1)\rangle|\theta^+(2)\rangle - |\eta^+(2)\rangle|\theta^+(1)\rangle\right] \right.$$
$$\left. + \frac{\sqrt{3}}{2}\left[|\eta^+(1)\rangle|\epsilon^+(2)\rangle - |\eta^+(2)\rangle|\epsilon^+(1)\rangle\right] \right\}$$
$$= -\tfrac{1}{2}|\eta^+\theta^+\rangle + \frac{\sqrt{3}}{2}|\eta^+\epsilon^+\rangle \tag{19.7.9}$$

As the reader may wish to verify, the $|\mathcal{Q}(t_2^5, e)1T_11x\rangle$ function given in (11.5.5) may also be found by this procedure.

19.8. The Simplest Off-Diagonal Case: Matrix Elements between a^n and $a^{n-1}b$

Spin-Independent Matrix Elements between a^n and $a^{n-1}b$

We now have the expansions (19.4.2) and (19.7.7) needed for the evaluation of a^n, $a^{n-1}b$ spin-independent matrix elements. Using them and (19.1.8) gives

$$\begin{aligned}
&\langle a^n \mathcal{S} h \mathcal{M} \theta | O_\phi^f | \mathcal{Q}\big(a^{n-1}(\mathcal{S}_1' h_1'), b\big) \mathcal{S}' h' \mathcal{M}' \theta' \rangle \\
&\quad = \delta_{\mathcal{S}\mathcal{S}'} \delta_{\mathcal{M}\mathcal{M}'} n \sum_{\mathcal{S}_1 h_1} (a^n \mathcal{S} h | a^{n-1} \mathcal{S}_1 h_1, a) \\
&\qquad \times \sum_{\mathcal{M}_1 m} (\mathcal{S}\mathcal{M} | \mathcal{S}_1 \mathcal{M}_1, \tfrac{1}{2} m) \sum_{\mathcal{M}_1' m'} (\mathcal{S}_1' \mathcal{M}_1', \tfrac{1}{2} m' | \mathcal{S}\mathcal{M}) \\
&\qquad \times \langle \big(a^{n-1}(\mathcal{S}_1 h_1 \mathcal{M}_1), a(m)\big) h\theta | o(n)_\phi^f | \\
&\qquad \times [(n-1)!n!]^{-1/2} \sum_{\mu=1}^{n!} (-1)^\mu P_\mu \big| \big(a^{n-1}(\mathcal{S}_1' h_1' \mathcal{M}_1'), b(m')\big) h'\theta' \rangle \rangle
\end{aligned} \tag{19.8.1}$$

Since all a and b functions are orthogonal, the integral is zero for all terms in which $o(n)_\phi^f$ does not connect the kets with the nth electron in $\langle a\, m\alpha|$ and $|b\, m'\beta\rangle$. Thus in the sum over μ, only those permutations P_μ which leave the nth electron in b contribute; those are the $(n-1)!$ permutations which permute the first $n-1$ electrons of the ket $|a^{n-1}\mathcal{S}_1' h_1' \mathcal{M}_1' \theta_1'\rangle$. These $(n-1)!$ permutations give $(n-1)!$ identical contributions to the integral [see (19.1.5)], and we therefore replace the sum over μ with multiplication by $(n-1)!$. The operator is spin-independent, so the same arguments we used to reach (19.4.5) apply. Thus (19.8.1) simplifies to

$$\begin{aligned}
&\delta_{\mathcal{S}\mathcal{S}'} \delta_{\mathcal{M}\mathcal{M}'} \sqrt{n}\, (a^n \mathcal{S} h | a^{n-1} \mathcal{S}_1' h_1', a) \\
&\qquad \times \langle \big(a^{n-1}(\mathcal{S}_1' h_1' \mathcal{M}_1'), a(m)\big) h\theta | o(n)_\phi^f | \big(a^{n-1}(\mathcal{S}_1' h_1' \mathcal{M}_1'), b(m)\big) h'\theta' \rangle
\end{aligned} \tag{19.8.2}$$

After reduction and use of (18.1.4) we obtain

$$\langle a^n \mathcal{S} h \mathcal{M} \| O^f \| \mathcal{Q}(a^{n-1}(\mathcal{S}_1' h_1'), b) \mathcal{S}' h' \mathcal{M}' \rangle$$
$$= \delta_{\mathcal{S}\mathcal{S}'} \delta_{\mathcal{M}\mathcal{M}'} \sqrt{n} \left(a^n \mathcal{S} h | a^{n-1} \mathcal{S}_1' h_1', a \right)$$
$$\times |h|^{1/2} |h'|^{1/2} \{h_1' b h'^*\} \{a^* f b\} \begin{Bmatrix} f & h' & h^* \\ h_1'^* & a & b \end{Bmatrix} \langle a \| o^f \| b \rangle \tag{19.8.3}$$

Note that the value of everything except $\langle a \| o^f \| b \rangle$ is completely symmetry-determined.

Derivation of the related $a^{n-1}b$, a^n equation proceeds in an analogous manner, with the result

$$\langle \mathcal{Q}(a^{n-1}(\mathcal{S}_1 h_1), b) \mathcal{S} h \mathcal{M} \| O^f \| a^n \mathcal{S}' h' \mathcal{M}' \rangle$$
$$= \delta_{\mathcal{S}\mathcal{S}'} \delta_{\mathcal{M}\mathcal{M}'} \sqrt{n} \left(a^{n-1} \mathcal{S}_1 h_1, a | a^n \mathcal{S}' h' \right)$$
$$\times |h|^{1/2} |h'|^{1/2} \{h_1 a h'^*\} \{b^* f a\} \begin{Bmatrix} f & h' & h^* \\ h_1^* & b & a \end{Bmatrix} \langle b \| o^f \| a \rangle \tag{19.8.4}$$

These equations are really just special cases of (20.2.10) and (20.2.9) in which $n = 1$, so the $6j$ and $9j$ coefficients simplify because some irreps are always A_1. The results also apply in the $|\mathcal{S} S M\rangle$ basis if only $\delta_{\mathcal{M}\mathcal{M}'}$ is replaced by $\delta_{SS'} \delta_{MM'}$.

Example: Calculation of an Orbital-Angular-Momentum Matrix Element between t_2^2 and $t_2 e$ States

We illustrate the use of (19.8.3) by calculating the orbital-angular-momentum matrix element between the octahedral states $|t_2^2\, 1T_1 1x\rangle$ and $|\mathcal{Q}(t_2, e) 1T_2 1\eta\rangle$. We use Sections C.7, C.11, C.12, C.15, and C.13 to obtain

$$\langle t_2^2\, 1T_1 1x | L_z | \mathcal{Q}(t_2(\tfrac{1}{2}T_2), e) 1T_2 1\eta \rangle$$
$$= \begin{pmatrix} T_1 \\ x \end{pmatrix} \begin{pmatrix} T_1 & T_1 & T_2 \\ x & z & \eta \end{pmatrix} \langle t_2^2\, 1T_1 1 \| L^{t_1} \| \mathcal{Q}(t_2(\tfrac{1}{2}T_2), e) 1T_2 1 \rangle$$
$$= (-1)\left(\frac{-1}{\sqrt{6}}\right)\left[\sqrt{2}(-1)\sqrt{3}\sqrt{3}(1)(-1)\left(\frac{-1}{2\sqrt{3}}\right)\langle t_2 \| l^{t_1} \| e \rangle\right]$$
$$= -\tfrac{1}{2}\langle t_2 \| l^{t_1} \| e \rangle \tag{19.8.5}$$

In this case the result may be easily checked, since we have already constructed the necessary functions. Using (19.7.9) and (19.3.5) for the functions and the methods of Section 13.2, we obtain

$$\left\langle \eta^+\zeta^+ |L_z| \left(-\tfrac{1}{2}|\eta^+\theta^+\rangle + \frac{\sqrt{3}}{2}|\eta^+\epsilon^+\rangle \right) \right\rangle$$

$$= -\tfrac{1}{2}\langle \eta^+\zeta^+ |L_z^{t_1}|\eta^+\theta^+\rangle + \frac{\sqrt{3}}{2}\langle \eta^+\zeta^+ |L_z^{t_1}|\eta^+\epsilon^+\rangle$$

$$= -\tfrac{1}{2}\langle \zeta^+ |l_z^{t_1}|\theta^+\rangle + \frac{\sqrt{3}}{2}\langle \zeta^+ |l_z^{t_1}|\epsilon^+\rangle$$

$$= \left[-\frac{1}{2}\begin{pmatrix} T_2 \\ \zeta \end{pmatrix}\begin{pmatrix} T_2 & T_1 & E \\ \zeta & z & \theta \end{pmatrix} + \frac{\sqrt{3}}{2}\begin{pmatrix} T_2 \\ \zeta \end{pmatrix}\begin{pmatrix} T_2 & T_1 & E \\ \zeta & z & \epsilon \end{pmatrix} \right]\langle t_2 \| l^{t_1} \| e \rangle$$

$$= \left[0 + \frac{\sqrt{3}}{2}(-1)\left(\frac{1}{\sqrt{3}} \right) \right]\langle t_2 \| l^{t_1} \| e \rangle$$

$$= -\tfrac{1}{2}\langle t_2 \| l^{t_1} \| e \rangle \tag{19.8.6}$$

The matrix element above was simple enough to evaluate easily by the methods of Section 13.2, but when the configurations become more complicated, formulas such as (19.8.3) are enormous time savers.

We must still evaluate the one-electron matrix element $\langle t_2 \| l^{t_1} \| e \rangle$. When the t_2 and e molecular orbitals may be approximated as pure d orbitals, the result is that given in (15.4.6).

Spin-Only Matrix Elements between a^n and $a^{n-1}b$

All matrix elements of spin-only operators between states of a configuration a^n and those of a configuration $a^{n-1}b$ are zero because of orthogonality of the a and b orbitals.

Spin–Orbit Matrix Elements between a^n and $a^{n-1}b$

Spin-orbit matrix elements between a^n and $a^{n-1}b$ are obtained much like the related matrix elements for O^f except that we use (19.7.1) rather than (19.7.7), (19.2.3) rather than (19.4.2), and (18.5.2) instead of (18.1.4). The reader can verify that we obtain

$$\langle a^n \mathbb{S}h \| \sum_k s^1 u^{f^*}(k) \| \mathcal{A}(a^{n-1}(\mathbb{S}_1' h_1'), b)\mathbb{S}' h' \rangle^{SO_3, G}$$

$$= \sqrt{n}\,(a^n \mathbb{S}h | a^{n-1} \mathbb{S}_1' h_1', a) H(\mathbb{S}_1' \tfrac{1}{2} \mathbb{S}) H(\mathbb{S}_1' \tfrac{1}{2} \mathbb{S}')$$

$$\times (|\mathbb{S}||\mathbb{S}'|)^{1/2} \{\mathbb{S}_1' \tfrac{1}{2} \mathbb{S}'\}\{\tfrac{1}{2}\}\{\tfrac{1}{2} 1 \tfrac{1}{2}\} \begin{Bmatrix} 1 & \mathbb{S}' & \mathbb{S} \\ \mathbb{S}_1' & \tfrac{1}{2} & \tfrac{1}{2} \end{Bmatrix}$$

$$\times (|h||h'|)^{1/2} \{h_1' b h'^*\}\{a^* f^* b\} \begin{Bmatrix} f^* & h' & h^* \\ h_1'^* & a & b \end{Bmatrix}$$

$$\times \frac{\sqrt{6}}{2} \langle a \| \xi l^{f^*} \| b \rangle^G \tag{19.8.7}$$

where in the last line we use (15.4.7) to give $\langle \tfrac{1}{2} \| s^1 \| \tfrac{1}{2} \rangle = \sqrt{6}/2$. For $a^{n-1}b$, a^n matrix elements the analgous equation is

$$\langle \mathcal{A}(a^{n-1}(\mathbb{S}_1 h_1), b)\mathbb{S}h \| \sum_k s^1 u^{f^*}(k) \| a^n \mathbb{S}' h' \rangle^{SO_3, G}$$

$$= \sqrt{n}\,(a^{n-1} \mathbb{S}_1 h_1, a | a^n \mathbb{S}' h') H(\mathbb{S}_1 \tfrac{1}{2} \mathbb{S}) H(\mathbb{S}_1 \tfrac{1}{2} \mathbb{S}')$$

$$\times (|\mathbb{S}||\mathbb{S}'|)^{1/2} \{\mathbb{S}_1 \tfrac{1}{2} \mathbb{S}'\}\{\tfrac{1}{2}\}\{\tfrac{1}{2} 1 \tfrac{1}{2}\} \begin{Bmatrix} 1 & \mathbb{S}' & \mathbb{S} \\ \mathbb{S}_1 & \tfrac{1}{2} & \tfrac{1}{2} \end{Bmatrix}$$

$$\times (|h||h'|)^{1/2} \{h_1 a h'^*\}\{b^* f^* a\} \begin{Bmatrix} f^* & h' & h^* \\ h_1^* & b & a \end{Bmatrix}$$

$$\times \frac{\sqrt{6}}{2} \langle b \| \xi l^{f^*} \| a \rangle^G \tag{19.8.8}$$

The above equations are just (20.2.14) and (20.2.13) respectively in which $n = 1$ and the $9j$ coefficients have been simplified because some irreps are necessarily A_1.

19.9. Calculation of cfp, $g(a^n, \mathbb{S}hfh')$, and $G(a^n, \mathbb{S}h\mathbb{S}'h')$ for the Pseudo-p^n Configurations of the Cubic Groups

In this section we illustrate more elegant methods than those of Section 19.3 (19.4.7), and (19.5.3) by which cfp, $g(a^n, \mathbb{S}hfh')$, and $G(a^n, \mathbb{S}h\mathbb{S}'h')$ may be calculated for the pseudo-p^n configurations, t_1^n and t_2^n, of the cubic groups. The methods, which were outlined for us by Butler [and are alluded to in Butler (1981), Chapters 7 and 9] use the Butler (1981) $3jm$ factors for $SO_3 \supset O$ and $O_3 \supset O_h \supset T_d$ to express these coefficients in terms of their p^n analogs, the values of which are given in Nielson and Koster (1963). The method is based on the fact that the values of these coefficients are independent of the nature of the group-O or group-T_d t_1^n and t_2^n configurations and may be calculated for any convenient choice. Since $l = p = 1$ transforms as $j = 1$ in SO_3 and as $j^- = 1^-$ in O_3, one can make use of the abstract isomorphism of $p^n(SO_3) \to t_1^n(O)$ to obtain the t_1^n coefficients and of $p^n(O_3) \to t_{1u}^n(O_h) \to t_2^n(T_d)$ to obtain the t_2^n coefficients.

Calculation of Coefficients of Fractional Parentage for t_1^n and t_2^n

A configuration a^n where $|a| = 3$ holds a maximum of six electrons. The cfp for such a six-particle configuration [Butler (1981), Chapter 7] may be viewed as a v.c.c. for a group–subgroup chain beginning with the unitary group for six particles, U_6. While the cfp are v.c.c. for a group chain not otherwise mentioned in this book, the important point for us is that they may be factored like other v.c.c. using the Racah factorization lemma (15.3.2). Thus for the t_1^n case we begin by writing our states as $\mathcal{S}(SO_3)\mathcal{L}(SO_3)h(O)$. We then use (15.3.2) and (15.3.3) and obtain

$$\begin{aligned}
&\left(t_1^{n-1}\mathcal{S}'h', t_1|t_1^n\mathcal{S}h\right)\\
&\quad= \left(t_1^{n-1}\mathcal{S}'\mathcal{L}'h', t_1\,\tfrac{1}{2}11|t_1^n\mathcal{S}\mathcal{L}h\right)^{O}\\
&\quad= (\mathcal{L}'h', 11|\mathcal{L}h)^{SO_3\supset O}\left(p^{n-1}\mathcal{S}'\mathcal{L}', p|p^n\,\mathcal{S}\mathcal{L}\right)^{SO_3}\\
&\quad= (-1)^{\mathcal{L}'-1+\mathcal{L}}\frac{|\mathcal{L}|^{1/2}}{|h|^{1/2}}\begin{pmatrix}\mathcal{L}' & 1 & \mathcal{L}\\ h' & 1 & h\end{pmatrix}^{SO_3}_{O}\left(p^{n-1}\mathcal{S}'\mathcal{L}', p|p^n\,\mathcal{S}\mathcal{L}\right)^{SO_3}
\end{aligned} \tag{19.9.1}$$

since the $2jm$ are all unity. The $(p^{n-1}\mathcal{S}'\mathcal{L}', p|p^n\,\mathcal{S}\mathcal{L})$ are the cfp for p^n given in Nielson and Koster (1963). Thus with the help of the p^n cfp and the Butler (1981) $SO_3 \supset O$ $3jm$ factors, Eq. (19.9.1) provides a very simple means for cfp calculation. A partial list of the $SO_3 \supset O$ $3jm$ factors is also given in Section B.5.

The method for t_2^n uses states labeled $\mathcal{S}^+(O_3)\mathcal{L}^\pm(O_3)h^\pm(O_h)h_1(T_d)$, much as in our (15.4.4) example, so the t_2^n equivalent of (19.9.1) involves $O_3 \supset O_h$

$\supset T_d$ $3jm$ factors which may themselves be factored into $O_3 \supset O_h$ and $O_h \supset T_d$ $3jm$ factors.

The cfp we tabulate for t_1^n and t_2^n in Section C.15 were determined as described above. The reader may find it an interesting exercise to verify some of the entries in Section C.15 by the procedure above.

Calculation of $g(a^n, \mathfrak{S}hfh')$ for t_1^n and t_2^n

The $g(a^n, \mathfrak{S}hfh')$ of (19.4.6) depend only on irrep transformation properties, so they may be calculated for t_1^n and t_2^n using the p^n isomorphism described above and any convenient operator with the desired transformation properties. For t_1^n we choose $O^f = O^{k(\mathrm{SO}_3)f(O)} = U^{k(\mathrm{SO}_3)f(O)}$ where U^k is a tensor operator transforming as $j = k$ with one-electron reduced matrix elements defined by Section B.4, Eq. (B.4.1). As in the cfp calculation above, we label our states as $\mathfrak{S}(\mathrm{SO}_3)\mathfrak{L}(\mathrm{SO}_3)h(O)$. Then using (15.4.3) and (19.4.6) (but leaving the $\mathfrak{M}$ labels implicit), we have

$$\langle t_1^n \mathfrak{S}\mathfrak{L}h \| U^{kf} \| t_1^n \mathfrak{S}'\mathfrak{L}'h' \rangle = \delta_{\mathfrak{S}\mathfrak{S}'} \begin{pmatrix} \mathfrak{L} & k & \mathfrak{L}' \\ h & f & h' \end{pmatrix}^{\mathrm{SO}_3}_{O} \langle p^n \mathfrak{S}\mathfrak{L} \| U^k \| p^n \mathfrak{S}\mathfrak{L}' \rangle$$

$$= \delta_{\mathfrak{S}\mathfrak{S}'}\, g(t_1^n, \mathfrak{S}hfh') \langle pt_1 \| u^{kf} \| pt_1 \rangle$$

$$= \delta_{\mathfrak{S}\mathfrak{S}'}\, g(t_1^n, \mathfrak{S}hfh') \begin{pmatrix} 1 & k & 1 \\ T_1 & f & T_1 \end{pmatrix}^{\mathrm{SO}_3}_{O} \langle p \| u^k \| p \rangle \tag{19.9.2}$$

where by definition [Eq. (B.4.1)]

$$\langle p \| u^k \| p \rangle = 1 \tag{19.9.3}$$

Thus

$$g(t_1^n, \mathfrak{S}hfh') = \frac{\begin{pmatrix} \mathfrak{L} & k & \mathfrak{L}' \\ h & f & h' \end{pmatrix}^{\mathrm{SO}_3}_{O}}{\begin{pmatrix} 1 & k & 1 \\ T_1 & f & T_1 \end{pmatrix}^{\mathrm{SO}_3}_{O}} \langle p^n \mathfrak{S}\mathfrak{L} \| U^k \| p^n \mathfrak{S}\mathfrak{L}' \rangle \tag{19.9.4}$$

The U^k reduced matrix elements are given in Section B.4 for $k = 0, 1$ and in Nielson and Koster (1963) for $k = 2$ to 6.

Note that the U^k reduced matrix elements for l^n are themselves simply $g(l^n, \mathfrak{S}\mathfrak{L}k\mathfrak{L}')$ for SO_3, since from (19.4.6) and (B.4.1) we obtain

$$\langle l^n\,\mathfrak{S}\mathfrak{L}\|U^k\|l^n\,\mathfrak{S}\mathfrak{L}'\rangle = g(l^n, \mathfrak{S}\mathfrak{L}k\mathfrak{L}')\langle l\|u^k\|l\rangle$$

$$= g(l^n, \mathfrak{S}\mathfrak{L}k\mathfrak{L}') \qquad (19.9.5)$$

[For SO_3, however, (19.4.7) would include $H(\mathfrak{L}_1 l\mathfrak{L})$-type factors, since they would be included in the equations in Section 18.1.]

An equation analogous to (19.9.4) defines the $g(t_2^n, \mathfrak{S}hfh')$ except that it, like the t_2^n cfp equation, involves both $O_3 \supset O_h$ and $O_h \supset T_d$ $3jm$ factors. As an exercise the reader might check that the $g(a^n, \mathfrak{S}hfh')$ for t_1^n (or t_2^n) of Section C.16 are consistent with both (19.9.4) (or its t_2^n equivalent) and with (19.4.7).

Calculation of the $G(a^n, \mathfrak{S}h\mathfrak{S}'h')$ for t_1^n and t_2^n

The calculation of the $G(t_i^n, \mathfrak{S}h\mathfrak{S}'h')$ for $i = 1, 2$ is similar to the $g(a^n, \mathfrak{S}hfh')$ calculation above. Here we note that the double tensor operator $U^{kkf} = U^{11f}$ transforms identically to $\Sigma_k s^1 u^{1f}(k)$ and may be used to determine the $G(t_1^n, \mathfrak{S}h\mathfrak{S}'h')$. Thus we may use the p^n-to-t_1^n isomorphism, (19.5.2), and (15.4.3) to write

$$\langle t_1^n\,\mathfrak{S}\mathfrak{L}h\|U^{11t_1}\|t_1^n\,\mathfrak{S}'\mathfrak{L}'h'\rangle$$

$$= \begin{pmatrix} \mathfrak{L} & 1 & \mathfrak{L}' \\ h & T_1 & h' \end{pmatrix}\begin{matrix} SO_3 \\ O \end{matrix} \langle p^n\,\mathfrak{S}\mathfrak{L}\|U^{11}\|p^n\,\mathfrak{S}'\mathfrak{L}'\rangle$$

$$= G(t_1^n, \mathfrak{S}h\mathfrak{S}'h')\langle \tfrac{1}{2}pt_1\|u^{11t_1}\|\tfrac{1}{2}pt_1\rangle$$

$$= G(t_1^n, \mathfrak{S}h\mathfrak{S}'h')\begin{pmatrix} 1 & 1 & 1 \\ T_1 & T_1 & T_1 \end{pmatrix}\begin{matrix} SO_3 \\ O \end{matrix} \langle \tfrac{1}{2}p\|u^{11}\|\tfrac{1}{2}p\rangle \qquad (19.9.6)$$

where by definition (Section B.4)

$$\langle \tfrac{1}{2}p\|u^{11}\|\tfrac{1}{2}p\rangle = 1 \qquad (19.9.7)$$

Thus

$$G(t_1^n, \mathfrak{S}h\mathfrak{S}'h') = \frac{\begin{pmatrix} \mathfrak{L} & 1 & \mathfrak{L}' \\ h & T_1 & h' \end{pmatrix}\begin{matrix} SO_3 \\ O \end{matrix}}{\begin{pmatrix} 1 & 1 & 1 \\ T_1 & T_1 & T_1 \end{pmatrix}\begin{matrix} SO_3 \\ O \end{matrix}} \langle p^n\,\mathfrak{S}\mathfrak{L}\|U^{11}\|p^n\,\mathfrak{S}'\mathfrak{L}'\rangle \qquad (19.9.8)$$

where

$$\langle p^n \mathbb{S}\mathbb{L} \| U^{11} \| p^n \mathbb{S}'\mathbb{L}' \rangle = \frac{\sqrt{2}}{\sqrt{3}} \langle p^n \mathbb{S}\mathbb{L} \| V^{11} \| p^n \mathbb{S}'\mathbb{L}' \rangle \qquad (19.9.9)$$

and the reduced matrix elements of l^n for V^{11} are given in Nielson and Koster (1963).

Note that it follows from the relations above and (19.5.2) that

$$\langle l^n \mathbb{S}\mathbb{L} \| U^{11} \| l^n \mathbb{S}'\mathbb{L}' \rangle = G(l^n, \mathbb{S}\mathbb{L}\mathbb{S}'\mathbb{L}') \qquad (19.9.10)$$

(Of course, for SO_3 (19.5.3) would include $H(\mathbb{L}_1 l \mathbb{L})$-type factors, since they would be included in the equations of Section 18.5.)

Again an equation analogous to (19.9.8) is easily derived for the t_2^n case.

20 Matrix Elements with Several Open Shells

20.1. Matrix Elements of $a^m b^n$

In this chapter we extend our analysis of matrix elements to include the general off-diagonal case and also matrix elements between two or more open shells. We begin by considering the matrix elements of an $a^m b^n$ configuration. As in Sections 19.4 and 19.5, we first wish to express the matrix elements in a form which allows the use of our equations in Sections 18.1 and 18.5. We consider matrix elements of a one-electron operator V and expand $|\mathcal{S}'h'\mathcal{M}'\theta'\rangle$ with (19.7.1):

$$\langle \mathcal{Q}(a^m(\mathcal{S}_1 h_1), b^n(\mathcal{S}_2 h_2))\mathcal{S}h\mathcal{M}\theta | V | \mathcal{Q}(a^m(\mathcal{S}_1' h_1'), b^n(\mathcal{S}_2' h_2'))\mathcal{S}'h'\mathcal{M}'\theta'\rangle$$

$$= \langle \mathcal{Q}(a^m(\mathcal{S}_1 h_1), b^n(\mathcal{S}_2 h_2))\mathcal{S}h\mathcal{M}\theta | V$$

$$\times \left| \left\{ [m!n!(m+n)!]^{-1/2} \sum_{\mu=1}^{(m+n)!} (-1)^{\mu} P_{\mu} \right.\right.$$

$$\left.\left. \times |(a^m(\mathcal{S}_1' h_1'), b^n(\mathcal{S}_2' h_2'))\mathcal{S}'h'\mathcal{M}'\theta'\rangle \right\}\right\rangle \tag{20.1.1}$$

Since our $|\mathcal{S}'h'\mathcal{M}'\theta'\rangle$ functions are (19.1.1)-type functions, all of the $(m+n)!$ terms in the sum over μ give the same contribution to the integral [here we use (19.1.10) via (19.7.6)]. Thus, after we expand $\langle \mathcal{S}h\mathcal{M}\theta|$ using (19.7.1) and collect the factorials, (20.1.1) becomes

$$(m!n!)^{-1} \left\langle \left[\sum_{\nu=1}^{(m+n)!} (-1)^{\nu} P_{\nu} \langle (a^m(\mathcal{S}_1 h_1), b^n(\mathcal{S}_2 h_2))\mathcal{S}h\mathcal{M}\theta | \right] \right|$$

$$\times V | (a^m(\mathcal{S}_1' h_1'), b^n(\mathcal{S}_2' h_2'))\mathcal{S}'h'\mathcal{M}'\theta'\rangle \tag{20.1.2}$$

Recall now that in the ket the notation indicates the first m electrons are confined to $a^m(\mathcal{S}_1'h_1')$ and the last n to $b^n(\mathcal{S}_2'h_2')$. Thus, since V is a one-electron operator, only $m!n!$ out of the $(m+n)!$ terms in the sum over ν contribute, since that is the number of permutations that keep the first m electrons in $a^m(\mathcal{S}_1h_1)$ and the last n electrons in $b^n(\mathcal{S}_2h_2)$. Moreover, since our a^m and b^n functions are individually antisymmetrized, we can use (19.1.5), which tells us these $m!n!$ contributions to the integral are identical. Replacing the sum over ν with multiplication by $m!n!$ gives the result we desire:

$$\langle \mathcal{Q}(a^m(\mathcal{S}_1h_1), b^n(\mathcal{S}_2h_2))\mathcal{S}h\mathfrak{M}\theta|V|\mathcal{Q}(a^m(S_1'h_1'), b^n(\mathcal{S}_2'h_2'))\mathcal{S}'h'\mathfrak{M}'\theta'\rangle$$

$$= \langle (a^m(\mathcal{S}_1h_1), b^n(\mathcal{S}_2h_2))\mathcal{S}h\mathfrak{M}\theta|V|(a^m(\mathcal{S}_1'h_1'), b^n(\mathcal{S}_2'h_2'))\mathcal{S}'h'\mathfrak{M}'\theta'\rangle \tag{20.1.3}$$

Spin-Independent Matrix Elements of a^mb^n

When V is a spin-independent operator it is convenient to use the expansion (19.7.7) rather than (19.7.1). Then, using the same logic which led earlier to (19.4.5), we sum over the spin quantum numbers $\mathfrak{M}_1$, $\mathfrak{M}_1'$, $\mathfrak{M}_2$, and $\mathfrak{M}_2'$ to obtain

$$\langle \mathcal{Q}(a^m(\mathcal{S}_1h_1), b^n(\mathcal{S}_2h_2))\mathcal{S}h\mathfrak{M}\theta|O_\phi^f|\mathcal{Q}(a^m(\mathcal{S}_1'h_1'), b^n(\mathcal{S}_2'h_2'))\mathcal{S}'h'\mathfrak{M}'\theta'\rangle$$

$$= \delta(\text{spin})\langle (a^m(\mathcal{S}_1h_1\mathfrak{M}_1), b^n(\mathcal{S}_2h_2\mathfrak{M}_2))\mathcal{S}h\theta|O_\phi^f$$

$$\times|(a^m(\mathcal{S}_1h_1'\mathfrak{M}_1), b^n(\mathcal{S}_2h_2'\mathfrak{M}_2))\mathcal{S}h'\theta'\rangle \tag{20.1.4}$$

where

$$\delta(\text{spin}) \equiv \delta_{\mathcal{S}_1\mathcal{S}_1'}\delta_{\mathcal{S}_2\mathcal{S}_2'}\delta_{\mathcal{S}\mathcal{S}'}\delta_{\mathfrak{M}\mathfrak{M}'} \tag{20.1.5}$$

Now that we have the general equations, (20.1.3) and (20.1.4), we are nearly finished. If we reduce (20.1.4) with respect to θ, ϕ, and θ', it is in precisely the form required for use of (18.1.2) and (18.1.4). We need only write our operator as $O_\phi^f = \Sigma_{i=1}^m o(i)_\phi^f + \Sigma_{j=m+1}^{m+n} o(j)_\phi^f$ and note that the first part operates only on the first m electrons, so we can use (18.1.2), while the second operates only on the last n, so (18.1.4) applies. Thus (20.1.4) becomes

$$\left\langle \mathcal{A}\left(a^m(\mathbb{S}_1 h_1), b^n(\mathbb{S}_2 h_2)\right)\mathbb{S}h\mathfrak{M}\right\| O^f \left\| \mathcal{A}\left(a^m(\mathbb{S}_1' h_1'), b^n(\mathbb{S}_2' h_2')\right)\mathbb{S}'h'\mathfrak{M}'\right\rangle$$

$$= \delta(\text{spin})\left[\delta_{h_2 h_2'} |h|^{1/2} |h'|^{1/2} \{h_1 h_2 h^*\}\{h_1^* f h_1'\} \begin{Bmatrix} f & h' & h^* \\ h_2^* & h_1 & h_1' \end{Bmatrix} \right.$$

$$\times \langle a^m\, \mathbb{S}_1 h_1 \mathfrak{M}_1 \| O^f \| a^m\, \mathbb{S}_1 h_1' \mathfrak{M}_1 \rangle$$

$$+ \delta_{h_1 h_1'} |h|^{1/2} |h'|^{1/2} \{h_1' h_2' h'^*\}\{h_2^* f h_2'\} \begin{Bmatrix} f & h' & h^* \\ h_1^* & h_2 & h_2' \end{Bmatrix}$$

$$\left. \times \langle b^n\, \mathbb{S}_2 h_2 \mathfrak{M}_2 \| O^f \| b^n\, \mathbb{S}_2 h_2' \mathfrak{M}_2 \rangle \right] \qquad (20.1.6)$$

where $\delta(\text{spin})$ is defined as in (20.1.5).

Once again we find that everything in (20.1.6) but the reduced matrix elements is completely symmetry-determined. They may be further reduced using (19.4.6) to give

$$\left\langle \mathcal{A}\left(a^m(\mathbb{S}_1 h_1), b^n(\mathbb{S}_2 h_2)\right)\mathbb{S}h\mathfrak{M}\right\| O^f \left\| \mathcal{A}\left(a^m(\mathbb{S}_1' h_1'), b^n(\mathbb{S}_2' h_2')\right)\mathbb{S}'h'\mathfrak{M}'\right\rangle$$

$$= \delta(\text{spin})\left[\delta_{h_2 h_2'} |h|^{1/2} |h'|^{1/2} \{h_1 h_2 h^*\}\{h_1^* f h_1'\} \begin{Bmatrix} f & h' & h^* \\ h_2^* & h_1 & h_1' \end{Bmatrix} \right.$$

$$\times g(a^m, \mathbb{S}_1 h_1 f h_1') \langle a \| o^f \| a \rangle + \delta_{h_1 h_1'} |h|^{1/2} |h'|^{1/2} \{h_1' h_2' h'^*\}$$

$$\left. \times \{h_2^* f h_2'\} \begin{Bmatrix} f & h' & h^* \\ h_1^* & h_2 & h_2' \end{Bmatrix} g(b^n, \mathbb{S}_2 h_2 f h_2') \langle b \| o^f \| b \rangle \right]$$

$$(20.1.7)$$

Since the matrix elements above are spin-independent, their values are independent of the spin basis used. Thus our results above are equally valid in the $|\mathbb{S}SM\rangle$ spin basis, if only $\delta_{\mathfrak{M}\mathfrak{M}'}$ is replaced by $\delta_{SS'}\delta_{MM'}$ in $\delta(\text{spin})$.

Spin-Only Matrix Elements of $a^m b^n$

Spin-only matrix elements of $a^m b^n$ are evaluated as discussed in Section 19.4. They are diagonal in all orbital irrep labels and all SO_3 spin irrep labels. Thus for spin-only matrix elements reduced in a subgroup of SO_3 [as in (18.1.9)] we have

$$\left\langle \mathcal{A}\left(a^m(\mathbb{S}_1h_1), b^n(\mathbb{S}_2h_2)\right)\mathbb{S}Sh\theta\middle\|S^f\middle\|\mathcal{A}\left(a^m(\mathbb{S}_1'h_1'), b^n(\mathbb{S}_2'h_2')\right)\mathbb{S}'S'h'\theta'\right\rangle_q$$

$$= \delta(\text{space})\delta_{\mathbb{S}_1\mathbb{S}_1'}\,\delta_{\mathbb{S}_2\mathbb{S}_2'}\delta_{\mathbb{S}\mathbb{S}'}\langle\mathbb{S}S\|S^f\|\mathbb{S}S'\rangle_q \tag{20.1.8}$$

where

$$\delta(\text{space}) \equiv \delta_{h_1h_1'}\delta_{h_2h_2'}\delta_{hh'}\delta_{\theta\theta'} \tag{20.1.9}$$

and $\langle\mathbb{S}S\|S^f\|\mathbb{S}S'\rangle_q$ is evaluated as described in Section 15.4. Spin-only matrix elements reduced in SO_3 are also nonzero only if all delta functions in (20.1.8) and (20.1.9) above are unity; the nonzero values are then calculated via (15.4.7).

Spin–Orbit Matrix Elements of a^mb^n

Spin–orbit matrix elements of a^mb^n are calculated by first applying (18.4.2) and (18.4.3), or (18.4.6). Then the $\Sigma su(k)$ matrix elements are evaluated using (20.1.3) together with (18.5.1), (18.5.2), and (19.5.5). We find

$$\Big\langle \mathcal{A}\left(a^m(\mathbb{S}_1h_1), b^n(\mathbb{S}_2h_2)\right)\mathbb{S}h\Big\|\sum_k s^1u^{f^*}(k)$$

$$\Big\|\mathcal{A}\left(a^m(\mathbb{S}_1'h_1'), b^n(\mathbb{S}_2'h_2')\right)\mathbb{S}'h'\Big\rangle^{SO_3,\,G}$$

$$= H(\mathbb{S}_1\mathbb{S}_2\mathbb{S})H(\mathbb{S}_1'\mathbb{S}_2'\mathbb{S}')(|\mathbb{S}|\,|\mathbb{S}'|\,|h|\,|h'|)^{1/2}$$

$$\times\Bigg[\delta_{\mathbb{S}_2\mathbb{S}_2'}\delta_{h_2h_2'}\{\mathbb{S}_1\mathbb{S}_2\mathbb{S}\}\{\mathbb{S}_1'\}\{\mathbb{S}_1 1\mathbb{S}_1'\}\begin{Bmatrix}1 & \mathbb{S}' & \mathbb{S}\\ \mathbb{S}_2 & \mathbb{S}_1 & \mathbb{S}_1'\end{Bmatrix}\{h_1h_2h^*\}$$

$$\times\{h_1^*f^*h_1'\}\begin{Bmatrix}f^* & h' & h^*\\ h_2^* & h_1 & h_1'\end{Bmatrix}\frac{\sqrt{6}}{2}G(a^n,\mathbb{S}_1h_1f\mathbb{S}_1'h_1')\langle a\|\xi l^{f^*}\|a\rangle^G$$

$$+\delta_{\mathbb{S}_1\mathbb{S}_1'}\delta_{h_1h_1'}\{\mathbb{S}_1'\mathbb{S}_2'\mathbb{S}'\}\{\mathbb{S}_2'\}\{\mathbb{S}_2 1\mathbb{S}_2'\}\begin{Bmatrix}1 & \mathbb{S}' & \mathbb{S}\\ \mathbb{S}_1 & \mathbb{S}_2 & \mathbb{S}_2'\end{Bmatrix}\{h_1'h_2'h'^*\}$$

$$\times\{h_2^*f^*h_2'\}\begin{Bmatrix}f^* & h' & h^*\\ h_1^* & h_2 & h_2'\end{Bmatrix}\frac{\sqrt{6}}{2}G(b^n,\mathbb{S}_2h_2f\mathbb{S}_2'h_2')\langle b\|\xi l^{f^*}\|b\rangle^G\Bigg]$$

$$\tag{20.1.10}$$

20.2. The Off-Diagonal Case: Matrix Elements Between $a^m b^{n-1}$ and $a^{m-1} b^n$

The Initial Procedure

Unlike all the matrix elements we have considered previously in this chapter, these off-diagonal matrix elements cannot be simplified using our equations in Section 18.1 or 18.5, since the coupling is more complex. We could, as before, first express them in terms of matrix elements between functions of the type (19.7.2), but then we would still have to work out the analogs of the equations in Section 18.1 or 18.5 to exploit fully the group theory of the problem. Instead, we merge the stages of the analysis in Chapters 18 and 19. Only at the end [see (20.2.16)] do we give a general equation in terms of functions of the type (19.7.2) (that is, an equation like those in Chapter 18), as an illustration of the difference between these matrix elements and our earlier ones.

We consider the derivation of the $a^{m-1}b^n$, $a^m b^{n-1}$ formulas in detail and simply present the result for the $a^m b^{n-1}$, $a^{m-1}b^n$ case. The analysis is lengthy but proceeds with similar logic to that employed in the derivation of (19.8.3), (20.1.3), and our equations in Sections 18.1 and 18.5. The first step is to expand $|\mathbb{S}'h'\mathfrak{M}'\theta'\rangle$ using (19.7.1) and to replace the sum over μ with multiplication by $(m+n-1)!$ [we use (19.1.10) via (19.7.6)]:

$$\left\langle \mathcal{Q}\left(a^{m-1}(\mathbb{S}_1 h_1), b^n(\mathbb{S}_2 h_2)\right)\mathbb{S}h\mathfrak{M}\theta\right| V$$

$$\times \left|\mathcal{Q}\left(a^m(\mathbb{S}_1' h_1'), b^{n-1}(\mathbb{S}_2' h_2')\right)\mathbb{S}'h'\mathfrak{M}'\theta'\right\rangle$$

$$= \left[\frac{(m+n-1)!}{m!(n-1)!}\right]^{1/2} \left\langle \mathcal{Q}\left(a^{m-1}(\mathbb{S}_1 h_1), b^n(\mathbb{S}_2 h_2)\right)\mathbb{S}h\mathfrak{M}\theta\right| V$$

$$\times \left|\left(a^m(\mathbb{S}_1' h_1'), b^{n-1}(\mathbb{S}_2' h_2')\right)\mathbb{S}'h'\mathfrak{M}'\theta'\right\rangle \qquad (20.2.1)$$

The a and b orbitals are orthogonal, so only those terms in V which connect a b orbital on the left with an a orbital on the right are nonzero. The notation of the right-hand ket confines the first m electrons to the a orbitals. Thus, we replace $V = \sum_{i=1}^{m+n-1} v(i)$ with $V = \sum_{i=1}^{m} v(i)$ and finally use (19.1.8) with $V = mv(m)$. We next expand $|(a^m, b^{n-1})\mathbb{S}'h'\mathfrak{M}'\theta'\rangle$ with

(19.7.2) and expand $\langle \mathcal{S} h \mathcal{M} \theta |$ with (19.7.1) to obtain

$$m[m!(n-1)!(m-1)!n!]^{-1/2} \sum_{\mathcal{M}_1'\mathcal{M}_2'} (\mathcal{S}_1'\mathcal{M}_1', \mathcal{S}_2'\mathcal{M}_2' | \mathcal{S}'\mathcal{M}')$$

$$\times \sum_{\theta_1'\theta_2'} (h_1'\theta_1', h_2'\theta_2' | h'\theta')$$

$$\times \left\langle \left[\sum_\nu (-1)^\nu P_\nu \left\langle (a^{m-1}(\mathcal{S}_1 h_1), b^n(\mathcal{S}_2 h_2)) \mathcal{S} h \mathcal{M} \theta \right| \right] \right|$$

$$\times v(m) \left| \left[|a^m \mathcal{S}_1' h_1' \mathcal{M}_1' \theta_1' \rangle |b^{n-1} \mathcal{S}_2' h_2' \mathcal{M}_2' \theta_2' \rangle \right] \right\rangle \qquad (20.2.2)$$

Then $|a^m \mathcal{S}_1' h_1' \mathcal{M}_1' \theta_1' \rangle$ is expressed in terms of $|a^{m-1} \mathcal{S}_1'' h_1'' \mathcal{M}_1'' \theta_1'' \rangle | a\, m_1' \alpha \rangle$ using (19.2.3) and (19.2.4), and $\langle (a^{m-1}, b^n) \mathcal{S} h \mathcal{M} \theta |$ is expressed in terms of $\langle a^{m-1} \mathcal{S}_1 h_1 \mathcal{M}_1 \theta_1 | \langle b^n \mathcal{S}_2 h_2 \mathcal{M}_2 \theta_2 |$ using (19.7.2). After simplifying the factorial we obtain

$$\sqrt{\frac{m}{n}}\, [(n-1)!(m-1)!]^{-1} \sum_{\mathcal{S}_1'' h_1''} (a^{m-1} \mathcal{S}_1'' h_1'', a | a^m \mathcal{S}_1' h_1')$$

$$\times \sum_{\mathcal{M}_1 \mathcal{M}_2} (\mathcal{S} \mathcal{M} | \mathcal{S}_1 \mathcal{M}_1, \mathcal{S}_2 \mathcal{M}_2) \sum_{\mathcal{M}_1'\mathcal{M}_2'} (\mathcal{S}_1'\mathcal{M}_1', \mathcal{S}_2'\mathcal{M}_2' | \mathcal{S}'\mathcal{M}')$$

$$\times \sum_{\mathcal{M}_1'' m_1'} (\mathcal{S}_1''\mathcal{M}_1'', \tfrac{1}{2} m_1' | \mathcal{S}_1'\mathcal{M}_1') \sum_{\theta_1\theta_2} (h\theta | h_1\theta_1, h_2\theta_2)$$

$$\times \sum_{\theta_1'\theta_2'} (h_1'\theta_1', h_2'\theta_2' | h'\theta') \sum_{\theta_1''\alpha} (h_1''\theta_1'', a\alpha | h_1'\theta_1')$$

$$\times \left\langle \sum_\nu (-1)^\nu P_\nu \left[\langle a^{m-1} \mathcal{S}_1 h_1 \mathcal{M}_1 \theta_1 | \langle b^n \mathcal{S}_2 h_2 \mathcal{M}_2 \theta_2 | \right] \right| v(m)$$

$$\times \left| \left[|a^{m-1} \mathcal{S}_1'' h_1'' \mathcal{M}_1'' \theta_1'' \rangle | a\, m_1' \alpha \rangle |b^{n-1} \mathcal{S}_2' h_2' \mathcal{M}_2' \theta_2' \rangle \right] \right\rangle \quad (20.2.3)$$

We now simplify the final integral by summing over ν. In the ket we always have the first $m-1$ electrons in $|a^{m-1} \mathcal{S}_1'' h_1'' \mathcal{M}_1'' \theta_1'' \rangle$, the mth in $|a\, m_1' \alpha \rangle$, and the last $n-1$ in $|b^{n-1} \mathcal{S}_2' h_2' \mathcal{M}_2' \theta_2' \rangle$. Since the operator operates only on $|a\, m_1' \alpha \rangle$, only those $(m-1)!n!$ permutations in the sum over ν which keep the first $m-1$ electrons in $\langle a^{m-1} \mathcal{S}_1 h_1 \mathcal{M}_1 \theta_1 |$ and the last n in $\langle b^n \mathcal{S}_2 h_2 \mathcal{M}_2 \theta_2 |$ contribute, and they all contribute equally [see (19.1.5)]. We must also have $\mathcal{S}_1 = \mathcal{S}_1''$, $h_1 = h_1''$, $\mathcal{M}_1 = \mathcal{M}_1''$, and $\theta_1 = \theta_1''$ in the a^{m-1} kets. At this point, after integrating over the coordinates of the first $m-1$

electrons, we have for the final integral in (20.2.3) above

$$(m-1)!n!\delta_{\mathcal{S}_1\mathcal{S}_1''}\delta_{h_1h_1''}\delta_{\mathfrak{M}_1\mathfrak{M}_1''}\delta_{\theta_1\theta_1''}$$
$$\times\langle b^n\,\mathcal{S}_2h_2\mathfrak{M}_2\theta_2|v(m)\big|\big[|a\,m_1'\alpha\rangle|b^{n-1}\,\mathcal{S}_2'h_2'\mathfrak{M}_2'\theta_2'\rangle\big]\big\rangle \quad (20.2.4)$$

We now use (19.3.13) and (19.3.14) to expand $\langle b^n\,\mathcal{S}_2h_2\mathfrak{M}_2\theta_2|$ in terms of $\langle b\,m_1\beta|\langle b^{n-1}\,\mathcal{S}_2''h_2''\mathfrak{M}_2''\theta_2''|$ and integrate over the coordinates of the last $n-1$ electrons. The final integration requires $\mathcal{S}_2''=\mathcal{S}_2'$, $h_2''=h_2'$, $\mathfrak{M}_2''=\mathfrak{M}_2'$ and $\theta_2''=\theta_2'$. Putting all the above steps together and summing over all double primed indices gives the result

$$\Big\langle\mathcal{Q}\big(a^{m-1}(\mathcal{S}_1h_1),\,b^n(S_2h_2)\big)\mathcal{S}h\mathfrak{M}\theta\Big|V$$
$$\times\Big|\mathcal{Q}\big(a^m(\mathcal{S}_1'h_1'),\,b^{n-1}(\mathcal{S}_2'h_2')\big)\mathcal{S}'h'\mathfrak{M}'\theta'\Big\rangle$$
$$=\sqrt{mn}\,\big(b^n\,\mathcal{S}_2h_2|b,\,b^{n-1}\,\mathcal{S}_2'h_2'\big)\big(a^{m-1}\,\mathcal{S}_1h_1,\,a|a^m\,\mathcal{S}_1'h_1'\big)$$
$$\times\sum_{\mathfrak{M}_1\mathfrak{M}_2m_1\mathfrak{M}_1'\mathfrak{M}_2'm_1'}(\mathcal{S}\mathfrak{M}|\mathcal{S}_1\mathfrak{M}_1,\,\mathcal{S}_2\mathfrak{M}_2)(\mathcal{S}_1'\mathfrak{M}_1',\,\mathcal{S}_2'\mathfrak{M}_2'|\mathcal{S}'\mathfrak{M}')$$
$$\times\big(\mathcal{S}_2\mathfrak{M}_2|\tfrac{1}{2}m_1,\,\mathcal{S}_2'\mathfrak{M}_2'\big)\big(\mathcal{S}_1\mathfrak{M}_1,\,\tfrac{1}{2}m_1'|\mathcal{S}_1'\mathfrak{M}_1'\big)$$
$$\times\sum_{\theta_1\theta_2\beta\theta_1'\theta_2'\alpha}(h\theta|h_1\theta_1,\,h_2\theta_2)(h_1'\theta_1',\,h_2'\theta_2'|h'\theta')$$
$$\times(h_2\theta_2|b\beta,\,h_2'\theta_2')(h_1\theta_1,\,a\alpha|h_1'\theta_1')\langle b\,m_1\beta|v|a\,m_1'\alpha\rangle \quad (20.2.5)$$

The equivalent formula for the a^mb^{n-1}, $a^{m-1}b^n$ case is

$$\Big\langle\big(a^m(\mathcal{S}_1h_1),\,b^{n-1}(\mathcal{S}_2h_2)\big)\mathcal{S}h\mathfrak{M}\theta\Big|V$$
$$\times\Big|\mathcal{Q}\big(a^{m-1}(\mathcal{S}_1'h_1'),\,b^n(\mathcal{S}_2'h_2')\big)\mathcal{S}'h'\mathfrak{M}'\theta'\Big\rangle$$
$$=\sqrt{mn}\,\big(a^m\,\mathcal{S}_1h_1|a^{m-1}\,\mathcal{S}_1'h_1',\,a\big)\big(b,\,b^{n-1}\,\mathcal{S}_2h_2|b^n\mathcal{S}_2'h_2'\big)$$
$$\times\sum_{\mathfrak{M}_1\mathfrak{M}_2m_1\mathfrak{M}_1'\mathfrak{M}_2'm_1'}(\mathcal{S}\mathfrak{M}|\mathcal{S}_1\mathfrak{M}_1,\,\mathcal{S}_2\mathfrak{M}_2)(\mathcal{S}_1'\mathfrak{M}_1',\,\mathcal{S}_2'\mathfrak{M}_2'|\mathcal{S}'\mathfrak{M}')$$
$$\times\big(\mathcal{S}_1\mathfrak{M}_1|\mathcal{S}_1'\mathfrak{M}_1',\,\tfrac{1}{2}m_1\big)\big(\tfrac{1}{2}m_1',\,\mathcal{S}_2\mathfrak{M}_2|\mathcal{S}_2'\mathfrak{M}_2'\big)$$
$$\times\sum_{\theta_1\theta_2\alpha\theta_1'\theta_2'\beta}(h\theta|h_1\theta_1,\,h_2\theta_2)(h_1'\theta_1',\,h_2'\theta_2'|h'\theta')$$
$$\times(h_1\theta_1|h_1'\theta_1',\,a\alpha)(b\beta,\,h_2\theta_2|h_2'\theta_2')\langle a\,m_1\alpha|v|b\,m_1'\beta\rangle \quad (20.2.6)$$

Spin-Independent Matrix Elements between $a^m b^{n-1}$ and $a^{m-1} b^n$

From this point it is straightforward to proceed to the very useful general formulas for our $a^{m-1}b^n$, $a^m b^{n-1}$ and $a^m b^{n-1}$, $a^{m-1}b^n$ reduced matrix elements. For spin-independent operators we use (10.3.7) to give

$$\langle \mathcal{S} h \mathfrak{M} \| O^f \| \mathcal{S}' h' \mathfrak{M}' \rangle = \delta_{\mathcal{S}\mathcal{S}'} \delta_{\mathfrak{M}\mathfrak{M}'} \sum_{\theta\phi\theta'} \begin{pmatrix} h \\ \theta \end{pmatrix} \begin{pmatrix} h^* & f & h' \\ \theta^* & \phi & \theta' \end{pmatrix}^* \times \langle \mathcal{S} h \mathfrak{M} \theta | O^f_\phi | \mathcal{S} h' \mathfrak{M} \theta' \rangle \tag{20.2.7}$$

and note that

$$\langle b\, m_1 \beta | o^f_\phi | a\, m'_1 \alpha \rangle = \delta_{m_1 m'_1} \begin{pmatrix} b \\ \beta \end{pmatrix} \begin{pmatrix} b^* & f & a \\ \beta^* & \phi & \alpha \end{pmatrix} \langle b \| o^f \| a \rangle \tag{20.2.8}$$

These relations are used with (20.2.5), and the same equations, but with $a\alpha$ and $b\beta$ interchanged in (20.2.8), are used with (20.2.6). We then express all the v.c.c. in terms of $3jm$ coefficients, using (B.3.2) or (10.2.1), and sum the $3jm$, using (16.2.1) for the spin part and (16.4.1) for the space part, to obtain

$$\begin{aligned}
&\Big\langle \mathcal{Q}\big(a^{m-1}(\mathcal{S}_1 h_1), b^n(\mathcal{S}_2 h_2)\big) \mathcal{S} h \mathfrak{M} \Big\| O^f \\
&\quad \times \Big\| \mathcal{Q}\big(a^m(\mathcal{S}'_1 h'_1), b^{n-1}(\mathcal{S}'_2 h'_2)\big) \mathcal{S}' h' \mathfrak{M}' \Big\rangle \\
&= \delta_{\mathcal{S}\mathcal{S}'} \delta_{\mathfrak{M}\mathfrak{M}'} \big(b^n\, \mathcal{S}_2 h_2 | b, b^{n-1}\, \mathcal{S}'_2 h'_2\big) \big(a^{m-1}\, \mathcal{S}_1 h_1, a | a^m\, \mathcal{S}'_1 h'_1\big) \\
&\quad \times \sqrt{mn}\, H(\mathcal{S}_1 \mathcal{S}_2 \mathcal{S}) H(\mathcal{S}'_1 \mathcal{S}'_2 \mathcal{S}') H(\tfrac{1}{2} \mathcal{S}'_2 \mathcal{S}_2) H(\mathcal{S}_1 \tfrac{1}{2} \mathcal{S}'_1) \\
&\quad \times \big(|\mathcal{S}'_1| |\mathcal{S}_2|\big)^{1/2} \{\tfrac{1}{2}\} \{\mathcal{S}'_1 \mathcal{S}'_2 \mathcal{S}'\} \{\mathcal{S}_1 \tfrac{1}{2} \mathcal{S}'_1\} \begin{Bmatrix} \mathcal{S}_1 & \mathcal{S}_2 & \mathcal{S} \\ \mathcal{S}'_2 & \mathcal{S}'_1 & \tfrac{1}{2} \end{Bmatrix} \\
&\quad \times \big(|h'_1| |h_2| |h| |h'|\big)^{1/2} \begin{Bmatrix} h_1 & h_2 & h^* \\ a & b^* & f \\ h'^*_1 & h'^*_2 & h' \end{Bmatrix} \langle b \| o^f \| a \rangle
\end{aligned} \tag{20.2.9}$$

and

$$\left\langle \mathcal{Q}\left(a^m(\mathcal{S}_1h_1), b^{n-1}(\mathcal{S}_2h_2)\right)\mathcal{S}h\mathfrak{M}\middle\|O^f\right.$$
$$\times\left\|\mathcal{Q}\left(a^{m-1}(\mathcal{S}_1'h_1'), b^n(\mathcal{S}_2'h_2')\right)\mathcal{S}'h'\mathfrak{M}'\right\rangle$$
$$= \delta_{\mathcal{S}\mathcal{S}'}\delta_{\mathfrak{M}\mathfrak{M}'}\left(a^m\,\mathcal{S}_1h_1|a^{m-1}\,\mathcal{S}_1'h_1', a\right)\left(b, b^{n-1}\,\mathcal{S}_2h_2|b^n\,\mathcal{S}_2'h_2'\right)$$
$$\times\sqrt{mn}\,H(\mathcal{S}_1\mathcal{S}_2\mathcal{S})H(\mathcal{S}_1'\mathcal{S}_2'\mathcal{S}')H(\mathcal{S}_1'\tfrac{1}{2}\mathcal{S}_1)H(\tfrac{1}{2}\mathcal{S}_2\mathcal{S}_2')$$
$$\times\ (|\mathcal{S}_1||\mathcal{S}_2'|)^{1/2}\{\mathcal{S}'\}\{\mathcal{S}_1'\mathcal{S}_2'\mathcal{S}'\}\{\tfrac{1}{2}\mathcal{S}_2\mathcal{S}_2'\}\begin{Bmatrix} \mathcal{S}_1 & \mathcal{S}_2 & \mathcal{S} \\ \mathcal{S}_2' & \mathcal{S}_1' & \tfrac{1}{2} \end{Bmatrix}$$
$$\times\ (|h_1||h_2'||h||h'|)^{1/2}\{h^*fh'\}\{a^*fb\}$$
$$\times\begin{Bmatrix} h_1 & h_2 & h^* \\ h_1'^* & h_2'^* & h' \\ a^* & b & f \end{Bmatrix}\langle a\|o^f\|b\rangle \tag{20.2.10}$$

Note that the spin phase factors above may be simplified considerably by using the relations in Section B.3. Note also that we generally give cfp explicitly in the $(b^{n-1}, b|b^n)$ form only. Thus to evaluate the $(b, b^{n-1}|b^n)$ cfp in (20.2.9) and (20.2.10), Eq. (19.3.15) must first be applied.

Spin-Only Matrix Elements Between a^mb^{n-1} and $a^{m-1}b^n$

Matrix elements of spin-only operators between a^mb^{n-1} and $a^{m-1}b^n$ are, of course, zero because of the orthogonality of the a and b spatial orbitals.

Spin–Orbit Matrix Elements Between a^mb^{n-1} and $a^{m-1}b^n$

As in our previous spin–orbit matrix-element calculations, we first apply (18.4.2)–(18.4.3) or (18.4.6). The resulting $\Sigma_k su(k)$ reduced matrix elements are then expressed in terms of matrix elements such as (20.2.5) or (20.2.6), using the double-tensor analog of (10.3.7):

$$\langle\mathcal{S}h\|\sum_k s^1u^{f^*}(k)\|\mathcal{S}'h'\rangle^{SO_3,G} = \sum_{\mathfrak{M}m\mathfrak{M}'}\begin{pmatrix}\mathcal{S}\\ \mathfrak{M}\end{pmatrix}\begin{pmatrix}\mathcal{S} & 1 & \mathcal{S}' \\ -\mathfrak{M} & m & \mathfrak{M}'\end{pmatrix}^*$$
$$\times\sum_{\theta\phi^*\theta'}\begin{pmatrix}h\\ \theta\end{pmatrix}\begin{pmatrix}h^* & f^* & h' \\ \theta^* & \phi^* & \theta'\end{pmatrix}^*\langle\mathcal{S}h\mathfrak{M}\theta|\sum_k s_m^1u_{\phi^*}^{f^*}(k)|\mathcal{S}'h'\mathfrak{M}'\theta'\rangle \tag{20.2.11}$$

Then (13.3.10) and (19.5.4) are used to reduce and simplify the one-electron matrix elements of (20.2.5):

$$\langle b\, m_1 \beta | s^1_m u^{f^*}_{\phi^*} | a\, m'_1 \alpha \rangle$$

$$= \begin{pmatrix} \frac{1}{2} \\ m_1 \end{pmatrix} \begin{pmatrix} \frac{1}{2} & 1 & \frac{1}{2} \\ -m_1 & m & m'_1 \end{pmatrix} \begin{pmatrix} b \\ \beta \end{pmatrix} \begin{pmatrix} b^* & f^* & a \\ \beta^* & \phi^* & \alpha \end{pmatrix} \frac{\sqrt{6}}{2} \langle b \| \xi l^{f^*} \| a \rangle^G \tag{20.2.12}$$

The same equation but with $a\alpha$ and $b\beta$ interchanged applies for the analogous one-electron matrix element in (20.2.6). In each case we then express all v.c.c. in terms of $3jm$ using (B.3.2) for the spin v.c.c. and (10.2.1) for the orbital v.c.c., and sum the $3jm$ using (16.4.1). We obtain

$$\Big\langle \mathcal{a}\big(a^{m-1}(\mathbb{S}_1 h_1), b^n(\mathbb{S}_2 h_2)\big)\mathbb{S}h \Big\| \sum_k s^1 u^{f^*}(k)$$

$$\times \Big\| \mathcal{a}\big(a^m(\mathbb{S}'_1 h'_1), b^{n-1}(\mathbb{S}'_2 h'_2)\big)\mathbb{S}'h' \Big\rangle^{SO_3,G}$$

$$= \big(b^n\, \mathbb{S}_2 h_2 | b, b^{n-1}\, \mathbb{S}'_2 h'_2\big)\big(a^{m-1}\, \mathbb{S}_1 h_1, a | a^m\, \mathbb{S}'_1 h'_1\big)$$

$$\times \sqrt{mn}\, H(\mathbb{S}_1 \mathbb{S}_2 \mathbb{S})\, H(\mathbb{S}'_1 \mathbb{S}'_2 \mathbb{S}')\, H(\tfrac{1}{2} \mathbb{S}'_2 \mathbb{S}_2)\, H(\mathbb{S}_1 \tfrac{1}{2} \mathbb{S}'_1)$$

$$\times \big(|\mathbb{S}'_1||\mathbb{S}_2||\mathbb{S}||\mathbb{S}'|\big)^{1/2} \{\mathbb{S}_2\}\{\mathbb{S}'\} \begin{Bmatrix} \mathbb{S}_1 & \mathbb{S}_2 & \mathbb{S} \\ \frac{1}{2} & \frac{1}{2} & 1 \\ \mathbb{S}'_1 & \mathbb{S}'_2 & \mathbb{S}' \end{Bmatrix}$$

$$\times \big(|h'_1||h_2||h||h'|\big)^{1/2} \begin{Bmatrix} h_1 & h_2 & h^* \\ a & b^* & f^* \\ h'^*_1 & h'^*_2 & h' \end{Bmatrix} \frac{\sqrt{6}}{2} \langle b \| \xi l^{f^*} \| a \rangle^G \tag{20.2.13}$$

and

$$\Big\langle \mathcal{a}\big(a^m(\mathbb{S}_1 h_1), b^{n-1}(\mathbb{S}_2 h_2)\big)\mathbb{S}h \Big\| \sum_k s^1 u^{f^*}(k)$$

$$\times \Big\| \mathcal{a}\big(a^{m-1}(\mathbb{S}'_1 h'_1), b^n(\mathbb{S}'_2 h'_2)\big)\mathbb{S}'h' \Big\rangle^{SO_3,G}$$

$$= \left(a^m \mathbb{S}_1 h_1 | a^{m-1} \mathbb{S}_1' h_1', a\right)\left(b, b^{n-1} \mathbb{S}_2 h_2 | b^n \mathbb{S}_2' h_2'\right)$$

$$\times \sqrt{mn}\, H(\mathbb{S}_1 \mathbb{S}_2 \mathbb{S}) H(\mathbb{S}_1' \mathbb{S}_2' \mathbb{S}') H(\mathbb{S}_1' \tfrac{1}{2} \mathbb{S}_1) H(\tfrac{1}{2} \mathbb{S}_2 \mathbb{S}_2')$$

$$\times \left(|\mathbb{S}_1||\mathbb{S}_2'||\mathbb{S}||\mathbb{S}'|\right)^{1/2} \{\mathbb{S}_1\}\{\mathbb{S}'\}\{\mathbb{S} 1 \mathbb{S}'\}\{\tfrac{1}{2} 1 \tfrac{1}{2}\} \begin{Bmatrix} \mathbb{S}_1 & \mathbb{S}_2 & \mathbb{S} \\ \mathbb{S}_1' & \mathbb{S}_2' & \mathbb{S}' \\ \tfrac{1}{2} & \tfrac{1}{2} & 1 \end{Bmatrix}$$

$$\times \left(|h_1||h_2'||h||h'|\right)^{1/2} \{h^* f^* h'\}\{a^* f^* b\} \begin{Bmatrix} h_1 & h_2 & h^* \\ h_1'^* & h_2'^* & h' \\ a^* & b & f^* \end{Bmatrix}$$

$$\times \frac{\sqrt{6}}{2} \langle a \| \xi l^{f^*} \| b \rangle^G \qquad (20.2.14)$$

Note that the comments following (20.2.10) also apply to the above equations.

Equations Analogous to those of Chapter 18 for Matrix Elements between $a^m b^{n-1}$ and $a^{m-1} b^n$

All of these equations could be written in the form of (19.4.5), (19.5.1), or (19.8.2) and then further reduced with equations in the form of Section 18.1 or 18.5. Thus, for example, (20.2.13) could be written

$$\left\langle \mathcal{Q}\left(a^{m-1}(\mathbb{S}_1 h_1), b^n(\mathbb{S}_2 h_2)\right) \mathbb{S} h \right\| \sum_k s^1 u^{f^*}(k)$$

$$\times \left\| \mathcal{Q}\left(a^m(\mathbb{S}_1' h_1'), b^{n-1}(\mathbb{S}_2' h_2')\right) \mathbb{S}' h' \right\rangle$$

$$= \sqrt{mn} \left(b^n \mathbb{S}_2 h_2 | b, b^{n-1} \mathbb{S}_2' h_2'\right)\left(a^{m-1} \mathbb{S}_1 h_1, a | a^m \mathbb{S}_1' h_1'\right)$$

$$\times \left\langle \left\{a^{m-1}(\mathbb{S}_1 h_1), \left[\left(b, b^{n-1}(\mathbb{S}_2' h_2')\right) b^n(\mathbb{S}_2 h_2)\right]\right\} \mathbb{S} h \right\| s^1 u^{f^*}(m)$$

$$\times \left\| \left\{\left[\left(a^{m-1}(\mathbb{S}_1 h_1), a\right) a^m(\mathbb{S}_1' h_1')\right], b^{n-1}(\mathbb{S}_2' h_2')\right\} \mathbb{S}' h' \right\rangle \qquad (20.2.15)$$

where in this case the equivalent of the equations in Section 18.5 is

$$\Big\langle\{a^{m-1}(\mathcal{S}_1h_1),\big[(b,b^{n-1}(\mathcal{S}_2'h_2'))b^n(\mathcal{S}_2h_2)\big]\}\mathcal{S}h\big\|s^1u^{f^*}(m)$$
$$\times\big\|\{\big[(a^{m-1}(\mathcal{S}_1h_1),a)a^m(\mathcal{S}_1'h_1')\big],b^{n-1}(\mathcal{S}_2'h_2')\}\mathcal{S}'h'\Big\rangle$$
$$=\big[\text{last three lines of (20.2.13) with factor }\sqrt{mn}\text{ omitted}\big] \tag{20.2.16}$$

There is not, however, much point in deriving general formulas of the type (20.2.16), since they do not have wide applicability.

Example: Calculation of the Spin–Orbit Matrix Elements between $U_u'(^2T_{1u})$ and $U_u'(^2T_{2u})$ for $IrCl_6^{2-}$

As an example of the use of (20.2.13) we calculate the off-diagonal spin–orbit matrix elements between the $U_u'(^2T_{1u})$ and $U_u'(^2T_{2u})$ states for $IrCl_6^{2-}$. The results have already been entered in Table 19.5.1. Equation (18.4.6) is first used to express the $\mathcal{H}_{SO}$ matrix elements in the $|(\mathcal{S}S,h)t\tau\rangle$ basis in terms of $\sum_k su(k)$ matrix elements in the $|\mathcal{S}h\mathfrak{M}\theta\rangle$ basis. Then (20.2.13) is employed to obtain the results in Table 19.5.1. The $9j$ are calculated with (16.4.8) followed by (16.1.11). These steps give

$$\Big\langle\big(\tfrac{1}{2}E_g',T_{1u}\big)U_u'\tau\big|\mathcal{H}_{SO}\big|\big(\tfrac{1}{2}E_g',T_{2u}\big)U_u'\tau\Big\rangle$$
$$=\frac{1}{2\sqrt{3}}\Big\langle\tfrac{1}{2}T_{1u}\Big\|\sum_k s^1u^{t_{1g}}(k)\Big\|\tfrac{1}{2}T_{2u}\Big\rangle^{SO_3,O}$$
$$=\frac{1}{2\sqrt{3}}\Big\langle\mathcal{Q}\big(t_{1u}^5(\tfrac{1}{2}T_{1u}),t_{2u}^6(0A_1)\big)\tfrac{1}{2}T_{1u}\Big\|\sum_k s^1u^{t_{1g}}(k)$$
$$\times\Big\|\mathcal{Q}\big(t_{1u}^6(0A_1),t_{2u}^5(\tfrac{1}{2}T_{2u})\big)\tfrac{1}{2}T_{2u}\Big\rangle^{SO_3,O}$$
$$=\left(\frac{1}{2\sqrt{3}}\right)\left(\frac{\sqrt{6}}{2}\right)\langle t_{2u}\|\xi l^{t_{1g}}\|t_{1u}\rangle$$
$$=\frac{1}{2\sqrt{2}}\langle t_{2u}\|\xi l^{t_{1g}}\|t_{1u}\rangle \tag{20.2.17}$$

Then

$$\Big\langle\big(\tfrac{1}{2}E_g',T_{2u}\big)U_u'\tau\big|\mathcal{H}_{SO}\big|\big(\tfrac{1}{2}E_g',T_{1u}\big)U_u'\tau\Big\rangle=\frac{1}{2\sqrt{2}}\langle t_{2u}\|\xi l^{t_{1g}}\|t_{1u}\rangle^* \tag{20.2.18}$$

since $\mathcal{H}_{SO}$ is a Hermitian operator and (20.2.18) is the Hermitian conjugate of (20.2.17).

20.3. More Than Two Open Shells

Configurationally Diagonal Matrix Elements for $(a^m b^n)c^l$, $a^m(b^n c^l)$, or $(a^m b^n)(c^l d^k)$ States

These matrix elements are very simple to evaluate by the following method which we illustrate only for $|\mathcal{A}\{[(a^m(\mathcal{S}_1 h_1), b^n(\mathcal{S}_2 h_2))\mathcal{S}_{12}h_{12}], c^l(\mathcal{S}_3 h_3)\}\mathcal{S}h\mathfrak{M}\theta\rangle$-type states and only for the spin-independent operator O^f. The method for $a^m(b^n c^l)$ or $(a^m b^n)(c^l d^k)$ matrix elements and for $\Sigma_k su(k)$ is analogous.

We first use (20.1.6) to obtain

$$\begin{aligned}
&\langle \mathcal{A}\{[(a^m(\mathcal{S}_1 h_1), b^n(\mathcal{S}_2 h_2))\mathcal{S}_{12}h_{12}], c^l(\mathcal{S}_3 h_3)\}\mathcal{S}h\mathfrak{M}\|O^f \\
&\quad\times\|\mathcal{A}\{[(a^m(\mathcal{S}_1' h_1'), b^n(\mathcal{S}_2' h_2'))\mathcal{S}_{12}'h_{12}'], c^l(\mathcal{S}_3' h_3')\}\mathcal{S}'h'\mathfrak{M}'\rangle \\
&= \delta(\text{spin})\Bigg[\delta_{h_3 h_3'}|h|^{1/2}|h'|^{1/2}\{h_{12}h_3h^*\}\{h_{12}^* f h_{12}'\}\begin{Bmatrix} f & h' & h^* \\ h_3^* & h_{12} & h_{12}' \end{Bmatrix} \\
&\quad\times\langle \mathcal{A}(a^m(\mathcal{S}_1 h_1), b^n(\mathcal{S}_2 h_2))\mathcal{S}_{12}h_{12}\mathfrak{M}_{12}\|O^f \\
&\quad\times\|\mathcal{A}(a^m(\mathcal{S}_1' h_1'), b^n(\mathcal{S}_2' h_2'))\mathcal{S}_{12}h_{12}'\mathfrak{M}_{12}\rangle \\
&\quad+\delta_{h_{12}h_{12}'}|h|^{1/2}|h'|^{1/2}\{h_{12}'h_3'h'^*\}\{h_3^* f h_3'\}\begin{Bmatrix} f & h' & h^* \\ h_{12}^* & h_3 & h_3' \end{Bmatrix} \\
&\quad\times\langle c^l\,\mathcal{S}_3 h_3\mathfrak{M}_3\|O^f\|c^l\,\mathcal{S}_3 h_3'\mathfrak{M}_3\rangle\Bigg]
\end{aligned} \tag{20.3.1}$$

where

$$\delta(\text{spin}) = \delta_{\mathcal{S}_{12}\mathcal{S}_{12}'}\delta_{\mathcal{S}_3\mathcal{S}_3'}\delta_{\mathcal{S}\mathcal{S}'}\delta_{\mathfrak{M}\mathfrak{M}'} \tag{20.3.2}$$

The $\langle \mathcal{S}_{12}h_{12}\mathfrak{M}_{12}\|O^f\|\mathcal{S}_{12}h_{12}'\mathfrak{M}_{12}\rangle$ matrix element is then itself reduced using (20.1.7), and $\langle c^l\,\mathcal{S}_3 h_3\mathfrak{M}_3\|O^f\|c^l\,\mathcal{S}_3 h_3'\mathfrak{M}_3\rangle$ is reduced with (19.4.6).

The above method may be easily extended to matrix elements between $a^m b^n c^l d^k \cdots$ states.

Matrix Elements between $(a^{m-1}b^n)c^l$ and $(a^mb^{n-1})c^l$ or between $a^m(b^{n-1}c^l)$ and $a^m(b^nc^{l-1})$

These matrix elements may be evaluated by methods very similar to those of the previous subsection. An illustration for the spin-independent operator O^f for the $(a^{m-1}b^n)c^l, (a^mb^{n-1})c^l$ case follows.

We again first use the "diagonal" formula (20.1.6) to obtain

$$\langle \mathcal{Q}\{[(a^{m-1}(\mathbb{S}_1h_1), b^n(\mathbb{S}_2h_2))\mathbb{S}_{12}h_{12}], c^l(\mathbb{S}_3h_3)\}\mathbb{S}h\mathfrak{M}\|O^f$$
$$\times\|\mathcal{Q}\{[(a^m(\mathbb{S}_1'h_1'), b^{n-1}(\mathbb{S}_2'h_2'))\mathbb{S}_{12}'h_{12}'], c^l(\mathbb{S}_3'h_3')\}\mathbb{S}'h'\mathfrak{M}'\rangle$$
$$= \delta(\text{spin})\bigg[\delta_{h_3h_3'}|h|^{1/2}|h'|^{1/2}\{h_{12}h_3h^*\}\{h_{12}^*fh_{12}'\}\begin{Bmatrix} f & h' & h^* \\ h_3^* & h_{12} & h_{12}' \end{Bmatrix}$$
$$\times\langle \mathcal{Q}(a^{m-1}(\mathbb{S}_1h_1), b^n(\mathbb{S}_2h_2))\mathbb{S}_{12}h_{12}\mathfrak{M}_{12}\|O^f$$
$$\times\|\mathcal{Q}(a^m(\mathbb{S}_1'h_1'), b^{n-1}(\mathbb{S}_2'h_2'))\mathbb{S}_{12}h_{12}'\mathfrak{M}_{12}\rangle + 0\bigg] \tag{20.3.3}$$

where $\delta(\text{spin})$ is again given by (20.3.2). The matrix element $\langle\mathbb{S}_{12}h_{12}\mathfrak{M}_{12}\|O^f\|\mathbb{S}_{12}h_{12}'\mathfrak{M}_{12}\rangle$ is here of the $a^{m-1}b^n$, a^mb^{n-1} type and so is evaluated using (20.2.9). The second term does not contribute, because of the orthogonality of the h_{12} and h_{12}' states.

More Complicated Cases

More complicated cases would include, for example, the calculation of matrix elements between $(a^{m-1}b^n)c^l$ and $(a^mb^n)c^{l-1}$ configurations. The above case is a more complex version of the $a^{m-1}b^n$, a^mb^{n-1} case and may be handled by methods analogous to those used to obtain (20.2.9) and (20.2.13). Often, however, by choosing a different order of coupling for the basis functions, matrix elements of this type may be avoided. Here, for example, if the states were constructed using the coupling sequences $(a^{m-1}c^l)b^n$ and $(a^mc^{l-1})b^n$, the very simple matrix-element methods of the last subsection would pertain.

If the basis consists of all possible $a^{m-1}b^nc^l$ and $a^mb^nc^{l-1}$ states, the coupling sequence should be chosen very carefully to enable the simplest possible calculations. Once a sequence is chosen, however, it must be used to calculate all the matrix elements required, or phase errors may result.

21 The jj-Coupling Approach to Molecules

21.1. Introduction to jj Coupling

In analogy with the atomic case, when the spin–orbit interactions become comparable to or larger than interelectronic repulsions V_{ee}, it is sometimes convenient to use a different coupling scheme called jj coupling. In this scheme the *one-electron* basis functions are chosen diagonal in $\mathcal{H}_{SO}$ and hence are diagonal in the total Hamiltonian if $V_{ee} = 0$. In the atomic case $\mathbf{j} = \mathbf{l} + \mathbf{s}$ is the total-angular-momentum operator of a single electron, and if $V_{ee} = 0$, j^2 and j_z commute with $\mathcal{H}_{SO}$ and hence with $\mathcal{H}$. Thus j and j_z are good quantum numbers. Since $s = \frac{1}{2}$ for a single electron, j can have the *two* possible values, $|\ell + \frac{1}{2}|$ and $|\ell - \frac{1}{2}|$ for each electron. One-electron $s\ell = \frac{1}{2}\ell$ basis functions, written $|\frac{1}{2}\ell\, m_s m_\ell\rangle$, can be coupled to form spin–orbit (jj) functions $|(\frac{1}{2}\ell)jm\rangle$ where (in units of $\hbar$)

$$
\begin{aligned}
j^2|(\tfrac{1}{2}\ell)jm\rangle &= j(j+1)|(\tfrac{1}{2}\ell)jm\rangle \\
j_z|(\tfrac{1}{2}\ell)jm\rangle &= m|(\tfrac{1}{2}\ell)jm\rangle
\end{aligned}
\qquad (21.1.1)
$$

The total energy of the system is the sum of the one-electron energies, and the corresponding eigenfunctions are antisymmetrized products of the $|(\frac{1}{2}\ell)jm\rangle$. These are assembled (i.e., are jj-coupled) to give eigenfunctions of J^2 and J_z, since $\mathcal{J}$ and $\mathfrak{M}_{\mathcal{J}}$ are always good quantum numbers.

We proceed in an analogous manner in the molecular case, using the notation of Section 12.2. The one-electron $s\ell$ eigenfunctions, written $|\frac{1}{2}sam\alpha\rangle$, are coupled to form one-electron spin–orbit functions $|(\frac{1}{2}s, a)p\pi\rangle$, where p and π designate irrep and irrep partner of the spin–orbit-coupled function. In the atomic case the $|\frac{1}{2}\ell\, m_s m_\ell\rangle$ are coupled by SO_3 $3jm$ to form $|(\frac{1}{2}\ell)jm\rangle$, and in the molecular case, in exact analogy, the symmetry-group $3jm$ are used to form $|(\frac{1}{2}s, a)p\pi\rangle$ from the $|\frac{1}{2}sam\alpha\rangle$. As always, we use lowercase letters to designate one-electron functions. Finally, in the atomic case we assemble (jj-couple) the one-electron functions into products (determinants) of appropriate $\mathcal{J}$ and $\mathcal{M}_{\mathcal{J}}$; likewise in the molecular case we assemble the $|(\frac{1}{2}s, a)p\pi\rangle$ into products belonging to $rt\tau$.

21.2 jj Coupling Using the OsX_6^{2-} t_{2g}^4 Configuration as an Example

The jj basis is particularly advantageous when it can be assumed to a first approximation that interelectronic repulsions are zero. As discussed in Sections 23.4–23.6, that is the case within the $t_{1u}^5 t_{2g}^5$ and $t_{2u}^5 t_{2g}^5$ charge-transfer configurations of OsX_6^{2-} (X = Cl, Br, or I). There the $A_{1g}(t_{2g}^4)$ ground state may also be described in the jj basis, and electric-dipole selection rules for states in that basis may be used. Here we use the example of OsX_6^{2-} to show how the states of a configuration may be described using the jj basis and how matrix elements may be calculated in such a basis using the irreducible-tensor method. We draw extensively on the results of Chapters 19 and 20, since the formalism developed there works equally well for antisymmetric functions and one-electron operators in other bases; no new ideas are really needed. Thus this chapter serves as a nice example of how irreducible-tensor methods may be applied to less conventional basis sets.

We begin our example by describing the fifteen $|rt\tau\rangle$ states of the t_{2g}^4 ground configuration of OsX_6^{2-} in both the $\mathcal{A}\ell$ and the jj bases and then calculating the A_{1g} spin–orbit matrix using them. We then calculate the spin–orbit energy and the magnetic moment of the lowest-energy T_{1u} state from the OsX_6^{2-} $t_{2u}^5 t_{2g}^5$ charge-transfer configuration. Finally we determine the dipole strength and $\mathcal{C}_1/\mathcal{D}_0$ ratio of the transition to that state. To simplify the equations we omit parity labels when they are unambiguous. A schematic MO diagram of OsX_6^{2-} is given in Fig. 23.2.1 (see Section 23.2).

Ground-State Basis of OsX_6^{2-} *Using* $\mathcal{SL}$ *Coupling*

The $\mathcal{SL}$ basis states are formed by coupling together one-electron spin–orbitals (designated $|\frac{1}{2}am\alpha\rangle$) to form $|a^n\,\mathcal{S}h\mathcal{M}\theta\rangle$ states. In our case $a^n = t_{2g}^4$ and there are four allowed terms. In the $^{2\mathcal{S}+1}h$ notation they are $^3T_1$, $^1A_1$, ^1E, and $^1T_2$. The $|a^n\,\mathcal{S}h\mathcal{M}\theta\rangle$ states may then be coupled to form spin–orbit basis states, the $|a^n(\mathcal{S}S, h)rt\tau\rangle$. This is the spin–orbit basis used in earlier chapters. In this example the index r need not be specified, since there are no repeated representations in $\mathcal{S} \otimes h$. The basis consists of:

1. The six $\mathcal{A}\ell$ or $|\frac{1}{2}am\alpha\rangle$ one-electron spin–orbitals (here $|1^+\rangle \equiv |\frac{1}{2}t_2\frac{1}{2}1\rangle$, etc.):

$$|1^+\rangle,\quad |1^-\rangle,\quad |0^+\rangle,\quad |0^-\rangle,\quad |-1^+\rangle,\quad |-1^-\rangle \qquad (21.2.1)$$

2. Fifteen $|a^n \mathbb{S} h \mathfrak{M} \theta\rangle$ states:

$$|t_2^4 1T_1 \mathfrak{M}\theta\rangle \quad \text{for} \quad \mathfrak{M} = 1, 0, -1 \text{ and } \theta = 1, 0, -1$$
$$|t_2^4 0A_1 0a_1\rangle$$
$$|t_2^4 0E0\theta\rangle \quad \text{and} \quad |t_2^4 0E0\epsilon'\rangle \tag{21.2.2}$$
$$|t_2^4 0T_2 01\rangle, \quad |t_2^4 0T_2 00\rangle, \quad \text{and} \quad |t_2^4 0T_2 0-1\rangle$$

3. Fifteen $|a^n t\tau\rangle = |a^n(\mathbb{S}S, h)t\tau\rangle$ states:

$$\left|t_2^4(1T_1, T_1)A_1a_1\right\rangle$$
$$\left|t_2^4(1T_1, T_1)E\tau\right\rangle, \qquad \tau = \theta, \epsilon'$$
$$\left|t_2^4(1T_1, T_1)T_1\tau\right\rangle, \qquad \tau = 1, 0, -1$$
$$\left|t_2^4(1T_1, T_1)T_2\tau\right\rangle, \qquad \tau = 1, 0, -1 \tag{21.2.3}$$
$$\left|t_2^4(0A_1, A_1)A_1a_1\right\rangle$$
$$\left|t_2^4(0A_1, E)E\tau\right\rangle, \qquad \tau = \theta, \epsilon'$$
$$\left|t_2^4(0A_1, T_2)T_2\tau\right\rangle, \qquad \tau = 1, 0, -1$$

Ground-State Basis of OsX_6^{2-} Using jj Coupling

The jj basis states are formed by first coupling together the spin and space parts of the $|\frac{1}{2}sam\alpha\rangle$ one-electron spin–orbitals to form $|(\frac{1}{2}s, a)p\pi\rangle$ one-electron jj orbitals. The $|(\frac{1}{2}s, a)p\pi\rangle$ functions are then coupled together to form antisymmetric $|p^l r_1 t_1 \tau_1\rangle$ and $|q^m r_2 t_2 \tau_2\rangle$ states, which are in turn coupled to form properly antisymmetrized $|a^n rt\tau\rangle = |\mathcal{A}(p^l(r_1t_1), q^m(r_2t_2))rt\tau\rangle$ states. The indices r_1, r_2, and r are specified only when repeated representations make the labels t_1, t_2, or t insufficient. The t_{2g}^4 basis consists of:

1. Six $|(\frac{1}{2}s, a)p\pi\rangle$ jj one-electron orbitals (here $|\kappa\rangle \equiv |(\frac{1}{2}e', t_2)u'\kappa\rangle$, $|\alpha''\rangle \equiv |(\frac{1}{2}e', t_2)e''\alpha''\rangle$, etc.):

$$|\kappa\rangle, \quad |\lambda\rangle, \quad |\mu\rangle, \quad |\nu\rangle, \quad |\alpha''\rangle, \quad |\beta''\rangle \tag{21.2.4}$$

2. Fifteen $|p^l\, r_1 t_1 \tau_1\rangle |q^m\, r_2 t_2 \tau_2\rangle$ states:

$$\left|(u')^4 A_1 a_1\right\rangle$$

$$\left|(u')^3 U'\tau_1\right\rangle |e''\, E''\tau_2\rangle \qquad \text{for} \quad \tau_1 = \kappa, \lambda, \mu, \nu \text{ and } \tau_2 = \alpha'', \beta''$$

$$\left|(u')^2 A_1 a_1\right\rangle \left|(e'')^2 A_1 a_1\right\rangle$$

$$\left|(u')^2 E\tau_1\right\rangle \left|(e'')^2 A_1 a_1\right\rangle \qquad \text{for} \quad \tau_1 = \theta, \epsilon'$$

$$\left|(u')^2 0T_2\tau_1\right\rangle \left|(e'')^2 A_1 a_1\right\rangle \qquad \text{for} \quad \tau_1 = 1, 0, -1$$

(21.2.5)

3. Fifteen $|a^n\, rt\tau\rangle = |\mathcal{Q}(p^l(r_1 t_1), q^m(r_2 t_2)) rt\tau\rangle$ states for $n = l + m$:

$$\left|(u')^4 A_1 a_1\right\rangle$$

$$\left|\mathcal{Q}\left((u')^3(U'), e''(E'')\right) E\tau\right\rangle \qquad \text{for} \quad \tau = \theta, \epsilon'$$

$$\left|\mathcal{Q}\left((u')^3(U'), e''(E'')\right) T_1\tau\right\rangle \qquad \text{for} \quad \tau = 1, 0, -1$$

$$\left|\mathcal{Q}\left((u')^3(U'), e''(E'')\right) T_2\tau\right\rangle \qquad \text{for} \quad \tau = 1, 0, -1 \qquad (21.2.6)$$

$$\left|\mathcal{Q}\left((u')^2(A_1), (e'')^2(A_1)\right) A_1 a_1\right\rangle$$

$$\left|\mathcal{Q}\left((u')^2(E), (e'')^2(A_1)\right) E\tau\right\rangle \qquad \text{for} \quad \tau = \theta, \epsilon'$$

$$\left|\mathcal{Q}\left((u')^2(0T_2), (e'')^2(A_1)\right) T_2\tau\right\rangle \qquad \text{for} \quad \tau = 1, 0, -1$$

Since the $q = e''$ orbital holds a maximum of two electrons, the minimum l value is 2 since $n = m + l = 4$. Note also that since the $|p^l\, r_1 t_1 \tau_1\rangle$ functions must be antisymmetric, $|(u')^2\, r_1 t_1 \tau_1\rangle$ states are allowed only for $r_1 t_1 = A_1$, E, and $0T_2$ (antisymmetric product of $U' \otimes U'$ in Section C.3).

Formal Construction of States of jj Configurations with One or Two Open Shells

The $|rt\tau\rangle$ states above may be defined by a set of equations of the same form as those for our $|\mathcal{Q}(a^m(\mathbb{S}_1 h_1), b^n(\mathbb{S}_2 h_2))\mathbb{S} h \mathfrak{M} \theta\rangle$ states [see (19.7.1) and (19.7.2)]. Thus

$$\left|\mathcal{A}\left(p^l(r_1t_1), q^m(r_2t_2)\right)rt\tau\right\rangle$$

$$= [l!m!(m+l)!]^{-1/2} \sum_{\nu=1}^{(l+m)!} (-1)^{\nu} P_\nu \left|\left(p^l(r_1t_1), q^m(r_2t_2)\right)rt\tau\right\rangle \tag{21.2.7}$$

where

$$\left|\left(p^l(r_1t_1), q^m(r_2t_2)\right)rt\tau\right\rangle = \sum_{\tau_1\tau_2} (t_1\tau_1, t_2\tau_2|rt\tau)|p^l\, r_1t_1\tau_1\rangle\,|q^m\, r_2t_2\tau_2\rangle \tag{21.2.8}$$

The $|p^l\, r_1t_1\tau_1\rangle$ states likewise have a form similar to that of our $|a^n\, \mathbb{S}h\mathfrak{M}\theta\rangle$ states [see (19.2.4) and (19.2.3)]. We write

$$|p^l\, r_1t_1\tau_1\rangle = \sum_{r_1't_1'} \left(p^{l-1} r_1't_1', p|p^l\, r_1t_1\right)\left|\left(p^{l-1}(r_1't_1'), p\right)r_1t_1\tau_1\right\rangle \tag{21.2.9}$$

and

$$\left|\left(p^{l-1}(r_1't_1'), p\right)r_1t_1\tau_1\right\rangle = \sum_{\tau_1'\pi} (t_1'\tau_1', p\pi|r_1t_1\tau_1)|p^{l-1}\, r_1't_1'\tau_1'\rangle\,|p\pi\rangle \tag{21.2.10}$$

In all cases the notation has the same meaning as in the referenced equations. Finally the $|p\pi\rangle = |(\frac{1}{2}s, a)p\pi\rangle$ orbitals are formed on a one-electron level very much like $|(\mathbb{S}S, h)t\tau\rangle$ functions [see (12.2.2)]:

$$\left|\left(\tfrac{1}{2}s, a\right)p\pi\right\rangle = \sum_{m\alpha} (sm, a\alpha|p\pi)|\tfrac{1}{2}sam\alpha\rangle \tag{21.2.11}$$

It is straightforward to construct $|p^l\, r_1t_1\tau_1\rangle$ functions and to determine the required coefficients of fractional parentage (cfp) using the same approach we used in Section 19.3 for the $|a^n\, \mathbb{S}h\mathfrak{M}\theta\rangle$ functions. We give the results in Section C.19. The cfp phases are chosen to follow the hole–particle convention of Section 9.7 in Griffith (1964)—see our Section 11.4.

Warning About Phases of jj States

When calculating off-diagonal matrix elements between jj configurations such as $p^lq^ms^{k-1}$ and $p^{l-1}q^ms^k$, it is in general necessary to include the q^m

configuration explicitly *even if q^m is a filled shell* ($m = 2|q|$). This is so because in point groups other than O_3 or SO_3 the phase of $|\mathcal{Q}\{[(p^l(r_1t_1), q^m(A_1))t_1], s^{k-1}(r_2t_2)\}rt\tau\rangle$ is the same as that for $|\mathcal{Q}(p^l(r_1t_1), s^{k-1}(r_2t_2))rt\tau\rangle$ when l is even, but opposite when l is odd. This difference, if ignored, can lead to phase errors in calculations. The reason for the phase difference is apparent from (10.2.1) and (10.3.1), which together give

$$(a\alpha, A_1a_1|a\alpha) = \{a\} \tag{21.2.12}$$

$$(A_1a_1, a\alpha|a\alpha) = 1 \tag{21.2.13}$$

and the fact that t_1 is doubled-valued and thus $\{t_1\} = -1$ only when l is odd. In SO_3 (or O_3) *both* v.c.c. above are unity, since the historical phase $H(j_1j_2j_3)$ enters into (10.2.1); thus closed-shell terms do not cause phase changes in spin–spin couplings in the $|\mathbb{S}h\mathfrak{M}\theta\rangle$ basis.

Phase problems in jj calculations involving off-diagonal matrix elements may be avoided most simply by converting each $p^lq^ms^k$-type configuration (where q^m is a closed shell) to the equivalent p^ls^k-type configuration multiplied by the appropriate phase from (21.2.12). The calculation may then proceed in the simpler p^ls^k-type basis. Examples of this procedure are given in Section 23.5.

21.3. Calculation of the Spin–Orbit Matrix for a^n: The Example of $|t_{2g}^4 A_1a_1\rangle$ in OsX_6^{2-}

The Matrix in the $\mathbb{S}\mathfrak{L}$ Basis

Calculation of spin–orbit matrix elements in the $\mathbb{S}\mathfrak{L}$ basis is described in detail in Chapters 19 and 20. In fact, the $|t_{2g}^4(\mathbb{S}S, h)A_1a_1\rangle$ spin–orbit matrix is calculated as an example in Section 19.5. Diagonalization of this matrix, which is given in (19.5.13), may be used to give a very rough approximation of the OsX_6^{2-} ground state since $\zeta_{Os^{4+}}$ is large (see Section 23.4 for further discussion). The resulting eigenvalues and eigenvectors are

$$\begin{aligned}
E_1 &= -2\zeta_{Os^{4+}},\\
\psi_1 &= \sqrt{\tfrac{2}{3}}\,|t_2^4(1T_1, T_1)A_1a_1\rangle + \frac{1}{\sqrt{3}}\,|t_2^4(0A_1, A_1)A_1a_1\rangle\\
E_2 &= \zeta_{Os^{4+}},\\
\psi_2 &= \frac{1}{\sqrt{3}}\,|t_2^4(1T_1, T_1)A_1a_1\rangle - \sqrt{\tfrac{2}{3}}\,|t_2^4(0A_1, A_1)A_1a_1\rangle
\end{aligned} \tag{21.3.1}$$

The Matrix in the jj Basis

The $|A_1a_1\rangle$ states in the $|\mathcal{Q}(p^l(r_1t_1), q^m(r_2t_2))rt\tau\rangle$ basis are $|(u')^4 A_1a_1\rangle$ and $|\mathcal{Q}((u')^2(A_1),(e'')^2(A_1))A_1a_1\rangle$. The spin–orbit matrix is diagonal in this basis, since $\mathcal{H}_{SO}$ is a one-electron operator transforming as A_1 and thus $\langle p\|\ell_{SO}\|q\rangle = \delta_{pq}\langle p\|\ell_{SO}\|p\rangle$. Moreover, the configurations differ by a two-electron jump, so no one-electron operator connects their states. Since the basis is diagonal, the *jj* states above have the same spin–orbit energies as ψ_1 and ψ_2 in (21.3.1). We demonstrate this by direct calculation.

The spin–orbit matrix element $\langle (u')^4 A_1a_1|\mathcal{H}_{SO}|(u')^4 A_1a_1\rangle$ is of the one-open-shell variety. Thus its form is very similar to that of the $\langle a^n \mathcal{S}h\mathfrak{M}\theta|O^f_\phi|a^n \mathcal{S}'h'\mathfrak{M}'\theta'\rangle$ matrix elements we calculated earlier in Section 19.4. $\mathcal{H}_{SO} = \Sigma^n_{k=1}\ell_{SO}(k)$ is a sum of one-electron operators; each $\ell_{SO}(k)$ operates on the $|p\pi\rangle$ function for the kth electron. This is analogous to $O^f_\phi = \Sigma^n_{k=1}o(k)^f_\phi$, since $o(k)^f_\phi$ operates on the $|a\alpha\rangle$ function for the kth electron. Thus using (19.2.2), (21.2.9), (10.2.2), (10.3.1), and (18.1.7), we obtain

$$\langle p^l rt\tau|(\mathcal{H}_{SO})^{A_1}_{a_1}|p^l r't'\tau'\rangle = \delta_{rr'}\delta_{tt'}\delta_{\tau\tau'}\left[l|p|^{-1/2}\langle p\|(\ell_{SO})^{A_1}\|p\rangle\right] \tag{21.3.2}$$

For our specific example, we find

$$\left\langle (u')^4A_1a_1\middle|\mathcal{H}_{SO}\middle|(u')^4A_1a_1\right\rangle = 2\langle u'\|\ell_{SO}\|u'\rangle \tag{21.3.3}$$

Similarly the form of $\langle\mathcal{Q}((u')^2(A_1),(e'')^2(A_1))A_1a_1|\mathcal{H}_{SO}|\mathcal{Q}((u')^2(A_1),$ $(e'')^2(A_1))A_1a_1\rangle$ is analogous to that of our $\langle\mathcal{Q}(a^m(\mathcal{S}_1h_1),\ b^n(\mathcal{S}_2h_2))$ $\mathcal{S}h\mathfrak{M}\theta|O^f_\phi|\mathcal{Q}(a^m(\mathcal{S}'_1h'_1),\ b^n(\mathcal{S}'_2h'_2))\mathcal{S}'h'\mathfrak{M}'\theta'\rangle$ matrix elements. Both are of the two-open-shell, diagonal-in-configuration variety. Thus we use (21.2.7) and the same method that took us from (20.1.1) to (20.1.3), together with other arguments like those in Section 20.1, and (21.3.2) above, to obtain the general formula

$$\begin{aligned}
&\left\langle\mathcal{Q}\left(p^l(r_1t_1), q^m(r_2t_2)\right)rt\tau\middle|(\mathcal{H}_{SO})^{A_1}_{a_1}\middle|\mathcal{Q}\left(p^l(r'_1t'_1), q^m(r'_2t'_2)\right)r't'\tau'\right\rangle\\
&\quad=\delta(rt\text{ labels})\Big[|t_1|^{-1/2}\langle p^l r_1t_1\|(\mathcal{H}_{SO})^{A_1}\|p^l r'_1t'_1\rangle\\
&\qquad\qquad+|t_2|^{-1/2}\langle q^m r_2t_2\|(\mathcal{H}_{SO})^{A_1}\|q^m r'_2t'_2\rangle\Big]\\
&\quad=\delta(rt\text{ labels})\Big[l|p|^{-1/2}\langle p\|(\ell_{SO})^{A_1}\|p\rangle\\
&\qquad\qquad+m|q|^{-1/2}\langle q\|(\ell_{SO})^{A_1}\|q\rangle\Big]
\end{aligned} \tag{21.3.4}$$

where

$$\delta(rt \text{ labels}) = \delta_{rr'}\delta_{tt'}\delta_{\tau\tau'}\delta_{r_1r_1'}\delta_{t_1t_1'}\delta_{r_2r_2'}\delta_{t_2t_2'} \tag{21.3.5}$$

Therefore we find

$$\left\langle \mathcal{C}\left((u')^2(A_1),(e'')^2(A_1)\right)A_1a_1|\mathcal{H}_{\text{SO}}|\mathcal{C}\left((u')^2(A_1),(e'')^2(A_1)\right)A_1a_1\right\rangle$$

$$= \langle u'\|h_{\text{SO}}\|u'\rangle + \sqrt{2}\,\langle e''\|h_{\text{SO}}\|e''\rangle \tag{21.3.6}$$

Equations (21.3.2) and (21.3.4) give us the very useful result that the spin–orbit matrix elements of a $|rt\tau\rangle$ state in the jj basis is *independent of the symmetry labels* ($rt\tau$, $r_1t_1\tau_1$, etc.) of the state. Thus within each strong-field configuration all the possible $|rt\tau\rangle$ (or $|\mathcal{C}((r_1t_1),(r_2t_2))rt\tau\rangle$) states of a given p^l (or p^lq^m) configuration have the same first-order spin–orbit energy.

It remains to evaluate the one-electron matrix elements $\langle p\|h_{\text{SO}}\|p\rangle$ above for $p = u'$ and e''. If we recall that $|p\pi\rangle = |(\frac{1}{2}s, a)p\pi\rangle$ and write $\langle p\|h_{\text{SO}}\|p\rangle$ in the unreduced form, $\langle(\frac{1}{2}s, a)p\pi|h_{\text{SO}}|(\frac{1}{2}s', a')p\pi\rangle$, we see it has the same structure as the multielectron matrix element $\langle(\mathbb{S}S, h)rt\tau|\mathcal{H}_{\text{SO}}|(\mathbb{S}'S', h')r't'\tau'\rangle$. Thus we may use (18.4.6) with the obvious substitutions $\mathbb{S} = \frac{1}{2}$, $S = s = e'$, $h = a$, $t = p$, $\tau = \pi$, and so on, as well as Section B.5 and (19.5.4), to obtain the cubic-group formula

$$\langle p\|h_{\text{SO}}\|p\rangle = |p|^{1/2}\left\langle(\tfrac{1}{2}e', a)p\pi\middle|h_{\text{SO}}\middle|(\tfrac{1}{2}e', a')p\pi\right\rangle$$

$$= (-1)|p|^{1/2}\{E'ap\}\{aT_1a'\}\begin{Bmatrix} E' & a' & p \\ a & E' & T_1 \end{Bmatrix}\frac{\sqrt{6}}{2}\langle a\|\xi l^{t_1}\|a'\rangle \tag{21.3.7}$$

This equation together with (19.5.12) gives

$$\langle u_g'\|h_{\text{SO}}\|u_g'\rangle = (-1)(2)(1)(-1)\left(\frac{1}{6}\right)\left(\frac{\sqrt{6}}{2}\right)\left(-\sqrt{6}\,\zeta_{\text{Os}^{4+}}\right) = -\zeta_{\text{Os}^{4+}} \tag{21.3.8}$$

$$\langle e_g''\|h_{\text{SO}}\|e_g''\rangle = (-1)(\sqrt{2})(-1)(-1)\left(\frac{1}{3}\right)\left(\frac{\sqrt{6}}{2}\right)\left(-\sqrt{6}\,\zeta_{\text{Os}^{4+}}\right) = \sqrt{2}\,\zeta_{\text{Os}^{4+}}$$

Thus our results in the $|\mathcal{C}(p^l(r_1t_1), q^m(r_2t_2))rt\tau\rangle$ basis are

$$\begin{aligned} E_1 &= -2\zeta_{\text{Os}^{4+}}, \qquad \psi_1 = \left|(u')^4A_1a_1\right\rangle \\ E_2 &= \zeta_{\text{Os}^{4+}}, \qquad \psi_2 = \left|\mathcal{C}\left((u')^2(A_1),(e'')^2(A_1)\right)A_1a_1\right\rangle \end{aligned} \tag{21.3.9}$$

Comparison of the Methods

The eigenvalues above agree, as required, with those obtained using the $|a^n(\mathcal{S}S, h)rt\tau\rangle$ basis [Eq. (21.3.1)]. Once one becomes familiar with the methods, it is equally easy to calculate these spin–orbit matrix elements in either basis. The jj basis becomes extremely advantageous, however, when $\mathcal{H}_{SO}$ is the *only* energy matrix needed to a first approximation; within each strong-field configuration $\mathcal{H}_{SO}$ is already diagonal in this basis, and furthermore the spin–orbit energy is independent of the symmetry labels ($rt\tau$, etc.) of the states [see (21.3.2) and (21.3.4)]. On the other hand, when interelectronic repulsions must be included, as in a d–d-type calculation, the $|a^n(\mathcal{S}S, h)rt\tau\rangle$ basis is usually preferred since the electrostatic matrices have been tabulated in the $|a^n\,\mathcal{S}h\mathfrak{M}\theta\rangle$ basis in terms of the Racah parameters A, B, and C—see Section C.9.

Our illustration of the calculation of the $|t_{2g}^4 A_1 a_1\rangle$ spin–orbit matrix using the jj basis shows how analogous methods to those we have used previously to derive formulas in the $|a^n(\mathcal{S}S, h)rt\tau\rangle$ or $\mathcal{S}\mathcal{L}$ basis may be used to derive formulas in the jj basis. We now give several additional examples of the approach.

21.4 Calculations for the T_{1u} States of the OsX_6^{2-} $t_{2u}^5 t_{2g}^5$ Configuration in the jj Basis

From the MO diagram in Fig. 23.2.1 and the group-O direct-product table, it follows that the $t_{2u}^5 t_{2g}^5$ charge-transfer configuration of OsX_6^{2-} gives rise to five T_{1u} states. These are described in the jj basis as

$$\begin{aligned}
\psi_1 &= \left|\mathcal{A}\left\{\left[\left((u_u')^4, e_u''\right)E_u''\right], \left[\left((u_g')^4, e_g''\right)E_g''\right]\right\} T_{1u}\tau\right\rangle \\
\psi_2 &= \left|\mathcal{A}\left\{\left[\left((u_u')^3, (e_u'')^2\right)U_u'\right], \left[\left((u_g')^4, e_g''\right)E_g''\right]\right\} T_{1u}\tau\right\rangle \\
\psi_3 &= \left|\mathcal{A}\left\{\left[\left((u_u')^4, e_u''\right)E_u''\right], \left[\left((u_g')^3, (e_g'')^2\right)U_g'\right]\right\} T_{1u}\tau\right\rangle \qquad (21.4.1)\\
\psi_4 &= \left|\mathcal{A}\left\{\left[\left((u_u')^3, (e_u'')^2\right)U_u'\right], \left[\left((u_g')^3, (e_g'')^2\right)U_g'\right]\right\} 0T_{1u}\tau\right\rangle \\
\psi_5 &= \left|\mathcal{A}\left\{\left[\left((u_u')^3, (e_u'')^2\right)U_u'\right], \left[\left((u_g')^3, (e_g'')^2\right)U_g'\right]\right\} 1T_{1u}\tau\right\rangle
\end{aligned}$$

The notation has been abbreviated somewhat here since each open jj shell gives rise to a state with the same symmetry as that shell.

Determination of the Spin–Orbit Energy of the Lowest T_{1u} State

Since the states above are already diagonal in $\mathcal{H}_{SO}$ we need only calculate their diagonal matrix elements to determine the spin–orbit energy of the lowest of the five possible T_{1u} states. We follow the method described in Section 20.3 for the calculation of $(a^m b^n)(c^l d^k)$ matrix elements. Thus our states have been written in the form $|\mathcal{A}\{[(p^l(r_1t_1), q^m(r_2t_2))r_{12}t_{12}], [(s^k(r_3t_3), u^n(r_4t_4))r_{34}t_{34}]\}rT_{1u}\tau\rangle$ (where we have $p = u'_u$, $q = e''_u$, $s = u'_g$, and $u = e''_g$). We use the general method of Section 20.3 together with (21.3.4) above to obtain the general formula

$$\begin{aligned}
&\big\langle \mathcal{A}\{[(p^l(r_1t_1), q^m(r_2t_2))r_{12}t_{12}], [(s^k(r_3t_3), u^n(r_4t_4))r_{34}t_{34}]\}rT_{1u}\tau\big| \\
&\quad \times \mathcal{H}_{SO}\big|\mathcal{A}\{[(p^l(r'_1t'_1), q^m(r'_2t'_2))r'_{12}t'_{12}], \\
&\quad [(s^k(r'_3t'_3), u^n(r'_4t'_4))r'_{34}t'_{34}]\}r'T_{1u}\tau'\big\rangle \\
&\quad = \delta_{\tau\tau'}\,\delta(\text{all } r\text{'s and } t\text{'s}) \\
&\quad \times \big[l|p|^{-1/2}\langle p\|h_{SO}\|p\rangle + m|q|^{-1/2}\langle q\|h_{SO}\|q\rangle \\
&\quad + k|s|^{-1/2}\langle s\|h_{SO}\|s\rangle + n|u|^{-1/2}\langle u\|h_{SO}\|u\rangle\big]
\end{aligned} \tag{21.4.2}$$

To complete the calculation we need only the one-electron h_{SO} matrix elements. From (21.3.8) we have

$$\begin{aligned}
\langle u'_g\|h_{SO}\|u'_g\rangle &= -\zeta_{Os^{4+}} \\
\langle e''_g\|h_{SO}\|e''_g\rangle &= \sqrt{2}\,\zeta_{Os^{4+}}
\end{aligned} \tag{21.4.3}$$

The remaining two h_{SO} matrix elements are calculated by first using (21.3.7) and then Table 19.6.1 to obtain $\langle t_{2u}\|\xi l^{t_1}\|t_{2u}\rangle$:

$$\begin{aligned}
\langle e''_u\|h_{SO}\|e''_u\rangle &= \left(\frac{-\sqrt{2}}{3}\right)\left(\frac{\sqrt{6}}{2}\right)\left(\frac{-\sqrt{6}}{2}\zeta_{X^-}\right) = \frac{1}{\sqrt{2}}\zeta_{X^-} \\
\langle u'_u\|h_{SO}\|u'_u\rangle &= \left(\frac{1}{3}\right)\left(\frac{\sqrt{6}}{2}\right)\left(\frac{-\sqrt{6}}{2}\zeta_{X^-}\right) = -\tfrac{1}{2}\zeta_{X^-}
\end{aligned} \tag{21.4.4}$$

The spin–orbit energies of our $(u'_u)^l(e''_u)^m(u'_g)^k(e''_g)^n$ states are therefore

$$E_1 = -\tfrac{1}{2}\zeta_{X^-} - \zeta_{Os^{4+}}$$

$$E_2 = \tfrac{1}{4}\zeta_{X^-} - \zeta_{Os^{4+}}$$

$$E_3 = -\tfrac{1}{2}\zeta_{X^-} + \tfrac{1}{2}\zeta_{Os^{4+}}$$

$$E_4 = E_5 = \tfrac{1}{4}\zeta_{X^-} + \tfrac{1}{2}\zeta_{Os^{4+}} \qquad (21.4.5)$$

and so the $\psi_1(T_{1u})$ state lies lowest.

The $(u'_u)^4 e''_u (u'_g)^4 e''_g$ configuration gives rise to an A_{1u} state in addition to the T_{1u} state. Equation (21.4.2) tells us that it is degenerate in spin–orbit energy with the T_{1u} state, and, as pointed out above, *all states for a given jj configuration from the same strong field configuration have the same spin–orbit energy*. Calculation of their energies in the *jj* basis is thus extremely simple compared to the same calculation in the $\mathfrak{SL}$ basis; the latter requires calculation and diagonalization of a 5×5 matrix to obtain our result (21.4.5).

The Magnetic Moment of the Lowest T_{1u} State

Calculation of the magnetic moment of the $|\mathfrak{A}\{[((u'_u)^4, e''_u)E''_u], [((u'_g)^4, e''_g)E''_g]\}T_{1u}\tau\rangle$ state is also straightforward. Closed shells [here $(u'_u)^4$ and $(u'_g)^4$] do not affect the values of diagonal matrix elements of non-totally-symmetric operators and may be neglected. We have two open shells remaining. Thus we expand the functions using (21.2.7) and then use an analogous procedure to that leading from (20.1.1) to (20.1.6) to derive the equation

$$\left\langle \mathfrak{A}\left(p^l(r_1t_1), q^m(r_2t_2)\right)rt\|O^f\|\mathfrak{A}\left(p^l(r'_1t'_1), q^m(r'_2t'_2)\right)r't'\right\rangle_k$$

$$= \delta_{r_2r'_2}\delta_{t_2t'_2}|t|^{1/2}|t'|^{1/2}\{t_1t_2t^*r\}\{t'_1\}$$

$$\times \sum_{k_1}\{t_1^*ft'_1k_1\}\begin{Bmatrix} f & t' & t^* \\ t_2^* & t_1 & t'_1 \end{Bmatrix}_{k_1r'rk}\langle p^l\, r_1t_1\|O^f\|p^l\, r'_1t'_1\rangle_{k_1}$$

$$+ \delta_{r_1r'_1}\delta_{t_1t'_1}|t|^{1/2}|t'|^{1/2}\{t'_1t'_2t'^*r'\}\{t'_2\}$$

$$\times \sum_{k_2}\{t_2^*ft'_2k_2\}\begin{Bmatrix} f & t' & t^* \\ t_1^* & t_2 & t'_2 \end{Bmatrix}_{k_2r'rk}\langle q^m\, r_2t_2\|O^f\|q^m\, r'_2t'_2\rangle_{k_2}$$

(21.4.6)

In our example $k = 0$, $O^f = \mu^{t_{1g}}$, and $l = m = 1$. Thus we obtain

$$\left\langle \mathcal{Q}\left(e_u''(E_u''), e_g''(E_g'')\right)T_{1u} \| \mu^{t_{1g}} \| \mathcal{Q}\left(e_u''(E_u''), e_g''(E_g'')\right)T_{1u}\right\rangle$$

$$= \langle e_u'' \, E_u'' \| \mu^{t_{1g}} \| e_u'' \, E_u'' \rangle + \langle e_g'' \, E_g'' \| \mu^{t_{1g}} \| e_g'' \, E_g'' \rangle$$

$$= \langle e_u'' \| \mu^{t_{1g}} \| e_u'' \rangle + \langle e_g'' \| \mu^{t_{1g}} \| e_g'' \rangle \qquad (21.4.7)$$

In the example above $p^l = e_u''$ and $q^m = e_g''$, so both l and m are one; thus the integrals in (21.4.6) are already one-electron integrals. For the general case, an equation analogous to (19.4.6) is required:

$$\langle p^l rt \| O^f \| p^l r't' \rangle_k$$

$$= l \sum_{r_1 t_1} \left(p^l rt | p^{l-1} r_1 t_1, p \right)\left(p^{l-1} r_1 t_1, p | p^l r't' \right)$$

$$\times |t|^{1/2} |t'|^{1/2} \{ t_1 p t'^* r' \} \{ p \}$$

$$\times \sum_{k_1} \{ p^* f p k_1 \} \begin{Bmatrix} f & t' & t^* \\ t_1^* & p & p \end{Bmatrix}_{k_1 r' r k} \langle p \| o^f \| p \rangle_{k_1}$$

$$(21.4.8)$$

With our hole–particle convention

$$\left\langle (u')^3 U' \| \mu^{t_1} \| (u')^3 U' \right\rangle_k = -\eta \langle u' \| \mu^{t_1} \| u' \rangle_k$$

$$= \langle u' \| \mu^{t_1} \| u' \rangle_k \qquad (21.4.9)$$

since $\eta = -1$ for μ (Section 14.10).

The one-electron integrals above are of the type

$$\langle p \| \mu^f \| q \rangle_k = \left\langle \left(\tfrac{1}{2}s, a\right) p \| (-\mu_B)(l^f + 2s^f) \| \left(\tfrac{1}{2}s, b\right) q \right\rangle_k \qquad (21.4.10)$$

and may be evaluated using (18.1.8) for the l^f part and (18.1.9) for the s^f part. Thus in our example these equations, together with Table 19.6.1 and

(15.4.9), give

$$\langle e_u''\|\mu^{t_{1g}}\|e_u''\rangle = \left\langle \left(\tfrac{1}{2}e_g', t_{2u}\right)e_u''\right\|(-\mu_B)(l^{t_{1g}} + 2s^{t_{1g}})\left\|\left(\tfrac{1}{2}e_g', t_{2u}\right)e_u''\right\rangle$$

$$= (-\mu_B)\left[\left(\frac{-2}{3}\right)\left(\frac{-\sqrt{6}}{2}\right) + 2\left(\frac{1}{3}\right)\left(\frac{\sqrt{6}}{2}\right)\right] = -\mu_B\frac{2\sqrt{6}}{3}$$

$$\langle e_g''\|\mu^{t_{1g}}\|e_g''\rangle = \left\langle \left(\tfrac{1}{2}e_g', t_{2g}\right)e_g''\right\|(-\mu_B)(l^{t_{1g}} + 2s^{t_{1g}})\left\|\left(\tfrac{1}{2}e_g', t_{2g}\right)e_g''\right\rangle$$

$$= (-\mu_B)\left[\left(\frac{-2}{3}\right)(-\sqrt{6}) + 2\left(\frac{1}{3}\right)\left(\frac{\sqrt{6}}{2}\right)\right] = -\mu_B\sqrt{6}$$

(21.4.11)

Substituting these results into (21.4.7), we obtain the magnetic moment of the lowest-energy T_{1u} state in reduced form:

$$\langle T_{1u}\|\mu^{t_{1g}}\|T_{1u}\rangle = -\mu_B\frac{5\sqrt{6}}{3} \tag{21.4.12}$$

Calculation of $\mathcal{A}_1/\mathcal{D}_0$ and $\mathcal{D}_0$ for the Transition to the Lowest T_{1u} State

Our final example is the calculation of $\mathcal{A}_1/\mathcal{D}_0$ and $\mathcal{D}_0$ for the $A_{1g} \to T_{1u}$ transition in OsX_6^{2-} in the *jj* approximation. The system is isotropic, so the equations of Section 17.2 apply. The T_{1u} state of interest is the lowest-lying T_{1u} state of the $t_{2u}^5 t_{2g}^5$ charge-transfer configuration which is well described by ψ_1 of (21.4.1). (See Section 23.4 for further discussion.) Since the ground state is A_{1g} and therefore has no magnetic moment, no information about it other than its symmetry is required to calculate $\mathcal{A}_1/\mathcal{D}_0$. To find $\mathcal{A}_1/\mathcal{D}_0$ we use (17.3.1) and then substitute $\langle J\|\mu^{t_{1g}}\|J\rangle$ using (21.4.12) above:

$$\frac{\mathcal{A}_1}{\mathcal{D}_0} = \frac{-\sqrt{6}}{\mu_B}\{A_1T_1T_1\}\begin{Bmatrix} T_1 & T_1 & T_1 \\ A_1 & T_1 & T_1 \end{Bmatrix}\langle T_{1u}\|\mu^{t_{1g}}\|T_{1u}\rangle$$

$$= \frac{-\sqrt{6}}{\mu_B}(1)\left(-\frac{1}{3}\right)\left(-\mu_B\frac{5\sqrt{6}}{3}\right) = -\frac{10}{3} \tag{21.4.13}$$

Note again that this result is independent of a detailed description of the A_{1g} ground state.

Calculation of $\mathfrak{D}_0$ requires us to specify further the A_{1g} ground state. It is *not* described particularly well by a single jj configuration (Section 23.4). Crystal-field calculations indicate that it is $\approx 80\%$ $|(u'_g)^4 A_{1g}a_{1g}\rangle$, but for the purposes of this example, we assume a $|(u'_g)^4 A_{1g}a_{1g}\rangle$ ground state. Thus we calculate $\mathfrak{D}_0$ for the transition from the A_{1g} state formed from the $(u'_u)^4(e''_u)^2(u'_g)^4$ configuration to the T_{1u} state from the $(u'_u)^4(e''_u)^1(u'_g)^4(e''_g)^1$ configuration. Shells which are closed in *both* configurations do not contribute to the magnitude of off-diagonal matrix elements of non-totally-symmetric operators, but may affect their phase (Section 21.2). However, since $\mathfrak{D}_0$ requires only the absolute valued squared of the electric-dipole matrix element $\langle A_{1g}\|m^{t_{1u}}\|T_{1u}\rangle$, in this case the phase will square to unity. Thus we may use the simpler $(e''_u)^2$ and $(e''_u)^1(e''_g)^1$ configurations in (17.2.19) to give

$$\begin{aligned}\mathfrak{D}_0 &= \tfrac{1}{3}\left|\langle A_{1g}\|m^{t_{1u}}\|T_{1u}\rangle\right|^2\\ &= \tfrac{1}{3}\left|\left\langle (e''_u)^2 A_{1g}\middle\|m^{t_{1u}}\middle\|\mathfrak{A}\left(e''_u(E''_u), e''_g(E''_g)\right)T_{1u}\right\rangle\right|^2 \qquad (21.4.14)\end{aligned}$$

General formulas for off-diagonal matrix elements in the jj basis are again easily derived by a process formally very similar to that used in the $\mathbb{SL}$ basis. Thus, analogous to (19.8.3), we obtain

$$\begin{aligned}&\left\langle p^l rt\|O^f\|\mathfrak{A}\left(p^{l-1}(r'_1t'_1), q\right)r't'\right\rangle_k\\ &= \sqrt{l}\left(p^l rt|p^{l-1}r'_1t'_1, p\right)|t|^{1/2}|t'|^{1/2}\{t'_1qt'^*r'\}\\ &\times \{q\}\sum_{k_2}\{p^*fqk_2\}\begin{Bmatrix} f & t' & t^*\\ t'^*_1 & p & q\end{Bmatrix}_{k_2r'rk}\langle p\|o^f\|q\rangle_{k_2} \qquad (21.4.15)\end{aligned}$$

In this case we have

$$\begin{aligned}&\left\langle (e''_u)^2 A_{1g}\|m^{t_{1u}}\|\mathfrak{A}\left(e''_u(E''_u), e''_g\right)T_{1u}\right\rangle\\ &= \sqrt{2}\,(1)(1)\sqrt{3}\,(1)(-1)(1)\left(\frac{1}{\sqrt{6}}\right)\langle e''_u\|m^{t_{1u}}\|e''_g\rangle\\ &= -\langle e''_u\|m^{t_{1u}}\|e''_g\rangle \qquad (21.4.16)\end{aligned}$$

The one-electron matrix element may be further simplified with (18.1.8):

$$\langle e_u''\|m^{t_{1u}}\|e_g''\rangle = \left\langle \left(\tfrac{1}{2}e_g', t_{2u}\right)e_u''\middle\|m^{t_{1u}}\middle\|\left(\tfrac{1}{2}e_g', t_{2g}\right)e_g''\right\rangle$$

$$= -\tfrac{2}{3}\langle t_{2u}\|m^{t_{1u}}\|t_{2g}\rangle \tag{21.4.17}$$

Combining the results of (21.4.14), (21.4.16), and (21.4.17), we find

$$\mathfrak{D}_0 = \left(\tfrac{1}{3}\right)\left(\tfrac{4}{9}\right)\left|\langle t_{2u}\|m^{t_{1u}}\|t_{2g}\rangle\right|^2$$

$$= \tfrac{4}{27}\left|\langle t_{2u}\|m^{t_{1u}}\|t_{2g}\rangle\right|^2 \tag{21.4.18}$$

The equation analogous to (21.4.15) for the $p^l q^{m-1}$, $p^{l-1}q^m$ case is similar to (20.2.10):

$$\left\langle \mathcal{A}\left(p^l(r_1t_1), q^{m-1}(r_2t_2)\right)rt\|O^f\|\mathcal{A}\left(p^{l-1}(r_1't_1'), q^m(r_2't_2')\right)r't'\right\rangle_k$$

$$= \left(p^l r_1 t_1|p^{l-1} r_1't_1', p\right)\left(q, q^{m-1} r_2 t_2|q^m r_2't_2'\right)$$

$$\times [lm|t_1|\,|t_2'|\,|t|\,|t'|]^{1/2}\{t_1\}\{t'\}\{t^*ft'k\}$$

$$\times \sum_{k_2}\{p^*fqk_2\}\begin{Bmatrix} t_1 & t_2 & t^* \\ t_1'^* & t_2'^* & t' \\ p^* & q & f \end{Bmatrix}\begin{matrix} r \\ r' \\ k_2 \end{matrix}\langle p\|o^f\|q\rangle_{k_2}$$
$$\begin{matrix} r_1 & r_2 & k \end{matrix} \tag{21.4.19}$$

When $l = |p|$ and $m = |q|$, it follows that $t_1 = A_1$, $t_2' = A_1$, $t = q$, and $t' = p$. Thus we may simplify (21.4.19) with (16.4.8) and (16.1.11) to obtain

$$\left\langle \mathcal{A}\left(p^{|p|}(A_1), q^{|q|-1}(q)\right)q\|O^f\|\mathcal{A}\left(p^{|p|-1}(p), q^{|q|}(A_1)\right)p\right\rangle_k$$

$$= \{q\}\{p^*fqk\}\langle p\|o^f\|q\rangle_k \tag{21.4.20}$$

22 MCD *Equations Using 3jm and 6j Coefficients: The Herzberg–Teller Case*

22.1. Introduction

In this chapter we show how group theory may be used in absorption and MCD calculations for transitions which are forbidden in the Franck–Condon (FC) approximation. We continue our Chapter 3 discussion of these electronically forbidden vibration-induced transitions, using the methods of Chapter 17. No new theoretical ideas need be introduced for our analysis, but compared to the equations in Chapter 17 for the electronically allowed case, the equations here are more complicated and lead to less conclusive predictions. Frequently a specific intensity-borrowing scheme must be specified before even the sign of $\mathcal{A}_1$ or $\mathcal{C}_0$ may be predicted. Nevertheless, in some common cases, such as octahedral systems with nondegenerate ground states, group theory allows valuable information to be obtained from MCD with very little effort.

We make the BO approximation and use harmonic-oscillator vibrational functions. The notation follows that of Chapters 3 and 7. The equations we derive may be used to calculate absorption and MCD parameters for both vibrationally induced bands and lines. The equations which relate the parameters calculated in this chapter to the experimental dispersion are those of (7.10.14). These equations apply to bands, but those for lines are closely analogous, as discussed in Section 7.11.

The equations in this chapter for $\mathcal{A}_1^{\text{vib}}$, $\mathcal{B}_0^{\text{vib}}$, $\mathcal{C}_0^{\text{vib}}$, and $\mathcal{D}_0^{\text{vib}}$ apply only for ambivalent groups. We have, however, kept the asterisk labels in the equations so they may be easily extended to nonambivalent groups, much as the ordinary MCD equations were in Section 17.6.

Statement of the Problem

We are interested here in transitions which are either forbidden or extremely weak in the FC approximation—that is, in transitions which are essentially electronically forbidden at $Q = Q_0$ but become allowed as the nuclei move away from Q_0 during molecular vibrations. Thus we cannot use the equations of Chapter 17, since the $Q = Q_0$ electric-dipole matrix elements which govern the intensity of $\mathcal{A}_1$, $\mathcal{B}_0$, $\mathcal{C}_0$, and $\mathcal{D}_0$ are zero or very weak; now in (3.6.5)

$$\langle A\alpha|m_\gamma|J\gamma\rangle^0 \approx 0 \tag{22.1.1}$$

but for at least one vibrational coordinate Q_η

$$\langle A\alpha|m_\gamma|J\gamma\rangle'_\eta \neq 0 \tag{22.1.2}$$

Electric-Dipole and Magnetic-Dipole Matrix Elements in the Herzberg–Teller Approximation

We use the Herzberg–Teller (HT) approximation as described in Section 3.7. Our vibronic basis states at $Q = Q_0$ are

$$|A\alpha\rangle^0|g\xi\rangle$$
$$|J\gamma\rangle^0|j\zeta\rangle \tag{22.1.3}$$

where $|A\alpha\rangle^0$ and $|J\gamma\rangle^0$ are electronic eigenfunctions and $|g\xi\rangle$ and $|j\zeta\rangle$ harmonic-oscillator vibrational functions as in Section 3.7. The first-order perturbed functions $|A\alpha\rangle$ and $|J\gamma\rangle$ are given in (3.7.3), and the electric-dipole matrix elements are those of (3.7.4) to (3.7.6).

To first order in Q_η, the magnetic-dipole matrix elements are the same as at $Q = Q_0$:

$$\begin{aligned}\langle A\alpha, g\xi|\mu_i|A\alpha', g\xi'\rangle &= \langle A\alpha|\mu_i|A\alpha'\rangle^0\delta_{\xi\xi'}\\ \langle J\gamma, j\zeta|\mu_i|J\gamma', j\zeta'\rangle &= \langle J\gamma|\mu_i|J\gamma'\rangle^0\delta_{\zeta\zeta'}\end{aligned} \tag{22.1.4}$$

22.2. The Dipole Strength of a Vibrationally Induced Transition

We derive equations for the oriented or isotropic case of Chapter 17. Later, however, we specify the changes to be made in the equations for the space-averaged case. Unlike the allowed case of Chapter 17, the vibrational-mode symmetry $h\theta$ plays a prominent role in the calculation. We assume (22.1.1), and also that intensity is induced by vibrations (the Q_η)

of symmetries $\eta = h\theta$. We define $\mathcal{D}_0^{\text{vib}}$ as

$$\mathcal{D}_0^{\text{vib}} = \frac{1}{|A|} \sum_{\alpha\gamma} \sum_{g\xi j\zeta} \frac{N_g}{N_G} |\langle A\alpha, g\xi | m_1 | J\gamma, j\zeta \rangle|^2 \tag{22.2.1}$$

This is essentially (17.2.17) plus vibrational functions and population factors. As in (7.3.15), $N_G \equiv \Sigma_g |g| N_g$. Next we make the harmonic approximation and use the HT approximation to expand the electric-dipole matrix elements as in Section 7.10:

$$\mathcal{D}_0^{\text{vib}} = \frac{1}{|A|} \sum_{\alpha\gamma} \sum_{h\theta} |\langle A\alpha | m_1 | J\gamma \rangle'_{h\theta}|^2 \frac{1}{N_G} \sum_{g\xi} N_g \sum_{j\zeta} |\langle g\xi | Q^\dagger_{h\theta} | j\zeta \rangle|^2 \tag{22.2.2}$$

The primed matrix elements are expanded according to (3.7.6), and operators are expressed in terms of group basis functions as discussed in Sections 9.8 and 17.2. We again use (17.2.9) for m_1 and μ_0. For $U_{h\theta}$ we write

$$U_{h\theta} = \langle h\theta | U_{h\theta} \rangle U^h_\theta \tag{22.2.3}$$

Since the operator transformation coefficients are phase factors, they multiply to unity in our $\mathcal{D}_0^{\text{vib}}$ expression. Thus, letting

$$\mathcal{D}_0^{\text{vib}} \equiv \sum_h \mathcal{D}_0^h \tag{22.2.4}$$

we have

$$\begin{aligned}
\mathcal{D}_0^h &= \sum_\theta \mathcal{D}_0(h\theta) \\
&\equiv \frac{1}{|A|} \sum_\theta \sum_{\alpha\gamma} \Bigg[\Bigg(\sum_{N\epsilon \neq A\alpha} \frac{\langle A\alpha | U^h_\theta | N\epsilon \rangle^0}{W^0_A - W^0_N} \langle N\epsilon | m^f_{\phi_1} | J\gamma \rangle^0 \\
&\quad + \sum_{N\epsilon \neq J\gamma} \langle A\alpha | m^f_{\phi_1} | N\epsilon \rangle^0 \frac{\langle N\epsilon | U^h_\theta | J\gamma \rangle^0}{W^0_J - W^0_N} \Bigg) \\
&\quad \times \Bigg(\sum_{N'\epsilon' \neq A\alpha} \frac{\langle A\alpha | U^h_\theta | N'\epsilon' \rangle^{0*}}{W^0_A - W^0_{N'}} \langle N'\epsilon' | m^f_{\phi_1} | J\gamma \rangle^{0*} \\
&\quad + \sum_{N'\epsilon' \neq J\gamma} \langle A\alpha | m^f_{\phi_1} | N'\epsilon' \rangle^{0*} \frac{\langle N'\epsilon' | U^h_\theta | J\gamma \rangle^{0*}}{W^0_J - W^0_{N'}} \Bigg) \Bigg] \overline{Q^2_{h\theta}}
\end{aligned} \tag{22.2.5}$$

where

$$\overline{Q_h^2} \equiv \overline{Q_{h\theta}^2} = \frac{1}{N_G} \sum_{g\xi} N_g \sum_{j\zeta} \left| \langle g\xi | (Q_\theta^h)^\dagger | j\zeta \rangle \right|^2 \qquad (22.2.6)$$

$\mathfrak{D}_0(h\theta)$ and $\overline{Q_{h\theta}^2}$ are equivalent to (7.10.7) and (7.10.10) respectively. The notation $\overline{Q_h^2}$ is introduced because (22.2.6) is easily shown to be independent of θ.

The electronic part of (22.2.5) contains three types of contributions. These are distinguished by whether the ground state A, the excited state J, or both A and J (as in the two cross-terms) are mixed with the N, N' set. We thus write

$$\begin{aligned} \mathfrak{D}_0^{\text{vib}} &= \sum_h \mathfrak{D}_0^h \\ &= \sum_h \left[\mathfrak{D}(J/N)_h + \mathfrak{D}(A/N)_h + \mathfrak{D}(C-T)_h \right] \end{aligned} \qquad (22.2.7)$$

$\mathfrak{D}(J/N)_h$ and $\mathfrak{D}(A/N)_h$ are easily simplified using (10.2.2) and (10.3.4). To obtain $\mathfrak{D}_0^{\text{vib}}$ in a form convenient for the MCD ratios, we next apply (18.3.12) and (18.3.14) to the electric dipole and U^h reduced matrix elements respectively. Our results are

$$\begin{aligned} \mathfrak{D}(J/N)_h = \frac{-1}{|A|} \sum_{N, N' \neq J} \Bigg[& \sum_{rs} \delta_{NN'} \delta(A^*fNr) \delta(N^*hJs) \\ & \times \frac{\{A^*fNr\}\{N^*hJs\}}{|f|\,|N|} \langle A \| m^f \| N \rangle_r^0 \frac{\langle N \| U^h \| J \rangle_s^0}{W_J^0 - W_N^0} \\ & \times \langle N' \| m^f \| A \rangle_r^0 \frac{\langle J \| U^{h^*} \| N' \rangle_s^0}{W_J^0 - W_{N'}^0} \Bigg] \overline{Q_h^2} \end{aligned} \qquad (22.2.8)$$

$$\begin{aligned} \mathfrak{D}(A/N)_h = \frac{-1}{|A|} \sum_{N, N' \neq A} \Bigg[& \sum_{rs} \delta_{NN'} \delta(A^*hNs) \delta(N^*fJr) \\ & \times \frac{\{A^*hNs\}\{N^*fJr\}}{|f|\,|N|} \frac{\langle A \| U^h \| N \rangle_s^0}{W_A^0 - W_N^0} \langle N \| m^f \| J \rangle_r^0 \\ & \times \frac{\langle N' \| U^{h^*} \| A \rangle_s^0}{W_A^0 - W_{N'}^0} \langle J \| m^f \| N' \rangle_r^0 \Bigg] \overline{Q_h^2} \end{aligned} \qquad (22.2.9)$$

In these equations, $\delta_{NN'}$ restricts the mixing states N and N' to the same symmetry, but N and N' need not be identical.

$\mathcal{D}(C - T)_h$ may be simplified with (16.2.1), but the resulting expression is still horrendous. We omit the $\mathcal{D}(C - T)_h$ result, since in practice (22.2.7) is nearly always used in an approximate form in which $\mathcal{D}(C - T)_h$ is presumed to be zero; the only exceptions are cases in which the entire sum in (22.2.7) cancels in an MCD ratio (Section 22.4).

Space-averaged equations are obtained from (22.2.8) and (22.2.9) above by first multiplying the equations by $\frac{1}{3}$ and then replacing f everywhere with f_j and summing over f_j [as in (17.4.14)].

Approximation of $\mathcal{D}_0^{\mathrm{vib}}$

An example where $\mathcal{D}_0^{\mathrm{vib}}$ may be simplified dramatically occurs when an excited state J lies very close in energy to another state K, and the transitions $A \to K$ are FC-allowed from the ground state. Then only the first term, $\Sigma_h \mathcal{D}(J/N)_h$, of (22.2.7) need be considered, and then only for $N = K$; ground-state mixing and cross-terms may be neglected, so symbolically,

$$\mathcal{D}_0^{\mathrm{vib}} \approx \sum_h \mathcal{D}(J/K)_h \tag{22.2.10}$$

Dobosh (1974) derives similar equations to those in this chapter for this special case for the octahedral group.

Less drastic but still useful simplification of (22.2.7) is possible when A and J are widely separated. Then intensity-providing mixing states may lie much closer in energy to J than to A, and ground-state mixing may be assumed negligible. Thus only the first term in (22.2.7) need be considered (but for all N). This model,

$$\mathcal{D}_0^{\mathrm{vib}} \approx \sum_h \mathcal{D}(J/N)_h \tag{22.2.11}$$

seems reasonable for parity-forbidden charge-transfer transitions in many systems. For example, see Section 23.8.

Finally, a particularly fortunate case is that of octahedral systems with nondegenerate ground states. In this case, all $\mathcal{D}_0^{\mathrm{vib}}$ reduced matrix elements cancel in $\mathcal{C}_1^{\mathrm{vib}}/\mathcal{D}_0^{\mathrm{vib}}$ ratios, so the complicated form of $\mathcal{D}_0^{\mathrm{vib}}$ is irrelevant—see Section 22.4.

22.3. $\mathcal{A}_1^{\text{vib}}$, $\mathcal{B}_0^{\text{vib}}$, and $\mathcal{C}_0^{\text{vib}}$

If the full $\mathcal{D}_0^{\text{vib}}$ equation seems intimidating, similar equations for $\mathcal{A}_1^{\text{vib}}$, $\mathcal{B}_0^{\text{vib}}$, and $\mathcal{C}_0^{\text{vib}}$ are even more so. In practice they are nearly always used in approximate form. We omit $\mathcal{B}_0^{\text{vib}}$ altogether. $\mathcal{A}_1^{\text{vib}}$ and $\mathcal{C}_0^{\text{vib}}$ are defined for the oriented or isotropic case as

$$\mathcal{A}_1^{\text{vib}} = \frac{-2}{\mu_B|A|} \sum_{\alpha\alpha'\gamma\gamma'} \sum_{g\xi\xi' j\zeta\zeta'} \frac{N_g}{N_G}$$

$$\times \left(\langle J\gamma,\, j\zeta|\mu_0|J\gamma',\, j\zeta'\rangle \delta_{\alpha\alpha'}\delta_{\xi\xi'} - \langle A\alpha,\, g\xi|\mu_0|A\alpha',\, g\xi'\rangle \delta_{\gamma\gamma'}\delta_{\zeta\zeta'} \right)$$

$$\times \langle A\alpha',\, g\xi'|m_1|J\gamma,\, j\zeta\rangle \langle J\gamma',\, j\zeta'|m_{-1}|A\alpha,\, g\xi\rangle \tag{22.3.1}$$

$$\mathcal{C}_0^{\text{vib}} = \frac{2}{\mu_B|A|} \sum_{\alpha\alpha'\gamma} \sum_{g\xi\xi' j\zeta} \frac{N_g}{N_G} \langle A\alpha,\, g\xi|\mu_0|A\alpha',\, g\xi'\rangle$$

$$\times \langle A\alpha',\, g\xi'|m_1|J\gamma,\, j\zeta\rangle \langle J\gamma,\, j\zeta|m_{-1}|A\alpha,\, g\xi\rangle \tag{22.3.2}$$

which except for the inclusion of the vibrational functions and population factors are the same as (17.2.2) and (17.2.6) respectively. Since

$$\mathcal{A}_1^{\text{vib}} = \mathcal{A}_1^{\text{vib}}(J) + \mathcal{A}_1^{\text{vib}}(A) \tag{22.3.3}$$

and

$$\mathcal{C}_0^{\text{vib}} = \mathcal{A}_1^{\text{vib}}(A) \tag{22.3.4}$$

it suffices to consider only $\mathcal{A}_1^{\text{vib}}$ in detail.

We first use the assumptions of Section 22.1, Section 17.2 (for operators), the HT approximation, and the harmonic approximation to obtain

$$\mathcal{A}_1^{\text{vib}} = \frac{-2}{\mu_B|A|} \sum_{\alpha\alpha'\gamma\gamma'} \left(\langle J\gamma|\mu_0|J\gamma'\rangle \delta_{\alpha\alpha'} - \langle A\alpha|\mu_0|A\alpha'\rangle \delta_{\gamma\gamma'} \right)$$

$$\times \sum_{h\theta} \langle A\alpha'|m_1|J\gamma\rangle'_{h\theta} \langle J\gamma'|m_{-1}|A\alpha\rangle'_{(h\theta)^\dagger} \overline{Q_h^2} \tag{22.3.5}$$

Electric-dipole matrix elements are expanded with (3.7.6). The resulting expression is simplified by the usual methods. We write our results as

$$\mathcal{A}_1^{\text{vib}} = \sum_h \left(\sum_\theta \mathcal{A}_1(h\theta) \right) \equiv \sum_h \mathcal{A}_1^h$$

$$\mathcal{A}_1^h = \mathcal{A}_1^h(J) + \mathcal{A}_1^h(A) \tag{22.3.6}$$

$$\mathcal{C}_0^{\text{vib}} = \sum_h \left(\sum_\theta \mathcal{C}_0(h\theta) \right) \equiv \sum_h \mathcal{C}_0^h = \sum_h \mathcal{A}_1^h(A)$$

where

$$\mathcal{A}_1^h(J) = \sum_h [\mathcal{A}(J, J/N)_h + \mathcal{A}(J, A/N)_h + \mathcal{A}(J, C-T)_h]$$

$$\mathcal{A}_1^h(A) = \sum_h [\mathcal{A}(A, J/N)_h + \mathcal{A}(A, A/N)_h + \mathcal{A}(A, C-T)_h] \tag{22.3.7}$$

$$= \mathcal{C}_0^h$$

The first two contributions to $\mathcal{A}_1^{\text{vib}}(J)$ are

$$\mathcal{A}(J, J/N)_h = (\text{MCD factor}) \frac{2}{\mu_B |A|} \sum_{qrr'ss'p} \sum_{N, N' \neq J} \{J\}\{J^* f' Jq\}$$

$$\times \begin{Bmatrix} f' & f & f \\ A^* & N'^* & N^* \end{Bmatrix}_{prr'0} \begin{Bmatrix} N^* & f' & N' \\ J & h^* & J \end{Bmatrix}_{sqs'p}$$

$$\times \langle J \| \mu^{f'} \| J \rangle_q^0 \langle A \| m^f \| N \rangle_r^0 \frac{\langle N \| U^h \| J \rangle_s^0}{W_J^0 - W_N^0} \langle N' \| m^f \| A \rangle_{r'}^0 \frac{\langle J \| U^{h^*} \| N' \rangle_{s'}^0}{W_J^0 - W_{N'}^0}$$

$$\times \overline{Q_h^2} \tag{22.3.8}$$

$$\mathcal{A}(J, A/N)_h = (\text{MCD factor}) \frac{2}{\mu_B |A|} \sum_{qrr's} \sum_{N, N' \neq A} \delta_{NN'} \delta(A^* h N s)$$

$$\times \frac{\{A^* h N s\}}{|N|} \begin{Bmatrix} f' & f & f \\ N^* & J^* & J^* \end{Bmatrix}_{qrr'0} \langle J \| \mu^{f'} \| J \rangle_q^0 \frac{\langle A \| U^h \| N \rangle_s^0}{W_A^0 - W_N^0}$$

$$\times \langle N \| m^f \| J \rangle_r^0 \frac{\langle N' \| U^{h^*} \| A \rangle_s^0}{W_A^0 - W_{N'}^0} \langle J \| m^f \| N' \rangle_{r'} \overline{Q_h^2} \tag{22.3.9}$$

and the first two contributions to $\mathcal{A}_1^{\text{vib}}(A) = \mathcal{C}_0^{\text{vib}}$ are

$$\mathcal{A}(A, J/N)_h = (\text{MCD factor})\frac{2}{\mu_B|A|}\sum_{qrr's}\sum_{N,N'\neq J}\delta_{NN'}\delta(N^*hJs)$$

$$\times\frac{\{N^*hJs\}}{|N|}\begin{Bmatrix} f' & f & f \\ N^* & A^* & A^* \end{Bmatrix}_{qr'r0}\langle A\|\mu^{f'}\|A\rangle_q^0$$

$$\times\langle A\|m^f\|N\rangle_r^0\frac{\langle N\|U^h\|J\rangle_s^0}{W_J^0 - W_N^0}\langle N'\|m^f\|A\rangle_{r'}^0\frac{\langle J\|U^{h^*}\|N'\rangle_s^0}{W_J^0 - W_{N'}^0}\overline{Q_h^2} \tag{22.3.10}$$

$$\mathcal{A}(A, A/N)_h = (\text{MCD factor})\frac{2}{\mu_B|A|}\sum_{qrr'ss'p}\sum_{N,N'\neq A}\{J\}\{A^*f'Aq\}$$

$$\times\begin{Bmatrix} f' & f & f \\ J^* & N^* & N'^* \end{Bmatrix}_{pr'r0}\begin{Bmatrix} N'^* & f' & N \\ A & h & A \end{Bmatrix}_{s'qsp}$$

$$\times\langle A\|\mu^{f'}\|A\rangle_q\frac{\langle A\|U^h\|N\rangle_s}{W_A^0 - W_N^0}\langle N\|m^f\|J\rangle_r\frac{\langle N'\|U^{h^*}\|A\rangle_{s'}}{W_A^0 - W_{N'}^0}$$

$$\times\langle J\|m^f\|N'\rangle_{r'}\overline{Q_h^2} \tag{22.3.11}$$

The MCD factor is defined in (17.2.12). We omit expressions for $\mathcal{A}(J, C - T)_h$ and $\mathcal{A}(A, C - T)_h$, since in practice they are little used.

To obtain space-averaged equations from the above, simply make analogous changes to those which change $\mathcal{A}_1$ in (17.2.14) into $\overline{\mathcal{A}}_1$ in (17.4.11).

Note that unlike the $\mathcal{D}_0^{\text{vib}}$ equations, (22.3.8)–(22.3.11) contain double sums over repeated representation indices in the electric-dipole terms.

Approximation of $\mathcal{A}_1^{\text{vib}}$ and $\mathcal{C}_0^{\text{vib}}$

$\mathcal{A}_1^{\text{vib}}$ and $\mathcal{C}_0^{\text{vib}}$ simplifications are analogous to those described in Section 22.2 for $\mathcal{D}_0^{\text{vib}}$. For excited-state mixing with only *one* nearby state, $N = K$,

$$\begin{aligned}\mathcal{A}_1^{\text{vib}} &\approx \sum_h \mathcal{A}(J, J/K)_h + \sum_h \mathcal{A}(A, J/K)_h \\ \mathcal{C}_0^{\text{vib}} &\approx \sum_h \mathcal{A}(A, J/K)_h\end{aligned} \tag{22.3.12}$$

and for excited-state mixing only, but with a set of states,

$$\mathcal{A}_1^{\text{vib}} \approx \sum_h \mathcal{A}(J, J/N)_h + \sum_h \mathcal{A}(A, J/N)_h$$
$$\mathcal{C}_0^{\text{vib}} \approx \sum_h \mathcal{A}(A, J/N)_h \tag{22.3.13}$$

22.4. Special Cases Where Vibration-Induced MCD Ratios Simplify

Octahedral Systems with A_{1g} or A_{2g} Ground States

All purely electronic $g \to g$ or $u \to u$ transitions, such as $d \to d$ or $f \to f$ transitions, are FC-forbidden in groups containing the inversion operation. Such parity-forbidden transitions in octahedral systems with A_{1g} or A_{2g} ground states are the most favorable ones for the study of vibration-induced MCD. We give a preliminary discussion of the absorption bandshape expected for some of these transitions in Section 3.7. In such systems $\mathcal{C}_0^{\text{vib}}$ is zero because the ground state is nondegenerate. And, since by symmetry only T_{1g} and T_{2g} excited states have magnetic moments, $\mathcal{A}_1^{\text{vib}}$ is nonzero only for excitations to states with these symmetries. Parity selection rules require coupling with odd-parity vibrations; in octahedral centers these have T_{1u} $[\nu_3(t_{1u}), \nu_4(t_{1u})]$ or T_{2u} $[\nu_6(t_{2u})]$ symmetry. With the above symmetries, group theory allows mixing states to have T_{1u} or T_{2u} symmetries only.

Thus N, N', h, f', f, and J in the $\mathcal{A}_1^{\text{vib}}$ and $\mathcal{D}_0^{\text{vib}}$ expansions must *all* have T_{1u}, T_{1g}, T_{2u}, or T_{2g} symmetries. And, since the group is simply reducible, all sums over q, r, r', s, s', and p disappear, and (10.2.5) holds. These considerations enable amazing simplification of expressions for vibration-induced MCD. For $A \to J + h$ where $A = A_{1g}$ or A_{2g} and $J = T_{1g}$ or T_{2g} we find, using Sections C.11 and C.13, that

$$\frac{\mathcal{A}_1^h}{\mathcal{D}_0^h} = \frac{\mathcal{A}_1^h(J)}{\mathcal{D}_0^h} = -\frac{1}{\mu_B\sqrt{6}}(-1)^{A+h}\langle J\|\mu^{T_1}\|J\rangle \tag{22.4.1}$$

The expression holds for an $A \to J + h$ vibronic line or for a vibronic band which couples to vibrations h of *one symmetry* only. Here the coupling is with vibrations of *either* $h = T_{1u}$ $[\nu_3(t_{1u})$ and/or $\nu_4(t_{1u})]$ *or* $h = T_{2u}$ $[\nu_6(t_{2u})]$. For a given excited-state magnetic moment, the sign of $\mathcal{A}_1^{\text{vib}}/\mathcal{D}_0^{\text{vib}}$ for $A \to J + h$ depends on the symmetry h of the coupling vibration. Or, conversely, for a given symmetry vibration, the sign depends on the sign of the magnetic moment. This result is independent of any assumption regarding mixing states such as those of Sections 22.2 and 22.3; ground-state mixing, excited-state mixing, and cross-term mixing are all included. Analogous results hold in some other point groups.

Unfortunately, if vibrations of both $h = T_{1u}$ $[(-1)^h = -1]$ and $h = T_{2u}$ $[(-1)^h = 1]$ couple into the same band (or line), we cannot write such a simple expression, since cancellation of $\mathcal{D}_0^{\text{vib}}$ reduced matrix elements is not possible. An example of the use of (22.4.1) is given in Piepho et al. (1971) for the UCl_6^{2-} $f \to f$ vibronic lines in $Cs_2ZrCl_6 : U^{4+}$ [the paper, however, uses the old MCD conventions and Griffith (1964) phases].

Systems with Excited-State Mixing Only and Mixing States of Only One Symmetry

Here we assume excited-state mixing only so that

$$\begin{aligned}
\mathcal{A}_1^{\text{vib}} &= \sum_h \mathcal{A}(J, J/N)_h + \sum_h \mathcal{A}(A, J/N)_h \\
\mathcal{C}_0^{\text{vib}} &= \sum_h \mathcal{A}(A, J/N)_h \qquad (22.4.2)\\
\mathcal{D}_0^{\text{vib}} &= \sum_h \mathcal{D}(J/N)_h
\end{aligned}$$

We further assume that all mixing states have the same symmetry, N. Sometimes, as in the cases for which (22.4.1) applies, group theory allows mixing states of only one symmetry. But generally the restriction of N to one symmetry must be rationalized by non-group-theoretic arguments. For example, if such an assumption leads to consistently good fits for a series of bands, it might be justified.

With the above assumptions, MCD ratios achieve simple forms except in a few cases where repeated representation indices cause trouble. Thus, as long as *both* N and A are not simultaneously U' states,

$$\begin{aligned}
\frac{\mathcal{C}_0^{\text{vib}}}{\mathcal{D}_0^{\text{vib}}} &= \frac{\sum_h \mathcal{A}(A, J/N)_h}{\sum_h \mathcal{D}(J/N)_h} \\
&= -\,(\text{MCD factor})\frac{2|f|}{\mu_B}\{A^*fN\}\begin{Bmatrix} f' & f & f \\ N^* & A^* & A^* \end{Bmatrix}_{q000} \\
&\quad \times \langle A\|\mu^{f'}\|A\rangle_q \qquad (22.4.3)
\end{aligned}$$

Note that all dependence on the mixing vibration symmetry h factors out. $\mathcal{C}_0^{\text{vib}}/\mathcal{D}_0^{\text{vib}}$ is *identical* to $\mathcal{C}_0/\mathcal{D}_0$ for the allowed transition $A \to N$ calculated from (17.2.25) or (17.3.6) with N substituted for J. If the sum over h is limited to h of one symmetry only, and if, for N a U' state, neither A nor J is U', then the following equation applies:

$$\frac{\mathcal{A}_1^{\text{vib}}}{\mathcal{D}_0^{\text{vib}}} = \frac{\sum_h \mathcal{A}(J, J/N)_h}{\sum_h \mathcal{D}(J/N)_h} + \frac{\mathcal{C}_0^{\text{vib}}}{\mathcal{D}_0^{\text{vib}}}$$

$$= \Bigg[-\text{(MCD factor)}\frac{2|f|}{\mu_B}|N|\{J\}\{A^*fN\}\{N^*hJ\}\sum_q\{J^*f'Jq\}$$

$$\times \sum_p \begin{Bmatrix} f' & f & f \\ A^* & N^* & N^* \end{Bmatrix}_{p000} \begin{Bmatrix} N^* & f' & N \\ J & h^* & J \end{Bmatrix}_{0q0p} \langle J\|\mu^{f'}\|J\rangle_q \Bigg]$$

$$+ \left[\frac{\mathcal{C}_0^{\text{vib}}}{\mathcal{D}_0^{\text{vib}}} \text{ from (22.4.3)} \right] \qquad (22.4.4)$$

Remember, however, that *both* (22.4.3) and (22.4.4) assume excited-state mixing only, and that mixing is restricted to states of a single symmetry N. The equations are similar to those derived in Dobosh (1974) for the octahedral group for the same special case.

Examples of the uses of the equations above are given in Section 23.8. Similar equations may be derived for other special cases from our general formulas.

23 *Analysis of Real Systems*

In this chapter we work from the point of view of the experimental spectroscopist rather than the theorist. We have included in earlier chapters many examples which illustrate theoretical points, and we refer back to these frequently. But our approach is to move from the system of interest to the theory, rather than vice versa.

23.1. Introduction

Our intention is not to give complete analyses of all spectral details of the systems discussed. Rather we aim to illustrate how one approaches the problems presented by the data. Once one knows what to look for and how to do the necessary calculations, detailed analysis becomes a tractable (though sometimes very time-consuming) procedure. More detailed discussions are available in the referenced literature.

Basic Considerations: Obtaining the Data

Before measuring the MCD and absorption spectrum of a system one should learn as much as possible about its electronic structure, symmetry, crystal structure, and vibrational (infrared and Raman) spectrum. All of this information is helpful in deciding at which temperatures and magnetic field strengths the spectrum should be obtained. It is particularly useful to know the symmetry of the ground state. Some important considerations are (see also Chapter 5):

1. Systems with ground-state magnetic moments give rise to $\mathcal{C}$ terms which vary inversely with the absolute temperature and tend to dominate even at room temperature. To separate the $\mathcal{C}$ term quantitatively from the $\mathcal{A}$ and $\mathcal{B}$ terms, which are also necessarily present, the MCD should be measured at a series of low temperatures such as 5, 10, and 20 K. But care must be taken to avoid $\mathcal{C}$-term saturation effects (Section 6.2) in quantitative work, as in moment analyses, if large magnetic fields are employed. For example, for a simple

Kramers-doublet ground state with $g = 2$, Eq. (6.3.4) shows that for $H = 5$ tesla and $T = 4.2$ K, the observed MCD will be about 17% lower than the linear limit value. [Alternatively, a study of the MCD saturation can give a quick and easy estimate of the ground-state g value (Section 6.3).]

2. The peak-to-trough amplitude for an $\mathcal{A}$ term is inversely proportional to the second power of the bandwidth. Thus $\mathcal{A}$ terms may show up much more clearly at low temperatures when linewidths are narrower than at room temperature (Section 5.3).
3. When lines are *very* narrow, the Zeeman splitting at high fields may no longer be much smaller than the linewidth as required for conventional MCD in the linear limit (Sections 6.4, 6.5), and the theory of Chapters 4, 5, 7, 17, and 22 is no longer applicable. Such a situation can always be avoided by using a lower field.
4. Vibronically induced, electronically forbidden transitions gain intensity with temperature according to the hyperbolic-cotangent law (Section 3.7). To check for these transitions it is desirable to measure the data at, for example, 10, 77, 200, and 300 K.
5. Low-temperature measurements are always desirable because bands narrow. Resolution of overlapping transitions and/or vibronic structure then makes detailed spectral analysis possible. Hot bands (Section 3.6) also convey information.
6. Both the absorption and the MCD spectrum should be measured in all regions accessible with the available instrumentation, since weak transitions with narrow linewidths may show up in the MCD even when they are undetectable in absorption.
7. The zero-field MCD should be carefully measured to provide an accurate baseline.
8. Depolarization measurements (Section 4.2) must be made if MCD results for solids are desired, since depolarization effects can significantly decrease the magnitude of the MCD dispersion or indeed render it completely meaningless.

Data Analysis: The Initial Stages

Once the raw data have been processed to obtain $\Delta A'$ and A (or $\Delta\epsilon'$ and ϵ) vs. cm^{-1}, it is usually easy to determine:

1. If the ground state is degenerate (from $\mathcal{C}$ terms).
2. Whether bands are allowed or forbidden (from absorption intensity data if sample concentration can be estimated and from hot-bands).

3. The presence of overlapping bands (assuming they have different MCD characteristics).
4. Excited-state degeneracy information (from $\mathcal{A}$ terms).
5. Vibrational progressions.
6. The presence of vibronically induced transitions (from temperature dependence and vibrational structure).

Further analysis allows the assignment of excitations to particular spectral features and in some cases brings an understanding of vibronic interactions, such as JT or HT coupling.

In the next section we begin the analysis of particular systems, proceeding from the simple to the complex. Here we assume the reader is familiar with molecular-orbital (MO) and ligand-field theory.

23.2. A Favorite Example: $IrCl_6^{2-}$

Perhaps the system with the most thoroughly studied absorption and MCD spectrum is the $d^5(t_{2g}^5)$ octahedral complex $IrCl_6^{2-}$. This ion may be studied either in solution or doped into a crystalline host such as Cs_2ZrCl_6. The $IrCl_6^{2-}$ spectra together with those for $OsCl_6^{2-}$ and $IrBr_6^{2-}$ illustrate many points in the analysis of MCD and absorption data. We therefore consider these species in detail.

It is now widely accepted that the Ir^{4+} spectrum in the visible–uv region arises from ligand-to-metal charge-transfer transitions. Since these transitions fill the t_{2g}^5 hole, excited-state configurations have a single open shell. This enormously simplifies calculations. These points are illustrated in the MO diagram of Fig. 23.2.1 and the state diagram of Fig. 23.2.2.

Sketching the Molecular-Orbital Diagram and Relating It to the Spectrum

An initial step in analyzing MCD data is to draw a preliminary MO diagram. This diagram is then used to predict the low-energy excitations for the system.

In transition-metal complexes three types of excitations must be considered: $d \rightarrow d$, ligand-to-metal charge-transfer ($\gamma \rightarrow d$), and metal-to-ligand charge-transfer ($d \rightarrow \gamma$) [Jørgensen (1962)]. The last of these fall too high in energy to account for the $IrCl_6^{2-}$ spectrum in solution, shown in Fig. 23.2.3. Moreover, since the group O_h contains the inversion operation, all $d \rightarrow d$ transitions are parity-forbidden. Such transitions generally have ϵ_{max} values well under 1000. Thus an explanation of the $IrCl_6^{2-}$ spectrum was sought in

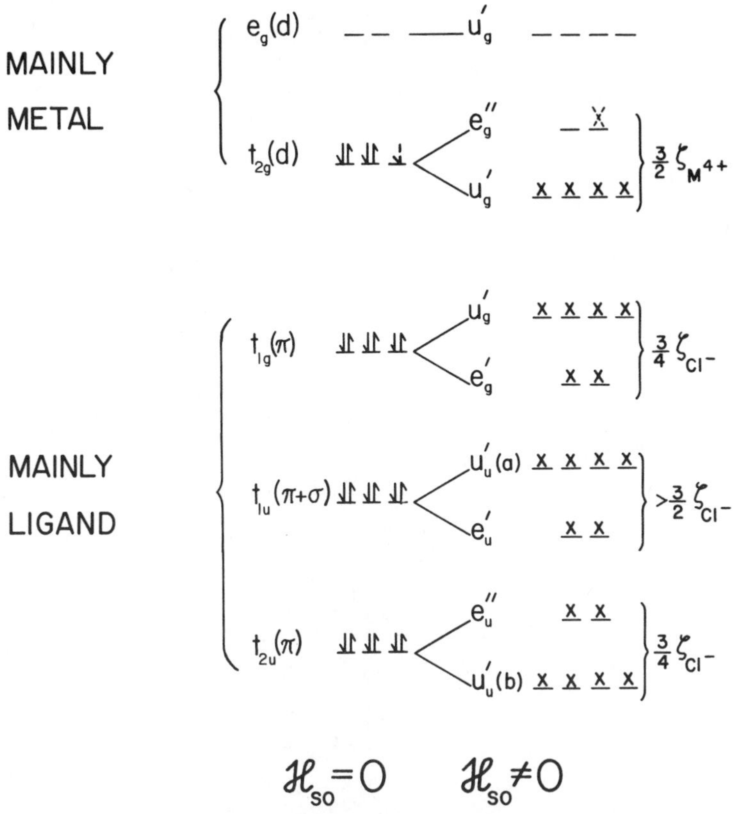

Figure 23.2.1. Schematic MO diagram in the absence and presence of first-order spin–orbit coupling in $IrCl_6^{2-}$ and $OsCl_6^{2-}$. The orbitals at the right are *jj* orbitals (Chapter 21). The $e_g''(t_{2g}(d))$ orbital is empty in the ground-state configuration for $M^{4+}= Os^{4+}$ and contains one electron when $M^{4+}= Ir^{4+}$. The relative orbital positions reflect the order of states assigned in the text.

terms of $\gamma \rightarrow d$ excitations. Again because of parity selection rules, the intense low-energy excitations of this type are $t_{1u}(\pi + \sigma) \rightarrow t_{2g}$, $t_{2u}(\pi) \rightarrow t_{2g}$, and $t_{1u}(\sigma + \pi) \rightarrow t_{2g}$.

MO calculations are not reliable in distinguishing the order of these excitations, and the establishment of the order illustrated in Fig. 23.2.1 was a major early achievement of MCD. It was first determined by analyzing the MCD of the $Fe(CN)_6^{3-}$ ion [Stephens (1965a), Schatz et al. (1966); see also Sections 13.2, 13.5, and 17.3] and was later shown by Henning et al. (1968) to hold for $IrCl_6^{2-}$ and related ions--see also Schatz (1982).

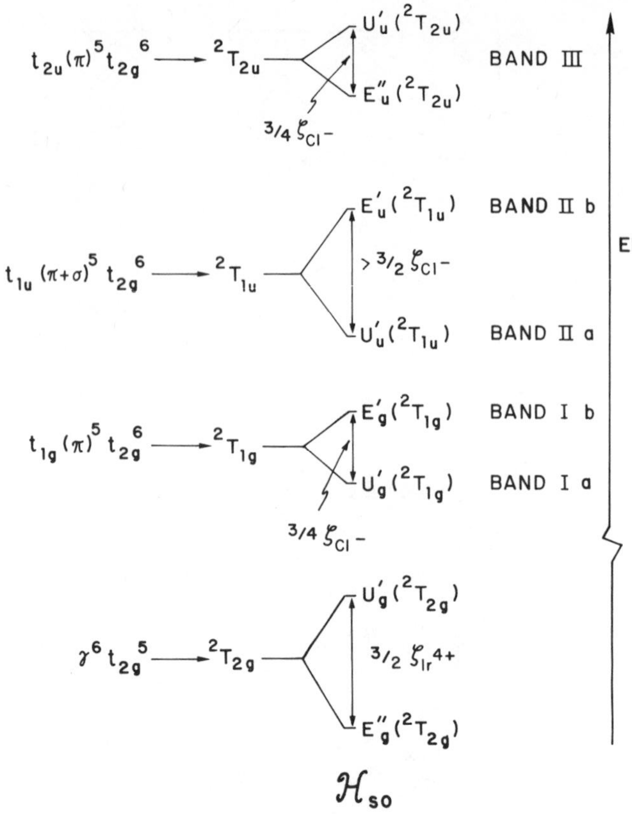

Figure 23.2.2. Energy-level diagram for the ground state and the $t_{1g}(\pi) \to t_{2g}$, $t_{1u}(\pi + \sigma) \to t_{2g}$, and $t_{2u}(\pi) \to t_{2g}$ charge-transfer states in $IrCl_6^{2-}$. The energy-level spacings for the charge-transfer states are roughly to scale, and the order corresponds to our assignments. The ${}^2T_{2g}$ splitting is larger than shown.

Arriving at an Electronic State Diagram

Once an MO diagram is proposed for a system, the configurations arising from low-lying excitations are easily determined. The next step is to establish the relative energies of the states within each configuration and to estimate the interactions between configurations. The end result is an electronic-state energy diagram such as Fig. 23.2.2, which may be related directly to the spectrum. $IrCl_6^{2-}$ presents a particularly simple example, since each configuration in Fig. 23.2.2 gives rise to a single state in the absence of spin–orbit interactions; interelectronic repulsions produce no additional splittings.

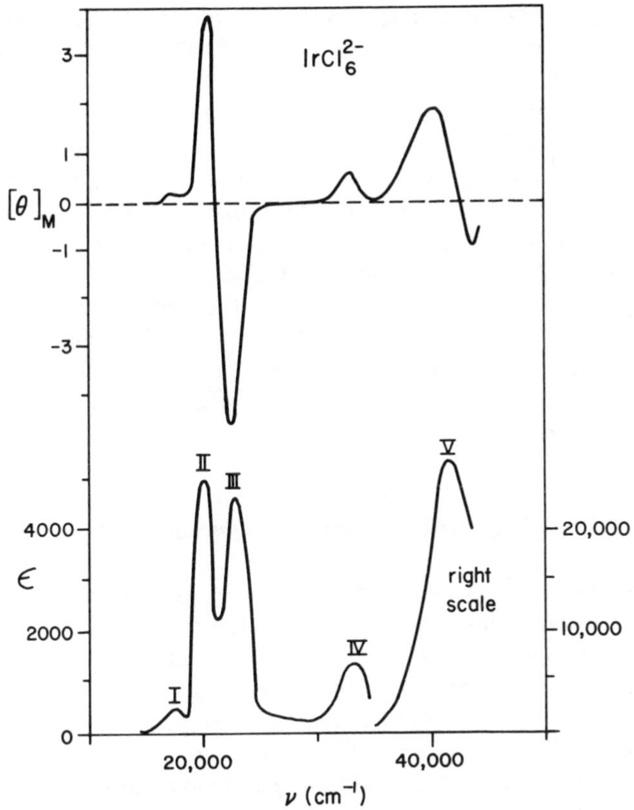

Figure 23.2.3. The absorption spectrum and MCD of $IrCl_6^{2-}$ in dichloroethane solution, from Henning et al. (1968). $[\theta]_M$ is the molar ellipticity (see section A.5). ϵ is the molar extinction coefficient.

First-order spin–orbit coupling splits the states of each configuration (Fig. 23.2.2), and second-order spin–orbit interactions mix states of the same symmetry belonging to different configurations. Spin–orbit matrix elements are calculated as illustrated in Sections 18.4, 19.5 and Chapter 20.

Relation of the State Diagram to the Absorption Spectrum: $IrCl_6^{2-}$ in Solution

Let us now relate Figs. 23.2.1 and 23.2.2 to the $IrCl_6^{2-}$ data of Fig. 23.2.3. Since the spin–orbit splittings of the $IrCl_6^{2-}$ excited states of Fig. 23.2.2 are less than the solution bandwidths (≈ 2000 cm^{-1}), we may, for the time being, disregard excited-state spin–orbit coupling. The ground-state spin–orbit splitting is much larger, about 5000 cm^{-1} [Keiderling et al. (1975)], since it depends on $\zeta_{Ir^{4+}} \approx 2800$ cm^{-1} rather than on $\zeta_{Cl^-} \approx 590$ cm^{-1}. Thus the ground state is $E_g''(^2T_{2g})$ to a good approximation.

The spectrum shows strong bands at 20,000 cm^{-1} (II) and 23,000 cm^{-1} (III) which are readily associated with the allowed $E_g''(^2T_{2g}) \rightarrow {}^2T_{1u}(\pi + \sigma)$ and $E_g''(^2T_{2g}) \rightarrow {}^2T_{2u}(\pi)$ charge-transfer transitions. Analysis of the MCD is required to establish the order of these transitions as discussed below. Comparison of Figs. 23.2.2 and 23.2.3 also suggests assignment of the weaker shoulder at $\approx$ 17,000 cm^{-1} (I) to the parity-forbidden $E_g''(^2T_{2g}) \rightarrow {}^2T_{1g}(\pi)$ charge-transfer transition. Intensity considerations then lead to assignment of the weaker band at 32,000 cm^{-1} (IV) to the $E_g''(^2T_{2g}) \rightarrow {}^2T_{1u}(\sigma + \pi)$ transition and the very intense band at $\approx$ 42,000 cm^{-1} (V) to ligand $\rightarrow e_g(d)$ excitations [Jørgensen (1962)].

Use of MCD *to Clarify* $IrCl_6^{2-}$ *Solution Assignments*

MCD provides a much more stringent test of assignments than intensity and energy criteria. This is because the MCD of each band has (in the linear limit) three separable contributions—the $\mathcal{A}$, $\mathcal{B}$, and $\mathcal{C}$ terms—each of which comes in two signs. Moreover, the $\mathcal{C}$ terms have a $1/kT$ temperature dependence, while the $\mathcal{A}$ and $\mathcal{B}$ terms are temperature-independent and have different dispersion forms. All of these terms may be calculated and compared with experiment. In addition, the Faraday ratios, $\mathcal{A}_1/\mathcal{D}_0$, $\mathcal{B}_0/\mathcal{D}_0$, and $\mathcal{C}_0/\mathcal{D}_0$, depend on magnetic moments which may be calculated much more reliably than the dipole strength $\mathcal{D}_0$, the measure of absorption intensity.

Since the $E_g''(^2T_{2g})$ ground state is paramagnetic and the solution bandwidth (Fig. 23.2.3) is large, the MCD of $IrCl_6^{2-}$ is dominated by $\mathcal{C}$ terms. The absorption and MCD band maxima coincide; $\mathcal{A}$ terms are undetectable in the solution spectra. (They become prominent in the $IrCl_6^{2-}$ crystal spectra at liquid-helium temperature, where the linewidth is much narrower.) $\mathcal{B}$ terms are neglected, since they depend inversely on the energy difference between states mixed by the magnetic field, and the states are well separated in this case. This assumption may be checked experimentally by studying the MCD as a function of temperature. [This was done by McCaffery, Schatz, and Lester (1969), in the solid state.] The experimental $\mathcal{C}$ terms for bands II and III have + and − signs respectively (new convention; see Appendix A). Calculations (see below) show that $\mathcal{C}_0/\mathcal{D}_0$ for $E_g'' \rightarrow U_u'(^2T_{1u}) + E_u'(^2T_{1u})$ is positive, and for $E_g'' \rightarrow E_u''(^2T_{2u}) + U_u'(^2T_{2u})$ is negative (new convention, so the signs of $\mathcal{C}$ terms are opposite to those quoted in the literature). Thus the MCD strongly supports the orbital and state diagrams of Figs. 23.2.1 and 23.2.2.

Calculation of $\mathcal{C}_0/\mathcal{D}_0$ for Bands II *and* III *of* $IrCl_6^{2-}$ *in Solution*

The calculations above are accomplished using a method similar to that of Section 18.2. Since $E_g'' \to E_u'$ is symmetry-forbidden, $\mathcal{C}_0/\mathcal{D}_0$ for $E_g'' \to U_u'(^2T_{1u}) + E_u'(^2T_{1u})$ is identical to $\mathcal{C}_0/\mathcal{D}_0$ for $E_g'' \to U_u'(^2T_{1u})$. For the latter, (17.3.6), Section C.11, and Section C.13 give

$$\frac{\mathcal{C}_0}{\mathcal{D}_0}\left[E_g'' \to U_u'(^2T_{1u})\right] = \frac{-\sqrt{6}}{\mu_B}\{E''T_1U'\}\begin{Bmatrix} T_1 & T_1 & T_1 \\ U' & E'' & E'' \end{Bmatrix}\langle E_g''\|\mu^{t_{1g}}\|E_g''\rangle$$

$$= \frac{-1}{\mu_B\sqrt{6}}\langle E_g''\|\mu^{t_{1g}}\|E_g''\rangle \tag{23.2.1}$$

Equations (18.2.4), (18.2.5), (18.2.8), and (15.4.9) then give

$$\langle E_g''\|\mu^{t_{1g}}\|E_g''\rangle = -\mu_B\left(\langle E_g''\|L^{t_{1g}}\|E_g''\rangle + 2\langle E_g''\|S^{t_{1g}}\|E_g''\rangle\right)$$

$$= -\sqrt{6}\,\mu_B \tag{23.2.2}$$

so we obtain

$$\frac{\mathcal{C}_0}{\mathcal{D}_0}\left[E_g'' \to U_u'(^2T_{1u})\right] = 1 \tag{23.2.3}$$

The calculation of $\mathcal{C}_0/\mathcal{D}_0$ for $E_g'' \to E_u''(^2T_{2u}) + U_u'(^2T_{2u})$ is more involved, since we must calculate the quantity

$$\frac{\mathcal{C}_0}{\mathcal{D}_0}\left[E_g'' \to E_u''(^2T_{2u}) + U_u'(^2T_{2u})\right] = \frac{\mathcal{C}_0\left[E_g'' \to E_u''\right] + \mathcal{C}_0\left[E_g'' \to U_u'\right]}{\mathcal{D}_0\left[E_g'' \to E_u''\right] + \mathcal{D}_0\left[E_g'' \to U_u'\right]} \tag{23.2.4}$$

using (17.2.13) and (17.2.20) rather than (17.3.6) (see Sections 7.6 and 7.8 for a discussion of the analysis of the MCD of overlapping bands using moments and Gaussian fitting). We obtain

$$\frac{\mathcal{C}_0}{\mathcal{D}_0} = \left[\frac{\dfrac{1}{\mu_B 3\sqrt{6}}\langle E_g''\|\mu^{t_{1g}}\|E_g''\rangle \times \left(\langle E_g''\|m^{t_{1u}}\|E_u''\rangle\langle E_u''\|m^{t_{1u}}\|E_g''\rangle + \frac{1}{2}\langle E_g''\|m^{t_{1u}}\|U_u'\rangle\langle U_u'\|m^{t_{1u}}\|E_g''\rangle\right)}{\frac{1}{6}\left(\langle E_g''\|m^{t_{1u}}\|E_u''\rangle\langle E_u''\|m^{t_{1u}}\|E_g''\rangle - \langle E_g''\|m^{t_{1u}}\|U_u'\rangle\langle U_u'\|m^{t_{1u}}\|E_g''\rangle\right)}\right] \tag{23.2.5}$$

To simplify further we use (18.2.3) to give

$$\langle E_g''\|m^{t_{1u}}\|E_u''\rangle = -\tfrac{2}{3}\langle {}^2T_{2g}\|m^{t_{1u}}\|{}^2T_{2u}\rangle \tag{23.2.6}$$

and (18.1.8) to obtain

$$\langle E_u''\|m^{t_{1u}}\|E_g''\rangle = -\tfrac{2}{3}\langle {}^2T_{2u}\|m^{t_{1u}}\|{}^2T_{2g}\rangle$$

$$\begin{aligned}\langle E_g''\|m^{t_{1u}}\|U_u'\rangle &= \left\langle \left(\tfrac{1}{2}E_g', T_{2g}\right)E_g''\middle\|m^{t_{1u}}\middle\|\left(\tfrac{1}{2}E_g', T_{2u}\right)U_u'\right\rangle \\ &= \frac{-\sqrt{2}}{3}\langle {}^2T_{2g}\|m^{t_{1u}}\|{}^2T_{2u}\rangle\end{aligned} \tag{23.2.7}$$

$$\langle U_u'\|m^{t_{1u}}\|E_g''\rangle = \frac{\sqrt{2}}{3}\langle {}^2T_{2u}\|m^{t_{1u}}\|{}^2T_{2g}\rangle$$

With these results and (23.2.2), Eq. (23.2.5) yields

$$\frac{\mathcal{C}_0}{\mathcal{D}_0}\left[E_g'' \to E_u''({}^2T_{2u}) + U_u'({}^2T_{2u})\right] = -1 \tag{23.2.8}$$

The results given in (23.2.3) and (23.2.8) are in very good agreement with the experimental values for bands II and III, which are 0.90 and -0.82 respectively.‡ Thus the MCD strongly supports the ordering of the ${}^2T_{1u}(\pi + \sigma)$ and ${}^2T_{2u}(\pi)$ states shown in Fig. 23.2.2 and thus the MO order of Fig. 23.2.1.

23.3. The $IrCl_6^{2-}$ Spectrum in the Solid State: $Cs_2ZrCl_6:Ir^{4+}$

When $IrCl_6^{2-}$ is studied at liquid-helium temperature doped into the cubic host Cs_2ZrCl_6, a really beautiful spectrum is obtained. This high-resolution spectrum, shown in Fig. 23.3.1, reveals a remarkable amount of detail completely absent in the solution spectrum. Individual vibronic transitions are now resolved, and $\mathcal{A}$ terms stand out very clearly for nearly all sharp lines, even though the $\mathcal{C}$-term contribution increases by a factor of ≈ 75 in going to low temperatures. (Compare ordinate scales in Fig. 23.3.1 et seq.

‡These experimental values are those of Henning et al. (1968) multiplied by the factor from (17.1.1) for $\mathcal{C}_0/\mathcal{D}_0$. They are actually values for $kT \times (\mathcal{B}_0 + \mathcal{C}_0/kT)/\mathcal{D}_0$, but $\mathcal{B}$ terms may be assumed very small, since the states are quite well separated.

with that in Fig. 23.2.3.) The $\mathcal{A}$ terms appear because their peak-to-trough deflection goes inversely as the second power of the linewidth.

We make the assumption in doped crystals that most of the prominent spectral features can be attributed to the isolated MX_6^{n-} moiety, and we focus on these features. Thus the crystal host is regarded as a spectroscopically "inert" solvent. Observation and analysis supports this view, but MX_6^{n-} is in fact coupled to the lattice, and vibronic structure due to lattice modes is certainly present. Bands I, II, and III of Fig. 23.3.1 should be compared with their counterparts in the solution spectrum of Fig. 23.2.3. While rough assignments can be made easily by analogy to solution results, assignment of individual vibronic lines is a much more challenging task. In fact the sharp lines found in the $Cs_2ZrCl_6 : Ir^{4+}$ spectrum and in the spectra of related systems were at first perplexing, since at the time the spectra were initially measured, sharp lines were thought to be typical of $d \rightarrow d$ or $f \rightarrow f$

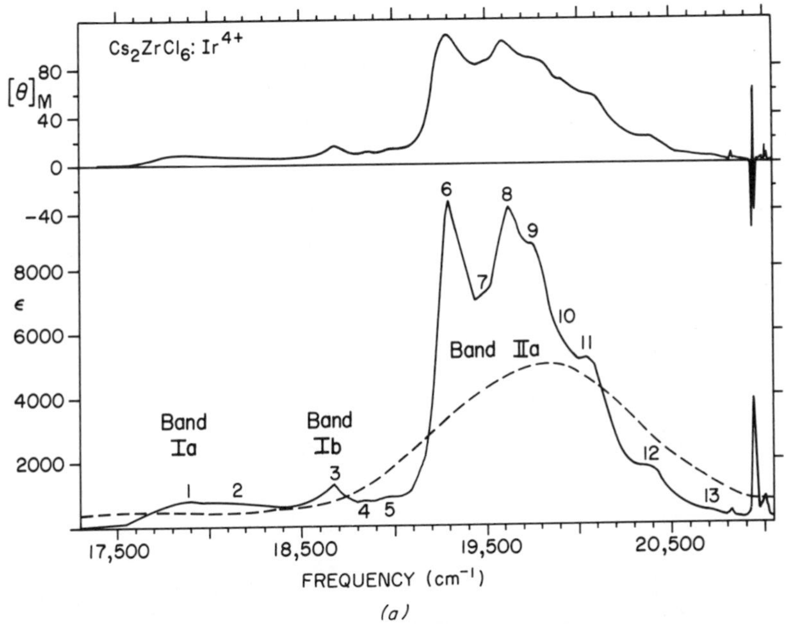

Figure 23.3.1. The absorption spectrum and MCD of $Cs_2ZrCl_6 : Ir^{4+}$ from Piepho et al. (1972b). $[\theta]_M$ is the MCD in molar ellipticity units (see Section A.5). ϵ is the molar extinction coefficient. The solid curves were run at $\approx$ 11 K, and the dashed curve [except in (d)] at room temperature. (a) Bands Ia, Ib, and IIa; (b) band IIb; (c) high-energy end of band IIb and band x; (d) band III (here the dashed line represents the 11-K absorption baseline).

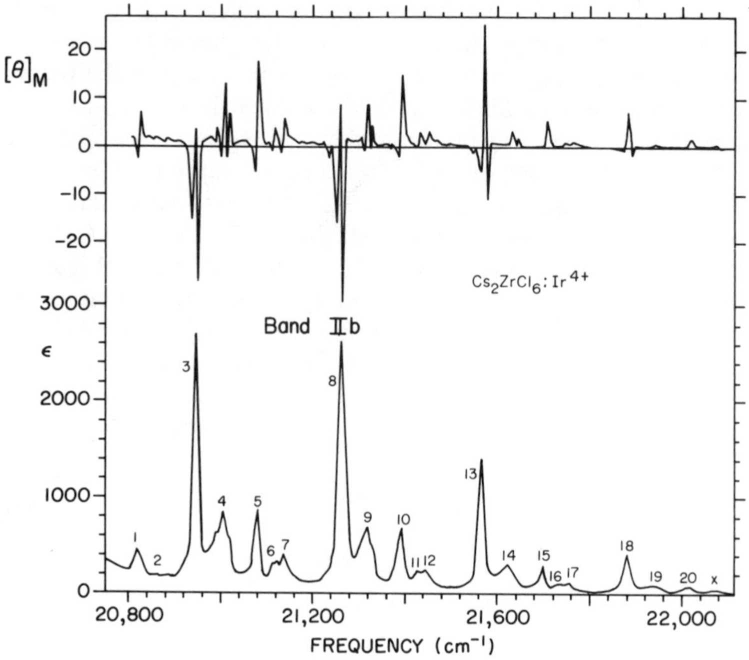

(b)

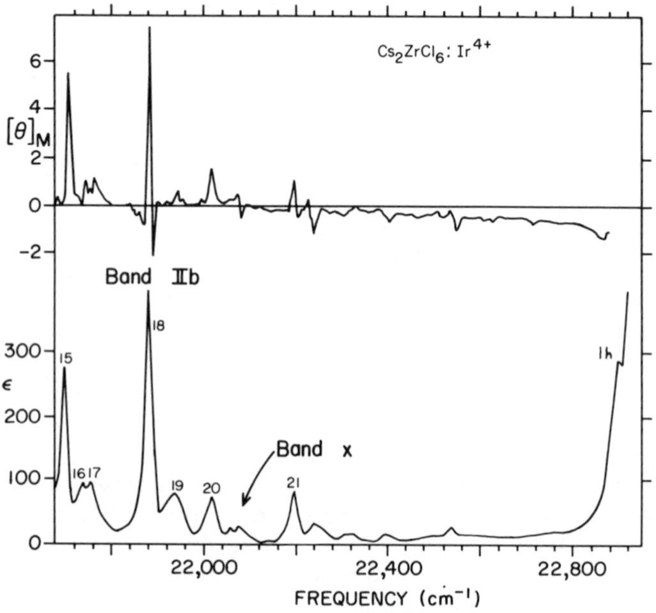

(c)

Figure 23.3.1. (*Continued*)

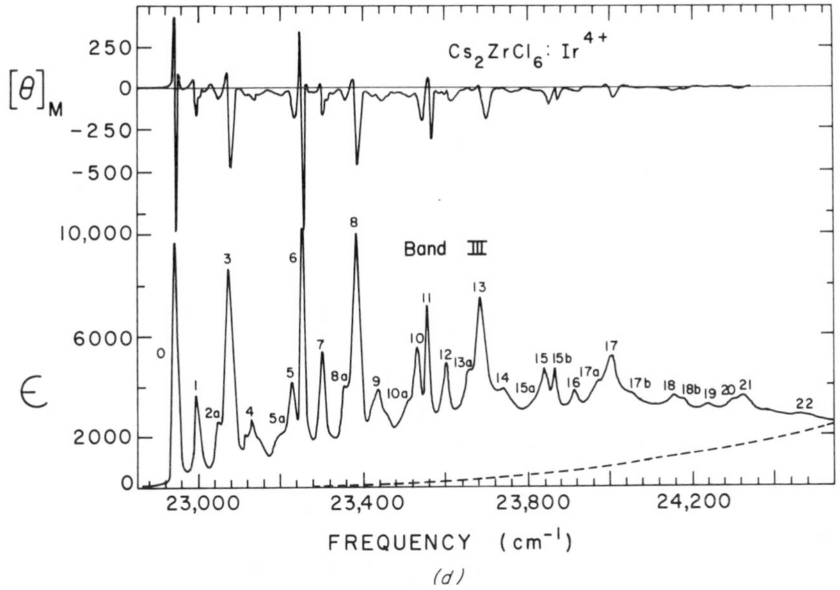

Figure 23.3.1. (*Continued*)

transitions, but not of $\gamma \rightarrow d$ transitions. Careful analyses of the high-resolution spectra of $Cs_2ZrCl_6 : Ir^{4+}$ and related systems showed unequivocally that virtually all lines in the visible and *uv* spectra arise from charge-transfer transitions.

$Cs_2ZrCl_6 : Ir^{4+}$, *Band* III: *The* $t_{2u}(\pi) \rightarrow t_{2g}$ *Band*

The most throughly analyzed band in the $Cs_2ZrCl_6 : Ir^{4+}$ spectrum is III, which is assigned to the $E_g'' \rightarrow E_u''(^2T_{2u})$ and $E_g'' \rightarrow U_u'(^2T_{2u})$ transitions. These arise from excitations from a spin–orbit-split $t_{2u}(\pi)$ orbital. Yeakel and Schatz (1974) demonstrated that the pattern of vibronic lines results from a strong Jahn–Teller (JT) effect in which the $^2T_{2u}$ excited state is coupled to the $\nu_5(t_{2g})$ and the $\nu_2(e_g)$ vibrational modes. We find that such vibronic coupling is quite pronounced in $OsCl_6^{2-}$ and $ReCl_6^{2-}$ as well. For these hexachlorides with a $t_{2u}(\pi)$ open shell, vibronic coupling is apparently of the same order of magnitude as spin–orbit coupling. In the hexabromides, spin–orbit coupling dominates, since ζ_{Br^-} is 2460 cm^{-1} as compared to $\zeta_{Cl^-} = 590$ cm^{-1}.

Since a complete analysis of the JT effect in band III in $Cs_2ZrCl_6 : Ir^{4+}$ is given by Yeakel and Schatz (1974), we do not repeat it here. We show, however, in Section 23.6 that virtually the same analysis also holds for the analogous band in $Cs_2ZrCl_6 : Os^{4+}$. The comparison both demonstrates the great power of group theory and lends support to the Yeakel–Schatz analysis for $Cs_2ZrCl_6 : Ir^{4+}$.

Cs_2ZrCl_6: Ir^{4+}, *Band II: The $t_{1u}(\pi + \sigma) \rightarrow t_{2g}$ Bands*

The band II region arises from excitations from the spin–orbit-split $t_{1u}(\pi + \sigma)$ orbital. The intense broad band IIa is assigned to the $E_g'' \rightarrow U_u'(^2T_{1u})$ transition with structure resulting from vibronic (JT) coupling. J. C. Collingwood has argued (private communication, 1974), as a result of detailed comparisons between $IrCl_6^{2-}$, $OsCl_6^{2-}$, and $ReCl_6^{2-}$, that the fine structure (IIb) between IIa and III must result from the forbidden $E_g'' \rightarrow E_u'(^2T_{1u})$ transition. While it is well documented from hot-band (and intensity) studies that the band is forbidden, the E_u' assignment requires a larger $^2T_{1u}(\pi + \sigma)$ spin–orbit splitting—$\approx 1500\ \text{cm}^{-1}$—than might seem reasonable. Collingwood, following Jørgensen (1962), explains the larger splitting by suggesting that the metal $6p$ orbital contributes significantly to the $t_{1u}(\pi + \sigma)$ orbital. Thus $|t_{1u}(\pi + \sigma)x\rangle$, for example, should be written

$$|t_{1u}(\pi + \sigma)x\rangle = -\left[a\left|\frac{1}{\sqrt{2}}(x_1 + x_3)\right\rangle + b\left|\tfrac{1}{2}(x_2 + x_4 + x_5 + x_6)\right\rangle + c\,|6p_x\rangle\right] \tag{23.3.1}$$

rather than as in (19.6.1), where the metal $6p_x$ contribution is absent. Thus instead of the result in Table 19.6.1, we have

$$\langle t_{1u}(\pi + \sigma)\|u\|t_{1u}(\pi + \sigma)\rangle = \frac{-\sqrt{6}}{2}\left[\zeta_{\text{Cl}^-}(b^2 + 2\sqrt{2}\,ab) + \zeta_{6p}(2c^2)\right] \tag{23.3.2}$$

Jørgensen estimates ζ_{6p} for Ir^{4+} as $\approx 5000\ \text{cm}^{-1}$, so even a small coefficient c increases the magnitude of $\langle t_{1u}(\pi + \sigma)\|u\|t_{1u}(\pi + \sigma)\rangle$ and hence the $^2T_{1u}$ spin–orbit splitting. In this manner a $\approx 1500\text{-cm}^{-1}$ separation of the $U_u'(^2T_{1u})$ and $E_u'(^2T_{1u})$ transitions could be accounted for. Strong support for the assignment comes from the presence of bands closely analogous to IIa and IIb in $OsCl_6^{2-}$ (Section 23.4) and $ReCl_6^{2-}$ [Collingwood et al. (1975)]. This similarity is predicted if the bands in each complex arise from excitations from a spin–orbit-split $t_{1u}(\pi + \sigma)$ orbital, but otherwise seems inexplicable.

Detailed assignments of individual lines in IIa and IIb have not been made. Analysis of the shape of the $U_u'(^2T_{1u})$ band (IIa) would require a JT calculation. [Our present suggestion is that the assignment of IIa by Piepho et al. (1972b) to *both* $^2T_{1u}$ transitions is incorrect, since we now believe IIb is the $E_u'(^2T_{1u})$ band.] Likewise, the structure of IIb remains a puzzle. While we can assign lines 3 and 4 to $E_u' + \nu_5(t_{2g})$ and $E_u' + \nu_2(e_g)$ respectively, line 5, for example, cannot be readily assigned on the basis of an isolated $IrCl_6^{2-}$ moiety. No doubt vibronic interactions are important in this region too, and vibronic coupling to the $U_u'(^2T_{1u})$ state (and perhaps also to the $^2T_{2u}$ states) is responsible for the observed pattern. Very similar problems in assignment exist in the analogous band in $OsCl_6^{2-}$.

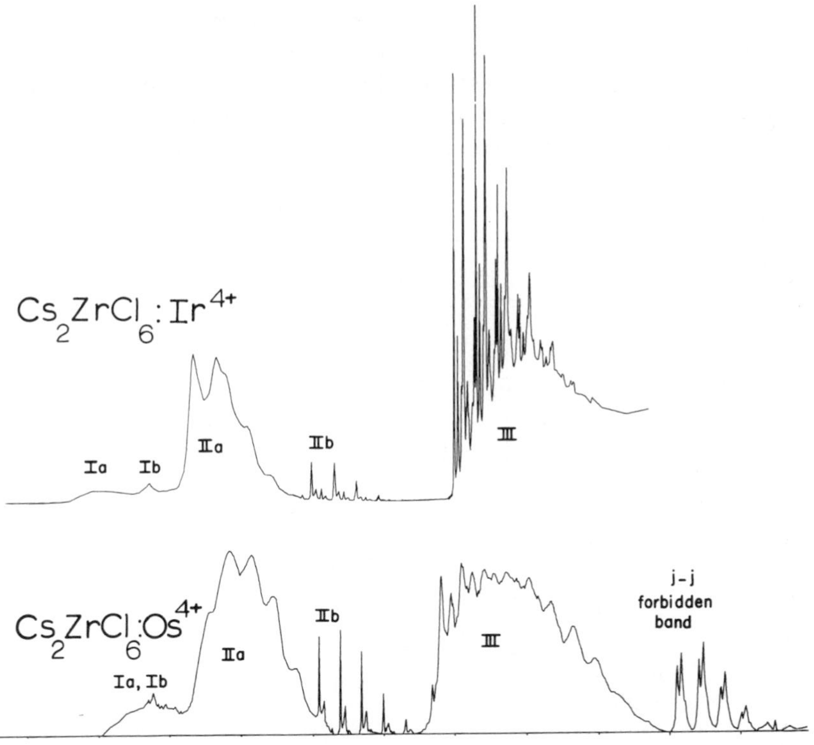

Figure 23.4.1. The absorption spectrum of $Cs_2ZrCl_6 : Ir^{4+}$ and $Cs_2ZrCl_6 : Os^{4+}$. The $OsCl_6^{2-}$ spectrum has been displaced downward in energy to emphasize the close correspondence between the charge-transfer bands.

23.4. $OsCl_6^{2-}$ in the Solid-State Host Cs_2ZrCl_6

As with Ir^{4+}, Os^{4+} doped into Cs_2ZrCl_6 gives a beautiful spectrum. And surprisingly, the absorption spectra of the two ions in the visible–uv region are *nearly identical* if the $OsCl_6^{2-}$ spectrum is uniformly shifted about 6000 cm^{-1} to the red. This is shown in Fig. 23.4.1. Figure 23.4.2 gives a more detailed comparison of the two spectra in the band-III region to stress how similar even the vibronic structure is in the two cases. The data strongly suggest that the similarity is no accident, and we now show how group-theoretic arguments can very elegantly explain the similarities using a jj-coupling model (Chapter 21). The complete $Cs_2ZrCl_6 : Os^{4+}$ absorption and MCD spectrum in the visible–uv region is given in Figure 23.4.3.

The jj Approximation in $OsCl_6^{2-}$

The close correspondence between the charge-transfer absorption spectra of $OsCl_6^{2-}$ and $IrCl_6^{2-}$ was first pointed out many years ago by Jørgensen (1959) when he studied the broad-band solution data for these systems. He suggested then and in Jørgensen and Preetz (1967) that the correspondence could be explained if calculations for $OsCl_6^{2-}$ were made in the jj limit. The jj designation means that within any strong-field configuration, spin–orbit

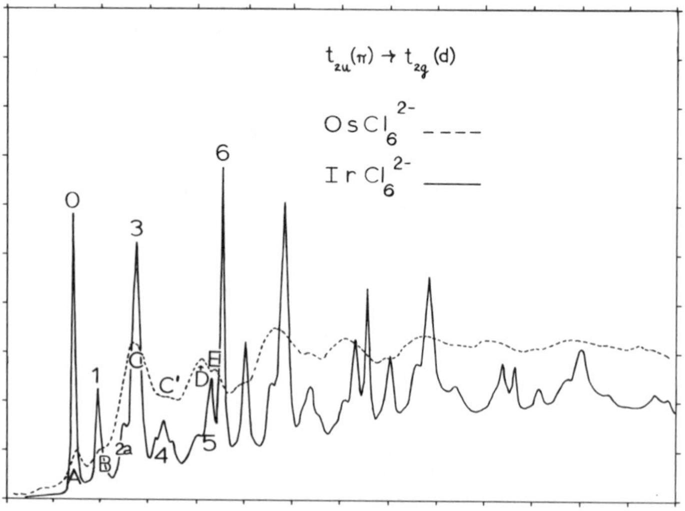

Figure 23.4.2. A detailed comparison of the vibronic lines of band III in $OsCl_6^{2-}$ and $IrCl_6^{2-}$. The $OsCl_6^{2-}$ spectrum has been displaced downward in energy as in Figure 23.4.1.

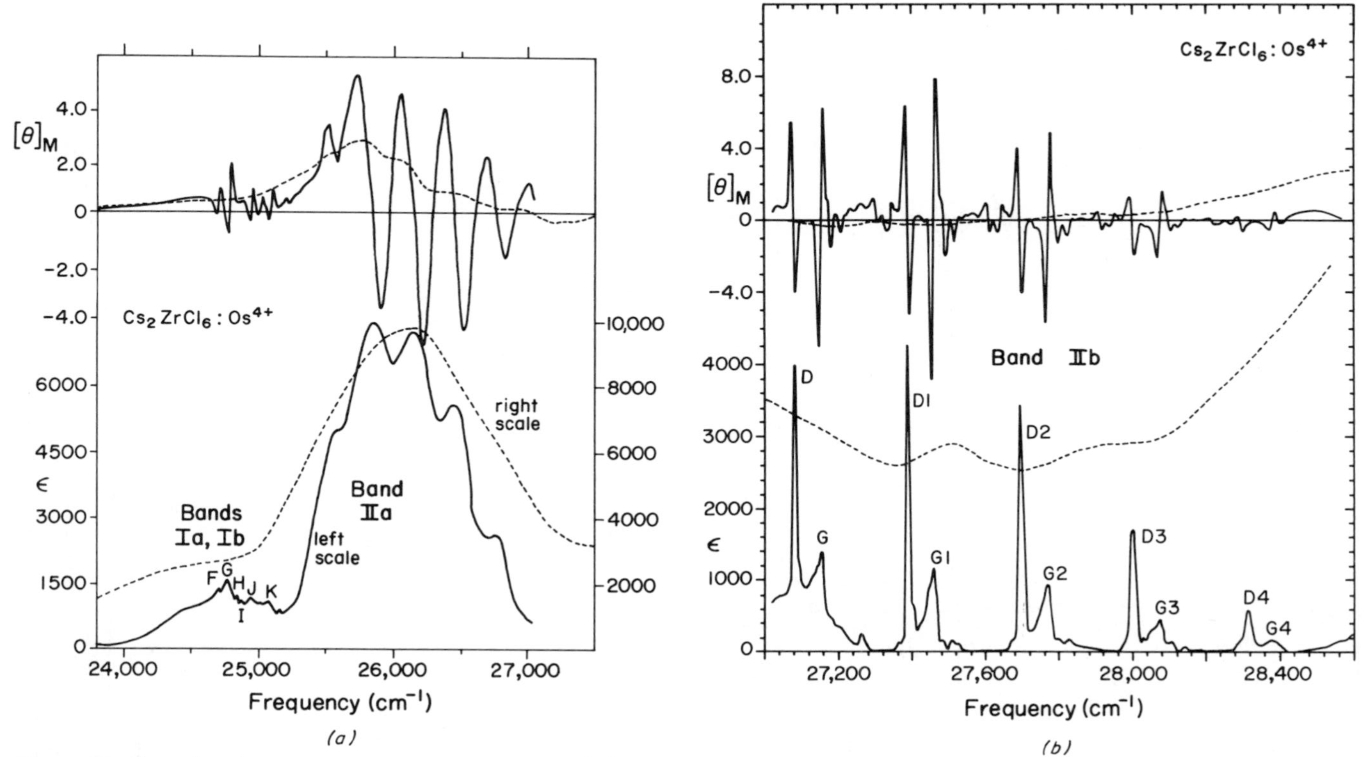

Figure 23.4.3. The absorption and MCD spectrum of $Cs_2ZrCl_6 : Os^{4+}$ from Piepho et al. (1972a). $[\theta]_M$ is the MCD in molar ellipticity units (see Section A.5). ϵ is the molar extinction coefficient normalized to a solution value. (*a*) Bands Ia, Ib, and IIa at room temperature (dashed line) and at liquid helium temperature (solid line); (*b*) band IIb; (*c*) band III; (*d*) *jj* forbidden band(s).

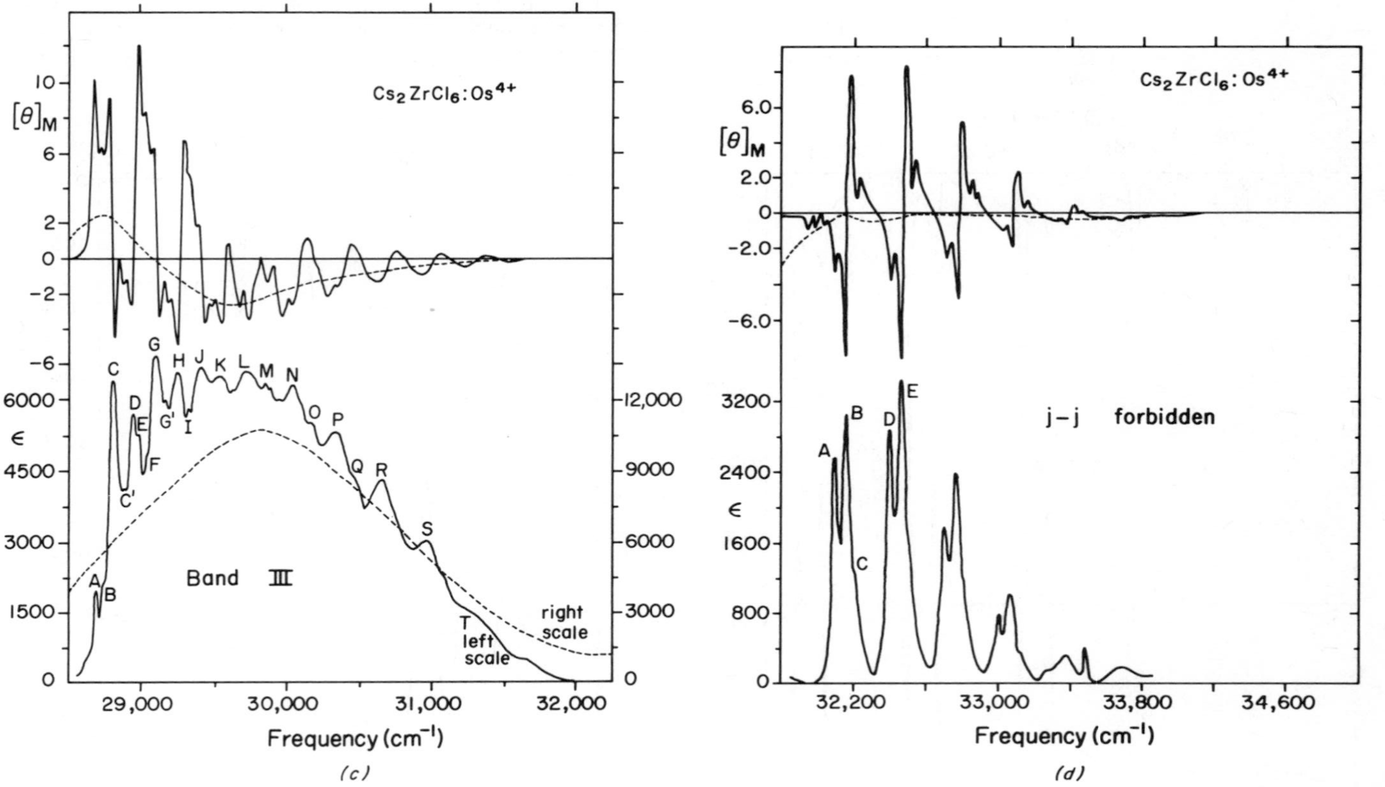

Figure 23.4.3. (*Continued*)

coupling dominates the interelectronic repulsions, so that to a first approximation the latter may be neglected.

The approach is most easily described with reference to Fig. 23.2.1 and Tables 23.4.1–23.4.2. Figure 23.2.1 shows the orbitals involved in the most intense low-energy charge-transfer band in $OsCl_6^{2-}$ and $IrCl_6^{2-}$, while Table 23.4.1 shows the ground- and excited-state configurations for the two complex ions based on this MO diagram in the jj limit.

So far electrostatic interactions (interelectronic repulsions) have been neglected. Relative energies of the states may therefore be calculated in the jj limit as described in Chapter 21. The results, which are given in Table 23.4.2, are almost obvious from the MO diagram of Fig. 23.2.1.

The diagrams and tables we have introduced have nearly all the information necessary to explain qualitatively the similarity of $OsCl_6^{2-}$ and $IrCl_6^{2-}$. We need only consider the selection rules for electric-dipole-allowed transitions in this limit. Those for $IrCl_6^{2-}$ have been introduced previously; using the jj nomenclature of Fig. 23.2.1 we find that $u'_u(a) \rightarrow e''_g$, $e''_u \rightarrow e''_g$, and $u'_u(b) \rightarrow e''_g$ are the only allowed excitations into the e''_g orbital, and they give rise to IIa and III.

The surprising result for $OsCl_6^{2-}$ is that in the jj limit identical excitations to those in $IrCl_6^{2-}$ are allowed. In $OsCl_6^{2-}$ they are the transitions between the ground state and the T_{1u} group-(a) excited states of Table 23.4.1. The latter have the same relative energies as the $IrCl_6^{2-}$ U'_u, E''_u, and U'_u states (Fig. 23.2.2 and Table 23.4.2). Transitions to the $OsCl_6^{2-}$ $E'_u(^2T_{1u}) \otimes E''_g(^2T_{2g})$ states are symmetry-forbidden, just like the transition to the $E'_u(^2T_{1u})$ state in $IrCl_6^{2-}$. All transitions to the remaining states—the $\Gamma_u \otimes U'_g(^2T_{2g})$ or group-(b) states—in Table 23.4.2 or 23.4.1 are forbidden because they describe two-electron jumps in the jj limit, and the electric-dipole operator is a one-electron operator. For this reason, for example, the $OsCl_6^{2-}$ $(u'_u)^4(e''_u)^2(u'_g)^4 \rightarrow (u'_u)^3(e''_u)^2(u'_g)^3(e''_g)^2$ excitation is forbidden.

Thus in the jj approximation the same number of allowed transitions is expected in $OsCl_6^{2-}$ as in $IrCl_6^{2-}$ with the same relative energies. We show in Section 23.5 that the transitions have identical relative intensities, and in Section 23.6 that nearly identical vibronic structure is to be expected for the two complex ions. Thus the jj approximation very nicely explains the spectral similarity in Fig. 23.4.1. Now we consider what we have omitted in working in the jj limit—the electrostatic interactions.

Electrostatic Interactions in $OsCl_6^{2-}$ *and* $IrCl_6^{2-}$

Interelectronic repulsions couple the jj states of Table 23.4.1 with states of the same symmetry from the same strong-field configuration. We see from

Table 23.4.1. A Comparison of $IrCl_6^{2-}$ and $OsCl_6^{2-}$ in the jj Approximation

(a) $IrCl_6^{2-}$

Strong-Field Configuration	jj Configuration[a]	Symmetry of States[b]
$t_{2u}(\pi)^5 t_{2g}^6$	$(u'_u)^3(e''_u)^2$	$U'_u(^2T_{2u}) = U'_u$ *
	$(u'_u)^4 e''_u$	$E''_u(^2T_{2u}) = E''_u$ *
$t_{1u}(\pi + \sigma)^5 t_{2g}^6$	$e'_u(u'_u)^4$	$E'_u(^2T_{1u}) = E'_u$
	$(e'_u)^2(u'_u)^3$	$U'_u(^2T_{1u}) = U'_u$ *
Ground state	$(u'_g)^3(e''_g)^2$	$U'_g(^2T_{2g}) = U'_g$
$\gamma_u^6 t_{2g}^5$	$(u'_g)^4 e''_g$	$E''_g(^2T_{2g}) = E''_g$

(b) $OsCl_6^{2-}$

Strong-Field Configuration	jj Configuration[c]	Symmetry of States[b]	Group
$t_{2u}(\pi)^5 t_{2g}^5$	$(u'_u)^3(e''_u)^2(u'_g)^3(e''_g)^2$	$U'_u(^2T_{2u}) \otimes U'_g(^2T_{2g})$ $= A_{1u} + A_{2u} + E_u$ $+2T_{1u} + 2T_{2u}$	(b)
	$(u'_u)^4 e''_u(u'_g)^3(e''_g)^2$	$E''_u(^2T_{2u}) \otimes U'_g(^2T_{2g})$ $= E_u + T_{1u} + T_{2u}$	(b)
	$(u'_u)^3(e''_u)^2(u'_g)^4 e''_g$	$U'_u(^2T_{2u}) \otimes E''_g(^2T_{2g})$ $= E_u + T_{1u}^* + T_{2u}$	(a)
	$(u'_u)^4 e''_u(u'_g)^4 e''_g$	$E''_u(^2T_{2u}) \otimes E''_g(^2T_{2g})$ $= A_{1u} + T_{1u}^*$	(a)
$t_{1u}(\pi + \sigma)^5 t_{2g}^5$	$e'_u(u'_u)^4(u'_g)^3(e''_g)^2$	$E'_u(^2T_{1u}) \otimes U'_g(^2T_{2g})$ $= E_u + T_{1u} + T_{2u}$	(b)
	$(e'_u)^2(u'_u)^3(u'_g)^3(e''_g)^2$	$U'_u(^2T_{1u}) \otimes U'_g(^2T_{2g})$ $= A_{1u} + A_{2u} + E_u$ $+2T_{1u} + 2T_{2u}$	(b)
	$e'_u(u'_u)^4(u'_g)^4 e''_g$	$E'_u(^2T_{1u}) \otimes E''_g(^2T_{2g})$ $= A_{2u} + T_{2u}$	(a)
	$(e'_u)^2(u'_u)^3(u'_g)^4 e''_g$	$U'_u(^2T_{1u}) \otimes E''_g(^2T_{2g})$ $= E_u + T_{1u}^* + T_{2u}$	(a)
Ground state $\gamma_u^6 t_{2g}^6$	$(u'_g)^2(e''_g)^2$	A_{1g}	
	$(u'_g)^3 e''_g$	$E_g + T_{1g} + T_{2g}$	
	$(u'_g)^4$	$A_{1g} = (\sqrt{2}/\sqrt{3})\lvert A_{1g}(^3T_{1g})\rangle$ $+(1/\sqrt{3})\lvert A_{1g}(^1A_{1g})\rangle$	

[a] The t_{2g}^6 and γ_u^6 parts are omitted.

[b] Transitions to states marked with an asterisk are strongly allowed.

[c] The γ_u^6 parts are omitted from the ground-state configurations.

Table 23.4.2. Relative Energies of $IrCl_6^{2-}$ and $OsCl_6^{2-}$ *jj* States within Configurations

	$IrCl_6^{2-}$	Relative Energy[a]	$OsCl_6^{2-}$	Relative Energy[a]
$t_{2u}(\pi)$			$U_u'(^2T_{2u}) \otimes U_g'(^2T_{2g})$	$\frac{1}{4}\zeta_{Cl^-} + \frac{1}{2}\zeta_{Os^{4+}}$
$\rightarrow t_{2g}$			$E_u''(^2T_{2u}) \otimes U_g'(^2T_{2g})$	$-\frac{1}{2}\zeta_{Cl^-} + \frac{1}{2}\zeta_{Os^{4+}}$
	$U_u'(^2T_{2u})$	$\frac{1}{4}\zeta_{Cl^-}$	$U_u'(^2T_{2u}) \otimes E_g''(^2T_{2g})$	$\frac{1}{4}\zeta_{Cl^-} - \zeta_{Os^{4+}}$
	$E_u''(^2T_{2u})$	$-\frac{1}{2}\zeta_{Cl^-}$	$E_u''(^2T_{2u}) \otimes E_g''(^2T_{2g})$	$-\frac{1}{2}\zeta_{Cl^-} - \zeta_{Os^{4+}}$
$t_{1u}(\pi + \sigma)$			$E_u'(^2T_{1u}) \otimes U_g'(^2T_{2g})$	$\frac{1}{2}k + \frac{1}{2}\zeta_{Os^{4+}}$
$\rightarrow t_{2g}$			$U_u'(^2T_{1u}) \otimes U_g'(^2T_{2g})$	$-\frac{1}{4}k + \frac{1}{2}\zeta_{Os^{4+}}$
	$E_u'(^2T_{1u})$	$\frac{1}{2}k$	$E_u'(^2T_{1u}) \otimes E_g''(^2T_{2g})$	$\frac{1}{2}k - \zeta_{Os^{4+}}$
	$U_u'(^2T_{1u})$	$-\frac{1}{4}k$	$U_u'(^2T_{1u}) \otimes E_g''(^2T_{2g})$	$-\frac{1}{4}k - \zeta_{Os^{4+}}$

[a] Here $k = \zeta_{Cl^-}(b^2 + 2\sqrt{2}\,ab) + \zeta_{M^{4+}}(2c^2)$, where $M^{4+} = Ir^{4+}$ or Os^{4+}.

the table that for $IrCl_6^{2-}$ there is only one state of a given symmetry in each configuration, so in fact no electrostatic interactions have been neglected.

In $OsCl_6^{2-}$ the electrostatic interactions mix the four T_{1u} states in the $t_{1u}(\pi + \sigma)^5 t_{2g}^5$ configuration; likewise the five T_{1u} states in the $t_{2u}(\pi)^5 t_{2g}^5$ configuration are mixed. Clearly this mixing makes all of these T_{1u} states formally allowed. Since the mixing necessarily involves "ligand–metal" electrostatic interactions, this mixing is expected to be relatively small.

The *jj* limit is more drastic with respect to the $OsCl_6^{2-}$ ground configuration, since the interelectronic repulsions can mix the two A_{1g} states arising from the t_{2g}^4 $OsCl_6^{2-}$ ground configuration. These are metal *d–d* electrostatic interactions and are known to be significant. Crystal-field calculations show that inclusion of these interactions results in a ground state of $\approx 80\%$ $A_{1g}(u_g')^4$ character, so the *jj* limit is still valid as a first approximation. The $A_{1g}[(u_g')^2(e_g'')^2]$ character mixed into the ground state makes transitions to the $\Gamma_u \otimes U_g'(^2T_{2g})$ levels allowed, since, for example,

$$(u_u')^4(e_u'')^2(u_g')^2(e_g'')^2 \rightarrow (u_u')^3(e_u'')^2(u_g')^3(e_g'')^2 \qquad (23.4.1)$$

is a one-electron jump.

23.5. MCD and Absorption Calculations for $IrCl_6^{2-}$ and $OsCl_6^{2-}$

In this section and those which follow we neglect vibronic interactions (JT effects) and calculate $\mathfrak{D}_0$, $\mathcal{C}_0/\mathfrak{D}_0$, and $\mathcal{A}_1/\mathfrak{D}_0$ for the $u_u'(a) \rightarrow e_g''$, $e_u'' \rightarrow e_g''$, and $u'_u(b) \rightarrow e''_g$ excitations in $IrCl_6^{2-}$ and $OsCl_6^{2-}$. The $OsCl_6^{2-}$ calculation is done in the jj limit (Section 23.4). We find in both compounds that $\mathcal{A}_1/\mathcal{D}_0$ values for corresponding excitations are identical. Likewise, the ratio of $\mathcal{D}_0$ values for successive excitations are the same. However, since $OsCl_6^{2-}$ has a nondegenerate ground state, it has no $\mathcal{C}$ terms, while $IrCl_6^{2-}$ has a $E''_g(^2T_{2g})$ ground state and so $\mathcal{C}$ terms are important. Equations (17.2.19), (17.3.5), and (17.3.6) are used to obtain $\mathcal{D}_0$, $\mathcal{A}_1/\mathcal{D}_0$, and $\mathcal{C}_0/\mathcal{D}_0$; alternatively (17.3.1) and (17.3.2) may be used instead of (17.3.5) and (17.3.6) for $OsCl_6^{2-}$.

The only symmetry-allowed transitions in $OsCl_6^{2-}$ are $A_{1g} \rightarrow T_{1u}$. The MCD equations yield the following results:

$IrCl_6^{2-}$

1. Excitation $u'_u(a) \rightarrow e''_g$, transition $E''_g(^2T_{2g}) \rightarrow U'_u(^2T_{1u})$:

$$\mathcal{D}_0 = \tfrac{1}{6}\left|\langle E''_g \| m^{t_{1u}} \| U'_u \rangle\right|^2$$

$$\frac{\mathcal{C}_0}{\mathcal{D}_0} = \frac{-1}{\sqrt{6}\,\mu_B}\langle E''_g \| \mu^{t_{1g}} \| E''_g \rangle \tag{23.5.1}$$

$$\frac{\mathcal{A}_1}{\mathcal{D}_0} = \frac{\sqrt{6}}{\mu_B}\left(\frac{-1}{6\sqrt{2}}\langle U'_u \| \mu^{t_{1g}} \| U'_u \rangle_0 + \frac{1}{3\sqrt{2}}\langle U'_u \| \mu^{t_{1g}} \| U'_u \rangle_1 - \tfrac{1}{6}\langle E''_g \| \mu^{t_{1g}} \| E''_g \rangle\right)$$

2. Excitation $e''_u \rightarrow e''_g$, transition $E''_g(^2T_{2g}) \rightarrow E''_u(^2T_{2u})$:

$$\mathcal{D}_0 = \tfrac{1}{6}\left|\langle E''_g \| m^{t_{1u}} \| E''_u \rangle\right|^2$$

$$\frac{\mathcal{C}_0}{\mathcal{D}_0} = \frac{\sqrt{6}}{3\mu_B}\langle E''_g \| \mu^{t_{1g}} \| E''_g \rangle \tag{23.5.2}$$

$$\frac{\mathcal{A}_1}{\mathcal{D}_0} = \frac{\sqrt{6}}{3\mu_B}\left(\langle E''_u \| \mu^{t_{1g}} \| E''_u \rangle + \langle E''_g \| \mu^{t_{1g}} \| E''_g \rangle\right)$$

3. Excitation $u'_u(b) \rightarrow e''_g$: identical to $u'_u(a) \rightarrow e''_g$ except that the transition is $E''_g(^2T_{2g}) \rightarrow U'_u(^2T_{2u})$.

$OsCl_6^{2-}$

1. Excitation $u'_u(a) \rightarrow e''_g$, transition $A_{1g} \rightarrow T_{1u}$ with T_{1u} from $U'_u(^2T_{1u}) \otimes E''_g(^2T_{2g})$:

$$\mathfrak{D}_0 = \tfrac{1}{3}\left|\langle A_{1g}\|m^{t_{1u}}\|T_{1u}\rangle\right|^2$$

$$\frac{\mathcal{C}_0}{\mathfrak{D}_0} = 0 \tag{23.5.3}$$

$$\frac{\mathcal{A}_1}{\mathfrak{D}_0} = \frac{\sqrt{6}}{3\mu_B}\langle T_{1u}\|\mu^{t_{1g}}\|T_{1u}\rangle$$

2. Excitation $e''_u \rightarrow e''_g$, transition $A_{1g} \rightarrow T_{1u}$: same as above except T_{1u} is from $E''_u(^2T_{2u}) \otimes E''_g(^2T_{2g})$.
3. Excitation $u'_u(b) \rightarrow e''_g$, transition $A_{1g} \rightarrow T_{1u}$: same as above except T_{1u} is from $U'_u(^2T_{2u}) \otimes E''_g(^2T_{2g})$.

We are now ready to evaluate the reduced matrix elements using the conventional methods of Chapters 19 and 20 or the $\mathcal{jj}$ methods of Chapter 21. For $IrCl_6^{2-}$ the $\mathcal{jj}$ method is less familiar but somewhat shorter; for $OsCl_6^{2-}$, the $\mathcal{jj}$ approach has distinct advantages. The T_{1u} states required for either method are those which are diagonal to first-order in $\mathcal{H}_{SO}$; they are simply the states with the $\mathcal{jj}$ configurations of Table 23.4.1. However, to define the $OsCl_6^{2-}$ T_{1u} eigenstates of $\mathcal{H}_{SO}$ in the more conventional $|(\mathbb{S}S, h)rt\tau\rangle$ basis, a 5×5 spin–orbit matrix must be diagonalized for the five $t_{2u}(\pi) \rightarrow t_{2g}$ T_{1u} states, and a 4×4 for the $t_{1u}(\pi + \sigma)$ case.

We present the calculation using both methods for $IrCl_6^{2-}$, but use only the $\mathcal{jj}$ method for $OsCl_6^{2-}$.

Calculation of $IrCl_6^{2-}$ *Reduced Matrix Elements: The Method of Chapters 19–20*

Equation (23.2.2) gives

$$\langle E''_g\|\mu^{t_{1g}}\|E''_g\rangle = -\sqrt{6}\,\mu_B \tag{23.5.4}$$

Then using (18.1.8), (18.1.9), (19.4.6), Table 19.6.1, and (15.4.9), we have

$$\langle E_u''\|\mu^{t_{1g}}\|E_u''\rangle = \left\langle\left(\tfrac{1}{2}^+ E_g', T_{2u}\right)E_u''\right\|-\mu_B(L^{t_{1g}}+2S^{t_{1g}})\left\|\left(\tfrac{1}{2}^+ E_g', T_{2u}\right)E_u''\right\rangle$$

$$= -\mu_B\left(-\tfrac{2}{3}\langle t_{2u}\|l^{t_{1g}}\|t_{2u}\rangle + \frac{\sqrt{6}}{3}\right)$$

$$= \frac{-2\sqrt{6}}{3}\mu_B \qquad (23.5.5)$$

Next, for both $U_u'(^2T_{1u})$ and $U_u'(^2T_{2u})$

$$\langle U_u'\|\mu^{t_{1g}}\|U_u'\rangle_p = -\mu_B\left(\langle U_u'\|L^{t_{1g}}\|U_u'\rangle_p + 2\langle U_u'\|S^{t_{1g}}\|U_u'\rangle_p\right) \qquad (23.5.6)$$

As above, we use (18.1.8) for the space-only parts and find

$$\left\langle\left(\tfrac{1}{2}^+ E_g', T_{1u}\right)U_u'\right\|L^{t_{1g}}\left\|\left(\tfrac{1}{2}^+ E_g', T_{1u}\right)U_u'\right\rangle_p$$

$$= \begin{cases} -(\sqrt{2}/3)\langle T_{1u}\|L^{t_{1g}}\|T_{1u}\rangle & \text{for } p=0 \\ -(2\sqrt{2}/3)\langle T_{1u}\|L^{t_{1g}}\|T_{1u}\rangle & \text{for } p=1 \end{cases} \qquad (23.5.7)$$

$$\left\langle\left(\tfrac{1}{2}^+ E_g', T_{2u}\right)U_u'\right\|L^{t_{1g}}\left\|\left(\tfrac{1}{2}^+ E_g', T_{2u}\right)U_u'\right\rangle_p$$

$$= \begin{cases} (\sqrt{2}/3)\langle T_{2u}\|L^{t_{1g}}\|T_{2u}\rangle & \text{for } p=0 \\ -(2\sqrt{2}/3)\langle T_{2u}\|L^{t_{1g}}\|T_{2u}\rangle & \text{for } p=1 \end{cases} \qquad (23.5.8)$$

For the spin-only parts (18.1.9) and (15.4.9) give

$$\left\langle\left(\tfrac{1}{2}^+ E_g', T_{1u}\right)U_u'\right\|S^{t_{1g}}\left\|\left(\tfrac{1}{2}^+ E_g', T_{1u}\right)U_u'\right\rangle_p = \begin{cases} 1/\sqrt{3} & \text{for } p=0 \\ 2/\sqrt{3} & \text{for } p=1 \end{cases} \qquad (23.5.9)$$

$$\left\langle\left(\tfrac{1}{2}^+ E_g', T_{2u}\right)U_u'\right\|S^{t_{1g}}\left\|\left(\tfrac{1}{2}^+ E_g', T_{2u}\right)U_u'\right\rangle_p = \begin{cases} 1/\sqrt{3} & \text{for } p=0 \\ -2/\sqrt{3} & \text{for } p=1 \end{cases} \qquad (23.5.10)$$

We next use (19.4.6) and Table 19.6.1 to obtain

$$\langle T_{1u} \| L^{t_{1g}} \| T_{1u} \rangle = \langle t_{1u}(\pi + \sigma) \| l^{t_{1g}} \| t_{1u}(\pi + \sigma) \rangle$$

$$= \frac{-\sqrt{6}}{2}(b^2 + 2\sqrt{2}\,ab) \tag{23.5.11}$$

$$\langle T_{2u} \| L^{t_{1g}} \| T_{2u} \rangle = \langle t_{2u}(\pi) \| l^{t_{1g}} \| t_{2u}(\pi) \rangle$$

$$= \frac{-\sqrt{6}}{2} \tag{23.5.12}$$

When the metal $6p$ orbital is assumed to contribute to $t_{1u}(\pi + \sigma)$ as in (23.3.1), Eq. (23.5.11) becomes

$$\langle T_{1u} \| L^{t_{1g}} \| T_{1u} \rangle = \frac{-\sqrt{6}}{2}(b^2 + 2\sqrt{2}\,ab + 2c^2)$$

$$\equiv \frac{-\sqrt{6}}{2}\bar{z} \tag{23.5.13}$$

Inserting these results into (23.5.6) yields

$$\left\langle \left(\tfrac{1}{2}^+ E_g', T_{1u}\right) U_u' \middle\| \mu^{t_{1g}} \middle\| \left(\tfrac{1}{2}^+ E_g', T_{1u}\right) U_u' \right\rangle_p$$

$$= \begin{cases} -\dfrac{(\bar{z}+2)}{\sqrt{3}}\mu_B & \text{for} \quad p = 0 \\ -\dfrac{2(\bar{z}+2)}{\sqrt{3}}\mu_B & \text{for} \quad p = 1 \end{cases} \tag{23.5.14}$$

$$\left\langle \left(\tfrac{1}{2}^+ E_g', T_{2u}\right) U_u' \middle\| \mu^{t_{1g}} \middle\| \left(\tfrac{1}{2}^+ E_g', T_{2u}\right) U_u' \right\rangle_p$$

$$= \begin{cases} -\dfrac{\mu_B}{\sqrt{3}} & \text{for} \quad p = 0 \\ \dfrac{2}{\sqrt{3}}\mu_B & \text{for} \quad p = 1 \end{cases}$$

where $\bar{z}$ is defined by (23.5.13).

The electric-dipole reduced matrix elements are found using (18.1.8):

$$\left\langle\left(\tfrac{1}{2}^{+}E'_g, T_{2g}\right)E''_g\middle\| m^{t_{1u}}\middle\|\left(\tfrac{1}{2}^{+}E'_g, T_{1u}\right)U'_u\right\rangle = \frac{-\sqrt{2}}{\sqrt{3}}\langle T_{2g}\|m^{t_{1u}}\|T_{1u}\rangle$$

$$\left\langle\left(\tfrac{1}{2}^{+}E'_g, T_{2g}\right)E''_g\middle\| m^{t_{1u}}\middle\|\left(\tfrac{1}{2}^{+}E'_g, T_{2u}\right)E''_u\right\rangle = -\tfrac{2}{3}\langle T_{2g}\|m^{t_{1u}}\|T_{2u}\rangle$$

$$\left\langle\left(\tfrac{1}{2}^{+}E'_g, T_{2g}\right)E''_g\middle\| m^{t_{1u}}\middle\|\left(\tfrac{1}{2}^{+}E'_g, T_{2u}\right)U'_u\right\rangle = \frac{-\sqrt{2}}{3}\langle T_{2g}\|m^{t_{1u}}\|T_{2u}\rangle$$

(23.5.15)

Finally (20.2.10) gives

$$\langle T_{2g}\|m^{t_{1u}}\|T_{1u}\rangle$$

$$= \left\langle \mathcal{A}\left(t_{1u}^6(0A_{1g}), t_{2g}^5\left(\tfrac{1}{2}T_{2g}\right)\right)\tfrac{1}{2}T_{2g}\middle\| m^{t_{1u}}\middle\|\mathcal{A}\left(t_{1u}^5\left(\tfrac{1}{2}T_{1u}\right), t_{2g}^6(0A_{1g})\right)\tfrac{1}{2}T_{1u}\right\rangle$$

$$= 18\begin{Bmatrix} 0 & \tfrac{1}{2} & \tfrac{1}{2} \\ 0 & \tfrac{1}{2} & \tfrac{1}{2} \end{Bmatrix}\begin{Bmatrix} A_1 & T_2 & T_2 \\ T_1 & A_1 & T_1 \\ T_1 & T_2 & T_1 \end{Bmatrix}\langle t_{1u}\|m^{t_{1u}}\|t_{2g}\rangle$$

$$= (18)\left(-\tfrac{1}{2}\right)\left(\tfrac{1}{9}\right)\langle t_{1u}\|m^{t_{1u}}\|t_{2g}\rangle = -\langle t_{1u}\|m^{t_{1u}}\|t_{2g}\rangle \qquad (23.5.16)$$

and, as illustrated in (18.2.10),

$$\langle T_{2g}\|m^{t_{1u}}\|T_{2u}\rangle = \langle t_{2u}\|m^{t_{1u}}\|t_{2g}\rangle \qquad (23.5.17)$$

Substituting these results back into $\mathcal{D}_0$, $\mathcal{A}_1/\mathcal{D}_0$, and $\mathcal{C}_0/\mathcal{D}_0$ of (23.5.1) and (23.5.2), we obtain for $IrCl_6^{2-}$:

1. $u'_u(a) \to e''_g$:

$$\mathcal{D}_0 = \tfrac{1}{9}\left|\langle t_{1u}\|m^{t_{1u}}\|t_{2g}\rangle\right|^2$$

$$\frac{\mathcal{C}_0}{\mathcal{D}_0} = 1$$

$$\frac{\mathcal{A}_1}{\mathcal{D}_0} = -\frac{1}{2}\bar{z}$$

[where $\bar{z}$ is defined in (23.5.13)].

2. $e_u'' \rightarrow e_g''$:

$$\mathcal{D}_0 = \tfrac{2}{27}\left|\langle t_{2u}\|m^{t_{1u}}\|t_{2g}\rangle\right|^2$$

$$\frac{\mathcal{C}_0}{\mathcal{D}_0} = -2$$

$$\frac{\mathcal{A}_1}{\mathcal{D}_0} = -\frac{10}{3}$$

(23.5.18)

3. $u_u'(b) \rightarrow e_g''$:

$$\mathcal{D}_0 = \tfrac{1}{27}\left|\langle t_{2u}\|m^{t_{1u}}\|t_{2g}\rangle\right|^2$$

$$\frac{\mathcal{C}_0}{\mathcal{D}_0} = 1$$

$$\frac{\mathcal{A}_1}{\mathcal{D}_0} = \frac{11}{6}$$

We are ready to compare our results with experiment. Neglecting $\mathcal{B}_0$ terms, moment analysis of IIa yields $\mathcal{C}_0/\mathcal{D}_0$ values of 0.6 to 1.2 [results of Piepho et al. (1972b) multiplied by the factors from (17.1.1)], which agrees reasonably well with the rough theoretical value, $\mathcal{C}_0/\mathcal{D}_0 = 1$. $\mathcal{C}_0/\mathcal{D}_0$ values for the $E_g'' \rightarrow E_u''(^2T_{2u})$ and $E_g'' \rightarrow U_u'(^2T_{2u})$ transitions cannot be obtained separately from experiment by moment analysis, since the $^2T_{2u}$ spin–orbit splitting of $\frac{3}{4}\zeta_{\mathrm{Cl}^-} \approx 440\ \mathrm{cm}^{-1}$ is much less than the width of III. From the discussion in Section 7.6 we see that if $\mathcal{B}_0$ terms due to mixing with states outside the $^2T_{2u}$ manifold are neglected, the ratio of the zeroth moment of the MCD to the zeroth moment of absorption for III is $\mu_B B/kT$ times

$$\frac{\mathcal{C}_0}{\mathcal{D}_0}\left[E_g'' \rightarrow E_u'' + U_u'\right] = \frac{\mathcal{C}_0\left(E_g'' \rightarrow E_u''\right) + \mathcal{C}_0\left(E_g'' \rightarrow U_u'(^2T_{2u})\right)}{\mathcal{D}_0\left(E_g'' \rightarrow E_u''\right) + \mathcal{D}_0\left(E_g'' \rightarrow U_u'(^2T_{2u})\right)}$$

(23.5.19)

We then use (23.5.18) in (23.5.19) to obtain

$$\frac{\mathcal{C}_0}{\mathcal{D}_0}\left[E_g'' \rightarrow E_u''(^2T_{2u}) + U_u'(^2T_{2u})\right] = -1 \qquad (23.5.20)$$

which is identical to (23.2.8). This is in satisfactory agreement with results of -0.6 to -1.0 obtained by moment analysis for the band [results of Piepho et al. (1972b) multiplied by the factors from (17.1.1)].

Comparison of theoretical and experimental MCD parameters based on first-moment analysis for the same bands is far trickier, since the experimental parameter involves $\mathcal{B}_1$ and $\mathcal{C}_1$ in addition to $\mathcal{A}_1$. Things become particularly complex when JT effects are significant or when several states are near-degenerate. Conversely, $\mathcal{B}_1$ and $\mathcal{C}_1$ are zero when the BO approximation can be made for both ground and excited states and no overlapping bands are present. Unfortunately this is not the case here, since both the $^2T_{1u}$ and $^2T_{2u}$ states show evidence of strong JT effects, and the small $^2T_{2u}$ spin–orbit splitting leaves $E_u''(^2T_{2u})$ and $U_u'(^2T_{2u})$ nearly degenerate. Thus to obtain information from first-moment data for these bands, a vibronic (JT) calculation must be undertaken so that $\mathcal{H}_T$ eigenfunctions may be used in the moment calculations (Section 3.8).

Neither experimental nor theoretical first moments have been determined for IIa in $Cs_2ZrCl_6:Ir^{4+}$ (or for the analogous band in $OsCl_6^{2-}$). Band III, however, has been analyzed in some detail by Yeakel and Schatz (1974). They obtained $\mathcal{H}_T$ eigenfunctions using a truncated vibronic basis and used them to calculate the MCD of individual vibronic lines. Gaussian fits of the lowest-energy (mostly no-phonon) lines agreed well with experiment. These results may also be used to analyze the MCD of the analogous lines in $Cs_2ZrCl_6:Os^{4+}$ as discussed in Section 23.6.

Calculation of $IrCl_6^{2-}$ *Reduced Matrix Elements: The* jj *Approach*

In the jj approximation, the basis states are those of Table 23.4.1. Closed shells do not affect values of magnetic moments for states diagonal in configuration. Thus

$$\langle E_g''\|\mu^{t_{1g}}\|E_g''\rangle = \langle e_g''\, E_g''\|\mu^{t_{1g}}\|e_g''\, E_g''\rangle$$

$$= \langle e_g''\|\mu^{t_{1g}}\|e_g''\rangle \tag{23.5.21}$$

Likewise

$$\langle E_u''\|\mu^{t_{1g}}\|E_u''\rangle = \langle e_u''\|\mu^{t_{1g}}\|e_u''\rangle \tag{23.5.22}$$

and for both U_u' states of Table 23.4.1, Eq. (21.4.9) yields

$$\langle U_u'\|\mu^{t_{1g}}\|U_u'\rangle_p = \langle u_u'\|\mu^{t_{1g}}\|u_u'\rangle_p \tag{23.5.23}$$

The one-electron integrals are easy to evaluate since the jj orbitals are simply $|(\mathbb{S}S, h)rt\tau\rangle$-type orbitals. Thus (18.1.8) and (18.1.9) apply as illustrated in Section 21.4. From (21.4.11) we have

$$\langle e_g''\|\mu^{t_{1g}}\|e_g''\rangle = -\sqrt{6}\,\mu_B \tag{23.5.24}$$

For the $|(\frac{1}{2}e_g', t_{1u})u_u'\tau\rangle = |u_u'(a)\tau\rangle$ orbitals

$$\left\langle u_u'(a)\|\mu^{t_{1g}}\|u_u'(a)\right\rangle_p = \begin{cases} -\dfrac{\bar{z}+2}{\sqrt{3}}\mu_B & \text{for } p=0 \\ -\dfrac{2(\bar{z}+2)}{\sqrt{3}}\mu_B & \text{for } p=1 \end{cases} \tag{23.5.25}$$

where $\bar{z}$ is defined as in (23.5.13). Finally for the $|(\frac{1}{2}e_g', t_{2u})u_u'(b)\tau\rangle$ orbitals

$$\left\langle u_u'(b)\|\mu^{t_{1g}}\|u_u'(b)\right\rangle_p = \begin{cases} -\dfrac{\mu_B}{\sqrt{3}} & \text{for } p=0 \\ \dfrac{2}{\sqrt{3}}\mu_B & \text{for } p=1 \end{cases} \tag{23.5.26}$$

Only electric-dipole matrix elements remain to be calculated. The same configurational basis must be used for all off-diagonal-in-configuration matrix elements for a given MCD parameter; otherwise phase errors may occur in jj calculations which involve, for example, pseudo-$\mathcal{A}$ terms such as $\mathcal{A}_1'(T_{1u}, T_{1u}')$ in (23.6.6) below. In our present case the configurational basis has the form $p^l q^m s^k u^n = (u_u')^l(e_u'')^m(u_g')^k(e_g'')^n$ which may be immediately simplified as discussed in Section 21.2; then (21.4.20) may be used. (The order of p^l, q^m, s^k, and u^n may be chosen for convenience, but should be kept fixed for all matrix elements in the calculation of a given observable.) For example

$$\begin{aligned}
&\langle E_g''\|m^{t_{1u}}\|E_u''\rangle \\
&\quad= \Big\langle \mathcal{A}\Big\{\big[\big((u_u')^4, (e_u'')^2\big)A_{1g}\big], \big[\big((u_g')^4, e_g''\big)E_g''\big]\Big\}E_g''\Big\|m^{t_{1u}} \\
&\qquad\times\Big\|\mathcal{A}\Big\{\big[\big((u_u')^4, e_u''\big)E_u''\big], \big[\big((u_g')^4, (e_g'')^2\big)A_{1g}\big]\Big\}E_u''\Big\rangle \\
&\quad= \Big\langle \mathcal{A}\big\{(e_u'')^2(A_{1g}), e_g''(E_g'')\big\}E_g''\Big\|m^{t_{1u}}\Big\|\mathcal{A}\big\{e_u''(E_u''), (e_g'')^2(A_{1g})\big\}E_u''\Big\rangle \\
&\quad= \{E''\}\{E''T_1E''\}\langle e_u''\|m^{t_{1u}}\|e_g''\rangle \\
&\quad= -\langle e_u''\|m^{t_{1u}}\|e_g''\rangle
\end{aligned} \tag{23.5.27}$$

where the phase which enters in the second equality is $+1$ because $(u'_u)^4$ and $(u'_g)^4$ are uncoupled from $(e''_u)^m$ and $(e''_g)^n$ respectively via (21.2.13). For the U'_u state from the $u'_u(b) \to e''_g$ excitation

$$\langle E''_g \| m^{t_{1u}} \| U'_u \rangle$$

$$= \Big\langle \mathcal{Q}\big\{\big[\big((u'_u)^4, (e''_u)^2\big)A_{1g}\big], \big[\big((u'_g)^4, e''_g\big)E''_g\big]\big\} E''_g \Big\| m^{t_{1u}}$$

$$\times \Big\| \mathcal{Q}\big\{\big[\big((u'_u)^3, (e''_u)^2\big)U'_u\big], \big[\big((u'_g)^4, (e''_g)^2\big)A_{1g}\big]\big\} U'_u \Big\rangle$$

$$= -\Big\langle \mathcal{Q}\big\{(u'_u)^4(A_{1g}), e''_g(E''_g)\big\} E''_g \Big\| m^{t_{1u}} \Big\| \mathcal{Q}\big\{(u'_u)^3(U'_u), (e''_g)^2(A_{1g})\big\} U'_u \Big\rangle$$

$$= -\langle u'_u(b) \| m^{t_{1u}} \| e''_g \rangle \tag{23.5.28}$$

The phase factor in the second equality is now -1. Here $(u'_g)^4$ is again uncoupled from $(e''_g)^n$ via (21.2.13), but in this case (21.2.12) is required to uncouple $(e''_u)^2$ from $(u'_u)^l$. When l is odd (as it is in the ket) a phase factor is introduced (see discussion in Section 21.2). For $u'_u(a) \to e''_g$, using the configurational basis of Table 23.4.1, we find $\langle E''_g \| m^{t_{1u}} \| U'_u \rangle = \langle u'_u(a) \| m^{t_{1u}} \| e''_g \rangle$.

Once again we need to use (18.1.8) to obtain the remaining one-electron jj matrix elements in terms of orbital functions:

$$\langle u'_u(a) \| m^{t_{1u}} \| e''_g \rangle = \Big\langle \big(\tfrac{1}{2}e'_g, t_{1u}\big)u'_u \Big\| m^{t_{1u}} \Big\| \big(\tfrac{1}{2}e'_g, t_{2g}\big)e''_g \Big\rangle$$

$$= \frac{-\sqrt{2}}{\sqrt{3}} \langle t_{1u} \| m^{t_{1u}} \| t_{2g} \rangle$$

$$\langle e''_u \| m^{t_{1u}} \| e''_g \rangle = \Big\langle \big(\tfrac{1}{2}e'_g, t_{2u}\big)e''_u \Big\| m^{t_{1u}} \Big\| \big(\tfrac{1}{2}e'_g, t_{2g}\big)e''_g \Big\rangle$$

$$= -\tfrac{2}{3} \langle t_{2u} \| m^{t_{1u}} \| t_{2g} \rangle \tag{23.5.29}$$

$$\langle u'_u(b) \| m^{t_{1u}} \| e''_g \rangle = \Big\langle \big(\tfrac{1}{2}e'_g, t_{2u}\big)u'_u \Big\| m^{t_{1u}} \Big\| \big(\tfrac{1}{2}e'_g, t_{2g}\big)e''_g \Big\rangle$$

$$= \frac{\sqrt{2}}{3} \langle t_{2u} \| m^{t_{1u}} \| t_{2g} \rangle$$

Substituting all of these results back into $\mathcal{D}_0$, $\mathcal{A}_1/\mathcal{D}_0$, and $\mathcal{C}_0/\mathcal{D}_0$ of (23.5.1) and (23.5.2), we again obtain the results given in (23.5.18).

Calculation of $OsCl_6^{2-}$ ***Reduced Matrix Elements in the*** jj ***Approximation***

In this subsection we demonstrate that the $OsCl_6^{2-}$ MCD calculation in the jj approximation is essentially finished once the $IrCl_6^{2-}$ MCD calculation has been completed. The $\mathcal{A}_1/\mathcal{D}_0$ expression for each excitation in $OsCl_6^{2-}$ reduces to the $IrCl_6^{2-}$ $\mathcal{A}_1/\mathcal{D}_0$ expression for the same excitation. Also $\mathcal{D}_0$ for each excitation in $OsCl_6^{2-}$ is exactly double that for the same excitation in $IrCl_6^{2-}$.

$\mathcal{C}_0$ is zero for all excitations in $OsCl_6^{2-}$, since the ground state is nondegenerate. For $\mathcal{A}_1$ and $\mathcal{D}_0$ we use the equations of Section 21.4 to calculate magnetic moments and dipole strengths in terms of jj one-electron functions as illustrated in Section 21.4. From (21.4.7) we have for the $e_u'' \to e_g''$ excitation

$$\begin{aligned}\langle T_{1u}\|\mu^{t_{1g}}\|T_{1u}\rangle &= \Big\langle \mathcal{A}\big(e_u''(E_u''), e_g''(E_g'')\big)T_{1u}\Big\|\mu^{t_{1g}}\\ &\quad\times\Big\|\mathcal{A}\big(e_u''(E_u''), e_g''(E_g'')\big)T_{1u}\Big\rangle\\ &= \langle e_u''\|\mu^{t_{1g}}\|e_u''\rangle + \langle e_g''\|\mu^{t_{1g}}\|e_g''\rangle \qquad (23.5.30)\end{aligned}$$

so (23.5.3) for $e_u'' \to e_g''$ gives

$$\frac{\mathcal{A}_1}{\mathcal{D}_0} = \frac{\sqrt{6}}{3\mu_B}\big(\langle e_u''\|\mu^{t_{1g}}\|e_u''\rangle + \langle e_g''\|\mu^{t_{1g}}\|e_g''\rangle\big) \qquad (23.5.31)$$

which is seen to be identical to the result obtained for the same excitation in $IrCl_6^{2-}$ by substituting (23.5.21) and (23.5.22) into (23.5.2). For the $u_u'(a) \to e_g''$ and $u_u'(b) \to e_g''$ excitations, we obtain, using (21.4.6) and (21.4.9)

$$\begin{aligned}&\langle T_{1u}\|\mu^{t_{1g}}\|T_{1u}\rangle\\ &= \Big\langle \mathcal{A}\big((u_u')^3(U_u'), e_g''(E_g'')\big)T_{1u}\Big\|\mu^{t_{1g}}\\ &\quad\times\Big\|\mathcal{A}\big((u_u')^3(U_u'), e_g''(E_g'')\big)T_{1u}\Big\rangle\\ &= \frac{-1}{2\sqrt{2}}\langle u_u'\|\mu^{t_{1g}}\|u_u'\rangle_0 + \frac{1}{\sqrt{2}}\langle u_u'\|\mu^{t_{1g}}\|u_u'\rangle_1\\ &\quad-\tfrac{1}{2}\langle e_g''\|\mu^{t_{1g}}\|e_g''\rangle \qquad (23.5.32)\end{aligned}$$

Thus for $u'_u(a) \to e''_g$ and $u'_u(b) \to e''_g$, (23.5.3) gives

$$\frac{\mathcal{A}_1}{\mathcal{D}_0} = \frac{-1}{2\sqrt{3}\,\mu_B}\langle u'_u\|\mu^{t_{1g}}\|u'_u\rangle_0 + \frac{1}{\sqrt{3}\,\mu_B}\langle u'_u\|\mu^{t_{1g}}\|u'_u\rangle_1 - \frac{1}{\sqrt{6}\,\mu_B}\langle e''_g\|\mu^{t_{1g}}\|e''_g\rangle \tag{23.5.33}$$

This result is identical to the $IrCl_6^{2-}$ result obtained by substituting (23.5.23) and (23.5.21) into (23.5.1). Thus upon evaluation of the one-electron jj reduced matrix elements the very same results are obtained.

The calculation of the electric-dipole matrix elements proceeds much as for $IrCl_6^{2-}$ [Eqs. (23.5.27)–(23.5.28)] except that (21.4.15) is used instead of (21.4.20). If we choose the same configuration order as for $IrCl_6^{2-}$, we obtain

$$\begin{aligned}\langle A_{1g}\|m^{t_{1u}}\|T_{1u}\rangle &= \pm\Big\langle (u'_u)^4 A_{1g}\Big\|m^{t_{1u}}\Big\|\mathcal{A}\big((u'_u)^3(U'_u), e''_g\big)T_{1u}\Big\rangle \\ &= \pm\langle u'_u\|m^{t_{1u}}\|e''_g\rangle\end{aligned} \tag{23.5.34}$$

where the + sign applies for $u'_u(a) \to e''_g$ and the $-$ sign for $u'_u(b) \to e''_g$. Thus for $u'_u(a) \to e''_g$ and $u'_u(b) \to e''_g$

$$\mathcal{D}_0 = \tfrac{1}{3}\left|\langle u'_u\|m^{t_{1u}}\|e''_g\rangle\right|^2 \tag{23.5.35}$$

Just as in (23.5.28), for the $u'_u(b) \to e''_g$ case a -1 phase enters when the $(e''_u)^2$ configuration is uncoupled to obtain simplified configurations. For the $e''_u \to e''_g$ excitation

$$\begin{aligned}\langle A_{1g}\|m^{t_{1u}}\|T_{1u}\rangle &= \Big\langle (e''_u)^2 A_{1g}\Big\|m^{t_{1u}}\Big\|\mathcal{A}\big(e''_u(E''_u), e''_g\big)T_{1u}\Big\rangle \\ &= -\langle e''_u\|m^{t_{1u}}\|e''_g\rangle\end{aligned} \tag{23.5.36}$$

so for $e''_u \to e''_g$

$$\mathcal{D}_0 = \tfrac{1}{3}\left|\langle e''_u\|m^{t_{1u}}\|e''_g\rangle\right|^2 \tag{23.5.37}$$

These results are exactly double the $IrCl_6^{2-}$ results found by substituting (23.5.28) and (23.5.27) into (23.5.1) and (23.5.2) respectively. Thus the dipole-strength expressions in $OsCl_6^{2-}$ have the same *relative* magnitudes as in $IrCl_6^{2-}$ but are twice as large.

Breakdown of the jj Approximation in $OsCl_6^{2-}$

Our study of a number of related hexahalides, including $RuCl_6^{2-}$, $OsCl_6^{2-}$, $ReCl_6^{2-}$, $OsBr_6^{2-}$, $ReBr_6^{2-}$, and $MoCl_6^{2-}$, suggests that "metal–ligand" electrostatic interactions are very small and that we need only consider interelectronic repulsions within the states of a metal (or ligand) configuration (or part of a configuration). In the case of $OsCl_6^{2-}$ this means that only those within the ground-state t_{2g}^4 configuration are important (since t_{2g}^5, t_{1u}^5, and t_{2u}^5 configurations give rise to only one state of a given symmetry). Inclusion of them, of course, makes transitions to the upper *jj* levels no longer two-electron jumps (Section 23.4), so these *jj*-forbidden transitions become weakly allowed. But this does not change the $\mathcal{C}_1/\mathcal{D}_0$ values for the allowed *jj* transitions; dipole strengths are reduced, but relative magnitudes remain constant.

23.6. Jahn–Teller Effects in a Series of Complex Ions: The General Approach

The JT effect is another perturbation which, like spin–orbit coupling, may be large for some orbitals but not for others. However, it is hard to predict which vibrations will lead to large couplings with particular orbitals. Also, calculations are difficult, since vibronic, rather than purely electronic, coupling is involved. However, once it is established that a given orbital and a given vibration are strongly coupled in one system, we may expect strong JT effects for the same orbital in similar systems.

Since the JT operator is one-electron, the effect can be parametrized using one-electron reduced matrix elements, much like spin–orbit coupling and other perturbations. In the following sections we illustrate this approach using $IrCl_6^{2-}$ and $OsCl_6^{2-}$ as examples.

The Jahn–Teller Effect in $IrCl_6^{2-}$

Since $IrCl_6^{2-}$ has only one open shell per configuration, it is an excellent compound for determining which orbitals give rise to significant JT effects. What we are really interested in is the relative magnitudes of the JT effect and spin–orbit coupling for a particular orbital, for when spin–orbit coupling is large compared to vibronic coupling, the latter may often be ignored to a first approximation.

From a study of the vibronic structure of the $E_g''(^2T_{2g}) \rightarrow U_g'(^2T_{2g})$ intraconfigurational t_{2g}^5 transition in $Cs_2ZrCl_6 : Ir^{4+}$, we have evidence that the JT effect for the $t_{2g}(5d)$ orbital is insignificant compared with the

$t_{2g}(5d)$ spin–orbit coupling. This is not surprising, since $\zeta_{Ir^{4+}} \approx 2800\ cm^{-1}$ [Keiderling et al. (1975)]. The $U_g'(^2T_{2g})$ band shows the type of structure expected for a parity-forbidden $d \to d$ transition made allowed by weak odd-mode vibronic coupling (HT type, Section 3.7). There is no evidence, for example, of lines built on more than one origin, such as would be expected if the JT effect were significant (Section 3.8).

In contrast, study of the charge-transfer bands in $Cs_2ZrCl_6{:}Ir^{4+}$ indicates significant JT interactions for both the $t_{1u}(\pi + \sigma)$ and the $t_{2u}(\pi)$ orbitals. One reason is that $\zeta_{Cl^-} \approx 590\ cm^{-1}$, which is much smaller than $\zeta_{Ir^{4+}}$. Yeakel and Schatz (1974) have studied the $t_{2u}(\pi) \to t_{2g}$ band (III, Fig. 23.3.1) in detail and have shown that the coupling is strong enough there to obscure almost completely the spin–orbit splitting of the $^2T_{2u}$ state (Section 3.8). The $t_{1u}(\pi + \sigma) \to t_{2g}$ transition gives rise to two bands (IIa and IIb, Fig. 23.3.1), as is expected when spin–orbit coupling splits the $^2T_{1u}$ state. Spin–orbit coupling is over twice as large for the $t_{1u}(\pi + \sigma)$ orbital in $IrCl_6^{2-}$ (Section 23.3) as for the $t_{2u}(\pi)$ orbital. However, large JT interactions are also clearly present, since unusual vibronic structure is observed for both the $u_u' \to e_g''$ (IIa) and $e_u' \to e_g''$ (IIb) transitions. While this structure remains unassigned, it is doubtful if it could arise without strong vibronic interactions, that is, a JT effect in the $^2T_{1u}$ state.

The Jahn–Teller Effect in $OsCl_6^{2-}$

Here we demonstrate, by way of the $t_{2u}(\pi) \to t_{2g}$ example, how the JT effect in $OsCl_6^{2-}$ may be understood using the JT analysis for $IrCl_6^{2-}$. We limit ourselves to linear JT coupling, so the vibronic terms which contribute to the JT Hamiltonian are (Sections 3.3, 3.7, and 3.8)

$$\mathcal{H}_{JT} \approx \sum_{h\theta} U_{h\theta}^{\dagger} Q_{h\theta} \tag{23.6.1}$$

where $h = t_{2g}$ and e_g, and $U_{h\theta} = (\partial V/\partial Q_{h\theta})_{Q_0}$ is a one-electron operator. Only the t_{2g} and e_g vibrational modes are JT-active.

To solve the $\mathcal{H}_{SO} + \mathcal{H}_{JT}$ problem it is convenient to begin with an electronic basis composed of the 36 $t_{2u}^5 t_{2g}^5$ determinantal functions, (that is, $|\xi_u^+ \eta_u^2 \zeta_u^2 \xi_g^+ \eta_g^2 \zeta_g^2\rangle$ and so on), which are defined as in (19.1.2). In this basis we may use a result which is derived much like (20.1.3) except that here the determinantal functions are uncoupled. The basis may be written using determinantal functions of the type

$$|a_i^m b_j^n\rangle = [m!n!(m+n)!]^{-1/2} \sum_{\nu=1}^{(m+n)!} (-1)^{\nu} P_{\nu} |a_i^m\rangle |b_j^n\rangle \tag{23.6.2}$$

where i and j refer to specific sets of orbitals of the a^m and b^n configurations. If a_i^m and b_j^n have no orbitals in common and V is a one-electron operator, then

$$\langle a_i^m b_j^n | V | a_{i'}^m b_{j'}^n \rangle = \langle a_i^m | V | a_{i'}^m \rangle + \langle b_j^n | V | b_{j'}^n \rangle \tag{23.6.3}$$

In our case this means

$$\left\langle (t_{2u}^5)_i (t_{2g}^5)_j \middle| \mathcal{H}_{SO} + U_{h\theta}^\dagger \middle| (t_{2u}^5)_{i'} (t_{2g}^5)_{j'} \right\rangle$$

$$= \left\langle (t_{2u}^5)_i \middle| \mathcal{H}_{SO} + U_{h\theta}^\dagger \middle| (t_{2u}^5)_{i'} \right\rangle + \left\langle (t_{2g}^5)_j \middle| \mathcal{H}_{SO} + U_{h\theta}^\dagger \middle| (t_{2g}^5)_{j'} \right\rangle \tag{23.6.4}$$

Thus the $\mathcal{H}_{SO} + \mathcal{H}_{JT}$ matrix for $OsCl_6^{2-}$ may be written in terms of matrix elements *identical* to those occurring in different parts of the $IrCl_6^{2-}$ problem. The t_{2u}^5 matrix elements are *exactly* those involved in the analysis by Yeakel and Schatz (1974) of the $t_{2u}(\pi) \to t_{2g}$ transition in $IrCl_6^{2-}$, while the t_{2g}^5 matrix elements are precisely those relevant to the low-energy intraconfigurational $E_g''(^2T_{2g}) \to U_g'(^2T_{2g})$ band in $IrCl_6^{2-}$. Analysis of the latter band encourages us to make the approximation that in $OsCl_6^{2-}$

$$\left\langle (t_{2g}^5)_j \middle| U_{h\theta}^\dagger \middle| (t_{2g}^5)_{j'} \right\rangle \approx 0 \qquad (\text{for} \quad h = t_{2g} \text{ and } e_g) \tag{23.6.5}$$

since no significant JT effect was apparent for the t_{2g}^5 band in $IrCl_6^{2-}$. We also assume that the t_{2u} one-electron JT parameters—$\langle t_{2u} \| u^{t_{2g}} \| t_{2u} \rangle$ and $\langle t_{2u} \| u^{e_g} \| t_{2u} \rangle$—for $OsCl_6^{2-}$ are identical to those for $IrCl_6^{2-}$.

We now need to diagonalize the electronic matrix in $\mathcal{H}_{SO}$ and then solve the vibronic problem. Actually we already have the answer without doing any further work. Since the t_{2g}^5 functions are diagonal in $U_{h\theta}$ ($h = t_{2g}$ and e_g) in our approximation, the eigenvalues fall into two groups as illustrated in Fig. 23.6.1. Each group consists of a set of vibronic levels with the same relative energies as given by the Yeakel–Schatz calculation. The groups differ in energy by the t_{2g}^5 spin–orbit splitting. In the jj approximation, all levels in the upper group (b) are jj-forbidden. In the lower group (a) the (vibronic E_u'') $\otimes$ $E_g''(^2T_{2g})$ and the (vibronic U_u') $\otimes$ $E_g''(^2T_{2g})$ are allowed, while the (vibronic E_u') $\otimes$ $E_g''(^2T_{2g})$ are forbidden. Thus the intensity pattern predicted for the lines of group (a) is exactly that of the Yeakel–Schatz analysis. In the section which follows, the $OsCl_6^{2-}$ data are interpreted using these theoretical results.

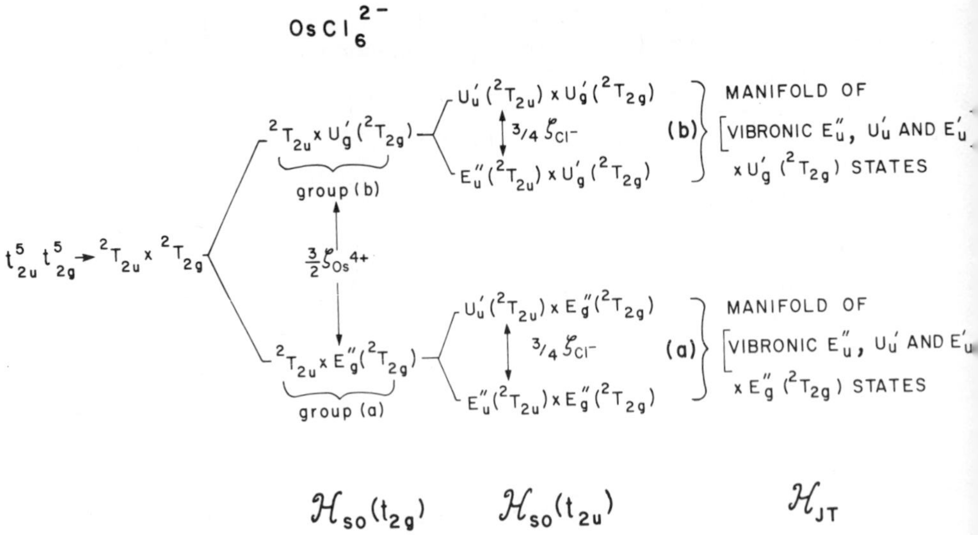

Figure 23.6.1. States of the $OsCl_6^{2-}$ $t_{2u}^5 t_{2g}^5$ configuration.

Analysis of the Absorption and MCD *Data for the* $OsCl_6^{2-}$ $t_{2u}(\pi) \rightarrow t_{2g}$ *Band in* $Cs_2ZrCl_6:Os^{4+}$

The 28,700-cm^{-1} absorption band in $OsCl_6^{2-}$ has a compelling similarity to the 22,900-cm^{-1} absorption band in $IrCl_6^{2-}$. Both band III's have an initial series of lines with approximately the same energy spacings (Fig. 23.4.2 and Table 23.6.1) and intensity pattern upon which are built totally symmetric progressions. Thus it seems highly reasonable to assign the initial lines following Yeakel and Schatz (1974) and to calculate the $OsCl_6^{2-}$ MCD parameters on that basis. The $OsCl_6^{2-}$ spectrum of Fig. 23.4.3 is first discussed in the context of the jj approximation. The effect of electrostatic interactions is then considered.

The $OsCl_6^{2-}$ MCD differs in several important respects from that for $IrCl_6^{2-}$. In the first place $OsCl_6^{2-}$ has an A_{1g} ground state and therefore can have no $\mathcal{C}$ terms. Secondly, the linewidths of the $Cs_2ZrCl_6:Ir^{4+}$ spectrum are far narrower than those of Fig. 23.4.3; thus $\mathcal{A}$ terms are much more prominent in $IrCl_6^{2-}$, since the peak-to-trough of an $\mathcal{A}$ term increases approximately as the inverse second power of the linewidth (Section 5.3). Thus while $\mathcal{A}_1/\mathcal{D}_0$ and $\mathcal{B}_0/\mathcal{D}_0$ for both individual vibronic lines and for the entire band are predicted in the jj approximation to have the same *magnitudes* for $IrCl_6^{2-}$ and $OsCl_6^{2-}$, the MCD spectra may *appear* quite different. In particular, $\mathcal{B}_0$ terms are important in the $Cs_2ZrCl_6:Os^{4+}$

Table 23.6.1. A Comparison of the Energies of the Observed Vibronic Lines within the Zeroth $\nu(a_{1g})$ Quantum for the $t_{2u}(\pi) \to t_{2g}$ Bands of $OsCl_6^{2-}$ and $IrCl_6^{2-}$

$IrCl_6^{2-}$		$OsCl_6^{2-}$	
Band III, Line[a]	Energy (cm^{-1})[a]	Band III, Line[a]	Energy (cm^{-1})[a]
0 0a	0 ≈ 5.8	A	0
1 1a 1b	51.1 59.2 71	B	54
2a 3	103.1 130.1	C	126
4a 4 4b	171.7 187.9 204.0	C′	192
5a 5	262.0 288.4	D	268

[a]The lines are numbered as in Fig. 23.4.2. Energies are given relative to the band origins.

spectrum, since linewidths are broad and no $\mathcal{C}$ terms are possible, while in $Cs_2ZrCl_6:Ir^{4+}$ $\mathcal{C}$ terms and $\mathcal{A}$ terms completely dominate the MCD.

We analyze the structure of the $OsCl_6^{2-}$ 28,700-cm^{-1} band, which is only semiresolved, in two ways: (1) the MCD first moment of the entire band is discussed, and (2) the MCD of the semiresolved vibronic lines is then considered. We do not discuss the zeroth moment of the MCD of the entire band, since it is experimentally very small, and it arises from interactions with states from all other configurations and so cannot be calculated accurately.

Moments may be easily extracted from experimental data when bands are isolated from one another (Chapter 7). They may then be compared with theoretical MCD parameters calculated for the entire band—in this case for the $t_{2u}(\pi) \to t_{2g}$ transition. The $\mathcal{A}_1$ contribution to the first moment for the band is independent of the vibronic details of the excited state (i.e. JT effects) and is independent of spin–orbit interactions so long as the ground state is diamagnetic. The discussion of Sections 7.4 and 7.5 indicates that here $\mathcal{B}_1$ contributions to the first moment are *not* independent of vibronic effects, since the spin–orbit coupling splits the Q_0 states. We have no reliable way of estimating such effects, since $\mathcal{B}_1$ involves magnetic mixing with states outside the band; thus we ignore $\mathcal{B}_1$ contributions in our calculations.

$\mathcal{A}_1$ and $\mathcal{D}_0$ are calculated as discussed in Section 7.6, since the $t_{2u}(\pi) \to t_{2g}$ jj-allowed excited states are nearly degenerate. In the absence of vibronic coupling they are separated by $\frac{3}{4}\zeta_{\mathrm{Cl}^-} \approx 440\ \mathrm{cm}^{-1}$ (see Fig. 23.6.1), and the spin–orbit splitting of the barycenters of these states is independent of vibronic interactions within the band. Since $440\ \mathrm{cm}^{-1}$ is considerably less than the band halfwidth of $1700\ \mathrm{cm}^{-1}$, the theory in Section 7.6 applies.

The equations applicable to this case (cf. Section 17.2), for $T_{1u} = \psi_1$ and $T'_{1u} = \psi_2$ of (21.4.1), give

$$\begin{aligned}
\mathcal{A}_1 &= \mathcal{A}_1(T_{1u}) + \mathcal{A}_1(T'_{1u}) + \mathcal{A}'_1(T_{1u}, T'_{1u}) \\
&= \frac{-2}{3\sqrt{6}\,\mu_B}\Big(\langle T_{1u}\|\mu^{t_{1g}}\|T_{1u}\rangle\langle A_{1g}\|m^{t_{1u}}\|T_{1u}\rangle\langle T_{1u}\|m^{t_{1u}}\|A_{1g}\rangle \\
&\quad + \langle T'_{1u}\|\mu^{t_{1g}}\|T'_{1u}\rangle\langle A_{1g}\|m^{t_{1u}}\|T'_{1u}\rangle\langle T'_{1u}\|m^{t_{1u}}\|A_{1g}\rangle \\
&\quad + 2\langle T'_{1u}\|\mu^{t_{1g}}\|T_{1u}\rangle\langle A_{1g}\|m^{t_{1u}}\|T'_{1u}\rangle\langle T_{1u}\|m^{t_{1u}}\|A_{1g}\rangle\Big) \\
\mathcal{D}_0 &= \mathcal{D}_0(T_{1u}) + \mathcal{D}_0(T'_{1u}) \\
&= \tfrac{1}{3}\Big(\left|\langle A_{1g}\|m^{t_{1u}}\|T_{1u}\rangle\right|^2 + \left|\langle A_{1g}\|m^{t_{1u}}\|T'_{1u}\rangle\right|^2\Big)
\end{aligned} \tag{23.6.6}$$

Many of these reduced matrix elements were calculated in Section 23.5. To complete the calculation we need $\langle T_{1u}\|m^{t_{1u}}\|A_{1g}\rangle$, $\langle T'_{1u}\|m^{t_{1u}}\|A_{1g}\rangle$, and $\langle T'_{1u}\|\mu^{t_{1g}}\|T_{1u}\rangle$. The electric-dipole reduced matrix elements may be found by equating (17.2.19) and (17.2.20) and using our earlier results. In this case

$$\begin{aligned}
\mathcal{D}_0 &= \tfrac{1}{3}|\langle A_{1g}\|m^{t_{1u}}\|T_{1u}\rangle|^2 \\
&= -\tfrac{1}{3}\langle A_{1g}\|m^{t_{1u}}\|T_{1u}\rangle\langle T_{1u}\|m^{t_{1u}}\|A_{1g}\rangle
\end{aligned} \tag{23.6.7}$$

so that (23.5.34)–(23.5.37) and (23.5.29) give

$$\begin{aligned}
\langle T_{1u}\|m^{t_{1u}}\|A_{1g}\rangle &= -\tfrac{2}{3}\langle t_{2u}\|m^{t_{1u}}\|t_{2g}\rangle^* \\
\langle T'_{1u}\|m^{t_{1u}}\|A_{1g}\rangle &= \frac{\sqrt{2}}{3}\langle t_{2u}\|m^{t_{1u}}\|t_{2g}\rangle^*
\end{aligned} \tag{23.6.8}$$

Next $\langle T'_{1u}\|\mu^{t_{1g}}\|T_{1u}\rangle$ is found using (21.4.6) [much as in (20.3.3)] and then the $p^{l-1}q^m$, p^lq^{m-1} analog of (21.4.19):

$$\begin{aligned}
&\langle T'_{1u}\|\mu^{t_{1g}}\|T_{1u}\rangle\\
&\quad=\Big\langle \mathcal{Q}\Big\{\Big[\big((u'_u)^3,(e''_u)^2\big)U'_u\Big],\Big[\big((u'_g)^4,e''_g\big)E''_g\Big]\Big\}T'_{1u}\Big\|\mu^{t_{1g}}\\
&\qquad\times\Big\|\mathcal{Q}\Big\{\Big[\big((u'_u)^4,e''_u\big)E''_u\Big],\Big[\big((u'_g)^4,e''_g\big)E''_g\Big]\Big\}T_{1u}\Big\rangle\\
&\quad=\tfrac{1}{2}\Big\langle \mathcal{Q}\big((u'_u)^3(U'_u),(e''_u)^2(A_{1g})\big)U'_u\Big\|\mu^{t_{1g}}\\
&\qquad\times\Big\|\mathcal{Q}\big((u'_u)^4(A_{1g}),e''_u(E''_u)\big)E''_u\Big\rangle\\
&\qquad+0\\
&\quad=\tfrac{1}{2}\langle e''_u\|\mu^{t_{1g}}\|u'_u\rangle
\end{aligned}\tag{23.6.9}$$

Finally (18.1.8), (18.1.9), Table 19.6.1, and (15.4.9) give

$$\begin{aligned}
\langle e''_u\|\mu^{t_{1g}}\|u'_u\rangle&=\Big\langle\big(\tfrac{1}{2}e'_g,t_{2u}\big)e''_u\|\mu^{t_{1g}}\|\big(\tfrac{1}{2}e'_g,t_{2u}\big)u'_u\Big\rangle\\
&=\frac{-5}{\sqrt{3}}\mu_B
\end{aligned}\tag{23.6.10}$$

Thus

$$\langle T'_{1u}\|\mu^{t_{1g}}\|T_{1u}\rangle=\frac{-5}{2\sqrt{3}}\mu_B\tag{23.6.11}$$

Substituting the above results and those of Section 23.5 into (23.6.6), we obtain

$$\begin{aligned}
\mathcal{Q}_1&=\frac{-2}{3\sqrt{6}\,\mu_B}\Bigg[\left(\frac{-5\sqrt{6}\,\mu_B}{3}\right)\left(\frac{2}{3}\right)\left(\frac{-2}{3}\right)+\left(\frac{11\mu_B}{2\sqrt{6}}\right)\left(\frac{-\sqrt{2}}{3}\right)\left(\frac{\sqrt{2}}{3}\right)\\
&\quad+2\left(\frac{-5\mu_B}{2\sqrt{3}}\right)\left(\frac{-\sqrt{2}}{3}\right)\left(\frac{-2}{3}\right)\Bigg]\big|\langle t_{2u}\|m^{t_{1u}}\|t_{2g}\rangle\big|^2
\end{aligned}\tag{23.6.12}$$

$$\quad=-\tfrac{1}{9}\big|\langle t_{2u}\|m^{t_{1u}}\|t_{2g}\rangle\big|^2$$

$$\mathcal{D}_0=\tfrac{2}{9}\big|\langle t_{2u}\|m^{t_{1u}}\|t_{2g}\rangle\big|^2$$

so

$$\frac{\mathcal{A}_1}{\mathcal{D}_0} = -\frac{1}{2} \qquad (23.6.13)$$

The $(\mathcal{A}_1 + \mathcal{B}_1)/\mathcal{D}_0$ value obtained experimentally from the ratio of the first moment of the MCD to the zeroth moment of absorption is -1.14, which has the same sign but about twice the magnitude of the theoretical $\mathcal{A}_1/\mathcal{D}_0$ value. The reason for this difference is not understood, but is probably in part due to $\mathcal{B}_1$. In general it is easier to imagine why an experimental MCD parameter is too *low*.

The MCD of the semiresolved vibronic structure cannot be analyzed in a rigorous way, but the general structure of the MCD can be explained qualitatively by reference to the $IrCl_6^{2-}$ results of Yeakel and Schatz (1974). We first consider the vibronic lines of Table 23.6.1. The Yeakel–Schatz calculations predict that $\approx$ 88% of the intensity of the lines in Table 23.6.1 arises from transitions to the E_u'' rather than the U_u' *vibronic* lines, or in $OsCl_6^{2-}$ to the (vibronic E_u'') $\otimes$ $E_g''(^2T_{2g})$ rather than the (vibronic U_u') $\otimes$ $E_g''(^2T_{2g})$ lines. All the $IrCl_6^{2-}$ E_u'' vibronic lines in this region are predicted to have negative $\mathcal{A}$ terms, the largest one being predicted for line 3, which is equivalent to our line C. In $OsCl_6^{2-}$ all the lines in Table 23.6.1 which show $\mathcal{A}$ terms do show negative ones, and the largest one is clearly that for line C [Fig. 23.4.3(*c*)]. Much of the MCD, however, is dominated by $\mathcal{B}$ terms, which for reasons discussed above are much more important in $OsCl_6^{2-}$ than in $IrCl_6^{2-}$. The pattern of $\mathcal{B}$ terms observed—line A, positive; line C, small; line D, negative—strongly suggests that an important portion of the $\mathcal{B}$-term intensity comes from interactions among these (vibronic E_u'') $\otimes$ $E_g''(^2T_{2g})$ lines. $\mathcal{B}$ terms are largest for magnetic mixing between neighboring lines, and they occur in equal and opposite pairs. If interaction between lines A and C produces the positive $\mathcal{B}$ term of line A, line C should show an equal and opposite negative $\mathcal{B}$ term. Then, if interaction between C and D produces a positive $\mathcal{B}$ term for line C, line D should have a negative $\mathcal{B}$ term. This would result in the pattern observed of a relatively small net $\mathcal{B}$ term (compared to the $\mathcal{A}$ term) for line C. As long as $\langle E_u''\alpha_i''|\mu_z|E_u''\alpha_j''\rangle > 0$, this pattern is predicted, since then the sign of $\mathcal{B}$ depends on the sign of $E_j - E_i$; calculation of the $\langle E_u''\alpha_i''|\mu_z|E_u''\alpha_j''\rangle$ for the $\mathcal{B}$ term above using the vibronic E_u'' vectors of Yeakel and Schatz confirms that they are positive and strongly supports our assignments.

The net positive MCD observed in the region of lines A–D is explained by magnetic interaction with additional (vibronic E_u'') $\otimes$ $E_g''(^2T_{2g})$ + [zero-quantum $\nu(a_{1g})$] lines to higher energy. Only $\approx$ 65% of the (vibronic

E_u'') ⊗ $E_g''(^2T_{2g})$ lines of the zero-quantum $\nu(a_{1g})$ set fall within the energy range of Table 23.6.1. It is all these interactions as well as those among the vibronic U_u' lines and the E_u'' and U_u' lines that add up to give the negative first moment observed for the entire band.

The structure of the remainder of the band can also be explained in a qualitative way by recalling that the barycenter of the (vibronic U_u') ⊗ $E_g''(^2T_{2g})$ lines is predicted to be 440 cm^{-1} to the blue of that for the (vibronic E_u'') ⊗ $E_g''(^2T_{2g})$ lines. Thus towards the middle of the band there are substantial contributions to the intensity from both $E_u'' \otimes E_g''(^2T_{2g})$ and $U_u' \otimes E_g''(^2T_{2g})$ lines, while the high-energy tail of the band arises from predominately $U_u' \otimes E_g''(^2T_{2g})$ lines; the overall ratio of $E_u'' \otimes E_g''(^2T_{2g})$ to $U_u' \otimes E_g''(^2T_{2g})$ intensity is 2 : 1. Thus the MCD of the higher-energy end of the band shows a rough progression in positive $\mathscr{A}$ terms, as expected for the $U_u' \otimes E_g''(^2T_{2g})$ lines.

A partial breakdown of the jj approximation in the $t_{2u}^5 t_{2g}^5$ levels should not modify the above picture substantially. In our treatment we have formally included all the vibronic interactions between the JT-active modes in the linear approximation and the $t_{2u}^5 t_{2g}^5$ electronic states. In so doing we produce a manifold of (vibronic E_u') ⊗ $E_g''(^2T_{2g})$, (vibronic E_u'') ⊗ $E_g''(^2T_{2g})$, and (vibronic U_u') ⊗ $E_g''(^2T_{2g})$ levels. Since $E_u' \otimes E_g'' = A_{2u} + T_{2u}$, $E_u'' \otimes E_g'' = A_{1u} + T_{1u}$, and $U_u' \otimes E_g'' = E_u + T_{1u} + T_{2u}$, each of these (vibronic Γ_u) ⊗ $E_g''(^2T_{2g})$ levels at this stage represents a set of exactly degenerate *vibronic* states. But since the ground state is A_{1g}, transitions are allowed *only* to the T_{1u} vibronic levels. There is no way the A_{1u}, A_{2u}, E_u, and T_{2u} vibronic levels may gain intensity directly from the $t_{2u}^5 t_{2g}^5$ vibronic T_{1u} levels, since electrostatic interactions do not mix them, and all important vibronic interactions within the band have already been taken into account. [Note that the $\nu(t_{1g})$ lattice mode, which formally couples the states, involves the librational motion of the MX_6^{2-} octahedron as a rigid unit and does not distort the octahedron; thus it is not expected to be important in the vibronic coupling.] Interactions with other states should enable the vibronic A_{1u}, A_{2u}, E_u, and T_{2u} states to gain intensity, but transitions to them are expected to be far weaker than those to the T_{1u} levels. Their presence, however, is probably one reason the $t_{2u}(\pi) \rightarrow t_{2g}$ linewidths are much greater in $Cs_2ZrCl_6 : Os^{4+}$ than in $Cs_2ZrCl_6 : Ir^{4+}$. Electrostatic interactions can mix the T_{1u} levels to some extent and redistribute intensity among the vibronic T_{1u} lines. The marked similarity between OsX_6^{2-} and IrX_6^{2-}, however, suggests that these electrostatic interactions are small in relation to spin–orbit and JT effects.

Inclusion of electrostatic interactions in calculating either the A_{1g} ground-state or the $t_{2u}^5 t_{2g}^5$ excited-state wavefunctions destroys the jj selection rule that forbids transitions to the (b) levels of Fig. 23.6.1. These (b) levels are expected in the *jj* approximation to be roughly $\frac{3}{2}\zeta_{Os^{4+}} \approx 3600$ cm^{-1} to the blue of the (a) levels, and so it is reasonable to assign the 32,100-cm^{-1} band, which lies $\approx 3400\ cm^{-1}$ from the 28,700-cm^{-1} band, to excitations to the (b) energy levels; as predicted, no comparable band is observed in the $IrCl_6^{2-}$ spectrum.

23.7. How to Treat Small Distortions from High Symmetry

Often the symmetry of a system is very close to that of a group G, but the true symmetry is known to be (or is suspected to be) that of a lower point group G_1. In such circumstances the distortion is easily handled theoretically using the methods of Butler (1981). The key step in such calculations is to employ a $G \supset \cdots \supset G_1$ basis (for notation, see Section 15.2). Thus the basis appropriate for a trigonally distorted octahedral complex is $O_h \supset D_{3d}$ (strong-field case) or $O_3 \supset O_h \supset D_{3d}$ (weak-field case). Use of such a group–subgroup basis enormously simplifies calculations, since matrix elements in the lower group may be expressed directly in terms of those for the higher group (Section 15.4). Those familiar with older methods for handling distortion calculations will appreciate the simplicity of this method, which we now illustrate with an example.

Example: The $t_{2u}(\pi) \rightarrow t_{2g}$ Transitions in Trigonally Distorted $IrCl_6^{2-}$

As an example we consider the $t_{2u}(\pi) \rightarrow t_{2g}$ transitions of $IrCl_6^{2-}$ (Sections 23.2, 23.3, 23.5) in a trigonally distorted crystalline environment. We calculate the t_{2g}^5 and t_{2u}^5 spin–orbit and distortion matrices and then $\mathcal{C}_0/\mathcal{D}_0$ for the $t_{2u}(\pi) \rightarrow t_{2g}$ transitions. The distortion is assumed small, so the ground and excited states may be described using octahedral t_{2g}^5 and $t_{2u}(\pi)^5$ configurations respectively. The t_{2g}^5 and t_{2u}^5 matrices have identical form except for parity labels; thus we first ignore parity and calculate t_2^5 matrices using an $O \supset D_3 \supset C_3$ spin–orbit basis. The results are then applied to both the t_{2g}^5 and $t_{2u}(\pi)^5$ configurations by affixing parity labels.

For each group-O t_2^5 configuration the spin–orbit basis consists of the $|(\mathbb{S}S, h)rt\tau\rangle = |(\mathbb{S}S, h)t(O)t_1(D_3)\tau_1\rangle$ states,

$$
\begin{aligned}
&\left|\left(\tfrac{1}{2}E', T_2\right)E''E'\tau_1\right\rangle, && \tau_1 = \alpha', \beta' \\
&\left|\left(\tfrac{1}{2}E', T_2\right)U'E'\tau_1\right\rangle, && \tau_1 = \alpha', \beta' \qquad (23.7.1) \\
&\left|\left(\tfrac{1}{2}E', T_2\right)U'P_1\rho_1\right\rangle, && \left|\left(\tfrac{1}{2}E', T_2\right)U'P_2\rho_2\right\rangle
\end{aligned}
$$

Here we use Butler (1981) chain notation to label the $O \supset D_3$ states (see Sections 15.2 and E.4). The basis is diagonal to first order in $\mathcal{H}_{SO}$, but not in the trigonal distortion $\mathcal{H}^{\text{trig}}$. A distortion Hamiltonian which splits degeneracies of the higher group (here O_h) transforms as A_{1g} or A_1 in the lower symmetry group (here D_{3d}), but not in the higher group (O_h). Here we see from Section E.5 that $\mathcal{H}^{\text{trig}}$ must transform as $T_{2g}(O_h)A_{1g}(D_{3d})$, or, dropping parity labels, as $T_2(O)A_1(D_3)$.

Since $\mathcal{H}^{\text{trig}}$ transforms as A_1 in the group D_3, it will only mix states with identical group-D_3 irrep and irrep-component labels. This is illustrated schematically in Fig. 23.7.1. The $\mathcal{H}^{\text{trig}} = V^{t_2 a_1 a_1} = V^{t_2 a_1}_{a_1} = V^{t_2}_{a_1 a_1}$ matrix elements are calculated using (15.4.1)–(15.4.2) followed by (18.1.8). Thus, for example,

$$\left\langle \left(\tfrac{1}{2}E', T_2\right)E''E'\alpha' \middle| V^{t_2}_{a_1 a_1} \middle| \left(\tfrac{1}{2}E', T_2\right)U'E'\alpha' \right\rangle^{O \supset D_3 \supset C_3}$$

$$= \begin{pmatrix} E'' \\ E' \end{pmatrix}^{O}_{D_3} \begin{pmatrix} E' \\ \alpha' \end{pmatrix}^{D_3}_{C_3} \begin{pmatrix} E'' & T_2 & U' \\ E' & A_1 & E' \end{pmatrix}^{O}_{D_3} \begin{pmatrix} E' & A_1 & E' \\ \beta' & a_1 & \alpha' \end{pmatrix}^{D_3}_{C_3}$$

$$\times \left\langle \left(\tfrac{1}{2}E', T_2\right)E'' \middle\| V^{t_2} \middle\| \left(\tfrac{1}{2}E', T_2\right)U' \right\rangle^{O}$$

$$= (1)(1)\left(\frac{-1}{\sqrt{3}}\right)\left(\frac{1}{\sqrt{2}}\right)\left[\sqrt{2}\,(2)(1)(1)\frac{1}{2\sqrt{3}}\langle T_2 \| V^{t_2} \| T_2 \rangle^{O}\right]$$

$$= -\tfrac{1}{3}\langle T_2 \| V^{t_2} \| T_2 \rangle \tag{23.7.2}$$

where we use Sections E.8 and E.9 for the $D_3 \supset C_3$ $2jm$ and $3jm$, Butler [(1981), pp. 242-243] for the $O \supset D_3$ $2jm$ and $3jm$ factors, and Sections

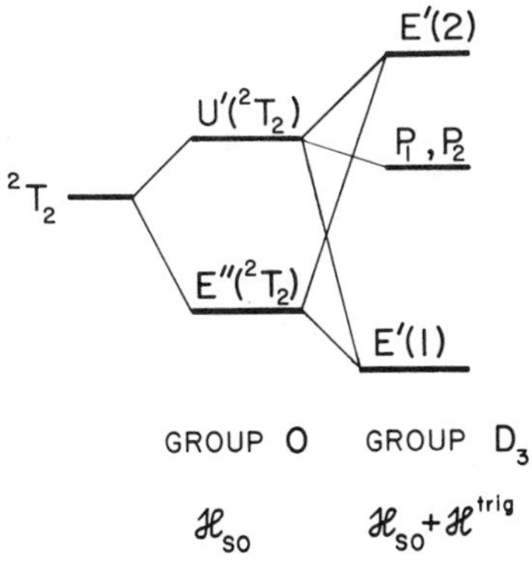

Figure 23.7.1. States of a t_2^5 configuration for a trigonally distorted octahedral complex.

C.11 and C.13 for the group-O coefficients needed in conjunction with (18.1.8). The spin–orbit matrix elements we require are given in Table 19.6.2. Thus the t_2^5 matrices are

$\mathcal{H}^{\text{trig}} + \mathcal{H}_{\text{SO}}$	$E''E'\tau_1$	$U'E'\tau_1$
$E''E'\tau_1$	$-k\zeta$	$-\frac{1}{3}\langle T_2\|V^{t_2}\|T_2\rangle$
$U'E'\tau_1$	$-\frac{1}{3}\langle T_2\|V^{t_2}\|T_2\rangle$	$\frac{-1}{3\sqrt{2}}\langle T_2\|V^{t_2}\|T_2\rangle + \frac{k}{2}\zeta$

(23.7.3)

$\mathcal{H}^{\text{trig}} + \mathcal{H}_{\text{SO}}$	$U'P_i\rho_i$
$U'P_i\rho_i$	$\frac{1}{3\sqrt{2}}\langle T_2\|V^{t_2}\|T_2\rangle + \frac{k}{2}\zeta$

(23.7.4)

where $k\zeta = \frac{1}{2}\zeta_{\text{Cl}^-}$ and $\zeta_{\text{Ir}^{4+}}$ for $t_{2u}(\pi)$ and $t_{2g}(d)$ respectively, and the $U'P_i\rho_i$ matrix applies for $i = 1, 2$.

With the trigonal distortion, the $t_{2u}(\pi) \to t_{2g}$ excitation gives rise to three allowed transitions. If we label the eigenvectors of (23.7.3) in the group D_3 as $E'_g(i)$, $i = 1, 2$, and $E'_u(i)$, $i = 1, 2$, these are

$$\begin{aligned} &E'_g(1) \to E'_u(1) \\ &E'_g(1) \to (P_{1u}, P_{2u}) \\ &E'_g(1) \to E'_u(2) \end{aligned} \tag{23.7.5}$$

where the $E'(i)$ are linear combinations of $|(\frac{1}{2}E', T_2)E''E'\tau_1\rangle$ and $|(\frac{1}{2}E', T_2)U'E'\tau_1\rangle$ basis states with the appropriate parity labels. For the first and third transitions, $\mathcal{C}_0/\mathcal{D}_0$ may be calculated with (17.2.25), since no complex irreps or repeated representations enter in the calculation. Using Table 17.2.1 for $D_3 \supset C_3$, Section E.8, and Section E.10, we find

$$\frac{\mathcal{C}_0}{\mathcal{D}_0}\left[E'_g(1) \to E'_u(i)\right] = \frac{-\sqrt{2}}{\mu_B}\langle E'_g(1)\|\mu^{a_{2g}}\|E'_g(1)\rangle^{D_{3d}} \tag{23.7.6}$$

for $i = 1, 2$. Then for the second transition of (23.7.5), the formalism of Section 17.6 applies. $\mathcal{C}_0/\mathcal{D}_0$ is calculated using (17.6.20) along with the same tables to obtain

$$\frac{\mathcal{C}_0}{\mathcal{D}_0}\left[E'_g(1) \to (P_{1u}, P_{2u})\right] = \frac{\sqrt{2}}{\mu_B}\left\langle E'_g(1)\middle\|\mu^{a_{2g}}\middle\|E'_g(1)\right\rangle^{D_{3d}} \tag{23.7.7}$$

The next step is to use the parameter-dependent eigenvector for the ground state $E'_g(1)$ of (23.7.3) to express the magnetic-moment reduced matrix element in (23.7.6)–(23.7.7) in terms of reduced matrix elements of our basis states. The latter are then easily evaluated. For example, if there were no trigonal distortion in the ground state, we would have $|E'_g(1)\tau_1\rangle = |(\frac{1}{2}{}^+E'_g, T_{2g})E''_g E'_g\tau_1\rangle$. Then (15.4.3), the $O \supset D_3$ $3jm$ factors from Butler [(1981), pp. 242–243], and (23.2.2) would give

$$\left\langle E'_g(1)\|\mu^{a_{2g}}\|E'_g(1)\right\rangle^{D_{3d}} = \langle E''_g E'_g\|\mu^{t_{1g}a_{2g}}\|E''_g E'_g\rangle^{O_h \supset D_{3d}}$$

$$= \begin{pmatrix} E'' & T_1 & E'' \\ E' & A_2 & E' \end{pmatrix}^{O}_{D_3} \langle E''_g\|\mu^{t_{1g}}\|E''_g\rangle^{O_h}$$

$$= \left(\frac{-1}{\sqrt{3}}\right)\left(-\sqrt{6}\,\mu_B\right) = \sqrt{2}\,\mu_B \qquad (23.7.8)$$

More often, the form of $E'_g(1)$ is more complicated, and a sum of matrix elements (which includes the one for $E''_g E'_g\tau_1$ above) must be calculated.

If the distortion is small, the three bands may overlap; it may then only be possible to obtain MCD parameters for the combined band. In such a case the MCD calculations above are inappropriate and $\mathcal{C}_0/\mathcal{D}_0$ for the combined band should be calculated (Section 7.6).

23.8. Analysis of the MCD of Vibronic Lines: The $IrBr_6^{2-}$ Spectrum

$IrBr_6^{2-}$ in the Cs_2ZrBr_6 host gives a very well-resolved spectrum [Fig. 23.8.1, from Dickenson et al. (1972)]. The analysis of the vibronic lines serves as an excellent example of the use of the theory of vibration-induced MCD developed in Chapter 22. $IrBr_6^{2-}$ differs from $IrCl_6^{2-}$ mainly in that $\zeta_{Br^-} \approx 2460\ cm^{-1}$ as compared to $\zeta_{Cl^-} \approx 590\ cm^{-1}$, so the excited-state spin–orbit splittings are larger. Results of MCD calculations for $IrCl_6^{2-}$ may otherwise be carried over to $IrBr_6^{2-}$. Figure 23.8.2 for $IrBr_6^{2-}$ is analogous to Fig. 23.2.2 for $IrCl_6^{2-}$, and the band numbers are chosen to stress the close correspondence. Thus, for example, bands IIa and IIb are assigned to $E''_g \to U'_u(^2T_{1u})$ and $E''_g \to E'_u(^2T_{1u})$ respectively, just as in $IrCl_6^{2-}$. In $IrBr_6^{2-}$ band III is now split into band IIIa, which we assign to $E''_g \to E''_u(^2T_{2u})$, and band IIIb, which we assign to $E''_g \to U'_u(^2T_{2u})$. Bands Ia and Ib are likewise assigned to $E''_g \to U'_g(^2T_{1g})$ and $E''_g \to E'_g(^2T_{1g})$ respectively.

As in $IrCl_6^{2-}$, the peculiar shape of the $U'_u(^2T_{1u})$ band (band IIa) is presumed to arise from Jahn–Teller (JT) coupling. The transitions giving

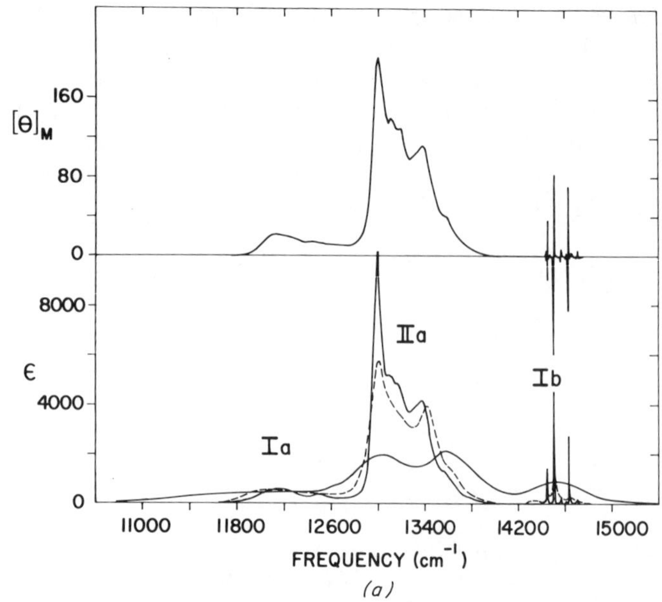

Figure 23.8.1. The absorption and MCD spectrum of $Cs_2ZrBr_6:Ir^{4+}$ from Dickinson et al. (1972). $[\theta]_M$ is the MCD in molar ellipticity units (see Section A5). ϵ is the molar extinction coefficient. (*a*) Bands Ia, IIa, and Ib. The MCD and sharpest absorption spectrum (solid line) were run at ≈ 8 K, the dashed one was run at liquid-nitrogen temperature, and the broad solid one at room temperature. (*b*) Bands IIIa, IIb, and IIIb. Temperatures are as in (*a*). (*c*) Detail of band Ib including hot bands. The cold spectrum was run at ≈ 8 K. The hot bands have been blown up along the ordinate for clarity, the factors being 10 and 2 for $[\theta]_M$ and ϵ respectively. (*d*) Detail of bands IIIa and IIb including hot bands. The cold spectrum was run at ≈ 8 K. The hot bands have been blown up along the ordinate for clarity, the factors being 40 and 3.75 for $[\theta]_M$ and ϵ respectively. (*e*) Detail of band IIIb including hot bands. The cold spectrum was run at ≈ 8 K. The hot bands have been blown up along the ordinate for clarity, the factors being 20 and 4.17 for $[\theta]_M$ and ϵ respectively.

rise to IIb and IIIa are so strongly coupled vibronically that the "forbidden" $E_u'(^2T_{1u})$ band has the intensity of an allowed transition. The detailed structure of these bands is not understood. Band IIIa seems to be dominated by a progression (lines 0, 1, 2, 3, and 4) in a low-energy mode, upon which is built the $\nu_1(a_{1g})$ progression.

JT coupling in band IIIb appears weak compared to that of band III in $IrCl_6^{2-}$. This is in part because spin–orbit coupling is so much larger in the hexabromides. The structure of band IIIb is very roughly a long $\nu_1(a_{1g})$

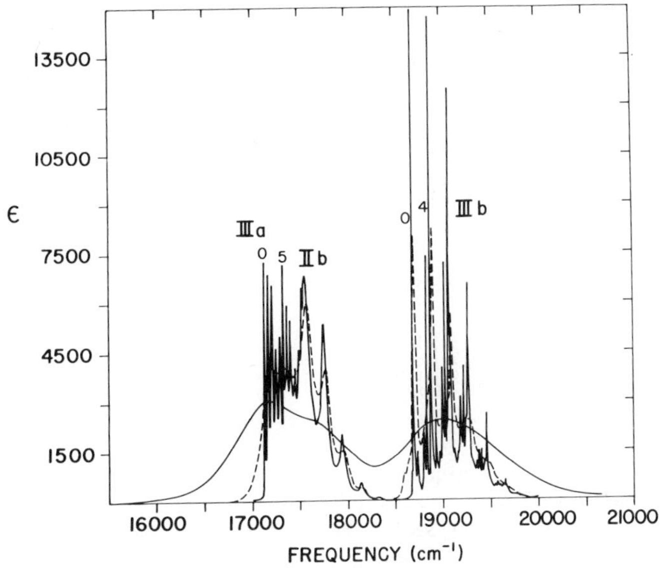

(b)

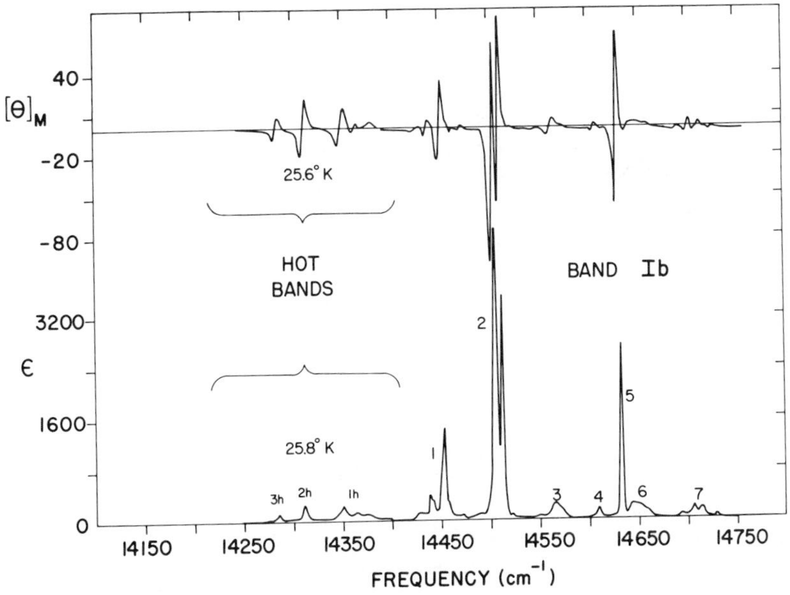

(c)

Figure 23.8.1. (*Continued*)

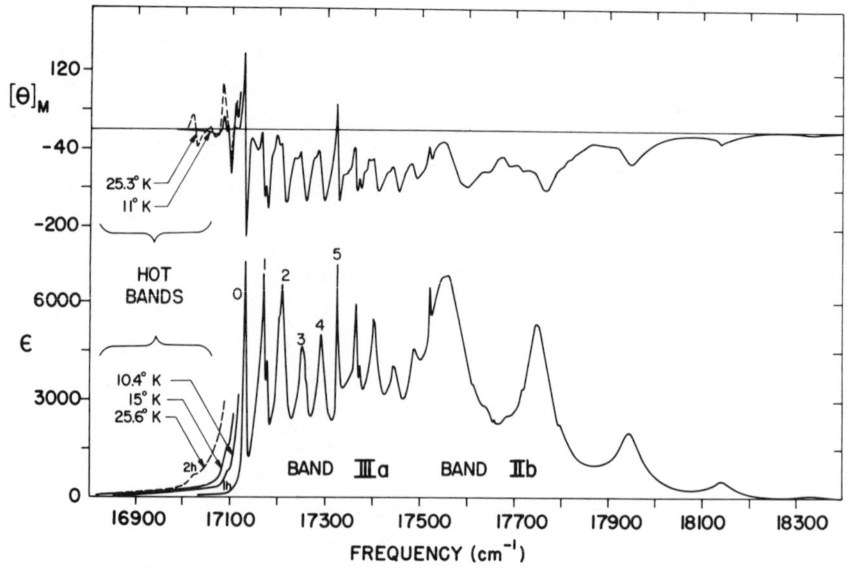

(d)

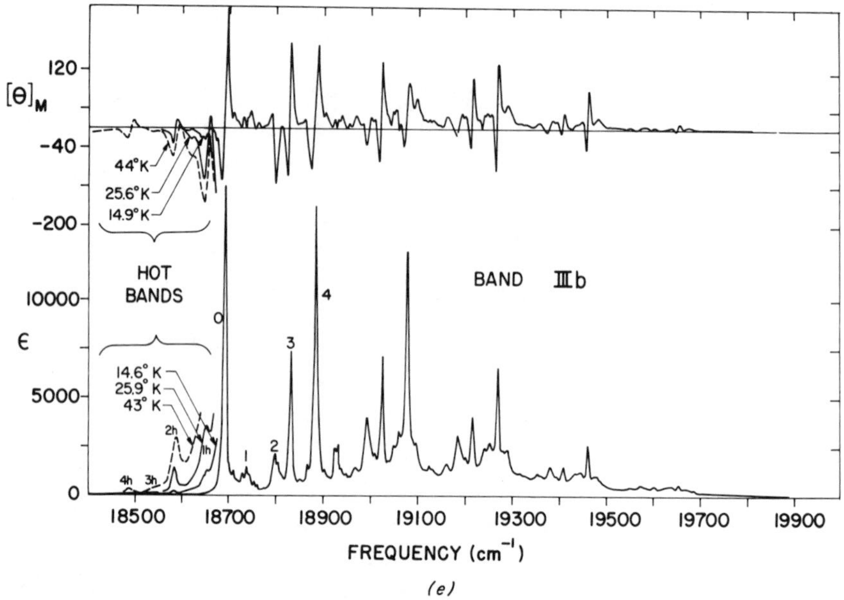

(e)

Figure 23.8.1. (*Continued*)

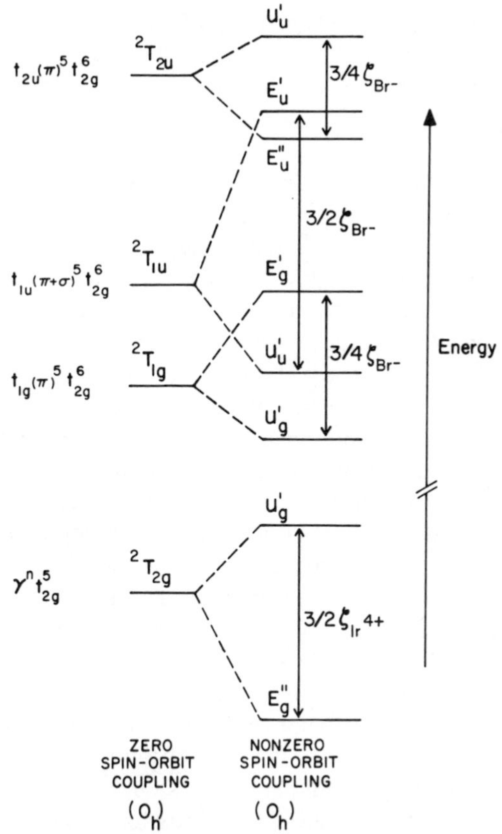

Figure 23.8.2. Energy-level diagram for $IrBr_6^{2-}$ for states of $\gamma^6 t_{2g}^5$ and low-lying $\gamma^5 t_{2g}^6$ configurations. The energy-level spacings are roughly to scale, and the order corresponds to our assignments.

progression built both upon a very strong no-phonon line (line 0) and on the weaker forbidden lines such as $E_g'' \rightarrow U_u' + \nu_2(e_g)$ (line 2) and $E_g'' \rightarrow U_u' + \nu_5(t_{2g})$ (line 3). The latter gain intensity through weak vibronic (JT) coupling to the no-phonon line.

As discussed below, studies of the absorption intensity as a function of temperature [see Fig. 23.8.1] support the assignment of bands IIa, IIb, IIIa, and IIIb to allowed transitions. In contrast, the temperature dependence of band Ib [Fig. 23.8.1(c)] is typical of an electric-dipole-forbidden transition which gains intensity through vibronic [Herzberg–Teller (HT)] coupling to other allowed states (Section 3.7). While band IIb is assigned to the $E_u'(^2T_{1u})$ transition, which is nominally electric-dipole-forbidden, the very strong coupling postulated between $E_u'(^2T_{1u})$ and $E_u''(^2T_{2u})$ states would redistribute intensity within the E_u' and E_u'' vibronic manifolds; thus we can understand why both bands show the temperature dependence typical of an allowed transition.

The Use of Hot-Band Data in $IrBr_6^{2-}$ *Assignments*

$Cs_2ZrBr_6:Ir^{4+}$ is extremely well suited for hot-band studies (Section 3.6), since the spectrum contains three bands with well-isolated origins, and sharp-line structure for two of the three persists even at liquid-nitrogen temperature. Hot-band studies give comparable information to that obtained from emission spectra, yet have the advantage that data may be obtained for any band with sufficiently sharp structure provided that initial lines are not appreciably overlapped by lower-energy bands. Hot-band data for bands Ib, IIIa, and IIIb of Fig. 23.8.1 are tabulated in Table 23.8.1.

The data strongly support the assignment of band Ib to a forbidden transition, since the hot and cold lines are reflected about an obvious gap in the spectrum at $\sim 14{,}400\ cm^{-1}$, where the no-phonon line would be expected for an allowed band. On the other hand, hot-band data for bands IIIa and IIIb show line 0 in both cases to be a strong no-phonon line, in keeping with their assignment to allowed transitions.

Table 23.8.1 gives assignments of hot bands to specific ground-state vibrational modes. Those for band Ib fit best with assignment to odd-parity vibrational modes, supporting assignment of the band to a *parity*-forbidden transition (as opposed to assignment, for example, to E_u', which is symmetry-forbidden but parity-allowed). Hot bands for IIIa and IIIb are assigned to even-parity modes.

The MCD *of* $Cs_2ZrBr_6:Ir^{4+}$

Since the amplitude of an $\mathcal{A}$ term to a $\mathcal{C}$ term goes as $kT/$(bandwidth), the MCD of narrow lines can be dominated by $\mathcal{A}$ terms while for broader bands the $\mathcal{A}$ terms are generally swamped by $\mathcal{C}$ terms. At low temperatures $\mathcal{B}$ terms are expected to be negligible compared to $\mathcal{C}$ terms, since their respective contributions are in the ratio $kT/\Delta W$.

The $\mathcal{C}$ *Terms of the Broader Bands*

Looking first at the MCD of the broader bands, we find that Ia and IIa have positive $\mathcal{C}$ terms, and IIIa and IIb have negative ones. The positive $\mathcal{C}$ term for IIa agrees with our earlier calculation for $u_u'(a) \rightarrow e_g''$ of (23.5.18), which predicts $\mathcal{C}_0/\mathcal{D}_0 = 1$ for the $E_g'' \rightarrow U_u'(^2T_{1u})$ electronically allowed

Table 23.8.1. $Cs_2ZrBr_6:Ir^{4+}$ **Hot-Band Data Near Liquid-Helium Temperature, from Dickinson et al. (1972), Compared with Ground-State Vibrational Energies**

Band Ib Fig. 23.8.1(*c*)		Band IIIa Fig. 23.8.1(*d*)		Band IIIb Fig. 23.8.1(*e*)		Hot-band assignment	Ground-State Vibrational Frequencies at Room Temperature		
Line	Energy (cm^{-1})[a]	Line	Energy (cm^{-1})	Line	Energy (cm^{-1})		Mode	K_2IrBr_6[c]	Cs_2SnBr_6[d]
1*h*	26	1*h*	36	1*h*	38	Lattice modes	$\nu_8(t_{1g})$		(46)[e]
	37		46		47		$\nu_9(t_{2g})$		(49)[e]
	51		(≈ 70)[b]		(≈ 70–75)[b]		$\nu_7(t_{1u})$	60	61
2*h*	89					ν_6	$\nu_6(t_{2u})$		(75)[e]
		2*h*	108	2*h*	108–112	ν_5	$\nu_5(t_{2g})$		109
3*h*	116					ν_4	$\nu_4(t_{1u})$	82	118
				3*h*	≈ 151	ν_2	$\nu_2(e_g)$		138
				4*h*	≈ 210	ν_1	$\nu_1(a_{1g})$		185
							$\nu_3(t_{1u})$	235	222
					≈ 258	ν_1 + lattice			

[a] Measured from the virtual origin estimated at 14,401 cm^{-1}.
[b] Very weak lines.
[c] Debeau, M. (1969), *Spectrochim. Acta* **25A**, 1311.
[d] Debeau, M., and H. Poulet (1969), *Spectrochim. Acta* **25A**, 1553.
[e] Parentheses designate calculated values.

Table 23.8.2. Energies of Initial Vibronic Lines and MCD Data for Bands Ib and IIIb of $Cs_2ZrBr_6:Ir^{4+}$ at Liquid-Helium Temperature, from Dickinson et al. (1972)

Band	Line	Energy (cm^{-1})	Assignment	$\mathcal{C}_1/\mathcal{D}_0$ Experimental	$\mathcal{C}_1/\mathcal{D}_0$ Theoretical
Ib		14,401[a]	$E_g'(^2T_{1g})$		
	1	27, 37, 51	+ lattice	0.70[b]	
	2	(≈ 88)[c], 105, 111, (≈ 122)[c]	$+\nu_6(t_{2u}), \nu_4(t_{1u})$	0.70[b,d]	1.0[e,f,g]
	3	166		> 0	
	4	208		> 0	
	5	231	$+\nu_3(t_{1u})$	0.70[b]	1.0[e,g]
	6	≈ 245	+ (lattice + $\nu_1(a_{1g})$)		
	7	291, 304, 312, 327	$+(\nu_6(t_{2u})+\nu_1(a_{1g}))$, $(\nu_4(t_{1u})+\nu_1(a_{1g}))$	0.50[b,d]	1.0[e,f,g]
IIIb	0	18,693	$U_u'(^2T_{2u})$ no-phonon	1.2[h]	$\frac{11}{6}$[g]
	1	(19)[c], 39, 47, (52)[c], (59)[c], 62, 68	+ lattice		
	2	107, 113	$+\nu_5(t_{2g})$	> 0	$\frac{7}{6}$[i,g]
	3	140	$+\nu_2(e_g)$	1.2[h]	$\frac{7}{6}$[j,g]
	4	194	$+\nu_1(a_{1g})$	> 0	$\frac{11}{6}$[g]
Entire band IIIb			$U_u'(^2T_{2u})$	≈ 2[b]	$\frac{11}{6}$[g]

[a] Estimated virtual origin.
[b] Using method of moments.
[c] Parentheses indicate shoulders.
[d] Total value for the four features.
[e] Assuming coupling to a U_u' state (see text for details) and using an orbital reduction factor of unity.
[f] Value for either vibration separately.
[g] Using an orbital reduction factor of unity and neglecting two-center integrals.
[h] By Gaussian fit.
[i] Assuming coupling to a $U_u'(^2T_{2u})$ state (see text).
[j] Assuming coupling to a U_u' state (see text).

Table 23.8.3. Theoretical Values of $\mathcal{A}_1/\mathcal{D}_0$ and $\mathcal{C}_0/\mathcal{D}_0$ for Some Vibronic Transitions[a]

Transition from $E_g''(^2T_{2g})$ Ground State	$\mathcal{A}_1/\mathcal{D}_0$: Symmetry of coupling vibration		$\mathcal{C}_0/\mathcal{D}_0$: Symmetry of coupling vibration		Symmetry of mixing state
$t_{1g}(\pi) \to t_{2g}$	$\nu(t_{1u})$	$\nu(t_{2u})$	$\nu(t_{1u})$	$\nu(t_{2u})$	
$U_g'(^2T_{1g})$	$-\frac{1}{10}$	$\frac{1}{6}$	1	1	U_u'[e]
	-3	$-\frac{1}{3}$	-2	-2	E_u''
$E_g'(^2T_{1g})$	1	1	1	1	U_u'
	Forbidden	-2	Forbidden	-2	E_u''
$t_{1u}(\pi+\sigma) \to t_{2g}$	$\nu(t_{2g})$	$\nu(e_g)$	$\nu(t_{2g})$	$\nu(e_g)$	
$E_u'(^2T_{1u})$	$(1 \text{ to } \frac{14}{9})$[b]	$(1 \text{ to } \frac{2}{3})$[b]	1	1	U_u'
	$(-2 \text{ to } -\frac{20}{9})$[b]	Forbidden	-2	Forbidden	E_u''
$t_{2u}(\pi) \to t_{2g}$	$\nu(t_{2g})$	$\nu(e_g)$	$\nu(t_{2g})$	$\nu(e_g)$	
$U_u'(^2T_{2u})$	$\frac{7}{6}$	$\frac{7}{6}$[c]	1	1	U_u'[f]
	$-\frac{7}{3}$	$-\frac{7}{3}$	-2	-2	E_u''
	No-phonon line[d]: $\mathcal{A}_1/\mathcal{D}_0 = \frac{11}{6}$		No-phonon line[d]: $\mathcal{C}_0/\mathcal{D}_0 = 1$		
$E_u''(^2T_{2u})$	$\frac{5}{3}$	$\frac{5}{3}$	1	1	U_u'
	0	0	0	0	E_u''
	No-phonon line[d]: $\mathcal{A}_1/\mathcal{D}_0 = -\frac{10}{3}$		No-phonon line[d]: $\mathcal{C}_0/\mathcal{D}_0 = -2$		

[a] Orbital reduction factor of unity assumed for simplicity, and two-center integrals neglected.
[b] Values depend on $\langle t_{1u}(\pi+\sigma)\|l\|t_{1u}(\pi+\sigma)\rangle$. Those given are for the cases of pure π ($a = 0$, $b = 1$) and maximum σ–π mixing ($a = 1/\sqrt{3}$, $b = \sqrt{2}/\sqrt{3}$) (see Table 19.6.1).
[c] Applies for any U_u' mixing state.
[d] Or entire band.
[e] For $\mathcal{A}_1/\mathcal{D}_0$ calculation assumes $U_u' = U_u'(^2T_{1u})$.
[f] For $\mathcal{A}_1/\mathcal{D}_0$ calculation assumes $U_u' = U_u'(^2T_{2u})$.

transition. If band Ia, the $E_g'' \to U_g'(^2T_{1g})$ transition, gains intensity by HT coupling with the nearby $U_u'(^2T_{1u})$ state, it too is predicted to have $\mathcal{C}_0/\mathcal{D}_0 = 1$; here we have used (22.4.3) assuming $N = U_u'$ and $h = t_{1u}$ and/or t_{2u}. Bands IIIa and IIb have the $\mathcal{C}$-term sign expected for the $E_g'' \to E_u''(^2T_{2u})$ electronically allowed transition [see (23.5.18) for $e_u'' \to e_g''$]. The MCD calculation here is not the HT type of Chapter 22, since we are very far indeed from the weak-coupling limit. Rather the applicable theory is that of Section 7.6 for near-degenerate E_u' and E_u'' states. However, since $E_g'' \to E_u'(^2T_{1u})$ is symmetry-forbidden, the zeroth-moment expression for the MCD or absorption for the entire band reduces to that of the isolated allowed $E_g'' \to E_u''(^2T_{2u})$ transition. Thus the $\mathcal{C}_0/\mathfrak{D}_0$ of (23.5.18) for $e_u'' \to e_g''$ is just that predicted for the combined IIIa + IIb band.

The $\mathcal{A}$ Terms of the Sharp Lines

The HT theory of Chapter 22 may be used to understand the MCD of the vibronic lines of the $E_g'' \to E_g'(^2T_{1g})$ band (Ib) and the $\nu_5(t_{2g})$ and $\nu_2(e_g)$ lines of the $E_g'' \to U_u'(^2T_{2u})$ band (IIIb). Results are given in Tables 23.8.2 and 23.8.3. With the approximations of Section 22.4 the calculation is straightforward from (22.4.3) and (22.4.4) except in the $E_g'' \to U_u' + \nu_i$ and $E_g'' \to U_g' + \nu_i$ cases, where the mixing state is of U' symmetry and ν_i belongs to a t_2 or t_1 irrep. In those cases repeated irrep labels occur in $\mathcal{A}(J, J/N)_h$ and prevent cancellation of electric-dipole and vibronic-mixing reduced matrix elements in the $|(\mathcal{S}S, h)rt\tau\rangle$ basis. Cancellation at a later stage is possible, however, if a specific U_u' mixing state is specified, such as $U_u'(^2T_{1u})$ or $U_u'(^2T_{2u})$ (Table 23.8.3). Then (18.1.8) may be used to reduce $|(\mathcal{S}S, h)rt\tau\rangle$ matrix elements to $|\mathcal{S}h\mathfrak{M}\theta\rangle$ matrix elements, as in our $CoCl_4^{2-}$ example of Section 18.2. The band-IIIb case is really that for a very weak JT effect, but such a calculation is identical to a HT calculation with $N = J$.

The allowed no-phonon line of band IIIb (and the totally symmetric progression built upon it) is predicted to have the same $\mathcal{A}_1/\mathfrak{D}_0$ as for the entire $E_g'' \to U_u'(^2T_{2u})$ band, Section 23.5—see (23.5.18) for $u_u'(b) \to e_g''$. This follows from the discussion in Section 7.9, since $\mathcal{B}_1$ and $\mathcal{C}_1$ should be close to zero because JT coupling for the band is small.

Finally, some of the lines of bands IIIa and IIb are sharp and show distinct $\mathcal{A}$ terms. However, calculation of $\mathcal{A}_1/\mathfrak{D}_0$ for these lines requires diagonalization of the vibronic matrix to obtain $\mathcal{H}_T$ eigenvectors, a lengthy calculation which we have not attempted. $\mathcal{A}_1/\mathfrak{D}_0$ could then be calculated much as it was for band III in $IrCl_6^{2-}$ in the Yeakel–Schatz treatment.

Appendix A *MCD Conventions and Notation*

Unfortunately, several different nomenclatures and conventions have appeared in the literature. In this book we have consistently followed the nomenclature and conventions of Stephens (1976). These are in turn an outgrowth of an infamous international gathering at the King's Arms* in September 1974. Stephens and Dr. Bernard Briat (Laboratoire d'Optique Physique EPCI, Paris) are the two individuals who have been most concerned with the question of notation and conventions.

In this appendix, we summarize these conventions and related matters. We also indicate where specific relevant formulas are introduced in this book.

A.1. Circularly Polarized Light

At a fixed point in space for light propagating in the $+Z$ direction,

$$\mathbf{E}_{R,L} \equiv \mathbf{E}_{\pm} = \frac{E^0}{\sqrt{2}}(\mathbf{e}_X \cos 2\pi\nu t \mp \mathbf{e}_Y \sin 2\pi\nu t) \tag{A.1.1}$$

This is (4.1.2) with $Z = 0$. Thus looking at the light source, we have

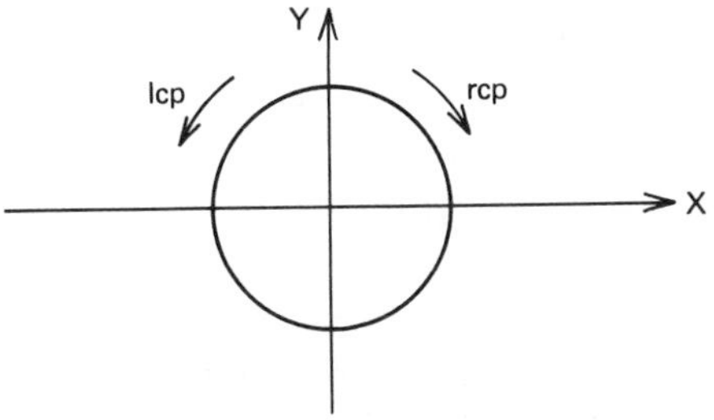

(A.1.2)

This is the conventional chemist's definition of rcp and lcp [Eyring, Walter,

*Parks Road and Hollywell Street, Oxford, England.

and Kimball (1944), Section 17c]. Physicists sometimes use the opposite convention.

A.2. (M)CD Sign

The CD sign is positive when absorption for lcp is greater than for rcp. Thus $\Delta A = A_{-} - A_{+} \equiv A_L - A_R$, the definition followed in Chapter 4.

A.3. Magnetic-Field Sign

The magnetic field is defined to be positive when it is parallel to the light propagation direction. Thus the magnetic field is positive when it points toward the detector [Eq. (4.3.2)].

A.4. MCD Equations Using the New [Stephens (1976)] Definitions of $\mathcal{A}_1$, $\mathcal{B}_0$, $\mathcal{C}_0$, and $\mathcal{D}_0$

The MCD has been commonly expressed in the literature in three different ways: as the absorbance difference ($\Delta A' = A'_L - A'_R$), as the absorption-coefficient difference ($\Delta k = k_L - k_R$), and as the molar ellipticity ($[\theta]_M$). We first interrelate these quantities. We then write the rigid-shift equations for $\Delta A'$, A, Δk, k, and $[\theta]_M$ using the new Stephens definitions of the Faraday parameters $\mathcal{A}_1$, $\mathcal{B}_0$, $\mathcal{C}_0$, and $\mathcal{D}_0$—the ones used exclusively throughout this book. Finally, in Section A.5 we relate these parameters and the equations containing them to several earlier definitions.

The absorbance A and the absorption coefficients k and κ may be related using (2.1.5), (2.1.6), and (4.2.3):

$$A \equiv \log\left(\frac{I_0}{I}\right) = \epsilon c l = \kappa l \log e \tag{A.4.1}$$

$$\kappa = \frac{4\pi\nu k}{c} \tag{A.4.2}$$

where $\log \equiv \log_{10}$, c is the sample concentration in moles/liter, l is the path length in cm, ν is the frequency, and c is the velocity of light.

The molar ellipticity $[\theta]_M$ enters in the following way. We start with the real part of (4.2.13) and note that ϕ/l is the (optical) rotation in rad cm^{-1}. In the traditional units of natural optical activity, the rotation in degrees per decimeter of path length for a solution of concentration 1 g cm^{-3} is designated as the specific rotation $[\alpha]$. The molar rotation $[\alpha]_M$ is the rotation in degrees per meter of path length for a 1 molar solution. Therefore

$$[\alpha]_M = [\alpha]\left(\frac{M}{100}\right) = \frac{18N_0\phi}{\pi Nl} = \frac{18{,}000\phi}{\pi cl} \tag{A.4.3}$$

where M is the molecular weight of the active species, N is its concentration in molecules cm^{-3}, and N_0 is Avogadro's number. In direct analogy, the molar ellipticity is defined using the imaginary part of (4.2.13):

$$[\theta]_M = \frac{18N_0\theta}{\pi Nl} = \frac{18{,}000\theta}{\pi cl} \tag{A.4.4}$$

where θ/l is in rad cm^{-1}.

To relate $[\theta]_M$ and Δk ($\equiv k_L - k_R$) to ΔA ($\equiv A_L - A_R$), we first note from (4.2.13) and (4.2.4) that

$$\frac{\theta}{l} = \frac{\pi\nu}{c}\Delta k \tag{A.4.5}$$

Then using (A.4.1), (A.4.2), and (A.4.4), we obtain

$$[\theta]_M = \left(\frac{4500}{\pi \log e}\right)\frac{\Delta A}{cl} = 3298.2\frac{\Delta A}{cl} \tag{A.4.6}$$

$$\Delta k = \frac{c}{4\pi\nu \log e}\frac{\Delta A}{l} \tag{A.4.7}$$

For a transition $a \to j$, the fundamental equations of MCD spectroscopy are given by (4.3.9) and (4.3.10):

$$\frac{\Delta A'}{\mathcal{E}} = \gamma\sum_{aj}\frac{N'_a - N'_j}{N}\left(|\langle a|m_-|j\rangle'|^2 - |\langle a|m_+|j\rangle'|^2\right)\rho'_{aj}(\mathcal{E}) \tag{A.4.8}$$

$$\frac{A}{\mathcal{E}} = \frac{\gamma}{2}\sum_{aj}\frac{N_a - N_j}{N}\left(|\langle a|m_-|j\rangle|^2 + |\langle a|m_+|j\rangle|^2\right)\rho_{aj}(\mathcal{E}) \tag{A.4.9}$$

where

$$\gamma \equiv \bar{\gamma} c l = \frac{2N_0 \pi^3 \alpha^2 c l \log e}{250 hcn} \tag{A.4.10}$$

$$\bar{\gamma} = 3.266 \times 10^{38} \frac{\alpha^2}{n} \tag{A.4.11}$$

$\mathcal{E} = h\nu$, N_a and N_j are the numbers of molecules per cm^3 in states a and j, $\rho_{aj}(\mathcal{E})$ is the lineshape function, and α^2/n is an effective-field correction (Section 2.11). In our notation, a prime designates a field-dependent quantity. [This differs slightly from Stephens (1976) and (1974), where primes are not used but superscript zeros designate zero-field quantities.] Equation (A.4.11) is in cgs units. (Note that our $\bar{\gamma}$ is Stephens's γ.) In addition

$$m_{\pm 1} \equiv \mp \frac{1}{\sqrt{2}}(m_x \pm i m_y) = \mp m_{\pm} \tag{A.4.12}$$

In the linear limit using the rigid-shift and BO approximations (Section 4.5), Eq. (A.4.8) becomes (4.5.14):

$$\frac{\Delta A'}{\mathcal{E}} = \frac{\Delta \epsilon' c l}{\mathcal{E}} = \gamma \mu_B B \left[\mathcal{A}_1 \left(\frac{-\partial f}{\partial \mathcal{E}} \right) + \left(\mathcal{B}_0 + \frac{\mathcal{C}_0}{kT} \right) f \right] \tag{A.4.13}$$

where $f \equiv f(\mathcal{E})$. The corresponding equations for $[\theta]_M$ and Δk follow immediately from (A.4.6) and (A.4.7). (Put a prime on A in these equations. We do not put primes on $[\theta]_M$ or Δk, to be consistent with the literature notation.)

Expressing the electric-dipole-moment operator in Debye units (10^{-18} esu cm) and using (A.4.11), Eq. (A.4.13) becomes

$$\frac{\Delta A'}{\mathcal{E}} = \frac{\Delta \epsilon' c l}{\mathcal{E}} = 326.6 \mu_B B c l \left[\mathcal{A}_1 \left(\frac{-\partial f}{\partial \mathcal{E}} \right) + \left(\mathcal{B}_0 + \frac{\mathcal{C}_0}{kT} \right) f \right] \tag{A.4.14}$$

In exactly the same way, using (7.8.11), we obtain for the corresponding moment equations

$$\left\langle \frac{\Delta A'}{\mathcal{E}} \right\rangle_0 = 326.6 \left(\mathcal{B}_0 + \frac{\mathcal{C}_0}{kT} \right) \mu_B B c l$$

$$\left\langle \frac{\Delta A'}{\mathcal{E}} \right\rangle_1^{\bar{\mathcal{E}}} = 326.6 \mathcal{A}_1 \mu_B B c l \tag{A.4.15}$$

[For simplicity we drop the factor (α^2/n) in (A.4.14), (A.4.15), and hereafter since we usually work with ratios, for example $\mathcal{A}_1/\mathcal{D}_0$, $\mathcal{B}_0/\mathcal{D}_0$, $\mathcal{C}_0/\mathcal{D}_0$, $\langle \Delta A'/\mathcal{E}\rangle_0/\langle A/\mathcal{E}\rangle_0$, $\langle \Delta A'/\mathcal{E}\rangle_1^{\mathcal{E}}/\langle A/\mathcal{E}\rangle_0$, which are independent of this factor.] Note that the right-hand sides of (A.4.14) and (A.4.15) are independent of energy units provided that *the same ones* are used for $\mu_B B$, $\mathcal{E}$, kT, and the energy denominators in $\mathcal{B}_0$. Finally, since $\mu_B = 0.4669\ \text{cm}^{-1}/\text{T}$, these equations can be written

$$\frac{\Delta A'}{\mathcal{E}} = \frac{\Delta\epsilon' c l}{\mathcal{E}} = 152.5 B c l\left[\mathcal{A}_1\left(\frac{-\partial f}{\partial \mathcal{E}}\right) + \left(\mathcal{B}_0 + \frac{\mathcal{C}_0}{kT}\right) f\right] \quad \text{(A.4.16)}$$

and

$$\left\langle \frac{\Delta A'}{\mathcal{E}} \right\rangle_0 = 152.5\left(\mathcal{B}_0 + \frac{\mathcal{C}_0}{kT}\right) B c l$$
$$\left\langle \frac{\Delta A'}{\mathcal{E}} \right\rangle_1^{\mathcal{E}} = 152.5 \mathcal{A}_1 B c l \quad \text{(A.4.17)}$$

if all energies are in cm^{-1} and B is in tesla. Similarly (A.4.9) becomes, in the rigid-shift and BO approximations [(4.5.22), (4.5.23)],

$$\frac{A}{\mathcal{E}} = \frac{\epsilon c l}{\mathcal{E}} = 326.6 \mathcal{D}_0 f c l \quad \text{(A.4.18)}$$

and the corresponding moment equation from (7.8.11) reads

$$\left\langle \frac{A}{\mathcal{E}} \right\rangle_0 = 326.6 \mathcal{D}_0 c l \quad \text{(A.4.19)}$$

We recommend that (A.4.16)–(A.4.19) *be the standard forms used in the literature* and reiterate that they apply for electric dipole moments in Debye units, that all energies are in cm^{-1}, and the magnetic field is in tesla.

Using (A.4.6), Eq. (A.4.16) translates to

$$\frac{[\theta]_M}{\mathcal{E}} = 5.028 \times 10^5 B\left[\mathcal{A}_1\left(\frac{-\partial f}{\partial \mathcal{E}}\right) + \left(\mathcal{B}_0 + \frac{\mathcal{C}_0}{kT}\right) f\right] \quad \text{(A.4.20)}$$

where Debye units, cm^{-1}, and tesla are required. Using (A.4.7), Eqs.

(A.4.13) and (A.4.18) translate to

$$\Delta k = 2\pi^2 N\mu_B B\left[\mathcal{A}_1\left(\frac{-\partial f}{\partial \mathcal{E}}\right) + \left(\mathcal{B}_0 + \frac{\mathcal{C}_0}{kT}\right)f\right] \tag{A.4.21}$$

$$k^0 = 2\pi^2 N\mathcal{D}_0 f \tag{A.4.22}$$

noting that $N_0 c = 10^3 N$. The form of the corresponding moment equations should be obvious [compare (A.4.14) and (A.4.15), (A.4.16) and (A.4.17), and (A.4.18) and (A.4.19)].

Note that (A.4.13)–(A.4.22) are applicable to several different cases. If $\mathcal{A}_1$ and $\mathcal{C}_0$ are given by (4.5.16), then Eqs. (A.4.13)–(A.4.17), (A.4.20), and (A.4.21) apply only for oriented or isotropic molecules and require that the eigenfunctions be diagonal in μ_z. *The operators are space-fixed* (Section 2.5). If (4.6.8) is used for $\mathcal{A}_1$ and $\mathcal{C}_0$, the eigenfunctions need not be diagonal in μ_z, but the operators are still space-fixed. [Equation (4.5.16) for $\mathcal{B}_0$ and (4.5.23) for $\mathcal{D}_0$ apply in both cases.] But (A.4.13)–(A.4.22) also apply for a collection of randomly oriented molecules if $\mathcal{A}_1, \mathcal{B}_0, \mathcal{C}_0, \mathcal{D}_0$ are simply replaced respectively by $\overline{\mathcal{A}}_1, \overline{\mathcal{B}}_0, \overline{\mathcal{C}}_0, \overline{\mathcal{D}}_0$ [Eq. (4.6.14)]. In that case, the operators are *molecule-fixed* (Section 2.5), and the eigenfunctions need not be diagonal in μ_z.

A.5. Earlier (and Obsolete) Definitions of the Faraday Parameters

At least two previous definitions of the Faraday parameters have appeared widely in the literature. Stephens (1970, 1974) used the rigid-shift BO MCD equations in the form

$$\frac{\Delta A}{\mathcal{E}} = -\tfrac{2}{3}\gamma\left[\mathcal{A}_1(\text{old})\left(-\frac{\partial f}{\partial \mathcal{E}}\right) + \left(\mathcal{B}_0(\text{old}) + \frac{\mathcal{C}_0(\text{old})}{kT}\right)f\right]H \tag{A.5.1}$$

$$\frac{A^0}{\mathcal{E}} = \frac{\gamma}{3}\mathcal{D}_0(\text{old})f \tag{A.5.2}$$

or equivalently

$$\Delta k = \frac{-4\pi^2 N}{3}\left[\mathcal{A}_1(\text{old})\left(-\frac{\partial f}{\partial \mathcal{E}}\right) + \left(\mathcal{B}_0(\text{old}) + \frac{\mathcal{C}_0(\text{old})}{kT}\right)\right]H \tag{A.5.3}$$

$$k^0 = \frac{2\pi^2 N}{3}\mathcal{D}_0(\text{old})f \tag{A.5.4}$$

Earlier still, Stephens et al. (1966) and Schatz et al. (1966) used the form

$$\frac{[\theta]_M}{\mathcal{E}} = \frac{-48\pi N_0 H}{hc}\left[A\frac{f_3}{\mathcal{E}} + \left(B + \frac{C}{kT}\right)\frac{f_4}{\mathcal{E}}\right] \tag{A.5.5}$$

$$= -21.346H\left[A\frac{f_3}{\mathcal{E}} + \left(B + \frac{C}{kT}\right)\frac{f_4}{\mathcal{E}}\right] \tag{A.5.6}$$

and the dipole strength was written D. [Yet earlier, Stephens (1965a, b) and Buckingham and Stephens (1966) defined A, B, and C as one-third the values given here.] In (A.5.5) and (A.5.6), $f_4/\mathcal{E} = (\pi/2)f(\mathcal{E}) \equiv (\pi/2)f$ and $(f_3/\mathcal{E}) = (\pi/2)\, df/d\mathcal{E}$. In (A.5.6), energies are in cm^{-1}, magnetic moments are in *Bohr magnetons*, and H is in gauss.

Note in (A.5.5) and (A.5.6) that the Faraday parameters are written without subscript as ordinary italic capitals (not script). In (A.5.1) and (A.5.2) identical symbols are used, as in the later definitions (Section A.4). We have therefore affixed the label "(old)" to emphasize the difference. If (A.5.1)–(A.5.6) are compared with the equations in Section A.4, the following relations emerge:

$$\begin{aligned}
\mathcal{A}_1(\text{new}) &= \frac{-2}{3\mu_B}\mathcal{A}_1(\text{old}) = \frac{2}{3\mu_B}A \\
\mathcal{B}_0(\text{new}) &= \frac{-2}{3\mu_B}\mathcal{B}_0(\text{old}) = \frac{-2}{3\mu_B}B \\
\mathcal{C}_0(\text{new}) &= \frac{-2}{3\mu_B}\mathcal{C}_0(\text{old}) = \frac{-2}{3\mu_B}C \\
\mathcal{D}_0(\text{new}) &= \tfrac{1}{3}\mathcal{D}_0(\text{old}) = \tfrac{1}{3}D
\end{aligned} \tag{A.5.7}$$

Exactly analogous relations apply for $n \geq 1$ in X_n where $X = \mathcal{A}, \mathcal{B}, \mathcal{C}$, or $\mathcal{D}$ —see (7.7.1).

For emphasis here we have added the label "(new)" to designate the parameters of Section A.4, which are used exclusively throughout this book. In dealing with the literature, we recommend first that the definition of parameters be ascertained by carefully comparing the literature MCD equations with those of this and the preceding subsection. It should then be easy, if required, to translate the Faraday parameters to the "new" ones via (A.5.7). Then all the equations of Section A.4 (and the entire book) are applicable.

A.6. Some Other Useful Equations

We conventionally express the CD as an absorbance difference (Section A.4), that is, in optical-density units (ODU), but millidegrees (mdeg) are sometimes used in the literature. The relation between these is readily established by combining (A.4.5) and (A.4.7):

$$\theta = \frac{\Delta A}{4 \log e} \tag{A.6.1}$$

where θ is in radians and ΔA is in absorbance (optical-density) units. Converting,

$$\theta = \left(\frac{\Delta A}{4 \log e}\right)\left(\frac{180}{\pi}\right) \times 10^3$$

$$= 32{,}982\, \Delta A \tag{A.6.2}$$

where θ is now in mdeg. We therefore obtain the conversion factor

$$1 \text{ mdeg} = 3.032 \times 10^{-5} \text{ ODU} \tag{A.6.3}$$

To obtain the dipole strength from an experimental absorption spectrum, we use (A.4.18) and write

$$\mathfrak{D}_0 = 3.062 \times 10^{-3} \int \frac{\epsilon}{\mathcal{E}}\, d\mathcal{E}$$

$$\approx \frac{3.062 \times 10^{-3}}{\nu_{\max}} \int \epsilon\, d\nu \tag{A.6.4}$$

where the latter is a good approximation for a symmetrical band. If the band can be approximated as a Gaussian [Eq. (7.8.4)],

$$\mathfrak{D}_0 \approx \frac{3.062 \times 10^{-3} \sqrt{\pi}\, \epsilon_{\max}\, \Delta}{\nu_{\max}} \tag{A.6.5}$$

which is actually (7.8.7).

Some authors use the oscillator strength f defined by

$$f = 4.319 \times 10^{-9} \int \epsilon\, d\nu \tag{A.6.6}$$

with ν in cm^{-1}. We do not use this unit.

Appendix B Tables for the Group of All Rotations, SO_3

B.1. Spherical Harmonics: The $SO_3 \supset SO_2$ Basis for Integer j

Spherical harmonics are the $|jm\rangle$ basis functions for SO_3 for integer j in the angular-momentum ($SO_3 \supset SO_2$) basis. Those symbolized by Y^j_m have the conventional Condon–Shortley phases and obey (14.3.8). In Table B.1.1 we give the Y^j_m in rectangular coordinates normalized to one. Unless noted otherwise, we use the Y^j_m as our standard $SO_3 \supset SO_2$ basis for integer j, and any atomic state $|jm\rangle$ for integer j is defined to transform identically to Y^j_m. The $\mathbf{D}^j(R)$ for the basis are those calculated according to Tinkham [(1964), Eq. (5-35)].

Another useful set of spherical harmonics is

$$\mathcal{Y}^j_m = i^j Y^j_m \tag{B.1.1}$$

The $\mathcal{Y}^j_m$ obey the simpler time-reversal equation (14.3.7).

B.2. Real Atomic Orbitals Defined in Terms of the Y^j_m

We tabulate in Table B.2.1 the angular parts of the real AOs for $l = 0$, 1, and 2. To obtain the complete AO, the functions should be multiplied by the appropriate atomic radial function $R_{nl}(r)$. The angular parts of AOs completely determine the orbital transformation properties, since the $R_{nl}(r)$ are spherically symmetric. The angular parts are independent of the quantum number n.

Table B.1.1. Spherical Harmonics

$$Y_0^0 = \sqrt{\frac{1}{4\pi}}$$

$$Y_{-1}^1 = \sqrt{\frac{3}{8\pi}}\,\frac{x - iy}{r}$$

$$Y_0^1 = \sqrt{\frac{3}{4\pi}}\,\frac{z}{r}$$

$$Y_1^1 = -\sqrt{\frac{3}{8\pi}}\,\frac{x + iy}{r}$$

$$Y_{-2}^2 = \sqrt{\frac{5}{4\pi}}\sqrt{\frac{3}{8}}\,\frac{(x - iy)^2}{r^2}$$

$$Y_{-1}^2 = \sqrt{\frac{5}{4\pi}}\sqrt{\frac{3}{2}}\,\frac{z(x - iy)}{r^2}$$

$$Y_0^2 = \sqrt{\frac{5}{4\pi}}\sqrt{\frac{1}{4}}\,\frac{3z^2 - r^2}{r^2}$$

$$Y_1^2 = -\sqrt{\frac{5}{4\pi}}\sqrt{\frac{3}{2}}\,\frac{z(x + iy)}{r^2}$$

$$Y_2^2 = \sqrt{\frac{5}{4\pi}}\sqrt{\frac{3}{8}}\,\frac{(x + iy)^2}{r^2}$$

$$Y_{-3}^3 = \sqrt{\frac{7}{4\pi}}\sqrt{\frac{5}{16}}\,\frac{(x - iy)^3}{r^3}$$

$$Y_{-2}^3 = \sqrt{\frac{7}{4\pi}}\sqrt{\frac{15}{8}}\,\frac{z(x - iy)^2}{r^3}$$

$$Y_{-1}^3 = \sqrt{\frac{7}{4\pi}}\sqrt{\frac{3}{16}}\,\frac{(x - iy)(5z^2 - r^2)}{r^3}$$

$$Y_0^3 = \sqrt{\frac{7}{4\pi}}\sqrt{\frac{1}{4}}\,\frac{z(5z^2 - 3r^2)}{r^3}$$

$$Y_1^3 = -\sqrt{\frac{7}{4\pi}}\sqrt{\frac{3}{16}}\,\frac{(x + iy)(5z^2 - r^2)}{r^3}$$

$$Y_2^3 = \sqrt{\frac{7}{4\pi}}\sqrt{\frac{15}{8}}\,\frac{z(x + iy)^2}{r^3}$$

$$Y_3^3 = -\sqrt{\frac{7}{4\pi}}\sqrt{\frac{5}{16}}\,\frac{(x + iy)^3}{r^3}$$

$$Y_{-4}^4 = \sqrt{\frac{9}{4\pi}}\sqrt{\frac{35}{128}}\,\frac{(x - iy)^4}{r^4}$$

$$Y_{-3}^4 = \sqrt{\frac{9}{4\pi}}\sqrt{\frac{35}{16}}\,\frac{z(x - iy)^3}{r^4}$$

$$Y_{-2}^4 = \sqrt{\frac{9}{4\pi}}\sqrt{\frac{5}{32}}\,\frac{(x - iy)^2}{r^4}(7z^2 - r^2)$$

$$Y_{-1}^4 = \sqrt{\frac{9}{4\pi}}\sqrt{\frac{5}{16}}\,\frac{(x - iy)}{r^4}(7z^3 - 3zr^2)$$

$$Y_0^4 = \sqrt{\frac{9}{4\pi}}\sqrt{\frac{1}{64}}\,\frac{35z^4 - 30z^2r^2 + 3r^4}{r^4}$$

$$Y_1^4 = -\sqrt{\frac{9}{4\pi}}\sqrt{\frac{5}{16}}\,\frac{(x + iy)}{r^4}(7z^3 - 3zr^2)$$

$$Y_2^4 = \sqrt{\frac{9}{4\pi}}\sqrt{\frac{5}{32}}\,\frac{(x + iy)^2}{r^4}(7z^2 - r^2)$$

$$Y_3^4 = -\sqrt{\frac{9}{4\pi}}\sqrt{\frac{35}{16}}\,\frac{z(x + iy)^3}{r^4}$$

$$Y_4^4 = \sqrt{\frac{9}{4\pi}}\sqrt{\frac{35}{128}}\,\frac{(x + iy)^4}{r^4}$$

$$Y_{-5}^5 = \sqrt{\frac{11}{4\pi}}\sqrt{\frac{63}{256}}\,\frac{(x - iy)^5}{r^5}$$

$$Y_{-4}^5 = \sqrt{\frac{11}{4\pi}}\sqrt{\frac{315}{128}}\,\frac{z(x - iy)^4}{r^5}$$

$$Y_{-3}^5 = \sqrt{\frac{11}{4\pi}}\sqrt{\frac{35}{256}}\,\frac{(x - iy)^3}{r^5}(9z^2 - r^2)$$

$$Y_{-2}^5 = \sqrt{\frac{11}{4\pi}}\sqrt{\frac{105}{32}}\,\frac{(x - iy)^2}{r^5}(3z^3 - zr^2)$$

$$Y_{-1}^5 = \sqrt{\frac{11}{4\pi}}\sqrt{\frac{15}{128}}\,\frac{(x - iy)}{r^5}(21z^4 - 14z^2r^2 + r^4)$$

$$Y_0^5 = \sqrt{\frac{11}{4\pi}}\,\frac{1}{8}\,\frac{63z^5 - 70z^3r^2 + 15zr^4}{r^5}$$

$$Y_1^5 = -\sqrt{\frac{11}{4\pi}}\sqrt{\frac{15}{128}}\,\frac{(x + iy)}{r^5}(21z^4 - 14z^2r^2 + r^4)$$

$$Y_2^5 = \sqrt{\frac{11}{4\pi}}\sqrt{\frac{105}{32}}\,\frac{(x + iy)^2}{r^5}(3z^3 - zr^2)$$

$$Y_3^5 = -\sqrt{\frac{11}{4\pi}}\sqrt{\frac{35}{256}}\,\frac{(x + iy)^3}{r^5}(9z^2 - r^2)$$

$$Y_4^5 = \sqrt{\frac{11}{4\pi}}\sqrt{\frac{315}{128}}\,\frac{z(x + iy)^4}{r^5}$$

$$Y_5^5 = -\sqrt{\frac{11}{4\pi}}\sqrt{\frac{63}{256}}\,\frac{(x + iy)^5}{r^5}$$

Table B.2.1

Orbital	Angular Part of the Orbital	Normalization N_l
s	$N_0 = 1$	$N_0 = 1$
p_x	$N_1 x = (1/\sqrt{2})(Y^1_{-1} - Y^1_1)$	
p_y	$N_1 y = (i/\sqrt{2})(Y^1_{-1} + Y^1_1)$	$N_1 = \frac{1}{r}\sqrt{\frac{3}{4\pi}}$
p_z	$N_1 z = Y^1_0$	
d_{yz}	$N_2\sqrt{2}\, yz = (i/\sqrt{2})(Y^2_{-1} + Y^2_1)$	
d_{zx}	$N_2\sqrt{2}\, zx = (1/\sqrt{2})(Y^2_{-1} - Y^2_1)$	
d_{xy}	$N_2\sqrt{2}\, xy = (i/\sqrt{2})(Y^2_{-2} - Y^2_2)$	$N_2 = \frac{1}{r^2}\sqrt{\frac{5}{4\pi}}\sqrt{\frac{3}{2}}$
d_{z^2}	$N_2(1/\sqrt{6})(3z^2 - r^2) = Y^2_0$	
$d_{x^2-y^2}$	$N_2(1/\sqrt{2})(x^2 - y^2) = (1/\sqrt{2})(Y^2_{-2} + Y^2_2)$	

B.3. High-Symmetry Coefficients and Phases for SO_3

All phase factors for the group of all rotations in the angular-momentum basis, $SO_3 \supset SO_2$, may be defined algebraically:

$$\begin{aligned}\begin{pmatrix} j \\ m \end{pmatrix} &= (-1)^{j-m} \\ \{j\} &= (-1)^{2j} \\ \{j_1 j_2 j_3\} &= (-1)^{j_1+j_2+j_3} \\ H(j_1 j_2 j_3) &= (-1)^{j_1-j_2+j_3} = (-1)^{-j_1+j_2-j_3}\end{aligned} \tag{B.3.1}$$

The group is ambivalent, so $j = j^*$ for all j. Also $m^* = -m$, $|j| = 2j + 1$, and all $3jm$ are real. The $3jm$, $6j$, and $9j$ are identical respectively to the familiar $3j$, $6j$, and $9j$ symbols. $3j$ and $6j$ symbols are tabulated in Rotenberg et al. (1959). We also give a partial list of $3jm$ and $6j$ below. $6j$ omitted include those with one or more of the j_i equal to 0, since they may be easily obtained using (16.1.11). Likewise $9j$ with one or more j_i equal to 0 are given by (16.4.8) or (16.4.9). Other $9j$ may be calculated as a sum of $6j$ using Butler [(1981), Eq. (3.3.37)]. Rotenberg et al. (1959) give references which tabulate $9j$.

Vector coupling coefficients (v.c.c.) may be obtained for $SO_3 \supset SO_2$ from $3jm$ using (10.2.1). However, if v.c.c. with the historical Condon–Shortley phases (i.e. the Wigner or Clebsch–Gordan coefficients) are desired, the phase factor $H(j_1 j_2 j_3)$ must be included in (10.2.1) and (10.2.3) as mentioned in Section 10.2. Thus

$$(j_1 m_1, j_2 m_2 | j_3 m_3) = (-1)^{j_1 - j_2 + j_3} |j_3|^{1/2} (-1)^{j_3 - m_3} \begin{pmatrix} j_1 & j_2 & j_3 \\ m_1 & m_2 & -m_3 \end{pmatrix} \tag{B.3.2}$$

and

$$\begin{aligned} |(j_1, j_2)\, j_3 m_3\rangle &= \sum_{m_1 m_2} (j_1 m_1, j_2 m_2 | j_3 m_3) | j_1 m_1\rangle | j_2 m_2\rangle \\ &= (-1)^{j_1 - j_2 + j_3} |j_3|^{1/2} (-1)^{j_3 - m_3} \\ &\quad \times \sum_{m_1 m_2} \begin{pmatrix} j_1 & j_2 & j_3 \\ m_1 & m_2 & -m_3 \end{pmatrix} | j_1 m_1\rangle | j_2 m_2\rangle \end{aligned} \tag{B.3.3}$$

Nielson and Koster (1963) use Condon–Shortley phases (and thus these relations) in deriving their equations.

Partial List of 3jm for $SO_3 \supset SO_2$

The $3jm$

$$\begin{pmatrix} j_1 & j_2 & j_3 \\ m_1 & m_2 & m_3 \end{pmatrix}$$

are ordered in Table B.3.1 so that $j_1 \geqslant j_2 \geqslant j_3$. The list is limited to $3jm$ with $j_1 \leqslant \frac{5}{2}$, $j_2 \leqslant \frac{3}{2}$, and $j_3 \leqslant 1$. Within this set, all $3jm$ neither found in the table nor related by (10.2.4) to those tabulated are zero. Recall that $3jm$ for $SO_3 \supset SO_2$ are zero if any of the following relations is not satisfied:

$$\begin{aligned} j_1 + j_2 - j_3 &\geqslant 0 \\ j_1 - j_2 + j_3 &\geqslant 0 \\ -j_1 + j_2 + j_3 &\geqslant 0 \\ j_1 + j_2 + j_3 &= \text{integer} \\ m_1 + m_2 + m_3 &= 0 \end{aligned} \tag{B.3.4}$$

Table B.3.1. Partial List of $3jm$ for $SO_3 \supset SO_2$

0	0	0	$3jm$
0	0	0	1

$\frac{1}{2}$	$\frac{1}{2}$	0	$3jm^\dagger$
$\frac{1}{2}$	$-\frac{1}{2}$	0	$1/\sqrt{2}$

1	$\frac{1}{2}$	$\frac{1}{2}$	$3jm$
-1	$\frac{1}{2}$	$\frac{1}{2}$	$-1/\sqrt{3}$
0	$-\frac{1}{2}$	$\frac{1}{2}$	$1/\sqrt{6}$
1	$-\frac{1}{2}$	$-\frac{1}{2}$	$-1/\sqrt{3}$

1	1	0	$3jm$
0	0	0	$-1/\sqrt{3}$
1	-1	0	$1/\sqrt{3}$

1	1	1	$3jm^\dagger$
-1	0	1	$1/\sqrt{6}$
0	0	0	0

$\frac{3}{2}$	1	$\frac{1}{2}$	$3jm^\dagger$
$-\frac{3}{2}$	1	$\frac{1}{2}$	$\frac{1}{2}$
$-\frac{1}{2}$	0	$\frac{1}{2}$	$-1/\sqrt{6}$
$-\frac{1}{2}$	1	$-\frac{1}{2}$	$-1/2\sqrt{3}$
$\frac{1}{2}$	-1	$\frac{1}{2}$	$1/2\sqrt{3}$
$\frac{1}{2}$	0	$-\frac{1}{2}$	$1/\sqrt{6}$
$\frac{3}{2}$	-1	$-\frac{1}{2}$	$-\frac{1}{2}$

$\frac{3}{2}$	$\frac{3}{2}$	0	$3jm^\dagger$
$\frac{1}{2}$	$-\frac{1}{2}$	0	$-\frac{1}{2}$
$\frac{3}{2}$	$-\frac{3}{2}$	0	$\frac{1}{2}$

$\frac{3}{2}$	$\frac{3}{2}$	1	$3jm$
$-\frac{1}{2}$	$-\frac{1}{2}$	1	$\sqrt{2}/\sqrt{15}$
$\frac{1}{2}$	$-\frac{3}{2}$	1	$-1/\sqrt{10}$
$\frac{1}{2}$	$-\frac{1}{2}$	0	$-1/2\sqrt{15}$
$\frac{1}{2}$	$\frac{1}{2}$	-1	$\sqrt{2}/\sqrt{15}$
$\frac{3}{2}$	$-\frac{3}{2}$	0	$\sqrt{3}/2\sqrt{5}$
$\frac{3}{2}$	$-\frac{1}{2}$	-1	$-1/\sqrt{10}$

2	1	1	$3jm$
-2	1	1	$1/\sqrt{5}$
-1	0	1	$-1/\sqrt{10}$
0	-1	1	$1/\sqrt{30}$
0	0	0	$\sqrt{2}/\sqrt{15}$
1	-1	0	$-1/\sqrt{10}$
2	-1	-1	$1/\sqrt{5}$

2	$\frac{3}{2}$	$\frac{1}{2}$	$3jm$
-2	$\frac{3}{2}$	$\frac{1}{2}$	$-1/\sqrt{5}$
-1	$\frac{1}{2}$	$\frac{1}{2}$	$\sqrt{3}/2\sqrt{5}$
-1	$\frac{3}{2}$	$-\frac{1}{2}$	$1/2\sqrt{5}$
0	$-\frac{1}{2}$	$\frac{1}{2}$	$-1/\sqrt{10}$
0	$\frac{1}{2}$	$-\frac{1}{2}$	$-1/\sqrt{10}$
1	$-\frac{3}{2}$	$\frac{1}{2}$	$1/2\sqrt{5}$
1	$-\frac{1}{2}$	$-\frac{1}{2}$	$\sqrt{3}/2\sqrt{5}$
2	$-\frac{3}{2}$	$-\frac{1}{2}$	$-1/\sqrt{5}$

$\frac{5}{2}$	$\frac{3}{2}$	1	$3jm^\dagger$
$-\frac{5}{2}$	$\frac{3}{2}$	1	$-1/\sqrt{6}$
$-\frac{3}{2}$	$\frac{1}{2}$	1	$1/\sqrt{10}$
$-\frac{3}{2}$	$\frac{3}{2}$	0	$1/\sqrt{15}$
$-\frac{1}{2}$	$-\frac{1}{2}$	1	$-1/2\sqrt{5}$
$-\frac{1}{2}$	$\frac{1}{2}$	0	$-1/\sqrt{10}$
$-\frac{1}{2}$	$\frac{3}{2}$	-1	$-1/2\sqrt{15}$
$\frac{1}{2}$	$-\frac{3}{2}$	1	$1/2\sqrt{15}$
$\frac{1}{2}$	$-\frac{1}{2}$	0	$1/\sqrt{10}$
$\frac{1}{2}$	$\frac{1}{2}$	-1	$1/2\sqrt{5}$
$\frac{3}{2}$	$-\frac{3}{2}$	0	$-1/\sqrt{15}$
$\frac{3}{2}$	$-\frac{1}{2}$	-1	$-1/\sqrt{10}$
$\frac{5}{2}$	$-\frac{3}{2}$	-1	$1/\sqrt{6}$

The table headings are j_1, j_2, and j_3, while entries beneath them are m_1, m_2, and m_3. Thus, for example,

$$\begin{pmatrix} \frac{3}{2} & \frac{3}{2} & 1 \\ \frac{1}{2} & -\frac{1}{2} & 0 \end{pmatrix} = \frac{-1}{2\sqrt{15}} \tag{B.3.5}$$

The $3jm$ in tables headed by $3jm^{\dagger}$ are multiplied by -1 for odd permutations of their columns.

Partial List of 6j for $\mathbf{SO_3}$

The $6j$

$$\begin{Bmatrix} j_1 & j_2 & j_3 \\ j_4 & j_5 & j_6 \end{Bmatrix}$$

are ordered in Table B.3.2 so that $j_1 \geqslant j_2 \geqslant j_3, j_1 \geqslant j_4$, and $j_2 \geqslant j_5$. If $j_1 = j_2$, $j_4 \geqslant j_5$. All $6j$ containing one or more j_i equal to 0 are given by (16.1.11) and are not tabulated. In SO_3 (16.1.11) may be written

$$\begin{Bmatrix} 0 & j_2 & j_3 \\ j_4 & j_5 & j_6 \end{Bmatrix} = \delta_{j_2 j_3} \delta_{j_5 j_6} \delta(j_4 j_2 j_6) \frac{(-1)^{j_4+j_2+j_6}}{(2j_2+1)^{1/2}(2j_5+1)^{1/2}} \tag{B.3.6}$$

The list is also limited to $6j$ with $j_1 \leqslant \frac{5}{2}$ and $j_2 \leqslant \frac{3}{2}$. Within this set, all $6j$ neither found in the table nor related to those tabulated by the symmetry rules of Section 16.1 are zero by (16.1.2). For the group SO_3, $\delta(j_1 j_2 j_3) \equiv 1$ if the first four relations of (B.3.4) are satisfied; otherwise $\delta(j_1 j_2 j_3) \equiv 0$.

The table headings are j_1, j_2, and j_3, while entries beneath them are j_4, j_5, and j_6. Thus, for example,

$$\begin{Bmatrix} 2 & \frac{3}{2} & \frac{1}{2} \\ \frac{3}{2} & 1 & 1 \end{Bmatrix} = \frac{1}{2\sqrt{30}} \tag{B.3.7}$$

B.4. Reduced Matrix Elements of SO_3: The Tables of Nielson and Koster

Nielson and Koster (1963) tabulate cfp and reduced matrix elements of tensor operators for the $^{2\mathcal{S}+1}\mathcal{L}$ terms of the atomic l^n configurations p^n, d^n,

Table B.3.2. Partial List of 6*j* for SO_3

1	$\frac{1}{2}$	$\frac{1}{2}$	$6j$
1	$\frac{1}{2}$	$\frac{1}{2}$	$\frac{1}{6}$

1	1	1	$6j$
$\frac{1}{2}$	$\frac{1}{2}$	$\frac{1}{2}$	$-\frac{1}{3}$
1	1	1	$\frac{1}{6}$

$\frac{3}{2}$	1	$\frac{1}{2}$	$6j$
$\frac{1}{2}$	1	$\frac{1}{2}$	$-\frac{1}{3}$
1	$\frac{1}{2}$	1	$-\frac{1}{6}$
$\frac{3}{2}$	1	$\frac{1}{2}$	$-\frac{1}{12}$

$\frac{3}{2}$	$\frac{3}{2}$	1	$6j$
$\frac{1}{2}$	$\frac{1}{2}$	1	$\sqrt{5}/6\sqrt{2}$
1	1	$\frac{1}{2}$	$\sqrt{5}/6\sqrt{2}$
1	1	$\frac{3}{2}$	$-1/3\sqrt{10}$
$\frac{3}{2}$	$\frac{1}{2}$	1	$\frac{1}{6}$
$\frac{3}{2}$	$\frac{3}{2}$	1	$\frac{11}{60}$

2	1	1	$6j$
1	1	1	$\frac{1}{6}$
2	1	1	$\frac{1}{30}$

2	$\frac{3}{2}$	$\frac{1}{2}$	$6j$
$\frac{1}{2}$	1	1	$1/2\sqrt{3}$
1	$\frac{1}{2}$	$\frac{3}{2}$	$1/2\sqrt{10}$
1	$\frac{3}{2}$	$\frac{1}{2}$	$\frac{1}{4}$
1	$\frac{3}{2}$	$\frac{3}{2}$	$-1/2\sqrt{5}$
$\frac{3}{2}$	1	1	$1/2\sqrt{30}$
2	$\frac{3}{2}$	$\frac{1}{2}$	$\frac{1}{20}$
2	$\frac{3}{2}$	$\frac{3}{2}$	$-\frac{1}{10}$

2	$\frac{3}{2}$	$\frac{3}{2}$	$6j$
$\frac{1}{2}$	1	1	$-1/2\sqrt{6}$
1	$\frac{1}{2}$	$\frac{3}{2}$	$-1/2\sqrt{5}$
1	$\frac{3}{2}$	$\frac{3}{2}$	$\frac{1}{20}$
$\frac{3}{2}$	1	1	$-\sqrt{2}/5\sqrt{3}$
2	$\frac{1}{2}$	$\frac{3}{2}$	$-\frac{1}{10}$
2	$\frac{3}{2}$	$\frac{3}{2}$	$\frac{3}{20}$

$\frac{5}{2}$	$\frac{3}{2}$	1	$6j$
$\frac{1}{2}$	$\frac{3}{2}$	1	$-\frac{1}{4}$
1	1	$\frac{3}{2}$	$-1/2\sqrt{10}$
$\frac{3}{2}$	$\frac{3}{2}$	1	$-\frac{1}{10}$
2	1	$\frac{3}{2}$	$-1/10\sqrt{6}$
$\frac{5}{2}$	$\frac{3}{2}$	1	$-\frac{1}{60}$

and f^n. These tables are very useful for crystal-field calculations in the weak-field basis and for the calculation of some of the coefficients used in strong-field calculations (Section 19.9). Free-ion electrostatic, spin–orbit, and magnetic-moment matrices are easily expressed in terms of these tabulated reduced matrix elements. Then, since the operator for the crystal-field potential may be written as a sum of tensor operators, it is straightforward to calculate the crystal-field matrix in the same basis. If the full basis is used for a given l^n configuration, a complete crystal-field calculation results. Nielson and Koster's tables may be used with either an atomic $|l^n(\mathcal{S}\mathcal{L})\mathcal{J}\mathcal{M}_{\mathcal{J}}\rangle$ (or $|l^n\,\mathcal{S}\mathcal{L}\mathcal{M}_{\mathcal{S}}\mathcal{M}_{\mathcal{L}}\rangle$) basis set or a weak-field $SO_3 \supset G \supset \cdots \supset C_n$ basis of the sort $|l^n(\mathcal{S}\mathcal{L})\mathcal{J}\Gamma\gamma\rangle$ (or $|l^n(\mathcal{S}S, \mathcal{L}h)r\Gamma\gamma\rangle$). In the latter cases $\Gamma\gamma$, S, and h are group-G irrep and irrep partner labels. If an atomic basis is used, matrices can be factored to some extent by $\mathcal{M}_{\mathcal{J}}$ values if a crystal field is present; a symmetry-group basis, however, allows for complete factoring.

Since Nielson and Koster (1963) tabulate values of reduced matrix elements for the ${}^{2\mathcal{S}+1}\mathcal{L}$ terms of l^n configurations, all desired matrix elements must be related to reduced matrix elements of the $|l^n\,\mathcal{S}\mathcal{L}\mathcal{M}_{\mathcal{S}}\mathcal{M}_{\mathcal{L}}\rangle$. If an $|l^n(\mathcal{S}\mathcal{L})\mathcal{J}\Gamma\gamma\rangle$ basis has been used, reduced matrix elements for the symmetry group G may be expressed in terms of reduced matrix elements of the $|l^n(\mathcal{S}\mathcal{L})\mathcal{J}\mathcal{M}_{\mathcal{J}}\rangle$ using (15.4.3) for $SO_3 \supset G$. Then $|l^n(\mathcal{S}\mathcal{L})\mathcal{J}\mathcal{M}_{\mathcal{J}}\rangle$ reduced matrix elements may be expressed in terms of $|l^n\,\mathcal{S}\mathcal{L}\mathcal{M}_{\mathcal{S}}\mathcal{M}_{\mathcal{L}}\rangle$ reduced matrix elements with the appropriate Butler (1981) equation, (4.3.8), (4.3.9), or (9.1.11); these are the analogs of our (18.1.8), (18.1.9), and (18.4.2), but were derived using the version of (10.2.1) which explicitly includes the factor $H(j_1 j_2 j_3)$. If a $|l^n(\mathcal{S}S, \mathcal{L}h)r\Gamma\gamma\rangle$ basis is employed, the appropriate equation, (18.1.8), (18.1.9), or (18.4.2), is first used to give reduced matrix elements in the $|l^n\,\mathcal{S}S\mathcal{L}hM\theta\rangle$ basis. These may then be expressed in terms of $|l^n\,\mathcal{S}\mathcal{L}\mathcal{M}_{\mathcal{S}}\mathcal{M}_{\mathcal{L}}\rangle$ reduced matrix elements as illustrated in our examples in Section 15.4.

Nielson and Koster tabulate values of reduced matrix elements for the ${}^{2\mathcal{S}+1}\mathcal{L}$ terms of atomic configurations as follows. Free-ion electrostatic matrices are given in terms of the familiar atomic F^k integrals (or for f^n, in terms of parameters which are linear combinations of them). Then matrix elements of the spin-independent one-electron tensor operators $U_q^k = \sum_{i=1}^n u(i)_q^k$, $k = 2$ to 6, and of the one-electron double-tensor operators $V_{qq}^{11} = \sum_{i=1}^n v(i)_{qq}^{11}$, are tabulated in terms of one-electron reduced matrix elements for u^k and v^{11} unit tensor operators. These are defined to have the values

$$
\begin{aligned}
\langle nl\|u^k\|n'l'\rangle &\equiv \delta_{nn'}\delta_{ll'} \\
\langle nsl\|v^{11}\|n's'l'\rangle &= \langle n\tfrac{1}{2}l\|v^{11}\|n'\tfrac{1}{2}l'\rangle \\
&= \frac{\sqrt{3}}{\sqrt{2}}\langle n\tfrac{1}{2}l\|u^{11}\|n'\tfrac{1}{2}l'\rangle \qquad \text{(B.4.1)} \\
&\equiv \frac{\sqrt{3}}{\sqrt{2}}\delta_{nn'}\delta_{ll'}
\end{aligned}
$$

Here n, s, and l are the usual atomic quantum numbers. The tensor operators, as always, transform as the $|jm\rangle$ of $SO_3 \supset SO_2$ (Sections 9.8 and 18.3). One-electron reduced matrix elements for angular-momentum operators in SO_3 are simply related to those of these unit tensor operators:

$$\langle nl\|l\|n'l'\rangle = [l(l+1)(2l+1)]^{1/2}\langle nl\|u^1\|n'l'\rangle$$
$$= \delta_{nn'}\delta_{ll'}[l(l+1)(2l+1)]^{1/2}$$
$$\langle s\|s\|s'\rangle = \langle\tfrac{1}{2}\|s\|\tfrac{1}{2}\rangle$$
$$= [\tfrac{1}{2}(\tfrac{1}{2}+1)(2\cdot\tfrac{1}{2}+1)]^{1/2} = \sqrt{\tfrac{3}{2}} \tag{B.4.2}$$
$$\langle n\tfrac{1}{2}l\|sl\|n'\tfrac{1}{2}l'\rangle = \sqrt{\tfrac{3}{2}}\,[l(l+1)(2l+1)]^{1/2}\langle n\tfrac{1}{2}l\|u^{11}\|n'\tfrac{1}{2}l'\rangle$$
$$= \delta_{nn'}\delta_{ll'}[\tfrac{3}{2}l(l+1)(2l+1)]^{1/2}$$
$$= \delta_{nn'}\delta_{ll'}[l(l+1)(2l+1)]^{1/2}\langle n\tfrac{1}{2}l\|v^{11}\|n'\tfrac{1}{2}l'\rangle$$

Nielson and Koster do not tabulate U^k reduced matrix elements for $k = 0$ or 1—the "trivial" cases. The U^0 and U^1 matrices are

$$\langle l^n\,\mathcal{S}\mathcal{L}\|U^0\|l^n\,\mathcal{S}'\mathcal{L}'\rangle = \delta_{\mathcal{S}\mathcal{S}'}\delta_{\mathcal{L}\mathcal{L}'}\,n|\mathcal{L}|^{1/2}|l|^{-1/2}$$
$$\langle l^n\,\mathcal{S}\mathcal{L}\|U^1\|l^n\,\mathcal{S}'\mathcal{L}'\rangle = \delta_{\mathcal{S}\mathcal{S}'}\delta_{\mathcal{L}\mathcal{L}'}\left[\frac{\mathcal{L}(\mathcal{L}+1)(2\mathcal{L}+1)}{l(l+1)(2l+1)}\right]^{1/2} \tag{B.4.3}$$

Note that the U^k are defined to be spin-independent tensor operators.

Other useful equations for the ${}^{2\mathcal{S}+1}\mathcal{L}$ terms of l^n are

$$\langle l^n\,\mathcal{S}\mathcal{L}\|S\|l^n\,\mathcal{S}'\mathcal{L}'\rangle = \delta_{\mathcal{L}\mathcal{L}'}\delta_{\mathcal{S}\mathcal{S}'}[\mathcal{S}(\mathcal{S}+1)(2\mathcal{S}+1)]^{1/2}$$
$$\langle l^n\,\mathcal{S}\mathcal{L}\|L\|l^n\,\mathcal{S}'\mathcal{L}'\rangle = \delta_{\mathcal{S}\mathcal{S}'}\delta_{\mathcal{L}\mathcal{L}'}[\mathcal{L}(\mathcal{L}+1)(2\mathcal{L}+1)]^{1/2} \tag{B.4.4}$$
$$\langle l^n\,\mathcal{S}\mathcal{L}\|\sum_{k=1}^{n} sl(k)\|l^n\,\mathcal{S}'\mathcal{L}'\rangle = [l(l+1)(2l+1)]^{1/2}\langle l^n\,\mathcal{S}\mathcal{L}\|V^{11}\|l^n\,\mathcal{S}'\mathcal{L}'\rangle$$

B.5. Partial List of $SO_3 \supset O$ 3*jm* Factors

The headings in Table B.5.1 are group-SO_3 irrep labels, while the entries beneath them are group-O irrep labels. Thus the table gives

$$\begin{pmatrix} 2 & 1 & 1 \\ T_2 & T_1 & T_1 \end{pmatrix}^{SO_3}_{O} = \frac{\sqrt{3}}{\sqrt{5}}$$

Permutation properties of the $3jm$ factors are given by (15.3.7). All $3jm$ factors below are even under column interchange except those listed as having a "phase" of -1. The $3jm$ factors are those of Butler [(1981), Chapter 13], but are expressed with Mulliken labels for the group-O irreps. The list is complete for j_1, j_2, and $j_3 \leqslant \frac{3}{2}$. For $3jm$ factors with one or more $j_i = \frac{5}{2}$, the list is limited to those containing one $j_i = 1$.

Table B.5.1. Partial List of $SO_3 \supset O$ $3jm$ Factors

0	0	0	$3jm$
A_1	A_1	A_1	1

1	1	0	$3jm$
T_1	T_1	A_1	1

1	1	1	$3jm$
T_1	T_1	T_1	-1

2	1	1	$3jm$
E	T_1	T_1	$\sqrt{2}/\sqrt{5}$
T_2	T_1	T_1	$\sqrt{3}/\sqrt{5}$

2	2	0	$3jm$
E	E	A_1	$\sqrt{2}/\sqrt{5}$
T_2	T_2	A_1	$\sqrt{3}/\sqrt{5}$

2	2	1	$3jm$
T_2	E	T_1	$-\sqrt{2}/\sqrt{5}$
T_2	T_2	T_1	$-1/\sqrt{5}$

2	2	2	$3jm$
E	E	E	$-2\sqrt{2}/\sqrt{5 \times 7}$
T_2	T_2	E	$-\sqrt{2 \times 3}/\sqrt{5 \times 7}$
T_2	T_2	T_2	$-3/\sqrt{5 \times 7}$

$\frac{1}{2}$	$\frac{1}{2}$	0	$3jm$
E'	E'	A_1	1

1	$\frac{1}{2}$	$\frac{1}{2}$	$3jm$
T_1	E'	E'	1

$\frac{3}{2}$	1	$\frac{1}{2}$	$3jm$
U'	T_1	E'	1

$\frac{3}{2}$	$\frac{3}{2}$	0	$3jm$
U'	U'	A_1	1

$\frac{3}{2}$	$\frac{3}{2}$	1	$3jm$
U'	U'	$T_1\ (r = 0)$	$1/\sqrt{5}$
U'	U'	$T_1\ (r = 1)$	$2/\sqrt{5}$

$\frac{5}{2}$	$\frac{3}{2}$	1	$3jm$	Phase
U'	U'	$T_1\ (r = 0)$	$2\sqrt{2}/\sqrt{3 \times 5}$	-1
U'	U'	$T_1\ (r = 1)$	$-\sqrt{2}/\sqrt{3 \times 5}$	-1
E''	U'	T_1	$-1/\sqrt{3}$	$+1$

$\frac{5}{2}$	$\frac{5}{2}$	1	$3jm$	Phase
U'	U'	$T_1\ (r = 0)$	$-\sqrt{2 \times 7}/3\sqrt{5}$	$+1$
U'	U'	$T_1\ (r = 1)$	$-4\sqrt{2}/3\sqrt{5 \times 7}$	$+1$
E''	U'	T_1	$-4/3\sqrt{7}$	-1
E''	E''	T_1	$\sqrt{5}/3\sqrt{7}$	$+1$

Appendix C *Tables for the Groups O and T_d*

C.1. Character Table for the Groups O and T_d

Irreps in Table C.1.1 are labeled using Mulliken (A_1, etc.), Bethe (Γ_n), and Butler (1981) notation. Transformation properties in the tetragonal basis of the coordinates x, y, and z and some of their products are indicated, as are those of the infinitesimal rotations R_x, R_y, and R_z. R represents the rotation through 360°, so $C_n^n = R$, not E. The transformation properties of individual partners of degenerate irreps depends on the basis used and on the way the coordinate system is chosen—see Sections C.5 and E.4.

Table C.1.1

T_d	O	T_d / O			E / E	R / R	$8C_3$ / $8C_3$	$8C_3R$ / $8C_3R$	$3C_2, 3C_2R$ / $3C_2, 3C_2R$	$6\sigma_d, 6\sigma_dR$ / $6C_2', 6C_2'R$	$6S_4$ / $6C_4$	$6S_4R$ / $6C_4R$
r^2	r^2	A_1	Γ_1	0	1	1	1	1	1	1	1	1
		A_2	Γ_2	$\tilde{0}$	1	1	1	1	1	−1	−1	−1
$(x^2 - y^2, \frac{3z^2 - r^2}{\sqrt{3}})$	$(x^2 - y^2, \frac{3z^2 - r^2}{\sqrt{3}})$	E	Γ_3	2	2	2	−1	−1	2	0	0	0
(R_x, R_y, R_z)	(R_x, R_y, R_z), (x, y, z)	T_1	Γ_4	1	3	3	0	0	−1	−1	1	1
(x, y, z), (yz, zx, xy)	(yz, zx, xy)	T_2	Γ_5	$\tilde{1}$	3	3	0	0	−1	1	−1	−1
		E'	Γ_6	$\frac{1}{2}$	2	−2	1	−1	0	0	$\sqrt{2}$	$-\sqrt{2}$
		E''	Γ_7	$\tilde{\frac{1}{2}}$	2	−2	1	−1	0	0	$-\sqrt{2}$	$\sqrt{2}$
		U'	Γ_8	$\frac{3}{2}$	4	−4	−1	1	0	0	0	0

C.2. Direct-Product Table for O and T_d

Table C.2.1

O, T_d	A_1	A_2	E	T_1	T_2	E'	E''	U'
A_1	A_1	A_2	E	T_1	T_2	E'	E''	U'
A_2	A_2	A_1	E	T_2	T_1	E''	E'	U'
E	E	E	$A_1 + A_2 + E$	$T_1 + T_2$	$T_1 + T_2$	U'	U'	$E' + E'' + U'$
T_1	T_1	T_2	$T_1 + T_2$	$A_1 + E + T_1 + T_2$	$A_2 + E + T_1 + T_2$	$E' + U'$	$E'' + U'$	$E' + E'' + 2U'$
T_2	T_2	T_1	$T_1 + T_2$	$A_2 + E + T_1 + T_2$	$A_1 + E + T_1 + T_2$	$E'' + U'$	$E' + U'$	$E' + E'' + 2U'$
E'	E'	E''	U'	$E' + U'$	$E'' + U'$	$A_1 + T_1$	$A_2 + T_2$	$E + T_1 + T_2$
E''	E''	E'	U'	$E'' + U'$	$E' + U'$	$A_2 + T_2$	$A_1 + T_1$	$E + T_1 + T_2$
U'	U'	U'	$E' + E'' + U'$	$E' + E'' + 2U'$	$E' + E'' + 2U'$	$E + T_1 + T_2$	$E + T_1 + T_2$	$A_1 + A_2 + E + 2T_1 + 2T_2$

C.3. Table of Symmetrized $[a^2]$ and Antisymmetrized (a^2) Squares

Table C.3.1

	O, T_d	
a	$[a^2]$	(a^2)
A_1	A_1	—
A_2	A_1	—
E	$A_1 + E$	A_2
T_1	$A_1 + E + T_2$	T_1
T_2	$A_1 + E + T_2$	T_1
E'	T_1	A_1
E''	T_1	A_1
U'	$A_2 + 2T_1 + T_2$	$A_1 + E + T_2$

C.4. Decomposition of O_h Relative to Its Subgroups, T_d, D_{4h}, and D_{3d}

This decomposition is given in Table C.4.1.

The decomposition of the group O with respect to D_4 and D_3 may be obtained from the table by dropping all u and g subscripts.

Table C.4.1

O_h	T_d	D_{4h}	D_{3d}
A_{1g}	A_1	A_{1g}	A_{1g}
A_{1u}	A_2	A_{1u}	A_{1u}
A_{2g}	A_2	B_{1g}	A_{2g}
A_{2u}	A_1	B_{1u}	A_{2u}
E_g	E	$A_{1g} + B_{1g}$	E_g
E_u	E	$A_{1u} + B_{1u}$	E_u
T_{1g}	T_1	$A_{2g} + E_g$	$A_{2g} + E_g$
T_{1u}	T_2	$A_{2u} + E_u$	$A_{2u} + E_u$
T_{2g}	T_2	$B_{2g} + E_g$	$A_{1g} + E_g$
T_{2u}	T_1	$B_{2u} + E_u$	$A_{1u} + E_u$
E'_g	E'	E'_g	E'_g
E'_u	E''	E'_u	E'_u
E''_g	E''	E''_g	E'_g
E''_u	E'	E''_u	E'_u
U'_g	U'	$E'_g + E''_g$	$E'_g + P_{1g} + P_{2g}$
U'_u	U'	$E'_u + E''_u$	$E'_u + P_{1u} + P_{2u}$

C.5. Tetragonal Bases for the Groups O and T_d

The standard basis functions defined below are identical to those of Butler (1981). Our real-O and complex-O (tetragonal) bases are respectively Butler's $O \supset D_4 \supset D_2 \supset C_2$ and $O \supset D_4 \supset C_4$ bases. Likewise our real-T_d and complex-T_d (tetragonal) bases are his $T_d \supset D_{2d} \supset D_2 \supset C_2$ and $T_d \supset D_{2d} \supset S_4$ bases. Basis chain nomenclature is discussed in Section 15.2. The basis differs in many respects from that of Griffith (1964), used previously by the present authors.

The behavior of the functions x, y, and z under the group generators (Section 9.4) is given in Table C.5.1(a). Recall that the active convention is used, so group operations rotate the functions, not the coordinate system (Section 8.2). While x, y, and z are in general not basis functions themselves, their transformations fix those of all basis functions expressible in rectangular coordinates (Section 8.7). These include basis functions for all single-valued irreps. The behavior of basis functions under C_4^Z and C_4^X alone defines the basis. The behavior of other group operations is fixed by these group generators (Section 9.4). For example, $C_3^{XYZ} = C_4^Z C_4^X$ (where C_4^X is performed first).

Table C.5.1(a). Behavior of x, y, and z Functions Under Group-O and Group-T_d Generators

	Group-O Generator		
Function	C_4^Z	C_4^X	C_3^{XYZ}
x	y	x	y
y	$-x$	z	z
z	z	$-y$	x
	Group-T_d Generator[a]		
Function	$(S_4^Z)^{-1}$	$(S_4^X)^{-1}$	C_3^{XYZ}
x	$-y$	$-x$	y
y	x	$-z$	z
z	$-z$	y	x

[a] The group generators for T_d were chosen to given irrep matrices isomorphic to those for the group-O generators. The superscript -1 designates "inverse," so that, for example, $(S_4^Z)^{-1} \equiv (S_4^Z)^3$.

Table C.5.1(b) defines the transformation properties of the real-O and real-T_d bases for single-valued irreps. Irrep components for T_2 are labeled

Table C.5.1(b). The Real-O and Real-T_d Bases for Single-Valued Irreps

Functions Which Transform: Identically to the Real-O Basis $(O \supset D_4 \supset D_2 \supset C_2)$	Functions Which Transform: Identically to the Real-T_d Basis $(T_d \supset D_{2d} \supset D_2 \supset C_2)$	Standard Basis: Real-T_d / Real-O	Group Generators: $(S_4^Z)^{-1}$ / C_4^Z	Group Generators: $(S_4^X)^{-1}$ / C_4^X	Group Generators: C_3^{XYZ} / C_3^{XYZ}
	$ixyz$	$\|A_1a_1\rangle \equiv \|0000\rangle$	a_1	a_1	a_1
$ixyz$		$\|A_2a_2\rangle \equiv \|\tilde{0}200\rangle$	$-a_2$	$-a_2$	a_2
$(-1/\sqrt{3})(3z^2 - r^2)$	$(-1/\sqrt{3})(3z^2 - r^2)$	$\|E\theta\rangle \equiv \|2000\rangle$	θ	$-\frac{1}{2}\theta + (\sqrt{3}/2)\epsilon$	$-\frac{1}{2}\theta - (\sqrt{3}/2)\epsilon$
$x^2 - y^2$	$x^2 - y^2$	$\|E\epsilon\rangle \equiv \|2200\rangle$	$-\epsilon$	$(\sqrt{3}/2)\theta + \frac{1}{2}\epsilon$	$(\sqrt{3}/2)\theta - \frac{1}{2}\epsilon$
$-x, -R_x$	$-R_x$	$\|T_1x\rangle \equiv \|11\tilde{1}1\rangle$	iT_1y	T_1x	iT_1y
iy, iR_y	iR_y	$\|T_1y\rangle \equiv \|1111\rangle$	iT_1x	iT_1z	iT_1z
z, R_z	R_z	$\|T_1z\rangle \equiv \|1\tilde{0}\tilde{0}0\rangle$	T_1z	iT_1y	$-T_1x$
$-iyz$	$-x, -iyz$	$\|T_2\xi\rangle \equiv \|\tilde{1}1\tilde{1}1\rangle$	$i\eta$	$-\xi$	$-i\eta$
zx	$-iy, zx$	$\|T_2\eta\rangle \equiv \|\tilde{1}111\rangle$	$i\xi$	$i\zeta$	$-i\zeta$
ixy	z, ixy	$\|T_2\zeta\rangle \equiv \|\tilde{1}\tilde{2}\tilde{0}0\rangle$	$-\zeta$	$i\eta$	$-\xi$

Table C.5.1(c). Relation between the Complex and Real Tetragonal Bases for Single-Valued Irreps

Complex-O $\lvert a\alpha\rangle \equiv \lvert a(O)a_1(D_4)a_2(C_4)\rangle$	Real-O Equivalent[a] $\lvert a\alpha\rangle \equiv \lvert a(O)a_1(D_4)a_2(D_2)a_3(C_2)\rangle$
Complex-T_d $\lvert a\alpha\rangle \equiv \lvert a(T_d)a_1(D_{2d})a_2(S_4)\rangle$	Real-T_d Equivalent[a] $\lvert a\alpha\rangle \equiv \lvert a(T_d)a_1(D_{2d})a_2(D_2)a_3(C_2)\rangle$
$\lvert A_1a_1\rangle = \lvert 000\rangle$	$\lvert A_1a_1\rangle = \lvert 0000\rangle$
$\lvert A_2a_2'\rangle = \lvert\tilde{0}22\rangle$	$-\lvert A_2a_2\rangle = -\lvert\tilde{0}200\rangle$
$\lvert E\theta\rangle = \lvert 200\rangle$	$\lvert E\theta\rangle = \lvert 2000\rangle$
$\lvert E\epsilon'\rangle = \lvert 222\rangle$	$-\lvert E\epsilon\rangle = -\lvert 2200\rangle$
$\lvert T_1 -1\rangle = \lvert 11-1\rangle$	$(1/\sqrt{2})(\lvert T_1x\rangle + \lvert T_1y\rangle) = (1/\sqrt{2})(\lvert 11\tilde{1}1\rangle + \lvert 1111\rangle)$
$\lvert T_10\rangle = \lvert 1\tilde{0}0\rangle$	$\lvert T_1z\rangle = \lvert 1\tilde{0}\tilde{0}0\rangle$
$\lvert T_11\rangle = \lvert 111\rangle$	$(1/\sqrt{2})(-\lvert T_1x\rangle + \lvert T_1y\rangle) = (1/\sqrt{2})(-\lvert 11\tilde{1}1\rangle + \lvert 1111\rangle)$
$\lvert T_2 -1\rangle = \lvert\tilde{1}1-1\rangle$	$(1/\sqrt{2})(\lvert T_2\xi\rangle + \lvert T_2\eta\rangle) = (1/\sqrt{2})(\lvert\tilde{1}1\tilde{1}1\rangle + \lvert\tilde{1}111\rangle)$
$\lvert T_20\rangle = \lvert\tilde{1}\tilde{2}2\rangle$	$\lvert T_2\zeta\rangle = \lvert\tilde{1}\tilde{2}\tilde{0}0\rangle$
$\lvert T_21\rangle = \lvert\tilde{1}11\rangle$	$(1/\sqrt{2})(-\lvert T_2\xi\rangle + \lvert T_2\eta\rangle) = (1/\sqrt{2})(-\lvert\tilde{1}1\tilde{1}1\rangle + \lvert\tilde{1}111\rangle)$

[a] For example, $\lvert T_11\rangle = (1/\sqrt{2})(-\lvert T_1x\rangle + \lvert T_1y\rangle)$

both by ξ, η, and ζ and by x, y, and z—the labels have identical meanings and are used interchangeably in the text. The table also indicates how common functions transform relative to the bases. Note carefully the differences for O and T_d. Further operator and function transformations are given in Section C.7.

In Table C.5.1(c) the complex-O and complex-T_d bases are given in terms of the real bases for single-valued irreps. Transformations may then be found using Table C.5.1(b).

Table C.5.1(d) defines the transformation properties of the complex-O and complex-T_d bases by giving the angular-momentum basis functions (the $\lvert jm\rangle$ of $SO_3 \supset SO_2$ and the $\lvert j^{\pm}m\rangle$ of $O_3 \supset SO_3 \supset SO_2$) which have the same transformation properties as our standard basis functions. Recall that SO_3 is the group of all rotations, while O_3 is the rotation–inversion group. Double-valued irreps (spinors) are not expressible in terms of rectangular coordinates. Transformations of the $\lvert jm\rangle$ [and hence of the Table C.5.1(d) spinors] are found as discussed in Section 9.4. Note that the $\lvert j^{-}m\rangle$ transform differently than the $\lvert j^{+}m\rangle$ in T_d—see notes in Section C.6.

Table C.5.1(d). **The Complex-O and Complex-T_d Bases for Double-Valued Irreps**

$O_3 \supset SO_3 \supset SO_2$ Equivalent[a]	Standard Basis	Group Generators	
For $\lvert j^+ m\rangle$ only	Complex-T_d ($T_d \supset D_{2d} \supset S_4$)	$(S_4^Z)^{-1}$	$(S_4^X)^{-1}$
For $\lvert j^+ m\rangle$ and $\lvert j^- m\rangle$	Complex-O ($O \supset D_4 \supset C_4$)	C_4^Z	C_4^X
$\lvert \frac{1}{2} -\frac{1}{2}\rangle$	$\lvert E'\beta'\rangle \equiv \lvert \frac{1}{2}\frac{1}{2} -\frac{1}{2}\rangle$	$(1/\sqrt{2})(1+i)\beta'$	$(1/\sqrt{2})(-i\alpha' + \beta')$
$\lvert \frac{1}{2}\frac{1}{2}\rangle$	$\lvert E'\alpha'\rangle \equiv \lvert \frac{1}{2}\frac{1}{2}\frac{1}{2}\rangle$	$(1/\sqrt{2})(1-i)\alpha'$	$(1/\sqrt{2})(\alpha' - i\beta')$
$(1/\sqrt{6})(-\sqrt{5}\lvert \frac{5}{2} -\frac{3}{2}\rangle + \lvert \frac{5}{2}\frac{5}{2}\rangle)$	$\lvert E''\beta''\rangle \equiv \lvert \tilde{\frac{1}{2}}\frac{3}{2} -\frac{3}{2}\rangle$	$(1/\sqrt{2})(-1+i)\beta''$	$(-1/\sqrt{2})(i\alpha'' + \beta'')$
$(1/\sqrt{6})(-\lvert \frac{5}{2} -\frac{5}{2}\rangle + \sqrt{5}\lvert \frac{5}{2}\frac{3}{2}\rangle)$	$\lvert E''\alpha''\rangle \equiv \lvert \tilde{\frac{1}{2}}\frac{3}{2}\frac{3}{2}\rangle$	$(1/\sqrt{2})(-1-i)\alpha''$	$(-1/\sqrt{2})(\alpha'' + i\beta'')$
$\lvert \frac{3}{2} -\frac{3}{2}\rangle$	$\lvert U'\nu\rangle \equiv \lvert \frac{3}{2}\frac{3}{2} -\frac{3}{2}\rangle$	$(1/\sqrt{2})(-1+i)\nu$	$(1/2\sqrt{2})(i\kappa - \sqrt{3}\lambda + i\sqrt{3}\mu + \nu)$
$-\lvert \frac{3}{2} -\frac{1}{2}\rangle$	$\lvert U'\mu\rangle \equiv \lvert \frac{3}{2}\frac{1}{2} -\frac{1}{2}\rangle$	$(1/\sqrt{2})(1+i)\mu$	$(1/2\sqrt{2})(\sqrt{3}\kappa + i\lambda - \mu + i\sqrt{3}\nu)$
$\lvert \frac{3}{2}\frac{1}{2}\rangle$	$\lvert U'\lambda\rangle \equiv \lvert \frac{3}{2}\frac{1}{2}\frac{1}{2}\rangle$	$(1/\sqrt{2})(1-i)\lambda$	$(1/2\sqrt{2})(-i\sqrt{3}\kappa - \lambda + i\mu - \sqrt{3}\nu)$
$\lvert \frac{3}{2}\frac{3}{2}\rangle$	$\lvert U'\kappa\rangle \equiv \lvert \frac{3}{2}\frac{3}{2}\frac{3}{2}\rangle$	$(1/\sqrt{2})(-1-i)\kappa$	$(1/2\sqrt{2})(\kappa - i\sqrt{3}\lambda + \sqrt{3}\mu + i\nu)$

[a] SO_3 is the full rotation group, and O_3 is the rotation–inversion group.

Irrep names are given in both the Griffith-type $|a\alpha\rangle$ notation and in the Butler chain notation (Section 15.2). In chain notation the real basis names are $|a(O)a_1(D_4)a_2(D_2)a_3(C_2)\rangle$ or $|a(T_d)a_1(D_{2d})a_2(D_2)a_3(C_2)\rangle$, while the complex-basis names are $|a(O)a_1(D_4)a_2(C_4)\rangle$ or $|a(T_d)a_1(D_{2d})a_2(S_4)\rangle$.

C.6. The Complex-O and Complex-T_d (Tetragonal) Bases Defined in Terms of the $O_3 \supset SO_3 \supset SO_2$ Basis

The $|j^- m\rangle$ transform identically to the $|j^+ m\rangle$ under O, but differently under T_d. Thus the $O_3 \supset O_h \supset T_d \supset D_{2d} \supset S_4$ chain is isomorphic with the $O_3 \supset O_h \supset O \supset D_4 \supset C_4$ chain for j^+ kets but not for j^- kets. In Table C.6.1 we give only the transformations for j^+. To obtain the j^- states in the complex-O chain, simply replace $|j^+a^+aa_1a_2\rangle$ with $|j^-a^-aa_1a_2\rangle$, $|j^+m\rangle$ with $|j^-m\rangle$, and the subscript g with u. The j^- basis for T_d may be obtained from the j^+ basis below using the method of Section 5.3 of Butler (1981). [See also Reid and Butler (1980, 1982).] In the notation of the table, one finds, for example, that in the complex-T_d chain

$$\begin{aligned} |T_2 -1\rangle &\sim |1^-1^-\tilde{1}1-1\rangle \equiv |1^-T_{1u}T_2E-1\rangle = |1^-1\rangle \\ |T_2 0\rangle &\sim |1^-1^-\tilde{1}\tilde{2}2\rangle \equiv |1^-T_{1u}T_2B_2b_2\rangle = |1^-0\rangle \\ |T_2 1\rangle &\sim |1^-1^-\tilde{1}11\rangle \equiv |1^-T_{1u}T_2E1\rangle = |1^- -1\rangle \end{aligned} \tag{C.6.1}$$

This result also follows from Tables C.5.1(b) and (c), and Eq. (C.7.4).

The numeric irrep notation for the chain is that of Butler (1981). The alphabetic notation for the chain gives the nearest Mulliken equivalent, with the exception that the last irrep in the chain (here a C_4 or S_4 irrep) is named using component labels from the next highest group in the chain (here D_4 or D_{2d}). The basis relations given follow directly from the tabulation of $SO_3 \supset O \supset D_4 \supset C_4$ partners in terms of $SO_3 \supset SO_2$ partners given in Chapter 16 of Butler (1981).

C.7. Function and Operator Transformation Coefficients

The coefficients are defined in Section 9.8 by the equations

$$\begin{aligned} \psi_i &= \sum_{a\alpha} \langle a\alpha|\psi_i\rangle |\psi_i a\alpha\rangle \\ \mathcal{O}_i &= \sum_{f\phi} \langle f\phi|\mathcal{O}_i\rangle O^f_\phi \end{aligned} \tag{C.7.1}$$

Table C.6.1

j	Complex-O / Complex-T_d	$O_3 \supset O_h \supset O \supset D_4 \supset C_4$ $\lvert j^+a^+aa_1a_2\rangle$ (For j^-a^-, $g \to u$) / $O_3 \supset O_h \supset T_d \supset D_{2d} \supset S_4$ $\lvert j^+a^+aa_1a_2\rangle$	$O_3 \supset SO_3 \supset SO_2$ $\lvert j^+m\rangle$ Equivalent
		True Irreps	
0	$\lvert A_1a_1\rangle$	$\lvert 0^+0^+000\rangle = \lvert 0^+A_{1g}A_1A_1a_1\rangle$	$\lvert 0^+0\rangle$
1	$\lvert T_1 -1\rangle$	$\lvert 1^+1^+11-1\rangle = \lvert 1^+T_{1g}T_1E-1\rangle$	$-\lvert 1^+-1\rangle$
	$\lvert T_10\rangle$	$\lvert 1^+1^+1\tilde{0}0\rangle = \lvert 1^+T_{1g}T_1A_2a_2\rangle$	$\lvert 1^+0\rangle$
	$\lvert T_11\rangle$	$\lvert 1^+1^+111\rangle = \lvert 1^+T_{1g}T_1E1\rangle$	$-\lvert 1^+1\rangle$
2	$\lvert E\theta\rangle$	$\lvert 2^+2^+200\rangle = \lvert 2^+E_gEA_1a_1\rangle$	$-\lvert 2^+0\rangle$
	$\lvert E\epsilon'\rangle$	$\lvert 2^+2^+222\rangle = \lvert 2^+E_gEB_1b_1'\rangle$	$(-1/\sqrt{2})(\lvert 2^+-2\rangle + \lvert 2^+2\rangle)$
	$\lvert T_2 -1\rangle$	$\lvert 2^+\tilde{1}^+\tilde{1}1-1\rangle = \lvert 2^+T_{2g}T_2E-1\rangle$	$\lvert 2^+-1\rangle$
	$\lvert T_20\rangle$	$\lvert 2^+\tilde{1}^+\tilde{1}\tilde{2}2\rangle = \lvert 2^+T_{2g}T_2B_2b_2\rangle$	$(1/\sqrt{2})(-\lvert 2^+-2\rangle + \lvert 2^+2\rangle)$
	$\lvert T_21\rangle$	$\lvert 2^+\tilde{1}^+\tilde{1}11\rangle = \lvert 2^+T_{2g}T_2E1\rangle$	$-\lvert 2^+1\rangle$
3	$\lvert A_2a_2'\rangle$	$\lvert 3^+\tilde{0}^+\tilde{0}22\rangle = \lvert 3^+A_{2g}A_2B_1b_1'\rangle$	$(1/\sqrt{2})(\lvert 3^+-2\rangle - \lvert 3^+2\rangle)$
	$\lvert T_1 -1\rangle$	$\lvert 3^+1^+11-1\rangle = \lvert 3^+T_{1g}T_1E-1\rangle$	$(-1/2\sqrt{2})(\sqrt{3}\lvert 3^+-1\rangle + \sqrt{5}\lvert 3^+3\rangle)$
	$\lvert T_10\rangle$	$\lvert 3^+1^+1\tilde{0}0\rangle = \lvert 3^+T_{1g}T_1A_2a_2\rangle$	$-\lvert 3^+0\rangle$
	$\lvert T_11\rangle$	$\lvert 3^+1^+111\rangle = \lvert 3^+T_{1g}T_1E1\rangle$	$(-1/2\sqrt{2})(\sqrt{5}\lvert 3^+-3\rangle + \sqrt{3}\lvert 3^+1\rangle)$
	$\lvert T_2 -1\rangle$	$\lvert 3^+\tilde{1}^+\tilde{1}1-1\rangle = \lvert 3^+T_{2g}T_2E-1\rangle$	$(1/2\sqrt{2})(-\sqrt{5}\lvert 3^+-1\rangle + \sqrt{3}\lvert 3^+3\rangle)$
	$\lvert T_20\rangle$	$\lvert 3^+\tilde{1}^+\tilde{1}\tilde{2}2\rangle = \lvert 3^+T_{2g}T_2B_2b_2\rangle$	$-(1/\sqrt{2})(\lvert 3^+-2\rangle + \lvert 3^+2\rangle)$
	$\lvert T_21\rangle$	$\lvert 3^+\tilde{1}^+\tilde{1}11\rangle = \lvert 3^+T_{2g}T_2E1\rangle$	$(1/2\sqrt{2})(\sqrt{3}\lvert 3^+-3\rangle - \sqrt{5}\lvert 3^+1\rangle)$
		Spin Irreps	
$\frac{1}{2}$	$\lvert E'\beta'\rangle$	$\lvert \frac{1}{2}^+\frac{1}{2}^+\frac{1}{2}\frac{1}{2}-\frac{1}{2}\rangle = \lvert \frac{1}{2}^+E_g'E'E'\beta'\rangle$	$\lvert \frac{1}{2}^+-\frac{1}{2}\rangle$
	$\lvert E'\alpha'\rangle$	$\lvert \frac{1}{2}^+\frac{1}{2}^+\frac{1}{2}\frac{1}{2}\frac{1}{2}\rangle = \lvert \frac{1}{2}^+E_g'E'E'\alpha'\rangle$	$\lvert \frac{1}{2}^+\frac{1}{2}\rangle$
$\frac{3}{2}$	$\lvert U'\nu\rangle$	$\lvert \frac{3}{2}^+\frac{3}{2}^+\frac{3}{2}\frac{3}{2}-\frac{3}{2}\rangle = \lvert \frac{3}{2}^+U_g'U'E''\beta''\rangle$	$\lvert \frac{3}{2}^+-\frac{3}{2}\rangle$
	$\lvert U'\mu\rangle$	$\lvert \frac{3}{2}^+\frac{3}{2}^+\frac{3}{2}\frac{1}{2}-\frac{1}{2}\rangle = \lvert \frac{3}{2}^+U_g'U'E'\beta'\rangle$	$-\lvert \frac{3}{2}^+-\frac{1}{2}\rangle$
	$\lvert U'\lambda\rangle$	$\lvert \frac{3}{2}^+\frac{3}{2}^+\frac{3}{2}\frac{1}{2}\frac{1}{2}\rangle = \lvert \frac{3}{2}^+U_g'U'E'\alpha'\rangle$	$\lvert \frac{3}{2}^+\frac{1}{2}\rangle$
	$\lvert U'\kappa\rangle$	$\lvert \frac{3}{2}^+\frac{3}{2}^+\frac{3}{2}\frac{3}{2}\frac{3}{2}\rangle = \lvert \frac{3}{2}^+U_g'U'E''\alpha''\rangle$	$\lvert \frac{3}{2}^+\frac{3}{2}\rangle$
$\frac{5}{2}$	$\lvert E''\beta''\rangle$	$\lvert \frac{5}{2}^+\widetilde{\tfrac{1}{2}}^+\widetilde{\tfrac{1}{2}}\frac{3}{2}-\frac{3}{2}\rangle = \lvert \frac{5}{2}^+E_g''E''E''\beta''\rangle$	$(1/\sqrt{6})(-\sqrt{5}\lvert \frac{5}{2}^+-\frac{3}{2}\rangle + \lvert \frac{5}{2}^+\frac{5}{2}\rangle)$
	$\lvert E''\alpha''\rangle$	$\lvert \frac{5}{2}^+\widetilde{\tfrac{1}{2}}^+\widetilde{\tfrac{1}{2}}\frac{3}{2}\frac{3}{2}\rangle = \lvert \frac{5}{2}^+E_g''E''E''\alpha''\rangle$	$(1/\sqrt{6})(-\lvert \frac{5}{2}^+-\frac{5}{2}\rangle + \sqrt{5}\lvert \frac{5}{2}^+\frac{3}{2}\rangle)$
	$\lvert U'\nu\rangle$	$\lvert \frac{5}{2}^+\frac{3}{2}^+\frac{3}{2}\frac{3}{2}-\frac{3}{2}\rangle = \lvert \frac{5}{2}^+U_g'U'E''\beta''\rangle$	$(1/\sqrt{6})(\lvert \frac{5}{2}^+-\frac{3}{2}\rangle + \sqrt{5}\lvert \frac{5}{2}^+\frac{5}{2}\rangle)$
	$\lvert U'\mu\rangle$	$\lvert \frac{5}{2}^+\frac{3}{2}^+\frac{3}{2}\frac{1}{2}-\frac{1}{2}\rangle = \lvert \frac{5}{2}^+U_g'U'E'\beta'\rangle$	$\lvert \frac{5}{2}^+-\frac{1}{2}\rangle$
	$\lvert U'\lambda\rangle$	$\lvert \frac{5}{2}^+\frac{3}{2}^+\frac{3}{2}\frac{1}{2}\frac{1}{2}\rangle = \lvert \frac{5}{2}^+U_g'U'E'\alpha'\rangle$	$\lvert \frac{5}{2}^+\frac{1}{2}\rangle$
	$\lvert U'\kappa\rangle$	$\lvert \frac{5}{2}^+\frac{3}{2}^+\frac{3}{2}\frac{3}{2}\frac{3}{2}\rangle = \lvert \frac{5}{2}^+U_g'U'E''\alpha''\rangle$	$(-1/\sqrt{6})(\sqrt{5}\lvert \frac{5}{2}^+-\frac{5}{2}\rangle + \lvert \frac{5}{2}^+\frac{3}{2}\rangle)$

Conversely

$$|\psi_i a\alpha\rangle = \sum_i \langle a\alpha|\psi_i\rangle^* \psi_i$$

$$O_\phi^f = \sum_i \langle f\phi|\mathcal{O}_i\rangle^* \mathcal{O}_i \qquad \text{(C.7.2)}$$

Note that the equations (C.7.2) require the complex conjugates of the coefficients in Table C.7.1. Whenever possible the basis is chosen so the sums above are unnecessary.

Table C.7.1. Function and Operator Transformation Coefficients

		$|\psi a\alpha\rangle$ or O_ϕ^f	
ψ or $\mathcal{O}$	$\langle a\alpha|\psi\rangle$ or $\langle f\phi|\mathcal{O}\rangle$	$a\alpha$ or $f\phi$	Equivalent-Chain Notation
Real-O Basis ($O \supset D_4 \supset D_2 \supset C_2$)			
x, p_x, V_x, R_x	-1	$T_1 x$	$11\tilde{1}1$
y, p_y, V_y, R_y	$-i$	$T_1 y$	1111
z, p_z, V_z, R_z	1	$T_1 z$	$1\tilde{0}\tilde{0}0$
d_{z^2}	-1	$E\theta$	2000
$d_{x^2-y^2}$	1	$E\epsilon$	2200
d_{yz}	i	$T_2\xi$	$\tilde{1}1\tilde{1}1$
d_{zx}	1	$T_2\eta$	$\tilde{1}111$
d_{xy}	$-i$	$T_2\zeta$	$\tilde{1}\tilde{2}\tilde{0}0$
Real-T_d Basis ($T_d \supset D_{2d} \supset D_2 \supset C_2$)[a]			
x, p_x, V_x	-1	$T_2\xi(x)$	$\tilde{1}1\tilde{1}1$
y, p_y, V_y	i	$T_2\eta(y)$	$\tilde{1}111$
z, p_z, V_z	1	$T_2\zeta(z)$	$\tilde{1}\tilde{2}\tilde{0}0$
Complex-O Basis ($O \supset D_4 \supset C_4$)			
$|1^{\pm}-1\rangle, V_{-1}, R_{-1}$	-1	$T_1 -1$	$11-1$
$|1^{\pm}0\rangle, V_0, R_0$	1	$T_1 0$	$1\tilde{0}0$
$|1^{\pm}1\rangle, V_1, R_1$	-1	$T_1 1$	111
Complex-T_d Basis ($T_{2d} \supset D_{2d} \supset S_4$)[b]			
$|1^{-}-1\rangle, V_{-1}$	1	$T_2 1$	$\tilde{1}11$
$|1^{-}0\rangle, V_0$	1	$T_2 0$	$\tilde{1}\tilde{2}2$
$|1^{-}1\rangle, V_1$	1	$T_2 -1$	$\tilde{1}1-1$

[a] **R** and the d functions transform as in the real-O basis.
[b] **R** and the $|1^{+}m\rangle$ for $m = -1, 0, 1$ transform as in the complex-O basis.

Vector operators which are odd under inversion transform as **V** (e.g. the electric dipole operator **m**) while those that are even under inversion transform as **R** (e.g. **l**, **s**, **j** and **μ**). Components of spherical vectors are defined in Section 9.8 to have transformation properties identical to the $|jm\rangle$ of $SO_3 \supset SO_2$ for $j = 1$. It follows that

$$R_{-1} \equiv \frac{1}{\sqrt{2}}(R_x - iR_y) \sim |1^+ -1\rangle$$

$$R_0 \equiv R_z \sim |1^+ 0\rangle \tag{C.7.3}$$

$$R_1 \equiv -\frac{1}{\sqrt{2}}(R_x + iR_y) \sim |1^+ 1\rangle$$

while

$$V_{-1} \equiv \frac{1}{\sqrt{2}}(V_x - iV_y) \sim |1^- -1\rangle$$

$$V_0 \equiv V_z \sim |1^- 0\rangle \tag{C.7.4}$$

$$V_1 \equiv -\frac{1}{\sqrt{2}}(V_x + iV_y) \sim |1^- 1\rangle$$

where the $|j^{\pm}m\rangle$ are the basis kets for the group chain $O_3 \supset SO_3 \supset SO_2$. Inverting (C.7.3) gives

$$R_x = \frac{1}{\sqrt{2}}(R_{-1} - R_1) \sim \frac{1}{\sqrt{2}}(|1^+ -1\rangle - |1^+ 1\rangle)$$

$$R_y = \frac{i}{\sqrt{2}}(R_{-1} + R_1) \sim \frac{i}{\sqrt{2}}(|1^+ -1\rangle + |1^+ 1\rangle) \tag{C.7.5}$$

$$R_z = R_0 \sim |1^+ 0\rangle$$

Inverting (C.7.4) also gives (C.7.5), but with the V substituted everywhere for R, and $|1^- m\rangle$ substituted for $|1^+ m\rangle$, $m = 0$, 1, and -1.

In Section 18.4 it is shown that the spin–orbit coupling operator transforms in the groups O and T_d as

$$\mathcal{H}_{SO} = -\sum_k \sum_\phi \begin{pmatrix} t_1 \\ \phi \end{pmatrix} s(k)^{1t_1}_{\phi} u(k)^{1t_1}_{\phi^*} \tag{C.7.6}$$

Thus, for example, in the complex-O basis

$$\mathcal{H}_{\mathrm{SO}} = \sum_k \left[-s(k)_1^{1t_1} u(k)_{-1}^{1t_1} + s(k)_0^{1t_1} u(k)_0^{1t_1} - s(k)_{-1}^{1t_1} u(k)_1^{1t_1} \right] \tag{C.7.7}$$

The transformation coefficients for the group O_h are identical to those for the group O, if only the proper parity labels are affixed. Thus x, y, and z, the p orbitals, and the components of **V**—all of which have odd (u) parity under inversion—go over to T_{1u} in the group O_h. Conversely, the d orbitals and the components of **R**, which have even (g) parity, transform as E_g, T_{2g} and T_{1g} respectively in the group O_h.

Note carefully that functions or operators which are *odd* under inversion transform *differently* in T_d and O (or O_h). Thus in the group T_d, x, y, and z, the p orbitals, and the components of **V** belong to T_2, but they belong to T_1 in the group O. *Even* parity functions or operators belong to the *same* irreps in groups T_d and O.

Transformation coefficients for d orbitals in the complex-O and complex-T_d bases follow directly from Section C.6, and so they are not tabulated here. The p and d atomic functions are defined in Sections B.1 and B.2.

The bases used in Table C.7.1 are those of Section C.5.

C.8. Functions for t_2^m and e^n Configurations

In writing the functions in Table C.8.1 we omit ket signs and use the abbreviation $\xi^2 = \xi^+\xi^-$, and so on. The functions were obtained as discussed in Sections 11.4 and 19.3 with the $3jm$ of Sections B.3 and C.12 and the cfp of Section C.15. The orbital basis used is defined in Table C.5.1(b).

Table C.8.1

a^n	$\lvert a^n \mathcal{S} h \mathcal{M} \theta \rangle$
t_2^1	$\lvert \tfrac{1}{2} T_2 \tfrac{1}{2} \xi \rangle = \lvert \xi^+ \rangle$
	$\lvert \tfrac{1}{2} T_2 \tfrac{1}{2} \eta \rangle = \lvert \eta^+ \rangle$
	$\lvert \tfrac{1}{2} T_2 \tfrac{1}{2} \zeta \rangle = \lvert \zeta^+ \rangle$
t_2^2	$\lvert 1 T_1 1 x \rangle = \lvert \eta^+ \zeta^+ \rangle$
	$\lvert 1 T_1 1 y \rangle = -\lvert \xi^+ \zeta^+ \rangle$
	$\lvert 1 T_1 1 z \rangle = \lvert \xi^+ \eta^+ \rangle$
	$\lvert 0 A_1 0 a_1 \rangle = (1/\sqrt{3})(-\lvert \xi^2 \rangle + \lvert \eta^2 \rangle - \lvert \zeta^2 \rangle)$
	$\lvert 0 E 0 \theta \rangle = (1/\sqrt{6})(\lvert \xi^2 \rangle - \lvert \eta^2 \rangle - 2\lvert \zeta^2 \rangle)$
	$\lvert 0 E 0 \epsilon \rangle = (1/\sqrt{2})(\lvert \xi^2 \rangle + \lvert \eta^2 \rangle)$
	$\lvert 0 T_2 0 \xi \rangle = (1/\sqrt{2})(\lvert \eta^+ \zeta^- \rangle - \lvert \eta^- \zeta^+ \rangle)$
	$\lvert 0 T_2 0 \eta \rangle = (1/\sqrt{2})(-\lvert \xi^+ \zeta^- \rangle + \lvert \xi^- \zeta^+ \rangle)$
	$\lvert 0 T_2 0 \zeta \rangle = (1/\sqrt{2})(\lvert \xi^+ \eta^- \rangle - \lvert \xi^- \eta^+ \rangle)$

Table C.8.1. Continued.

a^n	$\lvert a^n \mathfrak{S} h \mathfrak{M} \theta \rangle$
t_2^3	$\lvert \tfrac{3}{2} A_2 \tfrac{3}{2} a_2 \rangle = \lvert \xi^+ \eta^+ \zeta^+ \rangle$
	$\lvert \tfrac{1}{2} E \tfrac{1}{2} \theta \rangle = (1/\sqrt{2})(-\lvert \xi^+ \eta^- \zeta^+ \rangle + \lvert \xi^- \eta^+ \zeta^+ \rangle)$
	$\lvert \tfrac{1}{2} E \tfrac{1}{2} \epsilon \rangle = (1/\sqrt{6})(2\lvert \xi^+ \eta^+ \zeta^- \rangle - \lvert \xi^+ \eta^- \zeta^+ \rangle - \lvert \xi^- \eta^+ \zeta^+ \rangle)$
	$\lvert \tfrac{1}{2} T_1 \tfrac{1}{2} x \rangle = (1/\sqrt{2})(\lvert \xi^+ \eta^2 \rangle + \lvert \xi^+ \zeta^2 \rangle)$
	$\lvert \tfrac{1}{2} T_1 \tfrac{1}{2} y \rangle = (1/\sqrt{2})(-\lvert \xi^2 \eta^+ \rangle + \lvert \eta^+ \zeta^2 \rangle)$
	$\lvert \tfrac{1}{2} T_1 \tfrac{1}{2} z \rangle = (-1/\sqrt{2})(\lvert \xi^2 \zeta^+ \rangle + \lvert \eta^2 \zeta^+ \rangle)$
	$\lvert \tfrac{1}{2} T_2 \tfrac{1}{2} \xi \rangle = (1/\sqrt{2})(\lvert \xi^+ \eta^2 \rangle - \lvert \xi^+ \zeta^2 \rangle)$
	$\lvert \tfrac{1}{2} T_2 \tfrac{1}{2} \eta \rangle = (-1/\sqrt{2})(\lvert \xi^2 \eta^+ \rangle + \lvert \eta^+ \zeta^2 \rangle)$
	$\lvert \tfrac{1}{2} T_2 \tfrac{1}{2} \zeta \rangle = (1/\sqrt{2})(-\lvert \xi^2 \zeta^+ \rangle + \lvert \eta^2 \zeta^+ \rangle)$
t_2^4	$\lvert 1 T_1 1 x \rangle = -\lvert \xi^2 \eta^+ \zeta^+ \rangle$
	$\lvert 1 T_1 1 y \rangle = -\lvert \xi^+ \eta^2 \zeta^+ \rangle$
	$\lvert 1 T_1 1 z \rangle = -\lvert \xi^+ \eta^+ \zeta^2 \rangle$
	$\lvert 0 A_1 0 a_1 \rangle = (1/\sqrt{3})(-\lvert \eta^2 \zeta^2 \rangle + \lvert \xi^2 \zeta^2 \rangle - \lvert \xi^2 \eta^2 \rangle)$
	$\lvert 0 E 0 \theta \rangle = (1/\sqrt{6})(\lvert \eta^2 \zeta^2 \rangle - \lvert \xi^2 \zeta^2 \rangle - 2\lvert \xi^2 \eta^2 \rangle)$
	$\lvert 0 E 0 \epsilon \rangle = (1/\sqrt{2})(\lvert \xi^2 \zeta^2 \rangle + \lvert \eta^2 \zeta^2 \rangle)$
	$\lvert 0 T_2 0 \xi \rangle = (1/\sqrt{2})(\lvert \xi^2 \eta^+ \zeta^- \rangle - \lvert \xi^2 \eta^- \zeta^+ \rangle)$
	$\lvert 0 T_2 0 \eta \rangle = (1/\sqrt{2})(\lvert \xi^+ \eta^2 \zeta^- \rangle - \lvert \xi^- \eta^2 \zeta^+ \rangle)$
	$\lvert 0 T_2 0 \zeta \rangle = (1/\sqrt{2})(\lvert \xi^+ \eta^- \zeta^2 \rangle - \lvert \xi^- \eta^+ \zeta^2 \rangle)$
t_2^5	$\lvert \tfrac{1}{2} T_2 \tfrac{1}{2} \xi \rangle = -\lvert \xi^+ \eta^2 \zeta^2 \rangle$
	$\lvert \tfrac{1}{2} T_2 \tfrac{1}{2} \eta \rangle = \lvert \xi^2 \eta^+ \zeta^2 \rangle$
	$\lvert \tfrac{1}{2} T_2 \tfrac{1}{2} \zeta \rangle = -\lvert \xi^2 \eta^2 \zeta^+ \rangle$
t_2^6	$\lvert 0 A_1 0 a_1 \rangle = \lvert \xi^2 \eta^2 \zeta^2 \rangle$
e	$\lvert \tfrac{1}{2} E \tfrac{1}{2} \theta \rangle = \lvert \theta^+ \rangle$
	$\lvert \tfrac{1}{2} E \tfrac{1}{2} \epsilon \rangle = \lvert \epsilon^+ \rangle$
e^2	$\lvert 1 A_2 1 a_2 \rangle = -\lvert \theta^+ \epsilon^+ \rangle$
	$\lvert 0 A_1 0 a_1 \rangle = (1/\sqrt{2})(\lvert \theta^2 \rangle + \lvert \epsilon^2 \rangle)$
	$\lvert 0 E 0 \theta \rangle = (1/\sqrt{2})(-\lvert \theta^2 \rangle + \lvert \epsilon^2 \rangle)$
	$\lvert 0 E 0 \epsilon \rangle = (1/\sqrt{2})(\lvert \theta^+ \epsilon^- \rangle - \lvert \theta^- \epsilon^+ \rangle)$
e^3	$\lvert \tfrac{1}{2} E \tfrac{1}{2} \theta \rangle = \lvert \theta^+ \epsilon^2 \rangle$
	$\lvert \tfrac{1}{2} E \tfrac{1}{2} \epsilon \rangle = \lvert \theta^2 \epsilon^+ \rangle$
e^4	$\lvert 0 A_1 0 a_1 \rangle = \lvert \theta^2 \epsilon^2 \rangle$

C.9. Strong-Field Crystal-Field and Electrostatic Matrices for d^n Configurations

The strong-field crystal-field matrix is diagonal in configuration and independent of state symmetry labels. The crystal-field energies for all states of a group-O or group-T_d d^{m+n} configuration $t_2^m e^n$ are given by

$$E_{\text{crystal field}} = [m(-4) + n(6)]\,Dq \qquad \text{(C.9.1)}$$

Thus the $t_2^3 e$ states of a d^4 system have a crystal-field energy of $-6Dq$, and so on. Note that $Dq(O_h) = (-9/4)Dq(T_d) > 0$.

Strong-field electrostatic matrices for d^n configurations in the groups O and T_d are given in Tables A27–A30 of Griffith (1964). These matrices, however, cannot be combined with matrices obtained using our tables in Appendix C. They were calculated using Griffith [(1964), Table A24] t_2^m and e^n strong-field functions which differ from the strong-field functions obtained with our tables (Section C.8). The functions in Griffith's Table A24 are formed using his Table A39 for the cfp and his Table A20 for the real-O v.c.c. for orbital couplings. These cfp and v.c.c. differ in phase from ours. Moreover, irrep partners defined by Griffith [(1964), Section 6.8 and Table A16] differ from those of the real-O basis of Table C.5.1(b). Thus, while Griffith's spin couplings $|(\mathbb{S}_1, \mathbb{S}_2)\mathbb{S}\mathfrak{M}\rangle$ are identical to ours (both use Wigner coefficients), both his orbital basis and his orbital couplings $|(h_1, h_2)h\theta\rangle$ differ substantially.

It is possible to "Butlerize" Griffith's strong-field electrostatic matrices by multiplying each matrix element in Griffith's Tables A27–A30 by an appropriate correction (phase) factor. Such a procedure, although tedious, should be faster than recalculating the electrostatic matrices from first principles. We outline the steps below.

A typical ket is $|\mathcal{a}(t_2^m(\mathbb{S}_1 h_1), e^n(\mathbb{S}_2 h_2))\mathbb{S}h\mathfrak{M}\theta\rangle$. Since the spin–spin couplings in the two bases are identical, we may use functions of the type (19.7.8) for our discussion. Thus our functions (labeled B for Butler) are

$$\left|\left(t_2^m(\mathbb{S}_1 h_1 \mathfrak{M}_1), e^n(\mathbb{S}_2 h_2 \mathfrak{M}_2)\right) h\theta^B\right\rangle_B$$

$$= \sum_{\theta_1^B \theta_2^B} \left(h_1\theta_1^B, h_2\theta_2^B | h\theta^B\right)_B |t_2^m\, \mathbb{S}_1 h_1 \mathfrak{M}_1 \theta_1^B\rangle_B |e^n\, \mathbb{S}_2 h_2 \mathfrak{M}_2 \theta_2^B\rangle_B \qquad \text{(C.9.2)}$$

which, expressed in terms of the basis partners in Griffith [(1964), Table

A16] ($h_1\theta_1^G$ etc.), become

$$= \sum_{\theta_1^B\theta_2^B} \langle h_1\theta_1^B|h_1\theta_1^G\rangle\langle h_2\theta_2^B|h_2\theta_2^G\rangle\langle h\theta^G|h\theta^B\rangle\big(h_1\theta_1^G, h_2\theta_2^G|h\theta^G\big)_B$$

$$\times\langle h_1\theta_1^G|h_1\theta_1^B\rangle\langle h_2\theta_2^G|h_2\theta_2^B\rangle|t_2^m\,\mathcal{S}_1h_1\mathfrak{M}_1\theta_1^G\rangle_B|e^n\,\mathcal{S}_2h_2\mathfrak{M}_2\theta_2^G\rangle_B$$

$$= \langle h\theta^G|h\theta^B\rangle\sum_{\theta_1^G\theta_2^G}\big(h_1\theta_1^G, h_2\theta_2^G|h\theta^G\big)_B|t_2^m\,\mathcal{S}_1h_1\mathfrak{M}_1\theta_1^G\rangle_B|e^n\,\mathcal{S}_2h_2\mathfrak{M}_2\theta_2^G\rangle_B \tag{C.9.3}$$

The basis transformation coefficients correct for the differences between the real-O irrep-partner definitions of Griffith [(1964), Table A16] and those of Table C.5.1(b). For example, from his Table A16 and our Table C.5.1(b), respectively, we see that

$$|h_1\theta_1^B\rangle = |T_1y^B\rangle \sim iy$$

$$|h_1\theta_1^G\rangle = |T_1y^G\rangle \sim y \tag{C.9.4}$$

Thus

$$|T_1y^B\rangle = \langle T_1y^G|T_1y^B\rangle|T_1y^G\rangle$$

$$= i|T_1y^G\rangle \tag{C.9.5}$$

and so $\langle T_1y^G|T_1y^B\rangle = i$. Our task, however, is not complete, since our v.c.c. and t_2^m and e^n functions expressed in terms of Griffith's basis partners in (C.9.3) above do not necessarily have the same phases as those actually used by Griffith. We can write

$$|t_2^m\,\mathcal{S}_1h_1\mathfrak{M}_1\theta_1^G\rangle_B = p\big(t_2^m, \mathcal{S}_1h_1\big)|t_2^m\,\mathcal{S}_1h_1\mathfrak{M}_1\theta_1^G\rangle_G$$

$$|e^n\,\mathcal{S}_2h_2\mathfrak{M}_2\theta_2^G\rangle_B = p\big(e^n, \mathcal{S}_2h_2\big)|e^n\,\mathcal{S}_2h_2\mathfrak{M}_2\theta_2^G\rangle_G \tag{C.9.6}$$

$$\big(h_1\theta_1^G, h_2\theta_2^G|h\theta^G\big)_B = p(h_1h_2h)\big(h_1\theta_1^G, h_2\theta_2^G|h\theta^G\big)_G$$

where the t_2^m and e^n functions and the v.c.c. on the right are those of Table A24 and Table A20 respectively of Griffith (1964). Substituting these relations into (C.9.3) gives

$$\left|\left(t_2^m(\mathbb{S}_1h_1\mathfrak{M}_1),\, e^n(\mathbb{S}_2h_2\mathfrak{M}_2)\right)h\theta^B\right\rangle_B$$

$$= \langle h\theta^G|h\theta^B\rangle \sum_{\theta_1^G\theta_2^G} p(h_1h_2h)\left(h_1\theta_1^G,\, h_2\theta_2^G|h\theta^G\right)_G$$

$$\times p(t_2^m, \mathbb{S}_1h_1)|t_2^m\,\mathbb{S}_1h_1\mathfrak{M}_1\theta_1^G\rangle_G\, p(e^n, \mathbb{S}_2h_2)|e^n\,\mathbb{S}_2h_2\mathfrak{M}_2\theta_2^G\rangle_G$$

$$= \langle h\theta^G|h\theta^B\rangle p(h_1h_2h)p(t_2^m, \mathbb{S}_1h_1)p(e^n, \mathbb{S}_2h_2)$$

$$\times\left|\left(t_2^m(\mathbb{S}_1h_1\mathfrak{M}_1),\, e^n(\mathbb{S}_2h_2\mathfrak{M}_2)\right)h\theta^G\right\rangle_G \qquad \text{(C.9.7)}$$

where the final ket is that used by Griffith (1964) in the calculation of his electrostatic matrices in Tables A27–A30.

We now illustrate how the four phases in (C.9.7) are obtained. Suppose, for example, we wish to find the phase of the group-O function $|(t_2^2(0T_2\mathfrak{M}_1), e^2(1A_2\mathfrak{M}_2))T_1x^B\rangle_B$ relative to that used in Griffith [(1964), Table A29]. To find $p(t_2^2, 0T_2)$ we first note that Section C.8 gives

$$|t_2^2\,0T_20\xi^B\rangle_B = \frac{1}{\sqrt{2}}(|\eta_B^+\zeta_B^-\rangle - |\eta_B^-\zeta_B^+\rangle) \qquad \text{(C.9.8)}$$

Then, since the basis-partner definitions of Table C.5.1(b) and of Griffith [(1964), Table A16] give $|T_2\xi^B\rangle = -i|T_2\xi^G\rangle$, $|T_2\eta^B\rangle = |T_2\eta^G\rangle$, and $|T_2\zeta^B\rangle = i|T_2\zeta^G\rangle$, translating *both sides* of (C.9.8) into the Griffith basis leads to

$$-i|t_2^2\,0T_20\xi^G\rangle_B = \frac{i}{\sqrt{2}}(|\eta_G^+\zeta_G^-\rangle - |\eta_G^-\zeta_G^+\rangle) \qquad \text{(C.9.9)}$$

and so

$$|t_2^2\,0T_20\xi^G\rangle_B = \frac{-1}{\sqrt{2}}(|\eta_G^+\zeta_G^-\rangle - |\eta_G^-\zeta_G^+\rangle) \qquad \text{(C.9.10)}$$

Then Griffith [(1964), Table A24] gives

$$|t_2^2\,0T_20\xi^G\rangle_G = \frac{-1}{\sqrt{2}}(|\eta_G^+\zeta_G^-\rangle - |\eta_G^-\zeta_G^+\rangle) \qquad \text{(C.9.11)}$$

Comparison of the last two equations above defines $p(t_2^2, 0T_2) = 1$. Similarly, the same procedure gives the basis-partner relations $|A_2a_2^B\rangle = |A_2a_2^G\rangle$,

$|E\theta^B\rangle = -|E\theta^G\rangle$, and $|E\epsilon^B\rangle = |E\epsilon^G\rangle$, so that

$$|e^2 1A_2 1a_2^B\rangle_B = -|\theta_B^+\epsilon_B^+\rangle$$

$$= |e^2 1A_2 1a_2^G\rangle_B = |\theta_G^+\epsilon_G^+\rangle \qquad \text{(C.9.12)}$$

while Griffith [(1964), Table A24] gives

$$|e^2 1A_2 1a_2^G\rangle_G = |\theta_G^+\epsilon_G^+\rangle \qquad \text{(C.9.13)}$$

Thus $p(e^2, 1A_2) = 1$. We next find $p(T_2 A_2 T_1)$. From (10.2.1) and Sections C.11 and C.12 we have

$$\left(T_2\xi^B, A_2 a_2^B | T_1 x^B\right)_B = \sqrt{3}\,(-1)\left(\frac{1}{\sqrt{3}}\right) = -1 \qquad \text{(C.9.14)}$$

Table C.5.1(b) and Griffith [(1964), Table A16] give $|T_1 x^B\rangle = -|T_1 x^G\rangle$, and we already have the other basis relations required. Thus

$$\left(T_2\xi^B, A_2 a_2^B | T_1 x^B\right)_B = \langle T_2\xi^G | T_2\xi^B\rangle^* \langle A_2 a_2^G | A_2 a_2^B\rangle^* \langle T_1 x^G | T_1 x^B\rangle$$

$$\times \left(T_2\xi^G, A_2 a_2^G | T_1 x^G\right)_B$$

$$= (-i)^*(1)^*(-1)\left(T_2\xi^G, A_2 a_2^G | T_1 x^G\right)_B$$

$$= -i\, p(T_2 A_2 T_1)\left(T_2\xi^G, A_2 a_2^G | T_1 x^G\right)_G \qquad \text{(C.9.15)}$$

Then, since Griffith [(1964), Table A20] gives

$$\left(T_2\xi^G, A_2 a_2^G | T_1 x^G\right)_G = 1 \qquad \text{(C.9.16)}$$

it follows from (C.9.14)–(C.9.15) that $p(T_2 A_2 T_1) = -i$. Finally, substituting the above results into (C.9.7) and noting that $\langle T_1 x^G | T_1 x^B\rangle = -1$, we obtain

$$\left|\left(t_2^2(0T_2\mathfrak{M}_1), e^2(1A_2\mathfrak{M}_2)\right)T_1 x^B\right\rangle_B$$

$$= (-1)(-i)(1)(1)\left|\left(t_2^2(0T_2\mathfrak{M}_1), e^2(1A_2\mathfrak{M}_2)\right)T_1 x^G\right\rangle_G$$

$$= i\left|\left(t_2^2(0T_2\mathfrak{M}_1), e^2(1A_2\mathfrak{M}_2)\right)T_1 x^G\right\rangle_G \qquad \text{(C.9.17)}$$

The phases of other $t_2^m e^n$ states are found in a similar fashion, and since the phases needed in (C.9.7) recur, the calculation is not as lengthy as it might first appear.

C.10. Symmetry-Adapted Molecular Orbitals for Octahedral and Tetrahedral Complexes

The symmetry-adapted orbitals for an octahedral ML_6^{n-}-type metal–ligand complex, given in Table C.10.1(a), are defined in the right-handed coordinate system of Fig. C.10.1. Molecular orbitals are formed by taking linear combinations of symmetry orbitals belonging to the same irrep and irrep component. Ligand orbitals are centered at the ligands. When a ligand contains more than one atom (as in CN^-), x_1 should be interpreted as a linear combination of ligand orbitals which transform in the manner of an atomic p_x orbital centered at position 1, and so on. The basis is that of Table C.5.1(b). Parity labels are omitted in irrep-partner designations.

Table C.10.1(a). Octahedral Complex

Real-O_h Basis	Metal Orbitals	Ligand Symmetry Orbitals: σ-type	Ligand Symmetry Orbitals: π-type
$\vert A_{1g}a_1\rangle$	$(n+1)s$	$(1/\sqrt{6})(s_1+s_2+s_3+s_4+s_5+s_6)$, $(1/\sqrt{6})(x_1-x_3+y_2-y_4+z_5-z_6)$	
$\vert E_g\theta\rangle$	$-nd_{z^2}$	$(-1/2\sqrt{3})(2s_5+2s_6-s_1-s_2-s_3-s_4)$, $(-1/2\sqrt{3})(2z_5-2z_6-x_1+x_3-y_2+y_4)$	
$\vert E_g\epsilon\rangle$	$nd_{x^2-y^2}$	$\frac{1}{2}(s_1+s_3-s_2-s_4)$, $\frac{1}{2}(x_1-x_3-y_2+y_4)$	
$\vert T_{1g}x\rangle$			$-\frac{1}{2}(z_2-z_4-y_5+y_6)$
$\vert T_{1g}y\rangle$			$(i/2)(x_5-x_6-z_1+z_3)$
$\vert T_{1g}z\rangle$			$\frac{1}{2}(y_1-y_3-x_2+x_4)$
$\vert T_{1u}x\rangle$	$-(n+1)p_x$	$(-1/\sqrt{2})(s_1-s_3), (-1/\sqrt{2})(x_1+x_3)$	$-\frac{1}{2}(x_2+x_4+x_5+x_6)$
$\vert T_{1u}y\rangle$	$i(n+1)p_y$	$(i/\sqrt{2})(s_2-s_4), (i/\sqrt{2})(y_2+y_4)$	$(i/2)(y_5+y_6+y_1+y_3)$
$\vert T_{1u}z\rangle$	$(n+1)p_z$	$(1/\sqrt{2})(s_5-s_6), (1/\sqrt{2})(z_5+z_6)$	$\frac{1}{2}(z_1+z_3+z_2+z_4)$
$\vert T_{2g}\xi\rangle$	$-ind_{yz}$		$(-i/2)(y_5-y_6+z_2-z_4)$
$\vert T_{2g}\eta\rangle$	nd_{zx}		$\frac{1}{2}(z_1-z_3+x_5-x_6)$
$\vert T_{2g}\zeta\rangle$	ind_{xy}		$(i/2)(x_2-x_4+y_1-y_3)$
$\vert T_{2u}\xi\rangle$			$(-i/2)(x_2+x_4-x_5-x_6)$
$\vert T_{2u}\eta\rangle$			$\frac{1}{2}(y_5+y_6-y_1-y_3)$
$\vert T_{2u}\zeta\rangle$			$(i/2)(z_1+z_3-z_2-z_4)$

The coordinate system used for a tetrahedral ML_4^{n-} moiety [Ballhausen and Gray (1964)] is depicted in Fig. C.10.2. All coordinate systems are

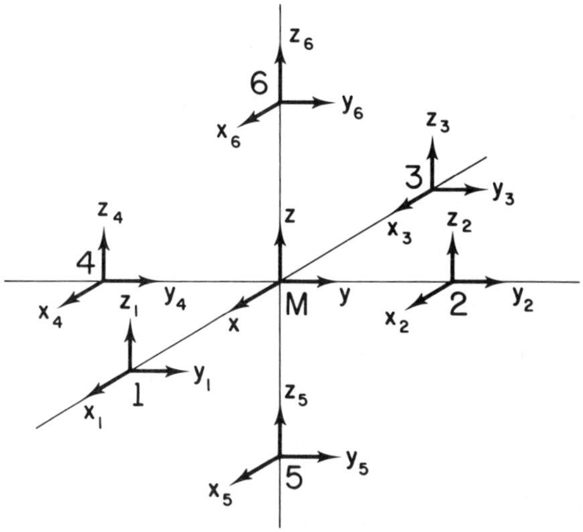

Figure C.10.1. Coordinate system for an octahedral complex.

right-handed, and Y_1–Y_4 are omitted for clarity. The Z_k ($k = 1, \ldots, 4$) point along ML_k; X_1 and X_4 are in the ML_1L_4 plane; X_2 and X_3 are in the ML_2L_3 plane. Y_1–Y_4 are perpendicular to Z.

Symmetry-adapted orbitals are given in Table C.10.1(b) using the real-T_d basis of Table C.5.1(b). The x_i, y_i, and so on should be interpreted as in Table C.10.1(a) above.

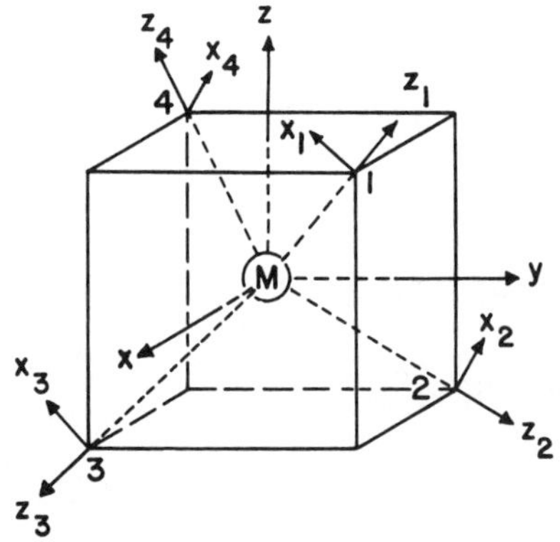

Figure C.10.2. Coordinate system for a tetrahedral compound.

Table C.10.1(b). Tetrahedral Complex

Real-T_d Basis	Metal Orbitals	Ligand Symmetry Orbitals
$\|A_1a_1\rangle$	$(n+1)s$	$\frac{1}{2}(s_1+s_2+s_3+s_4), \frac{1}{2}(z_1+z_2+z_3+z_4)$
$\|E\theta\rangle$	$-nd_{z^2}$	$-\frac{1}{2}(x_1-x_2-x_3+x_4)$
$\|E\epsilon\rangle$	$nd_{x^2-y^2}$	$\frac{1}{2}(y_1-y_2-y_3+y_4)$
$\|T_1x\rangle$		$-\frac{1}{4}\left[\sqrt{3}\,(-x_1-x_2+x_3+x_4)-y_1-y_2+y_3+y_4\right]$
$\|T_1y\rangle$		$(i/4)\left[\sqrt{3}\,(x_1-x_2+x_3-x_4)-y_1+y_2-y_3+y_4\right]$
$\|T_1z\rangle$		$\frac{1}{2}(y_1+y_2+y_3+y_4)$
$\|T_2\xi\rangle$	$-(n+1)p_x, -ind_{yz}$	$-\frac{1}{2}(z_1-z_2+z_3-z_4), -\frac{1}{2}(s_1-s_2+s_3-s_4),$ $-\frac{1}{4}\left[x_1+x_2-x_3-x_4+\sqrt{3}\,(-y_1-y_2+y_3+y_4)\right]$
$\|T_2\eta\rangle$	$-i(n+1)p_y, nd_{zx}$	$-(i/2)(z_1+z_2-z_3-z_4), (-i/2)(s_1+s_2-s_3-s_4),$ $(-i/4)\left[x_1-x_2+x_3-x_4+\sqrt{3}\,(y_1-y_2+y_3-y_4)\right]$
$\|T_2\zeta\rangle$	$(n+1)p_z, ind_{xy}$	$\frac{1}{2}(z_1-z_2-z_3+z_4), \frac{1}{2}(s_1-s_2-s_3+s_4),$ $\frac{1}{2}(-x_1-x_2-x_3-x_4)$

C.11. 3*j*, 2*j*, and 2*jm* Phases and Related Definitions

The bases used are defined in Section C.5.

The 3j Phase $\{abcr\}$

$3j$ phases are independent of the order of the irreps. With the exception of $\{U'U'T_21\}$, all $3j$ phases for the groups O and T_d are defined by

$$\{abcr\} = (-1)^a(-1)^b(-1)^c$$

where

$$(-1)^{A_1} = (-1)^E = (-1)^{T_2} = 1$$

$$(-1)^{A_2} = (-1)^{T_1} = -1$$

$$(-1)^{E'} = (-1)^{E''} = i$$

$$(-1)^{U'} = -i$$

The law-breaking phase has the value

$$\{U'U'T_2 1\} = 1$$

The 2j Phase $\{a\}$

We have

$$\{a\} = \begin{cases} 1 & \text{for } A_1, A_2, E, T_1, T_2 \\ -1 & \text{for } E', E'', U' \end{cases}$$

The 2jm $\begin{pmatrix} a \\ \alpha \end{pmatrix}$ *and Definitions of* a^* *and* α^*

These are given in Tables C.11.1 and C.11.2.

Table C.11.1. 2*jm* in the Real-*O* and Real-T_d Bases for Single-Valued Irreps[a]

$O \supset D_4 \supset D_2 \supset C_2$ $T_d \supset D_{2d} \supset D_2 \supset C_2$		
$\lvert a\alpha\rangle \equiv \lvert aa_1a_2a_3\rangle$		2*jm* $\begin{pmatrix} a \\ \alpha \end{pmatrix} \equiv \begin{pmatrix} a \\ a_1 \\ a_2 \\ a_3 \end{pmatrix}$
A_1a_1	0000	1
A_2a_2	$\tilde{0}\tilde{2}00$	1
$E\theta$	2000	1
$E\epsilon$	2200	1
T_1x	$11\tilde{1}1$	−1
T_1y	1111	1
T_1z	$1\tilde{0}\tilde{0}0$	−1
$T_2\xi$	$\tilde{1}1\tilde{1}1$	−1
$T_2\eta$	$\tilde{1}111$	1
$T_2\zeta$	$\tilde{1}\tilde{2}\tilde{0}0$	−1

[a] In these bases $a = a^*$, $\alpha = \alpha^*$, and $a_i = a_i^*$ ($i = 1$ to 3) for all a, α, and a_i.

Table C.11.2. 2*jm* in the Complex-O and Complex-T_d Bases

$$O \supset D_4 \supset C_4$$
$$T_d \supset D_{2d} \supset S_4$$

$\lvert a\alpha\rangle \equiv \lvert aa_1a_2\rangle$		2*jm* $\begin{pmatrix} a \\ \alpha \end{pmatrix} = \begin{pmatrix} a \\ a_1 \\ a_2 \end{pmatrix}$	$\lvert a^*\alpha^*\rangle \equiv \lvert a^*a_1^*a_2^*\rangle$	
A_1a_1	000	1	A_1a_1	000
A_2a_2'	$\tilde{0}22$	1	A_2a_2'	$\tilde{0}22$
$E\theta$	200	1	$E\theta$	200
$E\epsilon'$	222	1	$E\epsilon'$	222
$T_1 -1$	$11-1$	1	$T_1 1$	111
$T_1 0$	$1\tilde{0}0$	-1	$T_1 0$	$1\tilde{0}0$
$T_1 1$	111	1	$T_1 -1$	$11-1$
$T_2 -1$	$\tilde{1}1-1$	1	$T_2 1$	$\tilde{1}11$
$T_2 0$	$\tilde{1}\tilde{2}2$	-1	$T_2 0$	$\tilde{1}\tilde{2}2$
$T_2 1$	$\tilde{1}11$	1	$T_2 -1$	$\tilde{1}1-1$
$E'\beta'$	$\frac{1}{2}\frac{1}{2}-\frac{1}{2}$	-1	$E'\alpha'$	$\frac{1}{2}\frac{1}{2}\frac{1}{2}$
$E'\alpha'$	$\frac{1}{2}\frac{1}{2}\frac{1}{2}$	1	$E'\beta'$	$\frac{1}{2}\frac{1}{2}-\frac{1}{2}$
$E''\beta''$	$\tilde{\frac{1}{2}}\frac{3}{2}-\frac{3}{2}$	-1	$E''\alpha''$	$\tilde{\frac{1}{2}}\frac{3}{2}\frac{3}{2}$
$E''\alpha''$	$\tilde{\frac{1}{2}}\frac{3}{2}\frac{3}{2}$	1	$E''\beta''$	$\tilde{\frac{1}{2}}\frac{3}{2}-\frac{3}{2}$
$U'\nu$	$\frac{3}{2}\frac{3}{2}-\frac{3}{2}$	-1	$U'\kappa$	$\frac{3}{2}\frac{3}{2}\frac{3}{2}$
$U'\mu$	$\frac{3}{2}\frac{1}{2}-\frac{1}{2}$	-1	$U'\lambda$	$\frac{3}{2}\frac{1}{2}\frac{1}{2}$
$U'\lambda$	$\frac{3}{2}\frac{1}{2}\frac{1}{2}$	1	$U'\mu$	$\frac{3}{2}\frac{1}{2}-\frac{1}{2}$
$U'\kappa$	$\frac{3}{2}\frac{3}{2}\frac{3}{2}$	1	$U'\nu$	$\frac{3}{2}\frac{3}{2}-\frac{3}{2}$

C.12. 3*jm* for the Groups O and T_d in the Bases of Section C.5

The bases are defined in Section C.5. The headings in Tables C.12.1 and C.12.2 are group-O and group-T_d irrep labels, while the entries beneath them are irrep component labels for the specified basis in Section C.5. Thus, for example,

$$\begin{pmatrix} T_2 & T_1 & E \\ \eta & y & \theta \end{pmatrix} = -\frac{1}{2}$$

The 3*jm* in tables headed by 3*jm*† are multiplied by -1 for odd permutations of their columns. Thus the dagger indicates that $\{abcr\} = -1$ for the irreps of the 3*jm*.

$3jm$ which are neither found in the tables, nor related by (10.2.4) to those tabulated, are zero. All $3jm$ are consistent with the relations between the real and complex bases given in Table C.5.1(c). Thus the reduced matrix elements are identical in the two bases. The $3jm$ were calculated by taking products of Butler (1981) $3jm$ factors as described in Section 15.3.

Table C.12.1. $3jm$ in the Real-O and Real-T_d Bases for Single-Valued Irreps[a]

A_1	A_1	A_1	$3jm$
a_1	a_1	a_1	1

A_2	A_2	A_1	$3jm$
a_2	a_2	a_1	1

E	E	A_1	$3jm$
θ	θ	a_1	$1/\sqrt{2}$
ϵ	ϵ	a_1	$1/\sqrt{2}$

E	E	A_2	$3jm^\dagger$
θ	ϵ	a_2	$-1/\sqrt{2}$

E	E	E	$3jm$
θ	θ	θ	$-\frac{1}{2}$
θ	ϵ	ϵ	$\frac{1}{2}$

T_1	T_1	A_1	$3jm$
x	x	a_1	$-1/\sqrt{3}$
y	y	a_1	$1/\sqrt{3}$
z	z	a_1	$-1/\sqrt{3}$

T_1	T_1	E	$3jm$
x	x	θ	$1/2\sqrt{3}$
x	x	ϵ	$\frac{1}{2}$
y	y	θ	$-1/2\sqrt{3}$
y	y	ϵ	$\frac{1}{2}$
z	z	θ	$-1/\sqrt{3}$

T_1	T_1	T_1	$3jm^\dagger$
x	y	z	$1/\sqrt{6}$

T_2	T_1	A_2	$3jm$
ξ	x	a_2	$1/\sqrt{3}$
η	y	a_2	$1/\sqrt{3}$
ζ	z	a_2	$1/\sqrt{3}$

T_2	T_1	E	$3jm^\dagger$
ξ	x	θ	$\frac{1}{2}$
ξ	x	ϵ	$-1/2\sqrt{3}$
η	y	θ	$-\frac{1}{2}$
η	y	ϵ	$-1/2\sqrt{3}$
ζ	z	ϵ	$1/\sqrt{3}$

T_2	T_1	T_1	$3jm$
ξ	y	z	$1/\sqrt{6}$
η	z	x	$-1/\sqrt{6}$
ζ	x	y	$1/\sqrt{6}$

T_2	T_2	A_1	$3jm$
ξ	ξ	a_1	$-1/\sqrt{3}$
η	η	a_1	$1/\sqrt{3}$
ζ	ζ	a_1	$-1/\sqrt{3}$

T_2	T_2	E	$3jm$
ξ	ξ	θ	$1/2\sqrt{3}$
ξ	ξ	ϵ	$\frac{1}{2}$
η	η	θ	$-1/2\sqrt{3}$
η	η	ϵ	$\frac{1}{2}$
ζ	ζ	θ	$-1/\sqrt{3}$

T_2	T_2	T_1	$3jm^\dagger$
ξ	η	z	$1/\sqrt{6}$
η	ζ	x	$1/\sqrt{6}$
ζ	ξ	y	$-1/\sqrt{6}$

T_2	T_2	T_2	$3jm$
ξ	η	ζ	$1/\sqrt{6}$

[a] $r = 0$ for all $3jm$.

Table C.12.2. **$3jm$ in the Complex-O and Complex-T_d Bases**[a]

A_1	A_1	A_1	$3jm$
a_1	a_1	a_1	1

A_2	A_2	A_1	$3jm$
a_2'	a_2'	a_1	1

E	E	A_1	$3jm$
θ	θ	a_1	$1/\sqrt{2}$
ϵ'	ϵ'	a_1	$1/\sqrt{2}$

E	E	A_2	$3jm^\dagger$
θ	ϵ'	a_2'	$-1/\sqrt{2}$

E	E	E	$3jm$
θ	θ	θ	$-\frac{1}{2}$
θ	ϵ'	ϵ'	$\frac{1}{2}$

T_1	T_1	A_1	$3jm$
-1	1	a_1	$1/\sqrt{3}$
0	0	a_1	$-1/\sqrt{3}$

T_1	T_1	E	$3jm$
-1	-1	ϵ'	$-\frac{1}{2}$
-1	1	θ	$-1/2\sqrt{3}$
0	0	θ	$-1/\sqrt{3}$
1	1	ϵ'	$-\frac{1}{2}$

T_1	T_1	T_1	$3jm^\dagger$
-1	0	1	$-1/\sqrt{6}$

T_2	T_1	A_2	$3jm$
-1	-1	a_2'	$-1/\sqrt{3}$
0	0	a_2'	$-1/\sqrt{3}$
1	1	a_2'	$-1/\sqrt{3}$

T_2	T_1	E	$3jm^\dagger$
-1	-1	ϵ'	$1/2\sqrt{3}$
-1	1	θ	$-\frac{1}{2}$
0	0	ϵ'	$-1/\sqrt{3}$
1	-1	θ	$-\frac{1}{2}$
1	1	ϵ'	$1/2\sqrt{3}$

T_2	T_1	T_1	$3jm$
-1	1	0	$1/\sqrt{6}$
0	-1	-1	$1/\sqrt{6}$
0	1	1	$-1/\sqrt{6}$
1	0	-1	$-1/\sqrt{6}$

T_2	T_2	A_1	$3jm$
-1	1	a_1	$1/\sqrt{3}$
0	0	a_1	$-1/\sqrt{3}$

T_2	T_2	E	$3jm$
-1	-1	ϵ'	$-\frac{1}{2}$
-1	1	θ	$-1/2\sqrt{3}$
0	0	θ	$-1/\sqrt{3}$
1	1	ϵ'	$-\frac{1}{2}$

T_2	T_2	T_1	$3jm^\dagger$
-1	0	-1	$1/\sqrt{6}$
-1	1	0	$1/\sqrt{6}$
0	1	1	$1/\sqrt{6}$

T_2	T_2	T_2	$3jm$
-1	-1	0	$1/\sqrt{6}$
0	1	1	$-1/\sqrt{6}$

E'	E'	A_1	$3jm^\dagger$
β'	α'	a_1	$-1/\sqrt{2}$

E'	E'	T_1	$3jm$
β'	β'	1	$1/\sqrt{3}$
β'	α'	0	$1/\sqrt{6}$
α'	α'	-1	$1/\sqrt{3}$

E''	E'	A_2	$3jm$
β''	β'	a_2'	$1/\sqrt{2}$
α''	α'	a_2'	$1/\sqrt{2}$

E''	E'	T_2	$3jm^\dagger$
β''	β'	0	$-1/\sqrt{6}$
β''	α'	1	$1/\sqrt{3}$
α''	β'	-1	$-1/\sqrt{3}$
α''	α'	0	$1/\sqrt{6}$

E''	E''	A_1	$3jm^\dagger$
β''	α''	a_1	$-1/\sqrt{2}$

E''	E''	T_1	$3jm$
β''	β''	-1	$-1/\sqrt{3}$
β''	α''	0	$1/\sqrt{6}$
α''	α''	1	$-1/\sqrt{3}$

Table C.12.2. Continued.

U'	E'	E	$3jm$
ν	β'	ϵ'	$\frac{1}{2}$
μ	α'	θ	$-\frac{1}{2}$
λ	β'	θ	$\frac{1}{2}$
κ	α'	ϵ'	$\frac{1}{2}$

U'	E'	T_1	$3jm^\dagger$
ν	α'	1	$\frac{1}{2}$
μ	β'	1	$1/2\sqrt{3}$
μ	α'	0	$-1/\sqrt{6}$
λ	β'	0	$-1/\sqrt{6}$
λ	α'	-1	$1/2\sqrt{3}$
κ	β'	-1	$-\frac{1}{2}$

U'	E'	T_2	$3jm$
ν	β'	0	$-1/\sqrt{6}$
ν	α'	1	$-1/2\sqrt{3}$
μ	β'	1	$\frac{1}{2}$
λ	α'	-1	$\frac{1}{2}$
κ	β'	-1	$1/2\sqrt{3}$
κ	α'	0	$1/\sqrt{6}$

U'	E''	E	$3jm$
ν	α''	θ	$-\frac{1}{2}$
μ	β''	ϵ'	$\frac{1}{2}$
λ	α''	ϵ'	$\frac{1}{2}$
κ	β''	θ	$\frac{1}{2}$

U'	E''	T_1	$3jm^\dagger$
ν	β''	-1	$-1/2\sqrt{3}$
ν	α''	0	$-1/\sqrt{6}$
μ	α''	-1	$-\frac{1}{2}$
λ	β''	1	$\frac{1}{2}$
κ	β''	0	$-1/\sqrt{6}$
κ	α''	1	$-1/2\sqrt{3}$

U'	E''	T_2	$3jm$
ν	β''	-1	$-\frac{1}{2}$
μ	β''	0	$-1/\sqrt{6}$
μ	α''	-1	$1/2\sqrt{3}$
λ	β''	1	$-1/2\sqrt{3}$
λ	α''	0	$1/\sqrt{6}$
κ	α''	1	$-\frac{1}{2}$

U'	U'	A_1	$3jm^\dagger$
ν	κ	a_1	$-\frac{1}{2}$
μ	λ	a_1	$-\frac{1}{2}$

U'	U'	A_2	$3jm$
ν	μ	a_2'	$-\frac{1}{2}$
λ	κ	a_2'	$-\frac{1}{2}$

U'	U'	E	$3jm^\dagger$
ν	μ	ϵ'	$1/2\sqrt{2}$
ν	κ	θ	$-1/2\sqrt{2}$
μ	λ	θ	$1/2\sqrt{2}$
λ	κ	ϵ'	$-1/2\sqrt{2}$

U'	U'	T_1	$3jm$ $r = 0$	$3jm$ $r = 1$
ν	ν	-1	$-1/\sqrt{6}$	$1/2\sqrt{6}$
ν	λ	1	0	$1/2\sqrt{2}$
ν	κ	0	$1/2\sqrt{3}$	$1/2\sqrt{3}$
μ	μ	1	$-1/\sqrt{6}$	$-1/2\sqrt{6}$
μ	λ	0	$-1/2\sqrt{3}$	$1/2\sqrt{3}$
μ	κ	-1	0	$-1/2\sqrt{2}$
λ	λ	-1	$-1/\sqrt{6}$	$-1/2\sqrt{6}$
κ	κ	1	$-1/\sqrt{6}$	$1/2\sqrt{6}$

U'	U'	T_2	$3jm^\dagger$ $r = 0$	$3jm$ $r = 1$
ν	ν	-1	0	$1/2\sqrt{2}$
ν	μ	0	$-1/2\sqrt{3}$	$-1/2\sqrt{3}$
ν	λ	1	$1/\sqrt{6}$	$-1/2\sqrt{6}$
μ	μ	1	0	$-1/2\sqrt{2}$
μ	κ	-1	$1/\sqrt{6}$	$1/2\sqrt{6}$
λ	λ	-1	0	$-1/2\sqrt{2}$
λ	κ	0	$-1/2\sqrt{3}$	$1/2\sqrt{3}$
κ	κ	1	0	$1/2\sqrt{2}$

[a] Unless otherwise indicated, $r = 0$.

C.13. $6j$ for the Groups O and T_d

The $6j$ given in Table C.13.1 apply for all our group O and T_d bases (Sections C.5 and E.4). All $6j$ containing one or more A_1 irreps are given by (16.1.11) and are not tabulated:

$$\begin{Bmatrix} A_1 & b & c \\ d & e & f \end{Bmatrix}_{0rs0} = \delta_{bc}\delta_{ef}\delta_{rs}\,\delta(dbfr)|b|^{-1/2}|e|^{-1/2}\{dbfr\}$$

Other $6j$ which are neither found in the table, nor related to those tabulated by the symmetry rules of Section 16.1, are zero by (16.1.2).

The $6j$ are ordered by irreps according to the rules:

1. $a \leqslant b \leqslant c$.
2. $a \leqslant d$.
3. $b \leqslant e$, $b \leqslant f$.
4. If there is a choice, $c \leqslant f$ and $e \leqslant f$.

When no repeated irreps can occur for the particular irrep combinations in the $6j$, the r indices are all zero and are not listed. In other cases the r indices are ordered so:

1. r_1 and r_2 loop last (after f).
2. r_3 loops before f (thus the list gives $r_3 = 0$ for all f and then $r_3 = 1$ for all f).
3. r_4 loops together with a, b, and c (before d, e, and f).

Those $6j$ which change sign under odd permutation of their columns are followed by an asterisk (*). Thus the asterisk denotes that $(-1)^n = -1$ in (16.1.6).

The arrangement of each list is

a	b	c	
d	e	f	
d'	e'	f'	$r_1r_2r_3r_4$
. . . .			

Thus for example,

$$\begin{Bmatrix} E & T_1 & T_2 \\ U' & E' & U' \end{Bmatrix}_{0000} = \frac{1}{2\sqrt{6}}$$

Table C.13.1. 6*j* for the Groups *O* and T_d

A_2	E	E		6*j*
A_2	E	E		$\frac{1}{2}$
E	E	E		$\frac{1}{2}$
T_1	T_1	T_2		$1/\sqrt{6}$
T_2	T_1	T_2		$-1/\sqrt{6}$
E'	U'	U'		$1/2\sqrt{2}$
E''	U'	U'		$1/2\sqrt{2}$
U'	E'	E''		$\frac{1}{2}$
U'	U'	U'		$1/2\sqrt{2}$

A_2	T_1	T_2		6*j*
A_2	T_1	T_2		$\frac{1}{3}$
E	T_1	T_2		$\frac{1}{3}$
E	T_2	T_1		$\frac{1}{3}$
T_1	T_1	T_2		$\frac{1}{3}$
T_1	T_2	T_1		$\frac{1}{3}$
T_2	T_1	T_2		$-\frac{1}{3}$
T_2	T_2	T_1		$-\frac{1}{3}$
E'	E''	E'		$1/\sqrt{6}$
E'	U'	U'		$-1/2\sqrt{3}$
E''	E'	E''		$-1/\sqrt{6}$
E''	U'	U'		$-1/2\sqrt{3}$
U'	E'	E''		$-1/\sqrt{6}$
U'	E''	E'		$-1/\sqrt{6}$
U'	U'	U'	0000	$1/2\sqrt{3}$
U'	U'	U'	0100	0
U'	U'	U'	0010	0*
U'	U'	U'	0110	$1/2\sqrt{3}$ *

A_2	E'	E''	6*j*
A_2	E'	E''	$-\frac{1}{2}$
E	U'	U'	$1/2\sqrt{2}$
T_1	E''	E'	$\frac{1}{2}$
T_1	U'	U'	$1/2\sqrt{2}$
T_2	E'	E''	$\frac{1}{2}$
T_2	U'	U'	$-1/2\sqrt{2}$

A_2	U'	U'		6*j*
A_2	U'	U'		$-\frac{1}{4}$
E	U'	U'		$-\frac{1}{4}$
T_1	U'	U'	0000	$-\frac{1}{4}$
T_1	U'	U'	0100	0
T_1	U'	U'	0010	0
T_1	U'	U'	0110	$\frac{1}{4}$
T_2	U'	U'	0000	$\frac{1}{4}$
T_2	U'	U'	0100	0*
T_2	U'	U'	0010	0*
T_2	U'	U'	0110	$\frac{1}{4}$

E	E	E	6*j*
E	E	E	0
T_1	T_1	T_1	$-1/2\sqrt{3}$
T_1	T_1	T_2	$-1/2\sqrt{3}$
T_1	T_2	T_2	$-1/2\sqrt{3}$
T_2	T_2	T_2	$-1/2\sqrt{3}$
E'	U'	U'	$1/2\sqrt{2}$
E''	U'	U'	$-1/2\sqrt{2}$
U'	U'	U'	0

Table C.13.1. Continued.

E	T_1	T_1		$6j$
E	T_1	T_1		$\frac{1}{3}$
E	T_1	T_2		0
E	T_2	T_2		$-\frac{1}{3}$
T_1	T_1	T_1		$\frac{1}{6}$
T_1	T_1	T_2		$-1/2\sqrt{3}$
T_1	T_2	T_2		$-\frac{1}{6}$
T_2	T_1	T_1		$-\frac{1}{6}$
T_2	T_1	T_2		$-1/2\sqrt{3}$
T_2	T_2	T_2		$\frac{1}{6}$
E'	E'	U'		$1/2\sqrt{3}$
E'	U'	U'		$1/2\sqrt{6}$
E''	E''	U'		$1/2\sqrt{3}$
E''	U'	U'		$-1/2\sqrt{6}$
U'	E'	U'	0000	$-1/2\sqrt{6}$
U'	E'	U'	0100	$1/2\sqrt{6}$
U'	E''	U'	0000	$1/2\sqrt{6}$
U'	E''	U'	0100	$1/2\sqrt{6}$
U'	U'	U'	0000	0
U'	U'	U'	0100	$1/2\sqrt{6}$
U'	U'	U'	0010	$1/2\sqrt{6}$
U'	U'	U'	0110	0

E	T_1	T_2		$6j$
E	T_1	T_2		$\frac{1}{3}$
E	T_2	T_2		0
T_1	T_1	T_2		$-\frac{1}{6}$
T_1	T_2	T_1		$\frac{1}{6}$
T_1	T_2	T_2		$-1/2\sqrt{3}$
T_2	T_1	T_2		$\frac{1}{6}$
T_2	T_2	T_1		$-\frac{1}{6}$
T_2	T_2	T_2		$-1/2\sqrt{3}$
E'	E''	U'		$1/2\sqrt{3}$
E'	U'	E'		$1/2\sqrt{3}$
E'	U'	U'		$-1/2\sqrt{6}$
E''	E'	U'		$-1/2\sqrt{3}$
E''	U'	E''		$1/2\sqrt{3}$
E''	U'	U'		$1/2\sqrt{6}$
U'	E'	U'	0000	$1/2\sqrt{6}$

E	T_1	T_2 (*cont.*)		$6j$
U'	E'	U'	0100	$1/2\sqrt{6}$
U'	E''	U'	0000	$-1/2\sqrt{6}$
U'	E''	U'	0100	$1/2\sqrt{6}$
U'	U'	E'	0000	$-1/2\sqrt{6}$
U'	U'	E''	0000	$1/2\sqrt{6}$
U'	U'	U'	0000	0
U'	U'	U'	0100	$-1/2\sqrt{6}$
U'	U'	E'	0010	$1/2\sqrt{6}$ *
U'	U'	E''	0010	$1/2\sqrt{6}$ *
U'	U'	U'	0010	$-1/2\sqrt{6}$ *
U'	U'	U'	0110	0*

E	T_2	T_2		$6j$
E	T_2	T_2		$\frac{1}{3}$
T_1	T_2	T_2		$\frac{1}{6}$
T_2	T_2	T_2		$-\frac{1}{6}$
E'	E''	U'		$1/2\sqrt{3}$
E'	U'	U'		$1/2\sqrt{6}$
E''	E'	U'		$-1/2\sqrt{3}$
E''	U'	U'		$-1/2\sqrt{6}$
U'	E'	U'	0000	$1/2\sqrt{6}$
U'	E'	U'	0100	$-1/2\sqrt{6}$ *
U'	E''	U'	0000	$-1/2\sqrt{6}$
U'	E''	U'	0100	$-1/2\sqrt{6}$ *
U'	U'	U'	0000	0
U'	U'	U'	0100	$-1/2\sqrt{6}$ *
U'	U'	U'	0010	$1/2\sqrt{6}$ *
U'	U'	U'	0110	0

Table C.13.1. Continued.

E	E'	U'		$6j$
E	E'	U'		$-\frac{1}{4}$
E	E''	U'		$\frac{1}{4}$
E	U'	U'		$-\frac{1}{4}$
T_1	E'	U'		$\frac{1}{4}$
T_1	E''	U'		$\frac{1}{4}$
T_1	U'	E'	0000	$1/2\sqrt{2}$
T_1	U'	U'	0000	0
T_1	U'	E'	0010	0
T_1	U'	U'	0010	$\frac{1}{4}$
T_2	E'	U'		$\frac{1}{4}$
T_2	E''	U'		$\frac{1}{4}$
T_2	U'	E''	0000	$1/2\sqrt{2}$
T_2	U'	U'	0000	0
T_2	U'	E''	0010	0*
T_2	U'	U'	0010	$-\frac{1}{4}$*

E	E''	U'		$6j$
E	E''	U'		$-\frac{1}{4}$
E	U'	U'		$-\frac{1}{4}$
T_1	E''	U'		$\frac{1}{4}$
T_1	U'	E''	0000	$-1/2\sqrt{2}$
T_1	U'	U'	0000	0
T_1	U'	E''	0010	0
T_1	U'	U'	0010	$-\frac{1}{4}$
T_2	E''	U'		$\frac{1}{4}$
T_2	U'	U'	0000	0
T_2	U'	U'	0010	$\frac{1}{4}$*

E	U'	U'		$6j$
E	U'	U'		0
T_1	U'	U'	0000	$\frac{1}{4}$
T_1	U'	U'	0100	0
T_1	U'	U'	0010	0
T_1	U'	U'	0110	0
T_2	U'	U'	0000	$\frac{1}{4}$
T_2	U'	U'	0100	0*
T_2	U'	U'	0010	0*
T_2	U'	U'	0110	0

T_1	T_1	T_1		$6j$
T_1	T_1	T_1		$\frac{1}{6}$
T_1	T_1	T_2		$\frac{1}{6}$
T_1	T_2	T_2		$-\frac{1}{6}$
T_2	T_2	T_2		$-\frac{1}{6}$
E'	E'	E'		$\frac{1}{3}$
E'	E'	U'		$\frac{1}{6}$
E'	U'	U'	0000	$-1/6\sqrt{2}$
E'	U'	U'	1000	$-1/3\sqrt{2}$
E''	E''	E''		$-\frac{1}{3}$
E''	E''	U'		$-\frac{1}{6}$
E''	U'	U'	0000	$-1/6\sqrt{2}$
E''	U'	U'	1000	$1/3\sqrt{2}$
U'	U'	U'	0000	$-1/3\sqrt{2}$
U'	U'	U'	1000	0
U'	U'	U'	0100	0
U'	U'	U'	1100	$1/6\sqrt{2}$
U'	U'	U'	0010	0
U'	U'	U'	1010	$1/6\sqrt{2}$
U'	U'	U'	0110	$1/6\sqrt{2}$
U'	U'	U'	1110	0

Table C.13.1. Continued.

T_1	T_1	T_2		$6j$
T_1	T_1	T_2		$\frac{1}{6}$
T_1	T_2	T_2		$\frac{1}{6}$
T_2	T_2	T_1		$-\frac{1}{6}$
T_2	T_2	T_2		$-\frac{1}{6}$
E'	E''	U'		$1/2\sqrt{3}$
E'	U'	E'		$1/2\sqrt{3}$
E'	U'	U'	0000	$1/2\sqrt{6}$
E'	U'	U'	1000	0
E''	U'	E''		$-1/2\sqrt{3}$
E''	U'	U'	0000	$1/2\sqrt{6}$
E''	U'	U'	1000	0
U'	U'	E'	0000	$1/2\sqrt{6}$
U'	U'	E''	0000	$1/2\sqrt{6}$
U'	U'	U'	0000	0
U'	U'	U'	1000	0
U'	U'	U'	0100	0
U'	U'	U'	1100	$1/2\sqrt{6}$
U'	U'	E'	0010	0*
U'	U'	E''	0010	0*
U'	U'	U'	0010	0*
U'	U'	U'	1010	$-1/2\sqrt{6}$ *
U'	U'	U'	0110	$1/2\sqrt{6}$ *
U'	U'	U'	1110	0*

T_1	T_2	T_2		$6j$
T_1	T_2	T_2		$\frac{1}{6}$
T_2	T_2	T_2		$\frac{1}{6}$
E'	E''	E''		$-\frac{1}{3}$
E'	E''	U'		$\frac{1}{6}$
E'	U'	U'	0000	$-1/6\sqrt{2}$
E'	U'	U'	1000	$1/3\sqrt{2}$
E''	E'	E'		$\frac{1}{3}$
E''	E'	U'		$\frac{1}{6}$
E''	U'	U'	0000	$-1/6\sqrt{2}$
E''	U'	U'	1000	$-1/3\sqrt{2}$
U'	E'	E'		$\frac{1}{6}$
U'	E'	U'	0000	$-1/6\sqrt{2}$
U'	E'	U'	0100	$-1/3\sqrt{2}$ *
U'	E''	E''		$-\frac{1}{6}$

T_1	T_2	T_2 (*cont.*)		$6j$
U'	E''	U'	0000	$-1/6\sqrt{2}$
U'	E''	U'	0100	$1/3\sqrt{2}$ *
U'	U'	U'	0000	$-1/3\sqrt{2}$
U'	U'	U'	1000	0
U'	U'	U'	0100	0*
U'	U'	U'	1100	$1/6\sqrt{2}$ *
U'	U'	U'	0010	0*
U'	U'	U'	1010	$-1/6\sqrt{2}$ *
U'	U'	U'	0110	$-1/6\sqrt{2}$
U'	U'	U'	1110	0

T_1	E'	E'		$6j$
T_1	E'	E'		$\frac{1}{6}$
T_1	E'	U'		$-\frac{1}{3}$
T_1	U'	U'	0000	$1/6\sqrt{2}$
T_1	U'	U'	1000	$1/3\sqrt{2}$
T_2	E''	E''		$\frac{1}{6}$
T_2	E''	U'		$\frac{1}{3}$
T_2	U'	U'	0000	$-1/6\sqrt{2}$
T_2	U'	U'	1000	$1/3\sqrt{2}$

T_1	E'	U'		$6j$
T_1	E'	U'		$-\frac{1}{12}$
T_1	E''	U'		$-\frac{1}{4}$
T_1	U'	U'	0000	$-\frac{1}{6}$
T_1	U'	U'	1000	$\frac{1}{6}$
T_1	U'	U'	0010	$\frac{1}{6}$
T_1	U'	U'	1010	$\frac{1}{12}$
T_2	E'	U'		$\frac{1}{4}$
T_2	E''	E''		$-\frac{1}{3}$
T_2	E''	U'		$\frac{1}{12}$
T_2	U'	E''	0000	$-1/6\sqrt{2}$
T_2	U'	U'	0000	$\frac{1}{6}$
T_2	U'	U'	1000	$\frac{1}{6}$
T_2	U'	E''	0010	$1/3\sqrt{2}$ *
T_2	U'	U'	0010	$\frac{1}{6}$*
T_2	U'	U'	1010	$-\frac{1}{12}$*

Table C.13.1. Continued.

T_1	E''	E''		$6j$
T_1	E''	E''		$\frac{1}{6}$
T_1	E''	U'		$-\frac{1}{3}$
T_1	U'	U'	0000	$-1/6\sqrt{2}$
T_1	U'	U'	1000	$1/3\sqrt{2}$
T_2	U'	U'	0000	$1/6\sqrt{2}$
T_2	U'	U'	1000	$1/3\sqrt{2}$

T_1	E''	U'		$6j$
T_1	E''	U'		$-\frac{1}{12}$
T_1	U'	U'	0000	$-\frac{1}{6}$
T_1	U'	U'	1000	$-\frac{1}{6}$
T_1	U'	U'	0010	$-\frac{1}{6}$
T_1	U'	U'	1010	$\frac{1}{12}$
T_2	E''	U'		$\frac{1}{4}$
T_2	U'	U'	0000	$\frac{1}{6}$
T_2	U'	U'	1000	$-\frac{1}{6}$
T_2	U'	U'	0010	$-\frac{1}{6}$*
T_2	U'	U'	1010	$-\frac{1}{12}$*

T_1	U'	U'	$(r_4 = 0)$	$6j$
T_1	U'	U'	0000	$\frac{1}{12}$
T_1	U'	U'	1000	0
T_1	U'	U'	0100	0
T_1	U'	U'	1100	$-\frac{1}{6}$
T_1	U'	U'	0010	0
T_1	U'	U'	1010	$-\frac{1}{6}$
T_1	U'	U'	0110	$\frac{1}{12}$
T_1	U'	U'	1110	0
T_2	U'	U'	0000	$-\frac{1}{12}$
T_2	U'	U'	1000	0
T_2	U'	U'	0100	0*
T_2	U'	U'	1100	$-\frac{1}{6}$*
T_2	U'	U'	0010	0*
T_2	U'	U'	1010	$\frac{1}{6}$*
T_2	U'	U'	0110	$\frac{1}{12}$
T_2	U'	U'	1110	0

T_1	U'	U'	$(r_4 = 1)$	$6j$
T_1	U'	U'	0001	0
T_1	U'	U'	1001	$\frac{1}{12}$
T_1	U'	U'	0101	$-\frac{1}{6}$
T_1	U'	U'	1101	0
T_1	U'	U'	0011	$-\frac{1}{6}$
T_1	U'	U'	1011	0
T_1	U'	U'	0111	0
T_1	U'	U'	1111	$-\frac{1}{6}$
T_2	U'	U'	0001	0
T_2	U'	U'	1001	$\frac{1}{12}$
T_2	U'	U'	0101	$\frac{1}{6}$*
T_2	U'	U'	1101	0*
T_2	U'	U'	0011	$-\frac{1}{6}$*
T_2	U'	U'	1011	0*
T_2	U'	U'	0111	0
T_2	U'	U'	1111	$\frac{1}{6}$

T_2	T_2	T_2		$6j$
T_2	T_2	T_2		$\frac{1}{6}$
E'	E''	U'		$1/2\sqrt{3}$
E'	U'	U'	0000	$1/2\sqrt{6}$
E'	U'	U'	1000	0*
E''	U'	U'	0000	$1/2\sqrt{6}$
E''	U'	U'	1000	0*
U'	U'	U'	0000	0
U'	U'	U'	1000	0*
U'	U'	U'	0100	0*
U'	U'	U'	1100	$-1/2\sqrt{6}$
U'	U'	U'	0010	0*
U'	U'	U'	1010	$-1/2\sqrt{6}$
U'	U'	U'	0110	$-1/2\sqrt{6}$
U'	U'	U'	1110	0*

Table C.13.1. **Continued.**

T_2	E'	E''		$6j$
T_2	E'	E''		$\frac{1}{6}$
T_2	E'	U'		$-\frac{1}{3}$
T_2	U'	E''		$-\frac{1}{3}$
T_2	U'	U'	0000	$1/6\sqrt{2}$
T_2	U'	U'	1000	$1/3\sqrt{2}$ *

T_2	E'	U'		$6j$
T_2	E'	U'		$-\frac{1}{12}$
T_2	E''	U'		$-\frac{1}{4}$
T_2	U'	U'	0000	$-\frac{1}{6}$
T_2	U'	U'	1000	$\frac{1}{6}$*
T_2	U'	U'	0010	$\frac{1}{6}$*
T_2	U'	U'	1010	$\frac{1}{12}$

T_2	E''	U'		$6j$
T_2	E''	U'		$-\frac{1}{12}$
T_2	U'	U'	0000	$-\frac{1}{6}$
T_2	U'	U'	1000	$-\frac{1}{6}$*
T_2	U'	U'	0010	$-\frac{1}{6}$*
T_2	U'	U'	1010	$\frac{1}{12}$

T_2	U'	U'	$(r_4 = 0)$	$6j$
T_2	U'	U'	0000	$\frac{1}{12}$
T_2	U'	U'	1000	0*
T_2	U'	U'	0100	0*
T_2	U'	U'	1100	$-\frac{1}{6}$
T_2	U'	U'	0010	0*
T_2	U'	U'	1010	$-\frac{1}{6}$
T_2	U'	U'	0110	$\frac{1}{12}$
T_2	U'	U'	1110	0*

T_2	U'	U'	$(r_4 = 1)$	$6j$
T_2	U'	U'	0001	0*
T_2	U'	U'	1001	$\frac{1}{12}$
T_2	U'	U'	0101	$-\frac{1}{6}$
T_2	U'	U'	1101	0*
T_2	U'	U'	0011	$-\frac{1}{6}$
T_2	U'	U'	1011	0*
T_2	U'	U'	0111	0*
T_2	U'	U'	1111	$-\frac{1}{6}$

C.14. 9*j* for the Groups *O* and T_d

The majority of 9*j* needed in calculations contain one or more A_1 irreps and may be calculated using (16.4.8) and the symmetry rules of Section 16.4. A complete tabulation of the 9*j* for the groups *O* and T_d is given in Butler [(1981), Chapter 15, pp. 471–511].

C.15. Coefficients of Fractional Parentage for *O* and T_d

The coefficients of fractional parentage (cfp) are calculated as discussed in Sections 19.3 and 19.9. Only the $(a^{n-1}\mathfrak{S}'h', a|a^n\mathfrak{S}h)$ are given in Table

C.15.1. The $(a, a^{n-1}\mathbb{S}'h'|a^n \mathbb{S}h)$ may be calculated via (19.3.15). In all cases

$$\left(a^{n-1}\mathbb{S}'h', a|a^n \mathbb{S}h\right) = \left(a^n \mathbb{S}h|a^{n-1}\mathbb{S}'h', a\right)$$

$$\left(a^0, a|a^1\right) = \left(a, a^0|a^1\right) = 1$$

$$\left(a^{2|a|-1}, a|a^{2|a|}\right) = \left(a, a^{2|a|-1}|a^{2|a|}\right) = 1$$

The equations above define all the cfp for a_1^n and a_2^n to be unity. All $(a^{n-1}, a|a^n)$ cfp for e^n, t_1^n, and t_2^n for $n > 1$ are tabulated.

Table C.15.1

$(e^{n-1}\mathbb{S}'h', e|e^n \mathbb{S}h)$

$\mathbb{S}h$ \ $\mathbb{S}'h'$	e^1 Parent
e^2	$\frac{1}{2}E$
$0A_1$	1
$0E$	1
$1A_2$	1

$\mathbb{S}h$ \ $\mathbb{S}'h'$	e^2 Parents		
e^3	$0A_1$	$0E$	$1A_2$
$\frac{1}{2}E$	$1/\sqrt{6}$	$-1/\sqrt{3}$	$-1/\sqrt{2}$

$\mathbb{S}h$ \ $\mathbb{S}'h'$	e^3 Parent
e^4	$\frac{1}{2}E$
$0A_1$	1

$(t_1^{n-1}\mathbb{S}'h', t_1|t_1^n \mathbb{S}h)$

$\mathbb{S}h$ \ $\mathbb{S}'h'$	t_1^1 Parent
t_1^2	$\frac{1}{2}T_1$
$0A_1$	1
$0E$	1
$0T_2$	1
$1T_1$	1

$\mathbb{S}h$ \ $\mathbb{S}'h'$	t_1^2 Parents			
t_1^3	$0A_1$	$0E$	$0T_2$	$1T_1$
$\frac{1}{2}T_1$	$\sqrt{2}/3$	$-\frac{1}{3}$	$-1/\sqrt{6}$	$-1/\sqrt{2}$
$\frac{1}{2}E$	0	0	$-1/\sqrt{2}$	$1/\sqrt{2}$
$\frac{1}{2}T_2$	0	$-1/\sqrt{3}$	$-1/\sqrt{6}$	$1/\sqrt{2}$
$\frac{3}{2}A_1$	0	0	0	1

Table C.15.1. Continued.

$\mathbb{S}'h'$ \ $\mathbb{S}h$	t_1^3 Parents			
t_1^4	$\frac{1}{2}T_1$	$\frac{1}{2}E$	$\frac{1}{2}T_2$	$\frac{3}{2}A_1$
$0A_1$	1	0	0	0
$0E$	$-\frac{1}{2}$	0	$-\sqrt{3}/2$	0
$0T_2$	$-\frac{1}{2}$	$-1/\sqrt{2}$	$-\frac{1}{2}$	0
$1T_1$	$-\frac{1}{2}$	$1/\sqrt{6}$	$\frac{1}{2}$	$-1\sqrt{3}$

$\mathbb{S}'h'$ \ $\mathbb{S}h$	t_1^4 Parents			
t_1^5	$0A_1$	$0E$	$0T_2$	$1T_1$
$\frac{1}{2}T_1$	$1/\sqrt{15}$	$\sqrt{2}/\sqrt{15}$	$1/\sqrt{5}$	$\sqrt{3}/\sqrt{5}$

$\mathbb{S}'h'$ \ $\mathbb{S}h$	t_1^5 Parent
t_1^6	$\frac{1}{2}T_1$
$0A_1$	1

$(t_2^{n-1}\,\mathbb{S}'h', t_2|t_2^n\,\mathbb{S}h)$

$\mathbb{S}'h'$ \ $\mathbb{S}h$	t_2^1 Parent
t_2^2	$\frac{1}{2}T_2$
$0A_1$	1
$0E$	1
$0T_2$	-1
$1T_1$	-1

$\mathbb{S}'h'$ \ $\mathbb{S}h$	t_2^2 Parents			
t_2^3	$0A_1$	$0E$	$0T_2$	$1T_1$
$\frac{1}{2}T_2$	$\sqrt{2}/3$	$-\frac{1}{3}$	$1/\sqrt{6}$	$1/\sqrt{2}$
$\frac{1}{2}E$	0	0	$1/\sqrt{2}$	$-1/\sqrt{2}$
$\frac{1}{2}T_1$	0	$1/\sqrt{3}$	$-1/\sqrt{6}$	$1/\sqrt{2}$
$\frac{3}{2}A_2$	0	0	0	1

$\mathbb{S}'h'$ \ $\mathbb{S}h$	t_2^3 Parents			
t_2^4	$\frac{1}{2}T_2$	$\frac{1}{2}E$	$\frac{1}{2}T_1$	$\frac{3}{2}A_2$
$0A_1$	1	0	0	0
$0E$	$-\frac{1}{2}$	0	$\sqrt{3}/2$	0
$0T_2$	$\frac{1}{2}$	$-1/\sqrt{2}$	$-\frac{1}{2}$	0
$1T_1$	$\frac{1}{2}$	$1/\sqrt{6}$	$\frac{1}{2}$	$-1/\sqrt{3}$

$\mathbb{S}'h'$ \ $\mathbb{S}h$	t_2^4 Parents			
t_2^5	$0A_1$	$0E$	$0T_2$	$1T_1$
$\frac{1}{2}T_2$	$1/\sqrt{15}$	$\sqrt{2}/\sqrt{15}$	$-1/\sqrt{5}$	$-\sqrt{3}/\sqrt{5}$

$\mathbb{S}'h'$ \ $\mathbb{S}h$	t_2^5 Parent
t_2^6	$\frac{1}{2}T_2$
$0A_1$	1

C.16. Coefficients $g(a^n, \mathfrak{S}hfh')$ for O and T_d

The $g(a^n, \mathfrak{S}hfh')$ are defined in Section 19.4. In all cases when $f = A_1$

$$g(a^n, \mathfrak{S}hA_1h') = \delta_{hh'}n|h|^{1/2}|a|^{-1/2}$$

When $n = 2|a|$, we have $g(a^{2|a|}, 0hfh') = 0$ unless $f = A_1$. If $n = 1$,

$$g(a^1, \tfrac{1}{2}hfh') = \delta(a^*fa)$$

These equations define all $g(a^n, \mathfrak{S}hfh')$ for a_1^n and a_2^n. Those for e^n, t_1^n, and t_2^n for $1 < n \leqslant |a|$ are given in Table C.16.1. The results for $n > |a|$ follow from (19.4.12), which gives for $f = A_2$ and T_1

$$g(a^{2|a|-n}, \mathfrak{S}hfh') = g(a^n, \mathfrak{S}hfh')$$

and for $f = E$ and T_2

$$g(a^{2|a|-n}, \mathfrak{S}hfh') = -g(a^n, \mathfrak{S}hfh')$$

Table C.16.1

$g(e^n, \mathfrak{S}hfh')$ [a]

$e^2, f = A_1$

$\mathfrak{S}h$ \ $\mathfrak{S}'h'$	$0A_1$	$0E$	$1A_2$
$0A_1$	$\sqrt{2}$	0	0
$0E$	0	2	0
$1A_2$	0	0	$\sqrt{2}$

$e^2, f = A_2$

$\mathfrak{S}h$ \ $\mathfrak{S}'h'$	$0A_1$	$0E$	$1A_2$
$0A_1$	0	0	0
$0E$	0	-2	0
$1A_2$	0	0	0

$e^2, f = E$

$\mathfrak{S}h$ \ $\mathfrak{S}'h'$	$0A_1$	$0E$	$1A_2$
$0A_1$	0	$\sqrt{2}$	0
$0E$	$\sqrt{2}$	0	0
$1A_2$	0	0	0

Table C.16.1. Continued.

$g(t_1^n, \mathbb{S}hfh')$ and $g(t_2^n, \mathbb{S}hfh')$ [b]

t_1^2 and $t_2^2, f = A_1$

$\mathbb{S}h$ \ $\mathbb{S}'h'$	$0A_1$	$0E$	$0T_2$	$1T_1$
$0A_1$	$2/\sqrt{3}$	0	0	0
$0E$	0	$2\sqrt{2}/\sqrt{3}$	0	0
$0T_2$	0	0	2	0
$1T_1$	0	0	0	2

t_1^2 and $t_2^2, f = E$

$\mathbb{S}h$ \ $\mathbb{S}'h'$	$0A_1$	$0E$	$0T_2$	$1T_1$
$0A_1$	0	$2/\sqrt{3}$	0	0
$0E$	$2/\sqrt{3}$	$-2/\sqrt{3}$	0	0
$0T_2$	0	0	-1	0
$1T_1$	0	0	0	-1

t_1^2 and $t_2^2, f = T_1$

$\mathbb{S}h$ \ $\mathbb{S}'h'$	$0A_1$	$0E$	$0T_2$	$1T_1$
$0A_1$	0	0	0	0
$0E$	0	0	$\pm\sqrt{2}$	0
$0T_2$	0	$\pm\sqrt{2}$	± 1	0
$1T_1$	0	0	0	± 1

t_1^2 and $t_2^2, f = T_2$

$\mathbb{S}h$ \ $\mathbb{S}'h'$	$0A_1$	$0E$	$0T_2$	$1T_1$
$0A_1$	0	0	$\pm 2/\sqrt{3}$	0
$0E$	0	0	$\mp\sqrt{2}/\sqrt{3}$	0
$0T_2$	$\pm 2/\sqrt{3}$	$\mp\sqrt{2}/\sqrt{3}$	∓ 1	0
$1T_1$	0	0	0	∓ 1

t_1^3 and $t_2^3, f = A_1$

$\mathbb{S}h$ \ $\mathbb{S}'h'$	t_1^3	$\frac{1}{2}T_1$	$\frac{1}{2}E$	$\frac{1}{2}T_2$	$\frac{3}{2}A_1$
t_1^3	t_2^3	$\frac{1}{2}T_2$	$\frac{1}{2}E$	$\frac{1}{2}T_1$	$\frac{3}{2}A_2$
$\frac{1}{2}T_1$	$\frac{1}{2}T_2$	3	0	0	0
$\frac{1}{2}E$	$\frac{1}{2}E$	0	$\sqrt{6}$	0	0
$\frac{1}{2}T_2$	$\frac{1}{2}T_1$	0	0	3	0
$\frac{3}{2}A_1$	$\frac{3}{2}A_2$	0	0	0	$\sqrt{3}$

t_1^3 and $t_2^3, f = E$

$\mathbb{S}h$ \ $\mathbb{S}'h'$	t_1^3	$\frac{1}{2}T_1$	$\frac{1}{2}E$	$\frac{1}{2}T_2$	$\frac{3}{2}A_1$
t_1^3	t_2^3	$\frac{1}{2}T_2$	$\frac{1}{2}E$	$\frac{1}{2}T_1$	$\frac{3}{2}A_2$
$\frac{1}{2}T_1$	$\frac{1}{2}T_2$	0	0	$\pm\sqrt{3}$	0
$\frac{1}{2}E$	$\frac{1}{2}E$	0	0	0	0
$\frac{1}{2}T_2$	$\frac{1}{2}T_1$	$\mp\sqrt{3}$	0	0	0
$\frac{3}{2}A_1$	$\frac{3}{2}A_2$	0	0	0	0

t_1^3 and $t_2^3, f = T_1$

$\mathbb{S}h$ \ $\mathbb{S}'h'$	t_1^3	$\frac{1}{2}T_1$	$\frac{1}{2}E$	$\frac{1}{2}T_2$	$\frac{3}{2}A_1$
t_1^3	t_2^3	$\frac{1}{2}T_2$	$\frac{1}{2}E$	$\frac{1}{2}T_1$	$\frac{3}{2}A_2$
$\frac{1}{2}T_1$	$\frac{1}{2}T_2$	1	0	0	0
$\frac{1}{2}E$	$\frac{1}{2}E$	0	0	$\pm\sqrt{2}$	0
$\frac{1}{2}T_2$	$\frac{1}{2}T_1$	0	$\sqrt{2}$	1	0
$\frac{3}{2}A_1$	$\frac{3}{2}A_2$	0	0	0	0

t_1^3 and $t_2^3, f = T_2$

$\mathbb{S}h$ \ $\mathbb{S}'h'$	t_1^3	$\frac{1}{2}T_1$	$\frac{1}{2}E$	$\frac{1}{2}T_2$	$\frac{3}{2}A_1$
t_1^3	t_2^3	$\frac{1}{2}T_2$	$\frac{1}{2}E$	$\frac{1}{2}T_1$	$\frac{3}{2}A_2$
$\frac{1}{2}T_1$	$\frac{1}{2}T_2$	0	$\pm\sqrt{2}$	± 1	0
$\frac{1}{2}E$	$\frac{1}{2}E$	$-\sqrt{2}$	0	0	0
$\frac{1}{2}T_2$	$\frac{1}{2}T_1$	∓ 1	0	0	0
$\frac{3}{2}A_1$	$\frac{3}{2}A_2$	0	0	0	0

[a] All are zero for e^n when $f = T_1$ or T_2.

[b] The upper signs are for t_1^n and the lower signs for t_2^n. All are zero for t_1^n and t_2^n when $f = A_2$.

C.17. The Coefficients $G(a^n, \mathfrak{S}hf\mathfrak{S}'h')$ for O and T_d

The $G(a^n, \mathfrak{S}hf\mathfrak{S}'h')$ are defined in Section 19.5. For the groups O and T_d, they are zero unless $a = t_1$ or t_2, and $f = T_1$ only and is dropped.

For t_1^1 and t_2^1, we have

$$G\left(t_1^1, \tfrac{1}{2}T_1\tfrac{1}{2}T_1\right) = G\left(t_2^1, \tfrac{1}{2}T_2\tfrac{1}{2}T_2\right) = 1$$

For t_1^2 and t_2^2, the coefficients are shown in Table C.17.1; for t_1^3 and t_2^3, in Table C.17.2. In the tables, the upper signs are for t_1^n and the lower signs for t_2^n.

For t_1^4 and t_2^4 we have

$$G(a^4, \mathfrak{S}h\mathfrak{S}'h') = -G(a^2, \mathfrak{S}h\mathfrak{S}'h') \qquad \text{for} \quad a = t_1, t_2$$

For t_1^5 and t_2^5,

$$G\left(t_1^5, \tfrac{1}{2}T_1\tfrac{1}{2}T_1\right) = G\left(t_2^5, \tfrac{1}{2}T_2\tfrac{1}{2}T_2\right) = -1$$

Table C.17.1. The $G(t_i^2, \mathfrak{S}h\mathfrak{S}'h')$ for $i = 1, 2$.

$\mathfrak{S}h$ \ $\mathfrak{S}'h'$	$0A_1$	$0E$	$0T_2$	$1T_1$
$0A_1$	0	0	0	$\mp\sqrt{2}/\sqrt{3}$
$0E$	0	0	0	$\pm 1/\sqrt{3}$
$0T_2$	0	0	0	$\pm 1/\sqrt{2}$
$1T_1$	$\mp\sqrt{2}/\sqrt{3}$	$\pm 1/\sqrt{3}$	$\pm 1/\sqrt{2}$	± 1

Table C.17.2. The $G(t_i^3, \mathfrak{S}h\mathfrak{S}'h')$ for $i = 1, 2$.

$\mathfrak{S}h$ \ $\mathfrak{S}'h'$	t_1^3	$\frac{1}{2}T_1$	$\frac{1}{2}E$	$\frac{1}{2}T_2$	$\frac{3}{2}A_1$
t_1^3	t_2^3	$\frac{1}{2}T_2$	$\frac{1}{2}E$	$\frac{1}{2}T_1$	$\frac{3}{2}A_2$
$\frac{1}{2}T_1$	$\frac{1}{2}T_2$	0	$\mp\sqrt{2}/\sqrt{3}$	∓ 1	$\mp 2/\sqrt{3}$
$\frac{1}{2}E$	$\frac{1}{2}E$	$\sqrt{2}/\sqrt{3}$	0	0	0
$\frac{1}{2}T_2$	$\frac{1}{2}T_1$	± 1	0	0	0
$\frac{3}{2}A_1$	$\frac{3}{2}A_2$	$\mp 2/\sqrt{3}$	0	0	0

C.18. Functions for $(e')^n$, $(e'')^n$, and $(u')^n$ Configurations

The functions in Table C.18.1 were obtained using (21.2.9), (21.2.10), and (10.2.1) with the cfp of Section C.19 and the $3jm$ of Section C.12.

Table C.18.1

p^l	$\|p^l rt\tau\rangle$	p^l	$\|p^l rt\tau\rangle$
e'	$\|E'\beta'\rangle = \|\beta'\rangle$	$(u')^2$	$\|A_1a_1\rangle = (1/\sqrt{2})(\|\kappa\nu\rangle + \|\lambda\mu\rangle)$
	$\|E'\alpha'\rangle = \|\alpha'\rangle$		$\|E\theta\rangle = (1/\sqrt{2})(\|\kappa\nu\rangle - \|\lambda\mu\rangle)$
$(e')^2$	$\|A_1a_1\rangle = \|\alpha'\beta'\rangle$		$\|E\epsilon'\rangle = (1/\sqrt{2})(\|\kappa\lambda\rangle - \|\mu\nu\rangle)$
			$\|0T_2\,-1\rangle = -\|\lambda\nu\rangle$
e''	$\|E''\beta''\rangle = \|\beta''\rangle$		$\|0T_20\rangle = (-1/\sqrt{2})(\|\kappa\lambda\rangle + \|\mu\nu\rangle)$
	$\|E''\alpha''\rangle = \|\alpha''\rangle$		$\|0T_21\rangle = -\|\kappa\mu\rangle$
$(e'')^2$	$\|A_1a_1\rangle = \|\alpha''\beta''\rangle$	$(u')^3$	$\|0U'\nu\rangle = \|\lambda\mu\nu\rangle$
			$\|0U'\mu\rangle = -\|\kappa\mu\nu\rangle$
u'	$\|U'\nu\rangle = \|\nu\rangle$		$\|0U'\lambda\rangle = -\|\kappa\lambda\nu\rangle$
	$\|U'\mu\rangle = \|\mu\rangle$		$\|0U'\kappa\rangle = \|\kappa\lambda\mu\rangle$
	$\|U'\lambda\rangle = \|\lambda\rangle$	$(u')^4$	$\|A_1a_1\rangle = \|\kappa\lambda\mu\nu\rangle$
	$\|U'\kappa\rangle = \|\kappa\rangle$		

C.19. Coefficients of Fractional Parentage for $(e')^n$, $(e'')^n$, and $(u')^n$ jj Configurations

The cfp for jj configurations are defined in Section 21.2. Values are given below for group-O and group-T_d p^n configurations for $p = e'$, e'' and u'.

1. $(p^0, p|p) = 1$.
2. $(p, p|p^2\,rt) = 1$ when the antisymmetric square of $p \otimes p$ contains rt.
3. $(p^3, p|p^4) = 1$ for $p = u'$.
4. $(p^2(r't'), p|p^3\,rt)$ for $p = u'$:

rt \\ $r't'$	$(u')^2$ Parents		
$(u')^3$	A_1	E	$0T_2$
$0U'$	$1/\sqrt{6}$	$-1/\sqrt{3}$	$-1/\sqrt{2}$

Appendix D Tables for the Groups D_4 and D_{2d}

D.1. Character Table for D_4 and D_{2d}

See Table D.1.1.

For notation and comments, see Section C.1. We have

$$C_2 \equiv C_2^Z, \qquad 2C_2' \equiv \left(C_2^X, C_2^Y\right), \qquad 2C_2'' \equiv \left(C_2^{XY}, C_2^{-XY}\right)$$

Table D.1.1

		D_{2d}			E	R	$2S_4$	$2S_4R$	C_2, C_2R	$2C_2'$, $2C_2'R$	$2\sigma_d$, $2\sigma_dR$
D_{2d}	D_4	D_4			E	R	$2C_4$	$2C_4R$	C_2, C_2R	$2C_2'$, $2C_2'R$	$2C_2''$, $2C_2''R$
x^2+y^2, z^2	x^2+y^2, z^2	A_1	Γ_1	0	1	1	1	1	1	1	1
R_z	R_z, z	A_2	Γ_2	$\tilde{0}$	1	1	1	1	1	-1	-1
x^2-y^2	x^2-y^2	B_1	Γ_3	2	1	1	-1	-1	1	1	-1
xy, z	xy	B_2	Γ_4	$\tilde{2}$	1	1	-1	-1	1	-1	1
(x, y), (R_x, R_y), (yz, zx)	(x, y), (R_x, R_y), (yz, zx)	E	Γ_5	1	2	2	0	0	-2	0	0
		E'	Γ_6	$\frac{1}{2}$	2	-2	$\sqrt{2}$	$-\sqrt{2}$	0	0	0
		E''	Γ_7	$\frac{3}{2}$	2	-2	$-\sqrt{2}$	$\sqrt{2}$	0	0	0

D.2. Direct-Product Table for D_4 and D_{2d}

Table D.2.1

D_4, D_{2d}	A_1	A_2	B_1	B_2	E	E'	E''
A_1	A_1	A_2	B_1	B_2	E	E'	E''
A_2	A_2	A_1	B_2	B_1	E	E'	E''
B_1	B_1	B_2	A_1	A_2	E	E''	E'
B_2	B_2	B_1	A_2	A_1	E	E''	E'
E	E	E	E	E	$A_1 + A_2 + B_1 + B_2$	$E' + E''$	$E' + E''$
E'	E'	E'	E''	E''	$E' + E''$	$A_1 + A_2 + E$	$B_1 + B_2 + E$
E''	E''	E''	E'	E'	$E' + E''$	$B_1 + B_2 + E$	$A_1 + A_2 + E$

D.3. Table of Symmetrized $[a^2]$ and Antisymmetrized (a^2) Squares

Table D.3.1

	D_4, D_{2d}	
a	$[a^2]$	(a^2)
A_1	A_1	—
A_2	A_1	—
B_1	A_1	—
B_2	A_1	—
E	$A_1 + B_1 + B_2$	A_2
E'	$A_2 + E$	A_1
E''	$A_2 + E$	A_1

D.4. Bases for D_4 and D_{2d}

The nomenclature throughout this section [Tables D.4.1(a)–(d)] is analogous to that of Section C.5. The bases are identical to those of Butler (1981). Our real-D_4 and complex-D_4 bases are respectively Butler's $D_4 \supset D_2 \supset C_2$ and $D_4 \supset C_4$ bases. Likewise our real-D_{2d} and complex-D_{2d} bases are his $D_{2d} \supset D_2 \supset C_2$ and $D_{2d} \supset S_4$ bases.

Since the bases here are included in the chains in Section C.5, the table entries in the Sections D.4, D.5, and D.6 follow immediately from the analogous tables in Appendix C. The reader should compare Section C.5 with Section D.4, and Section C.7 with Section D.6, to appreciate the redundancy of the tables in Appendixes C and D when Butler bases are used. The only really new information in Sections D.4 and D.6 is the alphabetic irrep and irrep-component names in D_4 and D_{2d} for the basis chain [e.g. Table D.4.1(b) gives $|B_2 b_2\rangle \equiv |\tilde{2}\tilde{0}0\rangle$ and so on].

Table D.4.1(a). Behavior of x, y, and z Functions Under Group-D_4 and Group-D_{2d} Generators

	Group-D_4 Generator	
Function	C_4^Z	$C_2' = C_2^X$
x	y	x
y	$-x$	$-y$
z	z	$-z$
	Group-D_{2d} Generator	
Function	$(S_4^Z)^{-1}$	$C_2' = C_2^X$
x	$-y$	x
y	x	$-y$
z	$-z$	$-z$

Table D.4.1(b). The Real-D_4 and Real-D_{2d} Bases for Single-Valued Irreps[a]

Functions Which Transform		Standard Basis	Group Generators	
Identically to the Real-D_4 Basis ($D_4 \supset D_2 \supset C_2$)	Identically to the Real-D_{2d} Basis ($D_{2d} \supset D_2 \supset C_2$)	Real-D_{2d} Real-D_4	$(S_4^Z)^{-1}$ C_4^Z	$C_2' = C_2^X$ $C_2' = C_2^X$
$(-1/\sqrt{3})(3z^2 - r^2)$	$ixyz, (-1/\sqrt{3})(3z^2 - r^2)$	$\|A_1 a_1\rangle = \|000\rangle$	a_1	a_1
z, R_z	R_z	$\|A_2 a_2\rangle = \|\tilde{0}\tilde{0}0\rangle$	a_2	$-a_2$
$x^2 - y^2, ixyz$	$x^2 - y^2$	$\|B_1 b_1\rangle = \|200\rangle$	$-b_1$	b_1
ixy	z, ixy	$\|B_2 b_2\rangle = \|\tilde{2}\tilde{0}0\rangle$	$-b_2$	$-b_2$
$-x, -R_x, -iyz$	$-R_x, -x, -iyz$	$\|Ex\rangle = \|1\tilde{1}1\rangle$	iEy	Ex
iy, iR_y, zx	$iR_y, -iy, zx$	$\|Ey\rangle = \|111\rangle$	iEx	$-Ey$

[a] Irrep names in the Butler notation are $|a(D_4)a_1(D_2)a_2(C_2)\rangle$ and $|a(D_{2d})a_1(D_2)a_2(C_2)\rangle$ in the real-D_4 and real-D_{2d} bases respectively.

Table D.4.1(c). Relation between the Complex and Real Bases for Single-Valued Irreps

Complex-D_4 $\vert a\alpha\rangle \equiv \vert a(D_4)a_1(C_4)\rangle$ Complex-D_{2d} $\vert a\alpha\rangle \equiv \vert a(D_{2d})a_1(S_4)\rangle$	Real-D_4 Equivalent $\vert a\alpha\rangle \equiv \vert a(D_4)a_1(D_2)a_2(C_2)\rangle$ Real-D_{2d} Equivalent $\vert a\alpha\rangle = \vert a(D_{2d})a_1(D_2)a_2(C_2)\rangle$
$\vert A_1a_1\rangle = \vert 00\rangle$	$\vert A_1a_1\rangle = \vert 000\rangle$
$\vert A_2a_2\rangle = \vert\tilde{0}0\rangle$	$\vert A_2a_2\rangle = \vert\tilde{0}\tilde{0}0\rangle$
$\vert B_1b_1'\rangle = \vert 22\rangle$	$-\vert B_1b_1\rangle = -\vert 200\rangle$
$\vert B_2b_2\rangle = \vert\tilde{2}2\rangle$	$\vert B_2b_2\rangle = \vert\tilde{2}\tilde{0}0\rangle$
$\vert E-1\rangle = \vert 1-1\rangle$	$(1/\sqrt{2})(\vert Ex\rangle + \vert Ey\rangle) = (1/\sqrt{2})(\vert 1\tilde{1}1\rangle + \vert 111\rangle)$
$\vert E1\rangle = \vert 11\rangle$	$(1/\sqrt{2})(-\vert Ex\rangle + \vert Ey\rangle) = (1/\sqrt{2})(-\vert 1\tilde{1}1\rangle + \vert 111\rangle)$

Table D.4.1(d). The Complex-D_4 and Complex-D_{2d} Bases for Double-Valued Irreps

$O_3 \supset SO_3 \supset SO_2$ Equivalent	Standard Basis	Group Generators	
For $\vert j^+m\rangle$ only	Complex-D_{2d} $(D_{2d} \supset S_4)$	$(S_4^Z)^{-1}$	$C_2' = C_2^X$
For $\vert j^+m\rangle$ and $\vert j^-m\rangle$	Complex-D_4 $(D_4 \supset C_4)$	C_4^Z	$C_2' = C_2^X$
$\vert\frac{1}{2}-\frac{1}{2}\rangle$	$\vert E'\beta'\rangle \equiv \vert\frac{1}{2}-\frac{1}{2}\rangle$	$(1/\sqrt{2})(1+i)\beta'$	$-i\alpha'$
$\vert\frac{1}{2}\frac{1}{2}\rangle$	$\vert E'\alpha'\rangle \equiv \vert\frac{1}{2}\frac{1}{2}\rangle$	$(1/\sqrt{2})(1-i)\alpha'$	$-i\beta'$
$\vert\frac{3}{2}-\frac{3}{2}\rangle$	$\vert E''\beta''\rangle \equiv \vert\frac{3}{2}-\frac{3}{2}\rangle$	$(1/\sqrt{2})(-1+i)\beta''$	$i\alpha''$
$\vert\frac{3}{2}\frac{3}{2}\rangle$	$\vert E''\alpha''\rangle \equiv \vert\frac{3}{2}\frac{3}{2}\rangle$	$(1/\sqrt{2})(-1-i)\alpha''$	$i\beta''$

D.5. The Complex-D_4 and Complex-D_{2d} Bases Defined in Terms of the $O_3 \supset SO_3 \supset SO_2$ Basis

The complex-D_4 and complex-D_{2d} bases have already been defined in terms of the $O_3 \supset SO_3 \supset SO_2$ basis in Section C.6, since the central column of Table C.6.1 gives irrep labels for all groups in the chain.

D.6. Function and Operator Transformation Coefficients

See Table D.6.1.

The notation is analogous to that of Section C.7.

Table D.6.1

	$\langle a\alpha \vert \psi\rangle$	$\vert \psi a\alpha\rangle$ or O^f_ϕ	
ψ or Θ	or $\langle f\phi \vert \Theta\rangle$	$a\alpha$ or $f\phi$	Equivalent-Chain Notation
Real-D_4 Basis ($D_4 \supset D_2 \supset C_2$)			
x, p_x, V_x, R_x	-1	Ex	$1\tilde{1}1$
y, p_y, V_y, R_y	$-i$	Ey	111
z, p_z, V_z, R_z	1	A_2a_2	$\tilde{0}\tilde{0}0$
d_{z^2}	-1	A_1a_1	000
$d_{x^2-y^2}$	1	B_1b_1	200
d_{yz}	i	Ex	$1\tilde{1}1$
d_{zx}	1	Ey	111
d_{xy}	$-i$	B_2b_2	$\tilde{2}\tilde{0}0$
Real-D_{2d} Basis ($D_{2d} \supset D_2 \supset C_2$)[a]			
x, p_x, V_x	-1	Ex	$1\tilde{1}1$
y, p_y, V_y	i	Ey	111
z, p_z, V_z	1	B_2b_2	$\tilde{2}\tilde{0}0$
Complex-D_4 Basis ($D_4 \supset C_4$)			
$\vert 1^{\pm}-1\rangle, V_{-1}, R_{-1}$	-1	$E-1$	$1-1$
$\vert 1^{\pm}0\rangle, V_0, R_0$	1	A_2a_2	$\tilde{0}0$
$\vert 1^{\pm}1\rangle, V_1, R_1$	-1	$E1$	11
Complex-D_{2d} Basis ($D_{2d} \supset S_4$)[b]			
$\vert 1^{-}-1\rangle, V_{-1}$	1	$E1$	11
$\vert 1^{-}0\rangle, V_0$	1	B_2b_2	$\tilde{2}2$
$\vert 1^{-}1\rangle, V_1$	1	$E-1$	$1-1$

[a] **R** and d functions as in the real-D_4 basis.

[b] **R** and the $\vert 1^{+}m\rangle$ for $m = -1, 0, 1$ transform as in the complex-D_4 basis.

D.7. Symmetry-Adapted D_{4h} Molecular Orbitals for Square-Planar Compounds

In Table D.7.1 the notation is as in Section C.10. The coordinate system is identical to that of Fig. C.10.1 except that positions 5 and 6 are not occupied. We use the real-D_4 basis of Section D.4. Recall that $D_{4h} = D_4 \otimes C_i$.

Table D.7.1

Real-D_{4h} Basis	Metal Orbital	Ligand Orbitals: σ-type	Ligand Orbitals: π-type
$\lvert A_{1g}a_1\rangle$	$(n+1)s$, $-nd_{z^2}$	$\frac{1}{2}(s_1+s_2+s_3+s_4)$, $\frac{1}{2}(x_1-x_3+y_2-y_4)$	
$\lvert A_{2g}a_2\rangle$			$\frac{1}{2}(y_1-y_3-x_2+x_4)$
$\lvert A_{2u}a_2\rangle$	$(n+1)p_z$		$\frac{1}{2}(z_1+z_2+z_3+z_4)$
$\lvert B_{1g}b_1\rangle$	$nd_{x^2-y^2}$	$\frac{1}{2}(s_1+s_3-s_2-s_4)$, $\frac{1}{2}(x_1-x_3-y_2+y_4)$	
$\lvert B_{2g}b_2\rangle$	ind_{xy}		$(i/2)(x_2-x_4+y_1-y_3)$
$\lvert B_{2u}b_2\rangle$			$(i/2)(z_1+z_3-z_2-z_4)$
$\lvert E_gx\rangle$	$-ind_{yz}$		$(-i/\sqrt{2})(z_2-z_4)$
$\lvert E_gy\rangle$	nd_{zx}		$(1/\sqrt{2})(z_1-z_3)$
$\lvert E_ux\rangle$	$-(n+1)p_x$	$(-1/\sqrt{2})(s_1-s_3)$, $(-1/\sqrt{2})(x_1+x_3)$	$(-1/\sqrt{2})(x_2+x_4)$
$\lvert E_uy\rangle$	$i(n+1)p_y$	$(i/\sqrt{2})(s_2-s_4)$, $(i/\sqrt{2})(y_2+y_4)$	$(i/\sqrt{2})(y_1+y_3)$

D.8. $3j$, $2j$, and $2jm$ Phases and Related Definitions

The bases are defined in Section D.4.

1. The $3j$ phase $\{abc\}$: All $3j$ for D_4 and D_{2d} are given by

$$\{abc\} = (-1)^a(-1)^b(-1)^c$$

where

$$(-1)^{A_1} = (-1)^{B_1} = (-1)^{B_2} = 1$$

$$(-1)^{A_2} = (-1)^E = -1$$

$$(-1)^{E'} = i$$

$$(-1)^{E''} = -i$$

2. The $2j$ phase $\{a\}$:

$$\{a\} = \begin{cases} 1 & \text{for} \quad A_1, A_2, B_1, B_2 \text{ and } E \\ -1 & \text{for} \quad E' \text{ and } E'' \end{cases}$$

3. The $2jm$ $\begin{pmatrix} a \\ \alpha \end{pmatrix}$ and definitions of a^* and α^*: see Tables D.8.1 and D.8.2.

Table D.8.1. $2jm$ in the Real-D_4 and Real-D_{2d} Bases for Single-Valued Irreps[a]

$D_4 \supset D_2 \supset C_2$ $D_{2d} \supset D_2 \supset C_2$		
$\vert a\alpha\rangle \equiv \vert aa_1a_2\rangle$		$2jm$ $\begin{pmatrix} a \\ \alpha \end{pmatrix} \equiv \begin{pmatrix} a \\ a_1 \\ a_2 \end{pmatrix}$
A_1a_1	000	1
A_2a_2	$\tilde{0}\tilde{0}0$	−1
B_1b_1	200	1
B_2b_2	$\tilde{2}\tilde{0}0$	−1
Ex	$1\tilde{1}1$	−1
Ey	111	1

[a]In these bases $a = a^*$, $\alpha = \alpha^*$, and $a_i = a_i^*$ ($i = 1$ to 2) for all a, α, and a_i.

Table D.8.2. $2jm$ in the Complex-D_4 and Complex-D_{2d} Bases

$D_4 \supset C_4$ and $D_{2d} \supset S_4$				
$\vert a\alpha\rangle \equiv \vert aa_1\rangle$		$2jm$ $\begin{pmatrix} a \\ \alpha \end{pmatrix} \equiv \begin{pmatrix} a \\ a_1 \end{pmatrix}$	$\vert a^*\alpha^*\rangle \equiv \vert a^*a_1^*\rangle$	
A_1a_1	00	1	A_1a_1	00
A_2a_2	$\tilde{0}0$	−1	A_2a_2	$\tilde{0}0$
B_1b_1'	22	1	B_1b_1'	22
B_2b_2	$\tilde{2}2$	−1	B_2b_2	$\tilde{2}2$
$E-1$	$1-1$	1	$E1$	11
$E1$	11	1	$E-1$	$1-1$
$E'\beta'$	$\frac{1}{2}-\frac{1}{2}$	−1	$E'\alpha'$	$\frac{1}{2}\frac{1}{2}$
$E'\alpha'$	$\frac{1}{2}\frac{1}{2}$	1	$E'\beta'$	$\frac{1}{2}-\frac{1}{2}$
$E''\beta''$	$\frac{3}{2}-\frac{3}{2}$	−1	$E''\alpha''$	$\frac{3}{2}\frac{3}{2}$
$E''\alpha''$	$\frac{3}{2}\frac{3}{2}$	1	$E''\beta''$	$\frac{3}{2}-\frac{3}{2}$

D.9. $3jm$ for D_4 and D_{2d} in the Bases of Section D.4

See Tables D.9.1 and D.9.2. The bases are defined in Section D.4; otherwise the comments of Section C.12 also apply here.

Table D.9.1. $3jm$ in the Real-D_4 and Real-D_{2d} Bases for Single-Valued Irreps[a]

A_1	A_1	A_1	$3jm$
a_1	a_1	a_1	1

A_2	A_2	A_1	$3jm$
a_2	a_2	a_1	-1

B_1	B_1	A_1	$3jm$
b_1	b_1	a_1	1

B_2	B_1	A_2	$3jm^\dagger$
b_2	b_1	a_2	1

B_2	B_2	A_1	$3jm$
b_2	b_2	a_1	-1

E	E	A_1	$3jm$
x	x	a_1	$-1/\sqrt{2}$
y	y	a_1	$1/\sqrt{2}$

E	E	A_2	$3jm^\dagger$
x	y	a_2	$-1/\sqrt{2}$

E	E	B_1	$3jm$
x	x	b_1	$1/\sqrt{2}$
y	y	b_1	$1/\sqrt{2}$

E	E	B_2	$3jm$
x	y	b_2	$1/\sqrt{2}$

[a] $r = 0$ for all $3jm$.

Table D.9.2. $3jm$ in the Complex-D_4 and Complex-D_{2d} Bases[a]

A_1	A_1	A_1	$3jm$
a_1	a_1	a_1	1

A_2	A_2	A_1	$3jm$
a_2	a_2	a_1	-1

B_1	B_1	A_1	$3jm$
b_1'	b_1'	a_1	1

B_2	B_1	A_2	$3jm^\dagger$
b_2	b_1'	a_2	-1

B_2	B_2	A_1	$3jm$
b_2	b_2	a_1	-1

E	E	A_1	$3jm$
-1	1	a_1	$1/\sqrt{2}$

E	E	A_2	$3jm^\dagger$
-1	1	a_2	$-1/\sqrt{2}$

E	E	B_1	$3jm$
-1	-1	b_1'	$-1/\sqrt{2}$
1	1	b_1'	$-1/\sqrt{2}$

E	E	B_2	$3jm$
-1	-1	b_2	$1/\sqrt{2}$
1	1	b_2	$-1/\sqrt{2}$

E'	E'	A_1	$3jm^\dagger$
β'	α'	a_1	$-1/\sqrt{2}$

E'	E'	A_2	$3jm$
β'	α'	a_2	$1/\sqrt{2}$

E'	E'	E	$3jm$
β'	β'	1	$1/\sqrt{2}$
α'	α'	-1	$1/\sqrt{2}$

E''	E''	A_1	$3jm^\dagger$
β''	α''	a_1	$-1/\sqrt{2}$

E''	E''	A_2	$3jm$
β''	α''	a_2	$1/\sqrt{2}$

E''	E''	E	$3jm$
β''	β''	-1	$1/\sqrt{2}$
α''	α''	1	$1/\sqrt{2}$

E''	E'	B_1	$3jm$
β''	β'	b_1'	$1/\sqrt{2}$
α''	α'	b_1'	$1/\sqrt{2}$

E''	E'	B_2	$3jm$
β''	β'	b_2	$-1/\sqrt{2}$
α''	α'	b_2	$1/\sqrt{2}$

E''	E'	E	$3jm^\dagger$
β''	α'	1	$1/\sqrt{2}$
α''	β'	-1	$-1/\sqrt{2}$

[a] $r = 0$ for all $3jm$.

D.10. $6j$ for D_4 and D_{2d}

The $6j$ in Table D.10.1 apply for all bases in Section D.4. All $6j$ containing one or more A_1 irreps are given by (16.1.11) and are not tabulated:

$$\begin{Bmatrix} A_1 & b & c \\ d & e & f \end{Bmatrix} = \delta_{bc}\delta_{ef}\,\delta(dbf)|b|^{-1/2}|e|^{-1/2}\{dbf\}$$

Table D.10.1

A_2	B_1	B_2	$6j$
A_2	B_1	B_2	1
E	E	E	$-1/\sqrt{2}$
E'	E''	E''	$-1/\sqrt{2}$
E''	E'	E'	$-1/\sqrt{2}$

A_2	E	E	$6j$
A_2	E	E	$\frac{1}{2}$
B_1	E	E	$\frac{1}{2}$
B_2	E	E	$\frac{1}{2}$
E'	E'	E'	$-\frac{1}{2}$
E'	E''	E''	$\frac{1}{2}$
E''	E'	E'	$-\frac{1}{2}$
E''	E''	E''	$\frac{1}{2}$

A_2	E'	E'	$6j$
A_2	E'	E'	$-\frac{1}{2}$
B_1	E''	E''	$\frac{1}{2}$
B_2	E''	E''	$\frac{1}{2}$
E	E'	E'	$\frac{1}{2}$
E	E''	E''	$\frac{1}{2}$

A_2	E''	E''	$6j$
A_2	E''	E''	$-\frac{1}{2}$
E	E''	E''	$\frac{1}{2}$

B_1	E	E	$6j$
B_1	E	E	$\frac{1}{2}$
B_2	E	E	$-\frac{1}{2}$
E'	E'	E''	$\frac{1}{2}$
E''	E'	E''	$\frac{1}{2}$

B_1	E'	E''	$6j$
B_1	E'	E''	$-\frac{1}{2}$
B_2	E'	E''	$\frac{1}{2}$
E	E'	E''	$\frac{1}{2}$
E	E''	E'	$-\frac{1}{2}$

B_2	E	E	$6j$
B_2	E	E	$\frac{1}{2}$
E'	E'	E''	$\frac{1}{2}$
E''	E'	E''	$-\frac{1}{2}$

B_2	E'	E''	$6j$
B_2	E'	E''	$-\frac{1}{2}$
E	E'	E''	$\frac{1}{2}$
E	E''	E'	$\frac{1}{2}$

E	E'	E'	$6j$
E	E'	E'	0
E	E'	E''	$-\frac{1}{2}$
E	E''	E''	0

E	E'	E''	$6j$
E	E'	E''	0
E	E''	E''	$-\frac{1}{2}$

E	E''	E''	$6j$
E	E''	E''	0

Other $6j$ neither found in the table, nor related by the symmetry rules of Section 16.1, are zero by (16.1.2). The $6j$ are listed as in Section C.13. Since D_4 and D_{2d} are simply reducible, the simplified symmetry rules of Section 16.1 apply.

D.11. $9j$ for D_4 and D_{2d} for Single-Valued Irreps

$9j$ containing one or more A_1 irreps may be calculated with (16.4.8). Other $9j$ containing only single-valued irreps which are neither found below, nor related to those below, by the symmetry rules of Section 16.4 are zero:

$$\begin{Bmatrix} A_2 & B_1 & B_2 \\ B_2 & A_2 & B_1 \\ B_1 & B_2 & A_2 \end{Bmatrix} = -1, \qquad \begin{Bmatrix} E & E & A_2 \\ E & E & B_1 \\ A_2 & B_1 & B_2 \end{Bmatrix} = \frac{1}{2}$$

$$\begin{Bmatrix} E & E & B_1 \\ E & E & B_2 \\ B_1 & B_2 & A_2 \end{Bmatrix} = \frac{1}{2}, \qquad \begin{Bmatrix} E & E & B_2 \\ E & E & A_2 \\ B_2 & A_2 & B_1 \end{Bmatrix} = \frac{1}{2}$$

$$\begin{Bmatrix} A_2 & E & E \\ E & b & E \\ E & E & b \end{Bmatrix} = -\frac{1}{4} \quad \text{for } b = A_2, B_1, \text{ or } B_2$$

$$\begin{Bmatrix} A_2 & E & E \\ E & b & E \\ E & E & c \end{Bmatrix} = \frac{1}{4} \quad \text{for } \; b \neq c, \quad \text{and } b, c = A_2, B_1, \text{ or } B_2$$

$$\begin{Bmatrix} a & E & E \\ E & b & E \\ E & E & c \end{Bmatrix} = \frac{1}{4} \quad \text{for } \; a, b, c = B_1 \text{ or } B_2$$

D.12. $2jm$ and $3jm$ Factors for $O \supset D_4$ and $T_d \supset D_{2d}$

All $2jm$ factors for $O \supset D_4$ and $T_d \supset D_{2d}$ are unity. The notation for the $3jm$ factors in Table D.12.1 is that of Section B.5, except that here table headings are group-O (or T_d) irrep labels, and entries beneath them are group-D_4 (or D_{2d}) irrep labels. $3jm$ factors neither listed below nor related to those listed by (15.3.7) are zero. The $3jm$ factors are those of Butler [(1981), Chapter 13], but are expressed with Mulliken irrep labels.

Table D.12.1. 3*jm* Factors for $O \supset D_4$ and $T_d \supset D_{2d}$

A_1	A_1	A_1	$3jm$
A_1	A_1	A_1	1

A_1	A_2	A_2	$3jm$
A_1	B_1	B_1	1

A_1	E	E	$3jm$
A_1	A_1	A_1	$1/\sqrt{2}$
A_1	B_1	B_1	$1/\sqrt{2}$

A_1	T_1	T_1	$3jm$
A_1	A_2	A_2	$1/\sqrt{3}$
A_1	E	E	$\sqrt{2}/\sqrt{3}$

A_1	T_2	T_2	$3jm$
A_1	B_2	B_2	$1/\sqrt{3}$
A_1	E	E	$\sqrt{2}/\sqrt{3}$

A_2	E	E	$3jm$	Phase
B_1	A_1	B_1	$-1/\sqrt{2}$	-1

A_2	T_1	T_2	$3jm$	Phase
B_1	A_2	B_2	$1/\sqrt{3}$	-1
B_1	E	E	$\sqrt{2}/\sqrt{3}$	1

E	E	E	$3jm$
A_1	A_1	A_1	$-\frac{1}{2}$
A_1	B_1	B_1	$\frac{1}{2}$

E	T_1	T_1	$3jm$
A_1	A_2	A_2	$1/\sqrt{3}$
A_1	E	E	$-1/\sqrt{6}$
B_1	E	E	$1/\sqrt{2}$

E	T_1	T_2	$3jm$	Phase
A_1	E	E	$1/\sqrt{2}$	-1
B_1	A_2	B_2	$-1/\sqrt{3}$	1
B_1	E	E	$1/\sqrt{6}$	-1

E	T_2	T_2	$3jm$
A_1	B_2	B_2	$1/\sqrt{3}$
A_1	E	E	$-1/\sqrt{6}$
B_1	E	E	$1/\sqrt{2}$

T_1	T_1	T_1	$3jm$
A_2	E	E	$-1/\sqrt{3}$

T_1	T_1	T_2	$3jm$	Phase
A_2	E	E	$1/\sqrt{3}$	-1
E	E	B_2	$1/\sqrt{3}$	1

T_1	T_2	T_2	$3jm$	Phase
A_2	E	E	$-1/\sqrt{3}$	1
E	B_2	E	$-1/\sqrt{3}$	-1

T_2	T_2	T_2	$3jm$
B_2	E	E	$1/\sqrt{3}$

E'	E'	A_1	$3jm$
E'	E'	A_1	1

E'	E'	T_1	$3jm$
E'	E'	A_2	$1/\sqrt{3}$
E'	E'	E	$\sqrt{2}/\sqrt{3}$

E''	E'	A_2	$3jm$
E''	E'	B_1	1

E''	E'	T_2	$3jm$	Phase
E''	E'	B_2	$1/\sqrt{3}$	-1
E''	E'	E	$\sqrt{2}/\sqrt{3}$	1

E''	E''	A_1	$3jm$
E''	E''	A_1	1

E''	E''	T_1	$3jm$
E''	E''	A_2	$1/\sqrt{3}$
E''	E''	E	$-\sqrt{2}/\sqrt{3}$

U'	E'	E	$3jm$	Phase
E'	E'	A_1	$1/\sqrt{2}$	-1
E''	E'	B_1	$1/\sqrt{2}$	1

Table D.12.1. Continued.

U'	E'	T_1	$3jm$	Phase
E'	E'	A_2	$-1/\sqrt{3}$	-1
E'	E'	E	$1/\sqrt{6}$	-1
E''	E'	E	$1/\sqrt{2}$	1

U'	E'	T_2	$3jm$	Phase
E'	E'	E	$1/\sqrt{2}$	1
E''	E'	B_2	$1/\sqrt{3}$	1
E''	E'	E	$-1/\sqrt{6}$	-1

U'	E''	E	$3jm$	Phase
E''	E''	A_1	$1/\sqrt{2}$	-1
E'	E''	B_1	$1/\sqrt{2}$	1

U'	E''	T_1	$3jm$	Phase
E'	E''	E	$-1/\sqrt{2}$	1
E''	E''	A_2	$-1/\sqrt{3}$	-1
E''	E''	E	$-1/\sqrt{6}$	-1

U'	E''	T_2	$3jm$	Phase
E'	E''	B_2	$1/\sqrt{3}$	1
E'	E''	E	$1/\sqrt{6}$	-1
E''	E''	E	$-1/\sqrt{2}$	1

U'	U'	A_1	$3jm$
E'	E'	A_1	$1/\sqrt{2}$
E''	E''	A_1	$1/\sqrt{2}$

U'	U'	A_2	$3jm$
E'	E''	B_1	$-1/\sqrt{2}$

U'	U'	E	$3jm$	Phase
E'	E'	A_1	$-\frac{1}{2}$	1
E'	E''	B_1	$-\frac{1}{2}$	-1
E''	E''	A_1	$\frac{1}{2}$	1

			$r = 0$		$r = 1$	
U'	U'	T_1	$3jm$	Phase	$3jm$	Phase
E'	E'	A_2	$-1/\sqrt{6}$	1	$1/\sqrt{6}$	1
E'	E'	E	$-1/\sqrt{3}$	1	$-1/2\sqrt{3}$	1
E''	E'	E	0	-1	$1/2$	-1
E''	E''	A_2	$1/\sqrt{6}$	1	$1/\sqrt{6}$	1
E''	E''	E	$-1/\sqrt{3}$	1	$1/2\sqrt{3}$	1

			$r = 0$		$r = 1$	
U'	U'	T_2	$3jm$	Phase	$3jm$	Phase
E'	E'	E	0	-1	$-1/2$	1
E''	E'	B_2	$1/\sqrt{6}$	-1	$1/\sqrt{6}$	1
E''	E'	E	$1/\sqrt{3}$	1	$-1/2\sqrt{3}$	-1
E''	E''	E	0	-1	$1/2$	1

Appendix E Tables for the Groups D_3 and C_{3v} and for the Groups O and T_d Using Trigonal Bases

E.1. Character Table for D_3 and C_{3v}

See Tables E.1.1 and E.1.2. For notation and comments see Section C.1. The bases are those of Butler (1981). Note carefully that each of the four bases orients a triangle in a different way.

Table E.1.1

	$C_{3v} \supset C_3$ basis			E	R	$2C_3^Z$	$2C_3^ZR$	$3\sigma_{XZ}$	$3\sigma_{XZ}R$
	$C_{3v} \supset C_s$ basis			E	R	$2C_3^Y$	$2C_3^YR$	$3\sigma_{XY}$	$3\sigma_{XY}R$
	$D_3 \supset C_3$ basis			E	R	$2C_3^Z$	$2C_3^ZR$	$3C_2^Y$	$3C_2^YR$
	$D_3 \supset C_2$ basis			E	R	$2C_3^Y$	$2C_3^YR$	$3C_2^Z$	$3C_2^ZR$
	A_1	Γ_1	0	1	1	1	1	1	1
	A_2	Γ_2	$\tilde{0}$	1	1	1	1	-1	-1
	E	Γ_3	1	2	2	-1	-1	0	0
	E'	Γ_4	$\frac{1}{2}$	2	-2	1	-1	0	0
Mulliken	P_2	Γ_5	$\frac{3}{2}$	1	-1	-1	1	$-i$	i
E''	P_1	Γ_6	$-\frac{3}{2}$	1	-1	-1	1	i	$-i$

Table E.1.2

$C_{3v} \supset C_3$	$C_{3v} \supset C_s$	$D_3 \supset C_3$	$D_3 \supset C_2$	Irrep	
z, x^2+y^2, z^2	y, x^2+z^2, y^2	x^2+y^2, z^2	x^2+z^2, y^2	A_1	0
R_z	R_y	z, R_z	y, R_y	A_2	$\tilde{0}$
$(x, y), (R_x, R_y)$, (x^2-y^2, xy), (yz, zx)	$(x, z), (R_x, R_z)$, (x^2-z^2, zx), (yz, xy)	$(x, y), (R_x, R_y)$, (x^2-y^2, xy), (yz, zx)	$(x, z), (R_x, R_z)$, (x^2-z^2, zx), (yz, xy)	E	1

E.2. Direct-Product Table for D_3 and C_{3v}

Table E.2.1

D_3, C_{3v}	A_1	A_2	E	E'	P_1	P_2
A_1	A_1	A_2	E	E'	P_1	P_2
A_2	A_2	A_1	E	E'	P_2	P_1
E	E	E	$A_1 + A_2 + E$	$E' + P_1 + P_2$	E'	E'
E'	E'	E'	$E' + P_1 + P_2$	$A_1 + A_2 + E$	E	E
P_1	P_1	P_2	E'	E	A_2	A_1
P_2	P_2	P_1	E'	E	A_1	A_2

E.3. Table of Symmetrized $[a^2]$ and Antisymmetrized (a^2) Squares

Table E.3.1

	D_3, C_{3v}			D_3, C_{3v}	
a	$[a^2]$	(a^2)	a	$[a^2]$	(a^2)
A_1	A_1	—	E'	$A_2 + E$	A_1
A_2	A_1	—	P_1	A_2	—
E	$A_1 + E$	A_2	P_2	A_2	—

E.4. Bases for D_3 and C_{3v} and Trigonal Bases for O and T_d

The standard basis functions are identical to those of Butler (1981). They differ, however, in many respects from those of Griffith (1964) and those used previously by the present authors. In particular we note that the Butler coordinate systems differ for corresponding real and complex trigonal bases.

The coordinate systems for the octahedral and tetrahedral groups are those of Butler [(1981), Section 11.4]. Thus for $O \supset D_3 \supset C_3$ and $T_d \supset C_{3v} \supset C_3$ the Z and Y axes correspond respectively to the $C_3^{(-111)}$ and $C_2^{(110)}$ axes of $O \supset D_4 \supset C_4$ and $O \supset D_4 \supset D_2 \supset C_2$. The Z and Y axes for $O \supset D_3 \supset C_2$ and $T_d \supset C_{3v} \supset C_s$, however, differ; they correspond respectively to the $C_2^{(110)}$ and the $C_3^{(1-1-1)}$ axes for $O \supset D_4 \supset C_4$ and $O \supset D_4 \supset D_2 \supset C_2$.

The coordinate systems for the D_3 and C_{3v} bases may be deduced from the character table in Section E.1 and also from Table E.4.1(a), which shows the behavior of the functions x, y, and z under the group generators. For $D_3 \supset C_3$, the Z axis is the C_3 axis while the Y axis is a C_2 axis. In $C_{3v} \supset C_3$ the Z axis is again the C_3 axis, but σ_{XZ} (not σ_{YZ}) is one of the σ_v. Both $D_3 \supset C_2$ and $C_{3v} \supset C_s$ have the Y axis as the C_3 axis; however in $D_3 \supset C_2$, the Z axis is a C_2 axis, while in $C_{3v} \supset C_s$, σ_{XY} (not σ_{YZ}) is one of the σ_v. In all cases a right-handed coordinate system is used.

In Table E.4.1(b) we give representative functions which transform identically to our real-trigonal bases, $O \supset D_3 \supset C_2$ and $T_d \supset C_{3v} \supset C_s$. The same information for the $D_3 \supset C_2$ and $C_{3v} \supset C_s$ bases may be obtained from the table by dropping the first irrep label (a) from each $|aa_1a_2\rangle$ ket. Table E.4.1(c) defines the behavior of the complex-D_3 basis ($D_3 \supset C_3$) under the group generators. Then in Section E.5 the bases are further defined by specifying the $|j^{\pm}m\rangle$ of $O_3 \supset SO_3 \supset SO_2$ which have identical transformation properties to our standard basis functions.

We do not give a table relating the "real" and the "complex" trigonal bases for O, T_d, D_3, or C_{3v}. The reason is that the relations are complicated by the different coordinate systems used. (In this regard we note that the Butler [(1981), p. 544] table entitled "D_3–C_2 Partners as D_3–C_3 Partners" should not be used by itself to relate the "real" and "complex" trigonal bases, since it requires the axes in the two D_3 groups to be coincident.) We recommend in these cases that users chose either a real or complex basis for a given problem and stick to it; there is no compelling reason to move between bases.

E.5. Trigonal Bases Defined in Terms of the $O_3 \supset SO_3 \supset SO_2$ Basis

For an explanation of the notation and other comments, see Section C.6. As discussed there, $|j^+m\rangle$ and $|j^-m\rangle$ transform identically under group O but not under T_d. Thus the complex-trigonal basis chains (Table E.5.1), $O_3 \supset O_h \supset O \supset D_3 \supset C_3$ and $O_3 \supset O_h \supset T_d \supset C_{3v} \supset C_3$, are isomorphous for the j^+, but not for the j^-. In the complex-trigonal T_d chain

$$\begin{aligned} |1^-1^-\tilde{1}1-1\rangle &\equiv |1^-T_{1u}T_2E-1\rangle = -i|1^- -1\rangle \\ |1^-1^-\tilde{1}00\rangle &\equiv |1^-T_{1u}T_2A_1a_1\rangle = -i|1^-0\rangle \\ |1^-1^-\tilde{1}11\rangle &\equiv |1^-T_{1u}T_2E1\rangle = i|1^-1\rangle \end{aligned} \tag{E.5.1}$$

Table E.4.1(a). Behavior of x, y, and z Functions Under Group-D_3 and Group-C_{3v} Generators

	$D_3 \supset C_3$ Generator	
Function	C_3^Z	C_2^Y
x	$-\frac{1}{2}x + (\sqrt{3}/2)y$	$-x$
y	$-(\sqrt{3}/2)x - \frac{1}{2}y$	y
z	z	$-z$

	$D_3 \supset C_2$ Generator	
Function	C_3^Y	C_2^Z
x	$(-\sqrt{3}/2)z - \frac{1}{2}x$	$-x$
y	y	$-y$
z	$-\frac{1}{2}z + (\sqrt{3}/2)x$	z

	$C_{3v} \supset C_3$ Generator	
Function	C_3^Z	σ_{XZ}
x	$-\frac{1}{2}x + (\sqrt{3}/2)y$	x
y	$(-\sqrt{3}/2)x - \frac{1}{2}y$	$-y$
z	z	z

	$C_{3v} \supset C_s$ Generator	
Function	C_3^Y	σ_{XY}
x	$(-\sqrt{3}/2)z - \frac{1}{2}x$	x
y	y	y
z	$-\frac{1}{2}z + (\sqrt{3}/2)x$	$-z$

Table E.4.1(b). The Real-Trigonal Bases, $O_3 \supset D_3 \supset C_2$ and $T_d \supset C_{3v} \supset C_s$, for Single-Valued Irreps[a]

Functions Which Transform Identically to the		Real Trigonal Basis
$O \supset D_3 \supset C_2$ Basis	$T_d \supset C_{3v} \supset C_s$ Basis	$\|a(O)a_1(D_3)a_2(C_2)\rangle$ $\|a(T_d)a_1(C_{3v})a_2(C_s)\rangle$
		$\|000\rangle = \|A_1A_1a_1\rangle$
		$\|\tilde{0}\tilde{0}1\rangle = \|A_2A_2a_2\rangle$
		$\|211\rangle = \|EE\theta\rangle$
		$\|210\rangle = \|EE\epsilon\rangle$
$-x, -R_x$	$-R_x$	$\|111\rangle = \|T_1E\theta\rangle$
z, R_z	R_z	$\|110\rangle = \|T_1E\epsilon\rangle$
iy, iR_y	iR_y	$\|1\tilde{0}1\rangle = \|T_1A_2a_2\rangle$
	$-z$	$\|\tilde{1}11\rangle = \|T_2E\theta\rangle$
	$-x$	$\|\tilde{1}10\rangle = \|T_2E\epsilon\rangle$
	iy	$\|\tilde{1}00\rangle = \|T_2A_1a_1\rangle$

[a] The irrep notation is analogous to that of Section C.6. We label the partners of the E irrep of D_3 and C_{3v} as θ and ϵ (rather than as x and y) because they do not transform as x and y in the coordinate systems used. The transformation properties of the complex and real d orbitals may be easily determined from the relations in Sections E.5 and B.2.

Table E.4.1(c). Behavior of the Complex-D_3 Basis Under Group Generators

$D_3 \supset C_3$ Basis	Group Generators	
$\|a\alpha\rangle = \|a(D_3)a_1(C_3)\rangle$	C_3^Z	C_2^Y
$\|A_1a_1\rangle \equiv \|00\rangle$	a_1	a_1
$\|A_2a_2'\rangle \equiv \|\tilde{0}0\rangle$	a_2'	$-a_2'$
$\|E\,-1\rangle \equiv \|1\,-1\rangle$	$-\frac{1}{2}(1 - i\sqrt{3})\|E\,-1\rangle$	$\|E1\rangle$
$\|E1\rangle \equiv \|11\rangle$	$-\frac{1}{2}(1 + i\sqrt{3})\|E1\rangle$	$\|E\,-1\rangle$
$\|E'\beta'\rangle \equiv \|\frac{1}{2} - \frac{1}{2}\rangle$	$\frac{1}{2}(1 + i\sqrt{3})\beta'$	$-\alpha'$
$\|E'\alpha'\rangle \equiv \|\frac{1}{2}\frac{1}{2}\rangle$	$\frac{1}{2}(1 - i\sqrt{3})\alpha'$	β'
$\|P_1\rho_1\rangle \equiv \|-\frac{3}{2}\frac{3}{2}\rangle$	$-\rho_1$	$i\rho_1$
$\|P_2\rho_2\rangle \equiv \|\frac{3}{2}\frac{3}{2}\rangle$	$-\rho_2$	$-i\rho_2$

Likewise the real-trigonal bases (Table E.5.2), $O_3 \supset O_h \supset O \supset D_3 \supset C_2$ and $O_3 \supset O_h \supset T_d \supset C_{3v} \supset C_s$, are isomorphous for the j^+, but not for the j^- In the real-trigonal T_d chain

$$|1^-1^-\tilde{1}11\rangle \equiv |1^-T_{1u}T_2E\theta\rangle = -|1^-0\rangle = -z$$

$$|1^-1^-\tilde{1}10\rangle \equiv |1^-T_{1u}T_2E\epsilon\rangle = \frac{1}{\sqrt{2}}(|1^-1\rangle - |1^- -1\rangle) = -x \qquad \text{(E.5.2)}$$

$$|1^-1^-\tilde{1}00\rangle \equiv |1^-T_{1u}T_2A_1a_1\rangle = -\frac{1}{\sqrt{2}}(|1^-1\rangle + |1^- -1\rangle) = iy$$

Table E.5.1. Complex-Trigonal Bases

j	$\lvert j^+(O_3)a^+(O_h)a(O)a_1(D_3)a_2(C_3)\rangle$,[a] $\lvert j^+(O_3)a^+(O_h)a(T_d)a_1(C_{3v})a_2(C_3)\rangle$	$O_3 \supset SO_3 \supset SO_2$ $\lvert j^+m\rangle$ Equivalent
0	$\lvert 0^+0^+000\rangle = \lvert 0^+A_{1g}A_1A_1a_1\rangle$	$\lvert 0^+0\rangle$
1	$\lvert 1^+1^+11-1\rangle = \lvert 1^+T_{1g}T_1E-1\rangle$	$-\lvert 1^+-1\rangle$
	$\lvert 1^+1^+1\tilde{0}0\rangle = \lvert 1^+T_{1g}T_1A_2a_2'\rangle$	$\lvert 1^+0\rangle$
	$\lvert 1^+1^+111\rangle = \lvert 1^+T_{1g}T_1E1\rangle$	$-\lvert 1^+1\rangle$
2	$\lvert 2^+2^+21-1\rangle = \lvert 2^+E_gEE-1\rangle$	$(1/\sqrt{3})(\lvert 2^+2\rangle + \sqrt{2}\lvert 2^+-1\rangle)$
	$\lvert 2^+2^+211\rangle = \lvert 2^+E_gEE1\rangle$	$(1/\sqrt{3})(-\sqrt{2}\lvert 2^+1\rangle + \lvert 2^+-2\rangle)$
	$\lvert 2^+\tilde{1}^+\tilde{1}1-1\rangle = \lvert 2^+T_{2g}T_2E-1\rangle$	$(1/\sqrt{3})(-\sqrt{2}\lvert 2^+2\rangle + \lvert 2^+-1\rangle)$
	$\lvert 2^+\tilde{1}^+\tilde{1}00\rangle = \lvert 2^+T_{2g}T_2A_1a_1\rangle$	$-\lvert 2^+0\rangle$
	$\lvert 2^+\tilde{1}^+\tilde{1}11\rangle = \lvert 2^+T_{2g}T_2E1\rangle$	$(1/\sqrt{3})(-\lvert 2^+1\rangle - \sqrt{2}\lvert 2^+-2\rangle)$
3	$\lvert 3^+\tilde{0}^+\tilde{0}\tilde{0}0\rangle = \lvert 3^+A_{2g}A_2A_2a_2'\rangle$	$\frac{1}{3}(\sqrt{2}\lvert 3^+3\rangle - \sqrt{5}\lvert 3^+0\rangle - \sqrt{2}\lvert 3^+-3\rangle)$
	⋮	⋮
$\frac{1}{2}$	$\lvert \frac{1}{2}^+\frac{1}{2}^+\frac{1}{2}\frac{1}{2}-\frac{1}{2}\rangle = \lvert \frac{1}{2}^+E_g'E'E'\beta'\rangle$	$\lvert \frac{1}{2}^+-\frac{1}{2}\rangle$
	$\lvert \frac{1}{2}^+\frac{1}{2}^+\frac{1}{2}\frac{1}{2}\frac{1}{2}\rangle = \lvert \frac{1}{2}^+E_g'E'E'\alpha'\rangle$	$\lvert \frac{1}{2}^+\frac{1}{2}\rangle$
$\frac{3}{2}$	$\lvert \frac{3}{2}^+\frac{3}{2}^+\frac{3}{2}-\frac{3}{2}\frac{3}{2}\rangle = \lvert \frac{3}{2}^+U_g'U'P_1\rho_1\rangle$	$(1/\sqrt{2})(\lvert \frac{3}{2}^+\frac{3}{2}\rangle - i\lvert \frac{3}{2}^+-\frac{3}{2}\rangle)$
	$\lvert \frac{3}{2}^+\frac{3}{2}^+\frac{3}{2}\frac{1}{2}-\frac{1}{2}\rangle = \lvert \frac{3}{2}^+U_g'U'E'\beta'\rangle$	$-\lvert \frac{3}{2}^+-\frac{1}{2}\rangle$
	$\lvert \frac{3}{2}^+\frac{3}{2}^+\frac{3}{2}\frac{1}{2}\frac{1}{2}\rangle = \lvert \frac{3}{2}^+U_g'U'E'\alpha'\rangle$	$\lvert \frac{3}{2}^+\frac{1}{2}\rangle$
	$\lvert \frac{3}{2}^+\frac{3}{2}^+\frac{3}{2}\frac{3}{2}\frac{3}{2}\rangle = \lvert \frac{3}{2}^+U_g'U'P_2\rho_2\rangle$	$(1/\sqrt{2})(i\lvert \frac{3}{2}^+\frac{3}{2}\rangle - \lvert \frac{3}{2}^+-\frac{3}{2}\rangle)$
$\frac{5}{2}$	$\lvert \frac{5}{2}^+\tilde{\frac{1}{2}}^+\tilde{\frac{1}{2}}\frac{1}{2}-\frac{1}{2}\rangle = \lvert \frac{5}{2}^+E_g''E''E'\beta'\rangle$	$\frac{1}{3}(2\lvert \frac{5}{2}^+\frac{5}{2}\rangle - \sqrt{5}\lvert \frac{5}{2}^+-\frac{1}{2}\rangle)$
	$\lvert \frac{5}{2}^+\tilde{\frac{1}{2}}^+\tilde{\frac{1}{2}}\frac{1}{2}\frac{1}{2}\rangle = \lvert \frac{5}{2}^+E_g''E''E'\alpha'\rangle$	$\frac{1}{3}(-\sqrt{5}\lvert \frac{5}{2}^+\frac{1}{2}\rangle - 2\lvert \frac{5}{2}^+-\frac{5}{2}\rangle)$
	⋮	⋮

[a] For j^-a^-, replace g by u.

Table E.5.2. Real-Trigonal Bases

j	$\lvert j^{+}(\mathrm{O}_3)a^{+}(O_h)a(O)a_1(D_3)a_2(C_2)\rangle$,[a] $\lvert j^{+}(\mathrm{O}_3)a^{+}(O_h)a(T_d)a_1(C_{3v})a_2(C_s)\rangle$	$\mathrm{O}_3 \supset \mathrm{SO}_3 \supset \mathrm{SO}_2$ $\lvert j^{+}m\rangle$ Equivalent
0	$\lvert 0^{+}0^{+}000\rangle = \lvert 0^{+}A_{1g}A_1A_1a_1\rangle$	$\lvert 0^{+}0\rangle$
1	$\lvert 1^{+}1^{+}111\rangle = \lvert 1^{+}T_{1g}T_1E\theta\rangle$	$(1/\sqrt{2})(\lvert 1^{+}1\rangle - \lvert 1^{+}-1\rangle)$
	$\lvert 1^{+}1^{+}110\rangle = \lvert 1^{+}T_{1g}T_1E\epsilon\rangle$	$\lvert 1^{+}0\rangle$
	$\lvert 1^{+}1^{+}1\tilde{0}1\rangle = \lvert 1^{+}T_{1g}T_1A_2a_2\rangle$	$(-1/\sqrt{2})(\lvert 1^{+}1\rangle + \lvert 1^{+}-1\rangle)$
2	$\lvert 2^{+}2^{+}211\rangle = \lvert 2^{+}E_gEE\theta\rangle$	$(-1/\sqrt{3} - i/\sqrt{6})\lvert 2^{+}1\rangle + (-1/\sqrt{3} + i/\sqrt{6})\lvert 2^{+}-1\rangle$
	$\lvert 2^{+}2^{+}210\rangle = \lvert 2^{+}E_gEE\epsilon\rangle$	$(1/\sqrt{3} - i/2\sqrt{6})\lvert 2^{+}2\rangle + (i/2)\lvert 2^{+}0\rangle + (-1/\sqrt{3} - i/2\sqrt{6})\lvert 2^{+}-2\rangle$
	$\lvert 2^{+}\tilde{1}^{+}\tilde{1}11\rangle = \lvert 2^{+}T_{2g}T_2E\theta\rangle$	$(-1/\sqrt{6} + i/\sqrt{3})\lvert 2^{+}1\rangle + (-1/\sqrt{6} - i/\sqrt{3})\lvert 2^{+}-1\rangle$
	$\lvert 2^{+}\tilde{1}^{+}\tilde{1}10\rangle = \lvert 2^{+}T_{2g}T_2E\epsilon\rangle$	$(1/\sqrt{6} + i/2\sqrt{3})\lvert 2^{+}2\rangle - (i/\sqrt{2})\lvert 2^{+}0\rangle + (-1/\sqrt{6} + i/2\sqrt{3})\lvert 2^{+}-2\rangle$
	$\lvert 2^{+}\tilde{1}^{+}\tilde{1}00\rangle = \lvert 2^{+}T_{2g}T_2A_1a_1\rangle$	$(\sqrt{3}/2\sqrt{2})\lvert 2^{+}2\rangle + \frac{1}{2}\lvert 20\rangle + (\sqrt{3}/2\sqrt{2})\lvert 2^{+}-2\rangle$
3	$\lvert 3^{+}\tilde{0}^{+}\tilde{0}\tilde{0}1\rangle = \lvert 3^{+}A_{2g}A_2A_2a_2\rangle$	$(-\frac{5}{12} - i/6\sqrt{2})\lvert 3^{+}3\rangle + (-\sqrt{5}/4\sqrt{3} + i\sqrt{5}/2\sqrt{6})\lvert 3^{+}1\rangle$ $+(-\sqrt{5}/4\sqrt{3} - i\sqrt{5}/2\sqrt{6})\lvert 3^{+}-1\rangle + (-\frac{5}{12} + i/6\sqrt{2})\lvert 3^{+}-3\rangle$
	⋮	
$\frac{1}{2}$	$\lvert \frac{1}{2}^{+}\frac{1}{2}^{+}\frac{1}{2}\frac{1}{2}-\frac{1}{2}\rangle$	$\lvert \frac{1}{2}^{+}-\frac{1}{2}\rangle$
	$\lvert \frac{1}{2}^{+}\frac{1}{2}^{+}\frac{1}{2}\frac{1}{2}\frac{1}{2}\rangle$	$\lvert \frac{1}{2}^{+}\frac{1}{2}\rangle$
$\frac{3}{2}$	$\lvert \frac{3}{2}^{+}\frac{3}{2}^{+}\frac{3}{2}-\frac{3}{2}-\frac{1}{2}\rangle$	$\frac{1}{2}\lvert \frac{3}{2}^{+}\frac{3}{2}\rangle - (\sqrt{3}/2)\lvert \frac{3}{2}^{+}-\frac{1}{2}\rangle$
	$\lvert \frac{3}{2}^{+}\frac{3}{2}^{+}\frac{3}{2}\frac{1}{2}-\frac{1}{2}\rangle$	$(\sqrt{3}/2)\lvert \frac{3}{2}^{+}\frac{3}{2}\rangle + \frac{1}{2}\lvert \frac{3}{2}^{+}-\frac{1}{2}\rangle$
	$\lvert \frac{3}{2}^{+}\frac{3}{2}^{+}\frac{3}{2}\frac{1}{2}\frac{1}{2}\rangle$	$-\frac{1}{2}\lvert \frac{3}{2}^{+}\frac{1}{2}\rangle - (\sqrt{3}/2)\lvert \frac{3}{2}^{+}-\frac{3}{2}\rangle$
	$\lvert \frac{3}{2}^{+}\frac{3}{2}^{+}\frac{3}{2}\frac{3}{2}\frac{1}{2}\rangle$	$(\sqrt{3}/2)\lvert \frac{3}{2}^{+}\frac{1}{2}\rangle - \frac{1}{2}\lvert \frac{3}{2}^{+}-\frac{3}{2}\rangle$
$\frac{5}{2}$	$\lvert \frac{5}{2}^{+}\tilde{\frac{1}{2}}^{+}\tilde{\frac{1}{2}}\frac{1}{2}-\frac{1}{2}\rangle$	$(\sqrt{5}/6\sqrt{2} - i\sqrt{5}/6)\lvert \frac{5}{2}^{+}\frac{3}{2}\rangle + (\sqrt{5}/6 + i\sqrt{5}/3\sqrt{2})\lvert \frac{5}{2}^{+}-\frac{1}{2}\rangle$ $+(5/6\sqrt{2} - i/6)\lvert \frac{5}{2}^{+}-\frac{5}{2}\rangle$
	$\lvert \frac{5}{2}^{+}\tilde{\frac{1}{2}}^{+}\tilde{\frac{1}{2}}\frac{1}{2}\frac{1}{2}\rangle$	$(5/6\sqrt{2} + i/6)\lvert \frac{5}{2}^{+}\frac{5}{2}\rangle + (\sqrt{5}/6 - i\sqrt{5}/3\sqrt{2})\lvert \frac{5}{2}^{+}\frac{1}{2}\rangle$ $+(\sqrt{5}/6\sqrt{2} + i\sqrt{5}/6)\lvert \frac{5}{2}^{+}-\frac{3}{2}\rangle$
	⋮	

[a] For $j^{-}a^{-}$, replace g by u.

These relations are obtained using the method of Butler [(1981), Section 5.3], as further clarified in Reid and Butler (1982).

Table E.5.1 follows from Butler (1981), page 522, and Table E.5.2 from unpublished tables of Butler and Reid. Recall that the real and complex bases have different symmetry axes (see Section E.4).

E.6. Function and Operator Transformation Coefficients

The coefficients in Table E.6.1 are those defined in Section 9.8. The notation is similar to that of Section C.7, except that basis states are specified here only by chain notation (Section 15.2). To obtain the transformation coefficients for the D_3 and C_{3v} bases, simply drop the first irrep label (a or f) everywhere. Operators transform as $O^{f}_{f_1 f_2}$ in the O and T_d bases, and as $O^{f_1}_{f_2}$ in the D_3 and C_{3v} bases. For d-function transformations use Section E.5.

Table E.6.1

ψ or Θ	$\langle a a_1 a_2 \mid \psi \rangle$ or $\langle f f_1 f_2 \mid \Theta \rangle$	$\mid \psi a a_1 a_2 \rangle$, $O^{f}_{f_1 f_2}$, or $O^{f f_1}_{f_2}$ — $a a_1 a_2$ or $f f_1 f_2$
	Real-Trigonal Basis $O \supset D_3 \supset C_2$	
x, p_x, V_x, R_x	-1	$111 = T_1 E\theta$
y, p_y, V_y, R_y	$-i$	$1\tilde{0}1 = T_1 A_2 a_2$
z, p_z, V_z, R_z	1	$110 = T_1 E\epsilon$
	Real-Trigonal Basis $T_d \supset C_{3v} \supset C_s$[a]	
x, p_x, V_x	-1	$\tilde{1}10 = T_2 E\epsilon$
y, p_y, V_y	$-i$	$\tilde{1}00 = T_2 A_1 a_1$
z, p_z, V_z	-1	$\tilde{1}11 = T_2 E\theta$
	Complex-Trigonal Basis $O \supset D_3 \supset C_3$	
$\mid 1^{\pm} -1 \rangle, V_{-1}, R_{-1}$	-1	$11-1 = T_1 E -1$
$\mid 1^{\pm} 0 \rangle, V_0, R_0$	1	$1\tilde{0}0 = T_1 A_2 a_2'$
$\mid 1^{\pm} 1 \rangle, V_1, R_1$	-1	$111 = T_1 E1$
	Complex-Trigonal Basis $T_d \supset C_{3v} \supset C_3$[b]	
$\mid 1^{-} -1 \rangle, V_{-1}$	i	$\tilde{1}1-1 = T_2 E -1$
$\mid 1^{-} 0 \rangle, V_0$	i	$\tilde{1}00 = T_2 A_1 a_1$
$\mid 1^{-} 1 \rangle, V_1$	$-i$	$\tilde{1}11 = T_2 E1$

[a] **R** transforms as in $O \supset D_3 \supset C_2$.
[b] **R** and the $\mid 1^{+} m \rangle$ for $m = -1, 0, 1$ transform as in $O \supset D_3 \supset C_3$.

E.7. Symmetry-Adapted Molecular Orbitals for Trigonal Planar ML_3

In Table E.7.1 we give symmetry-adapted metal and ligand orbitals for a trigonal planar ML_3 complex which belongs to the group D_3. The D_3 coordinate system used has the Z axis as the C_3 axis and the Y axis as one of the three C_2 axes. The ligands are arranged symmetrically about the coordinate origin with ligand 1 in the $+X, -Y$ quadrant, ligand 2 along the $+Y$ axis, and ligand 3 in the $-X, -Y$ quadrant. In each case the ligand axes X_i, Y_i, and Z_i are parallel respectively to X, Y, and Z. The x_i, y_i, and so on should be interpreted as in Section C.10. d_0 transforms as Y_0^2 (Section B.1), p_0 as Y_0^1, and so on.

The $D_3 \supset C_3$ basis is that of Tables E.4.1(a) and E.4.1(c). We have taken advantage of the fact that in $D_3 \supset C_3$ (unlike $O \supset D_3 \supset C_3$—see Section E.5) each of the sets of $SO_3 \supset SO_2$ kets, $(|2\,-1\rangle, -|21\rangle)$ and $(|22\rangle, |2\,-2\rangle)$, transforms *independently* as the $D_3 \supset C_3$ basis set $(|E\,-1\rangle, |E1\rangle)$.

Table E.7.1

$D_3 \supset C_3$	Metal Orbitals	Ligand Orbitals
$\|A_1a_1\rangle$	$(n+1)s, -nd_0$	$(1/\sqrt{3})(s_1+s_2+s_3)$, $(1/2\sqrt{3})[-\sqrt{3}(x_1-x_3)+y_1-2y_2+y_3]$
$\|A_2a_2\rangle$	$(n+1)p_0$	$(1/\sqrt{3})(z_1+z_2+z_3)$, $(1/2\sqrt{3})[x_1-2x_2+x_3+\sqrt{3}(y_1-y_3)]$
$\|E\,-1\rangle$	$-(n+1)p_{-1}$, nd_{-1}, nd_2	$(-1/\sqrt{2})(\phi_1-i\phi_2)^a$
$\|E1\rangle$	$-(n+1)p_1$, $-nd_1, nd_{-2}$	$(1/\sqrt{2})(\phi_1+i\phi_2)^a$

ϕ_1	ϕ_2
$(1/\sqrt{2})(s_1-s_3)$	$(1/\sqrt{6})(-s_1+2s_2-s_3)$
$(1/\sqrt{3})(x_1+x_2+x_3)$	$(1/\sqrt{3})(y_1+y_2+y_3)$
$(1/2\sqrt{3})[x_1-2x_2+x_3-\sqrt{3}(y_1-y_3)]$	$(1/2\sqrt{3})[-\sqrt{3}(x_1-x_3)-y_1+2y_2-y_3]$
$(1/\sqrt{6})(z_1-2z_2+z_3)$	$(1/\sqrt{2})(z_1-z_3)$

[a] For each (ϕ_1, ϕ_2) set listed at the bottom.

E.8. $3j$, $2j$, and $2jm$ Phases for D_3 and C_{3v} and Related Definitions

The bases are defined in Section E.4.

1. The $3j$ phase $\{abc\}$: All $3j$ phases are by definition independent of the order of the irreps. With the exception of $\{EEE\}$ all $3j$ phases for D_3 are defined by

$$\{abc\} = (-1)^a(-1)^b(-1)^c$$

where

$$(-1)^{A_1} = 1$$

$$(-1)^{A_2} = (-1)^E = -1$$

$$(-1)^{E'} = i$$

$$(-1)^{P_1} = (-1)^{P_2} = -i$$

The law-breaking $3j$ phase has the value

$$\{EEE\} = 1$$

2. The $2j$ phase $\{a\}$:

$$\{a\} = \begin{cases} 1 & \text{for } A_1, A_2, \text{ and } E \\ -1 & \text{for } E', P_1, \text{ and } P_2 \end{cases}$$

3. The $2jm$ $\begin{pmatrix} a \\ \alpha \end{pmatrix}$ and definitions of a^* and α^*: See Tables E.8.1 and E.8.2.

Table E.8.1. $2jm$ in the Real-D_3 and Real-C_{3v} Bases for Single-Valued Irreps[a]

$D_3 \supset C_2$ and $C_{3v} \supset C_s$		
$\lvert a\alpha\rangle \equiv \lvert aa_1\rangle$		$\begin{pmatrix} a \\ \alpha \end{pmatrix} = \begin{pmatrix} a \\ a_1 \end{pmatrix}$
A_1a_1	00	1
A_2a_2	$\tilde{0}1$	1
$E\theta$	11	-1
$E\epsilon$	10	-1

[a] $a = a^*$, $\alpha = \alpha^*$, and $a_1 = a_1^*$ for all a, α, and a_1.

Table E.8.2. $2jm$ in the Complex-D_3 and Complex-C_{3v} Bases

$D_3 \supset C_3$ and $C_{3v} \supset C_3$				
$\|a\alpha\rangle \equiv \|aa_1\rangle$		$\begin{pmatrix} a \\ \alpha \end{pmatrix} = \begin{pmatrix} a \\ a_1 \end{pmatrix}$	$\|a^*\alpha^*\rangle \equiv \|a^*a_1^*\rangle$	
$A_1 a_1$	00	1	$A_1 a_1$	00
$A_2 a_2'$	$\tilde{0}0$	−1	$A_2 a_2'$	$\tilde{0}0$
$E-1$	$1-1$	1	$E1$	11
$E1$	11	1	$E-1$	$1-1$
$E'\beta'$	$\frac{1}{2}-\frac{1}{2}$	−1	$E'\alpha'$	$\frac{1}{2}\frac{1}{2}$
$E'\alpha'$	$\frac{1}{2}\frac{1}{2}$	1	$E'\beta'$	$\frac{1}{2}-\frac{1}{2}$
$P_1\rho_1$	$-\frac{3}{2}\frac{3}{2}$	−1	$P_2\rho_2$	$\frac{3}{2}\frac{3}{2}$
$P_2\rho_2$	$\frac{3}{2}\frac{3}{2}$	1	$P_1\rho_1$	$-\frac{3}{2}\frac{3}{2}$

E.9. $3jm$ for D_3 and C_{3v} in the Bases of Section E.4

See Tables E.9.1 and E.9.2. The bases are defined in Section E.4; otherwise the comments of Section C.12 also apply here.

Table E.9.1. Real Basis for Single-Valued Irreps: $3jm$ for $D_3 \supset C_2$ and $C_{3v} \supset C_s$

A_1	A_1	A_1	$3jm$
a_1	a_1	a_1	1

A_2	A_2	A_1	$3jm$
a_2	a_2	a_1	1

E	E	A_1	$3jm$
θ	θ	a_1	$-1/\sqrt{2}$
ϵ	ϵ	a_1	$-1/\sqrt{2}$

E	E	A_2	$3jm^{\dagger}$
θ	ϵ	a_2	$1/\sqrt{2}$

E	E	E	$3jm$
θ	θ	ϵ	$-i/2$
ϵ	ϵ	ϵ	$i/2$

E.10. $6j$ for D_3 and C_{3v}

The $6j$ in Table E.10.1 apply for both D_3 and C_{3v} bases defined in Section E.4. All $6j$ containing one or more A_1 irreps are given by (16.1.11) and are not tabulated:

$$\begin{Bmatrix} A_1 & b & c \\ d & e & f \end{Bmatrix} = \delta_{bc*}\delta_{ef}\,\delta(dbf^*)|b|^{-1/2}|e|^{-1/2}\{dbf^*\} \quad \text{(E.10.1)}$$

Table E.9.2. Complex Basis: 3*jm* for $D_3 \supset C_3$ and $C_{3v} \supset C_3$

A_1	A_1	A_1	$3jm$
a_1	a_1	a_1	1

A_2	A_2	A_1	$3jm$
a_2'	a_2'	a_1	-1

E	E	A_1	$3jm$
-1	1	a_1	$1/\sqrt{2}$

E	E	A_2	$3jm^\dagger$
-1	1	a_2'	$-1/\sqrt{2}$

E	E	E	$3jm$
-1	-1	-1	$1/\sqrt{2}$
1	1	1	$1/\sqrt{2}$

E'	E'	A_1	$3jm^\dagger$
β'	α'	a_1	$-1/\sqrt{2}$

E'	E'	A_2	$3jm$
β'	α'	a_2'	$1/\sqrt{2}$

E'	E'	E	$3jm$
β'	β'	1	$1/\sqrt{2}$
α'	α'	-1	$1/\sqrt{2}$

P_1	E'	E	$3jm^\dagger$
ρ_1	α'	1	$-i/\sqrt{2}$
ρ_1	β'	-1	$-1/\sqrt{2}$

P_1	P_1	A_2	$3jm$
ρ_1	ρ_1	a_2'	$-i$

P_2	E'	E	$3jm^\dagger$
ρ_2	α'	1	$-1/\sqrt{2}$
ρ_2	β'	-1	$-i/\sqrt{2}$

P_2	P_1	A_1	$3jm^\dagger$
ρ_2	ρ_1	a_1	1

P_2	P_2	A_2	$3jm$
ρ_2	ρ_2	a_2'	$-i$

Table E.10.1. 6*j* for D_3 and C_{3v}

A_2	E	E	$6j$
A_2	E	E	$\frac{1}{2}$
E	E	E	$\frac{1}{2}$
E'	E'	E'	$-\frac{1}{2}$
E'	P_1	P_2	$1/\sqrt{2}$
P_1	E'	E'	$-\frac{1}{2}$
P_2	E'	E'	$-\frac{1}{2}$

A_2	E'	E'	$6j$
A_2	E'	E'	$-\frac{1}{2}$
E	E'	E'	$\frac{1}{2}$
E	P_1	P_2	$1/\sqrt{2}$

A_2	P_1	P_1	$6j$
A_2	P_1	P_2	-1

E	E	E	$6j$
E	E	E	0
E'	E'	E'	$0^\dagger$
E'	E'	P_1	$-i/2^\dagger$
E'	E'	P_2	$i/2^\dagger$

E	E'	E'	$6j$
E	E'	E'	0
E	E'	P_1	$-\frac{1}{2}$
E	E'	P_2	$-\frac{1}{2}$

E	E'	P_1	$6j$
E	E'	P_1	$\frac{1}{2}$
E	E'	P_2	$-\frac{1}{2}$

E	E'	P_2	$6j$
E	E'	P_2	$\frac{1}{2}$

Other $6j$ which are neither found in the table nor related by the symmetry rules of Section 16.1 are zero by (16.1.2). The $6j$ are listed as in Section C.13. Those $6j$ which change sign under odd permutation of their columns are followed by a dagger (†). Thus the dagger denotes that $(-1)^n = -1$ in (16.1.6). Note that odd permutation of columns also changes d, e, and f to d^*, e^*, and f^*. Complex conjugation of irreps also accompanies other symmetry operations—see Section 16.1.

E.11. $9j$ for D_3 and C_{3v} for Single-Valued Irreps

$9j$ containing one or more A_1 irreps may be calculated with (16.4.8). Other $9j$ either are zero, or are given below, or are related by the symmetry rules of Section 16.4 to (16.4.8) or to those given below:

$$\begin{Bmatrix} A_2 & E & E \\ E & A_2 & E \\ E & E & A_2 \end{Bmatrix} = -\frac{1}{4} \qquad \begin{Bmatrix} A_2 & E & E \\ E & A_2 & E \\ E & E & E \end{Bmatrix} = \frac{1}{4} \qquad \begin{Bmatrix} A_2 & E & E \\ E & E & E \\ E & E & E \end{Bmatrix} = 0$$

$$\begin{Bmatrix} E & E & E \\ E & E & E \\ E & E & E \end{Bmatrix} = \frac{1}{4}$$

E.12. $2jm$, $3jm$, and Other Coefficients for O and T_d in the Trigonal Bases

$2jm$ and $3jm$ for the groups O and T_d in the trigonal bases are found as a product of $2jm$ and $3jm$ factors (Section 15.3). The $O \supset D_3$ (and $T_d \supset C_{3v}$) $2jm$ and $3jm$ factors are given in Butler [(1981), pp. 242–243], and those for $D_3 \supset C_2$ (and $C_{3v} \supset C_s$) and $D_3 \supset C_3$ (and $C_{3v} \supset C_3$) are given in Butler [(1981), pp. 221–222] (and also in our Sections E.8 and E.9). $2j$ and $3j$ phases (Section C.11) and all coefficients in Sections C.13–C.17 are basis-independent; thus they also apply for groups O and T_d in the trigonal basis.

Appendix F Tables for the Groups D_∞, $C_{\infty v}$, and D_6

The tables which follow are based on the equations of Butler and Reid (1979) and Butler (1981). Butler and Reid (1979) give a character table for D_∞. D_∞ is isomorphic with $C_{\infty v}$ but has symmetry operations $C_2(\phi)$ rather than $\sigma_v(\phi)$, where $0 \leqslant \phi \leqslant \pi$. The Butler D_∞ irrep notation is used: $0 \equiv \Sigma^+$, $\tilde{0} \equiv \Sigma^-$, $1 \equiv \Pi$, $2 \equiv \Delta$, $\frac{1}{2} = E_{1/2}$, $\frac{3}{2} = E_{3/2}$, and so on. All irreps except 0 and $\tilde{0}$ are two-dimensional. In the footnote to Table F.3.1 we follow Butler (1981) who defines $D_{\infty h}$ as $D_\infty \otimes C_i$. Note that the standard definition, $D_{\infty h} = C_{\infty v} \otimes C_i$, leads to *different basis labels* for the $|j^- m\rangle$.

F.1. Direct Products in D_∞ and $C_{\infty v}$

The irreps a and b are any two-dimensional irreps (1, 2, $\frac{1}{2}$, $\frac{3}{2}$, etc.). The direct product $a \otimes b$ is for $a \neq b$. In Table F.1.1, $(2a)$, $(a + b)$ and $|a - b|$ represent the numerical values calculated using the numerical irrep labels.

Table F.1.1

$D_\infty, C_{\infty v}$	0	$\tilde{0}$	a	b
0	0	$\tilde{0}$	a	b
$\tilde{0}$	$\tilde{0}$	0	a	b
a	a	a	$0 + \tilde{0} + (2a)$	$(a + b) + \|a - b\|$
b	b	b	$(a + b) + \|a - b\|$	$0 + \tilde{0} + (2a)$

F.2. Table of Symmetrized $[a^2]$ and Antisymmetrized (a^2) Squares for D_∞ and $C_{\infty v}$

In Table F.2.1 $(2a)$ represents the numerical value of $2 \times a$.

Table F.2.1

	$D_\infty, C_{\infty v}$	
a	$[a^2]$	(a^2)
0	0	—
$\tilde{0}$	0	—
a = integer	$0, (2a)$	$\tilde{0}$
a = half integer	$\tilde{0}, (2a)$	0

F.3. Basis for $D_\infty \supset C_\infty$ Defined in Terms of the $SO_3 \supset SO_2$ Basis

Table F.3.1

j	$\vert j(SO_3)\, j_1(D_\infty)\, a_1(C_\infty)\rangle$ $= \vert j(SO_3)\, j_1(D_\infty)\, m(SO_2)\rangle$	$SO_3 \supset SO_2$ $\vert jm\rangle$ Equivalent
0	$\vert 000\rangle$	$\vert 00\rangle$
1	$\vert 1\tilde{0}0\rangle$	$\vert 10\rangle$
	$\vert 11\,-1\rangle$	$\vert 1\,-1\rangle$
	$\vert 111\rangle$	$\vert 11\rangle$
2	$\vert 200\rangle$	$\vert 20\rangle$
	$\vert 21\,-1\rangle$	$\vert 2\,-1\rangle$
	$\vert 211\rangle$	$-\,\vert 21\rangle$
	$\vert 22\,-2\rangle$	$\vert 2\,-2\rangle$
	$\vert 222\rangle$	$\vert 22\rangle$
⋮	⋮[a]	⋮
$\frac{1}{2}$	$\vert \frac{1}{2}\frac{1}{2}-\frac{1}{2}\rangle$	$\vert \frac{1}{2}-\frac{1}{2}\rangle$
	$\vert \frac{1}{2}\frac{1}{2}\frac{1}{2}\rangle$	$\vert \frac{1}{2}\frac{1}{2}\rangle$
$\frac{3}{2}$	$\vert \frac{3}{2}\frac{1}{2}-\frac{1}{2}\rangle$	$\vert \frac{3}{2}-\frac{1}{2}\rangle$
	$\vert \frac{3}{2}\frac{1}{2}\frac{1}{2}\rangle$	$-\,\vert \frac{3}{2}\frac{1}{2}\rangle$
	$\vert \frac{3}{2}\frac{3}{2}-\frac{3}{2}\rangle$	$\vert \frac{3}{2}-\frac{3}{2}\rangle$
	$\vert \frac{3}{2}\frac{3}{2}\frac{3}{2}\rangle$	$\vert \frac{3}{2}\frac{3}{2}\rangle$
⋮	⋮[a]	⋮

[a] For even j, $j(SO_3) \to 0(D_\infty) + 1(D_\infty) + \cdots + j(D_\infty)$. For odd j, $j(SO_3) \to \tilde{0}(D_\infty) + 1(D_\infty) + \cdots + j(D_\infty)$. For half-integer j, $j \to \frac{1}{2} + \frac{3}{2} + \cdots + j$. Then $\vert j(SO_3)\ j_1(D_\infty) m(SO_2)\rangle = \langle jm \vert jj_1m\rangle \vert j(SO_3) m(SO_2)\rangle$, where $\langle \frac{1}{2}\frac{1}{2} \vert \frac{1}{2}\frac{1}{2}\frac{1}{2}\rangle = \langle j0 \vert j00\rangle = \langle j0 \vert j\tilde{0}0\rangle = 1$, $\langle jm \vert jj_1m\rangle = \delta_{mj_1}(-1)^{j-j_1}$ for $m > 0$, and $\langle jm \vert jj_1m\rangle = \delta_{-mj_1}$ for $m < 0$. These relations also relate the $\vert j^{\pm}j_1^{\pm}m\rangle$ of $O_3 \supset D_\infty \otimes C_i \supset C_\infty$ to the $\vert j^{\pm}m\rangle$ of $O_3 \supset SO_3 \supset SO_2$.

F.4. Phases and Coefficients for D_∞ and $C_{\infty v}$

1. $3j$ phase: In all cases $\{abc\} = (-1)^a(-1)^b(-1)^c$, where $(-1)^{\tilde{0}} \equiv -1$; otherwise $(-1)^a$ has its numerical value [i.e. $(-1)^0 = 1$, $(-1)^{1/2} = i$, $(-1)^1 = -1$, $(-1)^{3/2} = -i$, etc.].
2. $2jm$ phase:

$$\begin{pmatrix} \tilde{0} \\ 0 \end{pmatrix} = -1; \qquad \begin{pmatrix} a \\ \alpha \end{pmatrix} = (-1)^{a-\alpha} \quad \text{for all } a \neq \tilde{0},$$

where the exponent $a - \alpha$ has its numerical value. As in $SO_3 \supset SO_2$, we have $a^* = a$ and $\alpha^* = -\alpha$.

3. $3jm$ in the $D_\infty \supset SO_2 = D_\infty \supset C_\infty$ basis:

$$\begin{pmatrix} a & b & 0 \\ \alpha & \beta & 0 \end{pmatrix} = \delta_{ba}\delta_{\beta,-\alpha}|a|^{-1/2}\begin{pmatrix} a \\ \alpha \end{pmatrix}$$

$$\left.\begin{aligned} \begin{pmatrix} a & a & \tilde{0} \\ a & -a & 0 \end{pmatrix} &= \frac{1}{\sqrt{2}} \\ \begin{pmatrix} a+b & a & b \\ a+b & -a & -b \end{pmatrix} &= \frac{1}{\sqrt{2}}(-1)^{2b} \end{aligned}\right\} \quad \text{for} \quad a, b \neq 0 \text{ or } \tilde{0}$$

4. $6j$: Eq. (16.1.11) may be used to calculate $6j$ containing the 0 irrep. Other nonzero $6j$ may be found from the equations of Butler and Reid (1979) given below:

$$\begin{Bmatrix} 0 & \tilde{0} & \tilde{0} \\ 0 & \tilde{0} & \tilde{0} \end{Bmatrix} = \begin{Bmatrix} 0 & \tilde{0} & \tilde{0} \\ \tilde{0} & 0 & 0 \end{Bmatrix} = 1$$

For all $a, b, c \neq 0$ or $\tilde{0}$,

$$\begin{Bmatrix} a & b+\frac{1}{2} & a+b+\frac{1}{2} \\ b & a+\frac{1}{2} & \frac{1}{2} \end{Bmatrix} = \begin{Bmatrix} a+b & a & b \\ a & a+b & \tilde{0} \end{Bmatrix} = \tfrac{1}{2} \times (-1)^{2a+2b+1}$$

$$\begin{Bmatrix} a+b & a & b \\ a-c & a+b-c & c \end{Bmatrix} = \tfrac{1}{2} \times (-1)^{2a+2b}$$

$$\begin{Bmatrix} a+b & a & b \\ \tilde{0} & b & a \end{Bmatrix} = \tfrac{1}{2} \times (-1)^{2a+2b}$$

$$\begin{Bmatrix} a & a & \tilde{0} \\ a & a & \tilde{0} \end{Bmatrix} = \tfrac{1}{2} \times (-1)^{2a}$$

As always, (16.1.2) and the symmetry rules of Section 16.1 apply.

F.5. Tables for D_6

Butler (1981) gives tables for D_6 in a variety of bases, and we do not duplicate them here. The table entitled "SO_3–D_∞–D_6–C_6 Partners as *JM* Partners" of Butler [(1981), p. 541] relates the $SO_3 \supset D_\infty \supset D_6 \supset C_6$ basis to the $SO_3 \supset SO_2$ $|jm\rangle$. From this table (which is analogous to the tables of Sections C.6 and E.5) it is straightforward to determine function and operator transformation coefficients as discussed in Sections 9.8 and C.7. Then on p. 546 of Butler (1981) the table entitled "D_6–D_3–C_3 Partners as D_6–C_6 Partners" may be used to obtain basis information for the $D_6 \supset D_3 \supset C_3$ chain, since in this case coordinate systems have been chosen so that the two D_6 bases share common symmetry axes (compare Butler [(1981), Tables 11.8 and 11.11] for these bases).

References

Abragam, A., and B. Bleaney (1970), *Electron Paramagnetic Resonance of Transition Ions*, Clarendon Press, Oxford.

Andrews, D. L., and T. Thirunamachandran (1977), "On Three-Dimensional Rotational Averages," *J. Chem. Phys.* **67**, 5026.

Ballhausen, C. J. (1962), *Introduction to Ligand Field Theory*, McGraw-Hill, New York.

Ballhausen, C. J., and H. B. Gray (1964), *Molecular Orbital Theory*, W. A. Benjamin, New York.

Ballhausen, C. J., and A. E. Hansen (1972), "Electronic Spectra," *Ann. Rev. Phys. Chem.* **23**, 15.

Bishop, D. M. (1973), *Group Theory and Chemistry*, Clarendon Press, Oxford.

Born, M., and E. Wolf (1959), *Principles of Optics*, Pergamon Press, Oxford.

Böttcher, C. J. F. (1952) *Theory of Electric Polarization*, Elsevier, Amsterdam.

Brink, D. M., and G. R. Satchler (1968), *Angular Momentum*, 2nd edition, Clarendon Press, Oxford.

Buckingham, A. D., and P. J. Stephens (1966), "Magnetic Optical Activity," *Ann. Rev. Phys. Chem.* **17**, 399.

Bunker, P. R. (1979), *Molecular Symmetry and Spectroscopy*, Academic Press, New York.

Butler, P. H. (1975), "Coupling Coefficients and Tensor Operators for Chains of Groups," *Phil. Trans. R. Soc. London* **277**, 545.

Butler, P. H. (1981), *Point Group Symmetry Applications: Methods and Tables*, Plenum Press, New York.

Butler, P. H., and M. F. Reid (1979), "*j* Symbols and *jm* Factors for All Dihedral and Cyclic Groups," *J. Phys. A* **12**, 1655.

Ciardelli, F., and P. Salvadori, Eds. (1973), *Fundamental Aspects and Recent Developments in ORD and CD*; *Proceedings of the NATO Advanced Study Institute*, Heyden & Sons Ltd., New York.

Collingwood, J. C. (1974), private communication.

Collingwood, J. C., R. W. Schwartz, P. N. Schatz, and H. H. Patterson (1974), "Magnetic Circular Dichroism and Absorption Spectra of $Cs_2ZrCl_6 : Mo^{4+}$," *Mol. Phys.* **27**, 1291.

Collingwood, J. C., S. B. Piepho, R. W. Schwartz, P. A. Dobosh, J. R. Dickinson, and P. N. Schatz (1975), "The Optical Absorption and Magnetic Circular Dichroism Spectra of Re^{4+} in the Cubic Hosts, Cs_2ZrCl_6, Cs_2NaYCl_6, and Cs_2ZrBr_6," *Mol. Phys.* **29**, 793.

Cotton, F. A. (1971), *Chemical Applications of Group Theory*, 2nd edition, Wiley-Interscience, New York.

Day, P., Ed. (1975), *Electronic States of Inorganic Compounds: New Experimental Techniques*, *Proceedings of the NATO Advanced Study Institute*, D. Reidel, Dordrecht, Holland.

Denning, R. G. (1975), "The Technique of Magnetic Circular Dichroism," in P. Day, Ed., *Electronic States of Inorganic Compounds: New Experimental Techniques*, *Proceedings of the NATO Advanced Study Institute*, D. Reidel, Dordrecht, Holland, p. 157.

Denning, R. G., and J. A. Spencer (1969), "Magnetic Circular Dichroism of the Tetrachlorocobaltate Ion," *Symposia of the Faraday Society*, **1969**, 84.

Dickinson, J. R., S. B. Piepho, J. A. Spencer, and P. N. Schatz (1972), "High Resolution Magnetic Circular Dichroism and Absorption Spectra of $Cs_2ZrBr_6 : Ir^{4+}$," *J. Chem. Phys.* **56**, 2668.

Dobosh, P. A. (1972), "Irreducible-Tensor Theory for the Group O^*. I. V and W Coefficients," *Phys. Rev. A* **5**, 2376.

Dobosh, P. A. (1974), "Magnetic Circular Dichroism Calculations Using the Irreducible Tensor Method," *Mol. Phys.* **27**, 689.

Edmonds, A. R. (1963), *Angular Momentum in Quantum Mechanics*, Princeton University Press, Princeton, N.J.

Englman, R. (1972), *The Jahn–Teller Effect in Molecules and Crystals*, Wiley, New York.

Eyring, H., J. Walter, and G. E. Kimball (1944), *Quantum Chemistry*, Wiley, New York.

Fano, U., and G. Racah (1959), *Irreducible Tensorial Sets*, Academic Press, New York.

Gale, R., and A. J. McCaffery (1972), "Evidence for 'C' Terms in the Magnetic Circular Dichroism of $Fe(CN)_6^{3-}$," *J.C.S. Chem. Comm.* **1972**, 832.

Goldstein, H. (1950), *Classical Mechanics*, Addison-Wesley, Reading, Mass.

Griffith, J. S. (1962), *The Irreducible Tensor Method for Molecular Symmetry Groups*, Prentice-Hall, Englewood Cliffs, N.J.

Griffith, J. S. (1964), *The Theory of Transition-Metal Ions*, Cambridge University Press.

Ham, F. S. (1965), "Dynamical Jahn–Teller Effect in Paramagnetic Resonance Spectra: Orbital Reduction Factors and Partial Quenching of Spin–Orbit Interactions," *Phys. Rev.* **138**, A1727.

Harnung, S. E. (1973), "Irreducible Tensors in the Octahedral Spinor Group," *Mol. Phys.* **26**, 473.

Heitler, W. (1954), *The Quantum Theory of Radiation*, 3rd edition, Clarendon Press, Oxford.

Henning, G. N., A. J. McCaffery, P. N. Schatz, and P. J. Stephens (1968), "Magnetic Circular Dichroism of Charge-Transfer Transitions: Low-Spin d^5 Hexahalide Complexes," *J. Chem. Phys.* **48**, 5656.

Henry, C. H., S. E. Schnatterly, and C. P. Slichter (1965), "Effect of Applied Fields on the Optical Properties of Color Centers," *Phys. Rev.* **137**, A583.

Henry, C. H., and C. P. Slichter (1968), "Moments and Degeneracy in Optical Spectra," in W. B. Fowler, Ed., *Physics of Color Centers*, Academic Press, New York, p. 351.

Herzberg, Gerhard (1945), *Molecular Spectra and Molecular Structure II. Infrared and Raman Spectra of Polyatomic Molecules*, Van Nostrand, Princeton, N.J.

Herzberg, Gerhard (1966), *Molecular Spectra and Molecular Structure III. Electronic Spectra and Electronic Structure of Polyatomic Molecules*, Van Nostrand, Princeton, N.J.

Jensen, H. P., J. A. Schellman, and T. Troxell 1978), "Modulation Techniques in Polarization Spectroscopy," *Applied Spectroscopy* **32**, 192.

Jørgensen, C. K. (1959), "Electron Transfer Spectra of Hexahalide Complexes," *Mol. Phys.* **2**, 309.

Jørgensen, C. K. (1962), *Absorption Spectra and Chemical Bonding in Complexes*, Pergamon Press, Oxford.

Jørgensen, C. K., and W. Preetz (1967), "Interpretation of Electron Transfer Spectra of Iridium(IV) and Osmium(IV) Mixed Chloro–Bromo Complexes," *Z. Naturf. A* **22**, 945.

Kauzmann, W. (1957), *Quantum Chemistry*, Academic Press, New York.

Keiderling, T. A., P. J. Stephens, S. B. Piepho, J. L. Slater, and P. N. Schatz (1975), "Infrared Absorption and Magnetic Circular Dichroism of $Cs_2ZrCl_6:Ir^{4+}$," *Chemical Physics* **11**, 343.

Kobayashi, H., M. Shimizu, and Y. Kaizu (1970), "Magnetic Circular Dichroism of $Fe(CN)_6^{3-}$," *Bull. Chem. Soc. Japan* **43**, 2321.

Landau, L. D., and E. M. Lifshitz (1958), *Quantum Mechanics*, Addison-Wesley, Reading, Mass.

Landau, L. D., and E. M. Lifshitz (1960), *Electrodynamics of Continuous Media*, Addison-Wesley, Reading, Mass., Section 62.

Lax, M. (1952), "The Franck–Condon Principle and its Applications to Crystals," *J. Chem. Phys.* **20**, 1752.

Lohr, L. L. (1966), "Spin-Forbidden Electric-Dipole Transition Moments," *J. Chem. Phys.* **45**, 1362.

Longuet-Higgins, H. C. (1961), "Some Recent Developments in the Theory of Molecular Energy Levels," *Advances in Spectroscopy* **2**, 429.

Loudon, R. (1973), *The Quantum Theory of Light*, Clarendon Press, Oxford.

McCaffery, A. J., P. N. Schatz, and T. E. Lester (1969), "Magnetic Circular Dichroism of $IrCl_6^{2-}$ in Crystalline $(CH_3NH_3)_2SnCl_6$," *J. Chem. Phys.* **50**, 379.

Mason, S. F., Ed. (1979), *Optical Activity and Chiral Discrimination, Proceedings of the NATO Advanced Study Institute*, D. Reidel, Dordrecht, Holland.

Messiah, A. (1961), *Quantum Mechanics*, Vols. I and II, North-Holland, Amsterdam.

Nielson, C. W., and G. F. Koster (1963), *Spectroscopic Coefficients for the p^n, d^n, and f^n Configurations*, M.I.T. Press, Cambridge, Mass.

Orlandi, G., and W. Siebrand (1973), "Theory of Vibronic Intensity Borrowing. Comparison of Herzberg–Teller and Born–Oppenheimer Coupling," *J. Chem. Phys.* **58**, 4513.

Osborne, G. A., and P. J. Stephens (1972), "Magnetic Circular Dichroism of Impurities in Solids: Allowed Electronic Transitions and the LiF *F* Center," *J. Chem. Phys.* **56**, 609.

Pauling, L., and E. B. Wilson, Jr. (1935), *Introduction to Quantum Mechanics*, McGraw-Hill, New York.

Piepho, S. B. (1979), "Advanced Group Theoretical Techniques and Their Application to Magnetic Circular Dichroism Spectroscopy," in J. Donini, Ed., *Recent Advances in Group Theory and Their Application to Spectroscopy*, Plenum Press, New York, p. 405.

Piepho, S. B., J. R. Dickinson, and P. N. Schatz (1971), "Magnetic Circular Dichroism of Resolved Vibronic Lines in $Cs_2ZrCl_6:U^{4+}$," *Phys. Stat. Sol. (b)* **47**, 225.

Piepho, S. B., J. R. Dickinson, J. A. Spencer, and P. N. Schatz (1972a), "High Resolution Absorption and Magnetic Circular Dichroism Spectra of $Cs_2ZrCl_6:Os^{4+}$," *Mol. Phys.* **24**, 609.

Piepho, S. B., J. R. Dickinson, J. A. Spencer, and P. N. Schatz (1972b), "High Resolution Magnetic Circular Dichroism Spectrum of $Cs_2ZrCl_6:Ir^{4+}$," *J. Chem. Phys.* **57**, 982.

Piepho, S. B., W. H. Inskeep, P. N. Schatz, W. Preetz, and H. Homborg (1975), "M.C.D. Spectra of OsI_6^{2-} and the *trans*-Mixed Hexahalo Complexes of Ir^{4+} and Os^{4+}," *Mol. Phys.* **30**, 1569.

Piepho, S. B., E. R. Krausz, and P. N. Schatz (1978), "Vibronic Coupling Model for Calculation of Mixed Valence Absorption Profiles," *J. Am. Chem. Soc.* **100**, 2996.

Power, E. A., and T. Thirunamachandran (1974), "Circular Dichroism: A General

Theory Based on Quantum Electrodynamics." *J. Chem. Phys.* **60**, 3695.

Racah, G. (1943), "Theory of Complex Spectra. III," *Phys. Rev.* **63**, 367.

Racah, G. (1949), "Theory of Complex Spectra. IV," *Phys. Rev.* **76**, 1352.

Reid, M. F., and P. H. Butler (1980), "Orientations of Point Groups: Phase Choices in the Racah–Wigner Algebra," *J. Phys. A* **13**, 2889.

Reid, M. F., and P. H. Butler (1982), "Orientations of Reflection–Rotation Groups," *J. Phys. A* **15**, 2327.

Richardson, F. S. (1984), "Emission Circular Intensity Differentials," in F. Ciardelli, S. F. Mason, P. Salvadori, and G. Snatzke, Eds., *ORD and CD: Theory, Chemical Practice, and Biochemical Applications*, Wiley, New York.

Riehl, J. P., and F. S. Richardson (1977), "Theory of Magnetic Circularly Polarized Emission," *J. Chem. Phys.* **66**, 1988. [Note that $\mathcal{C}(180°)$ in their notation = $\mathcal{C}(0°)$ in ours. This paper uses the Stephens (1970) notation and conventions for $\mathcal{A}$, $\mathcal{B}$, and $\mathcal{C}$, rather than the nomenclature used in this book—see Section A.5. Note also that the authors use $\boldsymbol{\mu}$ and $\mathbf{m}$ with meanings the reverse of those in this book.]

Rotenberg, M., R. Bivins, N. Metropolis, and J. K. Wooten, Jr. (1959), *The 3-j and 6-j Symbols*, Technology Press, M.I.T., Cambridge, Mass.

Satten, R. A. (1971), "Vibronic Splitting and Zeeman Effect in Octahedral Molecules: Odd-Electron Systems," *Phys. Rev. A* **3**, 1246.

Satten, R. A., D. R. Johnston, and E. Y. Wong (1968), "Zeeman Splitting of Vibronic Levels for Octahedral Actinide and Lanthanide Complexes, Free and in Crystals," *Phys. Rev.* **171**, 370.

Schatz, P. N., A. J. MaCaffery, W. Suëtaka, G. N. Henning, A. B. Ritchie, and P. J. Stephens (1966), "Faraday Effect of Charge-Transfer Transitions in $Fe(CN)_6^{3-}$, MnO_4^-, and CrO_4^{2-}," *J. Chem. Phys.* **45**, 722.

Schatz, P. N., R. L. Mowery, and E. R. Krausz (1978), "M.C.D./M.C.P.L. Saturation Theory with Application to Molecules in $D_{\infty h}$ and its Subgroups," *Mol. Phys.* **35**, 1537.

Schatz, P. N. (1982), "Sign of the *g*-Value in $IrCl_6^{2-}$ Using Magnetic Circular Dichroism Data. A Correction of the Suggestion by Barron et al.," *Mol. Phys.* **47**, 673.

Schiff, L. I. (1968), *Quantum Mechanics*, 3rd edition, McGraw-Hill, New York.

Silver, B. L. (1976), *Irreducible Tensor Methods*, Academic Press, New York.

Slater, J. C. (1960), *Quantum Theory of Atomic Structure*, Vol. II, McGraw-Hill, New York.

Snyder, P. A., P. A. Lund, P. N. Schatz, and E. M. Rowe (1981), "Magnetic Circular Dichroism (MCD) of the Rydberg Transitions in Benzene Using Synchrotron Radiation," *Chem. Phys. Lett.* **82**, 546.

Stephens, P. J. (1965a), "The Faraday Rotation of Allowed Transitions: Charge-Transfer Transitions in $K_3Fe(CN)_6$," *Inorg. Chem.* **4**, 1690.

Stephens, P. J. (1965b), "Dispersion of the Faraday Effect in $CoCl_4^{2-}$," *J. Chem. Phys.* **43**, 4444.

Stephens, P. J. (1970), "Theory of Magnetic Circular Dichroism," *J. Chem. Phys.* **52**, 3489.

Stephens, P. J. (1974), "Magnetic Circular Dichroism," *Ann. Rev. Phys. Chem.* **25**, 201.

Stephens, P. J. (1976), "Magnetic Circular Dichroism," in I. Prigogine and S. A. Rice, Eds., *Adv. in Chem. Phys.* **35**, 197.

Stephens, P. J., W. Suëtaka, and P. N. Schatz (1966), "Magneto-optical Rotary Dispersion of Porphyrins and Phthalocyanines," *J. Chem. Phys.* **44**, 4592.

Stephens, P. J., R. L. Mowery, and P. N. Schatz (1971), "Moment Analysis of Magnetic Circular Dichroism: Diamagnetic Molecular Solutions," *J. Chem. Phys.* **55**, 224.

Tinkham, M. (1964), *Group Theory and Quantum Mechanics*, McGraw-Hill, New York.

Wigner, E. P. (1959), *Group Theory*, Academic Press, New York.

Wilson, E. B., Jr., J. C. Decius and Paul C. Cross (1955), *Molecular Vibrations. The Theory of Infrared and Raman Vibrational Spectra*, McGraw-Hill, New York.

Wong, K. Y., P. N. Schatz, and S. B. Piepho (1979), "Vibronic Coupling Model for Mixed Valence Compounds. Comparisons and Predictions," *J. Am. Chem. Soc.* **101**, 2793.

Wong, K. Y., and P. N. Schatz (1981), "A Dynamic Model for Mixed-Valence Compounds," in S. J. Lippard, Ed., *Prog. Inorg. Chem.* **28**, 369.

Yeakel, W. C. (1977), "Zeeman Anisotropy in the Spectra of Cubic Crystals," *Mol. Phys.* **33**, 1429.

Yeakel, W. C., and P. N. Schatz (1974), "Ham Effect in the $^2T_{2u}$ Charge Transfer Excited State of Octahedral $IrCl_6^{2-}$," *J. Chem. Phys.* **61**, 441.

Index

Absorbance, 7
Absorption, *see* $\mathfrak{D}_0$ parameter
Absorption coefficient:
  circularly polarized light, 67
  definition of, 6, 7, 61
  expression for, 18, 21, 25
  isotropic radiation, 19, 22
Absorption probability:
  derivation of, 7
  expression for, 12
Accidental degeneracy, 145, 185
Active convention, 166
$\mathfrak{A}_1/\mathfrak{D}_0$:
  groups *O* and T_d, 365
  $IrCl_6^{2-}$, 500
  isotropic case, 364
  oriented case, simply reducible groups, 364
  $OsCl_6^{2-}$, 501
  vibration-induced, $IrBr_6^{2-}$, 532
Adiabatic:
  functions, 30
  principle, 32
  surfaces, 31
Adjoint:
  operation, 303
  operators, 325
Allowed electronic transitions, 39. *See also* $\mathfrak{D}_0$ parameter
Alternating tensor, definition of, 87, 369
Ambivalent group, definition of, 305
Ambivalent groups, 231
  obtaining the 2jm, 310
Angular-momentum basis, 194, 212
  coupling coefficients and phases, 543
  time-reversal behavior, 308
  see also Group SO_3
Anharmonicity, 43
Antisymmetric product, 231
  groups D_3 and C_{3v}, 602
  groups D_4 and D_{2d}, 590
  groups D_∞ and $C_{\infty v}$, 613
  groups *O* and T_d, 533
Antisymmetrization, 255, 403, 406, 429
$\mathfrak{A}_1$ parameter:
  FC-forbidden case, HT approximation, 475
  $Fe(CN)_6^{3-}$ example, 246
  isotropic case, *see* $\mathfrak{A}_1$ parameter, oriented case
  oriented case:
    basis-invariant form, 85
    with complex irreps, 380
    diagonal basis, 79
    simplified form, 361
  properties of, 98
  simple example, 99
  space-averaged case, simplified form, 370
  space-averaged case ($\overline{\mathfrak{A}}_1$), 88
$\mathfrak{A}$ term:
  dispersion form, 149
  properties of, 114
  saturation analog of, 121
  simple example, 70
  temperature dependent, 108
  see also $\mathfrak{A}_1$ parameter
Atomic orbitals, angular parts, table of, 543

Bandshape function, definition of, 81
Basis:
  analogy to vectors, 172
  belonging to irreps, 189
  Butler basis, 344
  chain-of-groups definition, 334
  chain notation for, 335
  choice of, 180, 211
  compared to Griffith, 5, 212, 345
  complex, 221
  group D_∞, 614
  groups D_3 and C_{3v}, 602, 603
  groups D_4 and D_{2d}, 590

Basis (*Continued*)
groups *O* and T_d, tetragonal, 554, 558
trigonal, 602, 603
standard, 211
for transformation operators $\mathcal{R}$, 176, 180
transformations of, example, 177
used in this book, 344
vectors, 163
warning about, 5
see also Wavefunctions
Basis functions, *see* Basis
Beer-Lambert law, 7
Biaxial systems, 65
Birefringence, 64
Born-Oppenheimer (BO) approximation, 28
breakdown of, 49
electronic solutions, 33
vibronic solutions, 35
Born-Oppenheimer (BO) functions, 29
$\mathcal{B}_0$ parameter:
isotropic case, *see* $\mathcal{B}_0$ parameter, oriented case
oriented case, 79, 85
simplified form, 361
overlapping bands, 104
properties of, 102
simple example, 102
space-averaged case, simplified form, 371
space-averaged case ($\overline{\mathcal{B}}_0$), 88
$\mathcal{B}_1$ parameter:
in BO approximation, 138
cases where zero, 138
definition of, 136
Branching multiplicity, 336
$\mathcal{B}$ term:
overlapping bands, 104, 144
properties of, 115
saturation analog of, 121
simple example, 73
see also $\mathcal{B}_0$ parameter

$\mathcal{C}_0/\mathcal{D}_0$:
$CoCl_4^{2-}$ example, 388
$Fe(CN)_6^{3-}$ example, 366
groups *O* and T_d, 365
$IrBr_6^{2-}$ example, 386
$IrCl_6^{2-}$, 500
$IrCl_6^{2-}$ in solution, 486
isotropic case, 364
oriented case, simply reducible groups, 364
$OsCl_6^{2-}$, 501
vibration-induced, $IrBr_6^{2-}$, 528
Chain-of-groups method:
calculation of 3jm, 339
chain truncation, 334
factoring of 3jm and 2jm, 336
introduction to method, 333
notation for, 334
reduced-matrix-element evaluation, 340
Character, 186
Character analysis, 185, 186
of atomic basis sets, 194
basic equation for, 189
for normal coordinate determination, 191
significance of, 189
Character orthogonality relation, 186, 187
Character tables: form of, 190
groups D_3 and C_{3v}, 601
groups D_4 and D_{2d}, 589
groups *O* and T_d, 551
Circular dichroism:
electromagnetic origin, 60
magnetically-induced, 67
naturally occurring, 67
Classes, 187
Clebsch-Gordan coefficients, 204
$CoCl_4^{2-}$, *see* Group T_d
Coefficients of fractional parentage (cfp), 258, 260
calculation of, 407
pseudo-p^n case, 436
construction of a^n states, 407
general equations for, 411
groups *O* and T_d, table, 582
jj basis, groups *O* and T_d table, 588
Contragredience, 303
Coordinate system:
laboratory- or space-fixed, 14
definition, 16
molecule-fixed, 14
definition, 16
right-handed, 163
Coupling:
in another basis, 221
coefficients, *see* 3jm (coefficient); Vector coupling coefficients (v.c.c.)
of conjugate states, 235
definition of, 202
high-symmetry coefficients, 228
jj, 454
phase choice in tables, 220
$\mathcal{SL}$ (or SL or Russell-Saunders), 249
tables, how to construct, 214

with 3jm, 230
with v.c.c., 210
see also Wavefunctions
$\mathcal{C}_0$ parameter:
Fe(CN)$_6^{3-}$ example, 241, 296
isotropic case, *see* $\mathcal{C}_0$ parameter, oriented case
oriented case:
basis-invariant form, 85
with complex irreps, 377
diagonal basis, 79
simplified form, 361
properties of, 100
simple example, 101
space-averaged case, simplified form, 371
space-averaged case ($\bar{\mathcal{C}}_0$), 88
without spin-orbit coupling, 250
vibration-induced, HT approximation, 475
$\mathcal{C}_1$ parameter:
in BO approximation, 138
cases where zero, 138
definition of, 136
Crude adiabatic approximation, 34, 182
Crystal-field calculations:
crystal-field matrices, groups O and T_d, 564
electrostatic matrices, strong-field case, 564
Nielson and Koster tables, use of, 546
$\mathcal{C}$ term:
overlapping bands, 108, 145
properties of, 114
saturation analog of, 118
simple example, 72
see also $\mathcal{C}_0$ parameter

Delta symbol δ(abcr), definition of, 236
Depolarization, 65, 481
Determinantal functions, 256
permutation rules, 403
Diabatic surfaces, 31
Dipole-moment operator, *see* Electric-dipole transitions; Magnetic-dipole transitions
Dipole strength, 80. *See also* $\mathcal{D}_0$ parameter
Direct-product:
of functions, 202
groups, 198
examples of, 200
of matrices, 199
representations, 202
rules, 222, 243
table:
groups D_3 and C_{3v}, 602
groups D_4 and D_{2d}, 590
groups D_∞ and $C_{\infty v}$, 613
groups O and T_d, 552
Distortions, calculations with, 520
Dot product, 167
Double groups, 197, 267
$\mathcal{D}_0$ parameter:
Fe(CN)$_6^{3-}$ example, 245
isotropic case, *see* $\mathcal{D}_0$, oriented case
orientationally averaged case ($\bar{\mathcal{D}}_0$), 88
oriented case, 81, 85, 363
with complex irreps, 381
simplified form, 363, 364
space-averaged case, simplified form, 371
vibration-induced, HT approximation, 472
$\mathcal{D}_1$ parameter, definition of, 132

Electric-dipole transitions:
operator, 18
operator transformation coefficients, 224
transition moment, 18
see also $\mathcal{D}_0$ parameter
Electric-quadrupole:
for circularly polarized light, 67
transition moment, 21
Electrostatic interactions, 28
effect on jj model, 511, 519
ngelect of, 497
Electrostatic matrices:
from Griffith (1964) to Butler basis, how to convert, 564
from Nielson and Koster, 546
strong-field basis, groups O and T_d, 564
Ellipticity, 63
molar, 535
Emission:
induced, 12
magnetic circularly polarized, 70, 90
spontaneous, 12, 90
Emission probability:
induced, 12
spontaneous, 13
Equivalent representations, 170, 181, 184
ambivalent groups, 305
nonambivalent groups, 314

Faraday effect, 58
F-center:
method of moments, 144
use of higher moments, 147
Fe(CN)$_6^{3-}$, *see* Group O

Forbidden transitions:
  Herzberg-Teller (HT) approach to, 44
  MCD of, 156
  MCD parameters, 469
  moments of, 156
Franck-Condon (FC) approximation:
  definition of, 37
  deviations from, 38, 156
    MCD, 469
  selection rules in, 39
Franck-Condon (FC) overlap factor, 42
Functions, *see* Wavefunctions
Function transformation coefficients:
  definition and examples of, 223
  groups D_4 and D_{2d}, 593
  groups O and T_d, tetragonal, 558
  groups O, T_d, D_3 and C_{3v}, trigonal, 607

$g(a^n,\mathfrak{S}hfh')$ coefficients:
  calculation of, pseudo-p^n case, 437
  definition of, 416
  general equations for, 416
  groups O and T_d, table of, 585
$G(a^n,\mathfrak{S}hf\mathfrak{S}'h')$ coefficients:
  calculation of, pseudo-p^n case, 438
  definition of, 421
  general equations for, 422
  groups O and T_d, table of, 587
Gaussian:
  fitting:
    problems with, 152
    use of, 148
  lineshape, definition of, 148
  units, 8
Great Orthogonality Theorem, 185
Group C_i, 196
Group C_{3v}:
  2j and 3j phases, 609
  2jm:
    complex basis, 610
    real basis, 609
  3jm:
    complex basis, 611
    real basis, 610
  6j table, 611
  9j, single-valued irreps, 612
  basis, as O_3 subgroup, 603, 605
  basis definition, 602
  character table, 601
  direct-product table, 602
  function transformation coefficients, 607
  operator transformation coefficients, 607
Group $C_{\infty v}$:
  6j for, 615
  coupling coefficients and phases, 615
  direct-product table, 613
  see also Group D_∞
Group D_{2d}:
  2j phase, 595
  2jm, real and complex bases, 595
  3j phase, 594
  3jm, real and complex bases, 596
  6j table, 597
  9j, single-valued irreps, 598
  basis, as O_3 subgroup, 558
  basis definitions, 590
  character table, 589
  direct-product table, 590
  function transformation coefficients, 593
  operator transformation coefficients 593
Group D_3:
  2j and 3j phases, 609
  2jm:
    complex basis, 610
    real basis, 609
  3jm:
    complex basis, 611
    real basis, 610
  6j table, 611
  9j, single-valued irreps, 612
  basis, as SO_3 or O_3 subgroup, 603, 605
  basis definition, 602
  character table, 601
  direct-product table, 602
  example, 2jm for paired irreps, 314
  function transformation coefficients, 607
  molecular orbitals, trigonal planar case, 608
  near-octahedral case, $IrCl_6^{2-}$, 520
  operator transformation coefficients, 607
  procedure to obtain 3jm, 323
  v.c.c. for, 324
Group D_{3d}, *see* Group D_3
Group D_{3h}:
  BF_3 normal (and symmetry) coordinates, 192
  BF_3 representative, 171
  character table for, 200
Group D_4:
  2j phase, 595
  2jm, real and complex bases, 595
  3j phase, 594
  3jm, real and complex bases, 596
  6j table, 597
  9j, single-valued irreps, 598

basis, as SO_3 or O_3 subgroup, 558
basis definitions, 590
character table, 589
direct-product table, 590
function transformation coefficients, 593
MCD parameters, examples, 372
molecular orbitals, square planar case, 594
operator transformation coefficients, 593
spin-angular-momentum matrix, 418
spin-orbit matrix, reduction to $\mathbb{S}$h basis, 398
Group D_{4h}, *see* Group D_4
Group D_6, 616
Group D_∞:
6j for, 615
basis, as SO_3 subgroup, 614
coupling coefficients and phases, 615
direct-product table, 613
Group $D_{\infty h}$,
basis, as $D_\infty \otimes C_i$, 614
definition of, 613
see group D_∞
Group generators:
definition of, 212
groups D_3 and C_{3v}, 603
groups D_4 and D_{2d}, 591
groups *O* and T_d, 554
Group *O*:
2j phase, 571
2jm:
complex basis, tetragonal, 572
real basis, tetragonol, 571
3j phase, 570
3jm:
complex basis, tetragonal, 574
real basis, tetragonal, 573
3jm factors, $O \supset D_4$, 598
6j table, 576
9j, 582
basis:
as SO_3 or O_3 subgroup:
tetragonal, 558
trigonal, 603, 605
tetragonal, 554
trigonal, 602
cfp, table of, 582
chain notation, example of, 335
character table, 551
coupling table construction, 214
decomposition to subgroups, 553
direct-product table, 552
distortions, treatment of, 520
Group *O* (*Continued*)
$Fe(CN)_6^{3-}$ example, 239, 366
with s.o. coupling, 296
$Fe(CN)_6^{3-}$ matrix elements:
spin-independent, 278
spin-orbit, 282
function transformation coefficients:
tetragonal, 560
trigonal, 607
g(a^n,$\mathbb{S}$hfh′) table, 585
G(a^n,$\mathbb{S}$hf$\mathbb{S}$′h′) table, 587
$IrBr_6^{2-}$:
$\mathcal{C}$ term example, 386
solid state spectrum, 524
vibronic MCD, 523
$IrCl_6^{2-}$:
introductory concepts, 482
jj approach, 506
JT effect, 55, 511
MCD calculations, 499
$OsCl_6^{2-}$ comparison, 494
solid state spectrum, 488
solution MCD, 486
trigonally distorted, 520
magnetic moment matrix for t_{2g}^4, 419
molecular orbitals, octahedral complex, 568
off-diagonal:
orbital angular momentum, 433
spin-orbit, $IrCl_6^{2-}$, 451
one-electron matrix elements, multi-center, 425
operator transformation coefficients:
tetragonal, 560
trigonal, 607
orientation dependence of MCD, 374
$OsCl_6^{2-}$:
analysis of MCD, 514
breakdown of jj approximation, 511
jj coupling model, 494
JT effect, 512
MCD calculations, 499
solid state spectrum, 494
OsX_6^{2-}, *see* jj-coupling
parity-forbidden transitions, 48
with small perturbations, 143
spin-angular-momentum matrix elements, 343
spin-orbit coupling, high spin states, 274
spin-orbit matrix, reduction to $\mathbb{S}$h basis, 399
spin-orbit matrix for t_2^5 and t_1^5, 424
spin-orbit matrix for t_{2g}^4, 423
standard basis, significance of, 213

Group *O* (*Continued*)
  U′ multiplicity separation, 222, 273
  wavefunctions, t_2^m and e^n, 562
Group O_3, 196
  historical phase, 230
  transformations in O_3 subgroups, 226
  see also Group SO_3
Group O_h:
  decomposition to subgroups, 553
  definition of, 200
  as O_3 subgroup, 341
  see also Group *O*
Group SO_3, 194
  3jm, $SO_3 \supset SO_2$ basis (partial list), 544
  3jm factors, $SO_3 \supset O$ (partial list), 549
  6j (partial list), 546
  angular-momentum matrix elements, 549
  coupling coefficients and phases, 543
  group *O* and T_d basis chains:
    tetragonal, 558
    trigonal, 603, 605
  historical phase, 230
  Nielson and Koster tables, 546
  standard basis functions, 212
Group-subgroup chains, *see* Chain-of-groups method
Group T_d:
  2j phase, 571
  2jm:
    complex basis, tetragonal, 572
    real basis, tetragonal, 571
  3jm:
    complex basis, tetragonal, 574
    real basis, tetragonal, 573
  3jm factors, $T_d \supset D_{2d}$, 598
  3j phase, 570
  6j table, 576
  9j, 582
  basis:
    as O_3 subgroup:
      tetragonal, 558
      trigonal, 603, 605
    tetragonal, 554
    trigonal, 602
  cfp, table of, 582
  character table, 551
  $CoCl_4^{2-}$, $\mathcal{C}$ terms, 388
  direct-product table, 552
  function transformation coefficients:
    tetragonal, 560
    trigonal, 607
  g(a^n,$\mathcal{S}$hfh′) table, 585
  G(a^n,$\mathcal{S}$hf$\mathcal{S}$′h′) table, 587
  molecular orbitals, tetrahedral complex, 570
  operator transformation coefficients:
    tetragonal, 560
    trigonal, 607
  spin-angular-momentum matrix elements, 343
  wavefunctions, t_2^m and e^n, 562
Group theory and quantum mechanics, 181, 190, 249

Ham effect, 56, 155
Harmonic approximation, 35
Herzberg-Teller (HT) approximation:
  $\mathcal{A}_1$ and $\mathcal{C}_0$ parameters in, 475
  $\mathcal{D}_0$ parameter in, 472
  $IrBr_6^{2-}$ example, 523
  transition moment in, 46
Herzberg-Teller (HT) functions, 34
High symmetry coefficients, *see* 9j (coefficient or symbol); 6j (coefficient of symbol); 3jm (coefficient)
Historical phase H(abc),
  definition of, 230
  $SO_3 \supset SO_2$, 543
Hole-particle convention, 259
Hot bands,
  definition of, 43
  $IrBr_6^{2-}$ example, 528

Identity operation, 162
Interelectronic repulsions, *see* Electrostatic interactions
Invariant product:
  definition of, 304
  of three sets, 316
  of two sets, 315
$IrBr_6^{2-}$, *see* Group *O*
$IrCl_6^{2-}$, *see* Group *O*
Irreducible representations (irreps):
  complex (or paired), 305
  conjugate:
    definition of, 234
    reason for, 229
  definition of, 184
  dimension of, |a|, 206
  double-valued irreps, 198
  naming conventions, 191
  paired (or complex), 305
  partner (row), 184

single-valued irreps, 198
spin irreps, 198, 222
true irreps, 198
Irreducible tensor method:
equation simplification, example, 352
of Griffith, 4
guide to book, 3
matrix-element evaluation, introduction to, 382
simplification of MCD equations, 359, 370
see also Matrix-element evaluation
Isoscalar factor, 337

Jahn-Teller (JT) effect:
cause of, 30
Ham effect, 56, 155
$IrCl_6^{2-}$ example, 511
MCD of vibronic lines, 155
method of moments, 130, 137
$OsCl_6^{2-}$ example, 512
overlapping bands from, 144
in related systems, 511
static and dynamic, 54
treatment of, 49
use of higher moments, 147
jj coupling:
$\mathcal{A}_1/\mathcal{D}_0$ example, OsX_6^{2-}, 466
breakdown of *jj* approximation, 511, 519
cfp, groups *O* and T_d, 588
introduction, 454
$IrCl_6^{2-}$ matrix elements, 506
magnetic-moment matrix, OsX_6^{2-}, 465
$OsCl_6^{2-}$ example, 494
$OsCl_6^{2-}$ matrix elements, 509
spin-independent matrix elements, 464, 467
spin-orbit matrix, OsX_6^{2-}, 459, 463
wavefunctions:
general equations, 458
jj basis, groups *O* and T_d, 588
OsX_6^{2-}, 456

Kramers pair states, 101
properties of, 327

Least-squares fitting, 151
Length-preserving transformation, 168
Linear dichroism, 64
Linear limit of MCD, 75
Lineshape:
approximation of, 12
function, definition of, 154
Low-symmetry perturbations, 520
Luminescence, 72

Magnetic circular dichroism (MCD):
additivity, for overlapping transitions, 109
atomic example, 69
basic expression, 69
in biaxial systems, 64
conventions and notation, 533
data gathering, 480
depolarization effects, 65
derivation in linear limit, 74
deviations from linear limit, 116
dispersion forms, 110
electromagnetic origin, 60
examples, *see* Group *O*, etc.
experimental techniques, 65
$Fe(CN)_6^{3-}$ example, 239, 296
forbidden transitions, 156
guide to book, 2
with large Zeeman splitting, 122
limitations of simple model, 82
magnetic-dipole terms, 68
resolved vibronic lines, 153
saturation effects, 117, 122
without spin-orbit coupling, 247
use of time-reversal symmetry, 327
utility of, 114
see also Overlapping bands
Magnetic circularly polarized luminescence (MCPL), 90
with no photoselection, 93
with photoselection, 94
relation of MCPL and MCD parameters, 93
simple example, 72
Magnetic-dipole transitions:
for circularly polarized light, 67
operator, 23
selection rules, 46
transition moment, 21
Matrix-element evaluation:
a^mb^n configuration:
general procedure, 440
spin-angular-momentum, 443
spin-independent, 442
spin-orbit, 443
a^n configuration:
general procedure, 405
$IrCl_6^{2-}$, 501
magnetic moment example, 419
spin-angular-momentum, 418

Matrix-element evaluation (*Continued*)
spin-independent, 416
spin-orbit, 421
spin-orbit examples, 423
angular-momentum, group SO_3, 549
basic theorems, 279, 404
chain-of-groups method, 341
distortions, 521
distortions from high symmetry, 521
double tensor operators, 400
introduction to, 277
irreducible tensor method, fundamental equations, 382
more than two open shells, 452
Nielson and Koster tables, use of, 546
off-diagonal:
general procedure, 444
spin-angular-momentum, 434, 448
spin-independent, 388, 433, 447
spin-orbit, 449
spin-orbit coupling, 435
spin-orbit example, 451
one-electron, 287
multi-center, 290
group *O*, 425
single-center, 288, 342
orbital-angular-momentum, 341
repeated representation example, $CoCl_4^{2-}$, 388
spin-angular-momentum, 342
reduction to $\mathcal{S}$h basis, 385
spin-independent:
$Fe(CN)_6^{3-}$ example, 278
reduction to $\mathcal{S}$h basis, 385
spin-only, *see* Matrix-element evaluation, spin-angular-momentum
spin-orbit, reduction to $\mathcal{S}$h basis, 397
spin-orbit case, $Fe(CN)_6^{3-}$ example, 282
spin-orbit coupling, introduction, 281
see also *jj*-coupling
MCD factor:
definition:
oriented case, 361
space-averaged case, 370, 371
table of, oriented case, 361
MCD parameters, 56
with complex irreps, 376
cubic systems, 90
examples and calculation of, *see* Group *O*, etc.
former definitions, 538
from Gaussian fits, 148
isotropic case, *see* MCD parameters, oriented case
method of moments, allowed transitions, 131
neglecting spin-orbit coupling, 252
nonambivalent groups, 376
obtaining from experimental data, 82
obtaining in RS-BO-FC approximation, 148
orientation independence, cubic systems, 374
oriented case:
basis-invariant form, 84, 358
diagonal basis, 79
space-averaged case comparison, 372
oriented (or isotropic) case, defined, 356
properties of, 98, 114
quantitative comparisons of, 110
simplification using 3jm and 6j, 359
space-averaged case, 88, 367
defined, 356
examples, 372
time-reversal simplified expressions, 330
vibration-induced:
HT approximation, 469
$IrBr_6^{2-}$, 528
special cases, 477
see also $\mathcal{A}_1$ parameter; $\mathcal{B}_0$ parameter; $\mathcal{C}_0$ parameter; $\mathcal{D}_0$ parameter
Molar extinction coefficient, 7
Molecular orbital diagram, $IrCl_6^{2-}$, $OsCl_6^{2-}$, 483
Molecular orbitals:
octahedral complex, 568
square-planar complex, 594
tetrahedral complex, 570
trigonal planar molecule, 608
Molecular point group, 182
Moments, method of:
in BO approximation, 137
FC-allowed lines, 153
FC-forbidden bands, 156
$Fe(CN)_6^{3-}$ example, 300
forbidden lines, 159
ground-state near-degeneracy, 146
higher moments, 147
introduction, 124
moments of MCD, 127
moments of absorption, 126
in RS-BO-FC approximation, 152

use to obtain MCD parameters, allowed transitions, 131
vibronic effects, 137
see also Jahn-Teller (JT) effect; Overlapping bands
Multiplicity-free group, 231

Nielson and Koster tables, 342, 436
crystal-field calculations with, 546
9j (coefficient or symbol):
with A_1 irrep, 355
definition of, 353
groups D_3 and C_{3v}, single-valued irreps, 612
groups D_4 and D_{2d}, single-valued irreps, 598
groups *O* and T_d, 582
relation to recoupling coefficients, 355
symmetry rules:
general case, 354
simply reducible groups, 353
Nonadiabatic:
coupling, 57
functions, 51
Nonambivalent groups, 231
definition of, 305
obtaining 2jm, 313
No-phonon line, 41
Normal (and symmetry) coordinates, 191
BF_3 example, 192
definition of, 35
Notation,
for cfp, 260
for group-subgroup chain, 334
for irrep dimension, 206
for irreps, 191
MCD, 533
for transitions, 12

Octahedral, *see* Group *O*
Operators:
adjoint, 325
symmetry adapted tensor, 396
antilinear or antiunitary, 307
antisymmetrization, 429
dot product of tensors, 393
double tensor, 396
Hermitian, 327
standard vector operators, 225
step-up, step-down, 225, 284
time-reversed (complex-conjugate), 325
transformation to another center:
group *O* example, 292
group T_d example, 293
transformation coefficients, 223
see also specific operator names
Operator transformation coefficients:
definition and examples of, 223
groups D_4 and D_{2d}, 593
groups *O* and T_d, tetragonal, 558
groups *O*, T_d, D_3 and C_{3v}, trigonal, 607
Optical density, 7
Orbital reduction factor, 289
Order of group, 186
Orientational averaging, 86
and saturation, 120
Orthogonal matrix, 168
Orthonormal basis, 172
Oscillator strength, 540
$OsCl_6^{2-}$, *see* Group *O*
Overlapping bands:
accidental degeneracy, 145
$\mathcal{A}$ terms, $OsCl_6^{2-}$, 516
examples using moments, 143
Gaussian fitting of, 152
from ground-state near-degeneracy, 108, 146
$Fe(CN)_6^{3-}$ example, 296
method of moments, 141
pseudo-$\mathcal{A}$ terms, 104, 153
from spin-orbit splitting, 144
temperature-dependent $\mathcal{A}$ term, 145
from tetragonal distortion, 144

Parity:
of atomic orbitals, 196
definition, 200
direct-product groups, 200, 242
of spin states, 269
Parity-forbidden transitions, 46
Partners, of irreps, 184
Passive convention, 166
Pauli exclusion principle, 255
Phases:
of basis functions, 211
choice of, 220
compared to Griffith, 4, 212, 221
Condon-Shortley, 212, 230
effect of function order, 261
standardization of, v.c.c., 316, 321
of transformation coefficients, 227
warning about, 4
jj states, 458

Photoselection, 93, 94
Point group, 162
Polarization vector, for specific cases, 19
Polarized light, various types, 61
Potential energy surface:
BO approximation, 29
in Jahn-Teller case, 50
Progressions,
BO approximation, 41
Jahn-Teller case, 55
Pseudo-$\mathcal{A}$ term, 153
in BO-FC-RS approximation, 104
method of moments, 142
simple example, 105
Pseudo-Jahn Teller effect, cause of, 32

Quadrupole-moment operator, *see* Electric-quadrupole

Racah factorization lemma, 337
Radiation theory, semiclassical, 9
Recoupling coefficients:
relation to 6j, 351
relation to 9j, 355
Reduced matrix elements, 206, 208
adjoint:
symmetry operator, 397
vector operator, 396
in different subgroup bases, 222, 340
of double tensor operators, spin-orbit example, 284
inverted Wigner-Eckart equation, 236
repeated representation case, 276
see also Matrix-element evaluation
Repeated representations, 269, 271
3j phase, 232
$CoCl_4^{2-}$, $\mathcal{C}$ terms, 388
definition of, 209
significance of label r, 276
U′ multiplicity separation, 222, 273
Replacement theorem, 209
Representations:
definition of, 162
D(R), 176, 183
irreducible, 184
reducible, 188, 204
relation to standard basis, 213
see also Equivalent representations
Representative:
definition of, 162
Rigid shift (RS) approximation:
definition of, 77
moments in, 152
Rotational motion:
elimination of, 13

Saturation effects, 77
cause of, 117
treatment, 117
Scalar product, 167, 172, 205
of vectors, 225
Selection rules:
in Herzberg-Teller approximation, 46
in Jahn-Teller system, 54
parity, 200
vibronic, 39
Similarity transformation, 169
Simply reducible group, 231
6j (coefficient or symbol):
with A_1 irrep, 349
definition of, 346
groups D_3 and C_{3v}, 611
groups D_4 and D_{2d}, 597
groups D_∞ and $C_{\infty v}$, 615
group SO_3 (partial list), 546
groups O and T_d, 576
relation to recoupling coefficients, 351
symmetry rules:
general case, 347
simply reducible groups, 347
(3,1) equation, 350
Solvent effects:
effective field approximation, 24
$SO_3 \supset SO_2$, *see* Angular momentum basis; Group SO_3
Spatial averaging, 83
Spectroscopic stability, 83
Spherical harmonics, 212
Condon-Shortley, 308
table of, 542
parity of, 227
time-reversal behavior, 308
Spin-orbit coupling:
basis functions, 267
example with moments, 144
first-order effects, 273

Ham effect, 155
introduction to, 266
matrix-element evaluation, introduction, 281
operator, 267
symmetry-adapted, 395
spin quartets and sextets, group *O*, 274
wavefunction construction, 269, 271
ζ parameter, 267
see also Matrix-element evaluation
Spinors, 198, 212, 269
Strain birefringence, 65
Symmetrically equivalent sets, 197
Symmetric product, 231
groups D_3 and C_{3v}, 602
groups D_4 and D_{2d}, 590
groups D_∞ and $C_{\infty v}$, 613
groups *O* and T_d, 553
Symmetry operation, 162
Symmetry operators, 163. *See also* Transformation operators $\mathcal{R}$

Temperature dependence:
of absorption bands, 47, 158
of MCD parameters, 111
of pseudo-$\mathcal{A}$ terms, 108, 145
Tensors:
dot-product of, 393
double-tensor operator, 396
irreducible, 224
spherical, 224
unit tensor operators:
defined, 548
use of, 437
Term, 250
Term symbol, 250
3j phase {abcr}:
definition and properties of, 231
groups D_3 and C_{3v}, 609
groups D_4 and D_{2d}, 594
groups D_∞ and $C_{\infty v}$, 615
group SO_3, 543
groups *O* and T_d, 570
law-breaking factors, 231, 348
limitations on choice, 232
3jm (coefficient):
advantages of, 228
with A_1 irrep, 234
angular-momentum basis (partial list), 544
calculation from 3jm factors, example, 339
complex-D_4 and complex-D_{2d} bases, 596
complex-*O* and complex-T_d bases, tetragonal, 574
$D_\infty \supset C_\infty$ and $C_{\infty v} \supset C_\infty$, 615
definition of, 316
example of use, 236
factoring of, 338
groups D_3 and C_{3v}:
complex basis, 611
real basis, 610
groups *O* and T_d, trigonal basis, 612
orthonormality equations for, 235
with paired irreps, 319
parity rules, 243
permutation rules, 230
procedure to obtain, 321
properties, derivation of, 316
real-D_4 and real-D_{2d} bases, 596
real-*O* and real-T_d bases, tetragonal, 573
relation to Clebsch-Gordan coefficients, 230, 544
relation to 3jm*, 235
derivation, 320
relation to v.c.c., 230
3jm factor:
definition of, 337
$O \supset D_4$ and $T_d \supset D_{2d}$, 598
permutation rules, 338
$SO_3 \supset O$ (partial list), 549
Time-reversal operation, 305
use of, 307
Time-reversal symmetry:
use to simplify MCD equations, 330
Time-reversed operators, 325
Transformation operators $\mathcal{R}$:
definition of, 173
in double group space, 272
generation of **D**(R), 177
simple example, 174
Transformation properties:
definition, 185
Transition moment:
BO approximation, 36
electric dipole, 18
electric-quadrupole, 21
magnetic-dipole, 21
for polarized light, 19
relative magnitudes, 24
Transition probability:
absorption, 12
emission, 12

Translational motion, elimination of, 13
2jm (phase), 234
  angular-momentum basis, 543
  complex-D_3 and complex-C_{3v} bases, 610
  complex-D_4 and complex-D_{2d} bases, 595
  complex-O and complex-T_d bases, tetragonal, 572
  $D_\infty \supset C_\infty$ and $C_{\infty v} \supset C_\infty$, 615
  factoring of, 338
  groups O and T_d, trigonal basis, 612
  obtaining for ambivalent groups, 310
  obtaining for nonambivalent groups, 314
  real-D_3 and real-C_{3v} bases, 609
  real-D_4 and real-D_{2d} bases, 595
  real-O and real-T_d bases, tetragonal, 571
2jm factor:
  definition and properties, 337
  values in a chain, 338
2j phase {a}, 233
  groups D_3 and C_{3v}, 609
  groups D_4 and D_{2d}, 595
  group SO_3, 543
  groups O and T_d, 571

Uniaxial systems, 65
Unitary matrix, 169
Units:
  angular-momentum, 80
  Gaussian, 8
  MCD and absorption, 536, 537

V coefficients, 229, 319
Vector coupling coefficients (v.c.c.), 204, 207
  angular momentum basis, 544
  calculation from 3jm, 230
  coupling tables, how to construct, 214
  factoring of, 337
  for group D_3, 324
  for group O, sample tables, 218
  phase standardization, 316, 321
  warning about, 221
Vector coupling factor, 337
Vectors:
  definition, 163
  spherical vectors, 224
  standard vector operators, 225
  unit vectors, 225

Vibrations:
  symmetry classification of, 191
  see also Normal (and symmetry) coordinates
Vibronic spectroscopy:
  guide to book, 3
  see also Vibronic transitions
Vibronic transitions:
  FC-forbidden bands, 156
  FC-forbidden lines, 159
  Herzberg-Teller approximation, 44
  HT approximation, MCD, 469
  $IrBr_6^{2-}$ example, 523
  MCD in FC-allowed case, 153

Wavefunctions:
  a^n states, construction with cfp, 407
  determinantal properties, 403
  groups O and T_d, t_2^m and e^n, 562
  $\mathcal{jj}$ basis:
    general equations, 458
    groups O and T_d, 588
    OsX_6^{2-}, 456
    warning about phases, 458
  $|\mathcal{S}h\mathcal{M}\theta >$:
    construction of, 252, 261
    definition of, 250
  $\mathcal{SL}$, 249
  spin-orbit basis, 268
  spin-orbit states, construction of, 269, 271
  t_2e example, 431
  two-open-shell case, 429
Weak-field basis:
  matrix-element evaluation, 341, 548
Wigner-Eckart theorem:
  $Fe(CN)_6^{3-}$ example, 239
  with group-chain basis, 340
  group O example, 236
  repeated representation example, 276
  with 3jm, 230
  with v.c.c. (general form), 210
  with v.c.c. (restricted form), 208

X coefficients, 353

Zeeman:
  energy, 75
  Hamiltonian, 66
  splitting, 70